Narcolepsy

Christian R. Baumann
Claudio L. Bassetti
Thomas E. Scammell
Editors

Narcolepsy

Pathophysiology, Diagnosis, and Treatment

Editors
Christian R. Baumann, MD
Department of Neurology
University Hospital of Zurich
Zurich, Switzerland
christian.baumann@usz.ch

Claudio L. Bassetti, MD
Department of Neurology
Neurocenter EOC of Southern
Switzerland
Ospedale Civico
Lugano, Switzerland
claudio.bassetti@eoc.ch

Thomas E. Scammell, MD
Department of Neurology
Beth Israel Deaconess Medical Center
and Harvard Medical School
Boston, MA, USA
tscammel@bidmc.harvard.edu

ISBN 978-1-4419-8389-3 e-ISBN 978-1-4419-8390-9
DOI 10.1007/978-1-4419-8390-9
Springer New York Dordrecht Heidelberg London

Library of Congress Control Number: 2011926991

Printed on acid-free paper

Springer is part of Springer Science+Business Media (www.springer.com)

Preface

Narcolepsy is characterized by excessive daytime sleepiness, cataplexy, fragmented sleep, and other symptoms. It affects approximately 1 in 2,000 people and can have a huge impact on their ability to succeed in school and work. Narcolepsy was first recognized by clinicians over 125 years ago, yet until recently, its cause remained a mystery. In 2000, two research groups discovered that narcolepsy is caused by a selective loss of neurons in the hypothalamus that produce the hypocretin neuropeptides (also known as orexins). With this groundbreaking perspective, narcolepsy research has advanced in large steps, with new discoveries every year that have enhanced our understanding of the disorder.

In 1975, the First International Symposium on Narcolepsy was held in La Grande Motte in France, organized by William C. Dement, Christian Guilleminault, and Pierre Passouant. After a successful Fifth International Symposium on Monte Verità near Ascona (Switzerland) in 2004, many of the world's leading narcolepsy researchers – including the authors of this book – gathered again in this inspiring landscape for the Sixth International Symposium on Narcolepsy in 2009. In the course of the meeting, it became clear that researchers and clinicians have learned much about narcolepsy, yet many key questions remain unanswered, even in light of recent advances.

For instance, we still have no definite proof that narcolepsy is caused by an autoimmune attack or by another mechanism. The recent discovery that levels of specific antibodies are increased in some patients soon after the onset of narcolepsy provides some of the most compelling evidence for an autoimmune mechanism, but many questions remain unanswered. For example, it appears that narcolepsy is caused by a selective loss of the hypocretin-producing neurons, yet the target antigens are expressed by many non-hypocretin neurons. In addition, antibody titers appear normal in many narcolepsy patients. It remains possible that these antibodies are not pathogenic but are simply increased as a consequence of another process that kills the hypocretin neurons. Much more work is needed to determine what mechanism kills the hypocretin neurons in narcolepsy.

Many questions remain about the pathophysiology of cataplexy, hypnagogic hallucinations, and sleep paralysis. These symptoms have many similarities to rapid eye movement (REM) sleep, such as muscle atonia and dreaming, and they may represent the intrusion of fragments of REM sleep into wakefulness. Theories have proposed an increase in REM sleep pressure, a reduction in the threshold to transition into REM sleep, or dysregulation of the brainstem mechanisms that normally coordinate REM sleep phenomena. However, there is little evidence that these symptoms are influenced by manipulations of REM sleep, and gamma hydroxybutyrate strongly suppresses cataplexy yet it has no effect on REM sleep. Thus, the pathways underlying cataplexy and other narcolepsy symptoms remain elusive.

Furthermore, people and animals with narcolepsy transition frequently and rapidly between wakefulness and sleep. For over 20 years, this pattern has been referred to as behavioral state instability but its cause remains unknown. It is possible that hypocretin/orexin stabilizes the neural pathways that regulate sleep/wake transitions; so a loss of the hypocretin neurons would destabilize this mechanism, leading to frequent transitions between wakefulness and sleep. This could account for both excessive daytime sleepiness and fragmented nocturnal sleep in narcolepsy. However, this hypothesis is not yet proven, and the electrophysiological basis of this instability is still poorly understood.

Last but not least, narcolepsy is often accompanied by a variety of metabolic and psychiatric symptoms, including obesity and depression. These symptoms are unappreciated by many clinicians and their fundamental cause remains unknown. For instance, there is still no clear explanation why narcolepsy patients are often overweight. Hypocretin can enhance appetite, yet individuals with narcolepsy probably eat normal amounts. Their obesity may result from low physical activity or low basal metabolic rate.

Thus, despite much recent progress, many large questions remain about the causes, neurobiology, and physiology of narcolepsy. To provide a unified resource for clinicians and basic scientists, dozens of researchers with expertise in nearly all facets of narcolepsy have contributed to this book. Our intent is to provide a comprehensive and up-to-date overview on the pathophysiology and neurobiology of narcolepsy and to describe new clinical research on narcolepsy and the best approaches for treatment. The supplementary DVD offers a unique and large collection of movies displaying the symptoms of narcolepsy in people and animals. We have also highlighted many of the outstanding questions about narcolepsy, and hope this book will spark new perspectives and inspire new discoveries.

Finally, we thank the funders of the Sixth International Meeting on Narcolepsy, and above all the Centro Stefano Franscini on Monte Verità, the Swiss Federal Institute of Technology, Zurich, and also Actelion, Boehringer Ingelheim, Cephalon, and UCB Pharma. The production of the supplemental DVD was made possible by the funding from UCB. Special thanks go to Yvonne Fernandez and Sarah Eisenstein, the meeting secretaries.

Zurich, Switzerland — Christian R. Baumann
Lugano, Switzerland — Claudio L. Bassetti
Boston, MA, USA — Thomas E. Scammell

Contents

Contributors

Antoine Adamantidis, PhD
McGill University, Department of Psychiatry,
Douglas Mental Health University Institute,
Montréal, Canada

Chloé Alexandre, PhD
Department of Neurology, Beth Israel Deaconess Medical Center
and Harvard Medical School, Boston, MA, USA

Roberto Amici, MD
Department of Human and General Physiology,
University of Bologna, Bologna, Italy

Claudio L. Bassetti, MD
Department of Neurology, Neurocenter EOC of Southern Switzerland,
Ospedale Civico, Lugano, Switzerland

Christian R. Baumann, MD
Department of Neurology, University Hospital of Zurich,
Zurich, Switzerland

Cecilia Diniz Behn, PhD
Department of Mathematics, University of Michigan,
Ann Arbor, MI, USA

Thomas Birchler, MD
Institute of Experimental Immunology, University of Zurich,
Zurich, Switzerland

Stephanie L. Borgland, PhD
Department of Anesthesiology, Pharmacology and Therapeutics,
The University of British Columbia, Vancouver, BC, Canada

Denis Burdakov, PhD
Department of Pharmacology, University of Cambridge,
Cambridge, UK

Matthew E. Carter
Departments of Psychiatry and Behavioral Sciences,
Stanford University, Stanford, CA, USA

Norbert Dahmen, MD
Department of Psychiatry, University of Mainz,
Mainz, Germany

Yves Dauvilliers, MD
Centre de Référence Nationale Maladie Rare – Narcolepsie et
Hypersomnie Idiopathique, Service de Neurologie,
Hôpital Gui-de-Chauliac, Montpellier, France

Luis de Lecea, PhD
Departments of Psychiatry and Behavioral Sciences,
Stanford University, Stanford, CA, USA

Ralph J. DiLeone, PhD
Department of Psychiatry, Ribicoff Research Facilities,
Connecticut Mental Health Center, Yale University School
of Medicine, New Haven, CT, USA

Alice Engel
Department of Psychiatry, University of Mainz,
Mainz, Germany

Luigi Ferini-Strambi, MD, PhD
Department of Neuroscience, Sleep Disorders Center,
Università Vita-Salute San Raffaele, Milan, Italy

Adriano Fontana, MD
Institute of Experimental Immunology, University of Zurich,
Zurich, Switzerland

Patrice Fort, PhD
Institut Fédératif des Neurosciences de Lyon,
Université de Lyon, Lyon, France

Amanda A.H. Freeman, PhD
Department of Neurology, Emory University School of Medicine,
Atlanta, GA, USA

R. Fronczek, PhD
Department of Neurology, Leiden University Medical Centre,
Leiden, The Netherlands

Heidemarie Gast, MD
Department of Neurology, University Hospital Berne, Inselspital,
Berne, Switzerland

Junko Hara
Biosciences Division, SRI International, Menlo Park, CA, USA

Osamu Hayaishi, MD
Department of Molecular Behavioral Biology,
Osaka Bioscience Institute, Osaka, Japan

Alex Iranzo, MD, PhD
Hospital Clínic and Institut D'Investigació,
Neurology Service and Multidisciplinary Sleep Unit,
Barcelona, Spain

Mike Kalogiannis, DMD, DDS
Department of Physiology, New York Medical College,
Valhalla, NY, USA

Marietta Keckeis
Max Planck Institute of Psychiatry, Munich, Germany

Ramin Khatami, MD
Center of Sleep Medicine, Klinik Barmelweid AG,
Barmelweid, Switzerland

Thomas S. Kilduff, PhD
Biosciences Division, SRI International,
Menlo Park, CA, USA

Kristi A. Kohlmeier, PhD
Department of Physiology, New York Medical College,
Valhalla, NY, USA

G.J. Lammers, MD, PhD
Department of Neurology, Leiden University Medical Center,
Leiden, The Netherlands

Hans-Peter Landolt, PhD
Institute of Pharmacology & Toxicology,
Zürich Center for Integrative Human Physiology (ZIHP),
University of Zürich, Zürich, Switzerland

Michel Lecendreux, MD
Pediatric Sleep Center and Narcoleptic Reference Center,
Hospital Robert Debré, Paris, France

Christopher S. Leonard, PhD
Department of Physiology, New York Medical College, Valhalla, NY, USA

Régis Lopez
Centre de Référence Nationale Maladie Rare – Narcolepsie et
Hypersomnie Idiopathique, Service de Neurologie,
Hôpital Gui-de-Chauliac, Montpellier, France

Pierre-Hervé Luppi, PhD
Institut Fédératif des Neurosciences de Lyon, Université de Lyon,
Lyon, France

Michael Lutter, MD, PhD
Department of Psychiatry, University of Texas
Southwestern Medical Center, Dallas, TX, USA

Johannes Mathis, MD
Sleep Disorders Centre and Department of Neurology, Inselspital, Bern University Hospital, and University of Bern, Switzerland

Geert Mayer, MD
Hephata Klinik, Schwalmstadt, Germany

Emmanuel Mignot, MD, PhD
Stanford Center for Sleep Sciences, Stanford University School of Medicine, Palo Alto, CA, USA

Seiji Nishino, MD, PhD
Stanford Center for Sleep Sciences, Stanford University School of Medicine, Palo Alto, CA, USA

Sebastiaan Overeem, MD, PhD
Department of Neurology, Donders Institute for Neuroscience, Radboud University Nijmegen Medical Centre, Nijmegen, The Netherlands

Christelle Peyron, PhD
Centre de Recherche en Neurosciences de Lyon, Faculté de Médecine Laennec, Université Lyon1, Lyon, France

Thomas Pollmächer, MD
Center of Mental Health, Klinikum Ingolstadt, Ingolstadt, Germany

Rositsa Poryazova, MD
Department of Neurology, University Hospital of Zurich, Zurich, Switzerland

David B. Rye, MD, PhD
Department of Neurology and Program in Sleep, Emory University School of Medicine, Atlanta, GA, USA

Takeshi Sakurai, MD, PhD
Department of Molecular Neuroscience and Integrative Physiology, Kanazawa University, Kanazawa, Japan

Clifford B. Saper, MD, PhD
Department of Neurology, Beth Israel Deaconess Medical Center and Harvard Medical School, Boston, MA, USA

Maysa Sarhan, PhD
Department of Psychiatry, Yale University School of Medicine, New Haven, CT, USA

Thomas E. Scammell, MD
Department of Neurology, Beth Israel Deaconess Medical Center and Harvard Medical School, Boston, MA, USA

Andreas Schuld, MD
Max Planck Institute of Psychiatry, Munich, Germany

Sophie Schwartz, PhD
Department of Neuroscience, University of Geneva, Geneva, Switzerland

Ruth Sharf
Department of Psychiatry, Yale University School of Medicine,
New Haven, CT, USA

Jerome Siegel, PhD
Department of Psychiatry, University of California at Los Angeles,
North Hills, CA, USA

Alessandro Silvani, MD
Department of Human and General Physiology,
University of Bologna, Bologna, Italy

Yoshihiro Urade, PhD
Department of Molecular Behavioral Biology,
Osaka Bioscience Institute, Osaka, Japan

Philipp O. Valko, MD
Department of Neurology, University Hospital of Zurich,
Zurich, Switzerland

Xinmin Xie, MD
AfaSci, Inc., Burlingame, CA, USA

Giovanna Zoccoli, MD
Department of Human and General Physiology,
University of Bologna, Bologna, Italy

Part I

Etiology of Narcolepsy

Etiology and Genetics of Human Narcolepsy

Emmanuel Mignot

Keywords

Human leukocyte antigen • Mutation • Autoimmunity • Genetics • Prevalence • Narcolepsy

Low levels of the neuropeptide hypocretin-1 (hcrt-1, also called orexin-A) are found in the cerebrospinal fluid (CSF) of most people with narcolepsy with cataplexy and in some without cataplexy [1–6]. As a result, in the most recent revision of the International Classification of Sleep Disorders (ICSD), narcolepsy with cataplexy and narcolepsy without cataplexy have been separated [7]. In this chapter, we will discuss the etiology of narcolepsy/hcrt deficiency, as there is a strong suggestion of homogeneity based on the very high association with human leukocyte antigen (HLA) DQB1*0602 and low CSF hcrt-1. References to narcolepsy without cataplexy, defined by sleepiness and a positive multiple sleep latency test (MSLT), will also be made, although the condition likely represents a constellation of problems and pathologies. We will also briefly discuss secondary narcolepsy cases.

E. Mignot (✉)
Stanford Center for Sleep Sciences, Stanford University School of Medicine,
701-B Welch Road, Basement, Room 145,
Palo Alto, CA 94304-5742, USA
e-mail: mignot@stanford.edu

Prevalence Studies

Population-based prevalence studies of narcolepsy–cataplexy have been performed in multiple countries. In Finland, 11,354 individual twins were asked to respond to a questionnaire. Subjects with answers suggestive of narcolepsy were contacted by phone and subjected to clinical interviews and polysomnography [8]. Three subjects with cataplexy and abnormal MSLT results were identified, leading to a prevalence of 0.026% [8]. Other studies have led to similar prevalence (0.013–0.067%) in Great Britain, France, Hong Kong, the Czech Republic, and in the USA [9–11]. A study performed in 1945 in African American Navy recruits also led to 0.02% in this ethnic group for narcolepsy–cataplexy, although this study concluded that narcolepsy was more frequent in this ethnic group because of "natural tendencies" [12]. Narcolepsy–cataplexy may be less frequent in Israel (0.002%) and more frequent in Japan (0.16–0.18%). It is of interest to note that DQB1*0602 is rare in Israel (4–6%), Japan (8%), and Korea (13%), but more common in most Caucasian (25%), Chinese (25%), and

C.R. Baumann et al. (eds.), *Narcolepsy: Pathophysiology, Diagnosis, and Treatment*,
DOI 10.1007/978-1-4419-8390-9_1, © Springer Science+Business Media, LLC 2011

African American populations (38%). Thus, a direct correspondence between the prevalence of narcolepsy and the frequency of DQB1*0602 is not evident.

The prevalence of narcolepsy without cataplexy is largely unknown, as a proper population-based study would require an MSLT of all subjects. In case series, narcolepsy without cataplexy represents 20–50% of cases [13]. Patients without cataplexy are, however, more likely to be underdiagnosed (e.g., narcolepsy plus sleep apnea), undiagnosed (no major complaint), or misdiagnosed (e.g., as depression or sleep apnea) [14]. Some studies have shown that 1–3% of the adult population have self-reported sleepiness and multiple SOREMPs during MSLT [15, 16]. A recent study identified all diagnosed narcoleptic patients in Olmsted County (MN, USA) using the medical records linkage system of the Rochester Epidemiology Project [17]. The study identified 0.036% of the population with narcolepsy–cataplexy and 0.021% with narcolepsy without cataplexy, suggesting a significant prevalence for narcolepsy without cataplexy [17]. In King County (WA, USA), a similarly designed recent study found 0.031% of the population with narcolepsy and only 0.009% without cataplexy (27% of DQB1*0602 positive) [11]. It is likely that registry-based estimations of prevalence of diagnosed cases underestimate, while population-based epidemiological studies that do not exclude other confounding factors overestimate the true population prevalence of narcolepsy without cataplexy, explaining the 300-fold range.

What percent of narcolepsy without cataplexy cases have hcrt deficiency also remains unclear. When all other causes of daytime sleepiness have been excluded, 5–30% of patients with this diagnosis are hcrt deficient [4–6, 18], with a mean of 15% overall and 31% of HLA DQB1*0602-positive subjects in a recent meta-analysis of 162 samples tested in our center [19]. This is also reflected by the % DQB1*0602 positivity in such samples, ranging from 27% (slightly above the 23% population frequency in Caucasians) to 40% in a large multicenter drug trial [13] and other samples [4, 19]. We have conducted systematic CSF hcrt-1 measurement in random samples of healthy individuals (approximately 250 subjects total) and have been unable to detect a single subject with CSF hcrt-1 below 110 pg/ml, the best diagnostic cut-off distinguishing narcolepsy–cataplexy vs. controls. Interestingly, using 162 patients without cataplexy tested in our center since 2000, we found that a cut-off of 200 pg/ml improved sensitivity for this test to 41% in the presence of HLA-DQB1*0602 and 9% in the absence of HLA-DQB1*0602 [19] (Table 1). It is, therefore, possible that some subjects without cataplexy have less pronounced hcrt cell loss, as reflected by intermediate (110–200 pg/ml) or normal CSF hcrt-1 [19, 20]. The notion is also supported by the slightly increased HLA frequency observed in narcolepsy without cataplexy subjects with normal CSF hcrt [21], although it is difficult in this case to exclude that some patients were diagnosed after HLA positivity was established, thus creating a bias.

Twin Studies and Environmental Factors in Narcolepsy

As mentioned above, the only systematic twin study available was performed by Hublin et al. in Finns [8], but the three twin pairs identified with narcolepsy were dizygotic and so are uninformative to establish concordance. Approximately 20 monozygotic twin reports are available in the literature (see [10] for review). Five to seven pairs are concordant for narcolepsy, depending on how strictly concordance is determined clinically [10, 22–24]. Most cases of human narcolepsy, therefore, require the influence of environmental factors for the pathology to develop. This is also substantiated by the fact that onset is not at birth but rather in adolescence, suggesting the existence of triggering factors.

The nature of the environmental factor(s) involved is still uncertain. Frequently cited factors have been head trauma [25–27], sudden change in sleep/wake habits [23, 28], or various infections [29, 30]. These factors may be involved, but these studies all used retrospective designs, limiting the value of any reported difference. A recent study found increased antistreptolysin O antibodies, a

Table 1 Sensitivity (SE) and specificity (SP) of various diagnostic tests for narcolepsy/cataplexy and narcolepsy without cataplexy

		Narcolepsy with cataplexy	Narcolepsy without cataplexy	Idiopathic hypersomnia
HLA	SE	89.3%(822/1291)	45.4% (306/1,291)	17.7% (62/1,291)
	SP	76.0% (1,291)	76.0%(1,291)	76.0%(1,291)
MSLT	SE	87.9% (964/1,095)	Not applicable	Not applicable
	SP	96.9% (1,095)	96.9%(1,095)	96.9% (1,995)
hcrt ≤ 110 pg/ml	SE	83.3% (233/182)	14.8%(162/182)	0.0%(49/182)
	SP	100%	100%	100%
hcrt ≤ 200 pg/ml	SE	85.0%(233/182)	22.8%(162/182)	6.1%(49/182)
	SP	98.9%	98.9%	98.9%

Not applicable because it is part of the clinical definition. Numbers in parentheses indicate the number of patients and corresponding number of controls used to calculate sensitivity (SE). For specificity (SP), only the number of controls is reported. Narcolepsy with atypical or no cataplexy is grouped as narcolepsy without cataplexy as per ICSD-2 [7]. Idiopathic hypersomnia includes both patients as defined by a positive MSLT or with prolonged sleep time independent of MSLT results. A positive MSLT is a mean sleep latency ≤8 min and ≥2SOREMP for narcolepsy without cataplexy, or a mean sleep latency ≤8 min and <2 SOREMP for idiopathic hypersomnia as per ICSD-2. hcrt: CSF hcrt-1. HLA data are from the Stanford Center for Narcolepsy Research database and ethnically matched controls [39]. MSLT data are from the Stanford Center for Narcolepsy Research database, and 1,095 random control measurements are from Mignot et al. [15] and Singh et al. [16]. CSF hcrt-1 measurements are from the Stanford Center for Narcolepsy Research database and 182 CSF samples from healthy control subjects (volunteers or subjects undergoing spinal surgery for back pain)

marker of recent *Streptococcus pyogenes* infection, in recent onset narcolepsy patients but not in cases with long-standing disease [31]. Strikingly, another population-based study looking at exposures in narcoleptic patients found that narcolepsy with cataplexy was more frequent in African Americans and in poorer households. Adjusting for these factors, the condition was also 5.4-fold more common among people reporting a physician-diagnosed strep throat before the age of 21 years [32]. These findings strongly suggest that *Streptococcus pyogenes* is involved in triggering narcolepsy, either as a principal factor (e.g., contributing a peptide involved in molecular mimicry), or as a cofactor (e.g., reactivating a dormant T-cell clone, or increased permeability of the blood–brain barrier).

Familial Aspects of Human Narcolepsy

In the recent past, narcolepsy was considered a familial disorder. In more recent studies, the risk of a first-degree relative to develop narcolepsy–cataplexy has been shown to be only 1–2% (see [10] for review). A larger portion of relatives (4–5%) may have isolated daytime sleepiness, when other causes of daytime sleepiness have been excluded [10]. These figures are important to keep in mind as they are helpful in reassuring patients regarding the risk to their children and relatives. A 1–2% risk is 10- to 40-fold higher than that in the general population, but remains manageable. A 4–5% risk for daytime sleepiness is not negligible, but similar values have been reported for excessive daytime sleepiness in the general population independent of narcolepsy [8, 33, 34].

As mentioned in the introduction, narcolepsy is not a purely genetic disorder since of 20 reported monozygotic twin pairs, only 7 (35%) are concordant. Also somewhat surprisingly, only three of five concordant twin pairs tested (60%) are HLA-DQB1*0602 positive [21]. This suggests that some non-HLA-DQB1*0602 cases may have a particularly high genetic predisposition. In support of this hypothesis, we analyzed data from our own database and report findings in 31 Caucasian multiplex families (two members with narcolepsy and definite cataplexy) (Table 2).

Table 2 HLA-DQB1*0602 in sporadic and familial Caucasian cases

Family structure	Clinical subgroups	DQB1*0602, *n* (%)
Sporadic cases	Narcolepsy–cataplexy	498/574 (87%)‡
	Narcolepsy without cataplexy	83/210 (40%)‡
	Unrelated controls	358/1,416 (25%)
Multiplex cases	Narcolepsy–cataplexy	51/74 (70%)**,‡
	Narcolepsy without cataplexy	21/39 (54%)
	Narcolepsy–cataplexy in families with ≤2 affected members	36/47 (77%)*,‡
	Narcolepsy–cataplexy families with >2 affected members	15/27 (56%)†,**
	Healthy relatives	78/164 (48%)

Sporadic cases are patients with narcolepsy without a family history. Data reported for multiplex cases include multiple cases in each multiplex family. Results are identical when only one proband per family ($n=35$ families) is included, data not shown. Typical cataplexy is defined as muscle weakness triggered at least sometimes by laughing or joking. *$P=0.05$ vs. sporadic cases; **$p<0.001$ vs. sporadic cases; †$p=0.05$ vs. narcolepsy in ≤2 affected members per family; ‡$p<0.001$ vs. unrelated controls or healthy relatives when appropriate

These results are compared to HLA typing data gathered from sporadic Caucasian narcoleptic subjects who lack a family history. HLA DQB1*0602 positivity was indeed significantly lower in familial cases (70%) than that in sporadic narcolepsy cases (87%), especially in families with more than two affected individuals (56%). HLA typing data in these non-DQB1*0602 families did not support the concept of linkage to other HLA subtypes [10]. This strongly suggests that non-HLA genetic factors may be involved in a subset of non-HLA-DQB1*0602 cases.

The pattern of extended HLA haplotype segregation was also examined in DQB1*0602-positive multiplex families. We found that in general, these families had a smaller number of affected members, most often only two cases (Table 2). Interestingly, in some cases, the extended HLA-DQB1*0602 haplotype was not linked with narcolepsy and may have come from different branches of the family. This suggests that in these cases, multiple DQB1*0602 haplotypes (if not all DQB1*0602 alleles in the general population) in the family were equally predisposing to narcolepsy [10]. These results are generally consistent with the notion that in many cases, similar etiologies cause narcolepsy in these families as in sporadic cases. As seen later, the result is also consistent with the observation of hcrt deficiency in these HLA-positive multiplex family cases.

Hypocretin (Orexin) Deficiency and Human Narcolepsy–Cataplexy

As expected from the observation that most cases of human narcolepsy are sporadic and not fully genetic as in dogs or rodent models, extensive screening studies did not identify preprohypocretin, Hcrtr1, and Hcrtr2 mutations in typical human narcolepsy cases [2, 35, 36]. Surprisingly, even familial cases of narcolepsy (some of which were HLA-DQB1*0602 negative) did not have any hcrt mutations, suggesting further heterogeneity in genetic cases [2]. Rather, only a single case with a signal peptide mutation in the preprohypocretin gene was identified. This case had an extremely early onset (6 months), severe narcolepsy–cataplexy, DQB1*0602 negativity, and undetectable Hcrt-1 in CSF [2]. This important observation indicates that hcrt system gene mutations can cause narcolepsy-like symptoms in humans as also seen in dogs and in rodent models.

After identifying mutations in the hypocretin receptor 2 gene underlying canine narcolepsy, we determined that most sporadic, HLA-DQB1*0602-positive, narcoleptic patients with cataplexy have undetectable hcrt-1 levels in the CSF [1–6, 37]. Follow-up neuropathological studies in ten narcoleptic patients also indicated dramatic loss of hcrt-1, hcrt-2, and preprohypocretin mRNA in the brain and hypothalami of narcoleptic patients [2, 3]. As mentioned above, these subjects have no hcrt gene mutations and a peri- or postpubertal disease onset [38], as opposed to the 6-month onset in the subject with a preprohypocretin mutation [2]. Together with the tight HLA association [13, 39], a likely pathophysiological mechanism in most narcolepsy cases could thus involve an autoimmune alteration of hcrt-containing cells in the central nervous system (CNS).

HLA-DR2, Narcolepsy, and Autoimmunity

The observation that narcolepsy is associated with HLA DR2 was first reported in Japan in 1983 [40, 41]. It was quickly confirmed in Europe and North America with 90–100% of all patients with cataplexy carrying the HLA DR2 subtype. Because many HLA (also called major histocompatibility complex or MHC)-associated diseases are known to be autoimmune, this discovery led to the hypothesis that narcolepsy may result from an autoimmune insult within the CNS. The finding of hcrt cell loss in human narcolepsy [2, 3] suggests that the autoimmune process could target this small population of hypothalamic neurons.

Attempts at verifying an autoimmune mediation have generally been disappointing [42]. Human narcolepsy is not associated with any striking pathological changes in the CNS and/or increased frequency in the occurrence of oligoclonal bands in the CSF. Gliosis in human narcolepsy brains has been reported [2, 3, 43] but remains controversial, as are imaging findings suggesting macroscopic hypothalamic changes [44–46]. Similarly, peripheral immunity does not seem to be altered even around disease onset [47, 48]. Recently, antibodies against tribbles homolog 2, a protein partially coexpressed with hcrt, have been reported [49]. These recent results bolster the autoimmune hypothesis, suggesting that tissue destruction may have been difficult to detect because of the small anatomical area involved. Furthermore, tissue destruction may be active only around disease onset.

DQB1*0602 and DQA1*0102 Are the Main HLA Narcolepsy Susceptibility Genes

The observation that narcolepsy is HLA associated but may not be a classical autoimmune disorder led to the hypothesis that HLA DR2 was only a marker for narcolepsy. To explore this hypothesis, investigators have isolated and tested novel markers in the HLA DR region and have studied neighboring HLA genes (e.g., HLA DQ). HLA testing techniques have also changed from serological, antibody-based technology to molecular typing at the DNA level, thus resulting in a further layer of complexity for the clinician. In order to facilitate the review of this nomenclature, the results are summarized in Fig. 1.

At the DR level, DR2 was first split into two subtypes, DR15 and DR16, using serological typing techniques. DR15 was then identified in DR2 narcoleptic subjects. Molecular subtypes of DR15 were further identified at the DRB1 level using DNA sequencing and oligotyping. The DR molecule is a heterodimer constituted of a polymorphic DR beta chain (encoded by the DRB genes) and a monomorphic DR alpha chain (encoded by the DRA gene), so all the diversity at the level of DR can be obtained by molecularly typing the DRB genes. DR15 subtypes recognized at the DNA level were identified as DRB1*1501 to DRB1*1514; note that most subtypes besides 1501–1503 are extremely rare. In Caucasians and Japanese, patients were found to carry the DRB1*1501, whereas most African American narcoleptic patients with the DR2 antigens were observed to have DRB1*1503. A significant number of African American patients were also found to be negative for DR2 and to generally carry the DR11 subtype DRB1*1101 [39].

DQ, another HLA antigen encoded by genes located 85 kb centromeric to DRB1, was also studied. Serologically, all patients were initially found to carry DQ1, a very frequent DQ antigen. DQ1 was then serologically split into DQ5 and DQ6, and all patients were found to have DQ6. At the molecular level, the DQA1 and DQB1 genes encoding the DQ molecule are both polymorphic,

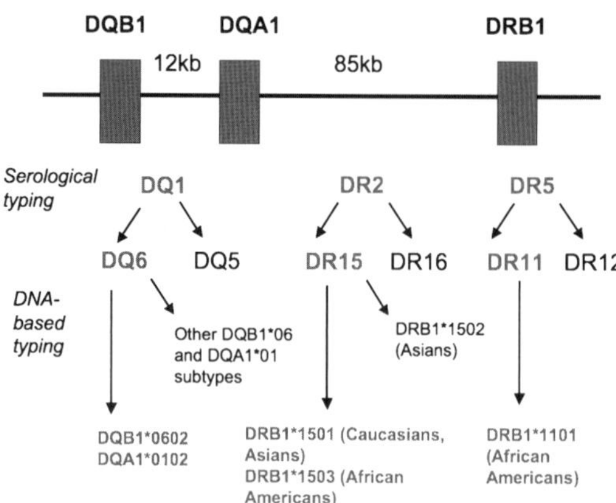

Fig. 1 Human leukocyte antigens (HLAs) DR and HLA DQ alleles typically observed in narcolepsy (*bold gray text*). The DR and DQ genes are located very close to each other on chromosome 6p21. These genes encode heterodimeric HLA proteins composed of an α and a β chain. In the DQ locus, both the DQα and DQβ chains have numerous polymorphic residues and are encoded by two polymorphic genes, DQA1 and DQB1, respectively. Polymorphism at the DR (α and β) level is mostly encoded by the DRB1 gene, so only this locus is depicted in this figure. DQB1*0602, a molecular subtype of the serologically defined DQ1 antigen, is the most specific marker for narcolepsy across all ethnic groups. It is always associated with the DQA1 subtype, DQA1*0102. In Caucasians and Asians, the associated DR2 subtype DRB1*1501 is typically observed with DQB1*0602 (and DQA1*0102) in narcoleptic patients. In African Americans, either DRB1*1503, a DNA-based subtype of DR2, or DRB1*1101, a DNA-based subtype of DR5, is most frequently observed together with DQB1*0602. Other DRB1 alleles (DRB1*0301, DRB1*0806, DRB1*08del, DRB1*12022, and DRB1*1602) have been observed together with DQB1*0602 in much rarer cases

so typing both the genes is theoretically required to identify the biologically active DQ antigen. DQB1 is, however, the most polymorphic of the two genes and usually determines the DQ serological specificity. Molecular subtypes of DQ6 are thus identified at the DQB1 level as DQB1*0601 to DQB1*0618 (most subtypes besides DQB1*0601, 0602, 0603, 0604, and 0609 are very rare). The DQ6 subtype identified in patients with narcolepsy was found to be DQB1*0602 [39].

Studies across ethnic groups have shown that DQB1*0602 is a better marker for narcolepsy. This is especially important in African Americans where many patients are DQB1*0602 positive but DR2 negative [39, 50]. All subjects were also found to be DQA1*0102 positive [50, 51]. Novel DNA markers developed in the HLA DQ region have been tested to further map the narcolepsy susceptibility region within the DQA1-DQB1 interval [52]. This segment was entirely sequenced and shown to contain no new genes. It is also worth noting that in all narcolepsy susceptibility DR-DQ haplotypes identified, both DQA1*0102 and DQB1*0602 are present [50], thus suggesting that the active DQA1*0102/DQB1*0602 heterodimer is necessary for disease predisposition. A number of other DR-DQ haplotypes in the population carry DQA1*0102 without DQB1*0602 and these do not predispose to narcolepsy [51]. Conversely, although DQB1*0602 subjects are almost always DQA1*0102 positive, rare haplotypes with DQB1*0602 but without DQA1*0102 are observed in the control population but not in the narcoleptic patients [51]. Both the DQA1*0102 and the DQB1*0602 alleles might thus be needed for disease predisposition [51].

Recent findings in families and in unrelated cases also suggest that most, if not all, of the DQB1*0602/DQA1*0102 alleles present in the population predispose equally to narcolepsy. One such finding comes from multiplex families where several patients are DQB1*0602 positive. In many cases, DQB1*0602 has been inherited from different branches of the family (e.g., in one case from the father and the other case from the mother) and are thus not “identical by descent” [10]. It was also recently shown that subjects homozygous for DQB1*0602 are at 2–4 times increased risk for developing narcolepsy when compared to DQB1*0602 heterozygous subjects [39, 53]. Finally, risk in DQB1*0602 heterozygous individuals is modulated by the other DQB1 allele. Most notably, risk is increased in DQB1*0602/DQB1*0301 heterozygotes and reduced in DQB1*0602/DQB1*0601 and DQB1*0602/DQB1*0501 heterozygotes [39]. These associations are remarkable as they are incredibly consistent across ethnic groups [54] (Table 3), again illustrating the likely etiological homogeneity of narcolepsy. Overall, the data accumulated to date strongly suggest that the HLA-DQ alleles themselves (most notably DQB1*0602 and

Table 3 Odds ratios (OR) of various heterozygotes across various ethnic groups

		Odds ratio (OR)			
DQA1-DQB1	DQA1-DQB1	USA Caucasians	Japan	Korea	USA Blacks
0102-0602	0102-0602	Reference	Reference	Reference	Reference
0102-0602	06-0301	Rare	1.2	1.0	Rare
0102-0602	05-0301	0.6	0.7	0.9	0.5
0102-0602	03-0301	0.8	Rare	Rare	Rare
0102-0602	0103-0603	0.3	Rare	Rare	0.8
0102-0602	Others	0.17	0.06	0.08	0.09
0102-0602	0101-0501	0.17	0.06	0.08	0.09
0102-0602	0103-0601	Rare	0.07	0.05	Rare

Data compiled from Mignot et al. [39] and Hong et al. [54]

DQA1*0102) rather than an unknown genetic factor in the region predispose to narcolepsy.

Usefulness of HLA Typing in Clinical Practice

The usefulness of HLA typing in clinical practice is limited by several factors. First, the HLA association is very high (>90%) only in narcoleptic patients with clear-cut cataplexy [13]. Clear-cut cataplexy is defined as episodes of muscle weakness triggered by laughter, joking, or anger. Muscle weakness episodes triggered by anger, stress, other negative emotions, or physical or sexual activity may not be cataplexy if joking or laughing is not mentioned as a triggering factor [55]. In patients without cataplexy or with doubtful cataplexy, HLA DQB1*0602 frequency is also increased (40–60%), but many patients are DQB1*0602 negative [13]. Second, a large number of control individuals (Table 1 and 2) have the HLA DQB1*0602 marker without having narcolepsy. Finally, a few rare patients with clear-cut cataplexy do not have the HLA DQB1*0602 marker [56].

Despite these limitations, HLA typing is probably most useful in atypical cases and in narcolepsy without definite cataplexy. A negative result should lead the clinician to be more cautious in excluding other possible causes of daytime sleepiness such as abnormal breathing during sleep or insufficient/disturbed nocturnal sleep. Practically, it is always more useful to request HLA DQ high-resolution typing rather than DR2 or DR15 typing to confirm the diagnosis of narcolepsy.

HLA DQB1*0602-negative subjects with typical and severe cataplexy have been reported, but these subjects are exceptionally rare [56]. An increase in DQB1*0301 has been suggested in these cases but needs further substantiation [39]. Most (but not all) of these patients have normal CSF hcrt-1, suggesting a different pathophysiology [4]. Interestingly, two partially concordant monozygotic twins reported in the literature were DQB1*0602 negative [10]. A number of DQB1*0602-negative families (with normal CSF hcrt-1) have been reported where narcolepsy and cataplexy seem to be transmitted as a highly penetrant autosomal dominant trait, with many patients experiencing narcolepsy–cataplexy, while other family members have sleepiness and documented REM abnormalities during the MSLT [10]. These results emphasize the fact that HLA typing and CSF hcrt-1 results should be interpreted in conjunction with a careful family history.

Genetic Factors Other than HLA

As mentioned above, genetic factors other than HLA-DQ and DR are likely to be involved in narcolepsy predisposition. The increased familial risk in first-degree relatives (10-fold in Japanese, 20–40-fold in Caucasians) cannot be solely explained by the sharing of HLA subtypes, estimated to explain two- to threefold increased risk [10]. Additionally, the existence of HLA-negative families suggests disease heterogeneity and the possible involvement of other genes. Linkage analysis in HLA-DQB1*0602-positive Japanese families has suggested the existence of a susceptibility gene on 4q13-23 [57]. A possible association with a TNF-alpha gene polymorphism (independent of HLA) has been suggested [58–60]. Other results indicate that polymorphisms in the catechol-*O*-methyltransferase and monoamine oxidase genes, key enzymes in the degradation of catecholamine, may also modulate disease severity [61, 62]. Additional studies are needed to identify non-HLA genetic factors.

The search for additional narcolepsy genetic factors is likely to accelerate, thanks to novel techniques allowing for genome-wide association studies (GWAS) of single nucleotide polymorphisms (SNPs). In a recent such study, Miyagawa et al. screened 222 narcoleptic patients and 289 Japanese controls [63] using a 500-K SNP microarray platform, with replication of top hits in 159 narcoleptics and 190 controls, followed by the testing of 424 Koreans, 785 individuals of European descent, and 184 African Americans. rs5770917, a SNP located between CPT1B and CHKB, was associated with narcolepsy in Japanese (rs5770917[C], odds ratio (OR) = 1.79, combined $P=4.4\times10^{-7}$) and other ancestry groups (OR = 1.40, $P=0.02$), although the association was primarily replicated in Koreans and was not significant in Caucasians. Real-time quantitative PCR assays in white blood cells indicated decreased CPT1B and CHKB expression in subjects with the C allele, suggesting that a genetic variant regulating CPT1B or CHKB expression is associated with narcolepsy. Either of these genes is a plausible candidate, as CPT1B regulates beta-oxidation, a pathway involved in regulating theta frequency during REM sleep, and CHKB is an enzyme involved in the metabolism of choline, a precursor of the REM- and wake-regulating neurotransmitter acetylcholine. In addition, the association was recently extended to hypersomnia cases without cataplexy [64] showing no interaction with HLA-DQB1*0602 status, suggesting that the polymorphism may be more involved in modulating REM sleep tendency in general, affecting REM sleep propensity in cases without hcrt deficiency, and increasing disease severity in cases with hcrt deficiency.

More recently, a larger GWAS compared 807 narcolepsy–cataplexy/hcrt deficient Caucasian subjects with 1,074 HLA-DQB1*0602-positive controls using the Affymetrix 500K and 6.0 platforms (Fig. 2). To ensure etiological homogeneity,

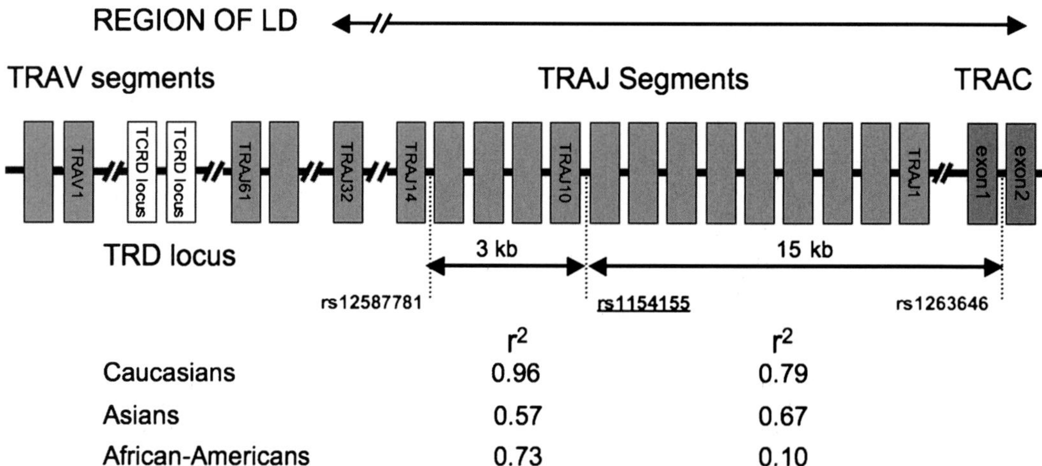

Fig. 2 Genomic region associated with narcolepsy within the TCRA locus. Linkage disequilibrium (LD) is high in Caucasians, but lower in Asians and African Americans. Association studies of these three markers in Asians and African Americans with narcolepsy indicate a strong association with rs1154155 (*underlined*) across all the three ethnic groups

patients were selected as having HLA-DQB1*0602 positivity, clear-cut cataplexy (~98% known to be hcrt deficient if a lumbar puncture was performed), or HLA-DQB1*0602 positivity with documented low CSF hcrt-1. This led to the discovery that a second major locus for narcolepsy is the T-cell receptor-alpha (TCRA) gene, as all three outlier SNPs mapped within this locus [65]. Using trans-ethnic mapping in Caucasians, Asians, and African Americans, the highest association was at rs1154155 within the J segment region (average allelic OR 1.69, genotype ORs 1.94 and 2.55, $p<10^{-21}$ based on 1,830 cases and 2,164 controls after replication [65]). Weaker associations within the T-cell receptor beta (TCRB) were also found, the strongest being at rs17231 (OR 0.59, $P=5.17\times10^{-7}$). This is the first documented involvement of the TCRA locus, the major receptor for HLA-peptide presentation, in any disease. It is still unclear how specific HLA alleles confer susceptibility to more than 100 known HLA-associated disorders, and narcolepsy may provide important new insights into how HLA–TCR interactions that may contribute to organ-specific autoimmune targeting.

CSF Hypocretin-1 as a Diagnostic Tool for Narcolepsy

The observation that CSF hcrt-1 levels are decreased in patients with narcolepsy provides a new test to diagnose this disorder [19]. Using a large sample of patients and controls, we recently conducted a quality receiver operating curve (QROC) analysis to determine the CSF hcrt-1 values most specific and sensitive to diagnose narcolepsy [4]. A cut-off value of 110 pg/mL (30% of mean control values) was the most predictive. In most subjects with cataplexy, hcrt-1 levels were undetectable (<40 pg/mL in most assays), while a few had detectable but very diminished levels. None of the patients with idiopathic hypersomnia, sleep apnea, restless leg syndrome, or insomnia had abnormal hcrt levels. The fact that the distribution is largely bimodal, with either very low or normal CSF hcrt-1, suggests that significant damage to hcrt cells and profound hcrt deficiency are needed before symptoms (especially cataplexy) can occur.

Using the 110 pg/mL cut-off, the measurement was especially predictive in cases with definite cataplexy (99% specificity and 87% sensitivity). Sensitivity and specificity are higher for this test than that for the MSLT. In most case series, approximately 15% of narcoleptic patients with cataplexy or hcrt deficiency do not have a positive MSLT. The trauma associated with the lumbar puncture must be balanced against the risk of mislabeling a patient with narcolepsy and possibly introducing lifelong treatment.

CSF hcrt-1 measurements have limited predictive power in cases with atypical or absent cataplexy. Of note, it is clear from HLA typing studies and especially from CSF hcrt-1 measurement studies that atypical cataplexy has no diagnostic value; thus, it is important to diagnose cataplexy only with typical presentations. Atypical presentations include very infrequent episodes (<1 per several months when untreated), only long-lasting episodes (usually >10 min), and most importantly, episodes that are never triggered by laughing or joking. In fact, episodes that occur mainly with stress or during sex suggest a psychogenic origin. A possible exception is that close to disease onset, episodes in young children can be atypical with tongue trusting or ill-defined weakness [66]. Specificity of the CSF hcrt-1 measurement is still extremely high (99%) but sensitivity is low (16%), with most cases having normal levels [4–6]. This is clearly a dilemma for the clinician as there is more often a need for a definitive diagnosis in these atypical cases.

The MSLT is generally a more useful first step as it will be more determinant for the diagnosis and possible treatment strategies. If a lumbar puncture is still required, HLA typing could be useful as a first step, as almost all cases of narcolepsy with low CSF hcrt-1 levels are also HLA-DQB1*0602 positive [4–6, 67]; only 3 exceptions have been reported (among several hundred patients with low hcrt-1), including one case with very mild and atypical cataplexy [2, 4, 67]. We estimate that the probability of observing low levels in HLA-negative cases without cataplexy is far less than 2%. Most (but not all) cases without any cataplexy and low CSF hcrt-1 levels have been children who develop cataplexy later in the course of the disease [4, 68, 69]. Therefore, we generally recommend CSF hcrt-1 testing in young children with excessive daytime sleepiness but without cataplexy, or in patients in whom there is a suspicion that cataplexy is present but not clearly reported.

While the diagnostic value of low CSF hcrt-1 (<110 pg/mL) has been well established, it is interesting to note that healthy control values have been shown to be above 200 pg/mL [4]. In rare cases of narcolepsy and hypersomnia, we have found hcrt-1 levels between these two values, raising the possibility of partial hcrt deficiency in these cases [4]. Similarly, in a recent meta-analysis, we found that 200 pg/ml is a better diagnostic cut-off for cases without cataplexy, suggesting milder deficiency in at least a portion of patients [19]. Such values should, however, be cautiously interpreted, as in a large series of individuals with various neurological disorders, we found that up to 15% had CSF hcrt-1 values within this intermediate range; most of these patients had severe brain pathology, most notably head trauma, encephalitis, and subarachnoid hemorrhage [37]. Decreased hcrt-1 levels in these cases may reflect damage to the hcrt system, or may be related to changes in CSF flow, as discussed below. Other authors have shown that CSF hcrt-1 increases with locomotor activity and decreases with treatment with serotonin reuptake inhibitors, though never to near-undetectable, narcolepsy-like levels [70]. Therefore, the finding of hcrt-1 levels in this intermediate range should alert the clinician to the possibility of an underlying brain pathology, which may require additional clinical evaluation, laboratory testing, or imaging. Whether genuine hcrt deficiency explains abnormal sleep in these neurological disorders is in need of further investigation.

Secondary Narcolepsy

Another potential application for CSF hcrt-1 testing lies in the complex field of narcolepsy and hypersomnia related to neurological disorders associated with trauma, tumors, infections, degenerative diseases, and genetic disorders. As discussed recently, however, this area is complex and the CSF test is likely to be hard to interpret when conducted in the context of an acutely ill patient, for example, immediately post head trauma, when comatose, or in the midst of an acute encephalitis [19]. Indeed, in many such cases, low to undetectable CSF hcrt can be observed and has been shown to improve with time, suggesting acute suppression of hcrt release (e.g., through the action of inflammatory factors or changes in internal milieu) or CSF flow dynamic changes. In such cases, low CSF hcrt-1 is thus less likely to reflect decreased hcrt cell count genuinely than in chronic, progressive neurological conditions. Conversely, some pathologies are associated with up to 50% decrease in hcrt cell counts, most notably Parkinson's disease [71, 72] and Huntington's chorea [73], yet with CSF hcrt-1 generally in the normal range [19]. A meaningful finding of low CSF hcrt as reflecting severe hcrt cell damage consistent with a narcolepsy impact is more likely in the presence of cataplexy, a more pathognomonic symptom, provided the observation is not coincidental, or in pathologies where there is no acute inflammation or trauma.

Von Economo was the first to suggest that narcolepsy may have its origins in the posterior hypothalamus and in some cases, a secondary etiology [74]. The cause of idiopathic narcolepsy that had been described some 50 years earlier was also speculated to involve this general area. This hypothesis was further refined by many authors who noted that tumors or other lesions located close to the third ventricle were also associated with secondary narcolepsy and hypothesized that the posterior hypothalamic region may be the culprit [25, 75]. A postulated hypothalamic cause of narcolepsy was widespread until the 1940s, but was then ignored during the psychoanalytic boom and replaced by hypotheses that the pathology lay in the brainstem [76].

Reports of lesions such as tumors near the third ventricle (hypothalamus and upper midbrain) in association with narcolepsy have been described for over 80 years; thus, it is clear that these tumors can cause or precipitate narcolepsy [76]. Interestingly, many secondary narcolepsy patients with cataplexy were HLA-DR2 or DQB1*0602 positive, suggesting that the association could be partially coincidental. In these cases, it is also possible that the tumor damaged the blood–brain barrier or induced an inflammatory response in the hypothalamus that could have triggered an autoimmune attack on the hcrt cells. In some of these cases, CSF hcrt was low or intermediate (110–200 pg/ml), although often the data are difficult to interpret as cataplexy was not clearly present and in some cases, measurements were made in very ill patients. Such intermediate levels may nonetheless reflect damage to nearby hcrt projection sites, with sufficient preservation of cell bodies to maintain detectable levels of hcrt-1. Alternatively (or additionally), other regions in the upper midbrain may also contribute to the symptomatology, especially sleepiness, as initially proposed by Von Economo.

The complex area of genetic or congenital disorders associated with primary central hypersomnolence is also of great interest. Specific familial syndromes combining HLA-negative narcolepsy–cataplexy with ataxia and deafness on the one hand [77] and obesity and type 2 diabetes on the other [78] are especially interesting, as both cataplexy and low to undetectable CSF hcrt-1 have been documented. Genetic disorders such as Coffin–Lowry syndrome [79], Moebius syndrome [80], Norrie's disease [80], Prader–Willi syndrome [4], Niemann–Pick disease type C [4, 81, 82], and myotonic dystrophy [83] have been reported to be associated with daytime sleepiness and/or cataplexy-like symptoms. CSF hcrt-1 has been measured in cases of Niemann–Pick disease type C, a condition where oculomotor symptoms are frequent, and intermediate levels have been found in some cases with cataplexy [4, 81, 82]. This condition is remarkable as cataplexy is often triggered by typical emotions (laughing) and responds to anticataplectic treatment. Some diseases are associated with the development of both narcolepsy and sleep-disordered breathing, such as myotonic dystrophy [83] and Prader–Willi syndrome [4]; in such cases, primary hypersomnia should be diagnosed only if excessive daytime sleepiness does not improve after adequate treatment of sleep-disordered breathing. We have explored CSF hcrt-1 levels in such cases and have found that some but not all of these patients have very low CSF hcrt-1 levels (<110 pg/mL), suggesting hcrt deficiency. Similarly, in one case of late-onset congenital hypoventilation syndrome, a disorder with reported hypothalamic abnormalities [84], we found very low CSF hcrt-1 levels in an individual with otherwise unexplained sleepiness and cataplexy-like episodes [4]. This patient also had an excellent response to anticataplectic therapy.

Conclusion and Perspectives

Narcolepsy–cataplexy is most commonly caused by a loss of hcrt-producing cells in the hypothalamus. Low CSF hcrt-1 can be used to diagnose the condition. The disorder is tightly associated with HLA-DQB1*0602, suggesting that the cause of most of these cases may be an autoimmune destruction of these cells. The hcrt system sends strong excitatory projections onto monoaminergic cells. The loss of hcrt is likely to create a cholinergic/monoaminergic imbalance. Abnormally sensitive cholinergic transmission in the forebrain and brainstem, and depressed dopaminergic and histaminergic transmission are believed to underlie abnormal REM sleep and daytime sleepiness in canine narcolepsy, and perhaps in human narcolepsy as well.

Whereas most cases with narcolepsy–cataplexy are caused by a ~95% hcrt cell loss, some cases with cataplexy and most cases without cataplexy have normal CSF hcrt-1 levels. This may reflect either disease heterogeneity or a partial loss of hcrt neurons without significant CSF hcrt-1 decrements. A critical area in need of further inquiry is the role of CSF hcrt-1 testing in predicting therapeutic response to medications already in use to treat the symptoms of narcolepsy. Developing an assay that could measure hcrt-1 in plasma reliably may be possible and would also be extremely useful if low levels are observed in narcolepsy. Measuring hcrt-1 levels may some day facilitate development of therapies that may interrupt or delay the development of disease. Experience suggests that a subset of patients without cataplexy (including those with idiopathic hypersomnia), or with an unclear clinical diagnosis of narcolepsy, may be more resistant to stimulant treatment, leading to management difficulties.

The recent discoveries of a TCRA association, anti-tribbles homolog antibodies, and a possible triggering by streptococcus infections have implications for our understanding of narcolepsy and for the study of autoimmune diseases in general, notably those affecting the brain. Interestingly, narcolepsy onset is not associated with a detectable inflammatory process, as exemplified by the measures of CRP or neuroimaging studies around disease onset. We thus suggest that in narcolepsy, the immune-mediated destruction of hcrt cells is a self-limited process, without significant epitope spreading. The limited epitope spreading may be a feature of the brain as an immune-protected organ, and could explain difficulties in detecting autoimmune abnormalities in narcolepsy and other brain-related autoimmune diseases. The specificity of this process may also explain the TCR genetic association. In this model, rs1154155 or a tightly linked marker within the TCR J region could favor the occurrence of specific V–J TCR pathogenic recombinants, resulting in a coding change within a J segment that alters TCR–peptide binding. The fact that narcolepsy, unlike other autoimmune diseases that have been subjected to GWAS, is associated at the genetic level with abnormalities in the TCR locus could reflect oligoclonal/monoclonal selectivity of the T-cell-mediated immune process in this pathology, in contrast to other autoimmune pathologies where complex polyclonal responses (and possibly multi-allelic TCR associations that would be difficult to detect) would be involved.

How would streptococcal infections be involved in this context? In a first model, destruction of hcrt neurons could occur through molecular mimicry of streptococcus-derived antigens with antigens expressed by hcrt cells, as suggested in rheumatic heart disease. In this case, if both the TCR idiotype and peptide could be identified, narcolepsy could offer the unique possibility of modeling a specific trimolecular HLA–peptide–TCR complex that leads to autoimmunity. TCR–peptide–HLA interactions have indeed been established in specific instances of autoimmunity, for example, after peptide injections in experimental mouse models, but have never been formally identified as causative in any "natural" autoimmune disease.

Alternatively, interactions could be mediated through streptococcus superantigen binding rather than peptide presentation. Superantigens (more often binding TCRB) are produced by various bacteria, notably Streptococcus and Staphylococcus, and are known to stimulate a large number of T cells. A TCR–peptide–HLA complex, however, seems more likely, considering the high target specificity (hcrt cells) in narcolepsy. Interestingly, superantigens are known to be involved in the mediation of toxic shock, and DRB1*1501-DQB1*0602 is protective against septic shock caused by *Streptococcus pyogenes*. Finally, these infections could simply make it permissive for other more specific factors to trigger narcolepsy, for example, by increasing blood–brain barrier permeability, or by reactivating a dormant pathogenic T-cell clone through superantigen activation.

Although still speculative at this juncture, a model is emerging in which narcolepsy is a uniquely selective autoimmune disease. The possibility that unique DQB1*0602-TCR narcolepsy-causing interactions would occur at the level of the brain independently of autoimmunity is, however, still possible, as the TCRB locus has been shown to be expressed in the brain [85]. In this case, however, only J–C unrecombined mRNA products were found without corresponding identification of coding products (TCRA was not studied). Even if expressed, these truncated proteins would not be likely to be functional enough to interact with HLA. This makes this model less likely, especially as neurons (in contrast to microglia and astrocytes) are unable to express HLA class II molecules such as HLA-DQ even under extreme stimulation.

If narcolepsy is found to be a selective autoimmune brain disease, it raises the possibility that other such diseases remain to be discovered. The fact that narcolepsy has a distinct phenotype may have facilitated the discovery that it is caused by loss of the hcrt cells. Indeed, considering brain plasticity, specific phenotypes are rarely the result of discrete lesions except at the cortical level, where specificity is ensured by circuit organization rather than by cellular/molecular diversity. In the sleep field, for example, almost all circuits are redundant, and phenotypic recovery is typical after lesions. It is thus possible that other brain areas are injured by selective autoimmune diseases, but result in nonspecific clinical effects, such as psychiatric symptoms. We hope that in time, narcolepsy will not only inform us about sleep, but also about other autoimmune diseases of the brain.

References

1. Nishino S, Ripley B, Overeem S, Lammers GJ, Mignot E. Hypocretin (orexin) deficiency in human narcolepsy. Lancet. 2000;355(9197):39–40.
2. Peyron C, Faraco J, Rogers W, et al. A mutation in a case of early onset narcolepsy and a generalized absence of hypocretin peptides in human narcoleptic brains. Nat Med. 2000;6(9):991–7.
3. Thannickal TC, Moore RY, Nienhuis R, et al. Reduced number of hypocretin neurons in human narcolepsy. Neuron. 2000;27(3):469–74.
4. Mignot E, Lammers GJ, Ripley B, et al. The role of cerebrospinal fluid hypocretin measurement in the diagnosis of narcolepsy and other hypersomnias. Arch Neurol. 2002;59(10):1553–62.
5. Kanbayashi T, Inoue Y, Chiba S, et al. CSF hypocretin-1 (orexin-A) concentrations in narcolepsy with and without cataplexy and idiopathic hypersomnia. J Sleep Res. 2002;11(1):91–3.
6. Krahn LE, Pankratz VS, Oliver L, Boeve BF, Silber MH. Hypocretin (orexin) levels in cerebrospinal fluid of patients with narcolepsy: relationship to cataplexy and HLA DQB1*0602 status. Sleep. 2002;25(7):733–6.
7. American Academy of sleep Medicine. ICSD-2. International classification of sleep disorders, Diagnostic and coding manual. 2nd ed. Westchester, IL: American Academy of Sleep Medicine; 2005.
8. Hublin C, Kaprio J, Partinen M, Heikkila K, Koskenvuo M. Daytime sleepiness in an adult, Finnish population. J Intern Med. 1996;239(5):417–23.
9. Dauvilliers Y, Billiard M, Montplaisir J. Clinical aspects and pathophysiology of narcolepsy. Clin Neurophysiol. 2003;114(11):2000–17.
10. Mignot E. Genetic and familial aspects of narcolepsy. Neurology. 1998;50(2 Suppl 1):S16–22.
11. Longstreth Jr WT, Ton TG, Koepsell T, Gersuk VH, Hendrickson A, Velde S. Prevalence of narcolepsy in King County, Washington, USA. Sleep Med. 2009; 10(4):422–6.
12. Solomon P. Narcolepsy in Negroes. Dis Nerv Syst. 1945;6:179–83.
13. Mignot E, Hayduk R, Black J, Grumet FC, Guilleminault C. HLA DQB1*0602 is associated with cataplexy in 509 narcoleptic patients. Sleep. 1997; 20(11):1012–20.
14. Chen W, Mignot E. Narcolepsy and hypersomnia of central origin: diagnosis, differential pearls, and management. In: Barkoukis T, Avidan A, editors. Review of sleep medicine. 2nd ed. Philadelphia: Butterworth Heinman Elsevier; 2007. p. 75–94.
15. Mignot E, Lin L, Finn L, et al. Correlates of sleep-onset REM periods during the Multiple Sleep Latency Test in community adults. Brain. 2006; 129(Pt 6):1609–23.
16. Singh M, Drake CL, Roth T. The prevalence of multiple sleep-onset REM periods in a population-based sample. Sleep. 2006;29(7):890–5.
17. Silber MH, Krahn LE, Olson EJ, Pankratz VS. The epidemiology of narcolepsy in Olmsted County, Minnesota: a population-based study. Sleep. 2002; 25(2):197–202.
18. Dauvilliers Y, Baumann CR, Carlander B, et al. CSF hypocretin-1 levels in narcolepsy, Kleine-Levin syndrome, and other hypersomnias and neurological conditions. J Neurol Neurosurg Psychiatry. 2003; 74(12):1667–73.
19. Bourgin P, Zeitzer JM, Mignot E. CSF hypocretin-1 assessment in sleep and neurological disorders. Lancet Neurol. 2008;7(7):649–62.
20. Thannickal TC, Nienhuis R, Siegel JM. Localized loss of hypocretin (orexin) cells in narcolepsy without cataplexy. Sleep. 2009;32(8):993–8.
21. Lin L, Mignot E. Human leukocyte antigen and narcolepsy: present status and relationship with familial history and hypocretin deficiency. In: Bassetti C, Billiard M, Mignot E, editors. Narcolepsy and hypersomnia, vol. 220. New York: Informa Health Care; 2007. p. 411–26.
22. Honda M, Honda Y, Uchida S, Miyazaki S, Tokunaga K. Monozygotic twins incompletely concordant for narcolepsy. Biol Psychiatry. 2001;49(11):943–7.
23. Honda Y. A monozygotic pair completely discordant for narcolepsy, with sleep deprivation as a possible precipitating factor. Sleep Biol Rhythm. 2003;1: 147–9.
24. Dauvilliers Y, Maret S, Bassetti C, et al. A monozygotic twin pair discordant for narcolepsy and CSF hypocretin-1. Neurology. 2004;62(11):2137–8.
25. Gill AW. Idopathic and traumatic narcolepsy. Lancet. 1941;1:474.
26. Guilleminault C, Faull KF, Miles L, van den Hoed J. Posttraumatic excessive daytime sleepiness: a review of 20 patients. Neurology. 1983;33(12):1584–9.

27. Lankford DA, Wellman JJ, O'Hara C. Posttraumatic narcolepsy in mild to moderate closed head injury. Sleep. 1994;17(8 Suppl):S25–8.
28. Orellana C, Villemin E, Tafti M, Carlander B, Besset A, Billiard M. Life events in the year preceding the onset of narcolepsy. Sleep. 1994;17(8 Suppl):S50–3.
29. Roth B. Narcolepsy and hypersomnia. Basel: Karger; 1980.
30. Mueller-Eckhardt G, Meier-Ewart K, Schiefer HG. Is there an infectious origin of narcolepsy? Lancet. 1990;335(8686):424.
31. Aran A, Lin L, Nevsimalova S, et al. Elevated anti-streptococcal antibodies in patients with recent narcolepsy onset. Sleep. 2009;32(8):979–83.
32. Koepsell TD, Longstreth WT, Ton TG. Medical exposures in youth and the frequency of narcolepsy with cataplexy: a population-based case-control study in genetically predisposed people. J Sleep Res. 2010; 19(1 Pt 1):80–6.
33. Young T, Palta M, Dempsey J, Skatrud J, Weber S, Badr S. The occurrence of sleep-disordered breathing among middle-aged adults. N Engl J Med. 1993; 328(17):1230–5.
34. D'Alessandro R, Rinaldi R, Cristina E, Gamberini G, Lugaresi E. Prevalence of excessive daytime sleepiness an open epidemiological problem. Sleep. 1995;18(5):389–91.
35. Hungs M, Lin L, Okun M, Mignot E. Polymorphisms in the vicinity of the hypocretin/orexin are not associated with human narcolepsy. Neurology. 2001;57(10):1893–5.
36. Olafsdottir BR, Rye DB, Scammell TE, Matheson JK, Stefansson K, Gulcher JR. Polymorphisms in hypocretin/orexin pathway genes and narcolepsy. Neurology. 2001;57(10):1896–9.
37. Ripley B, Overeem S, Fujiki N, et al. CSF hypocretin/orexin levels in narcolepsy and other neurological conditions. Neurology. 2001;57(12):2253–8.
38. Okun ML, Lin L, Pelin Z, Hong S, Mignot E. Clinical aspects of narcolepsy-cataplexy across ethnic groups. Sleep. 2002;25(1):27–35.
39. Mignot E, Lin L, Rogers W, et al. Complex HLA-DR and -DQ interactions confer risk of narcolepsy-cataplexy in three ethnic groups. Am J Hum Genet. 2001;68(3):686–99.
40. Honda Y, Asake A, Tanaka Y, Juji T. Discrimination of narcolepsy by using genetic markers and HLA. Sleep Res. 1983;12:254.
41. Juji T, Satake M, Honda Y, Doi Y. HLA antigens in Japanese patients with narcolepsy. All the patients were DR2 positive. Tissue Antigens. 1984;24:316–9.
42. Scammell TE. The frustrating and mostly fruitless search for an autoimmune cause of narcolepsy. Sleep. 2006;29(5):601–2.
43. Thannickal TC, Siegel JM, Nienhuis R, Moore RY. Pattern of hypocretin (orexin) soma and axon loss, and gliosis, in human narcolepsy. Brain Pathol. 2003;13(3):340–51.
44. Kaufmann C, Schuld A, Pollmacher T, Auer DP. Reduced cortical gray matter in narcolepsy: preliminary findings with voxel-based morphometry. Neurology. 2002;58(12):1852–5.
45. Draganski B, Geisler P, Hajak G, et al. Hypothalamic gray matter changes in narcoleptic patients. Nat Med. 2002;8(11):1186–8.
46. Overeem S, Steens SC, Good CD, et al. Voxel-based morphometry in hypocretin-deficient narcolepsy. Sleep. 2003;26(1):44–6.
47. Mignot E, Tafti M, Dement WC, Grumet FC. Narcolepsy and immunity. Adv Neuroimmunol. 1995;5(1):23–37.
48. Carlander B, Eliaou JF, Billiard M. Autoimmune hypothesis in narcolepsy. Neurophysiol Clin. 1993;23(1):15–22.
49. Cvetkovic-Lopes V, Bayer L, Dorsaz S, et al. Elevated Tribbles homolog 2-specific antibody levels in narcolepsy patients. J Clin Invest. 2010;120(3):713–9.
50. Mignot E, Lin X, Arrigoni J, et al. DQB1*0602 and DQA1*0102 (DQ1) are better markers than DR2 for narcolepsy in Caucasian and black Americans. Sleep. 1994;17(8 Suppl):S60–7.
51. Mignot E, Kimura A, Lattermann A, et al. Extensive HLA class II studies in 58 non-DRB1*15 (DR2) narcoleptic patients with cataplexy. Tissue Antigens. 1997;49(4):329–41.
52. Ellis MC, Hetisimer AH, Ruddy DA, et al. HLA class II haplotype and sequence analysis support a role for DQ in narcolepsy. Immunogenetics. 1997;46(5):410–7.
53. Pelin Z, Guilleminault C, Risch N, Grumet FC, Mignot E. HLA-DQB1*0602 homozygosity increases relative risk for narcolepsy but not disease severity in two ethnic groups. US Modafinil in Narcolepsy Multicenter Study Group. Tissue Antigens. 1998;51(1):96–100.
54. Hong SC, Lin L, Lo B, et al. DQB1*0301 and DQB1*0601 modulate narcolepsy susceptibility in Koreans. Hum Immunol. 2007;68(1):59–68.
55. Anic-Labat S, Guilleminault C, Kraemer HC, Meehan J, Arrigoni J, Mignot E. Validation of a cataplexy questionnaire in 983 sleep-disorders patients. Sleep. 1999;22(1):77–87.
56. Mignot E, Lin X, Kalil J, et al. DQB1-0602 (DQw1) is not present in most nonDR2 Caucasian narcoleptics. Sleep. 1992;15(5):415–22.
57. Nakayama J, Miura M, Honda M, Miki T, Honda Y, Arinami T. Linkage of human narcolepsy with HLA association to chromosome 4p13-q21. Genomics. 2000;65(1):84–6.
58. Hohjoh H, Terada N, Nakayama T, et al. Case-control study with narcoleptic patients and healthy controls who, like the patients, possess both HLA-DRB1*1501 and -DQB1*0602. Tissue Antigens. 2001;57(3):230–5.
59. Kato T, Honda M, Kuwata S, et al. Novel polymorphism in the promoter region of the tumor necrosis factor alpha gene: no association with narcolepsy. Am J Med Genet. 1999;88(4):301–4.
60. Wieczorek S, Gencik M, Rujescu D, et al. TNFA promoter polymorphisms and narcolepsy. Tissue Antigens. 2003;61(6):437–42.

61. Dauvilliers Y, Neidhart E, Billiard M, Tafti M. Sexual dimorphism of the catechol-*O*-methyltransferase gene in narcolepsy is associated with response to modafinil. Pharmacogenomics J. 2002;2(1):65–8.
62. Dauvilliers Y, Neidhart E, Lecendreux M, Billiard M, Tafti M. MAO-A and COMT polymorphisms and gene effects in narcolepsy. Mol Psychiatry. 2001;6(4):367–72.
63. Miyagawa T, Kawashima M, Nishida N, et al. Variant between CPT1B and CHKB associated with susceptibility to narcolepsy. Nat Genet. 2008;40(11):1324–8.
64. Miyagawa T, Honda M, Kawashima M, et al. Polymorphism located between CPT1B and CHKB, and HLA-DRB1*1501-DQB1*0602 haplotype confer susceptibility to CNS hypersomnias (essential hypersomnia). PLoS One. 2009;4(4):e5394.
65. Hallmayer J, Faraco J, Lin L, et al. Narcolepsy is strongly associated with the T-cell receptor alpha locus. Nat Genet. 2009;41(6):708–11.
66. Serra L, Montagna P, Mignot E, Lugaresi E, Plazzi G. Cataplexy features in childhood narcolepsy. Mov Disord. 2008;23(6):858–65.
67. Dalal MA, Schuld A, Pollmacher T. Undetectable CSF level of orexin A (hypocretin-1) in a HLA-DR2 negative patient with narcolepsy-cataplexy. J Sleep Res. 2002;11(3):273.
68. Hecht M, Lin L, Kushida CA, et al. Immunosuppression with prednisone in an 8-year-old boy with an acute onset of hypocretin deficiency/narcolepsy. Sleep. 2003;26(7):809–10.
69. Kubota H, Kanbayashi T, Tanabe Y, et al. Decreased cerebrospinal fluid hypocretin-1 levels near the onset of narcolepsy in 2 prepubertal children. Sleep. 2003;26(5):555–7.
70. Salomon RM, Ripley B, Kennedy JS, et al. Diurnal variation of cerebrospinal fluid hypocretin-1 (Orexin-A) levels in control and depressed subjects. Biol Psychiatry. 2003;54(2):96–104.
71. Thannickal TC, Lai YY, Siegel JM. Hypocretin (orexin) cell loss in Parkinson's disease. Brain. 2007;130(Pt 6):1586–95.
72. Fronczek R, Overeem S, Lee SY, et al. Hypocretin (orexin) loss in Parkinson's disease. Brain. 2007;130(Pt 6):1577–85.
73. Aziz A, Fronczek R, Maat-Schieman M, et al. Hypocretin and melanin-concentrating hormone in patients with Huntington disease. Brain Pathol. 2008;18(4):474–83.
74. von Economo C. Encephalitis lethargica ITS sequelae and treatment. London: Oxford University Press; 1931.
75. Daniels LE. Narcolepsy. Medicine. 1934;XIII(1):1–122.
76. Mignot E. A hundred years of narcolepsy research. Arch Ital Biol. 2001;139(3):207–20.
77. Melberg A, Ripley B, Lin L, Hetta J, Mignot E, Nishino S. Hypocretin deficiency in familial symptomatic narcolepsy. Ann Neurol. 2001;49(1):136–7.
78. Hor H, Vicário JL, Pfister C, Lammers GJ, Tafti M, Peraita-Adrados R. Familial narcolepsy, obesity, and type 2 diabetes with hypocretin deficiency. Eur J Med Sci. 2008;138(Supplementum 162):5S.
79. Nelson GB, Hahn JS. Stimulus-induced drop episodes in Coffin-Lowry syndrome. Pediatrics. 2003;111(3):197–202.
80. Parkes JD. Genetic factors in human sleep disorders with special reference to Norrie disease, Prader-Willi syndrome and Moebius syndrome. J Sleep Res. 1999;8 Suppl 1:14–22.
81. Kanbayashi T, Abe M, Fujimoto S, et al. Hypocretin deficiency in niemann-pick type C with cataplexy. Neuropediatrics. 2003;34(1):52–3.
82. Vankova J, Stepanova I, Jech R, et al. Sleep disturbances and hypocretin deficiency in Niemann-Pick disease type C. Sleep. 2003;26(4):427–30.
83. Martinez-Rodriguez JE, Lin L, Iranzo A, et al. Decreased hypocretin-1 (Orexin-A) levels in the cerebrospinal fluid of patients with myotonic dystrophy and excessive daytime sleepiness. Sleep. 2003;26(3):287–90.
84. Katz ES, McGrath S, Marcus CL. Late-onset central hypoventilation with hypothalamic dysfunction: a distinct clinical syndrome. Pediatr Pulmonol. 2000;29(1):62–8.
85. Syken J, Shatz CJ. Expression of T cell receptor beta locus in central nervous system neurons. Proc Natl Acad Sci USA. 2003;100(22):13048–53.

Narcolepsy: Autoimmunity or Secondary to Infection?

Adriano Fontana, Heidemarie Gast, and Thomas Birchler

Keywords

Autoimmunity • Infection • Tumor necrosis factor alpha • Humen leukocyte antigen

Narcolepsy is a sleep disorder that is characterized by excessive daytime sleepiness, cataplexy, hypnagogic hallucination, and sleep paralysis. In the review presented here, we aim at focusing on the immunological aspects of the disease.

Special attention will be given to the link between tumor necrosis factor-α (alpha) (TNF) and major histocompatibility class II (MHC II) antigens and on autoimmunity, autoinflammation, and neuronal cell death. The latter may affect mainly neurons that produce hypocretin peptides (*Hcrt-1* and *Hcrt-2*; also known as orexins A and B) and hypocretin receptors (*Hcrtr-1* and *Hcrtr-2*). The involvement of this neurotransmitter pathway in narcolepsy is extensively discussed within this book.

A mutation in the canine *Hcrtr-2* gene or disruption of the prepro-hypocretin gene in knockout mice causes narcolepsy. In humans, only one patient with early onset of disease in childhood has been reported to have a mutation in the hypocretin genes. However, hypocretin concentration in the cerebrospinal fluid (CSF) of narcolepsy patients is decreased. This finding points to an abnormal expression of the *Hcrt* gene, or release of hypocretin in the disease.

TNF and Its Receptors: Essential in the Pathogenesis of Narcolepsy?

A growing list of evidence supports a role of the cytokine TNF in sleep disorders including narcolepsy, daytime fatigue in infectious and autoimmune diseases, and sleep apnea. TNF is a homotrimeric cytokine that binds to two receptors: TNFRI and TNFRII. TNF is mainly produced by monocytes, macrophages, and dendritic cells. In the context of sleep disorders, it is of note that TNF is also produced in the central nervous system (CNS), mainly by microglial cells and astrocytes.

After synthesis in the endoplasmic reticulum, TNF is trafficked to the cell membrane. There the cytokine either acts as a membrane protein, or is cleaved by the TNF-converting enzyme TACE. Antibodies to TNF (infliximab or adalimumab) and soluble TNFRII (etanercept) prevent the binding of TNF to their membrane receptors and thereby interfere with the physiological function of TNF in the host response to infection and tumor defense. However, anti-TNF strategies have also

A. Fontana (✉)
Institute of Experimental Immunology, University of Zurich, Winterthurerstr. 190, 8057 Zurich, Switzerland
e-mail: adriano.fontana@uzh.ch

C.R. Baumann et al. (eds.), *Narcolepsy: Pathophysiology, Diagnosis, and Treatment*,
DOI 10.1007/978-1-4419-8390-9_2, © Springer Science+Business Media, LLC 2011

become a main line in the treatment of autoimmune diseases, including rheumatoid arthritis.

TNF is a pleiotropic inflammatory cytokine that acts on parenchymal cells in various organs including the CNS. There it modulates the function of microglia, oligodendrocytes, astrocytes, and neurons. In the context of sleep disorders, it is of note that TNF activates the production of glutamate by microglial cells [1, 2]. In cocultures of microglial and cerebellar granule cells, glutamate is the main effector molecule that leads to apoptosis of the neurons.

Detoxification of glutamate is achieved by astrocytes, the function being impaired by glucose oxidase, an enzyme that maintains steady-state levels of hydrogen peroxide. Thus, a possible mechanism whereby TNF alters neuronal function and causes neuronal injury is by its action on the production of excitatory amino acids by macrophages invading the CNS and resident microglia, and the prevention of the function of astrocytes in glutamate detoxification and balancing the electrolyte concentration. When discussing the potential role of TNF in sleep disorders, studies on overexpression of TNF in the CNS are of importance. Transgenic mice which overexpress TNF in the CNS lead to activation of macrophages/microglia, apoptosis of oligodendrocytes, demyelination, and axonal damage, the effect being mediated through the activation of TNFR1 [3].

Besides its role in inflammation and its response to infection, much interest has been gained in the role of TNF in sleep regulation (see review [4]). TNF has been identified to promote non-rapid eye movement sleep (NREM) and EEG delta (1/2–4 Hz) power, an index of sleep intensity. The same effect is seen with IL-1 when injected systemically or intracerebroventricularly. On the contrary, antibodies to these cytokines or their soluble receptor, as well as the IL-1 receptor antagonist (IL-1RA), interfere with spontaneous sleep [5].

These findings are of enormous importance for the understanding of dysfunction of sleep regulation in inflammatory and autoimmune disorders and in stress, because these conditions are associated with increased expression of TNF and IL-1. However, these cytokines are also claimed to play a role in normal, physiological sleep. The expression of both cytokines shows a circadian rhythm, with the highest levels in the brain and blood correlating with high sleep propensity. In rats, TNF bioactivity increases up to five times at 6 p.m. compared to that at 12 a.m. and 12 p.m. [6]. Failure of TNF signaling in TNFRI gene knockout mice and in TNFRI and RII double-deficient mice is associated with less NREM sleep [7, 8]. These findings in TNFRI knockout mice were not in agreement with another study [9]. The signals that lead to circadian regulation of TNF expression are not known. In this regard, it is interesting that ATP released at synapses during neurotransmission has been shown to act on purine Pz receptors and thereby promote the release of TNF from microglia [10]. Recently, a 72-h REM sleep deprivation in rats was associated with increased plasma levels of IL-1β (beta), TNF, and IL-17 [11]. Taken collectively, these data support a role of TNF and IL-1 in sleep regulation.

Studies on single nucleotide polymorphisms (SNPs) in cytokine genes in patients with autoimmune diseases, mainly type 1 diabetes and rheumatoid arthritis, point to an SNP in the TNF gene promoter position; that is, 308 G to A is found to be associated with increased TNF serum concentrations and high in vitro TNF transcription and expression [12, 13].

Since narcolepsy is strongly associated with the human leukocyte antigen (HLA) DQB1*0602, an autoimmune pathogenesis has been suggested. Together with the numerous reports on sleep regulation by cytokines, considerable interest has been paid to cytokines in narcoleptic patients. IL-1β (beta), IL-1RA, IL-2, TNF, and LTα (alpha) in plasma and in mitogen-stimulated monocytes and lymphocytes in narcoleptic patients were not found to differ from that in HLA-DR2-matched controls [14]. Only IL-6 was increased in LPS-activated monocytes. However, increased TNF and IL-6 serum levels compared to that in age- and gender-matched controls have been detected in a later study from Okun et al., who found TNF in patients' sera to be 13.9±1.39 pg/ml (control: 8.2±0.45 pg/ml) and IL-6 to be 6.7±1.45 pg/ml (control: 0.49±0.09 pg/ml) [15]. In the later study, stimulatory drugs were associated with lower TNF levels. Thus, stimulatory drugs may influence cytokine levels in narcoleptics. As outlined above, genetic polymorphism in the *Tnf* promoter may also influence TNF serum concentrations. The T-cell allele of the C-857T polymorphism was strongly associated in the subgroup of DRB1*15/16 (HLA-DR2 type)-negative patients [16]. This is interesting because elevations of both TNF and IL-6 have been reported in subjects who are sleep restricted, either experimentally or naturally by insomnia [17]. An increase in TNF is also seen in patients with sleep apnea, the TNF concentrations being normalized by continuous positive pressure [18]. In an acute animal model of obstructive sleep apnea, oxygen desaturation and respiratory effort were followed by an increase in TNF and IL-1 [19]. In a recent study on sleep apneics, TNFRI but not TNF serum concentrations correlated with different forms of arousal, namely, snore and spontaneous arousal, and periodic limb movement arousal (TNFRII was not assessed) [20].

Our laboratory has shown that subcutaneous infusion of TNF impairs locomotor activity of mice and lowers the expression of clock genes in the liver. TNF acts on the clock genes that are regulated by E-boxes in their promoters, namely, the PAR bZip clock-controlled genes *Dbp*, *Tef*, and *Hlf* and the period genes *Per1*, *Per2*, and *Per3*, but neither *Clock* nor *Bmal1* which do not have E-boxes in their regulatory DNA sequences [21]. Since clock genes are central in the sleep–wake cycle and mapped to mouse chromosome 5 within a region syntenic to the human chromosome 4g12, a region close to the narcolepsy susceptibility locus 4p/3-q21 identified recently, polymorphisms have been analyzed in the clock gene [22]. However, no differences in allelic and genotypic frequencies of two clock polymorphisms have been observed in narcoleptics compared to controls. One of the clock gene polymorphisms has been found to be associated with sleep genotypes [23].

In a well-controlled recent study, new information has been obtained by Himmerich et al. [24]. Whereas TNF was not increased, narcoleptic patients have higher TNFRII (but not sTNFRI) compared to controls. This may be explained by genetic polymorphisms. Positive correlations have been identified of TNF (−857T) and TNFRII (−196T) combination with narcolepsy (and DRB1*1501 and TNF (−857T) [25, 26]). Further studies should address the relationship of sTNFRII and HLA-DR2. Taken collectively, there is clear evidence that TNF and TNFR serum concentrations are influenced by many variables including obesity, stimulatory drugs, stress, oxygen saturation, *Tnf* gene polymorphism, and the HLA-DQB1*0602 allele. Thus, large population studies are required to control for these multivariable influences.

Anti-self T Lymphocytes and Activation of Macrophages/Microglial Cells: Key Factors in Narcolepsy?

Narcolepsy is genetically characterized by strong linkage to the HLA complex. More than 90% of the patients have the HLA-DR2 haplotype DQB1*0602. These data may point to an autoimmune mechanism. From a clinical point of view, a given autoimmune disease is often associated with other autoimmune diseases in the affected individual or in the family of the patient. However, there is no strong evidence for such a clustering of autoimmune diseases in narcolepsy. Unlike other autoimmune diseases including systemic lupus erythematosus, rheumatoid arthritis, or Sjögren syndrome, autoantibodies such as antinuclear antibodies, rheumatoid factor, and antibodies to nDNS, SS-A, Sm, and histone are not increased in narcolepsy [27]. Patient's markers indicating inflammation, such as increased blood sedimentation and C-reactive protein, are not found to be abnormal. Thus, these clinical data do not support a role for autoimmunity in narcolepsy.

Hallmarks of the T-cell system in autoimmune diseases are the demonstration of T cells sensitized to self-antigens, dysregulated CD4 effector T cells, e.g., CD4-TH17 cells, low regulatory T cells, and inflammation at sites of the autoimmune attack in distinct organs. The inflammation is characterized by local accumulation of CD4+ T cells and proinflammatory macrophages, with increased expression of MHC II and cytokines. None of these characteristic features of T-cell autoimmunity have been assessed in detail in narcolepsy.

In a highly interesting new study on T-cell receptor-alpha (TCRα), or -beta (β) subtypes, 807 narcolepsy patients positive for HLA-DQB1*0602 and hypocretin deficient, and 1,074 controls were selected for a genome-wide association study. The data identified an association between narcolepsy and polymorphisms in the T-cell receptor alpha [TRAα] locus. TRAα is expressed by T cells and interacts with HLA class I CD8 T cells and on CD4 cells with the HLA class II including the DQα β heterodimer denoted DQ 0602, which encodes for DQB1*0602 and DQA1*0102 alleles. Somatic cell recombination in the TRAα and TRBα loci leads to a diverse repertoire of distinct TCRα β idiotype-bearing T cells. Since narcolepsy is almost exclusively associated with a single HLA allele – DQB1*0602, the authors suggest that the polymorphism detected could influence VJ2 recombinations that bind DQ0602 and mediate autoimmunity to Hcrt neurons [28].

Since in autoimmune diseases histological examination reveals cellular infiltrates of lymphocytes, plasma cells, and macrophages in areas of tissue destruction, it is of importance whether this is seen in narcolepsy. Where should one observe? As outlined in this book, special attention should be given to Hcrt-1- and Hcrt-2-producing neurons in the hypothalamus and in the Hcrt projection fields. Since the disease has a good life expectancy, histological workup of brains is hardly available. In a patient with an *Hcrt* mutation and early-onset narcolepsy (age 6 months) and a long follow-up over many years, histological analysis did not show "obvious lesions" or gliosis in the perifornical area. Most importantly, immune histochemical staining of HLA-DR disclosed normally distributed resting microglia in both white and gray matter of two narcoleptic subjects. None of the cases were associated with activated, ameboid microglia. This is remarkable since upregulation of HLA-DR and transition from resting ramified microglia are hallmarks of immune-mediated inflammation in the CNS. In the context of the aforementioned discussion on dysregulated TNF expression, it is interesting that in situ hybridization of TNF did not produce significant signal in control and narcoleptic tissue [29]. Taken collectively, these findings do not support the idea of a T-cell/macrophage-mediated destruction of the reduced hypocretin neurons detected in postmortem studies. Histological analysis and HLA-DR staining in four adult and three young narcoleptic Dobermans did not reveal lymphocyte infiltration or inflammation in the CNS in canine narcolepsy. However, young dogs aged 3 and 8 months have a concomitant disease-onset diffuse increase in MHC II antigens on microglia. MHC II expression in older narcoleptic dogs did not differ from that in controls. With age, a general increase in MHC II molecules can be observed in microglia [30]. The observation in dogs may indicate that MHC II pathology is seen early at the time of loss of Hcrt neurons.

Are Anti-neuronal Antibodies Involved?

Loss of hypocretin neurotransmission may be due to either impairment of production and/or secretion of Hcrt by neurons, or loss of Hcrt neurons. Several studies have addressed the hypothesis that autoantibodies may lead to alterations in the Hcrt system. No increased IgG index or oligoclonal bands were detected in the CSF of 15 patients with narcolepsy. These data speak against an intrathecal synthesis of autoantibodies by local plasma cells [31]. Recent studies have failed to detect antibodies against orexin or orexin receptors [32, 33], or also against hypothalamic neurons [34]. Antibodies to hypothalamic neurons were claimed in only one of nine patients, with the antibody epitope not being characterized [35]. A new potential autoimmune target has been identified recently: Insulin-like growth factor-binding protein-3 (IGFBP3), which is expressed in hypocretin neurons and downregulated in narcolepsy. However, no anti-IGFBP3 antibodies were detected in human sera or CSF of patients. IGFBP3 concentration in CSF was not decreased [36].

Ex vivo mouse colonic migrating motor complex (CMMC) preparations are inhibited by IgG from narcoleptic patients. In this system, contractions migrating from proximal to the distal colon at 3- to 5-min intervals are recorded. The frequency of these contractions is severely interrupted or contractions even abolished by patient's IgG, but not by IgG from controls. The effect is not due to alterations of smooth muscle functions. Abrogation of contractions is followed by cholinergic myogenic hyperactivity. In the light of the effects found on the enteric nervous system, the authors wonder why there are no reports of abnormal colon movements in patients. The epitopes of the antibodies detected in the CMMC assay remain unclear and are unlikely to be Hcrt because orexin seems not to be present in the murine gut [37].

Even the detection of anti-neuronal antibodies in narcolepsy may not be indicative of immune-mediated damage of neurons. In paraneoplastic syndromes, autoantibodies are thought to be an epiphenomenon or footprint for autoimmunity, but are not directly involved in damage of the CNS. There are exceptions, such as autoantibodies directed against voltage-gated potassium or calcium channels located at nerve terminals, which may lead to paraneoplastic cerebellar degeneration and limbic encephalitis, respectively (for review, see [38]).

Conclusion and Hypothesis

The strong association of narcolepsy with HLA-DR2 and HLA-DQB1*0602 has provided strong interest in the hypothesis that narcolepsy is an autoimmune disease. However, several points about the aforementioned observations deserve emphasis. (1) Classical autoimmune diseases such as systemic lupus erythematosus, rheumatoid arthritis, or myasthenia gravis are not reported to be increased in narcoleptic patients and their families; (2) common autoantibodies against nuclear proteins are uncommon in sera of narcoleptic patients; (3) intrathecal synthesis and oligoclonal bands are not seen in the CSF; (4) there is no evidence of antibodies to hippocampal neurons, orexin, and orexin receptors in the disease; and (5) accumulation of T and B lymphocytes in the CNS, influx of monocytes from the blood, and activation of microglia in the tissue are – at least at late time points of the disease – not seen (Table 1).

Taking this into consideration, autoimmunity in narcolepsy becomes questionable. Difficulties in arriving at a suitable answer have been hampered by the fact that experiments demonstrating

Table 1 Autoimmunity in narcolepsy?

Pro	Contra
Association with HLA-DR2/DQB1*0602 Polymorphism in the T-cell receptor-α (alpha) locus Dysregulation of TNF/TNF receptor system	No association with autoimmune diseases in patients with narcolepsy in their family (e.g., SLE, rheumatoid arthritis, and thyroiditis)
	No autoantibodies to nuclear proteins (antinuclear antibodies, anti-nDNS) No increase in IgG index and no oligoclonal bands in CSF
Autoantibodies to Trib2	No narcolepsy-specific antibodies (anti-neuronal, anti-Hcrt, and anti-Hcrtr)

T-cell immunity to hippocampal neurons or orexin require stringent conditions, including availability of T cells from blood and CSF at the onset of disease. Further work should examine whether there is (1) a restricted usage of T-cell receptor genes, (2) whether T-cell activation and neurotoxic effects of T cells from patients on cocultures exist with immortalized hippocampal neurons transfected with HLA-DR2 genes, and (3) whether signs of narcolepsy develop in SCID mice that express human HLA-DR2 genes and are injected with reactivated CD4+ T cells from narcoleptic patients.

Very recent studies, however, provide new evidence that autoimmunity, superantigen-mediated T-cell activation, and non-T-cell-mediated activation by MHC II signaling could be involved in narcolepsy. New data identified an association between narcolepsy and polymorphisms in the TCR locus [39]. Because narcolepsy is almost exclusively associated with a single HLA allele – DQB1*0602 – the authors of this study hypothesized that the TCR polymorphism could contribute to autoimmunity directed against hypocretin neurons. Furthermore, in another recent study in narcolepsy patients, autoantibodies to Tribbles homolog 2 (Trib2), which is expressed by hypocretin neurons and by many other neurons, have been detected in a subset of patients [40]. The authors suggested that Trib2 is an autoantigen in patients with narcolepsy. In summary, whereas HLA-DQB1*0602 might select for recognition of self-antigens – and thereby lead to autoimmunity – the polymorphism of the TCR (alpha) gene might be crucial in superantigen-mediated T-cell activation. For the detection/confirmation of anti-CNS antibodies, future studies may want to concentrate on patients with very recent onset of disease.

However, besides the hypothesis of autoimmune mechanisms, other explanations for the association of HLA-DR2 with narcolepsy have to be taken into consideration. At disease onset, microglia of narcoleptic patients are reported to overexpress MHC II antigens. TNF acts synergistically with interferon gamma (IFN-γ) to upregulate MHC II expression on microglia. In T-cell-mediated diseases of the CNS, such as multiple sclerosis or experimental autoimmune encephalomyelitis, the function of MHC II is primarily that of antigen presentation by microglia and macrophages. However, in diseases such as Huntington's disease and Parkinson's disease or brain trauma, the aforementioned types of cells express MHC II, but no evidence for T-cell involvement has been observed. It has been suggested that an alternative role for MHC II is in signal transduction, which leads to activation, differentiation, and production of proinflammatory cytokines in vitro. Cuprizone-induced oligodendrocyte dysfunction with T-cell independent demyelination pathology is much less pronounced in MHC II $I\text{-}A_{\beta\,(beta)}^{-/-}$ mice or in mice with a truncated $I\text{-}A_{\beta\,(beta)}$ which lacks a cytoplasmic domain. This phenotype was associated with limited microglia/macrophage activation and reduced production of TNF, IL-1β (beta), and nitric oxide, molecules known to exert T-cell-independent toxic effects on oligodendrocytes. It is not clear yet how MHC II is being activated in the absence of T-cell function [41, 42]. These findings may be of relevance in narcolepsy. As a hypothesis, infectious pathogens may have a tropism to hypothalamic orexin neurons and, therefore, cause these neurons to activate microglia to increase signaling by their MHC II molecules. Necrotic neurons have been shown to activate microglia to upregulate MHC II, costimulatory molecules (CD40 and CD24), β-2 integrin, CD11b, iNOS, and cytokines including TNF [43]. In the MyD88-dependent step, activated microglia induce neurotoxicity through upregulation of glutaminase, an enzyme that produces glutamate, which is an NMDA receptor agonist [1, 2, 43]. As a consequence, MHC II may trigger the production of toxic molecules which destroy orexin neurons (Fig. 1a). Alternatively, infectious pathogens expressed in the hypothalamus may act as superantigens and bridge TCR on T cells with MHC II on microglia, which become activated to secrete neurotoxic molecules, e.g., via glutamate production (see [2]). The narcolepsy-associated polymorphism of the TCRα locus and DQB1*0602 on microglia may be required for the initiation of the superantigen-dependent process (Fig. 1b).

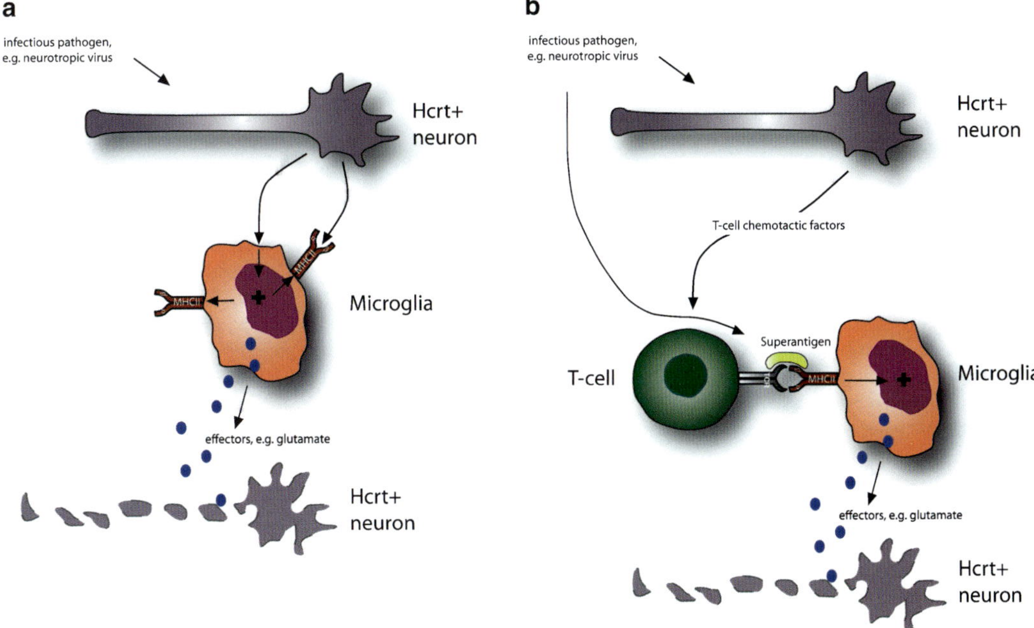

Fig. 1 Infectious pathogens lead to microglia-mediated toxicity of Hcrt-positive neurons. (**a**) Microglia-mediated T-cell-independent neurotoxicity. (**b**) Superantigen-induced microglia-mediated neurotoxicity

References

1. Piani D, Spranger M, Frei K, Schaffner A, Fontana A. Macrophage-induced cytotoxicity of *N*-methyl-D-aspartate receptor positive neurons involves excitatory amino acids rather than reactive oxygen intermediates and cytokines. Eur J Immunol. 1992;22(9):2429–36.
2. Piani D, Frei K, Do KQ, Cuenod M, Fontana A. Murine brain macrophages induced NMDA receptor mediated neurotoxicity in vitro by secreting glutamate. Neurosci Lett. 1991;133(2):159–62.
3. Probert L, Eugster HP, Akassoglou K, et al. TNFR1 signalling is critical for the development of demyelination and the limitation of T-cell responses during immune-mediated CNS disease. Brain. 2000;123 (Pt 10):2005–19.
4. Krueger JM. The role of cytokines in sleep regulation. Curr Pharm Des. 2008;14(32):3408–16.
5. Takahashi S, Kapas L, Fang J, Seyer JM, Wang Y, Krueger JM. An interleukin-1 receptor fragment inhibits spontaneous sleep and muramyl dipeptide-induced sleep in rabbits. Am J Physiol. 1996;271(1 Pt 2): R101–8.
6. Floyd RA, Krueger JM. Diurnal variation of TNF alpha in the rat brain. NeuroReport. 1997;8(4):915–8.
7. Deboer T, Fontana A, Tobler I. Tumor necrosis factor (TNF) ligand and TNF receptor deficiency affects sleep and the sleep EEG. J Neurophysiol. 2002;88(2): 839–46.
8. Kapas L, Bohnet SG, Traynor TR, et al. Spontaneous and influenza virus-induced sleep are altered in TNF-alpha double-receptor deficient mice. J Appl Physiol. 2008;105(4):1187–98.
9. Fang J, Wang Y, Krueger JM. Mice lacking the TNF 55 kDa receptor fail to sleep more after TNF alpha treatment. J Neurosci. 1997;17(15):5949–55.
10. Hide I, Tanaka M, Inoue A, et al. Extracellular ATP triggers tumor necrosis factor-alpha release from rat microglia. J Neurochem. 2000;75(3):965–72.
11. Yehuda S, Sredni B, Carasso RL, Kenigsbuch-Sredni D. REM sleep deprivation in rats results in inflammation and interleukin-17 elevation. J Interferon Cytokine Res. 2009;29(7):393–8.
12. Kumar R, Goswami R, Agarwal S, Israni N, Singh SK, Rani R. Association and interaction of the TNF-alpha gene with other pro- and anti-inflammatory cytokine genes and HLA genes in patients with type 1 diabetes from North India. Tissue Antigens. 2007; 69(6):557–67.
13. Khanna D, Wu H, Park G, et al. Association of tumor necrosis factor alpha polymorphism, but not the shared epitope, with increased radiographic progression in a seropositive rheumatoid arthritis inception cohort. Arthritis Rheum. 2006;54(4):1105–16.
14. Hinze-Selch D, Wetter TC, Zhang Y, et al. In vivo and in vitro immune variables in patients with narcolepsy and HLA-DR2 matched controls. Neurology. 1998; 50(4):1149–52.

15. Okun ML, Giese S, Lin L, Einen M, Mignot E, Coussons-Read ME. Exploring the cytokine and endocrine involvement in narcolepsy. Brain Behav Immun. 2004;18(4):326–32.
16. Wieczorek S, Gencik M, Rujescu D, et al. TNFA promoter polymorphisms and narcolepsy. Tissue Antigens. 2003;61(6):437–42.
17. Vgontzas AN, Papanicolaou DA, Bixler EO, Kales A, Tyson K, Chrousos GP. Elevation of plasma cytokines in disorders of excessive daytime sleepiness: role of sleep disturbance and obesity. J Clin Endocrinol Metab. 1997;82(5):1313–6.
18. Steiropoulos P, Kotsianidis I, Nena E, et al. Long-term effect of continuous positive airway pressure therapy on inflammation markers of patients with obstructive sleep apnea syndrome. Sleep. 2009;32(4):537–43.
19. Nacher M, Farre R, Montserrat JM, et al. Biological consequences of oxygen desaturation and respiratory effort in an acute animal model of obstructive sleep apnea (OSA). Sleep Med. 2009;10(8):892–7.
20. Yue HJ, Mills PJ, Ancoli-Israel S, Loredo JS, Ziegler MG, Dimsdale JE. The roles of TNF-alpha and the soluble TNF receptor I on sleep architecture in OSA. Sleep Breath. 2009;13(3):263–9.
21. Cavadini G, Petrzilka S, Kohler P, et al. TNF-alpha suppresses the expression of clock genes by interfering with E-box-mediated transcription. Proc Natl Acad Sci USA. 2007;104(31):12843–8.
22. Nakayama J, Miura M, Honda M, Miki T, Honda Y, Arinami T. Linkage of human narcolepsy with HLA association to chromosome 4p13-q21. Genomics. 2000;65(1):84–6.
23. Moreira F, Pedrazzoli M, Dos Santos Coelho FM, et al. Clock gene polymorphisms and narcolepsy in positive and negative HLA-DQB1*0602 patients. Brain Res Mol Brain Res. 2005;140(1–2):150–4.
24. Himmerich H, Beitinger PA, Fulda S, et al. Plasma levels of tumor necrosis factor alpha and soluble tumor necrosis factor receptors in patients with narcolepsy. Arch Intern Med. 2006;166(16):1739–43.
25. Hohjoh H, Terada N, Nakayama T, et al. Case-control study with narcoleptic patients and healthy controls who, like the patients, possess both HLA-DRB1*1501 and -DQB1*0602. Tissue Antigens. 2001;57(3):230–5.
26. Hohjoh H, Terada N, Miki T, Honda Y, Tokunaga K. Haplotype analyses with the human leucocyte antigen and tumour necrosis factor-alpha genes in narcolepsy families. Psychiatry Clin Neurosci. 2001;55(1):37–9.
27. Rubin RL, Hajdukovich RM, Mitler MM. HLA-DR2 association with excessive somnolence in narcolepsy does not generalize to sleep apnea and is not accompanied by systemic autoimmune abnormalities. Clin Immunol Immunopathol. 1988;49(1):149–58.
28. Hallmayer J, Faraco J, Lin L, et al. Narcolepsy is strongly associated with the T-cell receptor alpha locus. Nat Genet. 2009;41(6):708–11.
29. Peyron C, Faraco J, Rogers W, et al. A mutation in a case of early onset narcolepsy and a generalized absence of hypocretin peptides in human narcoleptic brains. Nat Med. 2000;6(9):991–7.
30. Tafti M, Nishino S, Aldrich MS, Liao W, Dement WC, Mignot E. Major histocompatibility class II molecules in the CNS: increased microglial expression at the onset of narcolepsy in canine model. J Neurosci. 1996;16(15):4588–95.
31. Fredrikson S, Carlander B, Billiard M, Link H. CSF immune variables in patients with narcolepsy. Acta Neurol Scand. 1990;81(3):253–4.
32. Black III JL, Silber MH, Krahn LE, et al. Analysis of hypocretin (orexin) antibodies in patients with narcolepsy. Sleep. 2005;28(4):427–31.
33. Tanaka S, Honda Y, Inoue Y, Honda M. Detection of autoantibodies against hypocretin, hcrtrl, and hcrtr2 in narcolepsy: anti-Hcrt system antibody in narcolepsy. Sleep. 2006;29(5):633–8.
34. Overeem S, Verschuuren JJ, Fronczek R, et al. Immunohistochemical screening for autoantibodies against lateral hypothalamic neurons in human narcolepsy. J Neuroimmunol. 2006;174(1–2):187–91.
35. Knudsen S, Mikkelsen JD, Jennum P. Antibodies in narcolepsy-cataplexy patient serum bind to rat hypocretin neurons. NeuroReport. 2007;18(1):77–9.
36. Honda M, Eriksson KS, Zhang S, et al. IGFBP3 colocalizes with and regulates hypocretin (orexin). PLoS One. 2009;4(1):e4254.
37. Baumann CR, Clark EL, Pedersen NP, Hecht JL, Scammell TE. Do enteric neurons make hypocretin? Regul Pept. 2008;147(1–3):1–3.
38. Dropcho EJ. Update on paraneoplastic syndromes. Curr Opin Neurol. 2005;18(3):331–6.
39. Hallmayer J, Faraco J, Lin L, Hesselson S, Winkelmann J, Kawashima M, et al. Narcolepsy is strongly associated with the T-cell receptor alpha locus. Nat Genet. 2009;41:708–11.
40. Cvetkovic-Lopes V, Bayer L, Dorsaz S, Maret S, Pradervand S, Dauvilliers Y, et al. Elevated Tribbles homolog 2-specific antibody levels in narcolepsy patients. J Clin Invest. 2010;120:713–9.
41. Matsushima GK, Taniike M, Glimcher LH, et al. Absence of MHC class II molecules reduces CNS demyelination, microglial/macrophage infiltration, and twitching in murine globoid cell leukodystrophy. Cell. 1994;78(4):645–56.
42. Hiremath MM, Chen VS, Suzuki K, Ting JP, Matsushima GK. MHC class II exacerbates demyelination in vivo independently of T cells. J Neuroimmunol. 2008;203(1):23–32.
43. Pais TF, Figueiredo C, Peixoto R, Braz MH, Chatterjee S. Necrotic neurons enhance microglial neurotoxicity through induction of glutaminase by a MyD88-dependent pathway. J Neuroinflammation. 2008;5:43.

Is Narcolepsy a Neurodegenerative Disorder?

Christelle Peyron

Keywords
Neurodegeneration • Hypocretin • Orexin • Dynorphin • Narcolepsy • Protein aggregates

Introduction

Human narcolepsy is a complex disorder characterized by a tetrad of symptoms: excessive daytime sleepiness, cataplexy, hypnagogic hallucinations, and sleep paralysis. Patients rarely have all four symptoms, and two main types of narcolepsy have been described by the presence or absence of cataplexy. The etiology of these different types is likely to be different. Narcolepsy with cataplexy (NC) is a chronic disorder, and in the large majority of patients it is a sporadic disorder. At least one specific genetic susceptibility marker is known since 85–95% of patients are positive for the human leukocyte antigen subtype DQB1*0602 (HLA+).

Shortly after the discovery of the hypocretin neuropeptides (also called orexin) [1, 2], it was established that an induced dysfunction in the hypocretin system (loss of the Hcrt-1 and Hcrt-2 neuropeptides or their receptors, Hcrt-R1 and Hcrt-R2) is sufficient to provoke narcolepsy symptoms in animal models [3–7]. Hypocretin loss is, however, not necessary since the CSF level of Hcrt-1 is usually in the normal range in narcoleptic patients without cataplexy [8].

Human narcolepsy is clearly not due to a genetic defect of the preprohypocretin and hypocretin receptor genes as only one case of mutation with unusual and extremely early onset (6 month of age) has been described among the 870 patients tested in all studies combined [9–13]. NC is, however, associated with a very low to undetectable level of CSF Hcrt-1 as measured by radioimmunoassay [8, 14]. This hypocretin deficiency is fairly specific to patients with NC and HLA+ status [8].

The etiology of human narcolepsy is still under debate. The current hypothesis is that NC may be of autoimmune origin due to its strong association with the HLA DQB1*0602 subtype (see Chap. "Narcolepsy: Autoimmunity or Secondary to Infection?" by Fontana). Furthermore, since CSF Hcrt-1 levels are low, it is hypothesized that the hypocretin neurons are targeted during the autoimmune attack, resulting in their destruction. As such, NC would be a neurodegenerative disease.

The definition of a neurodegenerative disease is debated but neurologists seem to agree on two main criteria: it involves (1) the death of neuronal cells (2) in a progressive manner. The cause of the cell

C. Peyron (✉)
Centre de Recherche en Neurosciences de Lyon, Faculté de Médecine Laennec, Université Lyon1, 7 rue Guillaume Paradin, 69372 Lyon cedex 08, France
e-mail: peyron@sommeil.univ-lyon1.fr

C.R. Baumann et al. (eds.), *Narcolepsy: Pathophysiology, Diagnosis, and Treatment*,
DOI 10.1007/978-1-4419-8390-9_3, © Springer Science+Business Media, LLC 2011

death can be very different from disease to disease. Accumulation of aberrant or misfolded proteins, ubiquitin-proteasome dysfunction, excitotoxic insult, autoimmune insult, oxidative and nitrosative stress, mitochondrial injury, synaptic failure, altered metal homeostasis, and failure of axonal and dendritic transport are events that have been described in progressive neurodegenerative disease.

Here, we review the arguments for and against the idea that NC might be a neurodegenerative disease considering only the NC/HLA+/Hcrt– subgroup of narcolepsies.

Tracking Signs of Neurodegeneration In Vivo

In vivo approaches were undertaken to determine whether hypocretin neurons die or are not functional in patients with narcolepsy, and to eventually develop a new diagnosis tool. A marked bilateral decrease in the gray matter content of the hypothalamus was detected in several studies [15–17] but this result was not replicated in others [18–20] using the same voxel-based morphometry MRI-technique. A FDG-PET study revealed significant hypometabolism in the posterior hypothalamus [21]. Some of these contradictory results likely depend on differences in methodology and clinical variabilities such as the use of small groups of patients, HLA typing, presence or not of cataplexy, co-morbidity or pharmacological history. As it is, these imaging experiments unfortunately do not provide backing for or against the neurodegenerative hypothesis.

The use of proton MR spectroscopy allows the measurement of several brain compounds such as the *N*-acetyl-aspartate (NAA) a neuronal marker that is reduced when neuronal loss or damage occurs in neurodegenerative or inflammatory disorders. A significant decrease in the NAA to creatine ratio limited to the hypothalamus was found in patients with NC/HLA+/Hcrt– [22–24]. Interestingly, the NAA to creatine ratio was not significantly different between controls and narcoleptic patients without cataplexy [22]. Although the CSF level of Hcrt-1 was not documented in this study, most narcoleptic patients without cataplexy have normal CSF concentration of Hcrt-1 [8]. These latest results support the hypothesis of a neuronal loss limited to the hypothalamus. It is tempting to attribute this to a loss of hypocretin neurons, but only histological studies may provide such information.

Are Hypocretin Neurons Missing in Patients with Narcolepsy-Cataplexy?

Several postmortem studies have examined the hypocretin neurons using in situ hybridization [9] or immunohistochemistry [25–29]. To this date, 14 brains from NC patients were processed for staining in all studies combined (Tables 1 and 2). In all of them, a lack or strong reduction in the number of hypocretin neurons was found (85–100%) compared to controls. No hypocretin cells were detected in NC brains when radioactive in situ hybridization was carried out on thin frozen sections collected on slides (Fig. 1) [9]. Furthermore, the estimation of the number of hypocretin neurons in the control population was significantly lower [9] compared to studies using immunohistochemistry on free-floating sections [25, 26]. It might be due to the methodology used, as staining on free-floating sections is considered to be a more sensitive technique. It is also possible that peptides accumulate in greater quantity or are more stable than RNA and thus were more easily detected. No correlation was found between the number of remaining hypocretin neurons and the postmortem delay or the time in fix for tissue [25]. A reduction in hypocretin axon density has also been reported [26]. The percentage of reduction in brain areas was positively correlated with the axon density found in the same areas of control brains ($r=0.75$, $t=3.37$, $p<0.0008$, $n=11$). Altogether, these data strongly suggest that patients with NC lack hypocretin neuropeptides. It does not mean, however, that there is an actual loss of neurons. Neurons usually expressing hypocretin could still be present and functionally active, while the expression of hypocretin peptides could be impaired.

To determine whether hypocretin neurons are dead or not, two groups of scientists looked at the presence of two other neuropeptides, namely dynorphin and NARP (neuronal activity regulated

Table 1 Demographic and clinical data regarding patients with narcolepsy and cataplexy in histological studies

Patient #	HLA+	Sex	Age at onset	Age at death	Cause of death	Treatment	PMI	References
1	+	F	NA	77	NA	NA	6.8	[9]
2	+	M	NA	67	NA	NA	17	[9]
3	NA	F	20	63	Adenocarcinoma	Clomipramine, viloxazine	5	[25–27]
4	NA	M	18	49	Cardiovascular	Dextroamphetamine, protriptyline, GHB	4.5	[25, 26]
5	NA	M	23	60	Sepsis	Methylphenidate, methamphetamine	4.2	[25–27]
6	NA	M	13	81	Cardiovascular	Dextroamphetamine	0.3	[26, 27]
7	NA	M	18	79	Atherosclerosis	NA	1	[27]
8	NA	M	13	80	Heart disease	Dextroamphetamine	<1	[27]
9	+	M	13	49	Viral cardiomyopathy	Methamphetamine	<24	[28]
10	+	M	<52	72	Cancer	Methylphenidate, clomipramine	<24	[28]
11	+	M	15	69	Epidural hematoma	NA	10	[29]
12	+	M	12	75	Spinal cord infarction, prostate cancer metastasis, sepsis (+AD)	NA	5.5	[29]
13	+	F	20	89	NA (+LBD)	NA	20	[29]
14	+	F	10	79	Chronic renal failure, heart failure, melena (+NFT)	NA	3	[29]

Patients were renumbered for clarity. Age at onset, at death, and the postmortem interval are reported in years
Abbreviations: *AD* Alzheimer's disease, *LBD* Lewy body disease, *NA* not available, *NC* narcolepsy with cataplexy, *NFT* senile dementia of the neurofibrillary tangle type, *PMI* postmortem interval

Table 2 Neuropeptides stained by in situ hybridization or immunohistochemistry in brains listed in Table 1

	Cell loss				Hcrt axon loss		
Patient #	Hcrt	PDYN	NARP	MCH	Hypothalamus	Pons	References
1	Y			N			[9]
2	Y			N			[9]
3	Y		Y	N	Y	Y	[25–27]
4	Y			N	Y	Y	[25, 26]
5	Y		Y	N	Y	Y	[25–27]
6	Y				Y	Y	[26, 27]
7	Y		Y				[27]
8	Y		Y				[27]
9	Y	Y	Y				[28]
10	Y		Y				[28]
11	Y						[29]
12	Y						[29]
13	Y						[29]
14	Y						[29]

Abbreviations: *Hcrt* hypocretin, *MCH* melanin-concentrating hormone, *N* no cell loss was found, *NARP* neuronal activity regulated pentraxin, *PDYN* pro-dynorphin, *Y* yes, cell loss was observed

pentraxin) that are known to be expressed specifically in hypocretin neurons in this brain area in rodents [30, 31]. In human control brains, 80% of hypocretin neurons contained pro-dynorphin and 91–99% contained NARP [27, 28]. Staining was processed on one NC brain only for pro-dynorphin

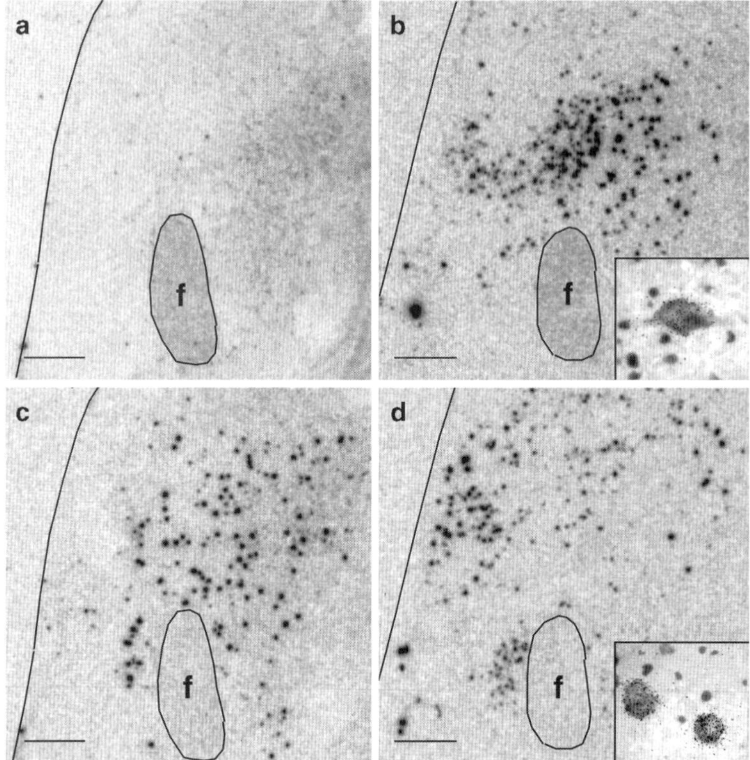

Fig. 1 Hypocretin and MCH expression studies in the hypothalamus of control and narcoleptic subjects. *Preprohypocretin* transcripts are detected in the hypothalamus of control (**b**) but not in narcoleptic (**a**) subjects. *MCH* transcripts are detected in the same region in both control (**d**) and narcoleptic (**c**) sections. *f* Fornix. Scale bar = 10 mm

using in situ hybridization [28] and on six NC brains for NARP using immunohistochemistry [27, 28] (Table 2). A 90% reduction in the number of pro-dynorphin neurons and a 89–95% decrease for NARP were found in NC brains compared to controls [27, 28]. No change in NARP staining was observed in brain areas that lack hypocretin neurons such as the paraventricular and supraoptic nuclei of the hypothalamus. It indicates that the loss of NARP peptides is limited to the neurons co-expressing NARP, pro-dynorphin, and hypocretins. Since three of the peptides expressed by the same cells are missing in the same proportion, it is likely that the hypocretin cells are actually missing or are very dysfunctional. These are the strongest arguments to date supporting the hypothesis that NC is a neurodegenerative disease harming the hypocretin neurons.

Interestingly, the intermingled but distinct neurons expressing melanin-concentrating hormone were found to be intact [9, 25] suggesting that the impairment is limited to the hypocretin neurons (Fig. 1, Table 2).

Looking for Signs of Inflammation or Gliosis

Additional experiments searching for signs of lesions, inflammation, or immune reactivity were undertaken on a limited number of postmortem brain samples from NC patients (Table 3), focusing mainly on the hypothalamic area containing the hypocretin neurons ($n = 11$, all studies combined) [9, 25, 26, 28, 29] and also on hypocretin projection sites such as the locus coeruleus

Table 3 Markers of gliosis or inflammation in the brains listed in Table 1

	GFAP+					Ubiquit	AIF1	Tau	Synuclein	Amyloid	TDP-43
Patient #	Hypothal	Thal	Pons	TNF+	HLA DR2+	In the remaining Hcrt cells					
1	N			N	N						
2	N			N	N						
3	Y	N	Y								
4	Y	N	Y								
5	Y	N	Y								
6			Y								
9	NQ										
10	NQ										
11	N					N	N	N	N	N	N
12	N					N	N	N	N	N	N
13	N					N	N	N	N	N	N
14	N					N	N	N	N	N	N

Abbreviations: *AIF1* allograft inflammatory factor 1 (a microglial activation marker), *Amyloid* ß (beta)-amyloid protein, *GFAP* glial fibrillary acidic protein, *NQ* no quantification; *synuclein* α (alpha)-synuclein; *Ubiquit* ubiquitin, *TNF* tumor necrosis factor

and the dorsal and median raphé nuclei in the pons ($n=4$) [25, 26]. The hypothalami contained no overt lesions or inflammation as assessed by simple cresyl violet counterstaining or staining for HLA-DR2 and AIF1, markers of microglial activation [9, 29]. The density of HLA-DR2 labeling was very variable from brain to brain and was not associated with narcolepsy [9]. AIF1 staining was unremarkable in NC tissue [29]. In situ hybridization of tumor necrosis factor (TNF)-alpha, a cytokine strongly expressed in many inflammatory CNS disorders produced no significant signal in narcoleptic and control brains [9].

An increase in GFAP staining density was, however, detected in the hypocretin hypothalamic area ($n=3$) by Thannickal and colleagues [25, 26]. Furthermore, an increase in the number of GFAP-stained astrocytes was observed in the locus coeruleus and raphe nuclei ($n=4$) but not in the dorsomedial thalamic nucleus ($n=3$) in NC versus control brain samples [25, 26]. In fact, regional gliosis in NC patients was highly correlated with the density of hypocretin axons in the normal brain ($r=0.87$, $p<0.0001$, $n=11$). A high density in axons was detected in the hypothalamic nuclei and the locus coeruleus, while a moderate density of axons was found in the raphe nuclei and no or occasional axons were seen in the dorsomedial nucleus of the thalamus, in human control brains [26]. A similar axonal distribution has been described in rats [32]. However, the presence of increased astrogliosis in the hypothalamus was not replicated by others on six brain samples [9, 29] as summarized in Table 3. Crocker and colleagues [28] found such high density of GFAP immunoreactivity in NC ($n=2$) and control hypothalami that they were not able to proceed in quantification. These data are quite puzzling considering that among the four brains studied by Honda and colleagues [29] three were obtained from patients diagnosed with comorbid neurodegenerative diseases such as Alzheimer's disease, Lewy body disease, and senile dementia of the neurofibrillary tangle type in whom gliosis would be expected. Brains from these patients contained signs of degeneration but not in the hypocretin field.

This controversy is difficult to disentangle. Although it seems unlikely, it may be accredited to the use of different antibodies directed against GFAP in the different studies and/or immunohistochemistry procedures. Thannickal and colleagues [25] did not find a correlation between the density of GFAP staining and the time in fix or the use of antigen retrieval or age at death. Since NC patients die at old age, it may rather reflect interindividual variance than a NC phenotype. Clinical and pharmacological history of patients and controls may need to be considered more carefully.

Testing the Neurodegenerative Hypothesis: Do the Remaining Hypocretin Neurons Contain Protein Aggregates?

Intracellular protein aggregates accumulate in specific neurons in some neurodegenerative diseases. Ubiquitin, tau, α (alpha)-synuclein, β (beta) amyloid, and TDP-43 are well-known components of protein aggregates. Examination of hypothalamic sections for possible protein aggregates revealed no inclusions in the remaining hypocretin neurons [29]. These results were consistent in all the studied NC cases ($n=4$) even though three of the four patients had comorbid dementia (Tables 1 and 3). Many ubiquitin inclusions were seen in the cortex of these patients in agreement with the diagnosis of dementia [29].

These data suggest that, if NC is due to the death of hypocretin neurons, the putative neurodegenerative process incriminated would not involve ubiquitinated inclusions, unless the remaining hypocretin cells are protected against it for some reason.

Progression of the Disease

By definition, neurodegenerative diseases are caused by progressive cell death. For example, in Parkinson's disease and Alzheimer's disease, neurodegeneration worsens over time, injuring many neuronal populations, and often resulting in dementia and death. In narcolepsy, symptoms appear around 15–25 years of age [33]. Once expressed, these symptoms are stable and may even improve slightly with age [34]. No neuronal population other than the hypocretin neurons is presumed to degenerate. However, cataplexy often appears several years after the onset of excessive daytime sleepiness, suggesting some progression.

Conclusion

Human narcolepsy with cataplexy is clearly associated with a lack of hypocretin neuropeptides. Data are still inconclusive and indirect on whether the absence of hypocretin neuropeptides is due to the death of these neurons. The main argument in favor of a neurodegenerative process in NC/HLA+/Hcrt– is carried by the fact that there is concomitant loss of several neuropeptides expressed by these same neurons (i.e., hypocretin, dynorphin, and NARP). However, this process does not seem to involve protein aggregates formed of ubiquitin, protein tau, α (alpha)-synuclein, β (beta)-amyloid, or TDP-43. As no signs of inflammation or microgliosis were found and the presence of astrogliosis is still debated, the mechanisms involved are still largely unknown. Furthermore, based on available data and clinical observations, the development of the disease is not clearly progressive, at least not in the proportion and with the outcome seen for other neurodegenerative disorders such as Parkinson's disease or Alzheimer's disease.

Acknowledgments Financial supports from the Centre National de la Recherche Scientifique and Claude Bernard University were acknowledged.

References

1. de Lecea L, Kilduff TS, Peyron C, et al. The hypocretins: hypothalamus-specific peptides with neuroexcitatory activity. Proc Natl Acad Sci U S A. 1998;95(1):322–7.
2. Sakurai T, Amemiya A, Ishii M, et al. Orexins and orexin receptors: a family of hypothalamic neuropeptides and G protein-coupled receptors that regulate feeding behavior. Cell. 1998;92(4):573–85.
3. Lin L, Faraco J, Li R, et al. The sleep disorder canine narcolepsy is caused by a mutation in the hypocretin (orexin) receptor 2 gene. Cell. 1999;98(3):365–76.
4. Chemelli RM, Willie JT, Sinton CM, et al. Narcolepsy in orexin knockout mice: molecular genetics of sleep regulation. Cell. 1999;98(4):437–51.
5. Hara J, Beuckmann CT, Nambu T, et al. Genetic ablation of orexin neurons in mice results in narcolepsy, hypophagia, and obesity. Neuron. 2001;30(2):345–54.
6. Willie JT, Chemelli RM, Sinton CM, et al. Distinct narcolepsy syndromes in Orexin receptor-2 and

Orexin null mice: molecular genetic dissection of Non-REM and REM sleep regulatory processes. Neuron. 2003;38(5):715–30.
7. Hondo M, Nagai K, Ohno K, et al. Histamine-1 receptor is not required as a downstream effector of orexin-2 receptor in maintenance of basal sleep/wake states. Acta Physiol (Oxf). 2010;198(3):287–94.
8. Mignot E, Lammers GJ, Ripley B, et al. The role of cerebrospinal fluid hypocretin measurement in the diagnosis of narcolepsy and other hypersomnias. Arch Neurol. 2002;59(10):1553–62.
9. Peyron C, Faraco J, Rogers W, et al. A mutation in a case of early onset narcolepsy and a generalized absence of hypocretin peptides in human narcoleptic brains. Nat Med. 2000;6(9):991–7.
10. Gencik M, Dahmen N, Wieczorek S, et al. A prepro-orexin gene polymorphism is associated with narcolepsy. Neurology. 2001;56(1):115–7.
11. Hungs M, Lin L, Okun M, Mignot E. Polymorphisms in the vicinity of the hypocretin/orexin are not associated with human narcolepsy. Neurology. 2001;57(10):1893–5.
12. Olafsdottir BR, Rye DB, Scammell TE, Matheson JK, Stefansson K, Gulcher JR. Polymorphisms in hypocretin/orexin pathway genes and narcolepsy. Neurology. 2001;57(10):1896–9.
13. Quinnell TG, Farooqi IS, Smith IE, Shneerson JM. Screening the human prepro-orexin gene in a single-centre narcolepsy cohort. Sleep Med. 2007;8(5):498–502.
14. Ripley B, Overeem S, Fujiki N, et al. CSF hypocretin/orexin levels in narcolepsy and other neurological conditions. Neurology. 2001;57(12):2253–8.
15. Draganski B, Geisler P, Hajak G, et al. Hypothalamic gray matter changes in narcoleptic patients. Nat Med. 2002;8(11):1186–8.
16. Buskova J, Vaneckova M, Sonka K, Seidl Z, Nevsimalova S. Reduced hypothalamic gray matter in narcolepsy with cataplexy. Neuro Endocrinol Lett. 2006;27(6):769–72.
17. Kim SJ, Lyoo IK, Lee YS, et al. Gray matter deficits in young adults with narcolepsy. Acta Neurol Scand. 2009;119(1):61–7.
18. Kaufmann C, Schuld A, Pollmacher T, Auer DP. Reduced cortical gray matter in narcolepsy: preliminary findings with voxel-based morphometry. Neurology. 2002;58(12):1852–5.
19. Overeem S, Steens SC, Good CD, et al. Voxel-based morphometry in hypocretin-deficient narcolepsy. Sleep. 2003;26(1):44–6.
20. Brenneis C, Brandauer E, Frauscher B, et al. Voxel-based morphometry in narcolepsy. Sleep Med. 2005;6(6):531–6.
21. Joo EY, Tae WS, Kim JH, Kim BT, Hong SB. Glucose hypometabolism of hypothalamus and thalamus in narcolepsy. Ann Neurol. 2004;56(3):437–40.
22. Lodi R, Tonon C, Vignatelli L, et al. In vivo evidence of neuronal loss in the hypothalamus of narcoleptic patients. Neurology. 2004;63(8):1513–5.
23. Plazzi G, Parmeggiani A, Mignot E, et al. Narcolepsy-cataplexy associated with precocious puberty. Neurology. 2006;66(10):1577–9.
24. Tonon C, Franceschini C, Testa C, et al. Distribution of neurochemical abnormalities in patients with narcolepsy with cataplexy: an in vivo brain proton MR spectroscopy study. Brain Res Bull. 2009;80(3):147–50.
25. Thannickal TC, Moore RY, Nienhuis R, et al. Reduced number of hypocretin neurons in human narcolepsy. Neuron. 2000;27(3):469–74.
26. Thannickal TC, Siegel JM, Nienhuis R, Moore RY. Pattern of hypocretin (orexin) soma and axon loss, and gliosis, in human narcolepsy. Brain Pathol. 2003;13(3):340–51.
27. Blouin AM, Thannickal TC, Worley PF, Baraban JM, Reti IM, Siegel JM. Narp immunostaining of human hypocretin (orexin) neurons: loss in narcolepsy. Neurology. 2005;65(8):1189–92.
28. Crocker A, Espana RA, Papadopoulou M, et al. Concomitant loss of dynorphin, NARP, and orexin in narcolepsy. Neurology. 2005;65(8):1184–8.
29. Honda M, Arai T, Fukazawa M, et al. Absence of ubiquitinated inclusions in hypocretin neurons of patients with narcolepsy. Neurology. 2009;73(7):511–7.
30. Chou TC, Lee CE, Lu J, et al. Orexin (hypocretin) neurons contain dynorphin. J Neurosci. 2001;21(19):RC168.
31. Reti IM, Reddy R, Worley PF, Baraban JM. Selective expression of Narp, a secreted neuronal pentraxin, in orexin neurons. J Neurochem. 2002;82(6):1561–5.
32. Peyron C, Tighe DK, van den Pol AN, et al. Neurons containing hypocretin (orexin) project to multiple neuronal systems. J Neurosci. 1998;18(23):9996–10015.
33. Dauvilliers Y, Montplaisir J, Molinari N, et al. Age at onset of narcolepsy in two large populations of patients in France and Quebec. Neurology. 2001;57(11):2029–33.
34. Dauvilliers Y, Gosselin A, Paquet J, Touchon J, Billiard M, Montplaisir J. Effect of age on MSLT results in patients with narcolepsy-cataplexy. Neurology. 2004;62(1):46–50.

Part II

Neurochemistry of Narcolepsy

The Roles of Hypocretin/Orexin in Narcolepsy, Parkinson's Disease, and Normal Behavior

Jerome Siegel

Keywords

Orexin • Hypocretin • Narcolepsy • Parkinson's disease • Hypothalamus • Sleep-wake regulation • Excessive daytime sleepiness

Despite the development of extremely useful rodent models [1–3], the best physiological model of human cataplexy remains the genetically narcoleptic dog [4]. In these animals, excitement induced by food or play produces abrupt loss of muscle tone which appears to be accompanied by maintained consciousness, as evidenced by tracking eye movements. Less severe attacks of partial weakness also mirror the human condition. Drugs effective in reversing human cataplexy are effective in blocking cataplexy in the narcoleptic dog [5]. In contrast, rodents lacking hypocretin (hcrt) have short periods of behavioral arrest, but it is not clear if they have tracking eye movements or other indications of maintained consciousness [6]. The generally positive emotions that trigger cataplexy in humans [7] and the feeding and play behavior that trigger it in dogs do not have clear parallels that allow behavioral arrest to be reliably triggered in mice. Clearly, murine models share key aspects of the underlying pathology of human narcolepsy, but the interaction of hcrt deficiency with a rodent limbic system produces differences in cataplexy elicitation, as should be expected. By comparing cataplexy in rodent, canine, and human subjects, each of which has a somewhat different underlying hypocretin pathology, we can most clearly identify the critical changes responsible for cataplexy and "behavioral arrest."

The reliable triggering of canine cataplexy and its behavioral similarities to human cataplexy have allowed an analysis of the neurophysiology of cataplexy at the neuronal level. We found that a majority of brainstem reticular formation cells abruptly cease discharge during cataplexy [8]. This is a striking difference between cataplexy and REM sleep, since these same cells have maximal activity in REM sleep. A small but critical subpopulation of cells, localized to the medial medullary region responsible for muscle tone suppression, is active only during cataplexy and REM sleep [9]. One may best understand this combination of cataplexy-on and cataplexy-off cells as indicating that in the normal animal, the generalized activation of medial reticular (presumably glutamatergic) neurons in REM sleep, which might be expected to cause motoneuron activation, is occluded by the activation of this

J. Siegel (✉)
Department of Psychiatry, Veterans Administration Greater Los Angeles Healthcare System, Neurobiology Research (151A3), University of California at Los Angeles, North Hills, CA 91343, USA
e-mail: jsiegel@ucla.edu

C.R. Baumann et al. (eds.), *Narcolepsy: Pathophysiology, Diagnosis, and Treatment*,
DOI 10.1007/978-1-4419-8390-9_4, © Springer Science+Business Media, LLC 2011

potent ponto-medullary inhibitory system releasing GABA and glycine onto motoneurons [10–13]. However, in cataplexy, not only is the inhibitory system activated, but most reticular neurons linked to movement are also inactivated, providing a simultaneous disfacilitation of muscle tone [8]. Recent work in hcrt KO mice has demonstrated an identical pattern [14].

We also examined the activity of noradrenergic [15], serotonergic [16], and histaminergic [17] neurons during cataplexy. These neurons all cease discharge during REM sleep in narcoleptic dogs, just as in normal cats, rats, and monkeys. We found that a more complex pattern occurred in cataplexy. Noradrenergic neurons ceased discharge in tight linkage to the onset of cataplexy and remained off during cataplexy. A burst of firing of these cells always immediately preceded the end of cataplexy. In contrast, presumed histaminergic REM sleep-off neurons always maintained discharge throughout cataplexy (Fig. 1). Studies have shown decreased monoamine release onto motoneuronal pools during muscle tone suppression triggered by medial medullary stimulation [18]. Noradrenergic agonists are the most effective treatment for cataplexy in humans and dogs [19], and clomipramine is effective in reducing behavioral arrests in mice [20]. It has been shown that the loss of norepinephrine signaling is more important in causing the loss of muscle tone in REM sleep than the increased release of GABA and glycine [21–23]. Therefore, both increased inhibition by glycine and GABA [9, 10] and decreased facilitation by norepinephrine contribute to the loss of muscle tone in cataplexy and REM sleep.

These studies illuminate the mechanisms responsible for muscle tone suppression in both normal REM sleep and cataplexy. They also clearly indicate that several changes characteristic of REM sleep are not required for muscle tone suppression. The changes not occurring in cataplexy include the activation of most medial reticular neurons and the cessation of activity in histamine and serotonin neurons. Conversely, it suggests the defining physiological differences between cataplexy and REM sleep. Loss of consciousness is linked to the cessation of activity in serotonergic and histaminergic neurons in slow wave sleep (SWS) and REM sleep. These neuronal groups are known to be important in maintaining alertness. The phasic motor activity in REM sleep is most likely linked to the activation of the majority of medial reticular cells, which have been shown to be related to specific lateralized movements of the eyes, neck, tongue, or limbs. These neurons are inactive during cataplexy and their activity is seen only during waking with motor activity and during REM sleep [8]. This pattern of neuronal activity that causes cataplexy is due to the abnormal operation of the hcrt system.

Human Narcolepsy with Cataplexy and Hypocretin

Human narcolepsy with cataplexy was shown to be characterized by low cerebrospinal hypocretin levels and loss of hypocretin cells [24–26]. We found that normal individuals have approximately 70,000 hcrt cells and that narcoleptics have on average a loss of 90% of these cells [26]. Since hcrt cells also contain Narp, we stained for this protein to determine if our inability to detect hcrt might be due to a cessation of hcrt production rather than death of these cells. We found that Narp-containing hcrt cells were lost to the same extent as the hcrt cells, but that Narp cells elsewhere in the brain were not lost [27]. This work indicated that Narp was not the antigen attacked in narcolepsy. Dynorphin is also contained in hcrt cells, and dynorphin cell staining in the hypothalamus is lost to the same extent as the hcrt and Narp staining [28]. Together, these results can most easily be explained by the loss of hcrt cells rather than by a specific metabolic dysfunction that prevents hcrt, Narp, and dynorphin synthesis. Although other work initially concluded that all hcrt cells are lost in narcolepsy (and that the normal number of hcrt cells was approximately 20,000), there now appears to be agreement that surviving hcrt cells remain and can be detected throughout the hcrt field and that the normal number is closer to 70,000. We find no obvious location or morphological difference between the surviving hcrt cells and the cells present in the same area in normal brains. No alpha synuclein inclusions have been observed in surviving hcrt cells [26, 29]. hcrt cells are intermixed with hypo-

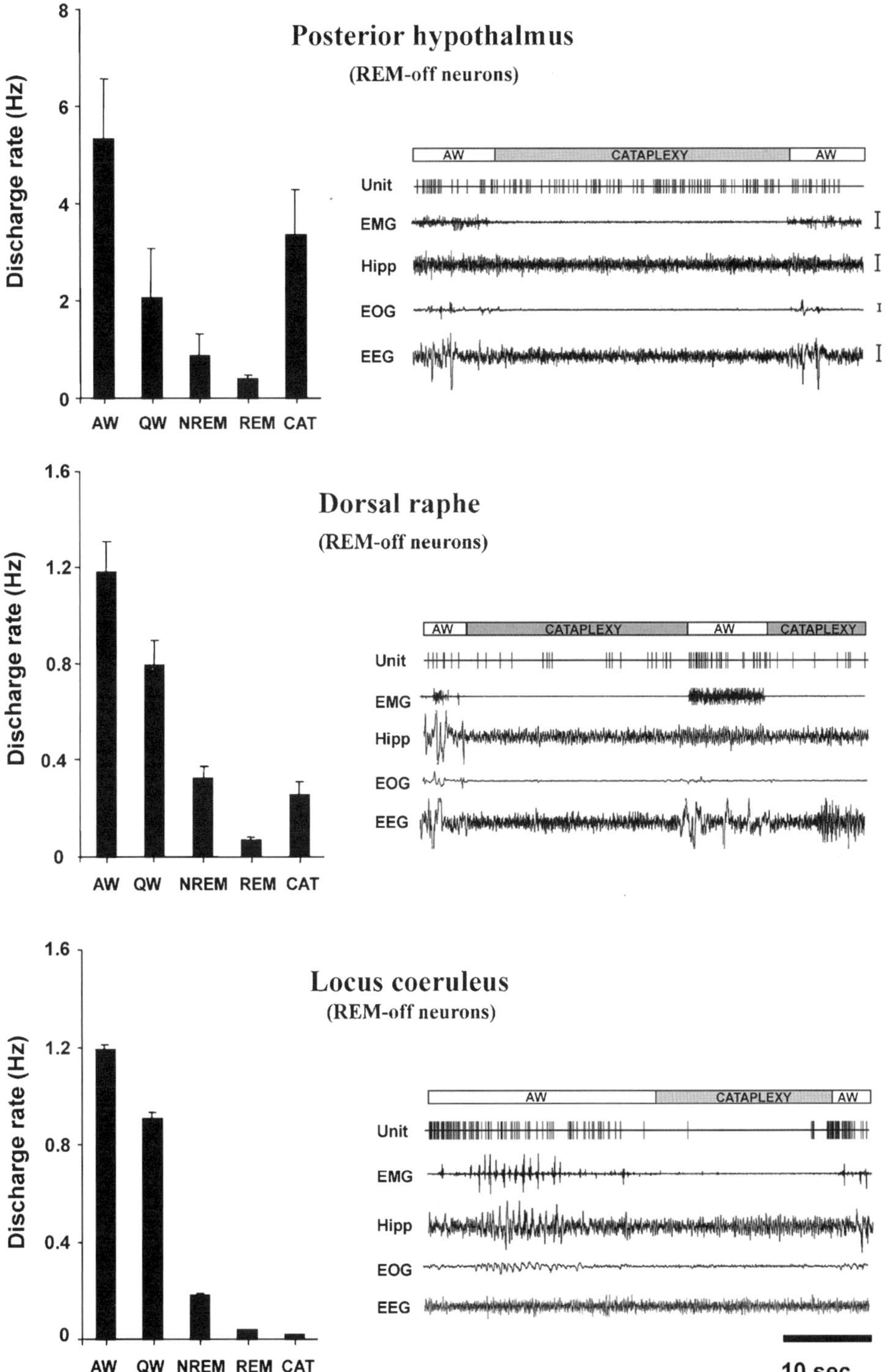

Fig. 1 Activity of noradrenergic, serotonergic, and histaminergic neurons during active waking (AW), quiet waking (QW), slow wave sleep (SWS), REM sleep (REM), and cataplexy (CAT) in the narcoleptic dog. In REM sleep, all three cell groups are silent in narcoleptics, as in normal animals. During cataplexy, noradrenergic activity decreases with loss of muscle tone, but histaminergic cell activity increases relative to REM sleep, SWS, and QW. Dorsal raphe serotonergic activity has a discharge pattern intermediate between that of noradrenergic and histaminergic cells during cataplexy. From [17] with permission

thalamic cells containing melanin-concentrating hormone (MCH). A quantitative analysis found that the number of these MCH cells is normal in narcolepsy with cataplexy [25, 26, 30]. This indicates that hcrt cell loss is not caused by a nonspecific loss of hypothalamic cells, although in rare cases such damage can cause narcoleptic symptomatology [31]. We reported hypothalamic gliosis in narcoleptic brains [26, 32]. A paper by Erlich and Itabashi [33] had reported gliosis in the hypothalamus of a narcoleptic patient in 1986, long before the discovery of hcrt brought the attention of narcolepsy researchers to that region. The most likely cause of such gliosis and the specific loss of hcrt cells is an autoimmune process [34–36]. We found that gliosis was present not only in the hcrt soma region but also in the terminal projection fields of hcrt cells, with the greatest concentration in regions containing hcrt receptor 2 (R2) [32]. These results suggest that hcrt cell loss may be related to an attack on the hcrtR2 or proteins associated with it. Additional evidence bearing on this is the reduced number of terminal varicosities in surviving hcrt axons in narcoleptics, suggesting that despite their normal soma morphology, their axonal innervation is reduced by the pathological process causing the disease [32].

The survival of hcrt cells in narcoleptics raises the question of whether these remaining cells modulate symptoms in narcoleptics. If remaining cells are functional and not maximally active, it might be possible to reduce narcoleptic symptomatology by increasing the activity of these cells by pharmacological, behavioral, or other interventions.

Narcolepsy Without Cataplexy

As many as one-third to one half of patients classified as narcoleptic by sleep latency and the appearance of sleep-onset REM sleep periods do not have cataplexy [37]. CSF studies indicate that some of these patients have decreased hcrt levels but most do not. (Conversely, most patients having narcolepsy with cataplexy have greatly reduced hcrt level, but some do not [34, 38].) An animal study suggests that CSF levels of hcrt do not decrease linearly with the loss of hcrt cells [39]. Most likely, CSF levels are a function not only of the number of cells lost, but also of the activity of surviving cells and the extent to which the hcrt they release enters the ventricular system.

The initial recruitment of patients to will their brain for postmortem analysis focused on patients with narcolepsy with cataplexy. However, we were able to secure two brains from patients having narcolepsy without cataplexy, one which was complete and one which contained only anterior hypothalamic tissue because a pathologist had removed more caudal regions. We found that the complete brain had gliosis and a loss of hcrt cells in the posterior hypothalamus but normal glial and hcrt cell levels in more anterior regions and that the partial brain showed a similarly intact rostral region [30]. It remains to be seen how typical of such patients this pattern is, but these findings raise the possibility that the approximately 90% hcrt cell loss in narcolepsy with cataplexy may represent one point on a continuum of hcrt cell loss causing sleepiness with a small loss, and the full spectrum of narcolepsy with cataplexy when 85% or more of hcrt cells are lost.

Parkinson's Disease and hcrt

Parkinson's disease can be preceded by excessive sleepiness [40, 41] and REM sleep behavior disorder [42] and is frequently accompanied by sleep attacks, nocturnal insomnia, REM sleep behavior disorder, and depression, symptoms that are common in narcolepsy. Therefore, we and others examined the hcrt system in these patients. We found a loss of hcrt cells that was progressive, though less complete than that seen in narcolepsy (Fig. 2). In contrast to narcolepsy, Parkinson's has long been known to be accompanied by a widespread loss of neurons that ultimately terminates in dementia. We found that unlike narcolepsy, MCH cells are lost in Parkinson's to a similar extent as hcrt cells [43, 44]. hcrt cell loss likely contributes to the symptoms of Parkinson's, and hcrt replacement may be valuable in treating Parkinson's as in narcolepsy. However, the symptoms of Parkinson's result from the interaction of

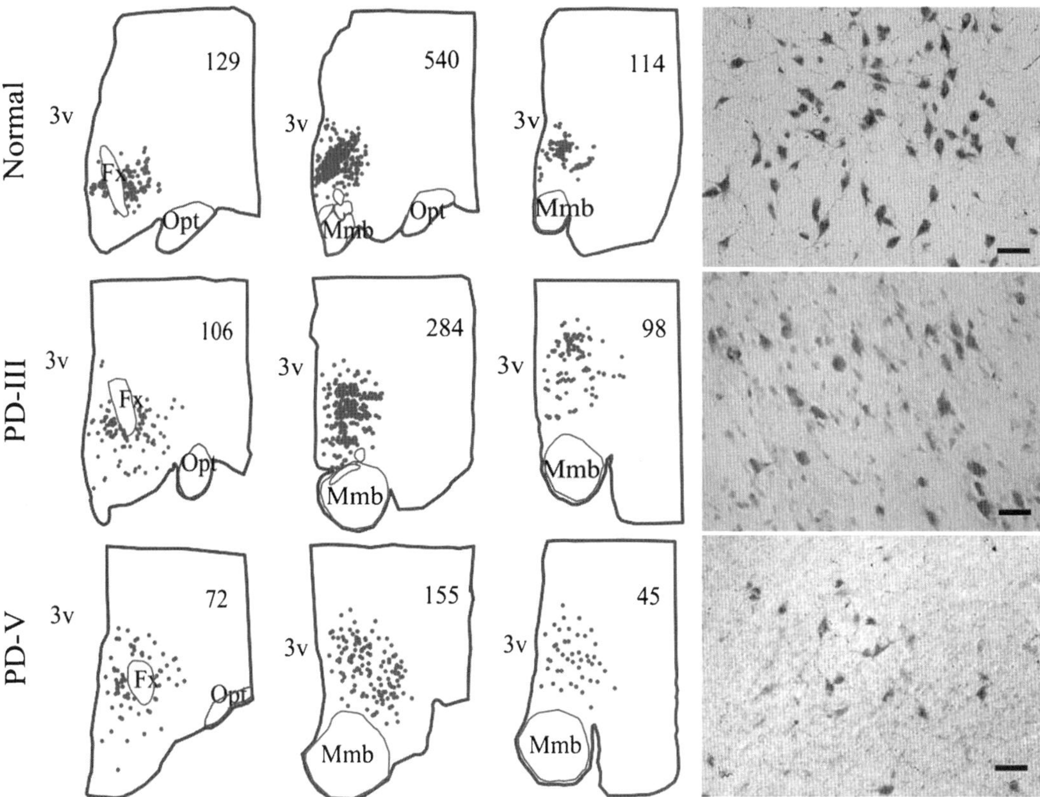

Fig. 2 Distribution of hcrt cells in normal subjects and across the stages of Parkinson's disease (PD). The clinical stages of PD are based on the Hoehn and Yahr criteria. The distribution of cells from sections in the anterior, middle, and posterior parts of the hypothalamus is illustrated from a normal subject, and from subjects with stage III (mild-moderate) and stage V (severe) PD. Cell counts are listed for each section. The number of hcrt cells is decreased with increasing severity of PD. *3v* third ventricle, *Fx* fornix, *Mmb* mammillary body, *Opt* optic tract. Scale bars, 50 mm. From [44] with permission

hcrt cell loss and the massive loss of other cell types, in contrast to the relatively localized hcrt cell loss in narcolepsy.

hcrt Administration

The identification of the loss of hcrt cells as the cause of narcolepsy naturally raises the possibility of hcrt administration as a treatment of narcolepsy. Narcoleptic dogs, while lacking a functional hcrtR2, contain a functional hcrtR1. hcrtR1 is the principal hcrt receptor present on locus coeruleus cells. Cessation of discharge in locus coeruleus cells is tightly linked to cataplexy [17] and the release of NE is tightly linked to the maintenance of muscle tone [15, 18, 45]. Because hcrt is known to cross the blood–brain barrier, we administered hcrt intravenously to narcoleptic dogs and found a reduction in cataplexy and normalization of sleep–wake patterns [46]. A subsequent study by Fujiki et al. saw similar changes but only at higher doses than we used [47, 48].

In a more recent work, the effect of hcrt administration on sleepiness has been studied in normal monkeys that had been sleep deprived. hcrt was administered intravenously and by nasal inhalation, which has been shown to be an effective and rapid method of delivering peptides to the brain. We found a rapid reversal of the behavioral deficits caused by sleepiness (Fig. 3) [49]. This behavioral reversal was correlated with a normalization of

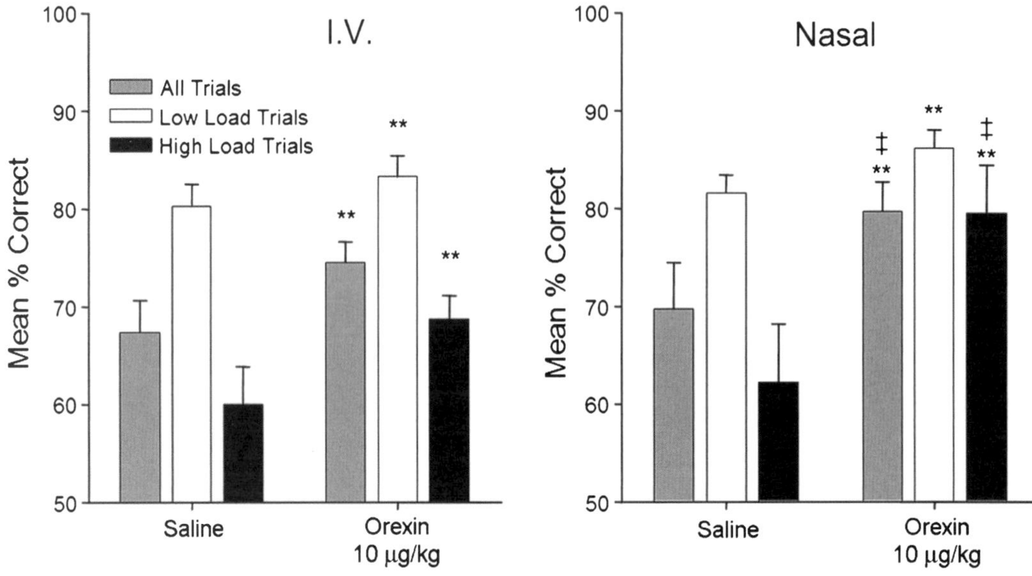

Fig. 3 Comparison of iv orexin-A and nasal orexin-A on reversal of the cognitive effects of sleep deprivation (SD) in rhesus monkeys. With either route of administration, orexin-A improved performance on a delayed matching to sample task. $^{**}p < 0.001$, difference relative to each respective saline condition; $^{‡}p < 0.001$, difference between iv versus nasal orexin A. From [49] with permission

the brain activity patterns caused by sleep deprivation, as evidenced by positron emission tomographic measurements. Preliminary studies using nasal administration of hcrt in human narcoleptics have been encouraging [50]; however, more work is necessary to demonstrate the safety and effectiveness of this approach.

Normal Role of hcrt

Early studies in which hcrt was injected into or adjacent to areas known to be involved in feeding led to the conclusion that hcrt was a feeding eliciting peptide and to the name "orexin" [51]. However, the non-physiological injection technique used may merely indicate that the injection of an excitatory peptide near a feeding-inducing brain region activates the region, not that the peptide is normally involved in this behavior. Therefore, we measured the levels of hcrt in CSF under baseline conditions, after 48 h of food deprivation and at various times after eating in normal dogs. We did not find an increase in hcrt levels in the CSF with hunger or a reduction with food consumption, as reported for well-established feeding-inducing transmitters. However, when the same dogs had their daily release into a large yard in which they played and interacted with each other (but did not eat), we saw a 50–100% increase in hcrt level in all dogs [52]. Therefore, it appears that motor activity and play behavior with their presumed emotional correlates have a much larger and more consistent effect on hcrt level than food deprivation or consumption [53]. These findings make sense in terms of the deficit seen in hcrt-deficient humans, i.e., narcoleptics. In narcoleptics, cataplexy is typically triggered by the sudden onset of positive emotions, most typically laughter, not by eating or pain [7].

In further studies, we developed techniques for identifying hcrt neurons in normal rats. By first performing a series of studies in anesthetized rats in which the recorded cells were identified by ejection of neurobiotin from a glass micropipette and subsequent immunohistochemical staining for hypocretin, we were able to determine the waveshape and conduction velocity that distinguished hcrt cells from adjacent hypothalamic cells. We then used these criteria to locate and record from

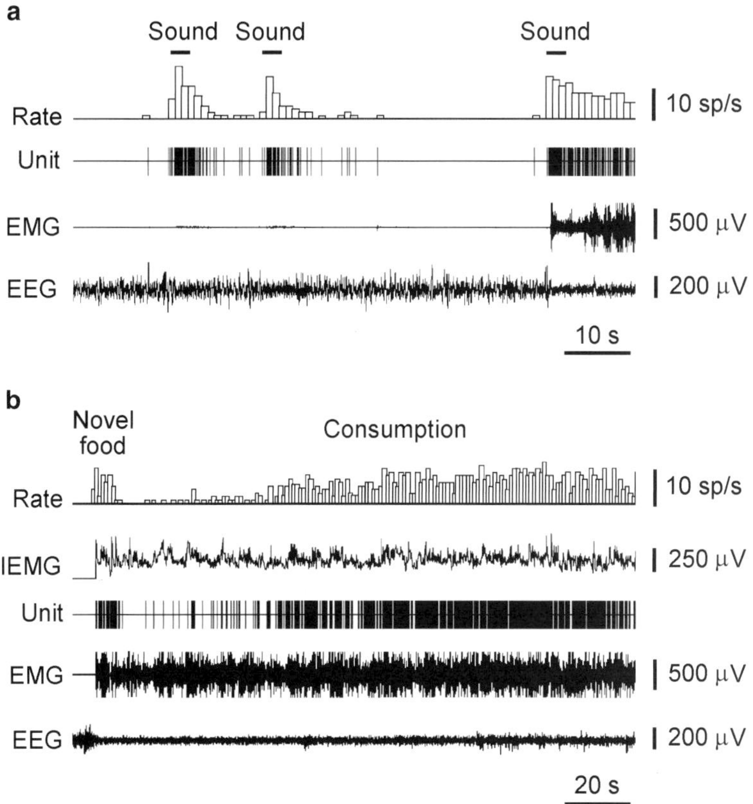

Fig. 4 Responses of a hcrt neuron to natural external stimuli. (**a**) Sound stimuli induce short-lasting hcrt cell excitation independent of marked neck muscle activation. (**b**) Transient decrease in hcrt cell activity in response to presentation of novel food (chicken) despite activated EEG. During decrease of firing rate, rat sniffed, tasted, and backed away from food. hcrt cell firing accelerated in conjunction with onset of "relaxed" food consumption. *IEMG* integrated neck muscle electromyogram, *EEG* electroencephalogram. From [54] with permission

these cells in freely moving animals using our microwire recording technique [54]. We and Lee et al. working in restrained animals found that these cells were maximally active in active waking and minimally active in sleep [54, 55]. hcrt unit discharge was correlated with motor activity. However, we found that these cells could be completely inactive even during periods of high arousal as indicated by EEG and behavioral criteria (Fig. 4). It appeared that anxiety, such as that caused by introducing a strange "new" food, from which the animal recoiled, caused a cessation of activity. This kind of data suggests, in accord with the data presented above and the human clinical experience, that hcrt activity is not simply arousal related, but has a particular role in mediating arousal in connection with positive experience. Other systems must mediate arousal in aversive situations since hcrt cells are silent at these times.

Summary

Animal studies have shown that cataplexy is correlated with a characteristic pattern of neuronal activity. There is strong activation of a restricted population of medial medullary neurons presumed to release GABA and glycine onto motoneurons. At the same time, most glutamatergic medial reticular neurons, which are active during REM sleep and waking movement, fall silent. Noradrenergic neurons which provide a tonic depolarizing input to motoneurons fall silent in tight temporal correlation with the loss of muscle tone. However, histamine neurons, which are silent in REM sleep, remain active during cataplexy. Most human narcolepsy with cataplexy is caused by a loss of hcrt cells and at least one case of narcolepsy without cataplexy has been characterized by a loss of posterior hypothalamic hcrt cells. Narcolepsy-like symptoms in Parkinson's disease may also be caused by hcrt cell loss. In intact animals, hcrt cell activity has little relation to feeding, but does appear to be correlated with arousal linked to positive emotions.

References

1. Hara J, Beuckmann CT, Nambu T, et al. Genetic ablation of orexin neurons in mice results in narcolepsy, hypophagia, and obesity. Neuron. 2001;30:345–54.
2. Beuckmann CT, Sinton CM, Williams SC, et al. Expression of a poly-glutamine-ataxin-3 transgene in orexin neurons induces narcolepsy-cataplexy in the rat. J Neurosci. 2004;24:4469–77.
3. Chemelli RM, Willie JT, Sinton CM, et al. Narcolepsy in orexin knockout mice: molecular genetics of sleep regulation. Cell. 1999;98:437–51.
4. Foutz AS, Mitler MM, Cavalli-Sforza LL, Dement WC. Genetic factors in canine narcolepsy. Sleep. 1979;1:413–21.
5. Nishino S, Mignot E. Pharmacological aspects of human and canine narcolepsy. Prog Neurobiol. 1997;52:27–78.
6. Scammell TE, Willie JT, Guilleminault C, Siegel JM. A consensus definition of cataplexy in mouse models of narcolepsy. Sleep. 2009;32:111–6.
7. Guilleminault C. Cataplexy. In: Guilleminault C, Dement WC, Passouant P, editors. Narcolepsy. 3rd ed. New York: Spectrum; 1976. p. 125–43.
8. Siegel JM, Nienhuis R, Fahringer HM, et al. Activity of medial mesopontine units during cataplexy and sleep-waking states in the narcoleptic dog. J Neurosci. 1992;12:1640–6.
9. Siegel JM, Nienhuis R, Fahringer H, et al. Neuronal activity in narcolepsy: identification of cataplexy related cells in the medial medulla. Science. 1991;252: 1315–8.
10. Kodama T, Lai YY, Siegel JM. Changes in inhibitory amino acid release linked to pontine-induced atonia: an in vivo microdialysis study. J Neurosci. 2003;23: 1548–54.
11. Lai YY, Siegel JM. Cardiovascular and muscle tone changes produced by microinjection of cholinergic and glutamatergic agonists in dorsolateral pons and medial medulla. Brain Res. 1990;514:27–36.
12. Lai YY, Siegel JM. Pontomedullary glutamate receptors mediating locomotion and muscle tone suppression. J Neurosci. 1991;11:2931–7.
13. Lai YY, Clements J, Siegel J. Glutamatergic and cholinergic projections to the pontine inhibitory area identified with horseradish peroxidase retrograde transport and immunohistochemistry. J Comp Neurol. 1993;336:321–30.
14. Thankachan S, Kaur S, Shiromani PJ. Activity of pontine neurons during sleep and cataplexy in hypocretin knock-out mice. J Neurosci. 2009;29:1580–5.
15. Wu MF, Gulyani S, Yao E, Mignot E, Phan B, Siegel JM. Locus coeruleus neurons: cessation of activity during cataplexy. Neuroscience. 1999;91:1389–99.
16. Wu MF, John J, Boehmer LN, Yau D, Nguyen GB, Siegel JM. Activity of dorsal raphe cells across the sleep-waking cycle and during cataplexy in narcoleptic dogs. J Physiol. 2004;554:202–15.
17. John J, Wu M-F, Boehmer LN, Siegel JM. Cataplexy-active neurons in the posterior hypothalamus: implications for the role of histamine in sleep and waking behavior. Neuron. 2004;42:619–34.
18. Lai YY, Kodama T, Siegel JM. Changes in monoamine release in the ventral horn and hypoglossal nucleus linked to pontine inhibition of muscle tone: an in vivo microdialysis study. J Neurosci. 2001;21:7384–91.
19. Mignot E, Renaud A, Arrigoni J, Guilleminault C, Dement WC. Canine cataplexy is preferentially controlled by adrenergic mechanisms: evidence using monoamine selective uptake inhibitors and release enhancers. Psychopharmacology. 1993;113:76–82.
20. Willie JT, Chemelli RM, Sinton CM, et al. Distinct narcolepsy syndromes in Orexin receptor-2 and Orexin null mice: molecular genetic dissection of Non-REM and REM sleep regulatory processes. Neuron. 2003;38:715–30.
21. Kubin L, Kimura H, Tojima H, Davies RO, Pack AI. Suppression of hypoglossal motoneurons during the carbachol-induced atonia of REM sleep is not caused by fast synaptic inhibition. Brain Res. 1993;611: 300–12.
22. Fenik VB, Davies RO, Kubin L. Noradrenergic, serotonergic and GABAergic antagonists injected together into the XII nucleus abolish the REM sleep-like depression of hypoglossal motoneuronal activity. J Sleep Res. 2005;14:419–29.
23. Horner RL. Neuromodulation of hypoglossal motoneurons during sleep. Respir Physiol Neurobiol. 2008;164:179–96.
24. Nishino S, Ripley B, Overeem S, Lammers GJ, Mignot E. Hypocretin (orexin) deficiency in human narcolepsy. Lancet. 2000;355:39–41.
25. Peyron C, Faraco J, Rogers W, et al. A mutation in a case of early onset narcolepsy and a generalized absence of hypocretin peptides in human narcoleptic brains. Nat Med. 2000;6:991–7.
26. Thannickal TC, Moore RY, Nienhuis R, et al. Reduced number of hypocretin neurons in human narcolepsy. Neuron. 2000;27:469–74.

27. Blouin AM, Thannickal TC, Worley PF, Baraban JM, Reti IM, Siegel JM. Narp immunostaining of human hypocretin (orexin) neurons: loss in narcolepsy. Neurology. 2005;65:1189–92.
28. Crocker A, Espana RA, Papadopoulou M, et al. Concomitant loss of dynorphin, NARP, and orexin in narcolepsy. Neurology. 2005;65:1184–8.
29. Honda M, Arai T, Fukazawa M, et al. Absence of ubiquitinated inclusions in hypocretin neurons of patients with narcolepsy. Neurology. 2009;73:511–7.
30. Thannickal TC, Nienhuis R, Siegel JM. Localized loss of hypocretin (orexin) cells in narcolepsy without cataplexy. Sleep. 2009;32:993–8.
31. Baumann CR, Bassetti CL, Valko PO, et al. Loss of hypocretin (orexin) neurons with traumatic brain injury. Ann Neurol. 2009;66:555–9.
32. Thannickal TC, Siegel JM, Moore RY. Pattern of hypocretin (orexin) soma and axon loss, and gliosis, in human narcolepsy. Brain Pathol. 2003;13:340–51.
33. Erlich SS, Itabashi HH. Narcolepsy: a neuropathologic study. Sleep. 1986;9:126–32.
34. Hallmayer J, Faraco J, Lin L, et al. Narcolepsy is strongly associated with the T-cell receptor alpha locus. Nat Genet. 2009;41:708–11.
35. Honda Y, Doi Y, Juji T, Satake M. Narcolepsy and HLA: positive DR2 as a prerequisite for the development of narcolepsy. Folia Psychiatr Neurol Jpn. 1984;38:360.
36. Mignot E, Young T, Lin L, Finn L. Nocturnal sleep and daytime sleepiness in normal subjects with HLA-DQB1*0602. Sleep. 1999;22:347–52.
37. Soldatos CR, Lugaresi E. Nosology and prevalence of sleep disorders. Semin Neurol. 1987;7:236–42.
38. Bassetti C, Gugger M, Bischof M, et al. The narcoleptic borderland: a multimodal diagnostic approach including cerebrospinal fluid levels of hypocretin-1 (orexin A). Sleep Med. 2003;4:7–12.
39. Gerashchenko D, Murillo-Rodriguez E, Lin L, et al. Relationship between CSF hypocretin levels and hypocretin neuronal loss. Exp Neurol. 2003;184: 1010–6.
40. Abbott RD, Ross GW, White LR, et al. Excessive daytime sleepiness and subsequent development of Parkinson disease. Neurology. 2005;65:1442–6.
41. Dhawan V, Healy DG, Pal S, Chaudhuri KR. Sleep-related problems of Parkinson's disease. Age Ageing. 2006;35:220–8.
42. Postuma RB, Gagnon JF, Rompre S, Montplaisir JY. Severity of REM atonia loss in idiopathic REM sleep behavior disorder predicts Parkinson disease. Neurology. 2010;74:239–44.
43. Fronczek R, Overeem S, Lee SY, et al. Hypocretin (orexin) loss in Parkinson's disease. Brain. 2007;130: 1577–85.
44. Thannickal TC, Lai YY, Siegel JM. Hypocretin (orexin) cell loss in Parkinson's disease. Brain. 2007;130:1586–95.
45. Schwarz PB, Yee N, Mir S, Peever JH. Noradrenaline triggers muscle tone by amplifying glutamate-driven excitation of somatic motoneurones in anaesthetized rats. J Physiol. 2008;586:5787–802.
46. John J, Wu MF, Siegel JM. Hypocretin-1 reduces cataplexy and normalizes sleep and waking durations in narcoleptic dogs. Sleep. 2000;23:A12.
47. Fujiki N, Yoshida Y, Ripley B, Mignot E, Nishino S. Effects of IV and ICV hypocretin-1 (orexin A) in hypocretin receptor-2 gene mutated narcoleptic dogs and IV hypocretin-1 replacement therapy in a hypocretin-ligand-deficient narcoleptic dog. Sleep. 2003; 26:953–9.
48. Siegel JM. Hypocretin administration as a treatment for human narcolepsy. Sleep. 2003;26:932–3.
49. Deadwyler SA, Porrino L, Siegel JM, Hampson RE. Systemic and nasal delivery of orexin-A (hypocretin-1) reduces the effects of sleep deprivation on cognitive performance in nonhuman primates. J Neurosci. 2007;27:14239–47.
50. Baier PC, Weinhold SL, Huth V, Gottwald B, Ferstl R, Hinze-Selch D. Olfactory dysfunction in patients with narcolepsy with cataplexy is restored by intranasal Orexin A (Hypocretin-1). Brain. 2008;131: 2734–41.
51. Sakurai T, Amemiya A, Ishii M, et al. Orexins and orexin receptors: a family of hypothalamic neuropeptides and G protein-coupled receptors that regulate feeding behavior. Cell. 1998;92:573–85.
52. Wu MF, John J, Maidment N, Lam HA, Siegel JM. Hypocretin release in normal and narcoleptic dogs after food and sleep deprivation, eating, and movement. Am J Physiol Regul Integr Comp Physiol. 2002;283:R1079–86.
53. Siegel JM, Boehmer LN. Narcolepsy and the hypocretin system-where motion meets emotion. Nat Clin Pract Neurol. 2006;2:548–56.
54. Mileykovskiy BY, Kiyashchenko LI, Siegel JM. Behavioral correlates of activity in identified hypocretin/orexin neurons. Neuron. 2005;46:787–98.
55. Lee MG, Hassani OK, Jones BE. Discharge of identified orexin/hypocretin neurons across the sleep-waking cycle. J Neurosci. 2005;25:6716–20.

Histamine in Narcolepsy and Excessive Daytime Sleepiness

Seiji Nishino

Keywords
Histamine • Tuberomammillary nucleus • H3 antagonists • Excessive daytime sleepiness • Narcolepsy

Introduction

Central histaminergic neurotransmission, originating from the tuberomammillary nucleus (TMN) in the posterior hypothalamus, constitutes one of the most important wake-active systems (see [1, 2] for review). Similar to other monoaminergic systems, histaminergic neurons consist of almost ubiquitous and long projections, a characteristic of neuronal circuits that are involved in vigilance control. Histaminergic neurons are active, and histamine release is high during wakefulness [3, 4]. There is evidence that histaminergic neurotransmission is interacting with other sleep–wake regulatory systems to mediate physiological wakefulness and other physiological functions during wakefulness [1, 2]. However, we do not well understand *how* the histaminergic system interacts with other sleep–wake regulatory systems, and how these systems are harmonized.

S. Nishino (✉)
Stanford Center for Sleep Sciences,
Stanford University School of Medicine,
1201 Welch Road, P213, Palo Alto, CA 94304, USA
e-mail: nishino@stanford.edu

Recently, the existence of inhibitory inputs from sleep-active neurons in the anterior hypothalamus (i.e., ventrolateral preoptic area, VLPO) on wake-active neurons has been emphasized [5, 6]. The critical pathophysiology of narcolepsy, a sleep–wake disorder characterized by excessive daytime sleepiness (EDS), cataplexy, and dissociated manifestations of REM sleep, is the dysfunction of hypocretin (orexin) neurotransmission because of a significant loss of hypocretin-producing cells, hypocretin ligands (in human narcolepsy with cataplexy), and loss of function of hypocretin-related genes (in animal narcolepsy) [7]. This discovery corroborated the role of the histaminergic system in physiological and pathophysiological aspects of vigilance control, since hypocretin neurons project downstream to the TMN and excite histaminergic neurons via hypocretin receptor 2 (hcrtr 2, one of the two hypocretin receptors critically involved in sleep–wake regulation and in the pathophysiology of narcolepsy) [8, 9]. These findings suggest that the histaminergic system belongs to the main executive systems for the wake-promoting hypocretin system. Therefore, altered histaminergic neurotransmission may be involved in the pathophysiology of narcolepsy.

In this chapter, recent progresses in the understanding of the role of the histaminergic system

C.R. Baumann et al. (eds.), *Narcolepsy: Pathophysiology, Diagnosis, and Treatment*,
DOI 10.1007/978-1-4419-8390-9_5, © Springer Science+Business Media, LLC 2011

in the regulation of wakefulness – with a special focus on its interaction with the hypocretin system and its involvement in narcolepsy and other disorders with EDS – are discussed. Also, the chapter discusses the development of histaminomimetic compounds for the treatment of narcolepsy and other hypersomnias in humans.

Physiology

Central Histaminergic Neurotransmission and the Control of Vigilance

Research evidence suggests that central histaminergic neurotransmission is involved in the control of vigilance (see [10] for review). Clinically, it is well known that histaminergic H1 blockers, such as promethazine or diphenhydramine, produce sedation, sleep, and disruptions of attention and cognition. These effects are less prominent with the second generation of H1 blockers with low central penetration [11]. In many countries, the first generation H1 blockers, such as diphenhydramine as well as doxylamine, are available as over-the-counter hypnotics.

Histamine neurons are located exclusively in the TMN of the posterior hypothalamus, from where they project to practically all brain regions, including areas important for vigilance control, such as the hypothalamus, basal forebrain, thalamus, cortex, and brainstem structures (see [1, 2] for review) (Fig. 1).

Histamine is synthesized in the brain from l-histidine by the enzyme histidine decarboxylase (HDC). To our knowledge, there is no high-affinity uptake system for histamine, and the termination of its action in the brain appears to require its catabolism to telemethyl histamine by the enzyme histamine *N*-methyl transferase. Inhibition of histamine synthesis can be achieved by the application of alpha-fluoromethylhistidine (α-FMH), an irreversible inhibitor of HDC, leading to a marked depression of histamine levels.

The intrinsic electrophysiological properties of TMN neurons (slow spontaneous firing, broad action potentials) are similar to those of other aminergic neurons, particularly dopaminergic neurons [1, 2]. Their firing rate varies across the sleep–wake cycle, reaching its maximum during waking and slowing down during light slow-wave sleep (SWS), and they are silent during deep SWS and REM sleep [3, 4]. Histamine release is enhanced under extreme conditions such as dehydration and hypoglycemia, and by a variety of stressors. Released histamine activates four types of receptors [1, 2]. H1 receptors are postsynaptic and are positively coupled to phospholipase C. High densities of H1 receptors are found in the hypothalamus and other limbic regions. H2 receptors are also postsynaptic and are positively coupled to adenylyl cyclase. These receptors are densely found in the hippocampus, amygdala, and basal ganglia. H3 receptors are found on cell bodies of the TMN neurons, and inhibit the firing of these same neurons. They have a presynaptic localization and are negatively coupled to adenylyl cyclase. Whereas H1 and H2 receptors are located both centrally and peripherally, H3 receptors are expressed exclusively in the central nervous system [12] (see Table 1), and high densities are found in the basal ganglia. These receptors mediate the presynaptic inhibition of histamine release and the release of other neurotransmitters. Finally, H4 receptors are mostly found in the bone marrow.

The central histamine system is also involved in many central nervous system functions other than vigilance control, such as anxiety, activation of the sympathetic nervous system, stress-related release of hormones from the pituitary and of central aminergic neurotransmitters, antinociception, water retention, and suppression of hunger.

Anatomical studies have shown that the TMN receives afferent inputs from the prefrontal/infralimbic cortex, all septal regions, and several cell groups of the hypothalamus, particularly the preoptic/anterior areas [13, 14] (Fig. 2). Monoaminergic inputs originate from the adrenergic cell groups C1–C3, the noradrenergic cell groups A1–A2, and the serotonergic cell groups B5–B9 [15]. Electrophysiological studies in brain slices revealed excitatory inputs mediated by AMPA and NMDA receptors from the lateral preoptic area and the lateral hypothalamus [16].

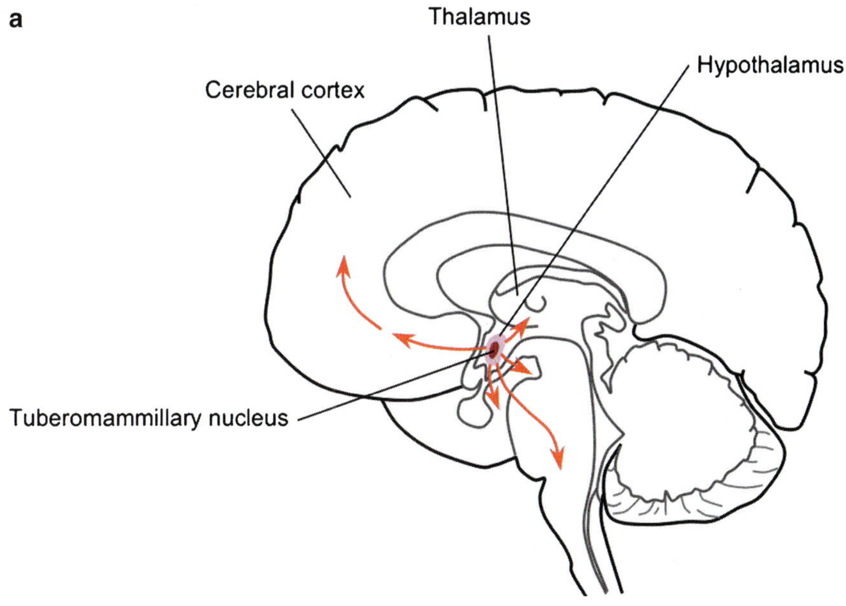

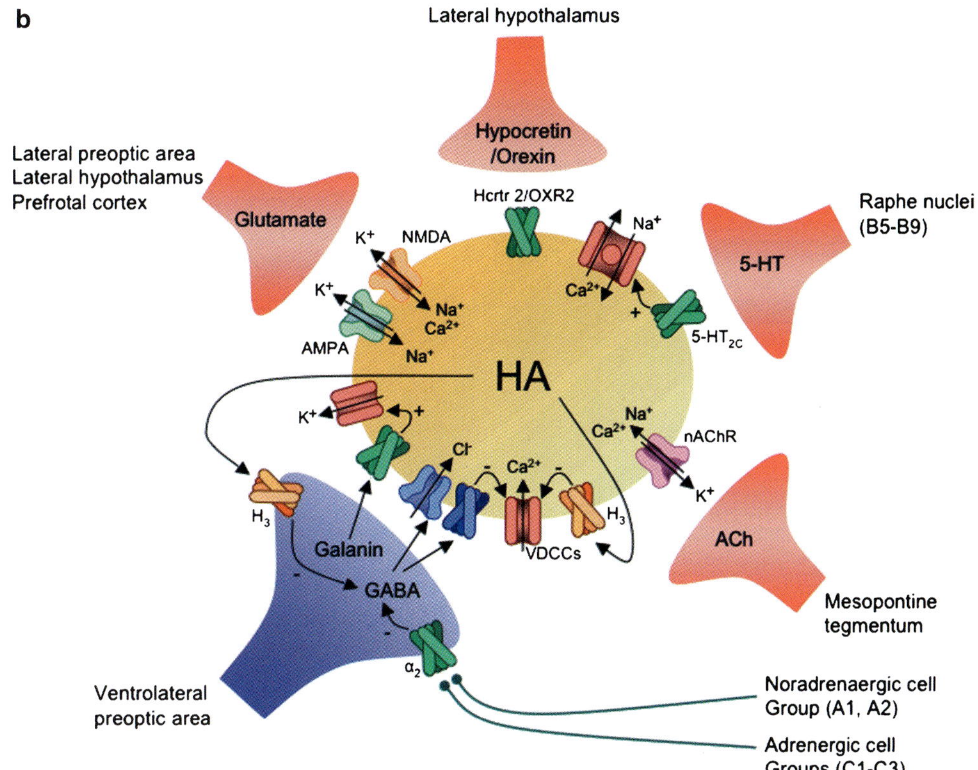

Fig. 1 (**a**) Th histaminergic system in the brain. In the CNS, histamine is produced exclusively in the tuberomammillary nucleus (TMN) of the hypothalamus. This small bilateral nucleus provides histaminergic input to most brain regions and to the spinal cord. (**b**) Afferent and autoregulation of histaminergic neurons. Abbreviations: *AMPA* α-amino-3-hydroxy-5-methyl-4-isoxazole propionic acid receptors, *nAChRs* nicotinic acetylcholine receptors, *NMDA N*-methyl-aspartate receptors, *Hcrtr* 2 hypocretin type 2 receptors, *VDCCs* voltage-dependent calcium channels

Table 1 Classification and characterizations of histamine receptors

Class	Coding sequence (amino acid)	Human chromosome	Location	Highest density in the brain	G protein	Autoreceptor	Specific agonist	Specific antagonist
H1	488	3p25	Brain/peripheral	Cerebellum, thalamus	Gi/o; Gq	No	2-(3-Trifluoromethy-lphenylhistamine)	Mepyramine
H2	359	5p35	Peripheral/brain	Striatum, cortex	Gs	No	Impromidine	Cimetidine
H3	445	20qTEL	Brain	Cortex, striatum	Gi/o	Yes	α-Methylhistamine	Thioperamide
H4	390	18q11.2	Bone marrow	Hardly detectable	Gi/o?	No	(Histamine)	JNJ 777120

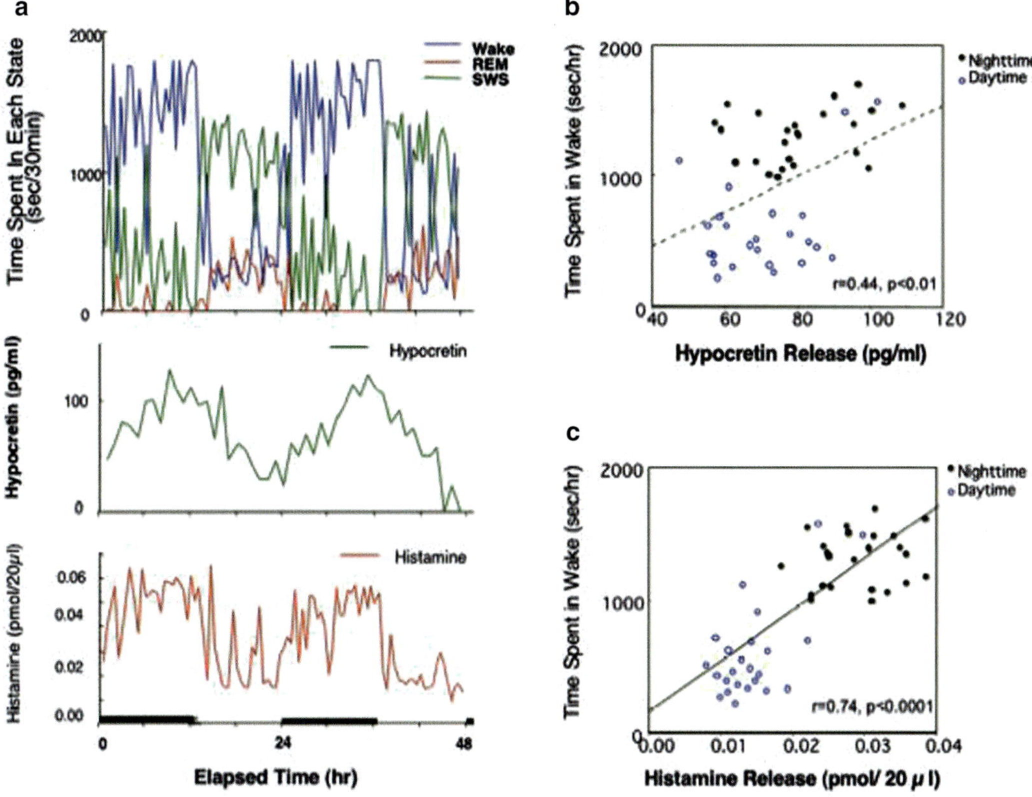

Fig. 2 Fluctuation of extracellular hypocretin and histamine levels in association with changes in the amount of wakefulness. Extracellular hypocretin and histamine levels simultaneously monitored in rats with sleep recordings under freely moving conditions using special floor-rotating rat chambers [30]. Changes in amount of wakefulness, hypocretin, and histamine levels were analyzed in 1 h time bins (the hypocretin levels were measured in 1 h time bins, while the mean of two consecutive 30 min bins were used for the histamine levels). The dark periods are indicated by *filled bars*. (**a**) Typical data from one animal. Amount of wakefulness is displayed as s/h, while hypocretin and histamine concentrations are shown as pg/ml and pmol/20 μl, respectively. (**b**, **c**) Correlations between time spent in wake and hypocretin/histamine releases in four rats. *Filled symbols* indicate data collected during dark periods. Extracellular histamine levels (**c**) are significantly (and positively) correlated with time spent in wakefulness across 24 h, while hypocretin levels (**b**) were clustered in light and dark periods, and the overall correlation between hypocretin levels and time spent in wakefulness is less significant

TMN neurons also receive ascending (presumed cholinergic) input from the mesopontine tegmentum [14] which is probably important for brain arousal mechanisms. TMN neurons receive a strong inhibitory GABAergic input that is responsible for decreasing their activity during sleep [5, 6, 17]. Anatomical and electrophysiological techniques have demonstrated GABAergic input to the TMN from the diagonal band of Broca, lateral hypothalamus, and from the most important sleep-promoting center, the VLPO [6]. In addition to these afferent systems, it has been shown that hypocretin fibers densely innervate the soma and proximal dendrites of TMN cells.

Parmentier and colleagues had shown that mice lacking histamine (i.e., HDC KO mice) show an increase in REM sleep [18]. Although no major difference was noted in the daily amount of spontaneous wake, HDC KO mice showed a deficit of wake at lights-off and signs of somnolence. These animals also showed decreased sleep latencies after various behavioral stimuli. This finding may indicate a reduced sensitivity to alerting stimuli, similar to the phenotype seen in human and canine narcolepsy [18]. Besides vigilance control, histamine is also involved in locomotor regulation [19]. Interestingly, both H1 receptor KO and HDC KO mice exhibit reduced activity during the active phase [19] (Yanai, personal communication), a pattern similar to the changes in rest/activity distribution seen in canine and human narcolepsy. These experiments indicate that genetic alterations in histaminergic transmission in animals can cause sleep abnormalities consistent with those observed in subjects with primary EDS.

Hypocretin–Histamine Interactions in Narcolepsy

After discovering the involvement of the hypocretin system in narcolepsy, several researchers studied how these deficits in neurotransmission induce narcolepsy. One of the keys to solving this question is revealing the functional differences between hypocretin receptor 1 (hcrtr 1) and hcrtr 2, since it is evident that hcrtr 2-mediated function plays more critical roles than hcrtr 1-mediated function in generating narcoleptic symptoms in animals [8, 20, 21]. In situ hybridization experiments in rats demonstrated that hcrtr 1 and hcrtr 2 mRNA exhibit marked differential distribution [22]. Hcrtr 1 is enriched in the ventromedial hypothalamic nucleus, tenia tecta, the hippocampal formation, dorsal raphe, and locus coeruleus (LC). In contrast, hcrtr 2 is enriched in the paraventricular nucleus, cerebral cortex, nucleus accumbens, ventral tegmental area, substantia nigra, and histaminergic TMN [22–24].

Of note, the TMN exclusively expresses hcrtr 2 [22], and a series of electrophysiological studies consistently demonstrated that hypocretin potently excites TMN histaminergic neurons through hcrtr 2 [9, 25, 26]. This is further confirmed by experiments using the brain slices of *hcrtr 2* KO mice. Willie et al. [8] reported that Ca^{2+} influx of TMN neurons by hypocretin-1 stimulation is abolished in these *hcrtr 2* KO mice. Furthermore, it has recently been demonstrated that the wake-promoting effects of hypocretins were totally abolished in histamine *H1* receptor KO mice, suggesting that the wake-promoting effects of hypocretin is dependent on the histaminergic neurotransmission [27].

Two groups have previously reported fluctuations of extracellular levels of histamine in the brains of animals using in vivo microdialysis techniques. Mochizuki et al. reported that histamine levels in the hypothalamus of rats show a clear diurnal variation: high during the active period and low during the resting period [28]. Strecker et al. demonstrated that histamine levels in the preoptic anterior hypothalamus in cats were high also during sleep deprivation and became lower during recovery sleep [29]. This pattern of fluctuations in histamine levels is very similar to those of extracellular hypocretin levels in rodent lateral hypothalamus and thalamus [30].

In order to study physiological correlations of the hypocretin and histamine systems in relation to the control of vigilance, we have simultaneously measured extracellular hypocretin and histamine levels in rats together with sleep recordings under freely moving conditions [31]. Two microdialysis probes were implanted in the brain (lateral hypothalamus for hypocretin measures and anterior preoptic area for histamine measures) of each rat. Microdialysis perfusates were collected in 30 min bins, and histamine and hypocretin assays were done on 1 h pooled bin samples. We found that both hypocretin-1 and histamine fluctuate significantly over 24 h periods, and are high during active periods and low during resting periods (Fig. 2). Histamine and hypocretin-1 levels appeared to fluctuate in correlation with the amount of wakefulness, but histamine changed more rapidly than hypocretin-1. Correlations between the amounts of wake, histamine levels, and hypocretin levels are presented in Fig. 2b, c. As demonstrated previously [30], and although hypocretin levels are positively correlated with the amount of wakefulness, the levels are rather clustered in two groups: the amounts of wakefulness in respective time periods during day and night are very different but the hypocretin levels are in a similar range. In contrast, histamine levels showed a better correlation with the amount of wakefulness over 24 h. By applying the stepwise, multiple regression analysis, a significant positive correlation between amounts of wakefulness and histamine levels are sustained, but not between wake amount and hypocretin levels. These results suggest that these two wake-promoting systems may have distinct roles for sleep–wake control, and the histaminergic system may be more tightly associated with the rapid state changes, while hypocretin tonus may govern much slower diurnal/circadian distribution of wake and sleep.

Pathophysiology

Decreased Histaminergic Neurotransmission has been Suggested to be Involved in Both hcrtr 2-Mutated Narcoleptic Dogs as well as in Hypocretin Ligand-Deficit Narcoleptic Dogs and Mice

The anatomical and functional interactions between the hypocretin and histaminergic systems suggest that altered histaminergic function is also involved in the generation of sleepiness in narcolepsy which is caused by a deficiency in hypocretin neurotransmission. To test this hypothesis, we measured histamine concentrations in the brains of narcoleptic ($n=9$) and control Dobermans ($n=9$). As a reference, contents of dopamine (DA), norephinephrine (NE), and serotonin (5HT) and their metabolites were also measured [32]. We found that histamine content in cortex and thalamus (the areas important in the control of wakefulness via histaminergic input [10]) was significantly lower in narcoleptic Dobermans compared to controls (Fig. 2). Brain tissue was collected during daytime, and separate sleep studies demonstrated that sleep amounts of narcoleptic dogs during the daytime did not significantly differ (besides more fragmentation) from those in control dogs [33], which suggests that these changes are not secondary to the change in sleep amount in these animals. Considering that hypocretins strongly excite TMN histaminergic neurons in vitro through hcrtr 2 stimulation [8, 9, 25, 26], the decrease in histaminergic content found in narcoleptic dogs may be due to the lack of excitatory input of hypocretin on TMN histaminergic neurons.

In contrast to histamine content, DA, NE, and 5HT contents in the brains of dogs with familial narcolepsy were high in these structures, and the increase of DA and NE in the cortex was statistically significant (Fig. 3). Two independent studies previously reported altered catecholamine contents in the brains of narcoleptic dogs [34, 35]. Both studies found an increase of DA and NE in many brain structures, especially of DA in

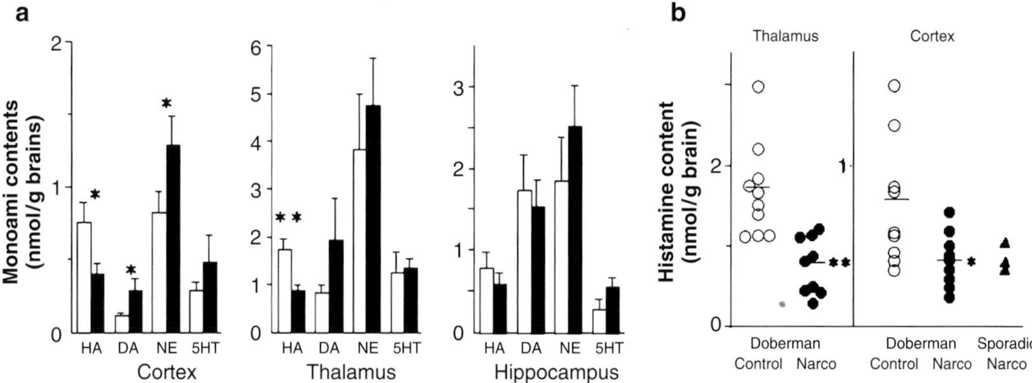

Fig. 3 Histamine, DA, NE, and 5HT contents in the cortex, thalamus, and hippocampus in *hcrtr 2*-mutated narcoleptic and control Dobermans as well as in three sporadic hypocretin-ligand deficient narcoleptic dogs. (**a**) Histamine (HA) content in the cortex and thalamus was significantly lower in narcoleptic Dobermans (*filled bars*) compared to controls (*open bars*), while DA and NE levels were higher in these structures. Increases were statistically significant in the cortex (**$p<0.01$, *$p<0.05$, by Student's t-test). (**b**) Scatter plots of histamine contents in the thalamus and cortex in each animal. Histamine contents in the cortex of three sporadic (ligand-deficient) narcoleptic dogs were also measured and these were lower than those of control Dobermans (adapted from [32])

the amygdala and of NE in the pontis reticularis oralis [34, 35]. These changes are not due to a reduction of turnover of these monoamines, since their turnover is rather high or at least not altered [32]. Compounds that enhance DA and NE transmission significantly improve symptoms of narcolepsy in animals (see Fig. 1), thus increases in DA and NE contents may be compensatory, either mediated by hcrtr 1, or by other neurotransmitter systems. Therefore, uncompensated low histamine levels in narcolepsy suggest that the hypocretin system may be the major excitatory input to histaminergic neurons.

We also measured histamine in the brains of three sporadic (ligand-deficient) narcoleptic dogs and found that histamine contents in these animals were equally low as in the *hcrtr 2*-mutated narcoleptic Dobermans (Fig. 3b) [32], thus suggesting that a decrease in histamine neurotransmission may also exist in ligand-deficient human narcolepsy. This finding was further confirmed in narcoleptic mice with hypocretin deficiency. Since our special floor-rotating rat chambers are powerful for in vivo microdialysis experiments with sleep recordings, we have developed mouse floor-rotating chambers (see Fig. 2). Using these mouse chambers, we have measured histamine release in the brain of hypocretin ligand deficient narcoleptic mice (i.e., orexin/ataxin-3 transgenic mice) over 3 days with simultaneous sleep recordings [36]. The histamine release in narcoleptic mice was compared with those in wild-type mice. Histamine exhibits a clear diurnal fluctuation pattern, and levels are high during dark (active) periods and low during light (resting) periods in both wild-type and narcoleptic mice. Histamine contents were positively correlated with amount of wakefulness in both narcoleptic and wild-type mice. We, however, found that mean histamine levels in the microdialysis perfusate in narcoleptic animals were much lower (about 50% reduced) compared to wild-type mice throughout the recoding period [36].

Altered histamine levels are unlikely to be secondary to the changes in sleep of narcoleptic mice considering the fact that low histamine levels were consistently observed in narcoleptic mice during the light period when both narcoleptic and wild-type mice spent a similar amount of time in sleep. Thus, altered histamine levels may play critical roles in the phenotype expression of the narcoleptic mice.

Human Narcolepsy and Other Hypersomnia Disorders of Central Origin Are Associated with Decreased Histaminergic Neurotransmission

Finally, we evaluated the histaminergic neurotransmission in human narcolepsy. Since the availability of the *postmortem* brain tissue in human narcolepsy is very limited, we examined CSF samples for the assessment of central histaminergic neurotransmission. However, the activity of both histaminergic neurons and (non-neuronal) brain mast cells affect histamine levels in the CSF. In order to characterize the physiology of histamine in the CSF, we performed first an animal study and measured the CSF histamine levels in rats across 24 h, after the administration of thioperamide (a H3 receptor antagonist), and after sleep deprivation [37].

We found that thioperamide (5 mg/kg, i.p.) significantly increased CSF histamine levels with little effects on locomotor activation. We also observed that the mean of the CSF histamine levels during the dark period was significantly higher than that during the light period. After 6 h sleep deprivation from ZT0, CSF histamine levels increased significantly in comparison with controls whose sleep were not deprived. Since H3 antagonist are known to enhance terminal histamine releases from histaminergic neurons, the activity of neuronal histamine is likely to be reflected in the CSF histamine levels. These results in rats suggest that CSF histamine levels at least partially reflect the central histamine neurotransmission and vigilance state changes.

We then conducted two clinical studies for evaluating CSF histamine in narcolepsy. In the first study, we included patients with narcolepsy with low CSF hypocretin-1 ($n=34$, all with cataplexy), narcolepsy without low CSF hypocretin-1 ($n=24$, 75% with cataplexy), and normal controls ($n=23$) [38]. Narcoleptic subjects with and without hypocretin deficiency were included in order to determine if histamine neurotransmission is dependent on the hypocretin status of each subject. A significant reduction of CSF histamine levels was found in the cases with low CSF hypocretin-1, and levels were intermediate in other narcolepsy cases. Mean CSF histamine levels were 133.2±20.1 pg/ml in narcoleptic subjects with low CSF hypocretin-1, 233.3±46.5 pg/ml in patients with normal CSF hypocretin-1, and 300.5±49.7 pg/ml in controls. Our results confirmed impaired histaminergic neurotransmission in human narcolepsy, but this is not entirely dependent on the hypocretin system.

We also examined CSF histamine levels in narcolepsy and other sleep disorders in a Japanese population (Fig. 4). This second clinical study included 67 narcolepsy subjects, 26 idiopathic hypersomnia (IHS) subjects, 16 obstructive sleep apnea syndrome (OSAS) subjects, and 73 neurological controls [39]. We found significant reductions in CSF histamine levels in hypocretin-deficient narcolepsy with cataplexy (mean±SEM; 176.0±25.8 pg/ml), hypocretin nondeficient narcolepsy with cataplexy (97.8± 38.4 pg/ml), hypocretin nondeficient narcolepsy without cataplexy (113.6±16.4 pg/ml), and idiopathic hypersomnia (161.0±29.3 pg/ml), while the levels in OSAS (259.3±46.6 pg/ml) did not statistically differ from those in the controls (333.8±22.0 pg/ml). Low CSF histamine levels were mostly observed in nonmedicated patients, and significant reductions in histamine levels were evident in nonmedicated patients with hypocretin-deficient narcolepsy with cataplexy (112.1±16.3 pg/ml) and

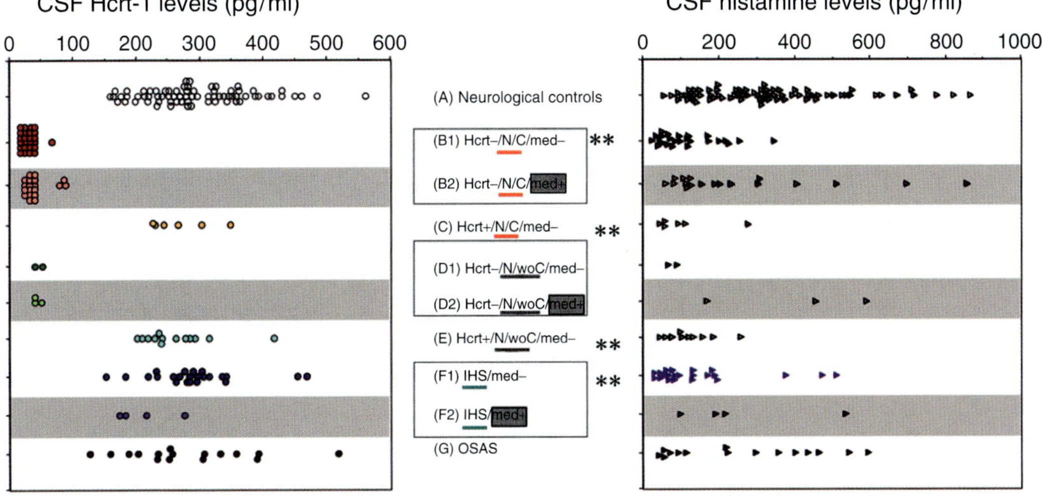

Fig. 4 CSF Hcrt-1 and histamine values for individuals with sleep–wake disorders. CSF Hcrt-1 (*left panel*) and histamine (*right panel*) values for each individual are plotted. The patient groups are indicated as Group A to Group G from above (see Table 1 for the group definition). The results of the subjects with CNS stimulants (*shadowed*) and without CNS stimulants medication are presented separately in the figure. The cutoff value of CSF hypocretin-1 level (less than or equal to 110 pg/ml) clearly segregated hypocretin deficiency from nondeficiency. None of the patients with idiopathic hypersomnia and OSAS showed hypocretin deficiency. We found significant reductions in CSF histamine levels in hypocretin deficient (B: 176±25.8 pg/ml) and nondeficient narcolepsy with cataplexy (C: 97.8±38.4 pg/ml), hypocretin nondeficient narcolepsy without cataplexy (E: 113.6±16.4 pg/ml) and idiopathic hypersomnia (F: 161.0±29.3 pg/ml), while those in hypocretin-deficient narcolepsy without cataplexy (D: 273.6±105 pg/ml) and OSAS (G: 259.3±46.6 pg/ml) were not statistically different from those in the control range (A: 333.8±22.0 pg/ml). The low CSF histamine levels were mostly observed in nonmedicated patients, and significant reductions in histamine levels were observed only in nonmedicated patients with hypocretin-deficient narcolepsy with cataplexy (B1: 112.1±16.3 pg/ml) and idiopathic hypersomnia (F1: 143.3±28.8 pg/ml). The levels in the medicated subjects are in the normal range (B2: 256.6±51.7 pg/ml and F2: 259.5±94.9 pg/ml). Nonmedicated subjects had a tendency for low CSF histamine levels in hypocretin-deficient narcolepsy without cataplexy (D1: 77.5±11.5 pg/ml) (adapted from [39])

idiopathic hypersomnia (143.3±28.8 pg/ml), while the levels in the medicated patients were in the normal range. Similar degrees of reduction, as seen in hypocretin-deficient narcolepsy with cataplexy were also observed in hypocretin nondeficient narcolepsy and in idiopathic hypersomnia, while those in OSAS (noncentral nervous system hypersomnia) were not altered. These results confirmed the result of the first study, but further suggest that an impaired histaminergic system may be involved in mediating sleepiness in a much broader category of patients with central EDS. The decrease of histamine in these subjects was more specifically observed in nonmedicated subjects, suggesting that CSF histamine is a biomarker reflecting the degree of hypersomnia of central origin. As histamine is a wake-promoting amine known to decrease during sleep, decreased histamine could either passively reflect or partially mediate daytime sleepiness in these pathologies.

Pharmacology

H3 Antagonists, New Wake-Promoting Compounds for the Treatment of Narcolepsy and Other Hypersomnia

If histamine deficiency is involved in hypersomnia of central origin, histaminomimetic compounds may have therapeutic applications for these conditions. The use of H1 agonists, a logical possibility, is made impossible by the lack of available CNS penetrating compounds and due to peripheral side effects. Current industry interest is therefore mostly focused on the H3 receptor, a receptor known to be, among other actions, an autoreceptor located on histaminergic cell bodies exclusively in the brain. Stimulation of this receptor is sedative while antagonists promote wakefulness (or reduce deep sleep) in rodents and dogs [10, 40]. Earlier experiments in narcoleptic canines have found anticataplectic and wake-promoting effects for H3 antagonists and H3 inverse agonists [40]. H3 antagonists enhance wakefulness in normal rats and cats [41]

We had also evaluated the effects of H3 antagonists on sleep in narcoleptic mice (i.e., orexin/ataxin-3 mice). Thioperamide (1 and 4 mg/kg, i.p.), a prototypical H3 antagonist, administered at ZT4 (during the normal sleep phase of the diurnal cycle) significantly enhanced wakefulness and reduced NREM in wild-type mice, and moderately reduced REM sleep [42]. The wake-promoting effects of thioperamide were very potent; the wake-promoting effect of 4 mg/kg of thioperamide was equipotent to that of 1 mg/kg of methamphetamine. We observed similar effects on wake and NREM in orexin/ataxin-3 mice. Wake-promoting effects of thioperamide were abolished in histamine H3 receptor KO mice [43], suggesting that the wake-promoting effects of thioperamide are at least mediated by H3 receptors.

Thioperamide possess an imidazoline ring (i.e., imidazoline compounds) and is likely to bind to imidazoline receptors. Imidazoline receptors are a newly discovered family of receptors, some of which (like alpha 2-adrenoceptors), have a presynaptic inhibitory effect on the release of norepinephrine, and thus imidazoline compounds may have an agonistic or antagonistic effect on these receptors, and the paradoxical effects on sleep seen in H3 receptor KO mice might be mediated by an interaction with the imidazol binding sites.

We had also tested one of the new H3 antagonists, RWJ 662733 (JNJ-10181457), and the effects of nighttime administration of the compound (0.3, 3, 10 mg/kg, i.p.) on sleep were evaluated [44]. The effects are similar but more striking than those of thioperamide [42], and this compound normalized the major sleep abnormalities seen in narcoleptic mice (increase in wakefulness, decrease in sleep and DREM, objective measures of abnormal REM sleep transitions seen in narcolepsy) to the level of wild-type animals, and the fragmentation of wakefulness is also normalized.

Lin and colleagues also tested tiprolisant, an inverse H3-receptor agonist in preprohypocretin KO mice and human narcoleptic subjects [45]. The authors found that tiprolisant (20 mg/kg, p.o.) promoted wakefulness and decreased abnormal DREMs in these animals. In addition, in a pilot single-blind trial on 22 patients receiving a placebo followed by tiprolisant (40 mg/day), both for 1 week, the Epworth Sleepiness Scale (ESS) score was reduced from a baseline value of 17.6 by 1.0 with the placebo ($p > 0.05$) and by 5.9 with tiprolisant ($p < 0.001$), suggesting H3 antagonists/reverse agonists may be useful to treat EDS in human narcolepsy.

Guo et al. also evaluated another novel H3 antagonist GSK189254 (3 and 10 mg/kg, p.o.) in preproorexin KO mice and found that acute administration of GSK189254 increased wake and decreased NREM sleep and REM sleep to a similar degree to modafinil (64 mg/kg), while it also reduced DREM in preproorexin KO mice [46]. After twice daily dosing for 8 days, the effect of GSK189254 (10 mg/kg) on wake in both preproorexin KO mice and wild-type mice was completely disappeared. Therefore, a rapid tolerance as has been generally observed with hypnotic H1 blockers may be the concern for clinical application of H3 antagonists. Similarly, long-term side effects of these compounds are also unknown at this moment.

Acute Deprivation of Histamine Does Not Modify Sleep–Wake Cycle and Hypocretin Release

As described in the physiology section, the rate-limiting enzyme of histamine synthesis is HDC, and a specific inhibitor of the HDC, α-FMH has been identified and used for various in vivo and in vitro experiments [47–49]. We therefore evaluated the effects of α-FMH on histamine release and sleep–wake cycle in rats and found that i.p. administration of α-FMH (100 mg/kg, i.p.) abolishes extracellular histamine levels by 66% over 24 h [31]. To our surprise, however, we did not observe any changes in the sleep parameters after administration of α-FMH [31]. This result that acute depletion of histamine synthesis does not have an impact on spontaneous sleep is difficult to reconcile if the histamine system is important for the regulation of physiological sleep. Interestingly, this large change in histamine release by α-FMH did not have a great impact on the diurnal fluctuation pattern of hypocretin release as similar hypocretin release patterns were continuously observed over 3 days. This suggests that the histamine system is located downstream of the hypocretin system and that the histaminergic system does not provide much feedback on hypocretin signaling. This is consistent with in vitro electrophysiological experiments showing that DA and NE inhibit hypocretin neuronal activity, but histamine does not [50].

One of the limitations of this experiment is the use of only one species; therefore, we also evaluated the effects of α-FMH in mice [51]. The animals were subjected to injection of three doses (50, 100, and 200 mg/kg i.p.) of α-FMH and a vehicle administration during dark periods (ZT12). In contrast to the results observed in rats, α-FMH doses of up to 100 mg/kg decreased wake by about −20% (6 h total-amount after injection) and enhanced NREM (400–700%) and REM sleep (40–300%) in a dose-dependent manner. These results (in contrast to those obtained in rats) may suggest that significant species differences likely exist in the roles of histamine in sleep–wake control.

Discussion

Multiple Wake-Promoting Systems

Based on our results that acute histamine deprivation does not alter sleep in rats, we hypothesize that histamine is critical for the physiology during wakefulness, but not for sleep–wake regulation by means of spontaneous EEG changes, at least for some tested animal species. Histamine release is tightly correlated with the amounts of wake, it declines at intermitted sleep episodes during active period, and it increases during wake in resting periods. Several authors proposed that histamine is critical for maintaining normal attention and cognition during wakefulness [52]. Impairing histamine signaling by H1 antagonists and due to neurological conditions induces reduced attention and cognition. It is also possible that histamine critically inhibits wakefulness under more specific conditions that we did not examine, e.g., in response to novelty [18] and other conditions of hyperarousal, but this notion does not contradict with our hypothesis that histamine may be more critical with attention or cognition during wakefulness.

In addition to histaminergic neurons in the TMN, there are several other neuronal groups implicated in central arousal. These include hypocretin neurons in the lateral hypothalamus, cholinergic neurons in the basal forebrain and mesopontine tegmentum, noradrenergic neurons in the LC, and serotonergic neurons in the dorsal raphe [53, 54]. VLPO neurons project to all of these brain areas, raising the possibility that spontaneous wake and sleep requires the simultaneous excitation and inhibition of multiple arousal systems [53]. Each of these multiple wake-promoting systems may have a distinct role. A good example is the dopaminergic system; not much attention has been paid to an involvement of dopaminergic system in the regulation of sleep–wake cycle. This is due to the fact that dopaminergic neurons in ventral tegmental area or substantia nigra do not change their firing rate across sleep cycles [55], yet there are changes in the firing pattern. Pharmacological studies, however, consistently reported that modes of action of most currently available wake-promoting agents, such as amphetamines and modafinil are mediated via stimulation of dopaminergic neurotransmission [56]. Since dopamine is involved in mood or motivation, the dopaminergic system may be more involved in forced wakefulness or motivated wakefulness that requires for more particular circumstances [54].

To be fully alert, each of these multiple wake-promoting systems with distinct roles may need to be activated and well coordinated.

References

1. Brown RE, Stevens DR, Haas HL. The physiology of brain histamine. Prog Neurobiol. 2001;63:637–72.
2. Haas H, Panula P. The role of histamine and the tuberomamillary nucleus in the nervous system. Nat Rev Neurosci. 2003;4:121–30.
3. Steininger TL, Alam MN, Gong H, Szymusiak R, McGinty D. Sleep-waking discharge of neurons in the posterior lateral hypothalamus of the albino rat. Brain Res. 1999;840:138–47.
4. Vanni-Mercier G, Sakai K, Jouvet M. Wake-state specific neurons for wakefulness in the posterior hypothalamus in the cat. C R Acad Sci III. 1984;298:195–200.
5. Sherin J, Shiromani P, McCarley R, Saper C. Activation of ventrolateral preoptic neurons during sleep. Science. 1996;271:216–20.
6. Sherin JE, Elmquist JK, Torrealba F, Saper CB. Innervation of histaminergic tuberomammillary neurons by GABAergic and galaninergic neurons in the ventrolateral preoptic nucleus of the rat. J Neurosci. 1998;18:4705–21.
7. Nishino S. Clinical and neurobiological aspects of narcolepsy. Sleep Med. 2007;8:373–99.
8. Willie JT, Chemelli RM, Sinton CM, Tokita S, Williams SC, Kisanuki YY, et al. Distinct narcolepsy syndromes in Orexin receptor-2 and Orexin null mice: molecular genetic dissection of Non-REM and REM sleep regulatory processes. Neuron. 2003;38:715–30.
9. Yamanaka A, Tsujino N, Funahashi H, Honda K, Guan JL, Wang QP, et al. Orexins activate histaminergic neurons via the orexin 2 receptor. Biochem Biophys Res Commun. 2002;290:1237–45.
10. Lin JS. Brain structures and mechanisms involved in the control of cortical activation and wakefulness, with emphasis on the posterior hypothalamus and histaminergic neurons. Sleep Med Rev. 2000;4:471–503.
11. Salmun LM. Antihistamines in late-phase clinical development for allergic disease. Expert Opin Investig Drugs. 2002;11:259–73.
12. Chen J, Liu C, Lovenberg TW. Molecular and pharmacological characterization of the mouse histamine H3 receptor. Eur J Pharmacol. 2003;467:57–65.
13. Wouterlood FG, Gaykema RP, Steinbusch HW, Watanabe T, Wada H. The connections between the septum-diagonal band complex and histaminergic neurons in the posterior hypothalamus of the rat. Anterograde tracing with *Phaseolus vulgaris*-leucoagglutinin combined with immunocytochemistry of histidine decarboxylase. Neuroscience. 1988;26:827–45.
14. Ericson H, Blomqvist A, Kohler C. Origin of neuronal inputs to the region of the tuberomammillary nucleus of the rat brain. J Comp Neurol. 1991;311:45–64.
15. Ericson H, Blomqvist A, Kohler C. Brainstem afferents to the tuberomammillary nucleus in the rat brain with special reference to monoaminergic innervation. J Comp Neurol. 1989;281:169–92.
16. Yang QZ, Hatton GI. Histamine mediates fast synaptic inhibition of rat supraoptic oxytocin neurons via chloride conductance activation. Neuroscience. 1994; 61:955–64.
17. Stevens DR, Kuramasu A, Haas HL. GABAB-receptor-mediated control of GABAergic inhibition in rat histaminergic neurons in vitro. Eur J Neurosci. 1999;11:1148–54.
18. Parmentier R, Ohtsu H, Djebbara-Hannas Z, Valatx JL, Watanabe T, Lin JS. Anatomical, physiological, and pharmacological characteristics of histidine decarboxylase knock-out mice: evidence for the role of brain histamine in behavioral and sleep-wake control. J Neurosci. 2002;22:7695–711.
19. Inoue I, Yanai K, Kitamura D, Taniuchi I, Kobayashi T, Niimura K, et al. Impaired locomotor activity and

exploratory behavior in mice lacking histamine H1 receptors. Proc Natl Acad Sci U S A. 1996;93: 13316–20.

20. Lin L, Faraco J, Li R, Kadotani H, Rogers W, Lin X, et al. The sleep disorder canine narcolepsy is caused by a mutation in the hypocretin (orexin) receptor 2 gene. Cell. 1999;98:365–76.
21. Ripley B, Fujiki N, Okura M, Mignot E, Nishino S. Hypocretin levels in sporadic and familial cases of canine narcolepsy. Neurobiol Dis. 2001;8: 525–34.
22. Marcus JN, Aschkenasi CJ, Lee CE, Chemelli RM, Saper CB, Yanagisawa M, et al. Differential expression of orexin receptors 1 and 2 in the rat brain. J Comp Neurol. 2001;435:6–25.
23. Trivedi P, Yu H, MacNeil DJ, Van der Ploeg LH, Guan XM. Distribution of orexin receptor mRNA in the rat brain. FEBS Lett. 1998;438:71–5.
24. Lu XY, Baganol D, Bagonol C, Lei FM, Burke S, Akil H, et al. Expression of orexin 1 and orexin 2 receptor mRNA are differently regulated in the rat brain by food deprivation. Abstr Soc Neurosci. 1999;25:958.
25. Bayer L, Eggermann E, Serafin M, Saint-Mleux B, Machard D, Jones B, et al. Orexins (hypocretins) directly excite tuberomammillary neurons. Eur J Neurosci. 2001;14:1571–5.
26. Eriksson KS, Sergeeva O, Brown RE, Haas HL. Orexin/hypocretin excites the histaminergic neurons of the tuberomammillary nucleus. J Neurosci. 2001;21:9273–9.
27. Huang ZL, Qu WM, Li WD, Mochizuki T, Eguchi N, Watanabe T, et al. Arousal effect of orexin A depends on activation of the histaminergic system. Proc Natl Acad Sci U S A. 2001;98:9965–70.
28. Mochizuki T, Yamatodani A, Okakura K, Horii A, Inagaki N, Wada H. Circadian rhythm of histamine release from the hypothalamus of freely moving rats. Physiol Behav. 1992;51:391–4.
29. Strecker RE, Nalwalk J, Dauphin LJ, Thakkar MM, Chen Y, Ramesh V, et al. Extracellular histamine levels in the feline preoptic/anterior hypothalamic area during natural sleep-wakefulness and prolonged wakefulness: an in vivo microdialysis study. Neuroscience. 2002;113:663–70.
30. Yoshida Y, Fujiki N, Nakajima T, Ripley B, Matsumura H, Yoneda H, et al. Fluctuation of extracellular hypocretin-1 (orexin A) levels in the rat in relation to the light-dark cycle and sleep-wake activities. Eur J Neurosci. 2001;14:1075–81.
31. Yoshida Y, Nishino S, Ishizuka T, Yamatodani A. Vigilance change, hypocretin and histamine release in rats before and after a histamine synthesis blocker (alpha-FMH) administration. Sleep. 2005;28:A18.
32. Nishino S, Fujiki N, Ripley B, Sakurai E, Kato M, Watanabe T, et al. Decreased brain histamine contents in hypocretin/orexin receptor-2 mutated narcoleptic dogs. Neurosci Lett. 2001;313:125–8.
33. Nishino S, Riehl J, Hong J, Kwan M, Reid M, Mignot E. Is narcolepsy REM sleep disorder? Analysis of sleep abnormalities in narcoleptic Dobermans. Neurosci Res. 2000;38:437–46.
34. Faull KF, Zeller-DeAmicis LC, Radde L, Bowersox SS, Baker TL, Kilduff TS, et al. Biogenic amine concentrations in the brains of normal and narcoleptic canines: current status. Sleep. 1986;9: 107–10.
35. Mefford IN, Baker TL, Boehme R, Foutz AS, Ciaranello RD, Barchas JD, et al. Narcolepsy: biogenic amine deficits in an animal model. Science. 1983;220:629–32.
36. Yoshida Y, Ishizuka T, Yamatodani A, Okuro M, Nishino S. Brain histamine release in freely moving narcoleptic and wild type mice. Sleep. 2009; 32:A12.
37. Soya S, Song YH, Kodama T, Honda Y, Fujiki N, Nishino S. CSF histamine levels in rats reflect the central histamine neurotransmission. Neurosci Lett. 2008;430:224–9.
38. Nishino S, Sakurai E, Nevsimalova S, Yoshida Y, Watanabe T, Yanai K, et al. Decreased CSF histamine in narcolepsy with and without low CSF hypocretin-1 in comparison to healthy controls. Sleep. 2009;32: 175–80.
39. Kanbayashi T, Kodama T, Kondo H, Satoh S, Inoue Y, Chiba S, et al. CSF histamine contents in narcolepsy, idiopathic hypersomnia and obstructive sleep apnea syndrome. Sleep. 2009;32:181–7.
40. Tedford CE, Edgar DM, Seidel WF, Mignot E, Nishino S, Pawlowski GP, et al. Effects of a novel, selective, and potent histamine H3 receptor antagonist, GT-2332, on rat sleep/wakefulness and canine cataplexy. Abstr Soc Neurosci. 1999;25:1134.
41. Lin JS, Sakai K, Vanni-Mercier G, Arrang JM, Garbarg M, Schwartz JC, et al. Involvement of histaminergic neurons in arousal mechanisms demonstrated with H3-receptor ligands in the cat. Brain Res. 1990;523:325–30.
42. Shiba T, Fujiki N, Wisor J, Edgar D, Sakurai T, Nishino S. Wake promoting effects of thioperamide, a histamine H3 antagonist in orexin/ataxin-3 narcoleptic mice. Sleep. 2004;27(suppl):A241–2.
43. Okuro M, Matsumura M, Fujiki N, Nishino S. Evaluations of wake promoting effects of histamine h3 antagonist in h3 receptor knockout mice. Sleep. 2008;31:A220–1.
44. Fujiki N, Yoshino F, Lovenberg TW, Nishino S. Wake promoting effects of non-imidazolin histamine H3 antagonist in orexin/ataxin-3 narcoleptic mice. Sleep. 2006;29:A230.
45. Lin JS, Dauvilliers Y, Arnulf I, Bastuji H, Anaclet C, Parmentier R, et al. An inverse agonist of the histamine H(3) receptor improves wakefulness in narcolepsy: studies in orexin–/– mice and patients. Neurobiol Dis. 2008;30:74–83.
46. Guo RX, Anaclet C, Roberts JC, Parmentier R, et al. Differential effects of acute and repeat dosing with the H3 antagonist GSK189254 on the sleep-wake cycle and narcoleptic episodes in Ox–/– mice. Br J Pharmacol. 2009;157:104–17.

47. Kiyono S, Seo ML, Shibagaki M, Watanabe T, Maeyama K, Wada H. Effects of alpha-fluoromethyl-histidine on sleep-waking parameters in rats. Physiol Behav. 1985;34:615–7.
48. Morimoto T, Yamamoto Y, Mobarakeh JI, Yanai K, Watanabe T, Yamatodani A. Involvement of the histaminergic system in leptin-induced suppression of food intake. Physiol Behav. 1999;67:679–83.
49. Kollonitsch J, Perkins LM, Patchett AA, Doldouras GA, Marburg S, Duggan DE, et al. Selective inhibitors of biosynthesis of aminergic neurotransmitters. Nature. 1978;274:906–8.
50. Yamanaka A, Muraki Y, Tsujino N, Goto K, Sakurai T. Regulation of orexin neurons by the monoaminergic and cholinergic systems. Biochem Biophys Res Commun. 2003;303:120–9.
51. Fujiki N, Yoshino F, Nishino S. Vigilance change by acute histamine depletion with a-FMH in orexin/ataxin-3 narcoleptic and wild type mice. Sleep. 2008; 31:A218–9.
52. Yanai K, Okamura N, Tagawa M, Itoh M, Watanabe T. New findings in pharmacological effects induced by antihistamines: from PET studies to knock-out mice. Clin Exp Allergy. 1999;29 Suppl 3:29–36. discussion 37–8.
53. Saper CB, Scammell TE, Lu J. Hypothalamic regulation of sleep and circadian rhythms. Nature. 2005;437:1257–63.
54. Mignot E, Taheri S, Nishino S. Sleeping with the hypothalamus: emerging therapeutic targets for sleep disorders. Nat Neurosci. 2002;5(Suppl):1–6.
55. Miller JD, Farber J, Gatz P, Roffwarg H, German DC. Activity of mesencephalic dopamine and non-dopamine neurons across stages of sleep and waking in the rat. Brain Res. 1983;273:133–41.
56. Nishino S, Mignot E. CNS stimulants in sleep medicine: basic mechanisms and pharmacology. In: Kryger MH, Roth T, Dement WC, editors. Principles and practice of sleep medicine. 4th ed. Philadelphia: Elsevier Saunders; 2005. p. 468–98.

Dopaminergic Substrates Underlying Hypersomnia, Sleepiness, and REM Sleep Expression

David B. Rye and Amanda A.H. Freeman

Keywords
Dopamine • Sleep-wake regulation • Narcolepsy • REM sleep • Mesotelencephalon

Dopamine is the most abundant monoamine, and it modulates diverse behaviors including movement, motivation/reward, cognition, and feeding that share one notable feature in common – viz., each occurs during wake. Dopamine's influence upon normal and pathological states of wake and sleep has only recently begun to receive widespread attention. This rebirth of interest bears directly upon the pathophysiology and treatment of impairments in arousal encountered in narcolepsy with cataplexy, narcolepsy lacking cataplexy, the primary hypersomnias, and several psychiatric and medical conditions in which narcolepsy-like phenotypes are encountered. A comprehensive accounting presented in a recent publication includes a more complete bibliography [1]. Here, we briefly summarize current knowledge of the functional anatomy of brain dopamine networks and how they modulate wake and REM-sleep.

D.B. Rye (✉)
Department of Neurology and Program in Sleep, Emory University School of Medicine, 101 Woodruff Circle, WMRB-Suite 6000, Atlanta, GA 30322, USA
e-mail: drye@emory.edu

Physiological Effects of Dopamine

Dopamine's physiological effects are best characterized as "neuromodulatory." Rather than eliciting excitatory or inhibitory postsynaptic potentials, dopamine gates inputs via altering membrane properties and specific ion conductances [2]. This manifests as different intensities, durations, and timings of neural output commensurate with environmental and homeostatic demands. Dopamine acts by way of five subtypes of seven transmembrane domain G-protein coupled receptors (D1–D5) which based upon similarities in pharmacology, biochemistry, and amino acid homology are divided in to two classes, $D_{1\text{-like}}$ (D_1, D_5) and $D_{2\text{-like}}$ (D_2, D_3, D_4) [3]. Ligand binding studies estimate the average K_d of dopamine for its receptors to be (in nM) D_3, 55; D_4, 209; D_2, 892; and D_1, 6,500 (and hence, the order of affinity for endogenous dopamine is $D_3 > D_4 > D_2 >> D_1$) [average of values obtained from references provided by a search of http://pdsp.cwru.edu/stestw.asp]. Both $D_{1\text{-like}}$ and $D_{2\text{-like}}$ receptors are found postsynaptically on neurons targeted by dopaminergic axons. $D_{2\text{-like}}$ receptors are also localized presynaptically on dendrites and soma, and the terminals of dopaminergic cells, functioning as autoreceptors that regulate neural firing and the synthesis and release of dopamine, respectively [4]. Each receptor subtype has a unique pattern of

C.R. Baumann et al. (eds.), *Narcolepsy: Pathophysiology, Diagnosis, and Treatment*, DOI 10.1007/978-1-4419-8390-9_6, © Springer Science+Business Media, LLC 2011

brain localization. Dopamine also promotes vasoconstriction of the brain's microvasculature via receptors present on endothelial cells.

Anatomy of Mesotelencephalic Dopamine Neurons in Relation to Wake–Sleep States

The mesocortical, mesolimbic, and mesostriatal systems are the most conspicuous of the brain's dopamine circuits, and are taught to govern cognition, emotion, and movement, respectively. Anatomical and physiological evidences highlight that midbrain dopamine neurons also mediate thalamocortical arousal (Fig. 1). Excitatory inputs originate from the bed nucleus of the stria terminals and central nucleus of the amygdala, hypothalamic hypocretin (i.e., orexin) neurons, glutamatergic and cholinergic pedunculopontine tegmental neurons (PPN), medullary, noradrenergic A1 and A2, and pontine A4–A5 (locus coeruleus; LC) neurons, and act via multiple glutamate receptor subtypes, hypocretin-2, muscarinic, nicotinic, and α_1-adrenoreceptors. Serotonergic inputs from the dorsal raphe have complex effects that are generally inhibitory and predominate in mesocorticolimbic versus nigrostriatal dopamine neurons. Dopaminergic "tone" is also influenced by local GABAergic neurons that themselves appear to modulate wake and REM-sleep [5].

Outputs from midbrain dopaminergic neurons are diverse and can affect thalamocortical arousal both directly and indirectly. An extensive set of thalamic collaterals originate from traditional nigrostriatal and mesocorticolimbic pathways. Because at least half of A8–A9–A10 neurons simultaneously target the thalamus and striatum, the spectrum of neuropsychiatric disorders thought to arise from midbrain dopamine neuron dysfunction should have an accompanying disturbance of thalamocortical arousal (i.e., sleep/wake state), albeit, yet to be universally appreciated [6]. The effects of dopamine upon thalamic phys-

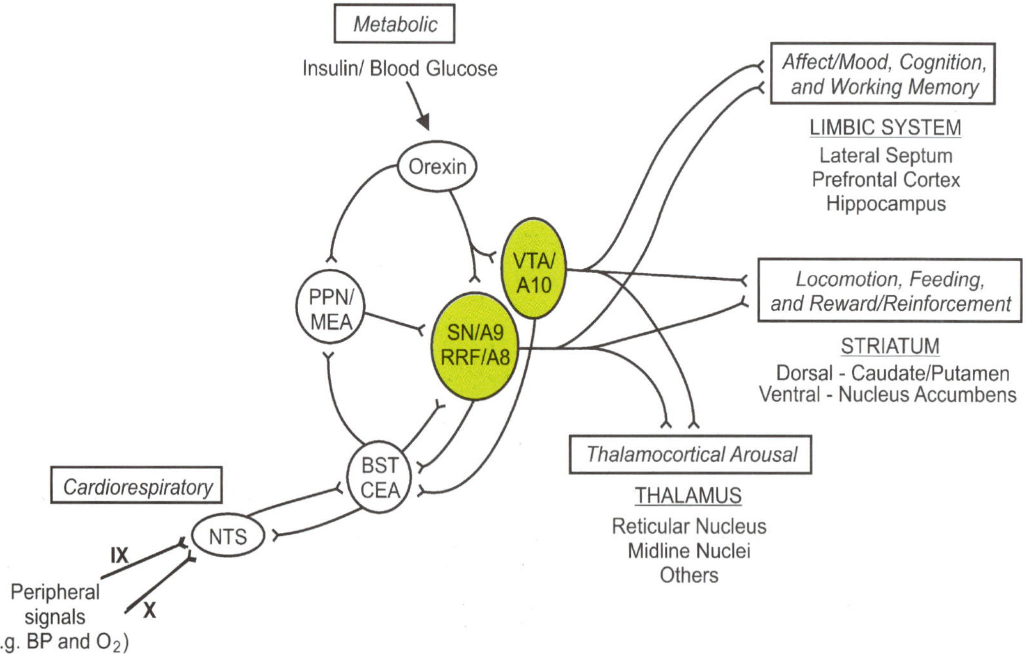

Fig. 1 The afferent and efferent connectivity of midbrain dopaminergic neurons (*highlighted in yellow*) is ideally suited to coordinate disparate behaviors inclusive of the organism's arousal state. Abbreviations: *BP* blood pressure, *BST* bed nucleus of the stria terminals, *CEA* central nucleus of the amygdala, *MEA* midbrain extrapyramidal area, *NTS* nucleus of the solitary tract, O_2 oxygen tension, *PPN* pedunculopontine tegmental nucleus, *RRF* retrorubral field, *SN* substantia nigra, *VTA* ventral tegmental area

iology are little studied and include both $D_{1\text{-like}}$ excitatory effects, $D_{2\text{-like}}$ receptor mediated presynaptic and GABAergic interneuron inhibition, and D_4-mediated inhibition of GABA release in the thalamic reticular nucleus critical to the gating of all thalamocortical transmission and spindle generation. Indirect effects of midbrain dopamine neurons upon thalamocortical arousal state might also occur by way of output pathways to the ventral forebrain, perifornical hypocretin/orexin and adjacent hypothalamic neurons (Fig. 2), and the LC (among others). The physiological and behavioral effects of these interactions are incompletely defined and include putative direct and indirect, excitatory and inhibitory effects upon magnocellular cholinergic and perifornical hypocretin/orexin neurons. Dopaminergic innervation of the LC originates from the hypothalamic A11 and A13, and midbrain VTA-A10 cell groups, and has been implicated in inhibiting REM-sleep and enhancing muscle tone in REM-sleep, as well as enhancing slow-wave and REM-sleep via α_2 adrenoreceptors. While there is experimental in vivo evidence from the preoptic area that dopamine can act via adrenoreceptors, the concentrations necessary appear to be supraphysiologic and unlikely to be encountered in nature [7].

Circadian, Homeostatic, and State Influences upon Dopamine Signaling

Diurnal rhythms in the content, turnover, release, and behavioral responsiveness of the brain's principal dopamine systems inclusive of retinal dopamine are well established. While we are far from a comprehensive understanding of what drives these rhythms, dopamine synthesis, release, and

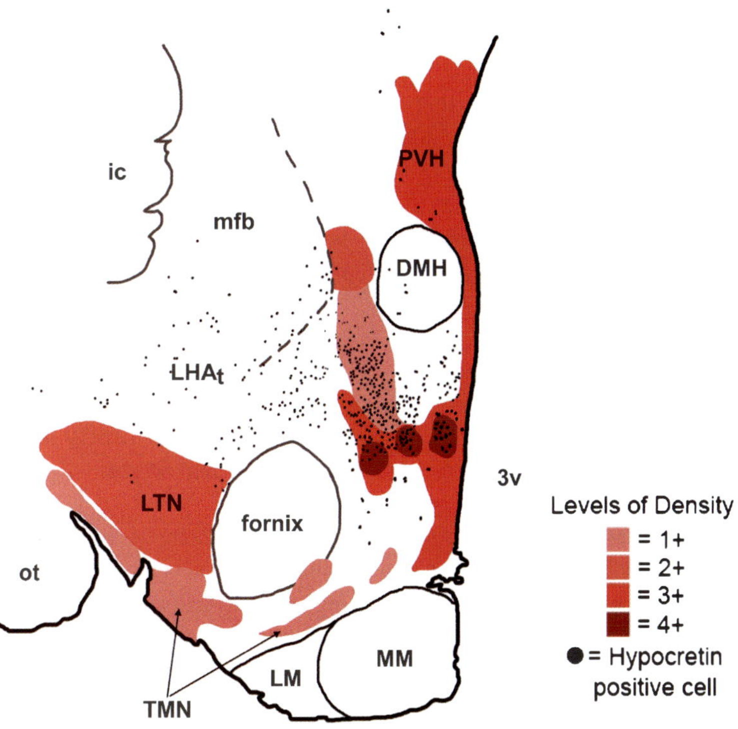

Fig. 2 Dopamine axons and terminal fields revealed by DAT immunohistochemistry in the human hypothalamus highlight their intimate relationship with hypocretin cell bodies (*black dots*), and other arousal-related nuclei such as the tuberomamillary nucleus (TMN). Abbreviations: *3v* third ventricle, *DMH* dorsomedial nucleus of the hypothalamus, *ic* internal capsule, *LHAt* tuberal region of the lateral hypothalamic area, *LM* lateral mammillary nucleus, *LTN* lateral tuberal nucleus, *mfb* medial forebrain bundle, *MM* medial mammillary nucleus, *ot* optic tract, *PVH* paraventricular nucleus of the hypothalamus

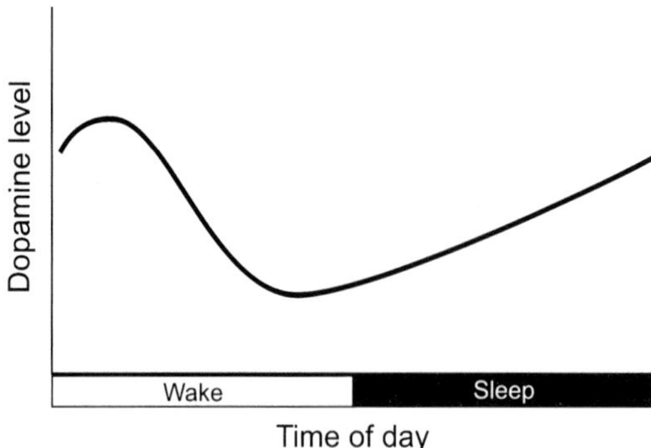

Fig. 3 Diurnal fluctuation of extracellular dopamine amounts across one 24-h activity-rest and wake–sleep cycle

signaling peak early during the active, wake period, and nadir early on in sleep (Fig. 3). Circadian influences are conveyed in part via melatonin as pinealectomy dampens striatal dopamine rhythms [8] as well as by circadian-related gene products. Increased expression and phosphorylation of tyrosine hydroxylase, and commensurate enhancements in A10 dopaminergic neural activity and associated behaviors, for example, are observed in mice expressing an inactive *Clock* protein [9]. Sleep and REM-sleep deprivation also lead to robust changes in the availability of dopamine and the density and affinity of its various receptors in mesotelencephalic circuits.

Diurnal Variations in Dopamine Signaling

Dopaminergic "tone" is most directly assessed from microdialysates collected from the behaving organism. Striatal and prefrontal cortical dopamine rise in anticipation of, and are maintained throughout the major wake period of the rat in concert with declines in dopamine metabolites (i.e., DOPAC, HVA, and 5-HIAA). This pattern is mirrored by tissue contents of (1) whole brain, striatal, cerebellar, cortical, hippocampal, midbrain, and brainstem dopamine contents; (2) tyrosine hydroxylase and tetrahydrobiopterin (both necessary for dopamine synthesis); and (3) the dopamine transporter (DAT) (a principal determinant of synaptic dopamine availability). These rhythms appear to be more pronounced in "limbic" as opposed to "motor" subcircuits although this remains somewhat controversial. Time of day also influences the kinetics of L-DOPA metabolism (clearance being highest early in wake), and receptor activation in species as diverse as fruit flies [10] and humans [11].

In nonhuman primates and humans, diurnal dopamine rhythms are evident in the brain, cerebrospinal fluid (CSF), plasma, and urine, and are generally concordant with those observed in rodents. Seminal studies of postmortem hypothalamic tissue reveal a peak in dopamine content during the active period, 1500–1800 h, with a nadir in the early morning hours. Dopamine content in the lumbar CSF [12], urine, and plasma peak at 0800–1000 h and nadir at 0400–0600 h. Similar patterns are evident for extracellular dopamine in dialysates from the amygdala and adjacent limbic cortex of humans and the putamen of nonhuman primates [13]. Afternoon and end of day dips in extracellular dopamine coincide with 10–20% reductions in $D_{1\text{-like}}$ receptor density in the putamen and frontal cortex in humans. The functional significance of these rhythms remains to be determined. More comprehensive analyses that include sleep deprivation and constant environmental conditions are needed in order to tease out the interrelationships between striatal, limbic, and cortical dopamine and homeostatic and circadian alerting signals.

Dopamine Availability in Relation to Behavioral State

There is a paucity of data on the release of dopamine during specific sleep states. Negative findings are difficult to interpret given the temporal and spatial limitations of microdialysis, and the real possibility for species and regional brain differences in dopamine release. Dopamine in the feline caudate assessed by voltammetry – whose temporal resolution is superior to that of microdialysis – fluctuates between sleep states, while microdialysates from the feline amygdala and locus coeruleus are unrevealing. Dopamine release in the more densely innervated medial prefrontal cortex and nucleus accumbens of the rat is highest in wake and REM-sleep. In the rhesus monkey putamen and human limbic brain regions, discernable fluctuations in dopamine in relation to specific sleep stages are undetectable by microdialysis.

Dopamine's Modulation of Wake–Sleep State

The regional and molecular substrates mediating dopamine's effects upon the sleep–wake continuum derive from (1) the systemic administration of receptor specific agents; (2) focal pharmacological manipulations and lesioning paradigms; (3) molecular engineering in flies and mice; and (4) animal models of disease. Hormonal release affected by hypothalamic dopamine is also relevant to this discussion since many hormones are known to modulate behavioral state. Complex feedback loops include not only the hypothalamus but also mesotelencephalic dopamine neurons themselves, since they, in turn, express receptors for many of these same hormones. The complexity of this reciprocity makes it difficult to reliably dissociate hormonal from dopaminergic influences upon state. This line of investigation is not presently being pursued, yet may ultimately have relevance to a comprehensive accounting of dopamine's modulation of behavioral state.

Exogenous Dopaminomimetic Effects on Wake–Sleep State

The proportion of time spent awake or asleep is greatly influenced by pharmacological agents that target molecules that modulate dopamine signaling. Because of differential regional expression in the central *and* peripheral nervous systems, and binding affinities of the dopamine receptor subtypes, and their pre- versus postsynaptic localizations, the effects of dopaminomimetics upon sleep–wake state are complex. Strategies to untangle this complexity have included systemic, intracerebroventricular, or focal brain delivery of pharmacological agents that activate or inactivate, more or less selectively, the individual pharmacologically and molecularly defined dopamine receptors. More recent paradigms include genetic engineering of some of the molecules involved in dopamine signaling. These studies tell a compelling story that dopamine is essential to normal arousal and rest-activity patterns.

Pharmacological agents that interfere with the primary route of removal of dopamine, namely, presynaptic reuptake via DAT, enhance wake and alpha power in the EEG in proportion to their binding affinities for DAT [14]. Genetically engineered mice lacking DAT exhibit increased activity levels and wake bout lengths, and are immune to the wake promoting effects of DAT blockers [15]. A genetically engineered lesion rendering DAT inactive in *Drosophila melanogaster* [16] yields a hyperactive mutant appropriately named fumin, whose literal translation from Japanese is "sleepless" [17]. These flies, as well as those with allelic DAT variants [18], spend an increased amount of time in the active (waking) phase. Blockade of dopamine synthesis in wild-type flies either pharmacologically or via genetic manipulation renders them unresponsive to methamphetamine-induced increases in wake-like activity [19].

The wake-promoting actions of dopamine signaling are further supported by consideration of the effects of systemic and focally applied dopaminomimetics upon sleep–wake state. These effects are dose- and receptor dependent and describe a biphasic dose–response relationship that also manifest in many other physiological and behavioral outputs such as locomotion, pain sensitivity, blood pressure, prolactin secretion, oxytocin release, and heart rate [20]. A more detailed discussion of the effects of receptor-specific dopaminomimetics is presented elsewhere [1]. In summary, $D_{1\text{-like}}$ receptors promote wake and suppress slow-wave and REM-sleep, while $D_{2\text{-like}}$ receptors promote sleep *and* wake, at low and high levels of endogenous dopamine, respectively. Sleep is mediated via $D_{2\text{-like}}$ inhibitory autoreceptors on the cell bodies or terminal axonal fields of dopaminergic VTA neurons. Local applications of $D_{2\text{-like}}$ receptor antagonists in the VTA, for example, block the sedation provoked by systemic agonists. Wake is promoted by the same VTA and contiguous dopaminergic neurons since amphetamine administered into their ventral forebrain targets initiates and maintains alert waking, and their selective lesioning disrupts wakefulness (see below).

Molecular engineering of murine dopamine receptors provides additional details. Dopamine by way of D_2 receptor modulation of nonvisual photic responses in the retina is necessary for light to disrupt circadian rest-activity rhythms (i.e., light "masking" of endogenous circadian rhythms) [21]. Mice lacking a functional D_3 receptor exhibit marked increases in wake (at the expense of slow-wave *and* REM-sleep), suggesting that this receptor is necessary for normal sleep and sleep stages [22]. This observation is particularly relevant because low doses of the D_3 preferring agonists promote sleep in rats and humans, and the D_3 receptor is most sensitive to endogenous dopamine (*vide supra*). Wakefulness attributable to dysfunctional D_3 receptors may alternatively be an epiphenomenon of a heightened response to novelty, or decrements in the gating of spinal sensorimotor and autonomic excitability. As wake is a prerequisite for locomotion, and a paucity of locomotion conducive to sleep, dissociating dopamine receptor effects upon arousal state as distinct from those upon locomotion will continue to necessitate simultaneous recording of related behaviors and careful interpretation.

Wake–Sleep-Related Physiology of Mesotelencephalic Dopamine Neurons

Historically, state-related, mean, discharge rates of mesencephalic dopaminergic neurons were believed to vary little across state and dealt a blow to considerations of dopamine as a lead player in arousal state regulation. As we have previously detailed [1], these studies failed to appreciate variances in interspike intervals (e.g., a burst firing pattern) recently rediscovered in VTA neurons prior to and during REM-sleep [23]. Such "burst firing" most efficiently drives synaptic dopamine release [24], and is modulated by afferents from brain regions intimately involved in thalamocortical arousal (i.e., wake and REM-sleep) [25].

Relevance of Dopamine to Arousal Disorders and Hypersomnia

Clinicians have long relied upon the wake-promoting effects of dopaminomimetics inclusive of traditional psychostimulants. Works in mice, rats, nonhuman primates, and humans are consistent in pointing to dopaminergic neurons of the medial aspects of the midbrain (i.e., the VTA and periaqueductal midline regions) as critical in modulating the quality, quantity, and timing of thalamacortical arousal states – both wake and REM-sleep. Hypersomnia with increases in both non-REM and REM-sleep lacking observable decrements in locomotion follow selective destruction of dopaminergic midbrain periaqueductal gray neurons [26], whereas lesions of the dopaminergic VTA and the adjacent substantia nigra yield a similar phenotype accompanied by a "non-dipping" heart rate and blood pressure pattern, absent the enhancement of REM-sleep [27]. We have further observed that these state changes are unique to mesolimbic versus nigrostriatal projections to the striatum (e.g., the nucleus accumbens vs. caudate/putamen) with the enhancement of REM-sleep being notably robust following selective lesioning of dopaminergic terminal fields [28, 29] (Fig. 4). A narcolepsy-like phenotype (i.e., sleepiness and REM-sleep intrusion into daytime naps) also occurs in methyl-4-phenyl-1,2,3, 6-tetrahydropyridine (MPTP)-induced parkinsonism in nonhuman primates, and in 20–50% of patients with idiopathic PD [30]. Distinct from genuine narcolepsy, however, decrements in alertness encountered with damage to midbrain dopamine cells manifests as a genuine hypersomnia. That is, sleep is excessive over a 24-h period, and cannot be

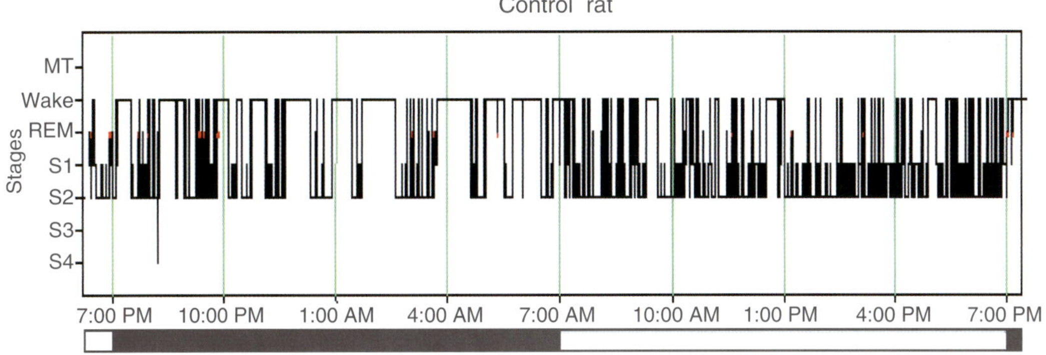

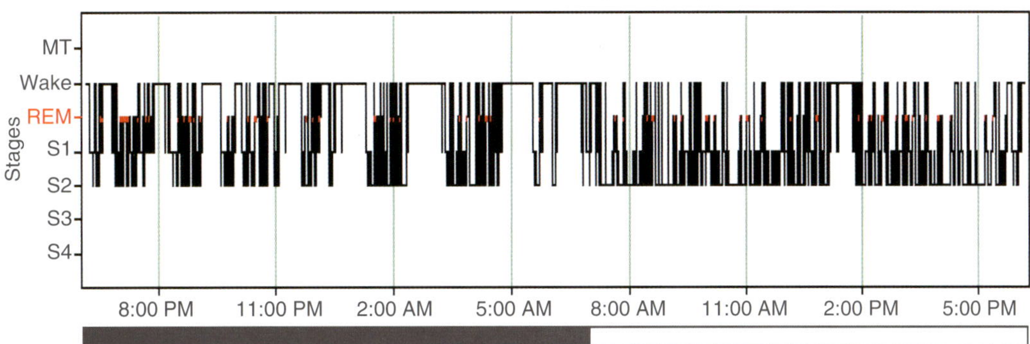

Fig. 4 Rat hypnogram observed 1 week after bilateral lesioning of dopaminergic ventral tegmental pathways made by direct injection of the dopaminergic toxin 6-hydroxydopamine into the nucleus accumbens [29]. Abundant non-REM *and* rapid eye movement sleep (REM-sleep) are evident during both the major wake (1900–0700) and sleep (0700–1900) periods. Maintenance of the major rest and activity periods to the 1200–1200, light–dark schedule (*indicated by the dark and light bars*), respectively, speaks to the elative preservation of circadian processes

attributed to medications or the quantity or quality of prior sleep. Temperature, activity, and neuroendocrine rhythms are generally unaffected by "chronic" parkinsonism induced by MPTP in mice and nonhuman primates, and PD, emphasizing that midbrain dopaminergic cell groups are intimately involved in homeostatic sleep–wake mechanisms (e.g., process S) as opposed to circadian (C) processes (see also, Fig. 4).

Narcolepsy with Cataplexy

In hypocretin-deficient mice, the absence of this excitatory input to dopaminergic neurons decreases dopamine turnover [31], while compensatory increases in dopamine signaling are preferred as an explanation for increases in dopamine metabolites in human narcoleptic CSF [32]. Modafinil, used to combat sleepiness in narcolepsy, may exert its benefits by way of enhancing dopamine signaling since it binds to DAT [33], and is ineffective in DAT-deficient mice [15] as well as those with impaired D_1 or D_2 receptor signaling [34].

Mesocorticolimbic emotive dopaminergic circuits also exacerbate cataplexy via actions at $D_{2\text{-like}}$ receptors consistent with the functional anatomy reviewed above [35, 36]. This dopaminergic effect upon cataplexy is upstream of coerulo-spinal, pro-locomotor circuits, since norepinephrine reuptake inhibitors retain their anticataplectic efficacy in the face of $D_{2\text{-like}}$ agonism.

Narcolepsy Without Cataplexy and the Idiopathic and Episodic Hypersomnias

There are several anecdotal observations and small case series that point to disruptions in dopamine signaling in additional clinical conditions with verifiable sleepiness. What remains unclear is whether these findings point to dopaminergic involvement in the core of the disease process, or more likely, as a final common endpoint somewhere downstream of the principal pathophysiology. Elevations in metabolites of dopamine in the CSF of idiopathic hypersomnics indicative of increased dopamine turnover have been reported by several groups [32, 37]. One recent imaging study of similar patients reveals increased D_2 receptor availability in mesotelencephalic dopamine circuits indicative of a disturbance in synaptic dopamine release [38]. In the episodic hypersomnias (e.g., Kleine–Levin syndrome), reduced hypothalamic dopaminergic tone during an ictal event [39] and nominal decreases in mesotelencephalic DAT binding in the asymptomatic phase [40] have been reported.

The Secondary Narcolepsies (e.g., Those Associated with a Medical or Psychiatric Condition)

More widespread use of multiple sleep latency testing reveals relatively high proportions (~3%) of the general population who meet criteria for narcolepsy without cataplexy and idiopathic hypersomnia [41, 42]. Much higher prevalences have been documented in PD and end-stage renal disease (ESRD) where hypofunctioning of dopamine is established, and appears causally related to PD.

Parkinson's Disease

By self-report, 10–75% of PD patients experience unintended sleep episodes or sleepiness that interfere with activities of daily living. Nearly one half of these patients meet criteria for narcolepsy lacking cataplexy. Pathologic sleepiness has been documented in 20–50% of clinic PD patients, and a narcolepsy-like phenotype in 15–50% of these. Impairments in arousal state bear little relationship to the primary motor manifestations of disease or polysomnographic measures. Dissociation of arousal state from the motor manifestations of disease and homeostatic sleep drives (i.e., sleep propensity should be inversely rather directly related to the quantity and quality of prior night sleep) emphasizes that endogenous dopamine or exogenous dopaminomimetics interact with homeostatic sleep mechanisms in important ways. It also points to the pathophysiologic basis of impaired thalamocortical arousal state residing outside of nigrostriatal pathways traditionally thought to underlie parkinsonian motor disabilities. A threshold of 60–90% of dopamine loss in the sensorimotor putamen is necessary for the emergence of waking clinical manifestations, and then proceeds in an orderly fashion through associative (i.e., caudate) and eventually limbic (i.e., nucleus accumbens) striatal subcircuits. Thus, it is dopamine loss in these latter circuits, most characteristic of advanced disease, which likely accounts for the narcolepsy-like phenotype discussed here.

The objective findings in a small number of unmedicated or young PD patients emphasize that the parkinsonian state itself is a major factor in the expression of sleepiness and SOREMs. The point is underscored by the supporting experimental lesion studies of ascending dopamine pathways from the medial midbrain (*vide supra*). Daytime sleepiness and SOREMs have also been reported in three rhesus monkeys following systemic delivery of the dopamine neurotoxin MPTP [30]. These effects were reversible in the sole animal tested with the dopamine precursor L-DOPA, the dopamine reuptake blocker bupropion, but not a $D_{2\text{-like}}$ agonist. Integrity of dopaminergic axons being necessary for the actions of L-DOPA and bupropion argues that their abilities to reverse sleepiness and SOREMs reflect actions upon surviving mesocorticolimbic dopaminergic circuits which are less vulnerable to MPTP and lost only later in the course of idiopathic PD.

Impairments in central release of hypocretins might seem an alternative, parsimonious explanation for the narcolepsy-like phenotype seen in a subset of PD patients; however, in most cases CSF hypocretin-1 levels are normal. Other plausible neural substrates worthy of investigation include targets of VTA dopamine neurons, such as the prefrontal cortex, the cholinergic magnocellular basal forebrain, and midline thalamic nuclei, and sites of extranigral pathology in PD, such as the dorsal raphe, LC, and PPN region.

End-Stage Renal Disease

Hypersomnia and excessive daytime sleepiness in ESRD have been well documented, but remain less well appreciated. Even in the absence of a nocturnal sleep disorder (i.e., sleep apnea, restless legs, and periodic leg movements of sleep) common in this population, one-third of ESRD patients meet physiological criteria for either narolepsy without cataplexy or idiopathic hypersomnia, and almost 40% of these exhibit extreme sleepiness (MSL <5 min) [43]. While hypofunctioning of dopamine versus other monoamines occurs in uremia [44], it is unknown whether this is causal to the documented disorders of arousal.

Psychoaffective Disorders

Hypersomnia, sleepiness, and SOREMs have been recognized in psychoaffective disorders; however, it remains controversial whether they occur at greater frequency than would be expected by chance. It has been our clinical experience that narcolepsy without cataplexy is common to the nonmanic state in bipolar disease and some atypical depressives. A single subject with narcolepsy lacking cataplexy and comborbid atypical depression, for example, experienced symptom relief and complete resolution of MSLT findings with the DAT and NET blocker, bupropion [45]. Converging lines of evidence and our personal experiences also highlight that the spectrum of vegetative symptoms observed with cytokine-induced sickness behavior often includes hypersomnia, sleepiness, and inappropriate intrusion of REM-sleep. In nonhuman primates, interferon-alpha is known to shorten REM latencies [46] and is associated with lower CSF concentrations of dopamine metabolites [47]. Parkinsonism is a known complication of cytokine treatment, and support for a causal role for impaired dopamine signaling comes from neuroimaging and ongoing physiological and interventional studies of cytokine-induced sickness behaviors in nonhuman primates and humans.

Summary

The integrity of mesotelencephalic dopamine systems, particularly the medial, mesocorticolimbic circuitry, is critical to the maintenance of arousal and normal expression of REM-sleep. The DAT, $D_{2\text{-like}}$ receptors, and *Clock*, are critical molecular determinants of thalamocortical arousal signaling in these circuits. Many additional genetic and environmental signals modulate brain dopamine signaling and thereby might account for the commonality of hypersomnia, sleepiness, and SOREMs. Dopaminergic neurons, for example, are exquisitely sensitive to blood pressure, hypoxia, metabolic state (e.g., kidney and liver failure), circadian and homeostatic processes, and humoral factors (e.g., estrogen, parathyroid hormone, insulin, and pro-inflammatory cytokines). Alterations in dopaminergic tone occur both acutely (within seconds of some physiologic challenges) and chronically, for example, as a consequence of intermittent hypoxia [48]. Dopamine neuron numbers are fewer in males, undergo attrition with age, differ between inbred strains of mice, and even between hemispheres within an animal. These attributes raise the intriguing possibility that they might underlie some of the sex, age-related, and genetic influences known to affect an individual's diathesis to sleep inclusive of the expression of REM-sleep.

Acknowledgments Special thanks are extended to our close colleagues Drs. Andy Miller, Glenda Keating, Michael Decker, Gillian Hue, and Jennifer Felger who contributed much to the body of this work. Dr. Rye is supported by USPHS grant NS-055015.

References

1. Freeman AA, Rye DB, editors. Dopamine in behavioral state control. Cambridge: Cambridge University Press; 2008.
2. Nicola S, Surmeier J, Malenka R. Dopaminergic modulation of neuronal excitability in the striatum and nucleus accumbens. Ann Rev Neurosci. 2000;23: 185–215.
3. Missale C, Nash SR, Robinson SW, Jaber M, Caron MG. Dopamine receptors: from structure to function. Physiol Rev. 1998;78:189–225.
4. Cragg S, Greenfield S. Differential autoreceptor control of somatodendritic and axon terminal dopamine release in substantia nigra, ventral tegmental area and striatum. J Neurosci. 1997;17:5738–46.
5. Lee RS, Steffensen SC, Henriksen SJ. Discharge profiles of ventral tegmental area GABA neurons during movement, anesthesia, and the sleep-wake cycle. J Neurosci. 2001;21:1757–66.
6. McCormick DA. Are thalamocortical rhythms the Rosetta Stone of a subset of neurological disorders? Nat Med. 1999;5:1349–51.
7. Keating G, Rye D. Where you least expect it: dopamine in the pons aion of sleep and REM-sleep. Sleep. 2003;26:788–9.
8. Khaldy H, Leon J, Escames G, Bikjdaouene L, Garcia JJ, Acuna-Castroviejo D. Circadian rhythms of dopamine and dihydroxyphenyl acetic acid in the mouse striatum: effects of pinealectomy and of melatonin treatment. Neuroendocrinology. 2002;75:201–8.
9. McClung C, Sidiropoulou K, Vitaterna M, et al. Regulation of dopaminergic transmission and cocaine reward by the *Clock* gene. Proc Natl Acad Sci U S A. 2005;102:9377–81.
10. Andretic R, Hirsh J. Circadian modulation of dopamine receptor responsiveness in *Drosophila melanogaster*. Proc Natl Acad Sci U S A. 2000;97:1873–8.
11. Garcia-Borreguero D, Larrosa O, Granizo JJ, de la Llave Y, Hening WA. Circadian variation in neuroendocrine response to L-dopa in patients with restless legs syndrome. Sleep. 2004;27:669–73.
12. Poceta JS, Parsons L, Engelland S, Kripke DF. Circadian rhythm of CSF monoamines and hypocretin-1 in restless legs syndrome and Parkinson's disease. Sleep Med. 2009;10:129–33.
13. Freeman A, Morales J, Beck J, et al. In vivo diurnal rhythm of dopamine measured in the putamen of non-human primates. Sleep. 2006;29:A69.
14. Nishino S, Mao J, Sampathkumaran R, Shelton J, Mignot E. Increased dopaminergic transmission mediates the wake-promoting effects of CNS stimulants. Sleep Res. 1998;1:49–61. Online.
15. Wisor J, Nishino S, Sora I, Uhl G, Mignot E, Edgar D. Dopaminergic role in stimulant-induced wakefulness. J Neurosci. 2001;21:1787–94.
16. Porzgen P, Park SK, Hirsh J, Sonders MS, Amara SG. The antidepressant-sensitive dopamine transporter in *Drosophila melanogaster*: a primordial carrier for catecholamines. Mol Pharmacol. 2001;59:83–95.
17. Kume K, Kume S, Park SK, Hirsh J, Jackson FR. Dopamine is a regulator of arousal in the fruit fly. J Neurosci. 2005;25:7377–84.
18. Wu MN, Koh K, Yue Z, Joiner WJ, Sehgal A. A genetic screen for sleep and circadian mutants reveals mechanisms underlying regulation of sleep in Drosophila. Sleep. 2008;31:465–72.
19. Andretic R, van Swinderen B, Greenspan RJ. Dopaminergic modulation of arousal in Drosophila. Curr Biol. 2005;15:1165–75.
20. Calabrese E. Dopamine: biphasic dose responses. Crit Rev Toxicol. 2001;31:563–83.
21. Doi M, Yujnovsky I, Hirayama J, et al. Impaired light masking in dopamine receptor-null mice. Nat Neurosci. 2006;9:732–4.
22. Hue G, Decker M, Solomon I, Rye D. Increased wakefulness and hyper-responsivity to novel environments in mice lacking functional dopamine D3 receptors. Soc Neurosci. 2003;616:16.
23. Dahan L, Astier B, Vautrelle N, Urbain N, Kocsis B, Chouvet G. Prominent burst firing of dopaminergic neurons in the ventral tegmental area during paradoxical sleep. Neuropsychopharmacology. 2007;32:1232–41.
24. Floresco SB, West AR, Ash B, Moore H, Grace AA. Afferent modulation of dopamine neuron firing differentially regulates tonic and phasic dopamine transmission. Nat Neurosci. 2003;6:968–73.
25. Kitai ST, Shepard PD, Callaway JC, Scroggs R. Afferent modulation of dopamine neuron firing patterns. Curr Opin Neurobiol. 1999;9:690–7.
26. Lu J, Jhou TC, Saper CB. Identification of wake-active dopaminergic neurons in the ventral periaqueductal gray matter. J Neurosci. 2006;26:193–202.
27. Sakata M, Sei H, Toida K, Fujihara H, Urushihara R, Morita Y. Mesolimbic dopaminergic system is involved in diurnal blood pressure regulation. Brain Res. 2002;928:194–201.
28. Decker M, Keating G, Freeman A, Rye D. Parkinsonian-like sleep-wake architecture in rats with bilateral striatal 6-OHDA lesions. Soc Neurosci Abstr. 2000;26:1514.
29. Decker MJ, Keating G, Hue GE, Freeman A, Rye DB. Mesolimbic dopamine's modulation of REM Sleep. J Sleep Res. 2002;51(Suppl):51–2.
30. Rye D, Daley J, Freeman A, Bliwise D. Daytime Sleepiness and Sleep Attacks in Idiopathic Parkinson's Disease. In: Bedard M-A, Agid Y, Chouinard S, Fahn S, Korcyzn A, Lesperance P, editors. Mental and behavioral dysfunction in movement disorders. Totawa, NJ: Humana; 2003. p. 527–38.
31. Mori T, Ito S, Kuwaki T, Yanagisawa M, Sawaguchi T. Monoaminergic neuronal changes in orexin deficient mice. Neuropharmacology. 2010;58:826–32.
32. Faull KF, Guilleminault C, Berger PA, Barchas JD. Cerebrospinal fluid monoamine metabolites in narcolepsy and hypersomnia. Ann Neurol. 1983;13: 258–63.

33. Volkow ND, Fowler JS, Logan J, et al. Effects of modafinil on dopamine and dopamine transporters in the male human brain: clinical implications. JAMA. 2009;301:1148–54.
34. Qu WM, Huang ZL, Xu XH, Matsumoto N, Urade Y. Dopaminergic D1 and D2 receptors are essential for the arousal effect of modafinil. J Neurosci. 2008; 28:8462–9.
35. Nishino S, Arrigoni J, Valtier D, et al. Dopamine D2 mechanisms in canine narcolepsy. J Neurosci. 1991;11:2666–71.
36. Okura M, Fujiki N, Kita I, et al. The roles of midbrain and diencephalic dopamine cell groups in the regulation of cataplexy in narcoleptic Dobermans. Neurobiol Dis. 2004;16:274–82.
37. Livrea P, Puca FM, Barnaba A, Di Reda L. Abnormal central monoamine metabolism in humans with "true hypersomnia" and "sub-wakefulness". Eur Neurol. 1977;15:71–6.
38. Bassetti C, Khatami R, Proyazova R, Buck F. Idiopathic Hypersomnia: a dopaminergic disorder? Sleep. 2009;32:A248–9.
39. Chesson Jr AL, Levine SN, Kong LS, Lee SC. Neuroendocrine evaluation in Kleine-Levin syndrome: evidence of reduced dopaminergic tone during periods of hypersomnolence. Sleep. 1991;14:226–32.
40. Hoexter MQ, Shih MC, Mendes DD, et al. Lower dopamine transporter density in an asymptomatic patient with Kleine-Levin syndrome. Acta Neurol Scand. 2008;117:370–3.
41. Mignot E, Lin L, Finn L, et al. Correlates of sleep-onset REM periods during the Multiple Sleep Latency Test in community adults. Brain. 2006;129: 1609–23.
42. Singh M, Drake CL, Roth T. The prevalence of multiple sleep-onset REM periods in a population-based sample. Sleep. 2006;29:890–5.
43. Parker KP, Bliwise DL, Bailey JL, Rye DB. Daytime sleepiness in stable hemodialysis patients. Am J Kidney Dis. 2003;41:394–402.
44. Adachi N, Lei B, Deshpande G, et al. Uraemia suppresses central dopaminergic metabolism and impairs motor activity in rats. Intensive Care Med. 2001;27: 1655–60.
45. Rye DB, Dihenia B, Bliwise DL. Reversal of atypical depression, sleepiness, and REM-sleep propensity in narcolepsy with bupropion. Depress Anxiety. 1998; 7:92–5.
46. Reite M, Laudenslager M, Jones J, Crnic L, Kaemingk K. Interferon decreases REM latency. Biol Psychiatry. 1987;22:104–7.
47. Felger JC, Alagbe O, Hu F, et al. Effects of interferon-alpha on rhesus monkeys: a nonhuman primate model of cytokine-induced depression. Biol Psychiatry. 2007;62:1324–33.
48. Decker MJ, Hue GE, Caudle WM, Miller GW, Keating GL, Rye DB. Episodic neonatal hypoxia evokes executive dysfunction and regionally specific alterations in markers of dopamine signaling. Neuroscience. 2003;117:417–25.

The Serotoninergic System in Sleep and Narcolepsy

Chloé Alexandre and Thomas E. Scammell

Keywords
Serotonin • 5-HT • Serotonin transporter • Orexin • Hypocretin • Narcolepsy • Dorsal raphe nucleus

Serotonin and Sleep–Wake Regulation

Over 60 years ago, researchers discovered that a blood-borne substance increased vascular tone and intestinal motility by constricting smooth muscle. They named this substance serotonin based on the Latin word *serum* and the Greek word *tonic*, and soon after, this factor was identified as 5-hydroxytryptamine (5-HT) [1, 2]. Decades of research have now shown that 5-HT plays a complex and multifaceted role in sleep–wake regulation, and altered 5-HT signaling may underlie some of the symptoms and pathophysiology of narcolepsy.

In humans, narcolepsy is caused by loss of the orexin/hypocretin-producing neurons in the lateral hypothalamus. Studies in rodents have demonstrated that orexin promotes arousal, inhibits REM sleep, and regulates transitions between behavioral states. Most importantly, disrupted orexin signaling in animals elicits narcolepsy-like symptoms including chronic sleepiness and cataplexy [3, 4]. The orexin neurons densely innervate most serotonergic nuclei, and in turn, they receive moderately heavy serotonergic inputs [5, 6]. Both orexin and 5-HT are thought to promote wakefulness and inhibit REM sleep, and destruction of either system can result in a narcolepsy-like behavior. For example, in cats, electrolytic lesions of the serotonergic neurons in the rostral raphe nuclei or suppression of neuronal activity in the dorsal raphe nucleus (DRN) can elicit direct transitions from wakefulness to REM sleep [7, 8]. These findings suggest that reduced 5-HT tone may result in symptoms of REM sleep dysregulation typical of narcolepsy, and conversely, increasing 5-HT levels with selective serotonin reuptake inhibitors (SSRIs) improves cataplexy [9]. This chapter reviews how 5-HT normally regulates sleep and wakefulness, and how abnormal 5-HT signaling may contribute to narcolepsy.

C. Alexandre (✉)
Department of Neurology, Beth Israel Deaconess Medical Center and Harvard Medical School, Room 717, 330 Brookline Ave., Boston, MA 02215, USA
e-mail: calexan1@bidmc.harvard.edu

C.R. Baumann et al. (eds.), *Narcolepsy: Pathophysiology, Diagnosis, and Treatment*, DOI 10.1007/978-1-4419-8390-9_7, © Springer Science+Business Media, LLC 2011

Brief Overview of the 5-HT System

In the central nervous system, 5-HT is exclusively synthesized from the essential amino acid L-tryptophan in small populations of neurons within the raphe nuclei that express the enzyme tryptophan hydroxylase-2. This enzyme adds a hydroxyl group to L-tryptophan to produce L-5-hydroxytryptophan (L-5-HTP), and this is then converted into 5-HT by aromatic L-amino acid decarboxylase. Hydroxylation is the rate-limiting step in 5-HT synthesis, and inhibition of this reaction with parachlorophenylalanine (pCPA) results in a long-lasting depletion of cerebral 5-HT.

The neurons that produce 5-HT extend along the midline from the caudal medulla to the midbrain and are clustered into nine cell groups (B1 to B9) within the raphe nuclei (Fig. 1) [10]. In the pons and midbrain, the dorsal and median raphe nuclei (DRN and MRN, respectively) mostly innervate the upper brainstem and the forebrain. More caudally, in the medulla, the nuclei raphe magnus, obscurus, and pallidus (NRM, NRO, and NRP, respectively) innervate the lower brainstem and spinal cord. Most serotonergic neurons lay close to the midline, but some lay in the lateral brainstem, including the B9 group (supralemniscal serotonergic group) [11]. The DRN contains about half of all 5-HT neurons in the rat, and the caudal nuclei contain many fewer 5-HT neurons. With their widespread projections throughout the entire brain, 5-HT neurons innervate many brain regions implicated in the control of sleep–wake behaviors and EEG activity, including the preoptic area, the basal forebrain, the lateral hypothalamus, the thalamus, and also brainstem structures that are key components of the REM sleep regulatory network, such as the locus coeruleus (LC), the mesopontine tegmentum, and the nucleus pontis oralis [12–14]. The more lateral B9 cells may also project to areas regulating behavioral states, such as the basal forebrain [15]. Neurons in the different serotonergic raphe nuclei have distinctive morphological properties and specific afferent and efferent connections [16–18]. In particular, Vertes and colleagues demonstrated that both the DRN and the MRN project to the forebrain, but to separate regions with very little overlap [12, 13]. Such differential patterns of innervation may reflect distinct functional roles for these two nuclei. Moreover, the patterns of neuronal activity differ across the raphe nuclei, with some regions exhibiting greater variations in activity as sleep–wake

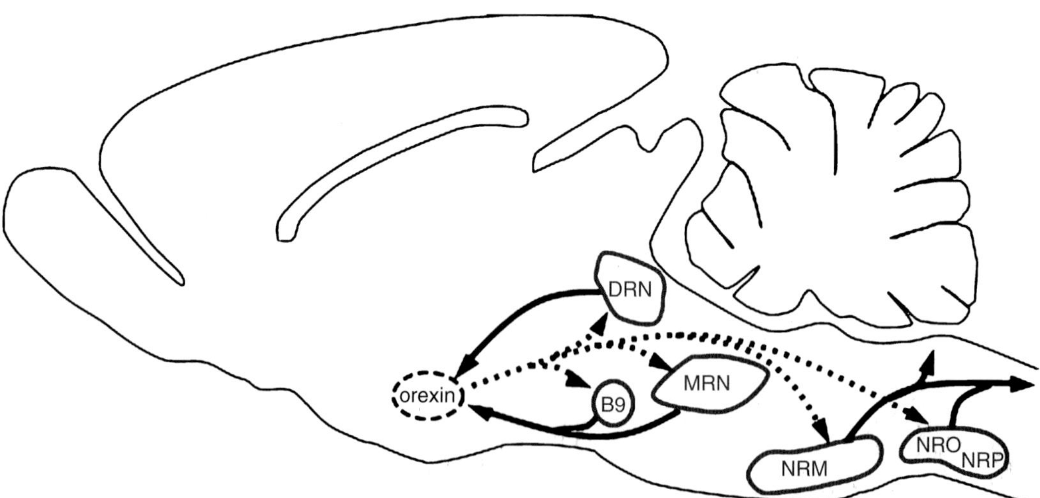

Fig. 1 The raphe nuclei and their interconnections with the orexin neurons. The *solid lines* show serotonergic projections, and the *dashed lines* show the orexinergic afferents to the raphe nuclei. The major serotonergic nuclei are the dorsal raphe nucleus (DRN; B6, B7); median raphe nucleus (MRN; B5, B8); nucleus raphe magnus (NRM; B3); nucleus raphe obscurus (NRO; B2); nucleus raphe pallidus (NRP; B1), and supralemniscal serotonergic neurons (B9)

behavior changes. It is likely then that the role of 5-HT in the control of wake–sleep behaviors may depend on the serotonergic subpopulation and the brain region involved.

Additionally, multiple 5-HT receptors add further complexity to the physiological effects of 5-HT. Thus far, 15 different 5-HT receptors in 7 major classes have been identified which are differentially distributed throughout the brain [19]. Except for the ionotropic 5-HT_3 receptor, all of these are G-protein-coupled receptors with diverse effects as they can couple with G_s (5-HT_4, 5-HT_6, and 5-HT_7), $G_{i/o}$ (5-HT_1 and 5-HT_5), or G_q (5-HT_2). 5-HT can inhibit postsynaptic neurons through the 5-HT_{1A} receptor, but serotonergic neurons also express 5-HT_{1A} autoreceptors. Serotonergic neurons also express inhibitory 5-HT_{1B} receptors on presynaptic serotonergic terminals. These autoreceptors form important negative feedback mechanisms to limit 5-HT signaling. Thus, 5-HT may exert net excitatory or inhibitory actions, depending on the receptor subtype and the transduction pathway activated.

Finally, it is important to note that the serotonin transporter (5-HTT) is an essential component for the regulation of the serotonergic neurotransmission, as it removes 5-HT from the synapse and thus regulates the levels of extracellular 5-HT. The 5-HTT is the main target of SSRIs, a major class of antidepressants that are also of clinical importance in the treatment of cataplexy.

Overall, this large diversity in the anatomical and physiological properties of 5-HT neurons suggests specific roles for the different serotonergic subpopulations in the regulation of sleep–wake behaviors.

Does Serotonin Promote Sleep or Wakefulness?

With such a variety of 5-HT nuclei, anatomical connections, and receptors, it is not surprising that the role of 5-HT in the regulation of sleep and wakefulness appears to be complex. In fact, one could conclude from some studies that 5-HT promotes wakefulness, while other experiments indicate that 5-HT promotes sleep. This tangled literature has baffled many students, and we will summarize it briefly below. In this review, we will focus mainly on the DRN and MRN, as their roles in the control of behavioral states have been extensively studied.

Early pharmacological studies reported that inhibition of 5-HT synthesis by pCPA in monkeys, cats, and rats decreased cerebral 5-HT levels and reduced both NREM and REM sleep 24 h after administration [20–24]. In the cat, short NREM sleep episodes reappeared about 60 h after pCPA injection, occasionally followed by REM sleep. However, REM sleep sometimes directly followed wakefulness, without any preceding NREM sleep. In the rat, Borbely and colleagues observed that slow-wave sleep and REM sleep were especially reduced during the phase of partial insomnia [23]. This insomnia could be transiently reversed by administration of 5-HT precursors via systemic or intraventricular routes, or when injected directly into the anterior hypothalamus [7, 22, 24–26]. In agreement with those pharmacologic studies, large electrolytic lesions of the raphe nuclei (80–90%) in the cat caused a severe and long-lasting insomnia with almost no sleep for 3–4 days [7, 27–29]. A small amount of NREM sleep then reappeared, but without any REM sleep. The sleep loss correlated positively with the size of the lesion and the depletion of brain 5-HT. More selective lesions limited to the rostral nuclei (DRN and MNR) markedly reduced NREM sleep with much less effect on REM sleep, though sometimes, direct transitions from wakefulness to REM sleep occurred, as seen in narcolepsy. From these experiments, Jouvet and coworkers hypothesized that the caudal part of the raphe system is required for the occurrence of both NREM and REM sleep, whereas the rostral part (DRN and MRN) mainly promotes NREM sleep. Similar experiments in rats using 5,7 DHT to destroy 5-HT neurons selectively also described a decrease in NREM sleep [30]. Altogether, these observations supported the concept of 5-HT as a sleep-promoting neurotransmitter.

However, other researchers dispute this hypothesis and found no significant changes in sleep or wakefulness after lesions of the DRN or MRN, or depletion of cerebral 5-HT using 5,7 DHT in rats [31–33]. Also, in apparent contradiction to the sleep-promoting effect of 5-HT, electrophysiological studies reveal that most putative 5-HT neurons fire tonically during wakefulness, then decrease their activity during NREM sleep, and are nearly silent during REM sleep [34–39]. These findings suggest that 5-HT promotes wakefulness and that the cessation of 5-HT neuronal firing is required for the appearance of REM sleep, much like the noradrenergic neurons of the LC or histaminergic neurons of the tuberomammillary nucleus.

Most putative 5-HT neurons in the DRN are wake active, slow their firing during NREM sleep, and are silent during REM sleep [34–39]. Interestingly, the discharge of these cells increases a few seconds before the transition from NREM or REM sleep to wakefulness, suggesting that they may be involved in the initiation of wakefulness. Still, it is a challenge to identify 5-HT neurons in vivo, and not all 5-HT neurons are wake active. Traditionally, 5-HT neurons have been identified by their broad spikes and slow, regular, firing pattern [39, 40]. However, recent studies challenge this classical profile and report great electrophysiological heterogeneity among DRN neurons [38, 41–44]. For instance, half of the slow-firing, seemingly serotonergic DRN neurons lack 5-HT, and some atypical 5-HT DRN neurons, identified by juxtacellular labeling, display fast firing rates [41, 43, 44]. Furthermore, though most putative serotonergic neurons fire during wakefulness and cease firing during REM sleep, some studies report 5-HT neurons with sustained activity during REM sleep or high discharge rates during NREM sleep [36–38, 44]. In part, these varied patterns of activity may be due to the fact that in addition to 5-HT neurons, the raphe nuclei also contain neurons that produce GABA and other neurotransmitters. Taken together, these studies highlight the anatomical, electrophysiological, and functional diversity among raphe neurons. Sorting out this heterogeneity will require much more work using juxtacellular labeling and other methods to identify neurons of the raphe nuclei definitively.

Still, most of the evidence suggests that the 5-HT neurons are mainly wake active and inhibit REM sleep. Microdialysis studies in cats and rats have shown that extracellular levels of 5-HT in the DRN are higher during wakefulness than during NREM and REM sleep, and similar patterns are also found in other brainstem areas and in the forebrain [45]. In addition, pharmacological studies support a prominent role for 5-HT in the inhibition of REM sleep and in the promotion of wakefulness [46, 47]. Increasing extracellular 5-HT with injection of SSRIs acutely increases wakefulness and markedly inhibits REM sleep in both humans and rodents [48–50]. In mice, this effect can be prevented by a 5-HT$_{1A}$ receptor (5-HT$_{1A}$R) antagonist, indicating that the suppression of REM sleep by SSRIs may be mediated by activation of 5-HT$_{1A}$R. Indeed, systemic administration of 5-HT$_{1A}$ or 5-HT$_{1B}$R agonists also causes an initial arousal and suppresses REM sleep [46]. Microdialysis of SSRIs or 5-HT$_{1A}$R agonists into the DRN increases REM sleep, perhaps through 5-HT$_{1A}$R-mediated inhibition of DRN neurons [47]. The tonic inhibition exerted by 5-HT on REM sleep may be mediated by postsynaptic 5-HT$_{1A}$R, since DRN lesion does not modify the REM sleep inhibition induced by systemic administration of 5-HT$_{1A}$R agonists [51]. On the contrary, these drugs decrease REM sleep when microdialyzed into REM sleep-promoting areas such as the LDT or the mesopontine reticular formation [47, 52, 53]. Others serotonergic receptors are thought to mediate the action of 5-HT on wakefulness and REM sleep, such as 5-HT$_2$ and 5-HT$_7$R [54, 55]. Thus, 5-HT, by acting on these receptors, may inhibit neuronal networks implicated in REM sleep generation. 5-HT may also promote wakefulness by inhibiting more rostral sleep-promoting neurons such as the GABAergic neurons in the ventrolateral preoptic area or in the basal forebrain and the cortex, and thus, facilitates wakefulness [56, 57].

These seemingly contradictory studies highlight the complexity of the 5-HT system, but overall, we believe that most 5-HT neurons are wake active and that 5-HT generally promotes arousal and suppresses REM sleep. Alternatively, and as suggested by Jouvet, 5-HT released during wakefulness may somehow prepare the brain for sleep, and thus large reductions in 5-HT interfere with the production of sleep [7]. Clearly, additional experiments are needed to characterize the precise roles of different 5-HT neurons in the regulation or modulation of behavioral states. Optogenetic techniques that enable stimulation or inhibition of specific raphe neurons would be especially helpful [58]. One could also use genetic techniques to eliminate 5-HT signaling focally in the raphe nuclei and examine the effects on sleep–wake behaviors. These powerful, new techniques should help define specific populations of 5-HT neurons and their roles in regulating behavioral states.

Interactions of the 5-HT and Orexin Systems

Both the orexin and 5-HT systems are thought to have important roles in promoting arousal and inhibiting REM sleep [59, 60], and both send projections to many brain areas linked to sleep–wake regulation. These two systems also share similar patterns of neuronal activity in relation to behavioral states, with the highest rate of discharge during active waking [39, 61]. These similarities may result in coordinated actions between 5-HT and orexin neurons in the modulation of arousal and REM sleep.

Effects of Orexins on Serotonergic Neurons

The orexin neurons project to many brain regions implicated in the regulation of behavioral states, including the raphe nuclei [5, 62, 63]. The DRN and MRN receive the densest inputs, and the NRM, NRP, and NRO also contain a moderate density of orexin terminals (Fig. 2) [5, 63, 64]. In the DRN, orexin terminals appose both 5-HT and GABAergic neurons [65]. At the electron microscopic level, orexin-containing terminals synapse on dendrites of DRN 5-HT neurons, and also on smaller neurons that are likely GABAergic interneurons [66, 67]. These findings indicate that orexins may influence 5-HT neurons directly and indirectly through interneurons at least in the DRN, but one can anticipate that this is also the case in other raphe nuclei.

Neurons of the raphe nuclei express mRNA for both orexin receptors (OX_1R and OX_2R) [68]. The DRN and MRN contain high levels of OX_1R mRNA, with labeling extending caudally into the NRM and the NRO. OX_2R mRNA is also found in the same nuclei except for the NRO. In the DRN, OX_1R immunoreactivity is present on both 5-HT and non-5-HT neurons [67], but in the MRN, OX_2R seems to be present only on 5-HT neurons, whereas OX_1R is probably expressed by local GABAergic neurons [69]. Immunostaining for orexin receptors may not be reliable, and more work is needed to establish just which raphe neurons express OX_1R or OX_2R and whether both receptors are expressed by individual neurons.

Electrophysiological recordings from the DRN have revealed that both orexin-A and -B directly depolarize 5-HT neurons by inducing a TTX-insensitive Na^+-dependent inward current [65, 70–72]. This effect may be mediated by OX_2R as this receptor responds well to both orexin-A and -B. The responses to orexins in the raphe are not simply excitatory. At higher concentrations, orexins may indirectly inhibit 5-HT neurons by exciting DRN GABA neurons that inhibit the 5-HT neurons [65]. Orexin-B can also reduce glutamate-mediated excitation of DRN 5-HT neurons, via a retrograde endocannabinoid (CB1) signal [73]. These mechanisms may help dampen orexin's excitatory effects on 5-HT neurons during periods of high orexin tone.

Overall, the major effect of orexins seems to be an enhancement of 5-HT release. In the DRN, microdialysis of orexin-A dose dependently increases extracellular 5-HT, although this response is not apparent in the MRN [69]. Orexin-B produces a smaller increase in 5-HT in both the DRN and MRN, suggesting that fewer serotonergic neurons respond to this peptide or perhaps orexin-B also triggers indirect effects that inhibit serotonergic neurons. Increased signaling by 5-HT and other monoamines may be the source of stereotypic behaviors induced by orexins, as 5-HT antagonists reduce the "wet dog shakes" induced by orexin-A or -B [74].

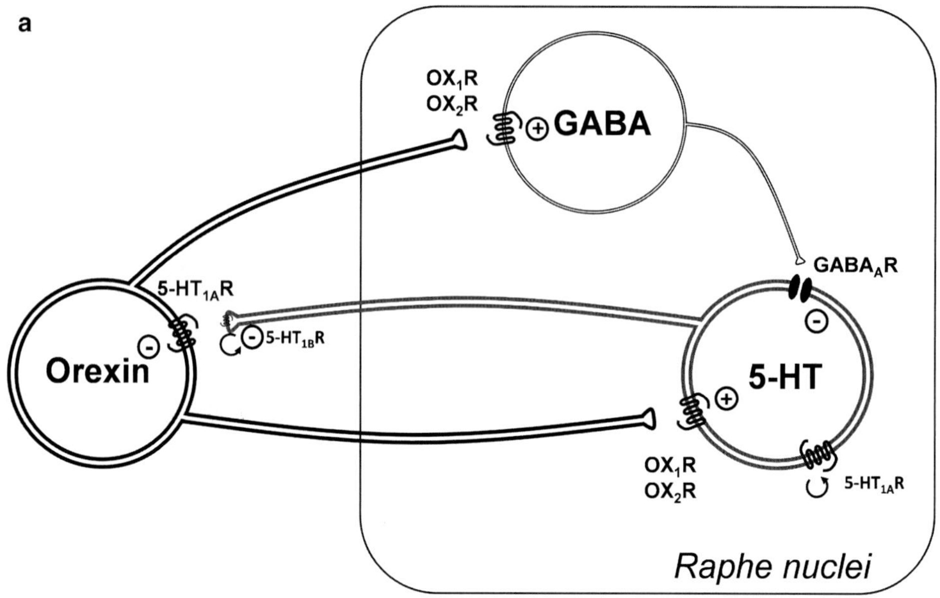

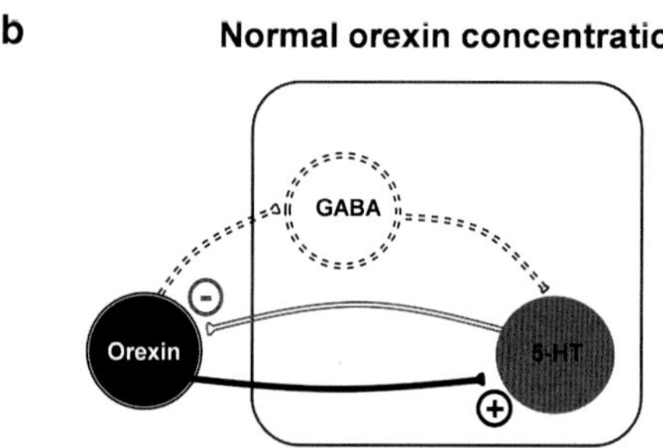

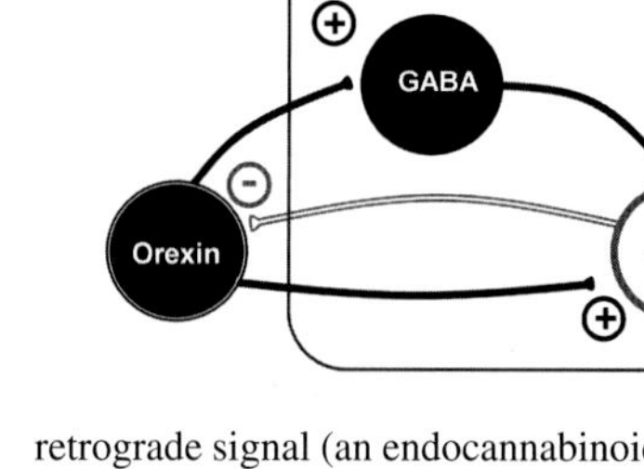

Fig. 2 A model of mutual interactions between orexin and 5-HT neurons. (**a**) Orexin may excite 5-HT neurons through both OX1- and OX2-R, and high concentrations of orexin may also activate local GABAergic neurons. 5-HT may inhibit orexin neurons through the activation of postsynaptic 5-HT_{1A}-R. Whether orexin neurons bear other serotonergic receptors or are indirectly regulated by other 5-HT receptors is still unknown. In addition, the activation of OX2-R may induce the release of a retrograde signal (an endocannabinoid) from serotonergic neurons that may inhibit the release of glutamate via the activation of presynaptic CB1-R (not shown). (**b**) Normal orexin concentrations may activate serotonergic but not GABAergic neurons. High concentrations of orexin may also activate GABAergic interneurons in the raphe that inhibit 5-HT neurons. In turn, 5-HT inhibits orexin neurons, but the conditions under which this occurs remain unknown

Altogether, these observations suggest that orexins enhance 5-HT signaling.

Effects of 5-HT on the Orexin Neurons

The orexin neurons are innervated by numerous serotonergic terminals [6, 75, 76]. Across the orexin field, about 60% of the orexin neurons are apposed by 5-HT terminals, and the number of appositions is especially high in the lateral part of the field [6]. However, the sources of this input are controversial. Using a genetic technique for retrograde labeling in mice, Sakurai and colleagues found that the orexin neurons are mainly innervated by serotonergic and non-serotonergic cells of the MRN but not by the DRN [76]. In contrast, Yoshida and collaborators found that the orexin field of rats is strongly innervated by the DRN and the MRN, with smaller input from the B9 group. Injections of conventional tracers in mice could establish whether these discrepancies are due to differences between species or technical aspects of the retrograde labeling.

This serotonergic input appears to inhibit the orexin neurons. In vitro, 5-HT hyperpolarizes and inhibits orexin neurons via postsynaptic 5-HT_{1A}R [75, 77]. In vivo, 5-HT may suppress orexin neuron activity during wakefulness as a selective 5-HT_{1A}R antagonist increases locomotor activity in wild-type mice but not in mice lacking the orexin neurons [75]. Along the same line, perfusion of 5-HT into the perifornical-lateral hypothalamic area in awake rats decreases the firing rate of putative orexinergic neurons and reduces c-fos expression in orexin neurons without changing the sleep–wake behavior [78]. In addition, the SSRI sertraline slightly decreases CSF orexin-A levels in depressed people [79].

Both orexin and 5-HT tone are high during active wakefulness and other behavioral conditions requiring a high level of arousal, such as stress [80–83]. Perhaps, negative feedback from the 5-HT system inhibits the orexin neurons to avoid excessively high levels of arousal.

Alternatively, the effects of 5-HT on the orexin neurons may be inhibitory or excitatory depending on the sleep–wake behavior of an animal, as has been shown with norepinephrine (NE) [84]. Like the 5-HT system, the noradrenergic neurons of the LC are densely innervated by the orexin neurons and in turn innervate the orexin neurons. After a period of sleep, NE excites the orexin neurons, but after 2 h of waking, NE inhibits the orexin neurons [84]. NE also inhibits the orexin neurons of mice killed during the day, just after the nocturnal active period [85]. Future studies could examine whether 5-HT has similar effects depending on an animal's sleep–wake behaviors.

5-HT and Narcolepsy

Early studies suggested that reduced signaling by 5-HT and other monoamines might be a fundamental aspect of narcolepsy [86, 87]. Specifically, many researchers hypothesized that low 5-HT tone could promote sleepiness and permit direct transitions from wakefulness into REM sleep-like states such as cataplexy. We now know that narcolepsy with cataplexy is caused by a marked reduction in orexins, yet whether this disrupts 5-HT signaling in people with narcolepsy remains poorly understood.

5-HT Signaling in Narcolepsy

Some supporting evidence for chronically reduced 5-HT signaling comes from studies of narcoleptic dogs with mutations in the OX2 receptor [88]. One early study reported significantly decreased 5-HIAA concentrations and reduced 5-HT turnover in narcoleptic dogs, suggesting a decrease in 5-HT signaling [89]. The dogs also had lower NE content and turnover that could also contribute to their sleepiness and REM sleep dysregulation. A subsequent study could not confirm the reduction in 5-HT signaling, although it did confirm the reduction in NE tone [90]. In narcoleptic dogs, signaling through OX1 receptors is intact, so it will be important to study 5-HT tone in narcoleptic mice, as complete loss of orexin signaling in such mice more closely parallels narcolepsy in people.

Acute reductions in 5-HT signaling may also contribute to the loss of muscle tone and paralysis of REM sleep and cataplexy. In narcoleptic dogs, DRN neurons fire much less during cataplexy than during wakefulness, and they become nearly silent during REM sleep [91]. Perhaps cataplexy does not require a complete absence of 5-HT tone, or perhaps other 5-HT neurons that affect muscle tone more directly, such as the medullary raphe nuclei that project to the spinal cord, would show larger reductions in firing rate during cataplexy. Indeed, during carbachol-induced REM sleep, the activity of medullary serotonergic neurons is almost fully suppressed [39]. In addition, chemical and electrical stimulation of the mesopontine area that induce muscle atonia in rats markedly reduces 5-HT release in the spinal cord and hypoglossal nucleus [92].

We are aware of only one study that directly examined 5-HT signaling in human narcolepsy, and surprisingly, this showed elevated concentrations of 5-HT and NE metabolites in cortical tissue [93]. This study is difficult to interpret because measurements of 5-HT in tissue do not provide much information on extracellular levels of 5-HT, and these patients were treated with amphetamines, which could increase monoamine levels. Clearly, more work is needed to examine 5-HT tone in untreated people with narcolepsy, both in CSF and in specific brain regions.

5-HT Pharmacology in Narcolepsy

Almost all drugs prescribed for the treatment of cataplexy increase monoamine signaling. Tricyclic antidepressants such as clomipramine and desipramine substantially improve cataplexy, most likely by inhibiting the reuptake of both NE and 5-HT [94, 95]. Because side effects are common with tricyclics, SSRIs and NE reuptake inhibitors have been used more frequently over the last 20 years. Many small series have shown clear improvements in cataplexy with SSRIs such as zimelidine, citalopram, fluvoxamine, and fluoxetine [9, 96–99]. Clinical experience suggests that SSRIs also improve hypnagogic hallucinations and sleep paralysis. Drugs that block NE uptake (e.g., atomoxetine) or the uptake of both 5-HT and NE (e.g., venlafaxine and duloxetine) may be even more effective [100, 101], but there have been no well-controlled clinical trials with any of these drugs or head-to-head comparisons to help guide treatment decisions.

In narcoleptic dogs, the evidence that cataplexy is improved by drugs that increase 5-HT signaling is less clear. $5\text{-HT}_{1A}R$ agonists reduce cataplexy, and this effect is blocked by coadministration of a $5\text{-HT}_{1A}R$ antagonist, suggesting that this improvement is mediated by $5\text{-HT}_{1A}R$ [102]. However, these results are difficult to interpret as the agonists produced abnormal behaviors that could have masked or suppressed cataplexy. Canine cataplexy is also improved by SSRIs, but the effects are modest and high doses are required [103, 104]. In fact, some of this improvement may be mediated by increases in NE tone, because while the parent compounds selectively block 5-HT reuptake, desmethyl metabolites of SSRIs can also block reuptake of NE and other monoamines [95]. Thus, in contrast to cataplexy in people, canine cataplexy is only modestly improved by drugs that increase 5-HT signaling, and some of the improvement may be due to increased NE signaling.

These discrepancies between narcoleptic people and dogs may be related to their underlying pathology; in people, narcolepsy is caused by a loss of the orexin neurons, whereas in dogs, it is caused by a null mutation in the OX_2R. Perhaps, residual signaling through the OX_1R is enough to maintain some 5-HT tone in narcoleptic dogs, but a complete lack of orexin signaling in narcoleptic humans may result in low 5-HT tone. The cause of this discrepancy could be determined by testing whether SSRIs and 5-HT agonists reduce cataplexy in mice lacking orexin peptides or the orexin neurons.

Basic research suggests that increasing 5-HT tone should promote wakefulness, but in the clinic, SSRIs generally produce only small improvements in sleepiness. Some SSRIs can disrupt sleep in healthy volunteers, with delayed sleep onset, less REM sleep, and reduced sleep continuity, but most show little improvement in the maintenance of wakefulness. Despite decades of research, researchers still debate whether 5-HT initiates wakefulness, helps maintain wakefulness, or somehow prepares the brain for sleep.

Conclusions

Compared to some other neurochemicals, many questions remain about how the 5-HT system regulates sleep and how 5-HT contributes to the symptoms of narcolepsy. The challenge of understanding 5-HT most likely arises from the complexity of the 5-HT system, with many sources of 5-HT, diverse 5-HT-responsive target regions, and numerous 5-HT receptors. Some recent work with knockout mice has added to the mystery by demonstrating how the 5-HT system is very adaptable, by adjusting to a constitutive lack of 5-HT reuptake [105–107]. The next generation of basic research should focus on understanding the roles of specific 5-HT nuclei and receptors while controlling for any compensatory mechanisms. For example, people with narcolepsy are frequently overweight, and low 5-HT promotes obesity in rats. SSRIs reduce food intake and obesity [108], and this effect is probably mediated, in part, through 5-HT_{2C} receptors in the hypothalamus [109]. Studies such as these that examine 5-HT signaling through specific pathways should shed light on how altered 5-HT signaling contributes to narcolepsy.

Despite these challenges, many of the ways in which 5-HT influences sleep–wake behaviors and narcolepsy are becoming clear. Most 5-HT neurons are active during wakefulness and probably suppress REM sleep and promote wakefulness. Normally, orexins may enhance 5-HT signaling, but in narcolepsy, loss of orexins may reduce 5-HT tone, which probably contributes to cataplexy, dysregulation of REM sleep, and possibly sleepiness.

Acknowledgments In part, writing of this chapter was supported by the National Institutes of Health Grant NS055367 to T.E.S.

References

1. Erspamer V, Asero B. Identification of enteramine, the specific hormone of the enterochromaffin cell system, as 5-hydroxytryptamine. Nature. 1952;169:800–1.
2. Rapport MM, Green AA, Page IH. Serum vasoconstrictor, serotonin; isolation and characterization. J Biol Chem. 1948;176:1243–51.
3. Nishino S. The hypothalamic peptidergic system, hypocretin/orexin and vigilance control. Neuropeptides. 2007;41:117–33.
4. Sakurai T. The neural circuit of orexin (hypocretin): maintaining sleep and wakefulness. Nat Rev Neurosci. 2007;8:171–81.
5. Peyron C, Tighe DK, van den Pol AN, et al. Neurons containing hypocretin (orexin) project to multiple neuronal systems. J Neurosci. 1998;18:9996–10015.
6. Yoshida K, McCormack S, Espana RA, et al. Afferents to the orexin neurons of the rat brain. J Comp Neurol. 2006;494:845–61.
7. Jouvet M. Biogenic amines and the states of sleep. Science. 1969;163:32–41.
8. Sakai K, Crochet S. Role of dorsal raphe neurons in paradoxical sleep generation in the cat: no evidence for a serotonergic mechanism. Eur J Neurosci. 2001;13:103–12.
9. Montplaisir J, Godbout R. Serotoninergic reuptake mechanisms in the control of cataplexy. Sleep. 1986; 9:280–4.
10. Dahlstrom A, Fuxe K. Localization of monoamines in the lower brain stem. Experientia. 1964;20:398–9.
11. Vertes RP, Crane AM. Distribution, quantification, and morphological characteristics of serotonin-immunoreactive cells of the supralemniscal nucleus (B9) and pontomesencephalic reticular formation in the rat. J Comp Neurol. 1997;378:411–24.
12. Vertes RP. A PHA-L analysis of ascending projections of the dorsal raphe nucleus in the rat. J Comp Neurol. 1991;313:643–68.
13. Vertes RP, Fortin WJ, Crane AM. Projections of the median raphe nucleus in the rat. J Comp Neurol. 1999;407:555–82.
14. Vertes RP, Kocsis B. Projections of the dorsal raphe nucleus to the brainstem: PHA-L analysis in the rat. J Comp Neurol. 1994;340:11–26.
15. Jones BE, Cuello AC. Afferents to the basal forebrain cholinergic cell area from pontomesencephalic–catecholamine, serotonin, and acetylcholine–neurons. Neuroscience. 1989;31:37–61.
16. Abrams JK, Johnson PL, Hollis JH, et al. Anatomic and functional topography of the dorsal raphe nucleus. Ann N Y Acad Sci. 2004;1018:46–57.
17. Imai H, Steindler DA, Kitai ST. The organization of divergent axonal projections from the midbrain raphe nuclei in the rat. J Comp Neurol. 1986;243:363–80.
18. Peyron C, Luppi PH, Fort P, et al. Lower brainstem catecholamine afferents to the rat dorsal raphe nucleus. J Comp Neurol. 1996;364:402–13.
19. Barnes NM, Sharp T. A review of central 5-HT receptors and their function. Neuropharmacology. 1999;38: 1083–152.
20. Weitzman ED, Rapport MM, McGregor P, et al. Sleep patterns of the monkey and brain serotonin concentration: effect of p-chlorophenylalanine. Science. 1968;160:1361–3.
21. Delorme F, Froment JL, Jouvet M. Suppression of sleep with p-chloromethamphetamine and p-chlorophenylalanine. C R Seances Soc Biol Fil. 1966;160:2347–51.
22. Koella WP, Feldstein A, Czicman JS. The effect of para-chlorophenylalanine on the sleep of cats. Electroencephalogr Clin Neurophysiol. 1968;25: 481–90.
23. Borbely AA, Neuhaus HU, Tobler I. Effect of p-chlorophenylalanine and tryptophan on sleep, EEG and motor activity in the rat. Behav Brain Res. 1981;2: 1–22.
24. Mouret J, Bobillier P, Jouvet M. Insomnia following parachlorophenylalanine in the rat. Eur J Pharmacol. 1968;5:17–22.
25. Denoyer M, Sallanon M, Kitahama K, et al. Reversibility of para-chlorophenylalanine-induced insomnia by intrahypothalamic microinjection of L-5-hydroxytryptophan. Neuroscience. 1989;28:83–94.
26. Petitjean F, Buda C, Janin M, et al. Insomnia caused by administration of para-chlorophenylalanine: reversibility by peripheral or central injection of

5-hydroxytryptophan and serotonin. Sleep. 1985;8: 56–67.

27. Jouvet M, Bobillier P, Pujol JF, et al. Effects of lesions of the raphe system on sleep and cerebral serotonin. C R Seances Soc Biol Fil. 1966;160:2343–6.
28. Jouvet M, Bobillier P, Pujol JF, et al. Suppression of sleep and decrease of cerebral serotonin caused by lesion of the raphe system in the cat. C R Acad Sci Hebd Seances Acad Sci D. 1967;264:360–2.
29. Jouvet M, Pujol JF. Effects of central alterations of serotoninergic neurons upon the sleep-waking cycle. Adv Biochem Psychopharmacol. 1974;11:199–209.
30. Kiianmaa K, Fuxe K. The effects of 5,7-dihydroxytryptamine-induced lesions of the ascending 5-hydroxytryptamine pathways on the sleep wakefulness cycle. Brain Res. 1977;131:287–301.
31. Bouhuys AL, Van Den Hoofdakker RH. Effects of midbrain raphe destruction on sleep and locomotor activity in rats. Physiol Behav. 1977;19:535–41.
32. Lu J, Sherman D, Devor M, et al. A putative flip-flop switch for control of REM sleep. Nature. 2006;441: 589–94.
33. Ross CA, Trulson ME, Jacobs BL. Depletion of brain serotonin following intraventricular 5,7-dihydroxytryptamine fails to disrupt sleep in the rat. Brain Res. 1976;114:517–23.
34. Gervasoni D, Peyron C, Rampon C, et al. Role and origin of the GABAergic innervation of dorsal raphe serotonergic neurons. J Neurosci. 2000;20:4217–25.
35. McGinty DJ, Harper RM. Dorsal raphe neurons: depression of firing during sleep in cats. Brain Res. 1976;101:569–75.
36. Rasmussen K, Heym J, Jacobs BL. Activity of serotonin-containing neurons in nucleus centralis superior of freely moving cats. Exp Neurol. 1984;83:302–17.
37. Sakai K, Crochet S. Differentiation of presumed serotonergic dorsal raphe neurons in relation to behavior and wake-sleep states. Neuroscience. 2001;104:1141–55.
38. Urbain N, Creamer K, Debonnel G. Electrophysiological diversity of the dorsal raphe cells across the sleep-wake cycle of the rat. J Physiol. 2006;573: 679–95.
39. Jacobs BL, Fornal CA. Activity of brain serotonergic neurons in the behaving animal. Pharmacol Rev. 1991;43:563–78.
40. Aghajanian GK, Wang RY, Baraban J. Serotonergic and non-serotonergic neurons of the dorsal raphe: reciprocal changes in firing induced by peripheral nerve stimulation. Brain Res. 1978;153:169–75.
41. Allers KA, Sharp T. Neurochemical and anatomical identification of fast- and slow-firing neurones in the rat dorsal raphe nucleus using juxtacellular labelling methods in vivo. Neuroscience. 2003;122:193–204.
42. Hajos M, Allers KA, Jennings K, et al. Neurochemical identification of stereotypic burst-firing neurons in the rat dorsal raphe nucleus using juxtacellular labelling methods. Eur J Neurosci. 2007;25:119–26.
43. Kirby LG, Pernar L, Valentino RJ, et al. Distinguishing characteristics of serotonin and non-serotonin-containing cells in the dorsal raphe nucleus: electrophysiological and immunohistochemical studies. Neuroscience. 2003;116:669–83.
44. Kocsis B, Varga V, Dahan L, et al. Serotonergic neuron diversity: identification of raphe neurons with discharges time-locked to the hippocampal theta rhythm. Proc Natl Acad Sci U S A. 2006;103:1059–64.
45. Portas CM, Bjorvatn B, Ursin R. Serotonin and the sleep/wake cycle: special emphasis on microdialysis studies. Prog Neurobiol. 2000;60:13–35.
46. Ursin R. Serotonin and sleep. Sleep Med Rev. 2002;6:55–69.
47. Monti JM, Jantos H. The roles of dopamine and serotonin, and of their receptors, in regulating sleep and waking. Prog Brain Res. 2008;172:625–46.
48. Maudhuit C, Jolas T, Lainey E, et al. Effects of acute and chronic treatment with amoxapine and cericlamine on the sleep-wakefulness cycle in the rat. Neuropharmacology. 1994;33:1017–25.
49. Monaca C, Boutrel B, Hen R, et al. 5-HT 1A/1B receptor-mediated effects of the selective serotonin reuptake inhibitor, citalopram, on sleep: studies in 5-HT 1A and 5-HT 1B knockout mice. Neuropsychopharmacology. 2003;28:850–6.
50. van Bemmel AL. The link between sleep and depression: the effects of antidepressants on EEG sleep. J Psychosom Res. 1997;42:555–64.
51. Tissier MH, Lainey E, Fattaccini CM, et al. Effects of ipsapirone, a 5-HT1A agonist, on sleep/wakefulness cycles: probable post-synaptic action. J Sleep Res. 1993;2:103–9.
52. Horner RL, Sanford LD, Annis D, et al. Serotonin at the laterodorsal tegmental nucleus suppresses rapid-eye-movement sleep in freely behaving rats. J Neurosci. 1997;17:7541–52.
53. Thakkar MM, Strecker RE, McCarley RW. Behavioral state control through differential serotonergic inhibition in the mesopontine cholinergic nuclei: a simultaneous unit recording and microdialysis study. J Neurosci. 1998;18:5490–7.
54. Popa D, Lena C, Fabre V, et al. Contribution of 5-HT2 receptor subtypes to sleep-wakefulness and respiratory control, and functional adaptations in knock-out mice lacking 5-HT2A receptors. J Neurosci. 2005;25: 11231–8.
55. Hedlund PB, Huitron-Resendiz S, Henriksen SJ, et al. 5-HT7 receptor inhibition and inactivation induce antidepressantlike behavior and sleep pattern. Biol Psychiatry. 2005;58:831–7.
56. Gallopin T, Fort P, Eggermann E, et al. Identification of sleep-promoting neurons in vitro. Nature. 2000;404:992–5.
57. Cape EG, Jones BE. Differential modulation of high-frequency gamma-electroencephalogram activity and sleep-wake state by noradrenaline and serotonin microinjections into the region of cholinergic basalis neurons. J Neurosci. 1998;18:2653–66.
58. Adamantidis AR, Zhang F, Aravanis AM, et al. Neural substrates of awakening probed with optogenetic control of hypocretin neurons. Nature. 2007;450:420–4.

59. McCarley RW. Neurobiology of REM and NREM sleep. Sleep Med. 2007;8:302–30.
60. Saper CB, Scammell TE, Lu J. Hypothalamic regulation of sleep and circadian rhythms. Nature. 2005;437:1257–63.
61. Lee MG, Hassani OK, Jones BE. Discharge of identified orexin/hypocretin neurons across the sleep-waking cycle. J Neurosci. 2005;25:6716–20.
62. Chemelli RM, Willie JT, Sinton CM, et al. Narcolepsy in orexin knockout mice: molecular genetics of sleep regulation. Cell. 1999;98:437–51.
63. Nambu T, Sakurai T, Mizukami K, et al. Distribution of orexin neurons in the adult rat brain. Brain Res. 1999;827:243–60.
64. Date Y, Ueta Y, Yamashita H, et al. Orexins, orexigenic hypothalamic peptides, interact with autonomic, neuroendocrine and neuroregulatory systems. Proc Natl Acad Sci U S A. 1999;96:748–53.
65. Liu RJ, van den Pol AN, Aghajanian GK. Hypocretins (orexins) regulate serotonin neurons in the dorsal raphe nucleus by excitatory direct and inhibitory indirect actions. J Neurosci. 2002;22:9453–64.
66. Wang QP, Guan JL, Matsuoka T, et al. Electron microscopic examination of the orexin immunoreactivity in the dorsal raphe nucleus. Peptides. 2003;24: 925–30.
67. Wang QP, Koyama Y, Guan JL, et al. The orexinergic synaptic innervation of serotonin- and orexin 1-receptor-containing neurons in the dorsal raphe nucleus. Regul Pept. 2005;126:35–42.
68. Marcus JN, Aschkenasi CJ, Lee CE, et al. Differential expression of orexin receptors 1 and 2 in the rat brain. J Comp Neurol. 2001;435:6–25.
69. Tao R, Ma Z, McKenna JT, et al. Differential effect of orexins (hypocretins) on serotonin release in the dorsal and median raphe nuclei of freely behaving rats. Neuroscience. 2006;141:1101–5.
70. Brown RE, Sergeeva OA, Eriksson KS, et al. Convergent excitation of dorsal raphe serotonin neurons by multiple arousal systems (orexin/hypocretin, histamine and noradrenaline). J Neurosci. 2002;22: 8850–9.
71. Brown RE, Sergeeva O, Eriksson KS, et al. Orexin A excites serotonergic neurons in the dorsal raphe nucleus of the rat. Neuropharmacology. 2001;40: 457–9.
72. Takahashi K, Wang QP, Guan JL, et al. State-dependent effects of orexins on the serotonergic dorsal raphe neurons in the rat. Regul Pept. 2005;126:43–7.
73. Haj-Dahmane S, Shen RY. The wake-promoting peptide orexin-B inhibits glutamatergic transmission to dorsal raphe nucleus serotonin neurons through retrograde endocannabinoid signaling. J Neurosci. 2005;25:896–905.
74. Matsuzaki I, Sakurai T, Kunii K, et al. Involvement of the serotonergic system in orexin-induced behavioral alterations in rats. Regul Pept. 2002;104:119–23.
75. Muraki Y, Yamanaka A, Tsujino N, et al. Serotonergic regulation of the orexin/hypocretin neurons through the 5-HT1A receptor. J Neurosci. 2004;24:7159–66.
76. Sakurai T, Nagata R, Yamanaka A, et al. Input of orexin/hypocretin neurons revealed by a genetically encoded tracer in mice. Neuron. 2005;46:297–308.
77. Li Y, Gao XB, Sakurai T, et al. Hypocretin/orexin excites hypocretin neurons via a local glutamate neuron-A potential mechanism for orchestrating the hypothalamic arousal system. Neuron. 2002;36: 1169–81.
78. Kumar S, Szymusiak R, Bashir T, et al. Effects of serotonin on perifornical-lateral hypothalamic area neurons in rat. Eur J Neurosci. 2007;25:201–12.
79. Salomon RM, Ripley B, Kennedy JS, et al. Diurnal variation of cerebrospinal fluid hypocretin-1 (Orexin-A) levels in control and depressed subjects. Biol Psychiatry. 2003;54:96–104.
80. Berridge CW, Espana RA, Vittoz NM. Hypocretin/orexin in arousal and stress. Brain Res. 2010;1314: 91–102.
81. Espana RA, Valentino RJ, Berridge CW. Fos immunoreactivity in hypocretin-synthesizing and hypocretin-1 receptor-expressing neurons: effects of diurnal and nocturnal spontaneous waking, stress and hypocretin-1 administration. Neuroscience. 2003;121:201–17.
82. Rachalski A, Alexandre C, Bernard JF, et al. Altered sleep homeostasis after restraint stress in 5-HTT knock-out male mice: a role for hypocretins. J Neurosci. 2009;29:15575–85.
83. Winsky-Sommerer R, Yamanaka A, Diano S, et al. Interaction between the corticotropin-releasing factor system and hypocretins (orexins): a novel circuit mediating stress response. J Neurosci. 2004;24:11439–48.
84. Grivel J, Cvetkovic V, Bayer L, et al. The wake-promoting hypocretin/orexin neurons change their response to noradrenaline after sleep deprivation. J Neurosci. 2005;25:4127–30.
85. Li Y, van den Pol AN. Direct and indirect inhibition by catecholamines of hypocretin/orexin neurons. J Neurosci. 2005;25:173–83.
86. Aldrich MS. The neurobiology of narcolepsy-cataplexy. Prog Neurobiol. 1993;41:533–41.
87. Nishino S. Clinical and neurobiological aspects of narcolepsy. Sleep Med. 2007;8:373–99.
88. Nishino S, Riehl J, Hong J, et al. Is narcolepsy a REM sleep disorder? Analysis of sleep abnormalities in narcoleptic Dobermans. Neurosci Res. 2000;38:437–46.
89. Faull KF, Barchas JD, Foutz AS, et al. Monoamine metabolite concentrations in the cerebrospinal fluid of normal and narcoleptic dogs. Brain Res. 1982;242: 137–43.
90. Miller JD, Faull KF, Bowersox SS, et al. CNS monoamines and their metabolites in canine narcolepsy: a replication study. Brain Res. 1990;509:169–71.
91. Wu MF, John J, Boehmer LN, et al. Activity of dorsal raphe cells across the sleep-waking cycle and during cataplexy in narcoleptic dogs. J Physiol. 2004;554: 202–15.
92. Lai YY, Kodama T, Siegel JM. Changes in monoamine release in the ventral horn and hypoglossal nucleus linked to pontine inhibition of muscle tone: an in vivo microdialysis study. J Neurosci. 2001;21:7384–91.

93. Kish SJ, Mamelak M, Slimovitch C, et al. Brain neurotransmitter changes in human narcolepsy. Neurology. 1992;42:229–34.
94. Guilleminault C, Raynal D, Takahashi S, et al. Evaluation of short-term and long-term treatment of the narcolepsy syndrome with clomipramine hydrochloride. Acta Neurol Scand. 1976;54:71–87.
95. Nishino S, Arrigoni J, Shelton J, et al. Desmethyl metabolites of serotonergic uptake inhibitors are more potent for suppressing canine cataplexy than their parent compounds. Sleep. 1993;16:706–12.
96. Guilleminault C, Pelayo R. Narcolepsy in children: a practical guide to its diagnosis, treatment and follow-up. Paediatr Drugs. 2000;2:1–9.
97. Thirumalai SS, Shubin RA. The use of citalopram in resistant cataplexy. Sleep Med. 2000;1:313–6.
98. Schachter M, Parkes JD. Fluvoxamine and clomipramine in the treatment of cataplexy. J Neurol Neurosurg Psychiatry. 1980;43:171–4.
99. Sonka K, Kemlink D, Pretl M. Cataplexy treated with escitalopram–clinical experience. Neuro Endocrinol Lett. 2006;27:174–6.
100. Izzi F, Placidi F, Marciani MG, et al. Effective treatment of narcolepsy-cataplexy with duloxetine: a report of three cases. Sleep Med. 2009;10:153–4.
101. Moller LR, Ostergaard JR. Treatment with venlafaxine in six cases of children with narcolepsy and with cataplexy and hypnagogic hallucinations. J Child Adolesc Psychopharmacol. 2009;19:197–201.
102. Nishino S, Shelton J, Renaud A, et al. Effect of 5-HT1A receptor agonists and antagonists on canine cataplexy. J Pharmacol Exp Ther. 1995;272:1170–5.
103. Babcock DA, Narver EL, Dement WC, et al. Effects of imipramine, chlorimipramine, and fluoxetine on cataplexy in dogs. Pharmacol Biochem Behav. 1976;5:599–602.
104. Foutz AS, Delashaw Jr JB, Guilleminault C, et al. Monoaminergic mechanisms and experimental cataplexy. Ann Neurol. 1981;10:369–76.
105. Adrien J, Alexandre C, Boutrel B, et al. Contribution of the "knock-out" technology to understanding the role of serotonin in sleep regulations. Arch Ital Biol. 2004;142:369–77.
106. Alexandre C, Popa D, Fabre V, et al. Early life blockade of 5-hydroxytryptamine 1A receptors normalizes sleep and depression-like behavior in adult knock-out mice lacking the serotonin transporter. J Neurosci. 2006;26:5554–64.
107. Popa D, Lena C, Alexandre C, et al. Lasting syndrome of depression produced by reduction in serotonin uptake during postnatal development: evidence from sleep, stress, and behavior. J Neurosci. 2008;28:3546–54.
108. Leibowitz SF, Alexander JT. Hypothalamic serotonin in control of eating behavior, meal size, and body weight. Biol Psychiatry. 1998;44:851–64.
109. Xu Y, Jones JE, Kohno D, et al. 5-HT2CRs expressed by pro-opiomelanocortin neurons regulate energy homeostasis. Neuron. 2008;60:582–9.

Sleep Homeostasis, Adenosine, Caffeine, and Narcolepsy

Hans-Peter Landolt

Keywords
Narcolepsy • Sleep homeostasis • Adenosine • Caffeine

Sleep Homeostasis Can Be Reliably Tracked in the Electroencephalogram

Many functional aspects of wakefulness and sleep, including excessive daytime sleepiness, vigilance, attention, sleep structure, and quantitative measures derived from the electroencephalogram (EEG) are tightly controlled by two interacting processes: (1) a circadian program providing temporal context to most physiological processes including sleep and (2) a homeostatic process keeping track of "sleep pressure." *Sleep homeostasis* is conceptualized in the two-process model of sleep regulation [1] as the build-up of sleep pressure (or "sleep need") during wakefulness and the dissipation of sleep pressure during sleep. The homeostatic regulation of rest/sleep is a common principle in invertebrates, fish, and mammals [2]. In humans, we think today that the circadian system opposes homeostatic changes in sleep pressure, to enable healthy people to stay awake and alert throughout a normal waking day despite accumulating sleep pressure associated with wakefulness [3]. Vice versa, circadian clock and sleep homeostasis interact to permit healthy individuals to remain asleep during the night despite the waning of sleep need. When wakefulness is prolonged ("sleep deprivation") and sleep pressure exceeds an average "reference value," subjective and objective measures of sleepiness increase, vigilance deteriorates, and attention is impaired. Moreover, theta activity in the waking EEG, as well as slow-wave sleep (SWS; nonREM sleep stages 3 and 4) and EEG slow-wave activity (SWA; spectral power within 0.75–4.5 Hz) are enhanced in recovery sleep. Particularly, SWA (or "delta activity") in nonREM sleep is predictably correlated with the duration of preceding wakefulness. This physiological measure constitutes the classical, highly reliable marker of sleep homeostasis, which served to delineate the basic concepts of the two-process model of sleep regulation [1, 4]. The neurobiological mechanisms underlying nonREM sleep homeostasis remain incompletely understood.

In this chapter, we review findings from recent studies in animals and humans which suggest

H.-P. Landolt (✉)
Institute of Pharmacology & Toxicology, University of Zürich, Winterthurerstrasse 190, 8057 Zürich, Switzerland
and
Zürich Center for Integrative Human Physiology (ZIHP), University of Zürich, Winterthurerstrasse 190, 8057 Zürich, Switzerland
e-mail: landolt@pharma.uzh.ch

C.R. Baumann et al. (eds.), *Narcolepsy: Pathophysiology, Diagnosis, and Treatment*,
DOI 10.1007/978-1-4419-8390-9_8, © Springer Science+Business Media, LLC 2011

that the adenosinergic neuromodulator/receptor system plays a key role in the homeostatic facet of sleep–wake regulation. We focus on the time course of EEG SWA in nonREM sleep. In addition, the question will be addressed whether adenosinergic mechanisms could contribute to disturbed sleep and wakefulness in narcolepsy patients.

The Neuromodulator Adenosine Is Involved in Sleep Homeostasis

Because systemic and local changes in adenosine levels, as well as agonistic and antagonistic interaction with adenosine receptors modulate the expression of wakefulness and sleep, in particular SWS and SWA in nonREM sleep, adenosine has long been thought to be importantly involved in sleep homeostasis (for recent reviews, see [5, 6]). Moreover, studies in rats and cats suggest that highest adenosine levels are found at the beginning of the circadian rest-period, whereas concentrations decline during sleep and sleep recovery [5].

Metabolism of energy-rich adenine nucleosides (e.g., adenosine-tri-phosphate, ATP) in response to increased energy demand leads to the formation of adenosine, which changes in the brain in activity-dependent manner. Different mechanisms contribute to the appearance of adenosine in the extracellular space. First, adenosine formed in neurons from adenosine-mono-phosphate (AMP) by cytosolic 5′-nucleotidase can be released through bidirectional, equilibrative nucleoside transporters. Elevated intracellular adenosine concentrations following increased utilization of ATP in conditions of high energy demand lead to release of adenosine. Second, extracellular adenosine is formed through ecto-nucleotidase-mediated hydrolysis of ATP. Release of ATP from synaptic vesicles occurs along with several neurotransmitters, including the brain's major excitatory neurotransmitter glutamate. Third, ATP and glutamate are also released from astrocytes by a recently established process referred to as gliotransmission. Molecular genetic manipulations in mice revealed that gliotransmission provides a significant source of extracellular adenosine in the brain [7].

Mice expressing a dominant-negative (dn) SNARE domain in astrocytes have reduced gliotransmission [8]. A recent study in these mice suggests that adenosine released from astrocytes is importantly involved in the homeostatic regulation of sleep pressure [9]. While both transgenic and wild-type mice spend similar proportions of the light-phase (i.e., the major sleep phase in mice) in nonREM sleep, REM sleep, and wakefulness, sleep pressure (or sleep intensity) as reflected in EEG SWA is significantly reduced when dnSNARE was expressed exclusively in astrocytes. The differences are specific to nonREM sleep and not present in REM sleep and wakefulness, and most pronounced in the 0.5–1.5-Hz range. Importantly, dnSNARE expression attenuates the sleep deprivation-induced increase in SWA in nonREM sleep, particularly in the low-frequency range and in 2–4 h of recovery sleep. Moreover, performance on a novel object recognition task is unimpaired after sleep deprivation in dnSNARE mice, while it is dramatically reduced in wild-type animals. Taken together, the data are consistent with an important role for glia-dependent regulation of adenosine in sleep homeostasis in mice.

Clearance of extracellular adenosine mostly occurs through the nonconcentrative nucleoside transporters [10]. The main intracellular metabolic pathways of adenosine are the formation of AMP by adenosine kinase, and the irreversible break down to inosine by adenosine deaminase (ADA). Ecto-ADA catalyzes also extracellular deamination of adenosine. Mainly due to the high activity of adenosine kinase, baseline levels of extracellular adenosine usually remain low. The action of ADA, which appears to be more abundantly expressed in astrocytes than in neurons [10], may be particularly important when large amounts of adenosine have to be cleared. Pharmacological inhibition of both, adenosine kinase and ADA, prolongs sleep and increases EEG SWA in rats (reviewed in [6]). The latter data are consistent with genetic findings in mice suggesting that a genomic region including the gene encoding ADA modifies the rate at which nonREM sleep need accumulates during wakefulness [11].

Also, support for a contribution of ADA in regulating nonREM sleep pressure comes from a recent genetic study in humans [12]. There are numerous allelic variants of *ADA* located on human chromosome 20 (Online Mendelian Inheritance in Man [OMIM] Database Accession No. 608958). As demonstrated in blood cells, healthy individuals carrying a G to A transition at nucleotide 22 (G/A genotype) exhibit reduced ADA activity when compared to individuals with the G/G genotype. This polymorphism may be expected to cause elevated extracellular adenosine levels. Consistent with a role for adenosine in sleep regulation, the polymorphism is associated in baseline condition with fewer perceived awakenings, prolonged SWS, and enhanced SWA in nonREM sleep [12]. The data demonstrate that genetic variation in *ADA* contributes to interindividual differences in subjective and objective markers of sleep homeostasis in humans. Studies are currently under way to examine the repercussions of sleep deprivation on waking performance, sleepiness, sleep, and the sleep EEG in healthy individuals with distinct *ADA* genotypes.

Adenosine A_1 and A_{2A} Receptors May Mediate Effects of Adenosine on Sleep Homeostasis

The cellular effects of adenosine are mediated via four subtypes of G-protein coupled adenosine receptors: A_1, A_{2A}, A_{2B}, and A_3 receptors. While in vitro studies indicate that physiological concentrations of endogenous adenosine can activate A_1, A_{2A}, and A_3 receptors, it is widely accepted that the high-affinity A_1 and A_{2A} receptors are primarily involved in mediating the effects of adenosine on vigilance and sleep in humans [13].

The stimulation of A_1 receptors inhibits adenylate cyclase through activation of G_i proteins. In humans, the A_1 receptor is ubiquitously but not homogenously expressed in the central nervous system [14]. In vivo imaging with the selective A_1 receptor antagonist, ^{18}F-CPFPX, recently developed for positron emission tomography (PET), reveals highest receptor occupancy in striatum and thalamus, as well as temporo-parietal and occipital cortex (Fig. 1). Because of the inhibition of excitatory neurotransmission after pre- and postsynaptic A_1 receptor activation, and because of the wide-spread distribution of this receptor subtype in the brain, it has been generally assumed that adenosine affects sleep and sleep homeostasis primarily via this receptor. Nevertheless, A_1 receptor knock-out mice have a normal homeostatic regulation of sleep, possibly because of compensation [15]. On the other hand, recent studies using conditional genetic deletions in mice and molecular imaging in rats and humans are more compatible with this assumption [16–19]. Inducible knock-out of adenosine A_1 receptors in neurons reduces SWA (3–4.5 Hz range) in nonREM sleep

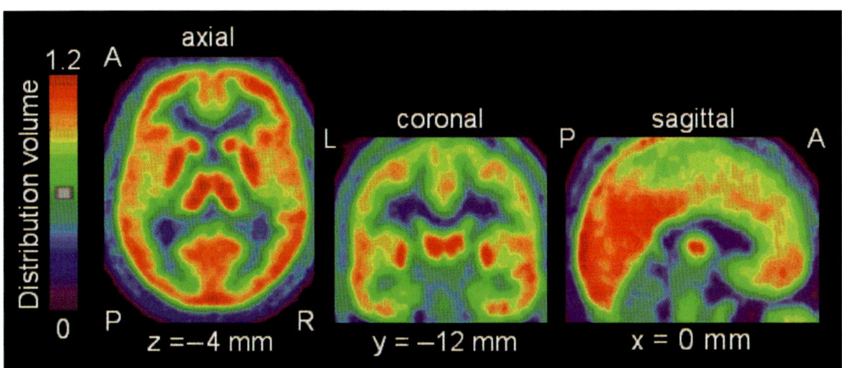

Fig. 1 Distribution of adenosine A_1 receptors in the human brain. Color-coded PET images illustrate average distribution volumes of the selective A_1 receptor radioligand ^{18}F-CPFPX in ten young healthy men. From left to right: axial (*A* anterior, *P* posterior, *R* right), coronal (*L* left), and sagittal (*P* posterior, *A* anterior) sections. Coordinates according to Montreal Neurological Institute brain atlas. Unpublished data from Geissler E, Ametamey SM, Buck A, and Landolt HP

under baseline conditions, and attenuates the homeostatically regulated rise in SWA following sleep restriction [17]. These knock-out mice also show reduced working memory performance when sleep is curtailed. In addition, A_1 receptor binding appears to be up-regulated in cortical and subcortical brain regions following prolonged wakefulness [18, 19]. Together, these data indicate a functional role for adenosine A_1 receptors in sleep homeostasis.

Stimulation of A_{2A} receptors increases adenylate cyclase activity through activation of G_s or G_{olf} (in striatum) proteins. Compared to the A_1 receptor, this adenosine receptor subtype is less widely distributed in the brain [14]. Highest expression in the human central nervous system is found in basal ganglia (particularly in putamen and caudate nucleus) and thalamus. Recent studies in rodents, including experiments in knock-out mice, suggest that also A_{2A} receptors contribute to the effects of adenosine on sleep. Local administration of the selective A_{2A} receptor agonist, CGS21680, to the subarachnoid space adjacent to basal forebrain and lateral preoptic area increases *c-fos* expression in the ventrolateral pre-optic (VLPO) area and promotes nonREM sleep [20]. Direct activation of sleep-promoting VLPO neurons upon stimulation of A_{2A} receptors may underlie this effect [21]. Interestingly, preliminary data indicate that CGS21680 is ineffective in mice lacking A_{2A} receptors and that a normal nonREM sleep rebound following sleep deprivation is absent in these knock-out mice [22].

Taken together, there is considerable evidence that both adenosine A_1 and A_{2A} receptor subtypes mediate effects of adenosine on sleep and sleep homeostasis, and these effects appear to be site- and receptor-dependent.

Caffeine Attenuates Markers of Sleep Homeostasis in Sleep and Wakefulness

Caffeine is the most widely consumed stimulant in the world. In the micromolar plasma concentrations reached after moderate consumption, caffeine acts as nonselective, competitive antagonist at both A_1 and A_{2A} receptors. Nevertheless, findings in knock-out mice indicate that the A_{2A} receptor is the main target for caffeine-induced wakefulness [23]. Consistent with this notion, a genetic polymorphism in the adenosine A_{2A} receptor gene (*ADORA2A*; OMIM Database Accession No. 102776) modulates individual sensitivity to subjective and objective measures of caffeine-induced sleep disturbance in humans [24].

Acute administration of caffeine in doses equivalent to one to two cups of coffee (100–200 mg) prolongs sleep latency, impairs sleep efficiency, and reduces the duration of SWS (for review, see [6]). The stimulant also induces changes in nonREM sleep, which are opposite to those of sleep deprivation. Specifically, spectral power in the low-delta range (~1–2.5 Hz) is decreased, whereas power in the spindle frequency range (~12–15 Hz) is increased. It is remarkable that these EEG alterations are still observed at night, even when caffeine is administered in the morning to either rested or sleep deprived subjects [6]. Thus, some caffeine-induced EEG changes in nonREM sleep may mimic the changes associated with physiological reduction in nonREM sleep pressure such as after a nap in the afternoon.

To further test this hypothesis, we examined the effects of caffeine on the evolution of sleep pressure during prolonged wakefulness. We found that compared to placebo, caffeine attenuates the increase of subjective sleepiness and EEG theta activity after sleep deprivation. Spectral power in the theta band (~5–8 Hz) of the waking EEG is considered an objective measure of sleep pressure. It increases more in an anterior EEG derivation than in a posterior derivation, primarily in self-rated caffeine-sensitive individuals [25]. Caffeine attenuates the waking-induced fronto-occipital power gradient in theta activity, and the effects of sleep deprivation and caffeine are negatively correlated. Moreover, the caffeine-induced reduction in theta power is more pronounced after 23 h than after 11 h of wakefulness [26]. Given that caffeine has no major effects on the circadian facet of sleep–wake regulation [27], these data suggest that the efficacy of caffeine to reduce theta activity in the waking EEG depends on the duration of prior wakefulness and, therefore, interacts with sleep homeostasis.

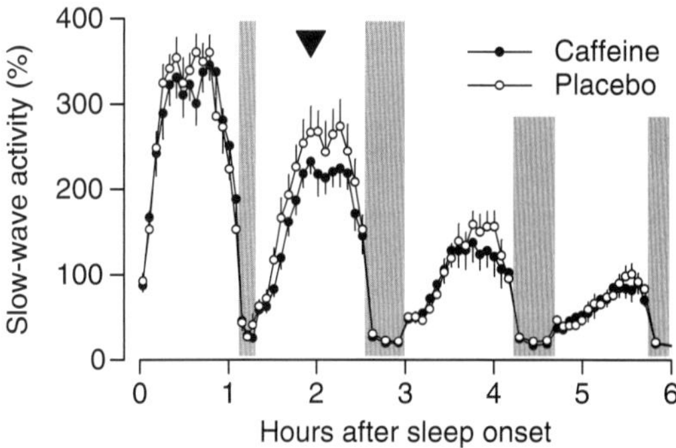

Fig. 2 Caffeine attenuates EEG slow-wave activity (SWA, 0.75–4.5 Hz) in recovery sleep following 40-h prolonged wakefulness. *Error bars* represent 1 SEM ($n=12$ young men). Power density values are expressed as a percentage of the mean nonREM sleep value in baseline, and plotted against the mean timing of nonREM and REM sleep episodes. *Gray shading* highlights REM sleep episodes. *The triangle* indicates significantly reduced SWA in the second nonREM sleep episode after caffeine when compared to placebo ($p<0.05$, paired, two-tailed t-test). Data from [26]

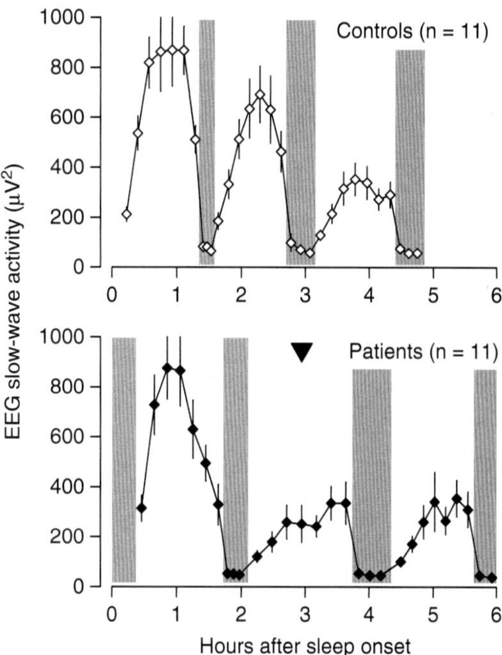

Fig. 3 Time course of EEG slow-wave activity SWA (0.75–4.5 Hz) in baseline sleep in 11 patients with narcolepsy-cataplexy (*black diamonds, bottom*) and 11 age- and sex-matched healthy controls (*open diamonds, top*). The data represent means ±1 SEM. *Gray shading* highlights REM sleep episodes. All patients showed a sleep onset REM sleep episode. *The triangle* indicates significantly reduced SWA in the second nonREM sleep episode in the patients when compared to the controls ($p<0.02$, unpaired, two-tailed t-test). Note that the duration of the second nonREM sleep episode is longer in the patients than in the controls. Slow-wave energy (i.e., SWA integrated over time) does not differ between the groups. Data re-plotted from [28]

This conclusion is further supported by the time course of caffeine-induced changes in recovery sleep after sleep deprivation. When compared to placebo, morning administration of the stimulant reduces nocturnal SWA in the second nonREM sleep episode, whereas the values in the beginning of the night do not differ (Fig. 2). Thus, a putative residual level of caffeine cannot readily account for the delayed reduction of SWA in recovery sleep. Taken together, pharmacological blockade of adenosine A_1 and A_{2A} receptors with caffeine interferes with the wakefulness-induced rise of sleep pressure. Caution is essential when comparing physiological mechanisms of sleep–wake regulation with pharmacological interventions. Nevertheless, the state-specific changes in predefined EEG markers of sleep homeostasis in sleep and wakefulness are highly unlikely to reflect an nonspecific action of caffeine on EEG generation.

Dysregulation of Sleep Homeostasis Involving Adenosinergic Mechanisms in Narcolepsy?

Only a few studies to date addressed the question whether changes in homeostatic sleep–wake regulation could contribute to the narcoleptic tetrad including disturbed nocturnal sleep and excessive daytime sleepiness. A most recent study in a series of investigation in carefully diagnosed patients with narcolepsy-cataplexy ($n=11$, 6 women) first focused on the dynamics of SWA in baseline sleep [28]. All-night power in nonREM and REM sleep does not differ from individually matched healthy volunteers, and mean SWA declines exponentially across the first three nonREM sleep episodes. Interestingly, the decline from the first to the second nonREM sleep episode is larger in the patients than in the controls (Fig. 3). Reduced SWA in the second nonREM sleep period is consistently found in narcolepsy-cataplexy [29]. Nevertheless, because frequent and long lasting intrusions of intermittent wakefulness preclude undisturbed evolution of SWA, it cannot be concluded that the dissipation of homeostatic sleep pressure is accelerated in the patients. Rather, the inability to consolidate sleep when nonREM sleep pressure has fallen underneath a hypothetical threshold value, may underlie the observed differences.

To test this hypothesis, two subgroups of six patients (three women) and six controls were kept awake for 40 h under continuous one-to-one supervision [30]. In both groups and in baseline and recovery nights, SWA was highest in the first nonREM sleep episode and declined in the course of sleep. The prolongation of wakefulness increased SWA to the same extent in both groups. Thus, the homeostatic facet of nonREM sleep regulation appears normal in the patients. The elevated sleep pressure following sleep deprivation also postponed the occurrence of wake intrusions when compared to baseline, supporting the hypothesis that sleep fragmentation in narcolepsy-cataplexy is associated with insufficient nonREM sleep intensity. The possible underlying mechanisms are presently unclear.

Although genetic studies in dogs suggest that narcolepsy may be transmitted as an autosomal recessive trait caused by a mutation in the canine *hypocretin (orexin) receptor 2* gene, no pathogenic mutations or polymorphisms in the genes encoding hypocretin and its two receptors (hypocretin-1 and hypocretin-2) were identified to date in human patients with narcolepsy [31]. Nevertheless, hypocretin-1 levels in cerebrospinal fluid (CSF) are undetectable in most patients with narcolepsy-cataplexy [32]. In conclusion, whereas polymorphisms may not play a role, CSF hypocretin deficiency could contribute to disturbed sleep and excessive daytime sleepiness in narcolepsy-cataplexy.

Hypocretin neurons have been proposed to fulfill an essential function in stabilizing sleep and wake states [33]. Consistent with this notion, hypocretin knock-out mice are unable to produce long consolidated periods of normal sleep and wakefulness, despite no evidence for homeostatic dysregulation of sleep [34]. These findings suggest that hypocretin neurotransmitter/receptor deficiency disrupts behavioral state control and results in fragmented sleep and wakefulness. Recent experiments indicate that adenosine can attenuate the activity of hypocretin neurons in vitro via both adenosine A_1 and A_{2A} receptor-mediated actions [35, 36].

In light of these findings, it would be worthwhile to study the effects of caffeine on sleep and the sleep EEG in patients with narcolepsy. Such studies are currently lacking. However, it is interesting to note that in a large sample of 530 patients, the inability to stay awake on the maintenance of wakefulness test is most severe in those narcolepsy patients who report moderate to heavy caffeine use [37] (Fig. 4). The therapeutic/adverse actions of acute and chronic caffeine can be quite different [13], and the data may be interpreted at least in

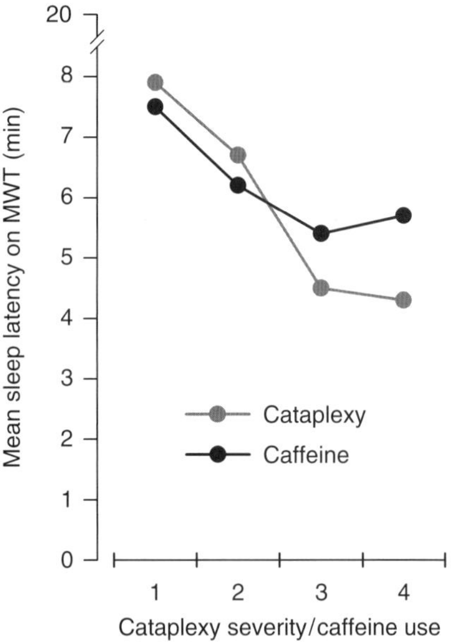

Fig. 4 Mean sleep latency over four trials of the maintenance of wakefulness test (MWT) in 530 patients with narcolepsy as a function of severity of cataplexy (physician rated, *gray dots*) and habitual caffeine use (*black dots*): **1** = no cataplexy (n = 101)/no caffeine use (n = 53); **2** = mild cataplexy (n = 219)/light caffeine use (one to two cups of coffee per day; n = 259); **3** = moderate cataplexy (n = 149)/moderate caffeine use (two to four cups; n = 186); **4** = severe cataplexy (n = 61)/heavy caffeine use (four to eight cups; n = 32). Mean values are plotted from published data [37]. Shorter sleep latency reflects higher sleepiness. Sleep latencies vary significantly as a function of cataplexy severity ($F_{3,526} = 15.9$, $p < 0.0001$, one-way ANOVA) and caffeine use ($F_{3,526} = 2.93$, $p < 0.03$)

two alternative ways. First, chronic high dietary caffeine consumption increases plasma concentrations of adenosine [38]. Elevated adenosine levels may inhibit the (remaining) hypocretin neurons and further exacerbate the symptoms of narcolepsy. Second, patients with narcolepsy may attempt to self-medicate excessive daytime sleepiness with caffeine. Those patients with the most severe symptoms (those with high sleepiness and severe cataplexy) may consume the highest amounts of caffeine. This reasoning, however, remains speculation at the moment.

Summary and Conclusions

Recent studies examined the effects of sleep deprivation on the time course of EEG SWA in nonREM sleep, after interference with adenosinergic neurotransmission by genetic reduction of adenosine release from astrocytes [9], knocking-out neuronal A_1 receptors [17], and blocking A_1 and A_{2A} receptors with caffeine [26]. All manipulations consistently reduce SWA in recovery sleep. Remarkably, the differences are not usually present at sleep initiation (i.e., when homeostatic sleep pressure is highest), but become significant after 2–4 h into the sleep episode. These data are consistent with the conclusion that adenosinergic mechanisms are involved in sleep homeostasis, yet other mechanisms must also contribute.

Sleep pressure in patients with narcolepsy may be comparable to the sleep pressure in healthy individuals after 2 days without sleep [39]. Interestingly, some differences in the evolution of SWA in nonREM sleep between patients and controls are reminiscent of the caffeine-induced changes in healthy volunteers in recovery sleep after sleep deprivation. The acute effects of caffeine on sleep and the sleep EEG in narcolepsy are unknown, but epidemiological studies indicate that the severity of narcolepsy-cataplexy and habitual caffeine use are associated. Further research is warranted to elucidate the possible relevance of this association for the pathophysiology of narcolepsy, and to investigate whether the adenosine neurotransmitter/receptor system could provide novel targets to treat disturbed sleep and wakefulness in narcolepsy-cataplexy or related sleep–wake disorders

Acknowledgments The author thanks Dr. Caroline Kopp for discussion and helpful comments on the manuscript. His research summarized in this chapter was supported by Swiss National Science Foundation, Center for Neuroscience Zürich (ZNZ), and Zürich Center for Integrative Human Physiology (ZIHP).

References

1. Borbély AA. A two process model of sleep regulation. Hum Neurobiol. 1982;1:195–204.
2. Cirelli C, Tononi G. Is sleep essential? PLoS Biol. 2008;6:1605–11.
3. Franken P, Dijk DJ. Circadian clock genes and sleep homeostasis. Eur J Neurosci. 2009;29:1820–9.
4. Daan S, Beersma DGM, Borbély AA. Timing of human sleep: recovery process gated by a circadian pacemaker. Am J Physiol. 1984;246:R161–78.
5. Basheer R, Strecker RE, Thakkar MM, McCarley RW. Adenosine and sleep-wake regulation. Prog Neurobiol. 2004;73:379–96.
6. Landolt HP. Sleep homeostasis: a role for adenosine in humans? Biochem Pharmacol. 2008;75:2070–9.
7. Haydon PG, Carmignoto G. Astrocyte control of synaptic transmission and neurovascular coupling. Physiol Rev. 2006;86:1009–31.
8. Pascual O, Casper KB, Kubera C, et al. Astrocytic purinergic signaling coordinates synaptic networks. Science. 2005;310:113–6.
9. Halassa MM, Florian C, Fellin T, et al. Astrocytic modulation of sleep homeostasis and cognitive consequences of sleep loss. Neuron. 2009;61:213–9.
10. Fredholm BB, Chen JF, Cunha RA, Svenningsson P, Vaugeois JM. Adenosine and brain function. Internatl Rev Neurobiol. 2005;63:191–270.
11. Franken P, Chollet D, Tafti M. The homeostatic regulation of sleep need is under genetic control. J Neurosci. 2001;21:2610–21.
12. Rétey JV, Adam M, Honegger E, et al. A functional genetic variation of adenosine deaminase affects the duration and intensity of deep sleep in humans. Proc Natl Acad Sci U S A. 2005;102:15676–81.
13. Sebastiao AM, Ribeiro JA. Adenosine receptors and the central nervous system. In: Wilson CN, Mustafa SJ, editors. Handbook of experimental pharmacology, vol. 193. Berlin, Heidelberg: Springer; 2009. p. 471–534.
14. Bauer A, Ishiwata K. Adenosine receptor ligands and PET imaging of the CNS. In: Wilson CN, Mustafa SJ, editors. Handbook of experimental pharmacology, vol. 193. Berlin, Heidelberg: Springer; 2009. p. 617–42.
15. Stenberg D, Litonius E, Halldner L, Johansson B, Fredholm BB, Porkka-Heiskanen T. Sleep and its homeostatic regulation in mice lacking the adenosine A1 receptor. J Sleep Res. 2003;12:283–90.

16. Thakkar MM, Winston S, McCarley RW. A1 receptor and adenosinergic homeostatic regulation of sleep-wakefulness: effects of antisense to the A1 receptor in the cholinergic basal forebrain. J Neurosci. 2003;23: 4278–87.
17. Bjorness TE, Kelly CL, Gao TS, Poffenberger V, Greene RW. Control and function of the homeostatic sleep response by adenosine A(1) receptors. J Neurosci. 2009;29:1267–76.
18. Elmenhorst D, Basheer R, McCarley RW, Bauer A. Sleep deprivation increases A(1) adenosine receptor density in the rat brain. Brain Res. 2009; 1258:53–8.
19. Elmenhorst D, Meyer PT, Winz OH, et al. Sleep deprivation increases A(1) adenosine receptor binding in the human brain: A positron emission tomography study. J Neurosci. 2007;27:2410–5.
20. Scammell TE, Gerashchenko DY, Mochizuki T, et al. An adenosine A2a agonist increases sleep and induces Fos in ventrolateral preoptic neurons. Neuroscience. 2001;107:653–63.
21. Gallopin T, Luppi PH, Cauli B, et al. The endogenous somnogen adenosine excites a subset of sleep-promoting neurons via A2A receptors in the ventrolateral preoptic nucleus. Neuroscience. 2005;134:1377–90.
22. Hayaishi O, Urade Y, Eguchi N, Huang Z-L. Genes for prostaglandin D synthase and receptor as well as adenosine A2A receptor are involved in the homeostatic regulation of NREM sleep. Arch Ital Biol. 2004;142:533–9.
23. Huang ZL, Qu WM, Eguchi N, et al. Adenosine A2A, but not A1, receptors mediate the arousal effect of caffeine. Nat Neurosci. 2005;8:858–9.
24. Rétey JV, Adam M, Khatami R, et al. A genetic variation in the adenosine A2A receptor gene (*ADORA2A*) contributes to individual sensitivity to caffeine effects on sleep. Clin Pharmacol Ther. 2007;81:692–8.
25. Rétey JV, Adam M, Gottselig JM, et al. Adenosinergic mechanisms contribute to individual differences in sleep-deprivation induced changes in neurobehavioral function and brain rhythmic activity. J Neurosci. 2006;26:10472–9.
26. Landolt HP, Rétey JV, Tönz K, et al. Caffeine attenuates waking and sleep electroencephalographic markers of sleep homeostasis in humans. Neuropsychopharmacology. 2004;29:1933–9.
27. Wyatt JK, Cajochen C, Ritz-De Cecco A, Czeisler CA, Dijk DJ. Low-dose repeated caffeine administration for circadian-phase-dependent performance degradation during extended wakefulness. Sleep. 2004;27:374–81.
28. Khatami R, Landolt HP, Achermann P, et al. Insufficient non-REM sleep intensity in narcolepsy-cataplexy. Sleep. 2007;30:980–9.
29. Besset A, Tafti M, Nobile L, Billiard M. Homeostasis and narcolepsy. Sleep. 1994;17:S29–34.
30. Khatami R, Landolt HP, Achermann P, et al. Challenging sleep homeostasis in narcolepsy-cataplexy: implications for non-REM and REM sleep regulation. Sleep. 2008;31:859–67.
31. Olafsdottir BR, Rye DB, Scammell TE, Matheson JK, Stefansson K, Gulcher JR. Polymorphisms in hypocretin/orexin pathway genes and narcolepsy. Neurology. 2001;57:1896–9.
32. Dauvilliers Y, Baumann CB, Carlander B, et al. CSF hypocretin-1 levels in narcolepsy, Kleine-Levin syndrome, and other hypersomnias and neurological conditions. J Neurol Neurosurg Psychiatry. 2003;74:1667–73.
33. Saper CB, Scammell TE, Lu J. Hypothalamic regulation of sleep and circadian rhythms. Nature. 2005;437:1257–63.
34. Mochizuki T, Crocker A, McCormack S, Yanagisawa M, Sakurai T, Scammell TE. Behavioral state instability in orexin knock-out mice. J Neurosci. 2004;24: 6291–300.
35. Liu ZW, Gao XB. Adenosine inhibits activity of hypocretin/orexin neurons by the A1 receptor in the lateral hypothalamus: A possible sleep-promoting effect. J Neurophysiol. 2007;97:837–48.
36. Satoh S, Matsumura H, Kanbayashi T, et al. Expression pattern of FOS in orexin neurons during sleep induced by an adenosine A(2A) receptor agonist. Behav Brain Res. 2006;170:277–86.
37. Mitler MM, Walsleben J, Sangal RB, Hirshkowitz M. Sleep latency on the maintenance of wakefulness test (MWT) for 530 patients with narcolepsy while free of psychoactive drugs. Electroencephalogr Clin Neurophysiol. 1998;107:33–8.
38. Conlay LA, Conant JA, deBros F, Wurtman R. Caffeine alters plasma adenosine levels. Nature. 1997; 389:136.
39. Siegel JM. Narcolepsy. Sci Am. 2000;282:76–81.

Prostaglandin D_2: An Endogenous Somnogen

Yoshihiro Urade and Osamu Hayaishi

Keywords
Prostaglandines • Prostaglandine D synthase • Adenosine • Narcolepsy • Ventrolateral preoptic area

Introduction

The humoral theory of sleep regulation, the concept that sleep and wakefulness are induced and regulated by a hormone-like chemical substance rather than by a neural network, was initially proposed by Kuniomi Ishimori of Nagoya, Japan, in 1909 and independently and concurrently, by a French neuroscientist, Henri Piéron of Paris, in 1913. They took samples of the brain and cerebrospinal fluid (CSF) from sleep-deprived dogs and infused them into the brains of normal dogs. The recipient dogs soon started to sleep. Thus these scientists became the pioneers of sleep research, demonstrating the existence of endogenous sleep-promoting substances. However, the chemical nature of their sleep substance(s) was not identified. During the twentieth century, numerous investigators have reported more than 30 so-called endogenous sleep- and wake-promoting substances in the brain, CSF, and urine of animals and humans. Among them, prostaglandin (PG) D_2 (Fig. 1) is now recognized as the most potent endogenous sleep-promoting substance; and its action mechanism to induce sleep is the best characterized.

Sleep Induction by PGD_2

PGs are a group of 20-carbon polyunsaturated fatty acids containing a unique 5-carbon ring structure. PGs of the two series, such as PGD_2, PGE_2, $PGF_{2\alpha}$, PGI_2 (prostacycline), and thromboxane A_2, are all produced from arachidonate ($C_{20:4}$ fatty acid) via a common intermediate, PGH_2, which is produced by the action of cyclooxygenase/PGH_2 synthase, a target of non-steroidal anti-inflammatory drugs such as aspirin and indomethacin. Each prostanoid is then produced from PGH_2 by the action of its terminal PG synthase, which in the case of PGD_2 is PGD synthase (PGDS). PGD_2 had long been considered as a minor and biologically inactive PG. However, in 1980s, PGD_2 was found to be the most abundant PG produced in the brains of rats [1] and other mammals including humans [2], suggesting that PGD_2 may play some important function in the CNS.

Y. Urade (✉)
Department of Molecular Behavioral Biology,
Osaka Bioscience Institute, 6-2-4, Furuedai,
Suita-shi, Osaka 565-0874, Japan
e-mail: uradey@obi.or.jp

C.R. Baumann et al. (eds.), *Narcolepsy: Pathophysiology, Diagnosis, and Treatment*,
DOI 10.1007/978-1-4419-8390-9_9, © Springer Science+Business Media, LLC 2011

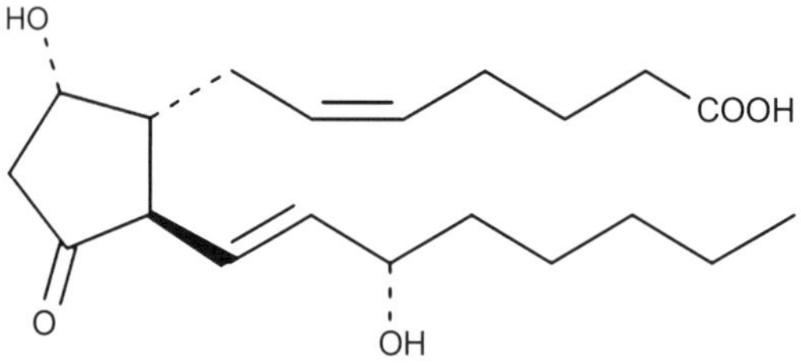

HO

COOH

O

OH

Fig. 1 Chemical structure of PGD_2

During the search for the neural function of PGD_2, its sleep-inducing activity was discovered; [3] i.e., the microinjection of nanomolar quantities of PGD_2 in the rat brain increased both non-rapid eye movement (non-REM, NREM) and REM sleep. By use of a sleep bioassay system based on electroencephalogram (EEG) and electromyogram (EMG) recordings during continuous intracerebroventricular (i.c.v.) infusion of drugs into freely moving rats, the somnogenic activity of PGD_2 was demonstrated to be both dose- and time-dependent [4]. The PGD_2-induced sleep was significant with as little as picomolar quantities per minute and indistinguishable from physiological sleep as judged by several electrophysiological and behavioral criteria. During the PGD_2 infusion, the rats were easily aroused by the sound of a clap; and their sleep was episodic, indicating that PGD_2 does not interfere with the minimum waking time for their survival. Essentially, the same sleep induction was demonstrated in the rhesus monkey *Macaca mulatta* during the i.c.v. infusion of PGD_2 [5].

The PGD_2 concentration in rat CSF showed a circadian fluctuation that paralleled the sleep–wake cycle [6] and became elevated with an increase in sleep propensity during sleep deprivation [7]. Also, the levels of PGD_2 were increased in mastocytosis, a disorder characterized by episodic and endogenous production of PGD_2 accompanied by deep sleep episodes [8]. In addition, PGD_2 concentrations were selectively and time-dependently elevated in the CSF of patients with African sleeping sickness, which is caused by infection with *Trypanosoma* [9]. These findings suggest that PGD_2 induces sleep in humans as well as in rodents and monkeys.

PGDS and PGD_2 Receptors in the CNS

There are two distinct types of PGDS (PGH_2 D-isomerase, EC.5.3.99.2); one is the lipocalin-type PGDS (L-PGDS) [10] and the other is hematopoietic PGDS (H-PGDS) [11, 12]. We purified L-PGDS and H-PGDS, isolated their cDNAs and chromosomal genes of the human and mouse enzymes, determined their X-ray crystallographic structures, and demonstrated that these two enzymes are quite different from each other in terms of their amino acid sequences, tertiary structures, evolutional origins, cellular distribution, etc. [13, 14]. We developed inhibitors selective for either L-PGDS or H-PGDS (AT-56 [17] and HQL-79 [18], respectively).

Two distinct subtypes of receptors for PGD_2 have been identified; one is the DP_1 (DP) receptor originally identified as a homolog of other PG receptors [15], and the other is the DP_2 (CRTH2) receptor, identified as a chemoattractant receptor for PGD_2 [16]. Selective agonists and antagonists for DP_1 or DP_2 were also made available, and these pharmacological tools are used for various studies including those on sleep.

L-PGDS and the DP_1 receptor are involved in the regulation of physiological sleep, as described later in detail, whereas H-PGDS is likely involved in pathological sleep, because H-PGDS is localized in microglial cells in the brain [19] and is upregulated in activated microglia in various neurodegenerative diseases. L-PGDS is localized in the leptomeninges, choroid plexus, and oligodendrocytes in the brain (Fig. 2) [20, 21]. DP_1 receptors are highly concentrated in the leptomeninges on the ventral surface of the rostral basal forebrain, whereas other brain areas are almost completely devoid of them. The DP_1-enriched leptomeninges was immunohistochemically defined as bilateral wings under the rostral basal forebrain lateral to the optic chiasm and extending back under the posterior hypothalamus, which are regions that contain sleep–wake regulatory neurons [22].

The lipocalin gene family comprises various secretory proteins that bind and transport small

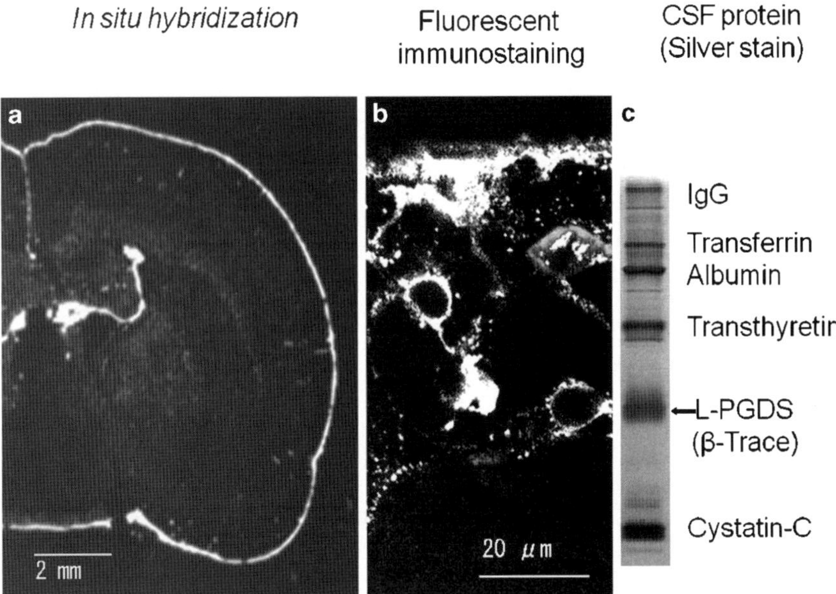

Fig. 2 Dominant expression of L-PGDS in leptomeninges and its secretion into CSF. (**a**) Negative film image of in situ hybridization of the mRNA for L-PGDS in a coronal section of the rat brain. Positive signals (*white spots*) are observed in the leptomeninges surrounding the brain, the choroid plexus in the ventricle, and oligodendrocytes of the brain parenchyma [21]. (**b**) Confocal immunofluorescence staining of L-PGDS in the rat brain. Positive fluorescence signal is found in the cytosol of arachnoid trabecular cells [20]. (**c**) SDS-polyacrylamide gel electrophoresis of rat CSF. L-PGDS (β-trace) is observed to be a broad band at a position of molecular weight of 26,000 Da [28]

hydrophobic substances, and L-PGDS is the only member associated with enzyme activity [23, 24] L-PGDS is a monomeric glycoprotein with a molecular weight of 26,000 Da and is composed of 189 and 190 amino acid residues in the mouse and human enzymes, respectively. L-PGDS is posttranslationally modified by the cleavage of an N-terminal hydrophobic signal peptide comprising 24 and 22 amino acid residues in the mouse and human enzymes, respectively, and by *N*-glycosylation at two positions, Asn51 and Asn78, in the mouse and human enzymes, with each glycosyl chain having a molecular weight of 3,000 Da. We recently determined the NMR solution structure of L-PGDS [25] and its X-ray crystallographic structures of two different conformers, i.e., one with an open calyx and the other, a closed calyx, at 2.1 Å resolution [26] and showed that L-PGDS possesses a typical lipocalin-fold and a β-barrel structure having a large hydrophobic pocket containing the catalytically essential activated thiol of the Cys-65 residue (Fig. 3).

L-PGDS is the same protein as β-trace (Fig. 2c) [27, 28], which was originally discovered in 1961 as a major protein in human CSF [29]. The serum L-PGDS/β-trace concentration shows a circadian change with a nocturnal increase that is suppressed during total sleep deprivation but is not affected by deprivation of REM sleep [30]. We recently found that L-PGDS/β-trace binds PGD_2 with a high affinity (Kd = 20 nM; Aritake K. and Y.U., unpublished results), suggesting that PGD_2 may circulate in CSF as a form bound to L-PGDS/β-trace and be transported to the DP_1 receptor to initiate NREM sleep.

The Molecular Mechanisms of PGD_2-Induced Sleep

We modified the above-mentioned sleep bioassay system to monitor the EEG/EMG of freely moving mice during continuous i.c.v. infusion of drugs, developed SLEEPSIGN software (Kissei Comtec, Nagano, Japan) for automatic

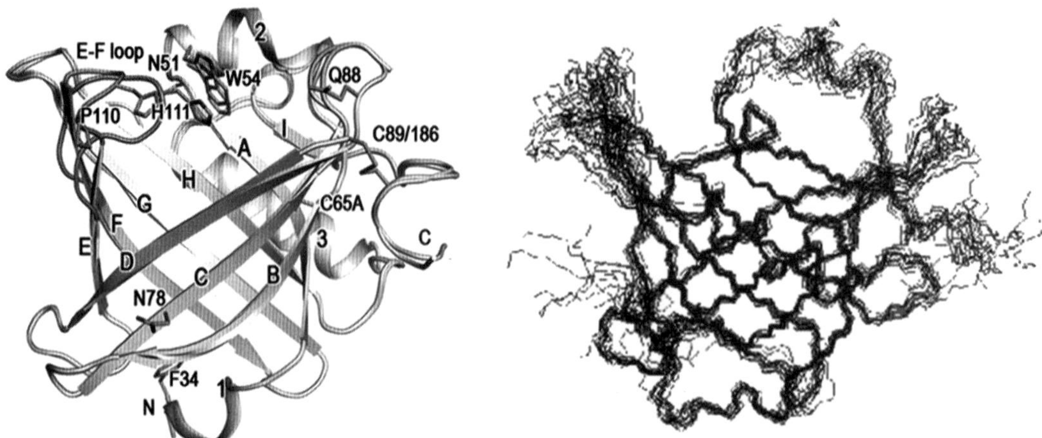

Fig. 3 X-ray crystallographic structure (*left*) and NMR solution structure (*right*) of L-PGDS. The X-ray crystallographic structure of mouse L-PGDS is shown by *ribbon diagrams*, in which β-strands and α-helices are labeled A-I and 1–3, respectively [26]. The EF-loop shows two different conformations between open and closed forms of L-PGDS. Stick models indicate two *N*-glycosylation sites (Asn [51] and Asn^{78}, denoted N51 and N78), the catalytically essential Cys^{65} substituted by Ala in the crystal (C65A), a disulfide bond between Cys^{89} and Cys^{186} (C89/186), and several residues with different conformations between the open and closed forms, Phe^{34} (F34), Gln^{88} (Q88), Trp^{54} (W54), Pro^{111} (P110), and His^{111} (H111). The flexibility of the EF-loop is clearly seen in the superposition of 15 lowest energy backbone conformers of L-PGDS determined by NMR [25]

scoring of the vigilance states of rats and mice [31], and used these systems and software to study the molecular mechanism of PGD_2-induced sleep.

When PGD_2 was infused into the subarachnoid space below the basal forebrain of wild-type mice, in which DP_1 receptors are remarkably abundant, the extracellular adenosine concentration increased dose-dependently. This PGD_2-induced increase in extracellular adenosine was not observed in KO mice for DP_1 receptors, indicating that the adenosine increase was dependent on the DP_1 receptors [22].

Adenosine has been proposed to be an endogenous sleep substance, because a number of stable adenosine analogues induce sleep when administered to rats and other experimental animals. When CGS21680, an A_{2A} receptor agonist, was infused into the lateral ventricle of WT mice, NREM sleep increased dose-dependently; whereas the A_1 receptor-selective agonist N^6-cyclopentyladenosine was totally inactive, indicating that A_{2A}, but not A_1, receptors are involved in NREM sleep regulation [32].

When PGD_2 or the A_{2A} receptor agonist CGS21680 was infused for 2 h into the PGD_2-sensitive zone of the subarachnoid space of the basal forebrain, the number of Fos-positive cells was remarkably increased in the leptomeningeal membrane as well as in the ventrolateral preoptic (VLPO) area; and concomitantly NREM sleep was induced [33, 34]. In contrast, the number of Fos-positive neurons decreased markedly in the tuberomammillary nucleus (TMN) of the posterior hypothalamus. The VLPO is known to send specific inhibitory GABAergic and galaninergic efferents to the TMN, the neurons of which contain the ascending histaminergic arousal system [35] (Fig. 4).

Intracellular recordings of VLPO neurons in rat brain slices demonstrated the existence of two distinct types of VLPO neurons in terms of their responses to serotonin and adenosine. VLPO neurons are inhibited uniformly by two arousal neurotransmitters, noradrenaline and acetylcholine, and mostly by an adenosine A_1 receptor agonist. Serotonin inhibits the type-1 neurons but excites the type-2 neurons. Also, an A_{2A} receptor

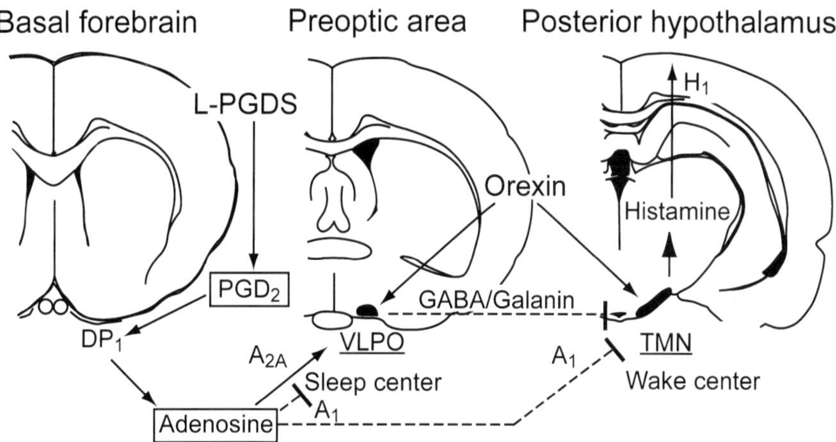

Fig. 4 Mechanisms of PGD_2-induced sleep. The endogenous somnogen PGD_2 is produced by L-PGDS, circulates within the CSF, stimulates DP_1 receptors below the basal forebrain and hypothalamus, and increases extracellular adenosine [22]. Adenosine diffuses into the brain parenchyma as the secondary somnogen and induces sleep by activating sleep-active VLPO neurons via A_{2A} receptors [36] and inhibiting arousal neurons in the TMN via A_1 receptors [43]. The flip-flop switch that helps regulate sleep and wakefulness [38, 39] is stabilized by their orexin-mediated activation and adenosine A_1 receptor-mediated suppression

agonist postsynaptically excites the type-2, but not the type-1, neurons. These results suggest that the type-2 VLPO neurons are involved in the initiation of sleep and that the type-1 neurons contribute to sleep consolidation, since they are activated only when released from inhibition by arousal systems [36].

In vivo microdialysis experiments revealed that infusion of the adenosine A_{2A} receptor agonist, CGS21680, into the basal forebrain inhibited the release of histamine in both the frontal cortex and medial preoptic area in a dose-dependent manner, and increased GABA release specifically in the TMN but not in the frontal cortex [37]. The CGS21680-induced inhibition of histamine release was antagonized by perfusion of the TMN with a $GABA_A$ antagonist, picrotoxin, suggesting that the A_{2A} agonist induced sleep by inhibiting the histaminergic system through an increase in GABA release in the TMN. These results support the original idea of the flip-flop mechanism, whereby sleep is driven by activity of the sleep-promoting neurons in the VLPO and inhibition of the histaminergic wake-promoting neurons in the TMN, as proposed by Saper and colleagues [38, 39].

Involvement of the Histaminergic Neurons in Sleep–Wake Regulation

The TMN is the sole source of histaminergic innervation of the mammalian CNS. The histaminergic system plays a central role in mediating wakefulness via H_1 receptors widely distributed in the CNS. The extracellular concentration of histamine in the rat cortex is positively correlated with the awake time more so than any other neurotransmitter [40]. The TMN is enriched in other receptors for various neuromodulators involved in the sleep–wake regulation, such as PGE_2, orexin, and adenosine. For example, EP_4 receptor mRNA was expressed in the TMN region, as revealed by in situ hybridization. Perfusion of the TMN with PGE_2 or an EP_4 receptor-agonist, ONO-AE1-329, but not with agonists for $EP_{1\text{-}3}$ receptors, significantly increased histamine release in the brain and induced wakefulness, indicating that PGE_2 induced wakefulness through activation of the histaminergic system via EP_4 receptors [41].

Orexin-2 receptors are also abundant in the TMN. Orexin neurons are exclusively localized in the lateral hypothalamic area and project their fibers densely to the TMN. The degeneration of orexin neurons causes narcolepsy. Perfusion of the rat TMN with orexin A (5–25 pmol/min) through a microdialysis probe promptly increased wakefulness and histamine release from both the medial preoptic area and the frontal cortex in a dose-dependent manner. Furthermore, the infusion of orexin A (1.5 pmol/min) into the lateral ventricle of wild-type mice increased wakefulness to the same level of the active period but did not change at all the sleep–wake cycle in KO mice for H_1 receptors [42]. These results indicate that orexin is a potent waking substance acting upon its receptor in the TMN and that the arousal effect of orexin A depends on the histaminergic neurotransmission mediated by H_1 receptors.

Adenosine deaminase, an enzyme that catabolizes adenosine to inosine, is strongly expressed in TMN neurons, and the TMN is also enriched in adenosine A_1 receptors. Therefore, the histaminergic arousal system in the TMN is likely regulated by adenosine. Recently, we showed that bilateral injection into the rat TMN of an A_1 receptor agonist, N^6-cyclopentyladenosine, significantly increased the amount of NREM sleep [43]. The bilateral injection of adenosine or an inhibitor of adenosine deaminase, coformycin, into the rat TMN also increased the amount of NREM sleep. This increase was completely abolished by coadministration of 1,3-dimethyl-8-cyclopenthylxanthine, a selective A_1R antagonist. These results indicate that endogenous adenosine in the TMN suppresses the histaminergic system via A_1 receptors to promote NREM sleep.

Sleep Abnormalities in Mice with Manipulations of the Genes Coding for L-PGDS, DP_1 Receptors, and A_{2A} Receptors

Using homologous recombination, we generated L-PGDS KO mice [44] and demonstrated that the KO mice grew normally but showed several functional abnormalities in their regulation of sleep [45]. The KO mice did not accumulate PGD_2 in their brain during sleep deprivation and lacked NREM sleep rebound after sleep deprivation; whereas the wild-type mice showed an increase in PGD_2 content in their brain during sleep deprivation, which induced NREM sleep rebound [45]. We also generated transgenic (TG) mice [46] that overexpressed human L-PGDS under the control of the β-actin promoter. We serendipitously discovered that these TG mice showed a transient increase in NREM sleep after their tails had been clipped for DNA sampling used for genetic analysis [46]. The noxious stimulation of tail clipping induced a remarkable increase in PGD_2 content in the brain of the TG mice but not in that of the wild-type ones, although we do not yet understand in detail the mechanism responsible for this increase.

KO mice for various receptors involved in the sleep–wake regulation, such as DP_1, A_1, A_{2A}, and H_1 receptors, have been already generated and their sleep–wake regulation examined. However, those KO mice showed essentially the same circadian profiles and daily amounts of sleep as wild-type mice, although a minor decrease in a short period of wakefulness was observed in H_1 KO mice [47]. These results raised questions about the involvement of those receptors in the regulation of physiological sleep. Considering that sleep is essential for life, it is likely that the sleep-regulatory system would be composed of a complicated network of systems, in which the deficiency of one system may be compensated by other systems during embryonic development. To minimize the effect of functional compensation on sleep–wake regulation, we used pharmacological tools, such as antagonists for those receptors, to examine the contribution of each system to physiological sleep, as described later.

On the other hand, after some stimulation, those KO mice showed abnormal sleep–wake regulation. Similar to L-PGDS KO mice, DP_1 KO mice did not exhibit NREM sleep rebound after sleep deprivation, indicating that the L-PGDS/

PGD_2/DP_1 receptor system is crucial for the homeostatic regulation of NREM sleep [45].

Contribution of PGD_2 to Physiological Sleep

We examined the effects of a DP_1 antagonist, ONO-4127Na, on the sleep of rats and that of $SeCl_4$, a selective inhibitor of PGDS, on the sleep of wild-type and KO mice for PGDS and DP_1 receptors [48].

ONO-4127Na is produced as a novel DP_1 antagonist by Ono Pharmaceutical Co. Ltd. (Osaka, Japan). This compound exhibited high specific binding affinity for DP_1 receptors (K_i=2.5 nM) and antagonism toward DP_1 receptors (pA_2=9.73). We infused this DP_1 antagonist into the subarachnoid space underlying the rostral basal forebrain of rats during their sleep period. ONO-4127Na infusion at 50 pmol/min had little effect on the sleep-stage distribution. However, ONO-4127Na given at 100 and 200 pmol/min reduced NREM sleep by 23 and 28%, respectively, and REM sleep by 49 and 63%, respectively, during perfusion for 6 h and postinfusion for 1 h. As shown in Fig. 5, ONO-4127Na infusion at 200 pmol/min decreased the amount of NREM sleep over a 7-h period by 30–40%, and reduced REM sleep by 60–90% commencing about 2 h after the beginning of ONO-4 127Na infusion, as compared with the baseline. We could not examine doses higher than 200 pmol/min, due to the low solubility of ONO-4127Na ($<5\times10^{-4}$ M in 5% sucrose, thus resulting in the maximum dosage for infusion of 200 pmol at the rate of 0.4 μl/min). These results clearly indicate that ONO-4127Na reduced NREM and REM sleep in a dose-dependent manner, and suggest that the stimulation of DP_1 receptors with endogenous PGD_2 is essential for the maintenance of physiological sleep.

Inorganic tetravalent selenium compounds are potent, relatively specific, and reversible inhibitors of PGDS [49]. When we examined the effect of $SeCl_4$ on the activities of mouse L-PGDS and H-PGDS in vitro, $SeCl_4$ inhibited both L-PGDS and H-PGDS efficiently in a concentration-dependent manner, giving IC_{50} values of 40 and 90 μM, respectively. We then determined the

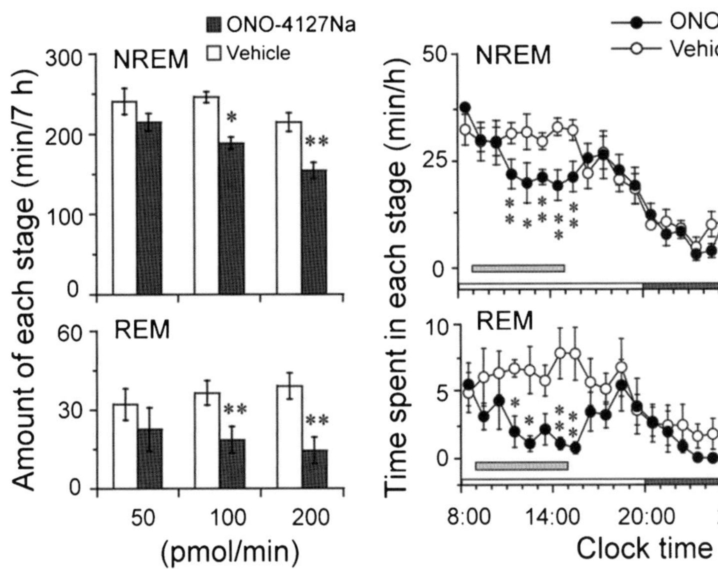

Fig. 5 Suppression of NREM and REM sleep by a DP_1 antagonist (ONO-4127Na). Dose-dependency (*left*) and time course (*right*) of DP_1 antagonist-induced inhibition of NREM and REM sleep are shown [48]. *$p<0.05$ and **$p<0.01$ vs. the vehicle-administrated group

PGD_2 content in the brains of wild-type mice 2 h after an intraperitoneal injection of $SeCl_4$ at a dose of 1.25–5 mg/kg body weight during the light period (when mice normally sleep). $SeCl_4$ given at 1.25 mg/kg had little effect on the PGD_2 content in the brain, but at 2.5 and at 5 mg/kg it reduced the PGD_2 content by 52 and 59%, respectively, without changing the PGE_2 and $PGF_{2\alpha}$ contents. These results indicated that the $SeCl_4$ administration selectively inhibited the production of PGD_2 in the brain in vivo without affecting the production of other PGs [48].

Next, we examined the sleep–wake pattern of wild-type mice before and after an intraperitoneal bolus injection of $SeCl_4$ at doses of 1.25–5 mg/kg at 11:00 h during the sleep period. In the case of 1.25 mg/kg $SeCl_4$, the sleep–wake pattern was almost identical before and after the injection, similar to the pattern of PGD_2 content in the brain. $SeCl_4$ at 2.5 mg/kg slightly reduced both NREM and REM sleep for 2–3 h after the injection and decreased the cumulative amounts of NREM and REM sleep for 5 h postinjection by 23 and 44%, respectively. At doses of 4 and 5 mg/kg $SeCl_4$, both NREM and REM sleep decreased promptly after the injection; and the $SeCl_4$-injected mice displayed almost complete insomnia within 1 h. The sleep suppression gradually decreased thereafter, lasting about 3 and 5 h after the injection of 4 and 5 mg/kg $SeCl_4$, respectively. $SeCl_4$ given at 4 and 5 mg/kg reduced the 5-h cumulative amount of NREM sleep during the daytime by 31 and 45%, respectively, and that of REM sleep by 63% and 81%, respectively (Fig. 6, data shown for 5 mg/kg dose only). It also induced a strong rebound-like reaction during the following nighttime to increase NREM and REM sleep. These results revealed that $SeCl_4$ given during the light period dose-dependently inhibited NREM and REM sleep of wild-type mice and increased both during the following dark period.

We then administered $SeCl_4$ at 5 mg/kg to various KO mice lacking H-PGDS, L-PGDS, or DP_1 receptors and compared the effect on their sleep–wake cycle with that on the wild-type mice (Fig. 6). In H-PGDS KO mice, $SeCl_4$ also inhibited NREM sleep by 80%, almost eliminated REM sleep, and increased the wake time about two-fold for 1 h after the injection. The hourly amounts of NREM sleep and wakefulness returned to the vehicle-injected level within 2–3 h after the administration, but the REM sleep suppression lasted for 6 h. The H-PGDS KO mice showed strong NREM and REM sleep induction and suppression of wakefulness during the following dark period, similar to the wild-type mice. Between the wild-type and H-PGDS KO mice, there were no statistically significant differences in the effects of $SeCl_4$ on wakefulness, NREM and REM sleep, suggesting that H-PGDS is most likely not involved in $SeCl_4$-induced insomnia.

In contrast, when L-PGDS KO mice were examined, $SeCl_4$ did not change the amounts of NREM and REM sleep or the time spent in wakefulness at all during the daytime, clearly indicating the sleep inhibition immediately after the administration of $SeCl_4$ is dependent on L-PGDS. When $SeCl_4$ was administered to DP_1 KO mice under the same experimental conditions, neither the inhibition of sleep immediately after administration of $SeCl_4$ nor the delayed increase in sleep during the night was observed. This indicates that $SeCl_4$ does not inhibit sleep through a nonspecific toxic mechanism but suppresses sleep by inhibiting the endogenous production of PGD_2 and signaling through DP_1 receptors. These results also suggest that sleep in wild-type mice is controlled by PGD_2 produced by L-PGDS, rather than H-PGDS, and that the PGD_2 produced is recognized by DP_1 receptors under physiological conditions.

Contribution of Adenosine to Physiological Sleep

Complete insomnia was also observed in wild-type mice for 2–3 h after an intraperitoneal injection of caffeine, a nonselective antagonist of adenosine A_1 and A_{2A} receptors, at a dose of 15 mg/kg, a dose corresponding to the intake of approximately three cups of coffee in humans. There was no disruption of sleep architecture after the 3-h period. We then used mice lacking A_1 or A_{2A} receptors and their respective wild-type

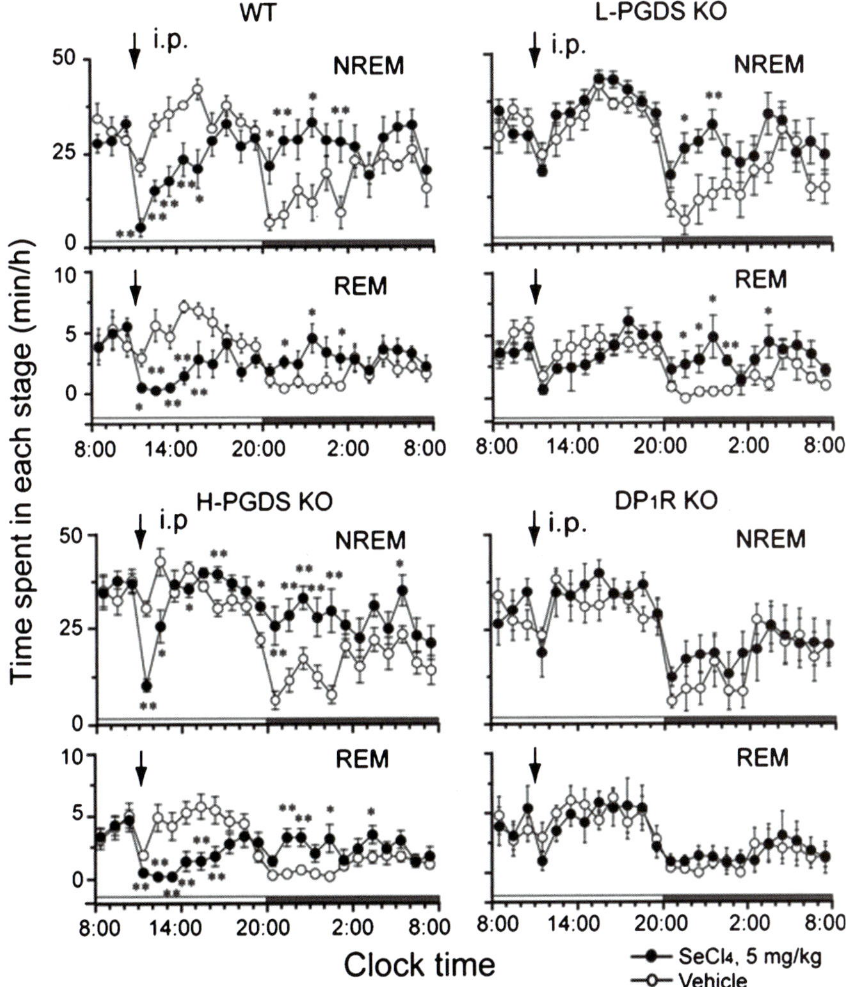

Fig. 6 $SeCl_4$-induced insomnia in wild-type and H-PGDS-KO mice, but not in L-PGDS- or DP_1 receptors (DP_1R)-KO mice. Both NREM and REM sleep were almost completely inhibited in wild-type (WT) and H-PGDS KO mice, at 1 h after an i.p. administration of $SeCl_4$ (5 mg/ml, at 10:00, *vertical arrows*) [48]. The $SeCl_4$-induced insomia was associated with delayed increases in both NREM and REM sleep during the dark period. The $SeCl_4$-induced sleep inhibition was not observed in L-PGDS- or DP_1-KO mice. $*p<0.05$ and $**p<0.01$ vs. the vehicle-administrated group

littermates of the inbred C57BL/6 strain to elucidate which subtype of receptors is involved in caffeine-induced wakefulness/insomnia [50]. The caffeine-induced insomnia in the A_1 receptor KO mice had the same intensity and duration as seen in the wild-type mice (Fig. 7a, b). In contrast, A_{2A} receptor KO mice did not show any change in the time spent in wakefulness after caffeine administration (Fig. 7), indicating that A_{2A}, but not A_1 receptors are crucial in caffeine-induced wakefulness. These results also indicate that the stimulation of A_{2A} receptors with endogenous adenosine is essential for the maintenance of physiological sleep. Alternatively, adenosine A_1 receptors are considered to regulate locally the sleep–wake cycle as in the case of the inhibition of the histaminergic arousal system, as described above [43].

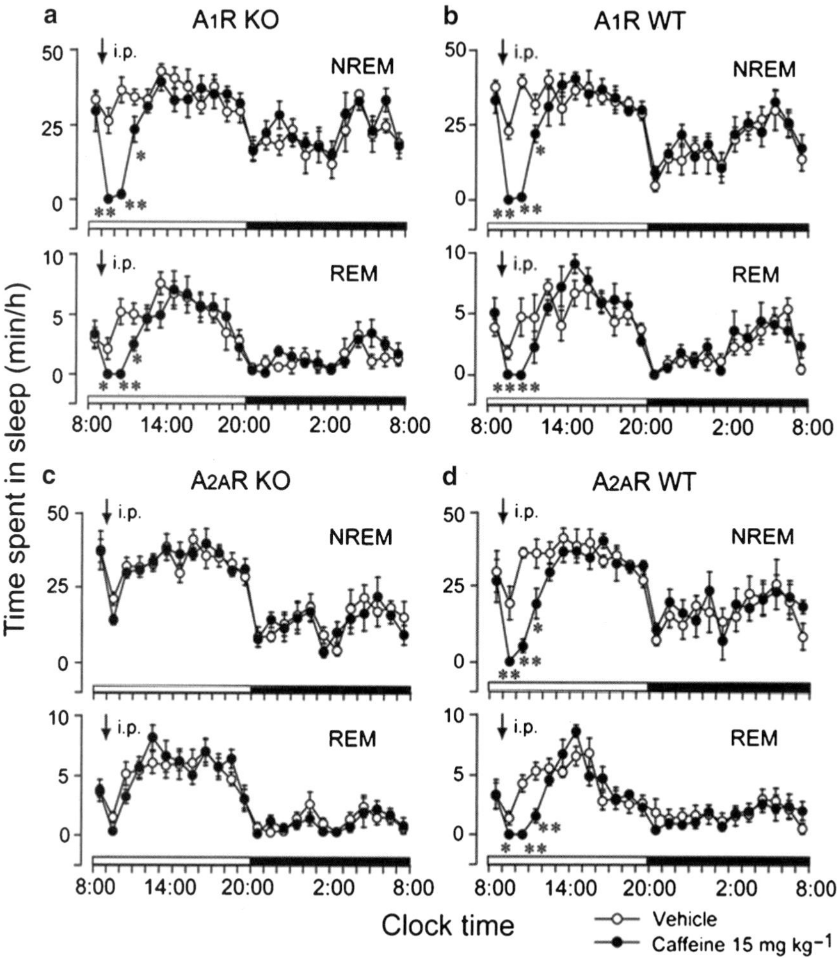

Fig. 7 Caffeine-induced insomnia in wild-type (WT) and A_1 receptor (R)-KO mice, but not in A_{2A} R-KO mice. Caffeine (15 mg/kg i.p., *arrows*) almost completely inhibited both NREM and REM sleep for 2 h in adenosine A_1 R KO mice (**a**) and their WT mice littermates (**b**). Caffeine did not reduce sleep in A_{2A} R KO mice (**c**), but it reduced sleep in their WT littermates (**d**) [50]. $^*p<0.05$ and $^{**}p<0.01$ vs. the vehicle-administrated group

Conclusions

In this review, we summarized recent progress on the molecular mechanisms of PGD_2-induced sleep. In the brain, PGD_2 is produced by L-PGDS in the leptomeninges, choroid plexus, and oligodendrocytes, and is secreted into the CSF as a sleep-promoting hormone. PGD_2 stimulates DP_1 receptors in the arachnoid membrane under the basal forebrain to release adenosine as a secondary sleep-promoting molecule. Adenosine then activates adenosine A_{2A} receptor-expressing neurons in the basal forebrain, which concomitantly excite sleep-active neurons in the VLPO and suppress the histaminergic arousal center in the TMN to induce sleep. The administration of an L-PGDS inhibitor ($SeCl_4$), a DP_1 antagonist (ONO-4127Na) or an adenosine A_2 receptor antagonist (caffeine) inhibits sleep in rats and wild-type mice, indicating that the PGD_2-adenosine system is crucial for the maintenance of physiological sleep. In narcoleptic patients, the serum L-PGDS concentration is significantly higher than that in healthy controls and correlated with excessive daytime sleepiness [51] and the

CSF L-PGDS level is reduced in patients with excessive daytime sleepiness [52]. Thus, the serum or CSF L-PGDS/β-trace concentration may be used as a marker of sleepiness in narcoleptic patients.

Acknowledgments Studies conducted in our laboratory were supported in part by the Program for Promotion of Basic Research Activities for Innovative Biosciences (PROBRAIN); the Program for Promotion of Fundamental Studies in Health Sciences of the National Institute of Biomedical Innovation (NIBIO); a Health and Labor Science Research Grant from the Ministry of Health, Labor, and Welfare, Japan; and by research grants from Takeda Science Foundation, Takeda Pharmaceutical Co. Ltd., Ono Pharmaceutical Co. Ltd., and Osaka city.

References

1. Narumiya S, Ogorochi T, Nakao K, et al. Prostaglandin D_2 in rat brain, spinal cord and pituitary: basal level and regional distribution. Life Sci. 1982;31:2093–103.
2. Ogorochi T, Narumiya S, Mizuno N, et al. Regional distribution of prostaglandins D_2, E_2, and $F_{2\ alpha}$ and related enzymes in postmortem human brain. J Neurochem. 1984;43:71–82.
3. Ueno R, Ishikawa Y, Nakayama T, et al. Prostaglandin D_2 induces sleep when microinjected into the preoptic area of conscious rats. Biochem Biophys Res Commun. 1982;109:576–82.
4. Ueno R, Honda K, Inoue S, et al. Prostaglandin D_2, a cerebral sleep-inducing substance in rats. Proc Natl Acad Sci U S A. 1983;80:1735–7.
5. Onoe H, Ueno R, Fujita I, et al. Prostaglandin D_2, a cerebral sleep-inducing substance in monkeys. Proc Natl Acad Sci U S A. 1988;85:4082–6.
6. Pandey HP, Ram A, Matsumura H, et al. Concentration of prostaglandin D_2 in cerebrospinal fluid exhibits a circadian alteration in conscious rats. Biochem Mol Biol Int. 1995;37:431–7.
7. Ram A, Pandey HP, Matsumura H, et al. CSF levels of prostaglandins, especially the level of prostaglandin D_2, are correlated with increasing propensity towards sleep in rats. Brain Res. 1997;751:81–9.
8. Roberts 2nd LJ, Sweetman BJ, Lewis RA, et al. Increased production of prostaglandin D_2 in patients with systemic mastocytosis. N Engl J Med. 1980;303: 1400–4.
9. Pentreath VW, Rees K, Owolabi OA, et al. The somnogenic T lymphocyte suppressor prostaglandin D_2 is selectively elevated in cerebrospinal fluid of advanced sleeping sickness patients. Trans R Soc Trop Med Hyg. 1990;84:795–9.
10. Urade Y, Fujimoto N, Hayaishi O. Purification and characterization of rat brain prostaglandin D synthetase. J Biol Chem. 1985;260:12410–5.
11. Christ-Hazelhof E, Nugteren DH. Purification and characterisation of prostaglandin endoperoxide D-isomerase, a cytoplasmic, glutathione-requiring enzyme. Biochim Biophys Acta. 1979;572:43–51.
12. Urade Y, Fujimoto N, Ujihara M, et al. Biochemical and immunological characterization of rat spleen prostaglandin D synthetase. J Biol Chem. 1987;262: 3820–5.
13. Urade Y, Eguchi N. Lipocalin-type and hematopoietic prostaglandin D synthases as a novel example of functional convergence. Prostaglandins Other Lipid Mediat. 2002;68–69:375–82.
14. Urade Y, Hayaishi O. Biochemical, structural, genetic, physiological, and pathophysiological features of lipocalin-type prostaglandin D synthase. Biochim Biophys Acta. 2000;1482:259–71.
15. Hirata M, Kakizuka A, Aizawa M, et al. Molecular characterization of a mouse prostaglandin D receptor and functional expression of the cloned gene. Proc Natl Acad Sci U S A. 1994;91:11192–6.
16. Hirai H, Tanaka K, Yoshie O, et al. Prostaglandin D_2 selectively induces chemotaxis in T helper type 2 cells, eosinophils, and basophils via seven-transmembrane receptor CRTH2. J Exp Med. 2001;193:255–61.
17. Irikura D, Aritake K, Nagata N, et al. Biochemical, functional, and pharmacological characterization of AT-56, an orally active and selective inhibitor of lipocalin-type prostaglandin D synthase. J Biol Chem. 2009;284:7623–30.
18. Aritake K, Kado Y, Inoue T, et al. Structural and functional characterization of HQL-79, an orally selective inhibitor of human hematopoietic prostaglandin D synthase. J Biol Chem. 2006;281:15277–86.
19. Mohri I, Eguchi N, Suzuki K, et al. Hematopoietic prostaglandin D synthase is expressed in microglia in the developing postnatal mouse brain. Glia. 2003;42: 263–74.
20. Beuckmann CT, Lazarus M, Gerashchenko D, et al. Cellular localization of lipocalin-type prostaglandin D synthase (beta-trace) in the central nervous system of the adult rat. J Comp Neurol. 2000;428:62–78.
21. Urade Y, Kitahama K, Ohishi H, et al. Dominant expression of mRNA for prostaglandin D synthase in leptomeninges, choroid plexus, and oligodendrocytes of the adult rat brain. Proc Natl Acad Sci U S A. 1993;90:9070–4.
22. Mizoguchi A, Eguchi N, Kimura K, et al. Dominant localization of prostaglandin D receptors on arachnoid trabecular cells in mouse basal forebrain and their involvement in the regulation of non-rapid eye movement sleep. Proc Natl Acad Sci U S A. 2001;98: 11674–9.
23. Urade Y, Eguchi N, Hayaishi O. Lipocalin-type prostaglandin D synthase as an enzymic lipocalin. Georgetown, TX: Landes Bioscience/Eurekah.com; 2006.
24. Urade Y, Hayaishi O. Prostaglandin D synthase: structure and function. Vitam Horm. 2000;58:89–120.
25. Shimamoto S, Yoshida T, Inui T, et al. NMR solution structure of lipocalin-type prostaglandin D synthase: evidence for partial overlapping of catalytic pocket

and retinoic acid-binding pocket within the central cavity. J Biol Chem. 2007;282:31373–9.
26. Kumasaka T, Aritake K, Ago H, et al. Structural basis of the catalytic mechanism operating in open-closed conformers of lipocalin type prostaglandin D synthase. J Biol Chem. 2009;284:22344–52.
27. Hoffmann A, Conradt HS, Gross G. Purification and chemical characterization of beta-trace protein from human cerebrospinal fluid: its identification as prostaglandin D synthase. J Neurochem. 1993;61:451–6.
28. Watanabe K, Urade Y, Mader M, et al. Identification of beta-trace as prostaglandin D synthase. Biochem Biophys Res Commun. 1994;203:1110–6.
29. Clausen J. Proteins in normal cerebrospinal fluid not found in serum. Proc Soc Exp Biol Med. 1961;107: 170–2.
30. Jordan W, Tumani H, Cohrs S, et al. Prostaglandin D synthase (beta-trace) in healthy human sleep. Sleep. 2004;27:867–74.
31. Kohtoh S, Taguchi Y, Matsumoto N, et al. Algorithm for sleep scoring in experimental animals based on fast Fourier transform power spectrum analysis of the electroencephalogram. Sleep Biol Rhythm. 2008;6: 163–71.
32. Urade Y, Eguchi N, Qu WM, et al. Sleep regulation in adenosine A_{2A} receptor-deficient mice. Neurology. 2003;61:S94–6.
33. Satoh S, Matsumura H, Koike N, et al. Region-dependent difference in the sleep-promoting potency of an adenosine A_{2A} receptor agonist. Eur J Neurosci. 1999;11:1587–97.
34. Scammell T, Gerashchenko D, Urade Y, et al. Activation of ventrolateral preoptic neurons by the somnogen prostaglandin D_2. Proc Natl Acad Sci U S A. 1998;95:7754–9.
35. Sherin JE, Elmquist JK, Torrealba F, et al. Innervation of histaminergic tuberomammillary neurons by GABAergic and galaninergic neurons in the ventrolateral preoptic nucleus of the rat. J Neurosci. 1998;18: 4705–21.
36. Gallopin T, Luppi PH, Cauli B, et al. The endogenous somnogen adenosine excites a subset of sleep-promoting neurons via A_{2A} receptors in the ventrolateral preoptic nucleus. Neuroscience. 2005;134: 1377–90.
37. Hong ZY, Huang ZL, Qu WM, et al. An adenosine A_{2A} receptor agonist induces sleep by increasing GABA release in the tuberomammillary nucleus to inhibit histaminergic systems in rats. J Neurochem. 2005;92:1542–9.
38. Saper CB, Chou TC, Scammell TE. The sleep switch: hypothalamic control of sleep and wakefulness. Trends Neurosci. 2001;24:726–31.
39. Saper CB, Scammell TE, Lu J. Hypothalamic regulation of sleep and circadian rhythms. Nature. 2005;437: 1257–63.
40. Chu M, Huang ZL, Qu WM, et al. Extracellular histamine level in the frontal cortex is positively correlated with the amount of wakefulness in rats. Neurosci Res. 2004;49:417–20.
41. Huang ZL, Sato Y, Mochizuki T, et al. Prostaglandin E_2 activates the histaminergic system via the EP_4 receptor to induce wakefulness in rats. J Neurosci. 2003;23:5975–83.
42. Huang ZL, Qu WM, Li WD, et al. Arousal effect of orexin A depends on activation of the histaminergic system. Proc Natl Acad Sci U S A. 2001;98:9965–70.
43. Oishi Y, Huang ZL, Fredholm BB, et al. Adenosine in the tuberomammillary nucleus inhibits the histaminergic system via A1 receptors and promotes non-rapid eye movement sleep. Proc Natl Acad Sci U S A. 2008;105:19992–7.
44. Eguchi N, Minami T, Shirafuji N, et al. Lack of tactile pain (allodynia) in lipocalin-type prostaglandin D synthase-deficient mice. Proc Natl Acad Sci U S A. 1999;96:726–30.
45. Hayaishi O, Urade Y, Eguchi N, et al. Genes for prostaglandin D synthase and receptor as well as adenosine A_{2A} receptor are involved in the homeostatic regulation of NREM sleep. Arch Ital Biol. 2004;142:533–9.
46. Pinzar E, Kanaoka Y, Inui T, et al. Prostaglandin D synthase gene is involved in the regulation of non-rapid eye movement sleep. Proc Natl Acad Sci U S A. 2000;97:4903–7.
47. Huang ZL, Mochizuki T, Qu WM, et al. Altered sleep-wake characteristics and lack of arousal response to H3 receptor antagonist in histamine H1 receptor knockout mice. Proc Natl Acad Sci U S A. 2006;103: 4687–92.
48. Qu WM, Huang ZL, Xu XH, et al. Lipocalin-type prostaglandin D synthase produces prostaglandin D_2 involved in regulation of physiological sleep. Proc Natl Acad Sci U S A. 2006;103:17949–54.
49. Islam F, Watanabe Y, Morii H, et al. Inhibition of rat brain prostaglandin D synthase by inorganic selenocompounds. Arch Biochem Biophys. 1991;289:161–6.
50. Huang ZL, Qu WM, Eguchi N, et al. Adenosine A_{2A}, but not A_1, receptors mediate the arousal effect of caffeine. Nat Neurosci. 2005;8:858–9.
51. Jordan W, Tumani H, Cohrs S, et al. Narcolepsy increased L-PGDS (β-trace) levels correlate with excessive daytime s;leepiness but not with cataplexy. J Neurol. 2005;252:1372–8.
52. Bassetti CL, Hersberger M, Baumann CR. CSF prostaglandin D synthase is reduced in excessive daytime sleepiness. J Neurol. 2006;253:1030–3.

Part III

The Role of the Hypocretins in Sleep–Wake Regulation

The Neurobiology of Sleep–Wake Systems: An Overview

Pierre-Hervé Luppi and Patrice Fort

Keywords
Sleep-wake regulation • Forebrain • Ventrolateral preoptic area • Hypothalamus • Brainstem • Orexin • hypocretin • Monoamines • Acetylcholine • GABA

Abbreviations

Ach	Acetylcholine
ADA	Adenosine
BF	Basal Forebrain
CTb	Cholera toxin b subunit
DPGi	Dorsal paragigantocellular reticular nucleus
dDpMe	Dorsal deep mesencephalic reticular nucleus
DRN	Dorsal raphe nucleus
Fos+	Neuron immunoreactive for Fos
GAD	Glutamate decarboxylase
GiA	Alpha gigantocellular nucleus
GiV	Ventral gigantocellular reticular nucleus
Gly	Glycine
Hcrt	Hypocretin
His	Histamine
LC	Locus coeruleus
LDT	Laterodorsal pontine nucleus
LHA	Lateral hypothalamic area
Mc	Nucleus reticularis magnocellularis
MCH	Melanin-concentrating hormone
MnPn	Median preoptic nucleus
NA	Norepinephrine
PeF	Perifornical hypothalamic area
peri-LC alpha	Peri-locus coeruleus alpha
PH	Posterior hypothalamus
PnC	Pontis caudalis nucleus
PnO	Pontis oralis nucleus
POA	Preoptic area
PPT	Pedunculopontine nucleus
SCN	Suprachiasmatic nucleus
SLD	Sublaterodorsal nucleus
TMN	Tuberomamillary nucleus
vlPAG	Ventrolateral periaqueductal gray
VLPO	Ventrolateral preoptic nucleus
ZI	Zona incerta

P.-H. Luppi (✉)
UMR5167 CNRS, Faculté de Médecine RTH Laennec, Institut Fédératif des Neurosciences de Lyon, 7 rue Guillaume Paradin, 69372 Lyon cedex 08, France
and
UMR5167 CNRS, Institut Fédératif des Neurosciences de Lyon (IFR 19), Université de Lyon, Lyon, France
e-mail: luppi@sommeil.univ-lyon1.fr

C.R. Baumann et al. (eds.), *Narcolepsy: Pathophysiology, Diagnosis, and Treatment*,
DOI 10.1007/978-1-4419-8390-9_10, © Springer Science+Business Media, LLC 2011

Neuronal Networks Responsible for Sleep Onset and Maintenance

The Forebrain Sleep Center

Following studies of patients with postinfluenza encephalitis, the neuropathologist von Economo reported that inflammatory lesions of the preoptic area (POA) were often associated with insomnia and therefore proposed that the POA was critical for the production of normal sleep [1]. Then, Ranson in monkeys, Nauta in rats, and McGinty in cats showed that POA lesions induce a profound and persistent insomnia [2–4]. It was later shown in cats that POA electrical stimulation induces EEG slow-wave activity and sleep (SWS) [5]. Finally, putative sleep-promoting neurons displaying an elevated discharge rate during SWS compared to waking (W), diffusely distributed within a large region encompassing the horizontal limb of the diagonal bands of Broca and the lateral preoptic area-substantia innominata were recorded in freely moving cats [6]. Altogether, these studies indicate that the POA is a unique brain structure containing neurons that directly promote sleep.

This simplicity highly contrasts with the complex network responsible for W, involving redundant neurotransmitter systems with populations of neurons disseminated from the upper brainstem, caudal hypothalamus to basal forebrain, such as neurons containing acetylcholine, norepinephrine, serotonin, histamine, and recently discovered hypocretin (Fig. 1). Collectively, through their widespread ascending projections to the cortical mantle and with their firing activity specific to W, these systems form the ascending reticular activating system (ARAS) that controls arousal with a low-voltage, high-frequency, cortical activation [7]. During the transition to sleep, the hypnogenic center would inhibit the multiple arousal systems of the ARAS via a sustained and coordinated inhibition (Fig. 2). The sleep pressure associated with drowsiness is believed to be regulated by the interaction of a homeostatic and a circadian processes, both able to modulate the hypnogenic center's activity to control the timing and duration of sleep [8].

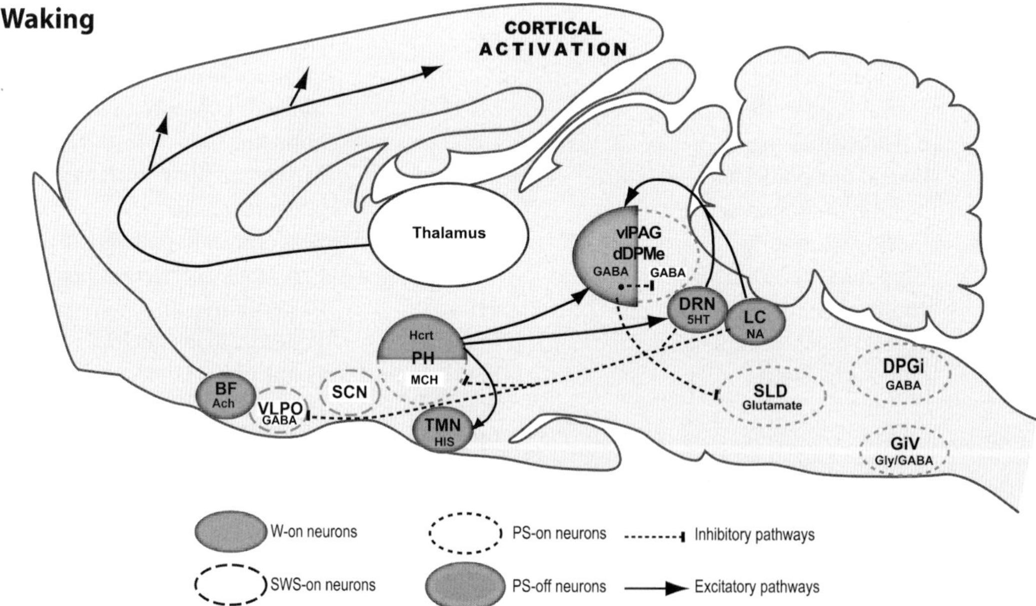

Fig. 1 Model of the network responsible for promoting wake. *Shaded nuclei* help promote wake and inhibit SWS and PS. Excitatory pathways are indicated by *solid lines*, and inhibitory pathways by *dashed lines*. Not shown on this figure are direct projections from monoaminergic and cholinergic nuclei to the cortex that may also contribute to the production of wake

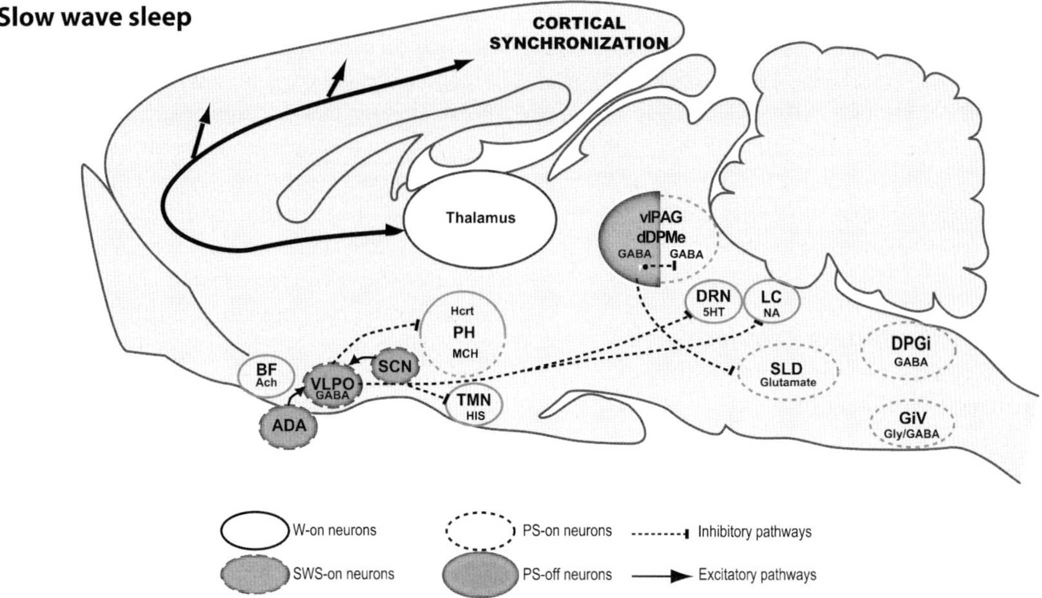

Fig. 2 Model of the network responsible for slow-wave sleep. During SWS, GABAergic neurons of the VLPO inhibit many wake-promoting regions. The VLPO neurons may be activated by adenosine or other homeostatic signals and by circadian signals from the SCN. Other GABAergic neurons in the vlPAG and dDPMe inhibit PS-promoting systems

Then, researchers identified specific sleep-active neurons by using Fos as a marker of neuronal activation in rats that had slept for a long period before sacrifice [9]. These neurons are distributed diffusely in the POA but are more densely packed in the median preoptic nucleus (MnPn) and the ventrolateral preoptic nucleus (VLPO). These studies showed that the number of Fos+ neurons in the VLPO and MnPn positively correlated with sleep quantity and sleep consolidation during the last hour preceding sacrifice. Numerous Fos+ neurons were also observed in sleep-deprived rats in the MnPn but not in VLPO [9, 10]. It appears, therefore, that VLPO neurons would be primarily responsible for the induction of sleep while MnPn neurons may also have a homeostatic role in sleep control. It was later demonstrated that VLPO and the suprachiasmatic nucleus (SCN) have synchronized activity [11]. Considering that both areas are interconnected and receive inputs from the retinal ganglion cells, it is thus possible that circadian- and photic-linked information may be conveyed to modulate VLPO activity [12] (Fig. 2).

Electrophysiology experiments in behaving rats have shown that neurons recorded in the VLPO and MnPn are active during SWS, generally anticipating its onset by several seconds. Further, their firing rate is positively correlated with sleep depth and duration. Some of these neurons are also active during PS with a higher firing frequency than during the preceding SWS [13]. In addition, VLPO and MnPn neurons display a firing pattern reciprocal to the wake-active neurons (see below). Functionally, bilateral neurotoxic destruction of VLPO is followed by a profound and long-lasting insomnia in rats [14]. Retrograde and anterograde tract-tracing studies indicate that VLPO and MnPn neurons are reciprocally connected with wake-active neurons such as those containing histamine in the tuberomammillary nucleus (TMN), hypocretin in the perifornical hypothalamic area (PeF), serotonin in the dorsal raphe nuclei (DRN), noradrenaline in the locus coeruleus (LC), and acetylcholine in the pontine (LDT/PPT) and basal forebrain nuclei. In these wake-promoting areas, extracellular levels of GABA increase during SWS compared to W.

It has also been shown that Fos+ neurons in the VLPO express galanin mRNA and 80% of VLPO neurons projecting to the TMN contain both galanin and GAD, the GABA-synthesizing enzyme. Finally, electrical stimulation of the VLPO area evokes a GABA-mediated inhibition of TMN neurons, suggesting that VLPO and MnPn efferents to the wake-promoting systems are inhibitory [12].

Neurotransmitters Regulating the Activity of Sleep-Promoting Neurons in the VLPO

Electrophysiological recordings of VLPO neurons in rat brain slices showed that it contains a homogeneous neuronal group with specific intrinsic membrane properties, a clear-cut chemomorphology and an inhibitory response to the major waking neurotransmitters. Their high proportion matching that of cells active during sleep and their pharmacological profile represent convincing arguments about their status as presumed sleep-promoting (PSP) neurons [15]. It was further shown that PSP neurons are GABAergic and galaninergic in nature, multipolar triangular shaped, and endowed with a potent low threshold calcium potential. These neurons are always inhibited by noradrenaline (NA), via postsynaptic alpha2-adrenoceptors. The wake-promoting drug modafinil was shown to increase the NA-mediated inhibition of VLPO neurons, perhaps by blocking NA reuptake by local noradrenergic terminals [16]. Interestingly, NA-inhibited neurons are also inhibited by acetylcholine, through muscarinic postsynaptic and nicotinic presynaptic actions on noradrenergic terminals. In contrast, histamine and hypocretin did not modulate PSP neurons [15, 17]. Finally, serotonin showed complex effects inducing either excitation (50%, Type 2) or inhibition (50%, Type 1) of the PSP neurons [12, 18].

Among processes that are likely to modulate the activity of PSP neurons, homeostatic mechanisms, involving natural sleep-promoting factors accumulating during waking, had long been thought to play a crucial role in triggering sleep. Among these factors, prostaglandin D2 and adenosine have been functionally implicated in sleep, although their neuronal targets and mechanisms of action remain largely unknown (see Chap. "Prostaglandin D_2: An Endogenous Somnogen" by Urade). In this context, we showed that application of an adenosine A_{2A} receptor ($A_{2A}R$) agonist evoked direct excitatory effects specifically in Type 2 PSP neurons that also activated by serotonin [12, 18]. Other results also suggested that adenosine may directly activate VLPO neurons at sleep onset via an action on postsynaptic $A_{2A}R$. Indeed, infusion of an $A_{2A}R$ agonist in the subarachnoid space rostral to the VLPO increases SWS and induces Fos expression in VLPO neurons [19] (Fig. 2).

In contrast, a number of studies showed that adenosine A_1 receptors (A_1R) promote sleep through inhibition of the wake-promoting neurons, in particular, cholinergic and hypocretin neurons [20, 21]. However, data using transgenic mice demonstrated that the lack of A_1R does not prevent the homeostatic regulation of sleep while the lack of $A_{2A}R$ does and blocks waking effect of caffeine, suggesting that the activation of $A_{2A}R$ is crucial in SWS [22–24].

An Updated Overview of the Neuronal Network Responsible for SWS

A line of evidence indicates that both the VLPO and the MnPn contain neurons responsible for sleep onset and maintenance. These neurons would be inhibited by noradrenergic and cholinergic inputs during waking. The majority of them would start firing at sleep onset (drowsiness) in response to excitatory, homeostatic (adenosine and serotonin), and circadian drives (suprachiasmatic inputs). These activated neurons, through the reciprocal GABAergic inhibition of all wake-promoting systems, would be in a position to suddenly unbalance the "flip-flop" network, as required for switching from W or drowsiness to SWS. Conversely, the slow removal of excitatory influences would result in a progressive firing decrease in VLPO neurons and therefore an activation of wake-promoting systems leading to awakening [18, 25].

Neuronal Network Responsible for Paradoxical Sleep

The Pontine Generator of PS and the Cholinergic Hypothesis

In 1959, Jouvet and Michel discovered PS in cats, a sleep phase characterized by a complete disappearance of muscle tone, paradoxically associated with cortical activation and rapid eye movements (REM) [26]. Rapidly, they demonstrated that the brainstem is necessary and sufficient to trigger and maintain PS in cats (pontine cat preparation). By using electrolytic and chemical lesions, it was then found that the dorsal part of the pontis oralis (PnO) and caudalis (PnC) nuclei contain the neurons responsible for PS onset [27]. Furthermore, large bilateral injections of a cholinergic agonist, carbachol into the PnO and PnC promotes PS in cats [28]. Later, PS induction with the shortest latencies was obtained with carbachol injection restricted to the dorsal area of the PnO and PnC, coined the peri-locus coeruleus alpha (peri-LC alpha). It was then shown by unit recordings in freely moving cats that many peri-LC alpha neurons show a tonic firing selective to PS (called PS-on neurons) [29]. Two types of PS-on neurons were identified. The first ones were inhibited by carbachol, an indication that they might be cholinergic. They were restricted to the rostro-dorsal peri-LC alpha and projected to forebrain including intralaminar thalamic nuclei, posterior hypothalamus, and basal forebrain. The second ones recorded over the whole peri-LC alpha were excited by carbachol and projected caudally to the nucleus reticularis magnocellularis (Mc) within the ventromedial medullary reticular formation [29, 30]. It has been then proposed that (1) the ascending PS-on neurons are cholinergic and are responsible for the cortical activation during PS and (2) the descending PS-on neurons are not cholinergic and generate muscle atonia during PS through excitatory projections to medullary glycinergic pre-motoneurons [12, 31–33].

In contrast to data in cats, carbachol iontophoresis into the rat sublaterodorsal tegmental nucleus (SLD), a very small area of the dorsolateral pontine tegmentum corresponding to the cat's peri-LC alpha, induces W with increased muscle activity [34]. Other studies using carbachol administration in rats described either a moderate PS enhancement or no effect [35–38]. It was first shown that the number of pedunculo-pontine (PPT) and laterodorsal (LDT) cholinergic neurons expressing Fos increases in rats during PS recovery following its selective deprivation by the flower-pot technique [39]. However, in our recent study reproducing these experiments (75 h PS deprivation; 3 h sleep recovery), we observed that only occasional cholinergic neurons stained for Fos in the PPT/LDT nuclei [40]. In conclusion, our results in rats are strongly against a role of pontine cholinergic neurons in PS genesis although unit recording combined with juxtacellular labeling is required to draw a more definitive conclusion.

SLD Neurons Triggering PS Are Glutamatergic

As described above, SLD neurons activated during PS are not cholinergic. We recently showed that they are not GABAergic, as well. Indeed, the small number of Fos+/GAD+ neurons in the SLD did not increase in rats displaying a PS rebound compared to control or PS-deprived rats [41]. It is more likely that the Fos+ neurons observed in the SLD specifically after PS recovery are glutamatergic since they express the vesicular glutamate transporter, vGlut2 [42]. Further, it has been shown that the SLD sends efferent projections to glycinergic neurons from the ventral and alpha gigantocellular nuclei (GiV and GiA, corresponding to the cat Mc) known to generate atonia during PS by direct projections to cranial and spinal motoneurons in the cat [12, 33]. These glycinergic neurons express Fos during a PS-like state inducted by the disinhibition of the SLD with bicuculline or gabazine (GABA-A receptor antagonists) [34]. Further, glutamate release in the GiV and GiA increases specifically during PS and injection of non-NMDA glutamate agonists in the GiV and GiA suppresses muscle tone. In contrast, an increased tonus is induced during PS after the lesion of these medullary areas in cats. Altogether, these results strongly suggest that the SLD neurons triggering PS are glutamatergic [12, 33].

It is likely that the glycinergic neurons of the GiA and Giv are also GABAergic since a large majority of the Fos+ neurons localized in these nuclei after PS recovery express GAD67 mRNA [41].

SLD Neurons Triggering PS Are Tonically Excited by Glutamate

In cats, the microdialysis administration of kainic acid, a glutamate agonist in the peri-LC alpha induces a PS-like state [43]. We reproduced these experiments in rats and also observed a firing activation of SLD neurons reliably associated with the induction of a PS-like state. Further, application of kynurenate, a glutamate antagonist, reversed the PS-like state induced by bicuculline [34]. These results suggest that SLD PS-on neurons are under a permanent glutamatergic barrage throughout the sleep–waking cycle, unmasked at the onset of PS by the removal of tonic GABAergic inputs. The best candidate structure for containing the glutamatergic neurons permanently activating SLD is the lateral and ventrolateral periaqueductal gray (vlPAG), containing numerous non-GABAergic, likely glutamatergic, efferent neurons to the SLD [44, 45]. Although Jouvet established that structures sufficient for PS are restricted to the brainstem, numerous non-GABAergic neurons projecting to the SLD located in the primary motor area of the frontal cortex, the bed nucleus of the stria terminalis or the central nucleus of the amygdala could also contribute to the activation of the SLD neurons during PS [45].

SLD Glutamatergic Neurons Are Inhibited by GABAergic Neurons During W and SWS

By early 2000, we observed that a long-lasting PS-like hypersomnia can be pharmacologically induced with a short latency in head-restrained rats by iontophoretic applications of bicuculline or gabazine into the SLD. Our results have been reproduced in freely moving rats and cats with pressure injection of bicuculline [46, 47]. During these experiments, we were also able to record neurons within the SLD that are specifically active during PS and excited following bicuculline or gabazine iontophoresis [48]. Taken together, our data indicate that the activation of SLD neurons is mainly due to the removal of GABAergic tone present during W and SWS. Combining retrograde tracing with cholera toxin b subunit (CTb) injected in SLD and immunostaining for glutamic acid decarboxylase 67 (GAD67, a synthetic enzyme for GABA), we thus identified neurons at the origin of these GABAergic inputs mainly within the pontine (including SLD *itself*) and the dorsal deep mesencephalic reticular nuclei (dDpMe) also named the lateral pontine tegmentum (LPT) [45, 49]. Supporting the contribution of local GABAergic neurons in the inhibition of PS-on neurons during SWS and W, a significant increase in PS is produced by the administration of antisense oligonucleotides against GAD67 mRNA centered to the cat nucleus peri-LC alpha [47]. In rats, the number of GABAergic Fos+ neurons in the rostral pontine reticular nucleus decreased following PS rebound, suggesting they are active during W and SWS and inactive during PS [50]. However, we recently demonstrated that the ventrolateral part of the PAG (vlPAG) and the dDPMe are the only brainstem structures containing a large number of Fos+ neurons expressing GAD67 mRNA after PS deprivation [41]. Further, injections of muscimol (a GABA-A receptor agonist) in the vlPAG and/or the dDpMe induce strong increases in PS in cats and rats [41, 51]. These congruent experimental data led us to propose that GABAergic neurons within the vlPAG and the dDpMe gate PS by tonically inhibiting PS-on neurons from the SLD during W and SWS.

Role of the Monoaminergic Neurons in the Control of SLD Glutamatergic Neurons

An important discovery in the course of the identification of PS mechanisms was the finding that serotonergic and noradrenergic neurons cease firing specifically during PS, i.e., show a PS-off firing activity reciprocal to that of PS-on neurons [12, 33]. Recently, it was shown that histaminergic and hypocretin neurons within PH also have a PS-off firing pattern [52–54]. These electrophysiological data were the basis for the classical "reciprocal interaction" model suggesting that PS onset is gated by reciprocal inhibitory interactions between PS-on and PS-off neurons [29, 55, 56]. Supporting this model, drugs enhancing serotonin and noradrenergic transmission (monoamine oxidase inhibitors and serotonin and norepinephrine reuptake blockers) specifically suppress PS. Further, applications of noradrenaline, adrenaline, or benoxathian (an alpha2 agonist) into the peri-LC alpha inhibit PS, but serotonin has no effect [57, 58]. In addition, noradrenaline via alpha2-adrenoceptors inhibits the noncholinergic PS-on neurons but has no effect on the putative cholinergic PS-on neurons from the peri-LC alpha while serotonin has no effect on both types of neurons [59]. Importantly, our recent data combining tyrosine hydroxylase and Fos staining after PS deprivation and recovery suggest that LC noradrenergic neurons are likely not involved in the inhibition of PS particularly during PS deprivation. Nevertheless, a substantial number of noradrenergic neurons from A1 and A2 cell groups displayed Fos after PS deprivation indicating that these medullary noradrenergic cell groups might contribute to PS inhibition [60].

GABAergic Neurons Responsible for the Inactivation of Monoaminergic Neurons During PS

According to the "reciprocal interaction" model, the inactivation of monoaminergic neurons at PS onset is due to a powerful inhibitory drive originating from cholinergic PS-on cells from peri-LC alpha [29, 56]. However, peri-LC alpha neurons do not send efferent projections to monoaminergic nuclei; acetylcholine excites noradrenergic LC neurons and is only weakly inhibitory on serotonergic DRN neurons [61, 62]. It has therefore been suggested that inhibitory amino acids such as GABA or glycine may be more powerful than acetylcholine [63, 64]. To test this hypothesis, effects of iontophoretic applications of bicuculline and strychnine (a glycine antagonist) were studied on the activity of LC and DRN cells in the head-restrained rat. Bicuculline or strychnine, applied during SWS or PS when monoaminergic neurons are silent, induces tonic firing in both types of neurons [65–67]. During W, the antagonists produce a sustained increase in discharge rate compared to control. These results indicate the existence of tonic GABA and glycinergic inputs to the LC and DRN that are active during all vigilance states. Importantly, when the strychnine effect occurred during transitions from PS to W, the discharge rate further increased at W onset. In the same situation with bicuculline, the discharge rate remained unchanged at the transition from PS to W. These distinctive electrophysiological responses strongly suggest that an increased GABA release is responsible for the PS-selective inactivation of monoaminergic neurons. This hypothesis is well supported by microdialysis experiments in cats measuring a significant increase in GABA release in the DRN and LC during PS as compared to W and SWS but no detectable changes in glycine concentration [68, 69]. It seems therefore that monoaminergic cells are under a tonic GABAergic inhibition with a gradually increased strength from W to PS resulting in their inactivation during sleep. GABAergic neurons located in the VLPO and MnPN areas may be involved in this inhibition during SWS while those located in the extended VLPO have been implicated in this inhibition during PS [70]. However, we could not confirm that Fos-positive neurons located in the VLPO and extended VLPO after PS rebound project to the LC [71]. Further, it is likely that the PS-related inhibition involves primarily GABAergic neurons within the brainstem, directly projecting to monoaminergic neurons and "turned on" specifically at the onset of and during PS. Indeed, PS-like episodes spontaneously occur in pontine cats and when pharmacologically induced by carbachol in decerebrate cats, PS episodes are still associated with the silencing of serotonergic neurons [72].

By combining retrograde tracing with CTb and GAD immunohistochemistry in rats, we found that the LC and DRN receive GABAergic inputs from neurons located in a large number of distant regions from the forebrain to medulla [67]. Two brainstem areas are common inputs to DRN and LC since they contain a substantial number of GABAergic/CTb+ neurons, and are thus candidates for mediating the PS-related inhibition of monoaminergic neurons: the vlPAG and the dorsal paragigantocellular nucleus (DPGi). Furthermore, we demonstrated by using Fos that both nuclei contain numerous LC-projecting neurons selectively activated during PS rebound following PS deprivation [71]. Since the DPGi has not previously been considered in sleep, we studied the firing activity of DPGi neurons across the sleep–wake cycle in head-restrained rats. In full agreement with our Fos data, the DPGi contains numerous PS-on neurons; they are silent during W and SWS but fire tonically during PS. They start discharging approximately 15 s before PS onset and become silent around 10 s before EEG signs of arousal. This singular electrophysiological pattern suggests their contribution to the generation of PS [73]. Indeed, local application of bicuculline blocked the DPGi-evoked inhibition of LC neurons and electrical DPGi stimulation induces an increase in PS quantities in rats [74, 75]. Taken together, these data support the presence within the DPGi of neurons responsible for the inactivation during PS of LC neurons. A contribution from the vlPAG in this inhibitory mechanism is also likely. In rat brainstem slices, iontophoretic NMDA application in the vlPAG-induced bicuculline-sensitive inhibitory postsynaptic potentials in serotonergic neurons [76]. Moreover, an increase in Fos+ GAD immunoreactive neurons has been reported in the vlPAG after PS rebound induced by deprivation in rats [39, 41].

Role of the Posterior Hypothalamus, in Particular MCH Neurons in PS

As reported above, numerous Fos+ cells were observed in structures implicated in PS, such as SLD, vlPAG, or DPGi in rats allowed to recover from a 75-h PS-selective deprivation. Surprisingly, we also observed a very large number of Fos+ cells in the PH including zona incerta (ZI), perifornical area (PeF), and the lateral hypothalamic area (LHA) [77]. Only a few experimental results already support the notion that the PH contributes to PS regulation. Indeed, bilateral injections of muscimol in the cat mammillary and tuberal hypothalamus induced a drastic inhibition of PS [78]. Further, neurons specifically active during PS were recorded in the PH of cats or head-restrained rats. By using double-immunostaining, we further showed that half of Fos+ cells (76% in the PeF) in PH are positive for the neuropeptide melanin-concentrating hormone (MCH) while almost none of the neighboring hypocretin neurons were Fos+. Almost 60% of the MCH-immunoreactive neurons counted in PH were Fos+ [77]. In support of our Fos data, it has been recently shown in head-restrained rats that hypothalamic neurons, identified ex vivo as containing MCH, fire quite exclusively during PS, with a slow tonic rate or phasically in doublets or burst of spikes [79]. In addition, rats receiving ICV administration of MCH showed a strong dose-dependent increase in PS and, to a minor extent, SWS quantities [77]. Further, subcutaneous injection of an MCH antagonist decreases SWS and PS quantities, and mice with genetically inactivated MCH signaling exhibit altered vigilance state architecture and sleep homeostasis [80–82]. Thus, growing evidence indicates that MCH neurons and/or MCH signaling play a key role in PS regulation/homeostasis. Since MCH is primarily an inhibitory peptide, we have proposed that MCH neurons might promote PS by inhibiting W active neurons permissive for PS like the vlPAG/dDpMe GABAergic, the monoaminergic PS-off neurons and the hypocretin neurons intermingled in the PH with them [12, 83]. Supporting this hypothesis, observations with electron microscopy indicate that MCH and Hcrt neurons are interconnected [84, 85] and recent data from MCH-R1 KO mice demonstrated interactions between Hcrt and MCH systems, implying that MCH may exert an inhibitory influence on Hcrt signaling [86]. In addition, GABA, present in MCH neurons, likely contributes to the cessation of activity of Hcrt neurons during SWS and PS since bicuculline application in the PH-induced W and Fos expression in hypocretin neurons [87, 88].

Projections of the Hypocretin Neurons Involved in PS Regulation and Cataplexy

Hypocretin neurons are specifically active during W [53, 54]. ICV or local injection of hypocretin in the LC or the TMN neurons induces W and inhibits PS [89–91]. Narcolepsy, a sleep disorder characterized by excessive daytime sleepiness and cataplexy, is caused by the lack of hypocretin (Hcrt) mRNA and peptides in humans or a disruption of the Hcrt-R2 or its ligands in dogs and mice [92–94]. Excessive daytime sleepiness is likely due to the absence of excitatory hypocretin inputs on the other waking systems such as the histaminergic and noradrenergic ones. The absence of these inputs might also be responsible for cataplexy. It is, however, more likely that the absence of an excitatory hypocretin input to the GABAergic PS-off dDPMe and vlPAG neurons tonically inhibiting SLD PS-on neurons during W is involved. Indeed, we have shown that the removal of the GABAergic inhibition of SLD neurons is sufficient to induce PS since these neurons are permanently excited by a glutamatergic input [45]. We therefore propose that normally during strong emotions, hypocretin neurons excite the PS-off GABAergic neurons to counteract a simultaneous increase in glutamatergic tone on SLD PS-on neurons that may be coming from the central amygala [45]. The absence of the excitatory hypocretin input on the PS-off GABAergic neurons in narcoleptics would allow the activation of the SLD PS-on neurons by the glutamate input.

A Network Model for PS Onset and Maintenance

Our model integrates all the experimental data described above. It also integrates the observation that PS episodes in the rat start from SWS after a relatively long intermediate state during which the EEG displays a mix of spindles and theta activity, and then terminate abruptly, associated with a short microarousal [95].

We propose that the induction of PS is due to the activation of glutamatergic PS-on neurons in the SLD. During W and SWS, the activity of these PS-on neurons would be inhibited by an inhibitory GABAergic tone originating from PS-off neurons localized in the vlPAG and dDpMe. These PS-off neurons would be activated during W by the Hcrt neurons and the monoaminergic neurons. The onset of PS would be due to the activation by intrinsic mechanisms of PS-on MCH/GABAergic neurons and PS-on GABAergic neurons localized in the DPGi and vlPAG. These neurons would inactivate the vlPAG/dDPMe PS-off neurons and the PS-off monoaminergic and Hcrt neurons during PS. The disinhibited ascending SLD PS-on neurons would in turn induce cortical activation via their projections to intralaminar thalamic relay neurons in collaboration with W/PS-on cholinergic and glutamatergic neurons from the LDT and PPT, mesencephalic and pontine reticular nuclei and the basal forebrain. Descending PS-on SLD neurons would induce muscle atonia and sensory inhibition via their excitatory projections to glycinergic/GABAergic premotoneurons localized in the alpha and ventral gigantocellular reticular nuclei and the nucleus raphe magnus. The exit from PS would be due to the activation of waking systems since PS episodes are almost always terminated by an arousal. The waking systems would inhibit the MCH/GABAergic and GABAergic PS-on neurons localized in the DPGi and vlPAG. Since the duration of PS is negatively coupled with metabolic rate, we propose that the activity of the waking systems is triggered to end PS to restore competing physiological parameters like thermoregulation (Fig. 3).

Regarding PS-related pathologies in humans, cataplectic episodes in narcoleptics would be due to the absence of an increase of the hypocretin excitation of GABAergic PS-off neurons in the dDPMe and the vlPAG during emotion and a concomitant excitation of the SLD by glutamatergic inputs from the lateral bed nucleus of the stria terminalis and the central amygdala.

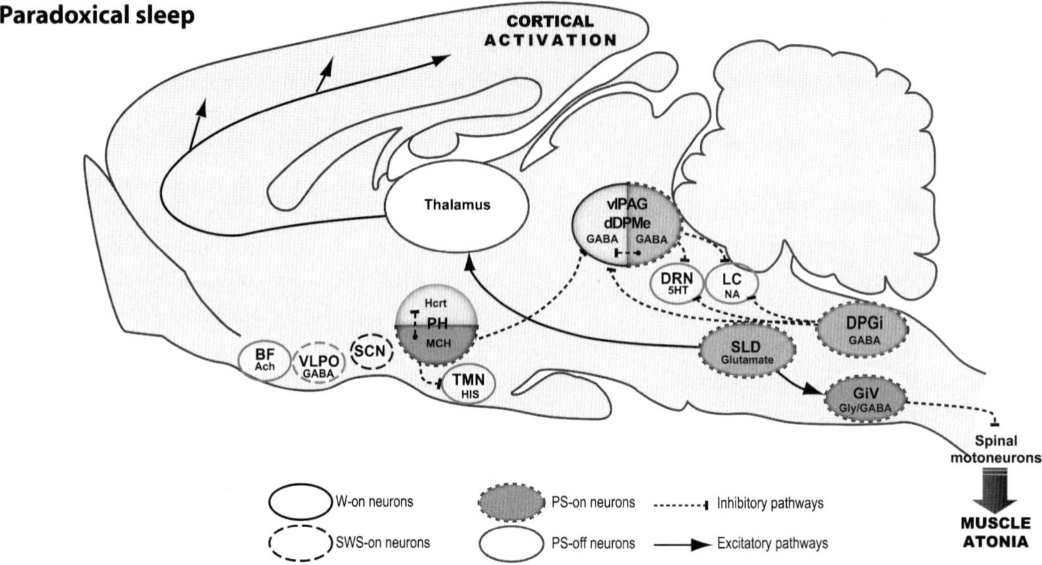

Fig. 3 Model of the network responsible for paradoxical sleep (REM sleep). During PS, glutamatergic neurons of the SLD may promote cortical activation through projections to the thalamus and promote atonia by activating GABA/glycinergic neurons in the GiV that inhibit spinal and cranial motor neurons. Monoaminergic neurons are inhibited during PS by GABAergic neurons in the DPGi, vlPAG, and by MCH neurons in the lateral hypothalamus

Finally, phasic motor movements during RBD would be due to a degeneration or destruction of at least part of the SLD PS-on neurons and/or the phasic activation of motor systems normally occurring during PS.

Acknowledgments This work was supported by CNRS and Université Claude Bernard Lyon.

References

1. von Economo C. Die pathologie des schlafes. In: Von Bethe A, Bergman GV, Embden G, Ellinger UA, editors. Handbuch des Normalen und Pathologischen Physiologie. Berlin: Springer; 1926. p. 591–610.
2. McGinty DJ, Sterman MB. Sleep suppression after basal forebrain lesions in the cat. Science. 1968;160(833): 1253–5.
3. Nauta W. Hypothalamic regulation of sleep in rats. An experimental study. J Neurophysiol. 1946;9:285–316.
4. Ranson SW. Somnolence caused by hypothalamic lesions in the monkey. Arch Neurol Psychiatry. 1939;41:1–23.
5. Sterman MB, Clemente CD. Forebrain inhibitory mechanisms: cortical synchronization induced by basal forebrain stimulation. Exp Neurol. 1962;6:91–102.
6. Szymusiak R, McGinty D. Sleep-related neuronal discharge in the basal forebrain of cats. Brain Res. 1986;370(1):82–92.
7. Moruzzi G. The sleep-waking cycle. Ergeb Physiol. 1972;64:1–165.
8. Borbely AA. From slow waves to sleep homeostasis: new perspectives. Arch Ital Biol. 2001;139(1–2):53–61.
9. Sherin JE, Shiromani PJ, McCarley RW, Saper CB. Activation of ventrolateral preoptic neurons during sleep. Science. 1996;271(5246):216–9.
10. Gvilia I, Turner A, McGinty D, Szymusiak R. Preoptic area neurons and the homeostatic regulation of rapid eye movement sleep. J Neurosci. 2006;26(11):3037–44.
11. Novak CM, Nunez AA. Daily rhythms in Fos activity in the rat ventrolateral preoptic area and midline thalamic nuclei. Am J Physiol. 1998;275(5 Pt 2):R1620–6.
12. Fort P, Bassetti CL, Luppi PH. Alternating vigilance states: new insights regarding neuronal networks and mechanisms. Eur J Neurosci. 2009;29(9):1741–53.
13. Szymusiak R, Alam N, Steininger TL, McGinty D. Sleep-waking discharge patterns of ventrolateral preoptic/anterior hypothalamic neurons in rats. Brain Res. 1998;803(1–2):178–88.
14. Lu J, Greco MA, Shiromani P, Saper CB. Effect of lesions of the ventrolateral preoptic nucleus on NREM and REM sleep. J Neurosci. 2000;20(10):3830–42.
15. Gallopin T, Fort P, Eggermann E, et al. Identification of sleep-promoting neurons in vitro. Nature. 2000;404(6781):992–5.
16. Gallopin T, Luppi PH, Rambert FA, Frydman A, Fort P. Effect of the wake-promoting agent modafinil on sleep-promoting neurons from the ventrolateral preoptic nucleus: an in vitro pharmacologic study. Sleep. 2004;27(1):19–25.

17. Eggermann E, Serafin M, Bayer L, et al. Orexins/hypocretins excite basal forebrain cholinergic neurones. Neuroscience. 2001;108(2):177–81.
18. Gallopin T, Luppi PH, Cauli B, et al. The endogenous somnogen adenosine excites a subset of sleep-promoting neurons via A2A receptors in the ventrolateral preoptic nucleus. Neuroscience. 2005;134(4):1377–90.
19. Scammell TE, Gerashchenko DY, Mochizuki T, et al. An adenosine A2a agonist increases sleep and induces Fos in ventrolateral preoptic neurons. Neuroscience. 2001;107(4):653–63.
20. Porkka-Heiskanen T, Strecker RE, McCarley RW. Brain site-specificity of extracellular adenosine concentration changes during sleep deprivation and spontaneous sleep: an in vivo microdialysis study. Neuroscience. 2000;99(3):507–17.
21. Rainnie DG, Grunze HC, McCarley RW, Greene RW. Adenosine inhibition of mesopontine cholinergic neurons: implications for EEG arousal. Science. 1994;263(5147):689–92.
22. Stenberg D, Litonius E, Halldner L, Johansson B, Fredholm BB, Porkka-Heiskanen T. Sleep and its homeostatic regulation in mice lacking the adenosine A1 receptor. J Sleep Res. 2003;12(4):283–90.
23. Urade Y, Eguchi N, Qu WM, et al. Minireview: sleep regulation in adenosine A(2A) receptor-deficient mice. Neurology. 2003;61(11 Suppl 6):S94–6.
24. Huang ZL, Qu WM, Eguchi N, et al. Adenosine A2A, but not A1, receptors mediate the arousal effect of caffeine. Nat Neurosci. 2005;8(7):858–9.
25. Fort P, Luppi PH, Gallopin T. In vitro identification of the presumed sleep-promoting neurons of the ventrolateral preoptic nucleus (VLPO). In: Luppi PH, editor. Sleep. Circuits and Functions: CRC Press; 2005. p. 43–64.
26. Jouvet M, Michel F. Corrélations électromyographiques du sommeil chez le chat décortiqué et mésencéphalique chronique. CR Soc Biol. 1959;153:422–5.
27. Jouvet M. The paradoxical phase of sleep. Int J Neurol. 1965;5(2):131–50.
28. George R, Haslett WL, Jenden DJ. A cholinergic mechanism in the brainstem reticular formation: induction of paradoxixal sleep. Int J Neuropharmacol. 1964;3:541–52.
29. Sakai K, Sastre JP, Kanamori N, Jouvet M. State-specific neurones in the ponto-medullary reticular formation with special reference to the postural atonia during paradoxical sleep in the cat. In: Pompeiano O, Aimone Marsan C, editors. Brain mechanisms of perceptual awareness and purposeful behavior. New York: Raven; 1981. p. 405–29.
30. Sakai K, Kanamori N, Jouvet M. Neuronal activity specific to paradoxical sleep in the bulbar reticular formation in the unrestrained cat. C R Seances Acad Sci D. 1979;289(6):557–61.
31. Sakai K, Crochet S, Onoe H. Pontine structures and mechanisms involved in the generation of paradoxical (REM) sleep. Arch Ital Biol. 2001;139(1–2):93–107.
32. Chase MH, Soja PJ, Morales FR. Evidence that glycine mediates the postsynaptic potentials that inhibit lumbar motoneurons during the atonia of active sleep. J Neurosci. 1989;9(3):743–51.
33. Luppi PH, Gervasoni D, Verret L, et al. Paradoxical (REM) sleep genesis: the switch from an aminergic-cholinergic to a GABAergic-glutamatergic hypothesis. J Physiol Paris. 2006;100(5–6):271–83.
34. Boissard R, Gervasoni D, Schmidt MH, Barbagli B, Fort P, Luppi PH. The rat ponto-medullary network responsible for paradoxical sleep onset and maintenance: a combined microinjection and functional neuroanatomical study. Eur J Neurosci. 2002;16(10):1959–73.
35. Gnadt JW, Pegram GV. Cholinergic brainstem mechanisms of REM sleep in the rat. Brain Res. 1986;384(1):29–41.
36. Shiromani PJ, Fishbein W. Continuous pontine cholinergic microinfusion via mini-pump induces sustained alterations in rapid eye movement (REM) sleep. Pharmacol Biochem Behav. 1986;25(6):1253–61.
37. Bourgin P, Escourrou P, Gaultier C, Adrien J. Induction of rapid eye movement sleep by carbachol infusion into the pontine reticular formation in the rat. Neuroreport. 1995;6(3):532–6.
38. Deurveilher S, Hars B, Hennevin E. Pontine microinjection of carbachol does not reliably enhance paradoxical sleep in rats. Sleep. 1997;20(8):593–607.
39. Maloney KJ, Mainville L, Jones BE. Differential c-Fos expression in cholinergic, monoaminergic, and GABAergic cell groups of the pontomesencephalic tegmentum after paradoxical sleep deprivation and recovery. J Neurosci. 1999;19(8):3057–72.
40. Verret L, Leger L, Fort P, Luppi PH. Cholinergic and noncholinergic brainstem neurons expressing Fos after paradoxical (REM) sleep deprivation and recovery. Eur J Neurosci. 2005;21(9):2488–504.
41. Sapin E, Lapray D, Berod A, et al. Localization of the brainstem GABAergic neurons controlling paradoxical (REM) sleep. PLoS ONE. 2009;4(1):e4272.
42. Clement O, Sapin E, Berod A, Gervasoni D, Fort P, Luppi P-H. Evidence that neurons of the sublaterodorsal tegmental nucleus triggering paradoxical (REM) sleep are glutamatergic *SfN* (Abstract). 2009.
43. Onoe H, Sakai K. Kainate receptors: a novel mechanism in paradoxical (REM) sleep generation. Neuroreport. 1995;6(2):353–6.
44. Beitz AJ. Relationship of glutamate and aspartate to the periaqueductal gray-raphe magnus projection: analysis using immunocytochemistry and microdialysis. J Histochem Cytochem. 1990;38(12):1755–65.
45. Boissard R, Fort P, Gervasoni D, Barbagli B, Luppi PH. Localization of the GABAergic and non-GABAergic neurons projecting to the sublaterodorsal nucleus and potentially gating paradoxical sleep onset. Eur J Neurosci. 2003;18(6):1627–39.
46. Pollock MS, Mistlberger RE. Rapid eye movement sleep induction by microinjection of the GABA-A antagonist bicuculline into the dorsal subcoeruleus area of the rat. Brain Res. 2003;962(1–2):68–77.
47. Xi MC, Morales FR, Chase MH. Evidence that wakefulness and REM sleep are controlled by a GABAergic pontine mechanism. J Neurophysiol. 1999;82(4):2015–9.

48. Boissard R, Gervasoni D, Fort P, Henninot V, Barbagli B, Luppi PH. Neuronal networks responsible for paradoxical sleep onset and maintenance in rats: a new hypothesis. Sleep. 2000;23(Suppl):107.
49. Lu J, Sherman D, Devor M, Saper CB. A putative flip-flop switch for control of REM sleep. Nature. 2006;441(7093):589–94.
50. Maloney KJ, Mainville L, Jones BE. c-Fos expression in GABAergic, serotonergic, and other neurons of the pontomedullary reticular formation and raphe after paradoxical sleep deprivation and recovery. J Neurosci. 2000;20(12):4669–79.
51. Sastre JP, Buda C, Kitahama K, Jouvet M. Importance of the ventrolateral region of the periaqueductal gray and adjacent tegmentum in the control of paradoxical sleep as studied by muscimol microinjections in the cat. Neuroscience. 1996;74(2):415–26.
52. Takahashi K, Lin JS, Sakai K. Neuronal activity of histaminergic tuberomammillary neurons during wake-sleep states in the mouse. J Neurosci. 2006; 26(40):10292–8.
53. Mileykovskiy BY, Kiyashchenko LI, Siegel JM. Behavioral correlates of activity in identified hypocretin/orexin neurons. Neuron. 2005;46(5):787–98.
54. Lee MG, Hassani OK, Jones BE. Discharge of identified orexin/hypocretin neurons across the sleep-waking cycle. J Neurosci. 2005;25(28):6716–20.
55. Hobson JA, McCarley RW, Wyzinski PW. Sleep cycle oscillation: reciprocal discharge by two brainstem neuronal groups. Science. 1975;189(4196):55–8.
56. McCarley RW, Hobson JA. Neuronal excitability modulation over the sleep cycle: a structural and mathematical model. Science. 1975;189(4196):58–60.
57. Tononi G, Pompeiano M, Cirelli C. Suppression of desynchronized sleep through microinjection of the alpha 2-adrenergic agonist clonidine in the dorsal pontine tegmentum of the cat. Pflugers Arch. 1991;418(5):512–8.
58. Crochet S, Sakai K. Effects of microdialysis application of monoamines on the EEG and behavioural states in the cat mesopontine tegmentum. Eur J Neurosci. 1999;11(10):3738–52.
59. Sakai K, Koyama Y. Are there cholinergic and non-cholinergic paradoxical sleep-on neurones in the pons? Neuroreport. 1996;7(15–17):2449–53.
60. Leger L, Goutagny R, Sapin E, Salvert D, Fort P, Luppi PH. Noradrenergic neurons expressing Fos during waking and paradoxical sleep deprivation in the rat. J Chem Neuroanat. 2008;37(3):139–57.
61. Guyenet PG, Aghajanian GK. ACh, substance P and met-enkephalin in the locus coeruleus: pharmacological evidence for independent sites of action. Eur J Pharmacol. 1979;53(4):319–28.
62. Koyama Y, Kayama Y. Mutual interactions among cholinergic, noradrenergic and serotonergic neurons studied by ionophoresis of these transmitters in rat brainstem nuclei. Neuroscience. 1993;55(4):1117–26.
63. Luppi PH, Charlety PJ, Fort P, Akaoka H, Chouvet G, Jouvet M. Anatomical and electrophysiological evidence for a glycinergic inhibitory innervation of the rat locus coeruleus. Neurosci Lett. 1991;128(1):33–6.
64. Jones BE. Noradrenergic locus coeruleus neurons: their distant connections and their relationship to neighboring (including cholinergic and GABAergic) neurons of the central gray and reticular formation. Prog Brain Res. 1991;88:15–30.
65. Darracq L, Gervasoni D, Souliere F, et al. Effect of strychnine on rat locus coeruleus neurones during sleep and wakefulness. Neuroreport. 1996;8(1):351–5.
66. Gervasoni D, Darracq L, Fort P, Souliere F, Chouvet G, Luppi PH. Electrophysiological evidence that noradrenergic neurons of the rat locus coeruleus are tonically inhibited by GABA during sleep. Eur J Neurosci. 1998;10(3):964–70.
67. Gervasoni D, Peyron C, Rampon C, et al. Role and origin of the GABAergic innervation of dorsal raphe serotonergic neurons. J Neurosci. 2000;20(11): 4217–25.
68. Nitz D, Siegel J. GABA release in the dorsal raphe nucleus: role in the control of REM sleep. Am J Physiol. 1997;273(1 Pt 2):R451–5.
69. Nitz D, Siegel JM. GABA release in the locus coeruleus as a function of sleep/wake state. Neuroscience. 1997;78(3):795–801.
70. Lu J, Bjorkum AA, Xu M, Gaus SE, Shiromani PJ, Saper CB. Selective activation of the extended ventrolateral preoptic nucleus during rapid eye movement sleep. J Neurosci. 2002;22(11):4568–76.
71. Verret L, Fort P, Gervasoni D, Leger L, Luppi PH. Localization of the neurons active during paradoxical (REM) sleep and projecting to the locus coeruleus noradrenergic neurons in the rat. J Comp Neurol. 2006;495(5):573–86.
72. Woch G, Davies RO, Pack AI, Kubin L. Behaviour of raphe cells projecting to the dorsomedial medulla during carbachol-induced atonia in the cat. J Physiol. 1996;490(Pt 3):745–58.
73. Goutagny R, Luppi PH, Salvert D, Lapray D, Gervasoni D, Fort P. Role of the dorsal paragigantocellular reticular nucleus in paradoxical (rapid eye movement) sleep generation: a combined electrophysiological and anatomical study in the rat. Neuroscience. 2008;152(3):849–57.
74. Ennis M, Aston-Jones G. GABA-mediated inhibition of locus coeruleus from the dorsomedial rostral medulla. J Neurosci. 1989;9(8):2973–81.
75. Kaur S, Saxena RN, Mallick BN. GABAergic neurons in prepositus hypoglossi regulate REM sleep by its action on locus coeruleus in freely moving rats. Synapse. 2001;42(3):141–50.
76. Liu R, Jolas T, Aghajanian G. Serotonin 5-HT(2) receptors activate local GABA inhibitory inputs to serotonergic neurons of the dorsal raphe nucleus. Brain Res. 2000;873(1):34–45.
77. Verret L, Goutagny R, Fort P, et al. A role of melanin-concentrating hormone producing neurons in the central regulation of paradoxical sleep. BMC Neurosci. 2003;4(1):19.
78. Lin JS, Sakai K, Vanni-Mercier G, Jouvet M. A critical role of the posterior hypothalamus in the mechanisms of wakefulness determined by microinjection

of muscimol in freely moving cats. Brain Res. 1989;479(2):225–40.

79. Hassani OK, Lee MG, Jones BE. Melanin-concentrating hormone neurons discharge in a reciprocal manner to orexin neurons across the sleep-wake cycle. Proc Natl Acad Sci U S A. 2009;106(7):2418–22.
80. Ahnaou A, Drinkenburg WH, Bouwknecht JA, Alcazar J, Steckler T, Dautzenberg FM. Blocking melanin-concentrating hormone MCH1 receptor affects rat sleep-wake architecture. Eur J Pharmacol. 2008;579(1–3):177–88.
81. Adamantidis A, Salvert D, Goutagny R, et al. Sleep architecture of the melanin-concentrating hormone receptor 1-knockout mice. Eur J Neurosci. 2008; 27(7):1793–800.
82. Willie JT, Sinton CM, Maratos-Flier E, Yanagisawa M. Abnormal response of melanin-concentrating hormone deficient mice to fasting: Hyperactivity and rapid eye movement sleep suppression. Neuroscience. 2008;156(4):819–29.
83. Peyron C, Sapin E, Leger L, Luppi PH, Fort P. Role of the melanin-concentrating hormone neuropeptide in sleep regulation. Peptides. 2009;30(11):2052–9.
84. Bayer L, Mairet-Coello G, Risold PY, Griffond B. Orexin/hypocretin neurons: chemical phenotype and possible interactions with melanin-concentrating hormone neurons. Regul Pept. 2002;104(1–3):33–9.
85. Guan JL, Uehara K, Lu S, et al. Reciprocal synaptic relationships between orexin- and melanin-concentrating hormone-containing neurons in the rat lateral hypothalamus: a novel circuit implicated in feeding regulation. Int J Obes Relat Metab Disord. 2002;26(12): 1523–32.
86. Rao Y, Lu M, Ge F, et al. Regulation of synaptic efficacy in hypocretin/orexin-containing neurons by melanin concentrating hormone in the lateral hypothalamus. J Neurosci. 2008;28(37):9101–10.
87. Goutagny R, Luppi PH, Salvert D, Gervasoni D, Fort P. GABAergic control of hypothalamic melanin-concentrating hormone-containing neurons across the sleep-waking cycle. Neuroreport. 2005;16(10):1069–73.
88. Alam MN, Kumar S, Bashir T, et al. GABA-mediated control of hypocretin- but not melanin-concentrating hormone-immunoreactive neurones during sleep in rats. J Physiol. 2005;563(Pt 2):569–82.
89. Espana RA, Baldo BA, Kelley AE, Berridge CW. Wake-promoting and sleep-suppressing actions of hypocretin (orexin): basal forebrain sites of action. Neuroscience. 2001;106(4):699–715.
90. Huang ZL, Qu WM, Li WD, et al. Arousal effect of orexin A depends on activation of the histaminergic system. Proc Natl Acad Sci U S A. 2001;98(17): 9965–70.
91. Hagan JJ, Leslie RA, Patel S, et al. Orexin A activates locus coeruleus cell firing and increases arousal in the rat. Proc Natl Acad Sci U S A. 1999;96(19):10911–6.
92. Peyron C, Faraco J, Rogers W, et al. A mutation in a case of early onset narcolepsy and a generalized absence of hypocretin peptides in human narcoleptic brains. Nat Med. 2000;6(9):991–7.
93. Lin L, Faraco J, Li R, et al. The sleep disorder canine narcolepsy is caused by a mutation in the hypocretin (orexin) receptor 2 gene. Cell. 1999;98(3):365–76.
94. Chemelli RM, Willie JT, Sinton CM, et al. Narcolepsy in orexin knockout mice: molecular genetics of sleep regulation. Cell. 1999;98(4):437–51.
95. Gervasoni D, Lin SC, Ribeiro S, Soares ES, Pantoja J, Nicolelis MA. Global forebrain dynamics predict rat behavioral states and their transitions. J Neurosci. 2004;24(49):11137–47.

The Hypocretins/Orexins: Master Regulators of Arousal and Hyperarousal

Matthew E. Carter, Antoine Adamantidis, and Luis de Lecea

Keywords
Orexin • Hypocretin • Arousal • Hyperarousal • Stress • Addiction • Optogenetics

Introduction

The hypocretins (Hcrts) – also known as "orexins" – are a pair of neuropeptides well known for their role in promoting wakefulness. Shortly after their discovery, multiple laboratories identified the crucial link between dysfunction of the Hcrt system and narcolepsy in mice, dogs, and humans [1–6]. Therefore, much research has focused on the wake-promoting properties of Hcrts and why their absence leads to a narcoleptic phenotype. Indeed, drugs that directly target the Hcrt system are currently under development to treat human patients with insomnia and other sleep disorders [7]. However, research also indicates that Hcrts play roles in behavioral phenotypes beyond wakefulness, regulating other arousal-related behaviors such as brain reward and motivation [8–11], stress/anxiety [12, 13], food intake [14, 15], sexual behavior [16], and attention [17]. In this chapter, we review the role of Hcrts in mediating arousal-related behaviors in addition to wakefulness, with a special emphasis on the recently discovered role of Hcrts in mediating the stress response and in regulating reward circuitry in the brain. These multiple roles of Hcrts in arousal-related behaviors should be considered when contemplating the role of Hcrts in narcolepsy, and also when targeting the Hcrt system to treat narcolepsy and other sleep disorders.

L. de Lecea (✉)
Departments of Psychiatry and Behavioral Sciences, Stanford University, 701B Welch Road, Stanford, CA 94304, USA
e-mail: llecea@stanford.edu

The Hcrts and their Receptors

The Hcrts were discovered independently by two groups in the late 1990s [14, 18]. They consist of a pair of secreted peptides, hypocretin-1 and hypocretin-2 (Hcrt1 and Hcrt2; also known as "Orexin A and Orexin B," respectively), that are produced from the same precursor, preprohypocretin (ppHcrt). These peptides are produced exclusively in a subset of neurons in the lateral hypothalamus (LH), distinct from cells that express melanin-concentrating hormone (MCH). Hcrt-expressing neurons receive widespread afferent projections from many brain regions including nuclei in the hypothalamus, allocortex, claustrum, bed nucleus of the stria terminalis,

C.R. Baumann et al. (eds.), *Narcolepsy: Pathophysiology, Diagnosis, and Treatment*, DOI 10.1007/978-1-4419-8390-9_11, © Springer Science+Business Media, LLC 2011

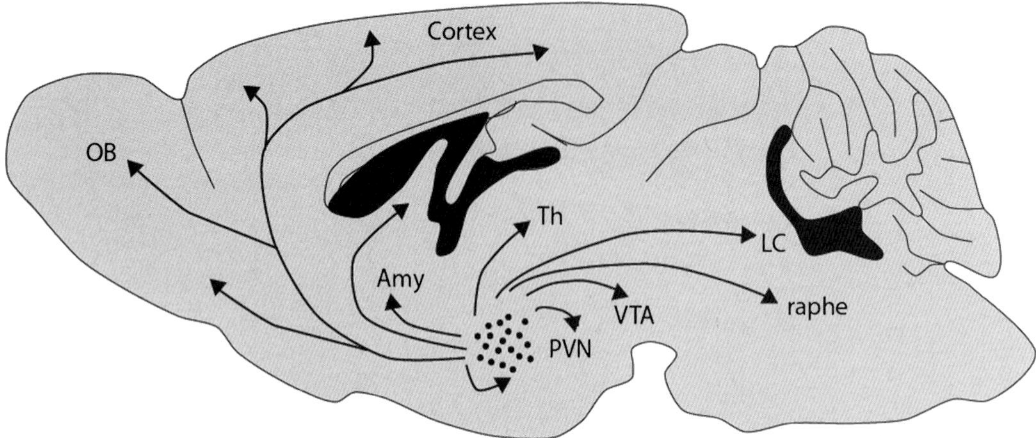

Fig. 1 Efferent projects of hypocretin-expressing neurons throughout the brain. Many targets of the hypocretin system express other neuromodulators and are capable of regulating diverse physiological functions and behaviors. *Amy* amygdala, *LC* locus coeruleus, *PVN* paraventricular nucleus, *OB* olfactory bulb, *Th* thalamus, *VTA* ventral tegmental area

periaquiductal gray, dorsal raphe nucleus, and lateral parabrachial nucleus [19, 20]. These neurons receive input from GABAergic, glutamatergic, and cholinergic neurons [21]. In vitro electrophysiology studies demonstrate that several neurotransmitters/neuromodulators excite Hcrt neurons (including corticotropin-releasing factor, ghrelin, neurotensin, oxytocin, and vasopressin) or inhibit Hcrt neurons (including serotonin, noradrenaline, dopamine, neuropeptide Y, and leptin) [22].

Hcrts excite their postsynaptic targets by binding to two G-coupled-protein receptors, hypocretin receptor-1 and -2 (Hcrt-r1 and Hcrt-r2, respectively) [14, 18]. Hcrt-r1 binds Hcrt1 with high affinity and binds Hcrt2 with 100- to 1,000-fold lower affinity [18, 23]. Hcrt-r2 has a high affinity for both Hcrt1 and Hcrt2. Hcrt-r1 mRNA levels are found within the hypothalamus, locus coeruleus (LC), the cerebral cortex, and several brainstem nuclei [24]. Hcrt-r2 mRNA is expressed in cholinergic nuclei in the brainstem, the ventral tegmental area (VTA), and histaminergic neurons in the tuberomammillary nucleus (TMN), as well as overlap expression with Hcrt-r1 in the hypothalamus [24]. The in situ hybridization pattern of these receptors is consistent with the projections of efferent fibers, including prominent projections to the noradrenergic LC, the histaminergic TMN, the serotonergic raphe nuclei, the dopaminergic VTA, the cholinergic pedunculopontine tegmental area (PPT) and laterodorsal tegmental area (LDT), and the galaninergic ventrolateral preoptic nucleus (VLPO) [25]. Hcrt neurons also project diffusely throughout the cerebral cortex (Fig. 1).

These anatomical and electrophysiological studies suggest that Hcrt neurons integrate a variety of homeostatic signals from the central nervous system and periphery, and project to numerous brain regions, many of which express other neuromodulators and are capable of regulating diverse physiological functions and behaviors. Because Hcrts receive and project dense projections to nuclei known to play a role in arousal and vigilance, initial studies of the physiological function of Hcrts involved the role of Hcrts in wakefulness.

Hcrts and Wakefulness

The role of Hcrts in promoting wakefulness is well known and is described more thoroughly in other excellent reviews [22, 26, 27]. Major evidence stems from the original finding that impairment of the Hcrt system causes the sleep disorder narcolepsy in mice [2, 3], dogs [1], and humans [4–6]. Most human narcoleptics have decreased levels of Hcrt in their cerebrospinal fluid [4], and postmortem analysis reveals a reduction in Hcrt neurons in human narcoleptic brains [6]. The naturally occurring canine model of narcolepsy is caused by a mutation in the Hcrt-r2 [1]. Mice with genetic ablation of the preprohypocretin precursor [2], Hcrt-r2 [28], or the hypocretin-expressing neurons themselves [3] exhibit a narcolepsy with cataplexy phenotype that is remarkably similar to that in humans. Interestingly, in addition to narcolepsy, a lack of Hcrt function also impairs normal emergence from general anesthesia [29].

In addition to loss-of-function studies that demonstrate Hcrts are necessary for normal wakefulness, gain-of-function studies indicate that Hcrts are sufficient to stimulate wakefulness and arousal. Intracerebroventricular (i.c.v.) injection of Hcrt1 and/or Hcrt2 increases the time spent awake and decreases the time spent in slow-wave and REM sleep in a variety of vertebrate species [30–33]. Hcrt-induced increases in time spent awake are correlated with a relative increase in arousal-related behaviors including eating, drinking, grooming, and locomotor activity [30, 34–38]. Artificial stimulation of Hcrt neurons using a light-activated cation channel, channelrhodopsin-2, increases the probability of transitions from sleep to wakefulness during slow-wave and REM sleep throughout the entire light/dark cycle (described in more detail below) [39, 40].

Finally, electrophysiological studies indicate that Hcrt neurons are indeed more active during wakefulness and arousal-related behaviors compared with during sleep. In vivo single-unit recordings of identified Hcrt neurons demonstrate a high discharge activity during arousal, especially behaviors accompanied with a strong locomotor activity [41–43]. Hcrt neurons are activated by neurotransmitters that promote arousal including glutamate [44, 45], corticotropin-releasing factor [46], ATP [47], noradrenaline [48], and carbachol [48]. Sleep-promoting neurotransmitters inhibit Hcrt neurons, including GABA (through $GABA_{a,b}$) [44, 45, 49] and adenosine [50].

Together, these studies clearly demonstrate a role for Hcrts in promoting and maintaining wakefulness. However, in recent years, many studies have established a role for Hcrt peptides beyond stabilizing an "awake" state, demonstrating a role for these peptides in generalized arousal and arousal-related behaviors. We describe these behaviors below.

Hcrts in the Context of Allostasis

The role of the Hcrt system in promoting wakefulness is often described as a role in "arousal." Generalized arousal is marked by increased motor activity and heightened responsiveness to sensory and emotionally salient stimuli [51–54]. Less often emphasized, however, is that arousal systems are involved in much more than just regulating sleep/wake cycles, such as stress/anxiety, motivation and reward, sexual behavior, and attention. Importantly, brain structures implicated in generalized arousal, including the reticular formation of the medulla and pons, midbrain, and the paraventricular, dorsomedial, and lateral hypothalamic nuclei, receive projections from Hcrt neurons [25, 55]. By appreciating the role arousal is known to play other than promoting an "awake" state, investigators may be able to make increasingly novel yet specific hypotheses about the function of Hcrts in nonsleep behaviors. For example, recent reports that Hcrts modulate behavior in human and murine models of depression [56, 57] are understandable and even anticipatable in the face of years of psychiatric research showing that arousal processing is impaired in humans with depression [58].

Hcrts seem to have their greatest impact when arousal is needed to regulate basic homeostatic pressures such as hunger, anxiety, or the drive for sex. Also, as discussed below, Hcrt signaling seems to have the strongest effects in conditions of addiction and stress, suggesting that these peptides are particularly important for allostasis. In contrast to homeostasis, allostasis maintains stability at levels outside the normal range and is achieved by varying the internal milieu to match perceived and anticipated environmental demands [59, 60]. For example, consider a recent study testing the effects of calorie restriction on stress and depression [61]. In mouse models of depression, calorie-restricted mice perform better in a forced swim test (have longer latencies to immobility and less total immobility) and do not exhibit social interaction deficits compared to ad libitum-fed mice. Fascinatingly, Hcrt null mice do not show either of these calorie-restriction benefits [61]. Moreover, the number of c-Fos-positive Hcrt neurons induced by calorie restriction correlates strongly with improvement on the social interaction test [61]. This suggests that Hcrt neurons mediate an allostatic generalized stress response to calorie restriction that allows an animal to overcome maladaptive depressive symptoms induced by chronic stress. Similarly, although Hcrts do not necessarily stimulate food intake under normal conditions, in situations of calorie restriction, Hcrts are necessary for adaptive increases in food-anticipatory behavior [62]. This study further demonstrates that Hcrt neurons mediate allostatic changes in behavior, in this case ensuring that animals will be awake and motivated to obtain food during the limited times it is available.

While more research is needed to understand the functions of Hcrts in various environmental challenges, recent work has illuminated significant roles for Hcrts in arousal-related behaviors, including brain reward and motivation, heightened arousal following acute stressors, sexual behavior, and attention.

Stress

Stimuli that increase arousal/wakefulness also often increase stress and anxiety. Therefore, the ability of Hcrts to promote wakefulness suggests that these peptides may play a role in increasing the behavioral and physiological hallmarks of stress. In support of this hypothesis are observations indicating that response to centrally administered Hcrts mimics the behavioral and physiological response to stress. For example, i.c.v. injection of Hcrt1 elicits a majority of stress-related behaviors, including grooming, chewing of inedible material, increased locomotor activity, and food consumption [30, 34–38]. Furthermore, an increase of Hcrts is also correlated with a variety of autonomic processes associated with stress or high levels of arousal, such as elevation of heart rate, body temperature, mean arterial blood pressure, and oxygen consumption [63–67]. Hypocretin cells also seem to be activated by environmental stressors. Studies show an increase in c-Fos, an immediate early gene and marker of neural activity, in Hcrt cells in response to acute stressful stimuli including a brightly lit novel environment, food deprivation, and cold exposure, in addition to more chronic stressors such as foot shock and immobilization stress [12, 13, 36, 68–70].

In addition to evidence suggesting that Hcrt cells play a role in the stress response, there is a clear anatomical connection between Hcrt cells and cells in the PVN that produce CRF, the peptide that initiates the hypothalamus–pituitary–adrenal (HPA) axis stress response. Hcrt fibers are located within close proximity to CRF neurons in the PVN [12, 13, 46]. Innervation of these structures by Hcrt efferents suggest the possibility that Hcrt may interact with central CRF systems to activate the HPA axis and other stress-related processes. Consistent with this hypothesis, bath application of Hcrt1 in vitro elicits depolarization and increased spike frequency of CRF-containing neurons in the PVN [71, 72].

Addiction

Recently, an exciting role for Hcrts has been established in reward seeking and addiction. Hcrt neurons project to the VTA, a critical site of synaptic plasticity induced by addictive drugs [8, 9]. Stimulation of Hcrt neurons or microinjection of Hcrt1 into the VTA or ventricles reinstates previously extinguished drug-seeking behaviors, and these effects are blocked by a Hcrtr1 antagonist [8, 9]. In vitro application of Hcrt1 and Hcrt2 induces potentiation of NMDA-mediated neurotransmission via insertion of NMDA receptors in VTA dopamine neuron synapses [10, 11]. In vitro application of Hcrt2 also increases presynaptic glutamate release in the VTA [11]. In vivo administration of a Hcrt1 antagonist blocks locomotion sensitization to cocaine and occludes cocaine-induced potentiation of excitatory currents in VTA dopamine neurons [10]. These seminal studies have sparked a rapidly growing body of research that repeatedly confirms that Hcrts modulate reward processing.

Hcrts in Other Arousal-Related Behaviors

Hcrts are implicated in many physiological functions other than maintaining wakefulness. For example, the alternate name of Hcrt, "orexin," was designated because i.c.v. infusion of Hcrts increased food intake in rodents [14]. These results are now considered to be an indirect effect of the wake-promoting effects of Hcrts, but this is still an active area of investigation. Microinjection of Hcrts into the arcuate nucleus stimulates orexigenic GABAergic neurons and inhibits anorexigenic POMC-expressing neurons [15]. Hcrts also inhibit neurons in the ventromedial hypothalamus, an established satiety center [15]. Thus, Hcrts act in a reciprocal manner to the satiety hormone leptin in important energy-homeostatic regions of the hypothalamus.

In addition to food intake, Hcrts have also been implicated in the heightened arousal during male sexual behavior [16] and attention [17]. Hcrts have also been hypothesized to play a role in the symptoms of Parkinson's disease [73], schizophrenia [74, 75], and depression [56, 57]. In sum, studies of the Hcrt system have progressed far beyond the initial discovery of the involvement of Hcrts in sleep and wakefulness.

New Methods to Study the Hcrt System

Our laboratory has recently employed optogenetic technology as a new method of studying the Hcrt system [39, 40]. This technology does not feature many of the spatial and temporal problems inherent in stimulation or inhibition of Hcrt neurons during experiments [76]. For example, the most common way of inducing electrical activity in neurons for the past century has been injecting current through a microelectrode. Although temporally precise, a microelectrode cannot distinguish between cell types in the stimulated area. Furthermore, a scientist may inadvertently stimulate other unintended regions that are adjacent to the electrode. Pharmacological techniques lack both spatial and temporal precision, as psychoactive substances can spread throughout the brain and require minutes to hours to clear the system.

Optogenetics can be thought of as a perfect combination of an electrode, which has high temporal precision, with a genetically encodable probe, which has a high spatial resolution [76]. To stimulate neural activity, scientists deliver an ion channel called channelrhodopsin-2 (ChR2) into specific populations of neurons. ChR2 is a nonspecific cation channel that depolarizes neurons (and thus stimulates neural activity) upon stimulation with blue light. This channel is naturally expressed in green algae and normally absent in animals. Because ChR2 is genetically encoded, scientists can use specific promoter and enhancer elements to express the channel in discrete populations of neurons in the brain. Once expressed, scientists can deliver blue light using thin fiber-optic cables, implanted into the brain for long-term experiments over days or weeks. To inhibit neural activity, scientists use a separate light-sensitive protein, halorhodopsin (NpHR), which is naturally expressed by archaebacteria and normally absent in animals. NpHR is a chloride pump that hyperpolarizes neurons (and thus inhibits neural activity) upon stimulation with yellow light. Importantly, ChR2 and NpHR both have a temporal resolution of milliseconds, allowing stimulation or silencing of single action potentials. Thus, optogenetic technology allows for temporally precise manipulation of specific populations of neurons in an awake, behaving animal.

Recently, we delivered ChR2 to Hcrt neurons using a well-characterized Hcrt-specific promoter in a lentiviral delivery system [39, 40]. We found that direct, deep brain optical stimulation of Hcrt neurons in the hypothalamus increased the probability of transitions to wakefulness from either slow-wave sleep (SWS) or rapid eye movement (REM) [39]. Photostimulation using 5–30-Hz light pulse trains reduced the latency to wakefulness, whereas 1-Hz trains did not [39]. More recently, we asked whether Hcrt-mediated sleep-to-wake transitions are affected by light/dark period and sleep pressure [40]. We found that stimulation of Hcrt neurons increased the probability of an awakening event throughout the entire light/dark period but that this effect was diminished with sleep pressure induced by 2 or 4 h of sleep deprivation [40]. These results suggest that the Hcrt system promotes wakefulness throughout the light/dark period by activating multiple downstream targets, which themselves are inhibited by increased sleep pressure.

The precise temporal and spatial specificity of the optogenetic technology, in combination with imaging and pharmacological techniques, will allow us to better understand not only the types of behaviors Hcrts regulate but the specific neural circuits underlying these behaviors.

Acknowledgments M.E.C. is supported by fellowships from the National Science Foundation and National Institutes of Health (Award #F31MH83439). L.d.L. is supported by grants from the National Institute on Drug Abuse, DARPA, and NARSAD.

References

1. Lin L, Faraco J, Li R, et al. The sleep disorder canine narcolepsy is caused by a mutation in the hypocretin (orexin) receptor 2 gene. Cell. 1999;98:365–76.
2. Chemelli RM, Willie JT, Sinton CM, et al. Narcolepsy in orexin knockout mice: molecular genetics of sleep regulation. Cell. 1999;98:437–51.
3. Hara J, Beuckmann CT, Willie JT, et al. Genetic ablation of orexin neurons in mice results in narcolepsy, hypophagia, and obesity. Neuron. 2001;30:345–54.
4. Nishino S, Ripley B, Overeem S, Lammers GJ, Mignot E. Hypocretin (orexin) deficiency in human narcolepsy. Lancet. 2000;355:39–40.
5. Peyron C, Faraco J, Rogers W, et al. A mutation in a case of early onset narcolepsy and a generalized absence of hypocretin peptides in human narcoleptic brains. Nat Med. 2000;6:991–7.
6. Thannickal TC, Moore RY, Nienhuis R, et al. Reduced number of hypocretin neurons in human narcolepsy. Neuron. 2000;27:469–74.
7. Brisbare-Roch C, Dingemanse J, Koberstein R, et al. Promotion of sleep by targeting the orexin system in rats, dogs, and humans. Nat Med. 2007;13:150–5.
8. Harris GC, Wimmer M, Aston-Jones G. A role for lateral hypothalamic orexin neurons in reward seeking. Nature. 2005;437:556–9.
9. Boutrel B, Kenny PJ, Specio SE, et al. Role for hypocretin in mediating stress-induced reinstatement of cocaine-seeking behavior. Proc Natl Acad Sci. 2005;102:19168–73.
10. Borgland SL, Taha SA, Sarti F, Fields HL, Bonci A. Orexin A in the VTA is critical for the induction of synaptic plasticity and behavioral sensitization to cocaine. Neuron. 2006;49:589–601.
11. Borgland SL, Storm E, Bonci A. Orexin B/hypocretin 2 increases glutamatergic transmission to ventral tegmental area neurons. Eur J Neurosci. 2008;28:1545–56.
12. Berridge CW, Espana RA. Hypocretin/orexin in stress and arousal. In: de Lecea L, Sutcliffe JG, editors. Hypocretins: integrators of physiological functions. New York, NY: Springer-Verlag; 2005.
13. Winsky-Sommerer R, Boutrel B, de Lecea L. Stress and arousal: the corticotropin releasing factor/hypocretin circuitry. Mol Neurobiol. 2005;32:285–94.
14. Sakurai T, Amemiya A, Ishii M, et al. Orexins and orexin receptors: a family of hypothalamic neuropeptides and G protein-coupled receptors that regulate feeding behavior. Cell. 1998;92:573–85.
15. Muroya S, Funahashi H, Yamanaka A, et al. Orexins (hypocretins) directly interact with neuropeptide Y, POMC, and glucose-responsive neurons to regulate Ca2+ signaling in a reciprocal manner to leptin: orexigenic neuronal pathways in the mediobasal hypothalamus. Eur J Neurosci. 2004;19:1524–34.
16. Muschamp JW, Dominguez JM, Sato SM, Shen R-Y, Hull EM. A role for hypocretin (orexin) in male sexual behavior. J Neurosci. 2007;27:2837–45.
17. Lambe EK, Olausson P, Horst NK, et al. Hypocretin and nicotine excite the same thalamocortical synapses in prefrontal cortex: correlation with improved attention in rat. J Neurosci. 2005;25:5225–9.
18. de Lecea L, Kilduff TS, Peyron C, et al. The hypocretins: hypothalamus-specific peptides with neuroexcitatory activity. Proc Natl Acad Sci USA. 1998;95:322–7.
19. Yoshida K, McCormack S, Espana RA, Crocker A, Scammel TE. Afferents to the orexin neurons of the rat brain. J Comp Neurol. 2006;494:845–61.
20. Sakurai T, Nagata R, Yamanak A, et al. Input of orexin/hypocretin neurons revealed by a genetically encoded tracer in mice. Neuron. 2005;46:297–308.
21. Henny P, Jones BE. Innervation of orexin/hypocretin neurons by GABAergic, glutamatergic or cholinergic basal forebrain terminals evidenced by immunostaining for presynaptic vesicular transporter and postsynaptic scaffolding proteins. J Comp Neurol. 2006;499:645–61.
22. Ohno K, Sakurai T. Orexin neuronal circuitry: role in the regulation of sleep and wakefulness. Front Neuroendocrinol. 2008;29:70–87.

23. Lang M, Bufe B, de Pol S, et al. Structural properties of orexins for activation of their receptors. J Pept Sci. 2006;12:258–66.
24. Trivedi P, Yu H, MacNeil DJ, Van der Ploeg LH, Guan XM. Distribution of orexin receptor mRNA in the rat brain. FEBS Lett. 1998;438:71–5.
25. Peyron C, Tighe DK, van den Pol AN, de Lecea L, Heller HC, Sutcliffe JG, et al. Neurons containing hypocretin (orexin) project to multiple neuronal systems. J Neurosci. 1998;18:9996–10015.
26. Sakurai T. The neural circuit of orexin (hypocretin): maintaining sleep and wakefulness. Nat Rev Neurosci. 2007;8:171–81.
27. Saper CB, Scammell TE, Lu J. Hypothalamic regulation of sleep and circadian rhythms. Nature. 2005;437: 1257–63.
28. Willie JT, Chemelli RM, Sinton CM, et al. Distinct narcolepsy syndromes in Orexin receptor-2 and Orexin null mice: molecular genetic dissection of non-REM and REM sleep regulatory processes. Neuron. 2003;38:715–30.
29. Kelz MB, Sun Y, Chen J, et al. An essential role for orexins in emergence from general anesthesia. Proc Natl Acad Sci USA. 2008;105:1309–14.
30. Espana RA, Baldo BA, Kelley AE, Berridge CW. Wake-promoting and sleep-suppressing actions of hypocretin (orexin): basal forebrain sites of action. Neuroscience. 2001;106:699–715.
31. Piper DC, Upton N, Smith MI, Hunter AJ. The novel brain neuropeptide, orexin-A, modulates the sleep-wake cycle of rats. Eur J Neurosci. 2000;12:726–30.
32. Prober DA, Rihel J, Onah AA, Sung RJ, Schier AF. Hypocretin/orexin overexpression induces an insomnia-like phenotype in zebrafish. J Neurosci. 2006;26: 13400–10.
33. ES da silva, dos Santos TV, Hoeller AA, et al. Behavioral and metabolic effects of central injections of orexins/hypocretins in pigeons (*Columba livia*). Regul Pept. 2008;147:9–18.
34. Espana RA, Plahn S, Berridge CW. Circadian-dependent and cidcadian-independent behavioral actions of hypocretin/orexin. Brain Res. 2002;943: 224–36.
35. Ida T, Nakahara K, Katayma T, Murakami N, Nakazato M. Effect of lateral cerebroventricular injection of the appetite-stimulating neuropeptide, orexin and neuropeptide Y, on the various behavioral activities of rats. Brain Res. 1999;821:526–9.
36. Ida T, Nakahara K, Murakami T, Hanada R, Makazato M, Murakami N. Possible involvement of orexin in the stress reaction in rats. Biochem Biophys Res Commun. 2000;270:318–23.
37. Espana RA, Valentino RJ, Berridge CW. Fos immunoreactivity in hypocretin synthesizing and hypocretin-1 receptor-expressing neurons: effects of diurnal and nocturnal spontaneous waking, stress, and hypocretin-1 administration. Neuroscience. 2003;121: 201–17.
38. Martins PJ, D'Almeida V, Pedrazzoli M, et al. Increased hypocretin-1 (orexin-a) levels in cerebrospinal fluid of rats after short-term forced activity. Regul Pept. 2004;117:155–8.
39. Adamantidis AR, Zhang F, Aravanis AM, Deisseroth K, de Lecea L. Neural substrates of awakening probed with optogenetic control of hypocretin neurons. Nature. 2007;450:420–4.
40. Carter ME, Adamantidis AR, Ohtsu H, Deisseroth K, de Lecea L. Sleep homeostasis modulates hypocretin-mediated sleep-to-wake transitions. J Neurosci. 2009;29:10939–49.
41. Lee MG, Hassani OK, Jones BE. Discharge of identified orexin/hypocretin neurons across the sleep-waking cycle. J Neurosci. 2005;25:6716–20.
42. Mileykovskiy BY, Kiyashchenko LI, Siegel JM. Behavioral correlates of activity in identified hypocretin/orexin neurons. Neuron. 2005;46:787–98.
43. Takahashi K, Lin JS, Sakai K. Neuronal activity of orexin and non-orexin waking-active neurons during wake-sleep states in the mouse. Neuroscience. 2008; 153:860–70.
44. Li Y, Gao XB, Sakurai T, van den Pol AN. Hypocretin/orexin excites hypocretin neurons via a local glutamate neuron-A potential mechanism for orchestrating the hypothalamic arousal system. Neuron. 2002;36:1169–81.
45. Yamanaka A, Beuckmann CT, Willie JT, et al. Hypothalamic orexin neurons regulate arousal according to energy balance in mice. Neuron. 2003;38:701–13.
46. Winsky-Sommerer R, Yamanaka A, Diano S, et al. Interactions between the corticotropin-releasing factor system and hypocretins (orexins): a novel circuit mediating stress response. J Neurosci. 2004;24: 11439–48.
47. Wolmann G, Acuna-Goycolea C, van den Pol AN. Direct excitation of hypocretin/orexin cells by extracellular ATP at P2X receptors. J Neurophysiol. 2005;94:2195–206.
48. Bayer L, Eggermann E, Serafin M, et al. Opposite effects of noradrenaline and acetylcholine upon hypocretin/orexin versus melanin concentrating hormone neurons in rat hypothalamic slices. Neuroscience. 2005;130:807–11.
49. Xie X, Crowder TL, Yamanaka A, et al. GABA(B) receptor-mediated modulation of hypocretin/orexin neurons in mouse hypothalamus. J Physiol. 2006;574: 399–414.
50. Liu ZW, Gao XB. Adenosine inhibits activity of hypocretin/orexin neurons by the A1 receptor in the lateral hypothalamus: a possible sleep-promoting effect. J Neurophysiol. 2007;97:837–48.
51. Pfaff D, Ribeiro A, Matthews J, Kow L-M. Concepts and mechanisms of generalized central nervous system arousal. Ann NY Acad Sci. 2008;1129:11–25.
52. Garey J, Goodwillie A, Frohlich J, et al. Genetic contributions to generalized arousal of brain and behavior. Proc Natl Acad Sci. 2003;100:11019–22.
53. Levenson RW. Autonomic nervous systems differences among emotions. Psychol Sci. 1992;3:23–7.
54. Levenson RW. Blood, sweat, and fears. Ann NY Acad Sci. 2003;1000:348–66.

55. Kerman I. Organization of brain somatomotor-sympathetic circuits. Exp Brain Res. 2008;187:1–16.
56. Brundin L, Björkqvist M, Petersén A, et al. Reduced orexin levels in the cerebrospinal fluid of suicidal patients with major depressive disorder. Eur Neuropsychopharmacol. 2007;17:573–9.
57. Feng P, Vurbic D, Wu Z, Hu Y, Strohl K. Changes in brain orexin levels in a rat model of depression induced by neonatal administration of clomipramine. J Psychopharmacol. 2008;22:784–91.
58. Moratti S, Rubio G, Campo P, Keil A, Ortiz T. Hypofunction of right temporoparietal cortex during emotional arousal in depression. Arch Gen Psychiatry. 2008;65:532–41.
59. McEwen B, Wingfield JC. The concept of allostasis in biology and biomedicine. Horm Behav. 2003;43: 2–15.
60. Roberts AJ, Heyser CJ, Cole M, Griffin P, Koob GF. Excessive ethanol drinking following a history of dependence: animal model of allostasis. Neuropsychopharmacology. 2000;22:581–94.
61. Lutter M, Krishnan V, Russo SJ, et al. Orexin signaling mediates the antidepressant-like effect of calorie restriction. J Neurosci. 2008;28:3071–5.
62. Akiyama M, Yuasa T, Hayasaka N, et al. Reduced food anticipatory activity in genetically orexin (hypocretin) neuron-ablated mice. Eur J Neurosci. 2004;20: 3054–62.
63. Chen CT, Hwang LL, Chang JK, Dun NJ. Pressor effects of orexins injected intracisternally and to rostral ventrolateral medulla of anesthetized rats. Am J Physiol Regul Integr Comp Physiol. 2000;278: R692–7.
64. Lubkin M, Stricker Krongrad A. Independent feeding and metabolic actions of orexins in mice. Biochem Biophys Res Commun. 1998;253:241–5.
65. Samson WK, Gosnell B, Chang JK, Resch ZT, Murphy TC. Cardiovascular regulatory actions of the hypocretins in brain. Brain Res. 1999;831:248–53.
66. Shirasaka T, Nakazato M, Matsukura S, Takasaki M, Kannan H. Sympathetic and cardiovascular actions of orexins in conscious rats. Am J Physiol. 1999;277: R1780–5.
67. Yoshimichi G, Yoshimatsu H, Masaki T, Sakata T. Orexin-A regulates body temperature in coordination with arousal status. Exp Biol Med. 2001;226: 468–76.
68. Sakamoto F, Yamada S, Ueta Y. Centrally administered orexin-A activates corticotropin-releasing factor-containing neurons in the hypothalamic paraventricular nucleus and central amygdaloid nucleus of rats: possible involvement of central orexins on stress-activated central CRF neurons. Regul Pept. 2004;118:183–91.
69. Sakurai T, Moriguchi T, Furuya K, et al. Structure and function of human prepro-orexin gene. J Biol Chem. 1999;274:17771–6.
70. Zhu L, Onaka T, Sakurai T, Yada T. Activation of orexin neurons after noxious but not conditioned fear stimuli in rats. NeuroReport. 2002;13:1351–3.
71. Samson WK, Taylar MM, Folwell M, Ferguson AV. Orexin actions in hypothalamic paraventricular nucleus: physiological consequences and cellular correlates. Regul Pept. 2002;104:97–103.
72. Shirasaka T, Miyahara S, Kunitake T, et al. Orexin depolarizes rat hypothalamic paraventricular nucleus neurons. Am J Physiol Regul Integr Comp Physiol. 2001;281:R1113–8.
73. Fronczek R, Overeem S, Lee SYY, et al. Hypocretin (orexin) loss in Parkinson's disease. Brain. 2007;130: 1577–85.
74. Salomon RM. Hypocretin measures in psychiatric disorders. In: Nishino S, Sakurai T, editors. The orexin/hypocretin system. New York, NY: Humana Press; 2005. p. 317–27.
75. Deutch AY, Bubser M. The orexins/hypocretins and schizophrenia. Schizophrenia Bull. 2007;33: 1277–83.
76. Zhang F, Aravanis AM, Adamantidis A, de Lecea L, Deisseroth K. Circuit breakers: technologies for probing neural signals and systems. Nat Rev Neurosci. 2007;8:577–81.

Optogenetic Probing of Hypocretins' Regulation of Wakefulness

Antoine Adamantidis and Luis de Lecea

Keywords
Hypothalamus • Sleep • Wakefulness • Optogenetics • Hcrts/Orexins

Introduction

In mammals, sleep is commonly defined as "a rapidly reversible state of (behavioral) immobility and greatly reduced sensory responsiveness to environmental stimuli" [1]. Sleep and wake states have been strongly conserved during evolution, and "sleep-like" states exist in most organisms, including worms, flies, and fish [2], suggesting common underlying neural circuits and endocrine systems. During the last decades, neural circuits that modulate the sleep–wake cycle have been identified using a combination of lesion, histological, pharmacological, genetic, and in vitro and in vivo electrophysiology techniques. Collectively, they support the "reciprocal interaction" and other computational models which describe the sleep–wake cycle as a complex, yet partially defined balance between subcortical excitatory and inhibitory neural circuits in the brain [3]. However, limitations of current techniques have hampered our understanding of their dynamics and functional connectivity. In this chapter, we summarize key experiments that led to the key hypothesis that the hypocretin (Hcrt; also known as orexin) system sets the arousal threshold. We discuss our implementation of in vivo optogenetic techniques to overcome previous techniques' limitations and establish causal links between Hcrt neuron activation and behavioral state transitions. Finally, we propose to use optogenetics as a tool to probe the necessity, sufficiency, and connectivity of defined neural circuits in the regulation of sleep and wakefulness.

In mammals, wakefulness, non-rapid eye movement (NREM sleep, also called "slow-wave sleep"), and rapid eye movement (REM) sleep are defined by electroencephalographic (EEG) and electromyographic (EMG) criteria. NREM sleep is characterized by slow oscillations (1–4 Hz) of high amplitude, which reflect increasing depth of sleep. REM sleep (also called "paradoxical" sleep [4]) is a singular behavioral state, characterized by fast oscillations (6–12 Hz) of low amplitude (frequently described as cortical (re)activation) and persistent muscle atonia. Due to this cortical activity, REM sleep has been linked to dreaming activity. Rapid eye movements, as well as fluctuation of the heart and breathing rates, are also associated with REM sleep events. Neuronal circuits that regulate the occurrence and maintenance of wakefulness, and NREM and REM

A. Adamantidis (✉)
McGill University, Department of Psychiatry, Douglas Mental Health University Institute, 6875 LaSalle Blvd, Montréal H4H 1R3, Canada
e-mail: antoine.adamantidis@mcgill.ca

C.R. Baumann et al. (eds.), *Narcolepsy: Pathophysiology, Diagnosis, and Treatment*,
DOI 10.1007/978-1-4419-8390-9_12, © Springer Science+Business Media, LLC 2011

sleep are described in the chapter "The Neurobiology of Sleep-Wake Systems: An Overview" by Luppi.

The Hcrt System as a Modulator of the Sleep–Wake Cycle

In 1998, two groups independently reported the identification of a pair of neuropeptides called the Hcrts exclusively expressed in a population of glutamatergic neurons in the lateral hypothalamus [5, 6]. These peptides bind to two Hcrt receptors distributed throughout multiple brain areas that match Hcrt-containing projections. According to this neuroanatomical distribution, the Hcrt system modulates multiple brain functions, including arousal, stress, and consummatory behaviors [7].

As for other sleep and wake systems, our understanding of the Hcrt modulatory functions stems from experiments that typically involve pharmacology – in which an agonist or antagonist (or Hcrt peptides themselves) is injected into the ventricular system or discrete brain regions – and in vitro and in vivo electrophysiology. Alternatively, many studies employed genetic manipulation of the Hcrt system using transgenic or knockout technologies in rodents. To illustrate the last decade of progress in narrowing down a role for the Hcrt system (Hcrt peptides and receptors) in "setting the arousal threshold" [8], we will summarize supporting experimental evidence and sort them into correlation, loss-of-function, and gain-of-function studies. The neuroanatomical description of the Hcrt system is reviewed in the chapter "The Neurobiology of Sleep-Wake Systems: An Overview" by Luppi.

Correlation Studies Between Hcrt Neural Activities and Behavioral States

A neuronal population that fires shortly before a specific behavior suggests that it is involved in the onset of that specific behavior, whereas a stable discharge during the behavior suggests a participation in its maintenance. In a first attempt to correlate the firing rates of neurons located in the Hcrt-immunoreactive field with state of vigilance, several groups proposed a classification of recorded neurons based on their activity during wake, NREM sleep, REM sleep, or both wake and REM sleep [9]. This suggested the existence of neuronal populations in the hypothalamus that either promote or maintain a particular behavioral state. More recently, technically challenging in vivo single-unit recordings of identified Hcrt neurons confirmed their high discharge activity during arousal elicited by environmental stimuli (e.g., tone) and behavior accompanied with a strong locomotor activity (e.g., goal-oriented behaviors) [10–12]. In contrast to their spontaneous firing in brain slices [13], Hcrt neurons are mostly silent during quiet wakefulness, NREM sleep, and REM sleep and are reactivated during REM sleep-to-wake transitions [10, 11].

Collectively, these results suggest that Hcrt neurons are involved in the transitions from REM sleep to wakefulness when the animal is asleep. When the animal is awake, their activation may participate in neurobiological mechanisms underlying alertness and goal-oriented behaviors. The principal limitation of single-unit recording is the low network resolution. Recording the activity of one cell may not allow extrapolation to neuronal circuit activity, mainly because subpopulations of Hcrt neurons may exist. Hopefully, long-term development of high-resolution bioimaging methods will overcome such issues.

Gain-of-Function Studies

Electrical stimulation of the lateral hypothalamic area decreased REM sleep duration in rats and cats [14], possibly through a Hcrt-mediated inhibition of neurons located in the oral nucleus of the pons neurons [15, 16], a structure that participates in the generation and maintenance of REM sleep. Disinhibition of LH cells by local injection of $GABA_A$ antagonists (bicuculline or gabazine) induced a continuous quiet waking state associated with robust muscle tone in head-restrained rats [17], possibly via activation of

Hcrt neurons [18]. However, none of these approaches have selectively targeted the Hcrt system. To better mimic physiological conditions, intracerebroventricular (icv) infusion of Hcrt peptides or Hcrt agonists [7], as well as local injection of Hcrt peptides in the LC, LH, laterodorsal tegmental nucleus, and basal forebrain structures, enhanced wakefulness and locomotor activity and markedly reduced REM and non-REM sleep in rodents and cats [7]. Local Hcrt-1 injections in the cholinoceptive neurons of the pons (nucleus pontis oralis) promoted wakefulness, and suppressed NREM sleep and REM sleep, whereas it directly inhibited REM sleep when injected in the ventral part of the NPO [7, 19, 20]. Disinhibition of Hcrt neurons using a genetically engineered mouse model (i.e., mice with a selective deletion of $GABA_B$ receptor in Hcrt neurons only) induced severe fragmentation of sleep–wake states during both the light and dark periods, without changes in total sleep time or signs of cataplexy [21].

Interestingly, results from gain-of-function approaches consistently support a role for the Hcrt system in promoting wakefulness and suppressing NREM and REM sleep. However, such methods possess low temporal and spatial resolution. The injected pharmacological compound stays active in the brain for minutes to hours before clearance and may act on different cell types in the vicinity of the injection site. It is, therefore, difficult to conclude a role for Hcrt in maintaining wakefulness from these studies.

Lack-of-Function Studies

Lack-of-function evidence is based on lesion, receptor blockade, and recombinant technologies and represents one of the most common approaches in neurobiology to determine the necessity of a particular system. Although electrical or chemical anatomical lesions of the LH are not specific to the arousal-promoting Hcrt neurons, such lesions were reported to induce prolonged aphagia and adipsia concomitant to a disorganized EEG (i.e., rapid low voltage activity and high voltage low frequency waves) [22]. Sleep and EEG parameters normalized after several days when rats progressively recovered. LH lesions also caused either insomnia [23] or transient hypersomnia [24] with disturbed hippocampal theta activity during both waking and REM sleep [25] in rodents and cats. In humans, lesions of the hypothalamus often disrupt sleep–wake architecture [26], and in some cases, they can increase total sleep time [27] and cause cataplexy [28].

To selectively target the Hcrt system in vivo, saporin-coupled Hcrt molecules have been used. Saporin is a ribosome-inactivating protein that, once it is internalized, kills target cells. Saporin was coupled to Hcrt peptides to kill Hcrt receptor-expressing neurons upon local brain infusion. When injected in the LH, it induced sleep [29] without altering adenosine levels in the basal forebrain [30]. In contrast, it provoked insomnia when injected in the VTA and substantia nigra [31]. Importantly, conjugated saporin showed dose-dependent effects, and higher doses may have nonselectively killed adjacent cells [32]. In such experiments, saline or vehicle is often used as the control condition, instead of nonconjugated saporin, limiting the interpretation of the results. Indeed, although saporin requires its internalization for toxicity, it remains unknown whether nonconjugated saporin has toxic properties when internalized by nontargeted cells.

Hcrt receptor antagonists have been extensively used as an alternative to silence the Hcrt system. Hcrt-R2 antagonist [33] and dual Hcrt receptor antagonists increased both non-REM and REM sleep in rats [34], caused somnolence and increased surrogate markers of REM sleep in dogs, and increased electrophysiological signs of sleep in humans [35]. Hcrt-R1 antagonist delays emergence from anesthesia, without changing anesthetic induction [36]. However, these antagonists differ in their affinities for both Hcrt-R1 and R2. Additionally, they may modulate other receptors when administered in vivo at very high concentrations. Thus, although this pharmacological approach is selective to the Hcrt neurotransmission, its limitations include low temporal resolution and receptor selectivity.

The most compelling lack-of-function evidence comes from the link between deficiency of the Hcrt system and the development of narcolepsy in humans, dogs, and mice [7]. Narcoleptic patients with cataplexy have a marked reduction in *prepro hcrt* gene transcripts in the hypothalamus and barely detectable levels of Hcrt peptides in the cerebrospinal fluid [7]. Doberman narcoleptic dogs with a mutation in the Hcrt-R2, and all genetically engineered rodents with either a deletion of the Hcrt gene, Hcrt-R2 gene, or Hcrt cells exhibit behavioral arrests that resemble cataplexy, the hallmark of narcolepsy [7]. Interestingly, Hcrt administration can reverse behavioral attacks in narcoleptic dogs. However, Hcrt-R2 KO mice are less affected than Hcrt KO mice, suggesting that the altered REM sleep control in narcolepsy–cataplexy syndrome emerges from loss of signaling through both Hcrt-R2-dependent and -independent pathways [37]. Alternative approaches to classic (i.e., nonconditional) genetic technologies include the use of short interfering RNAs (siRNA) targeting prepro-orexin mRNA that have a transient silencing action. Once injected into the rat LH, animals exhibited a transient increase in REM sleep (a few days) compared to scrambled siRNA-treated animals [38]. Although some animals even showed cataplexy-like episodes; wakefulness and NREM sleep were unaffected by the treatment.

Collectively, these lack-of-function studies strongly support a role for the Hcrt system in "lowering the arousal threshold" [8]. However, the variety of impact on the sleep–wake cycle parameters and behavior (e.g., animals presenting "cataplexy-like" attacks) may be due to the low temporal (e.g., mutant model) or spatial (e.g., Hcrt gene knockdown) resolutions and the possible compensatory mechanisms that may occur in KO animals.

Manipulating Hcrt Neuron Activity

Wake and sleep centers, in particular the lateral hypothalamus, encompass neuronal and nonneuronal populations that have distinct biochemical and electrical properties. These include neurons expressing GABA, glutamate, melanin-concentrating hormone (MCH), tyrotropin-releasing hormone (TRH), substance P, or neurotensin [39] as well as a bundle of fibers that cross the hypothalamus (fornix and medial forebrain bundle) (Fig. 1a). Some of these neuronal populations have recently been implicated in the regulation of arousal [40–44], adding a level of complexity to the functional deconstruction of the hypothalamic system. Indeed, electrical microstimulation or lesional approach experiments described above cannot distinguish between cell types in the targeted area and may inadvertently stimulate or damage adjacent cells or fibers. Furthermore, pharmacological infusion of agonists or antagonists (including Hcrt peptides) modulates Hcrt cells and surrounding cells that express a large variety of receptors, including receptors for neurotransmitters, neuromodulators, and neuropeptides, including Hcrt receptors [7].

In addition to the spatial resolution criteria, the temporal resolution is an important parameter to consider when modulating the activity of a given cell type. Lesional (electrical, physical, and saporin-based) approaches and traditional knockout mouse models are irreversible. Genetic engineering of rodent models is frequently associated with the development of compensatory mechanisms due to the absence of the targeted gene from the early embryonic stages. Also, genetic background inherent to the recombinant technologies may have deleterious consequences on the phenotype of interest, although conditional gene targeting and breeding strategies have been proposed to overcome these limitations [45].

Other pharmacological, saporin-based techniques and gene knockdown strategies show a transient modulation when infused in the brain. However, molecules can spread throughout the brain and remain active at the synapse where they may artificially prolong behavioral effect until they are cleared by extra- and intracellular machinery. This is the case for Hcrt peptide infusion strategy which, although reversible, does not reflect physiological release of the peptide by Hcrt neurons in vivo, and weaken conclusions drawn from these experiments.

a - Non-Selective Modulation of Hypothalamic Neurons **b - Optogenetic Stimulation of Hcrt Neurons**

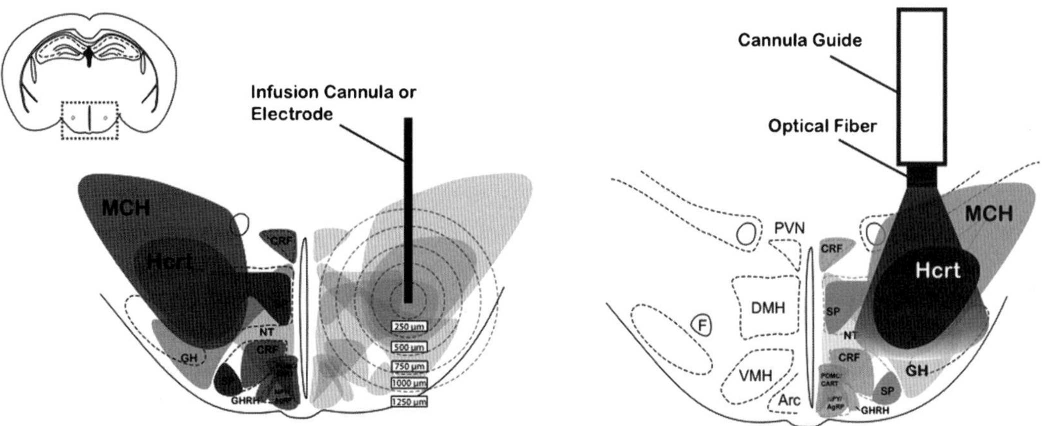

Fig. 1 Comparison between electrical/pharmacological activation or inhibition and optogenetic activation of Hcrt neurons in the lateral hypothalamus. Schematic drawings showing the heterogeneity of neuronal cell types in the vicinity of Hcrt neurons, including MCH, Hcrt, NPY/AgRP, POMC/CART, growth hormone releasing hormone (GHRH), growth hormone (GH), neurotensin (NT), and substance P. Note that, in addition to these defined cell types, the hypothalamus contain glutamatergic, GABAergic neurons and unidentified neurons. For clarity, neurons expressing nesfatin-1, thyrotropin-releasing hormone (TRH), vasopressin, and oxytocin are not represented. (**a**) Limitations of traditional microstimulation or pharmacological techniques to perform loss-of-function or gain-of-function studies include confounding effect on neuronal and nonneuronal cells surrounding the targeted cells. *Concentric circles* indicate the distance from the tip of the electrode or the cannula used for electrical or pharmacological stimulation and inhibition of lateral hypothalamic neurons. For instance, infusion of Hcrt peptide or non-peptide agonists or antagonists can spread up to 1,000 μm away from the infusion site, thereby activating or inhibiting several neuronal populations in addition to the Hcrt neurons. Alternatively, temporally precise microelectrode stimulation cannot distinguish between cell types in the stimulated area, and thus may inadvertently stimulate other, unintended regions that are adjacent to the electrode. (**b**) Optogenetic technology allows selective stimulation (using ChR2) or inhibition (using NpHR) of genetically targeted Hcrt neurons, with no confounding modulation of surrounding cells, or fiber of passage, that may regulate the same brain function

Modern biochemical techniques, including caged-compound, photoactivatable genetically encodable probes, and delivery of ligand-gated ion channel not expressed by the target cells [46], improve the spatial and temporal properties of traditional methods with variable efficiency.

The heterogeneity of cells in the hypothalamus has imposed limitations in interpreting results from traditional loss-of-function or gain-of function studies. Thus, selectively stimulating or inhibiting Hcrt neurons without affecting surrounding cells represents a challenge for scientists interested in hypothalamic functions. Therefore, we have recently used a new technology called optogenetics to overcome this problem and selectively stimulate Hcrt neurons in vivo in freely moving animals to probe their function in sleep. Optogenetic tools combine the high temporal precision of optical stimulation (millisecond timescale) with the spatial precision of a genetically encodable probe [47, 48], allowing the stimulation or inhibition of the complete Hcrt neuron population (Fig. 1b) [49].

Optogenetics uses expression of an ion channel called channelrhodopsin-2 (ChR2) to manipulate the membrane potential and thus induce action potentials in specific populations of neurons. ChR2 is a nonspecific cation channel from the algae *Chlamydomonas reinhardtii* that opens upon optical stimulation with blue light (473 nm) [47, 50]. The spatial resolution of optogenetics is defined by the specific promoter and enhancer elements used to express the ChR2 channel in discrete neuronal populations of the brain. Once functionally expressed, ChR2 can be activated in brain slices or in vivo using a blue light source coupled to optical fibers and implanted into the brain for long-term experiments over days or weeks [49]. A second microbial-based opsin from *Volvox carteri*, called VChR1, now permits the use of a red-shifted wavelength (589 nm) to increase neuronal activity [51]. A third distinct light-sensitive protein from *Natronomonas pharaonis*, called halorhodopsin (NpHR), can be used to inhibit neuronal activity [50]. NpHR is a chloride pump that hyperpolarizes neurons (and thus inhibits neural activity) upon stimulation with yellow light (580 nm).

In addition to a genetically defined spatial resolution, ChR2, VChR1, and NpHR have a millisecond-timescale temporal resolution that allows stimulation or silencing of single action potentials. Hence, optogenetic technology allows for bimodal manipulation of defined neuronal populations with temporal resolutions relevant to in vivo physiological conditions in an awake, behaving animal [49, 52]. One drawback of the optogenetic technique is that cell specificity requires identification of specific promoters or selective Cre recombinase driver mouse lines.

In an attempt to better understand the role of the Hcrt system on arousal, we studied the consequences of Hcrt neuronal activation on NREM and REM sleep transitions to wakefulness using in vivo optogenetics. First, we genetically delivered the light-activated channel ChR2 to Hcrt neurons using a specific Hcrt promoter and a lentivirus approach [49]. Although the lentivirus infected multiple hypothalamic cell types, only the cells with the endogenous cellular machinery to express the Hcrt gene were found to express ChR2. Therefore, deep brain optical stimulation activated only the Hcrt neurons and not the surrounding cells. Second, the fast on–off kinetics of ChR2 was found to induce action potential in brain slices in less than a millisecond in ChR2-expressing Hcrt neurons. Furthermore, deep brain delivery of high frequency light pulse trains activated Hcrt neurons in vivo (as measured by c-Fos immunohistochemistry).

The main breakthrough of using optogenetic strategy is that we can now mimic the physiological range of Hcrt neuronal spiking rate, and thus overcome related limitations of previous techniques (e.g., uncontrolled persistence of Hcrt peptide in the brain after local infusion). In addition, it allows us to probe the behavioral consequences of different optical stimulation patterns (e.g., frequency of firing) of Hcrt neurons. We found that direct, unilateral deep brain optical stimulation of Hcrt neurons in the hypothalamus increased the probability of transition to wakefulness from either NREM or REM sleep [49]. Interestingly, photostimulation using 5–30-Hz light pulse trains reduced latency to wakefulness, whereas 1-Hz trains did not (Fig. 2). These results suggest that (1) Hcrt peptides and other neurotransmitters/modulators (glutamate and dynorphin) are released when Hcrt neurons fire at frequencies higher than 5 Hz, and (2) their release

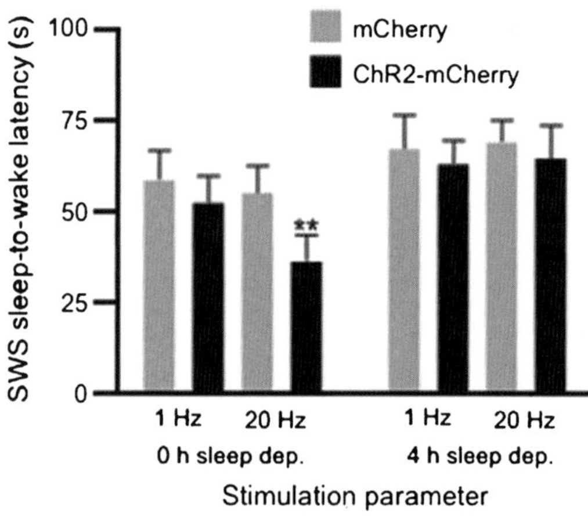

Fig. 2 Optogenetic probing of Hcrt-mediated sleep-to-wake transitions in baseline and sleep pressure conditions. After photostimulation at 20 Hz, the latency to awaken from slow-wave sleep (SWS) was shorter in mice transduced with ppHcrt::ChR2-mCherry ($n=6$) than in controls transduced with ppHcrt::ChR2-mCherry ($n=6$), but this response was absent after 4-h sleep deprivation. Photostimulation was administered as a single 10-s bout at 1 or 20 Hz (15-ms light pulses) under baseline conditions and during the first hour after 4 h of sleep deprivation. Data analysis is based on an average of 30 stimulations per frequency and per mouse during SWS. Latencies are represented as mean ± SEM. $**p<0.0001$ using a two-tailed Student's t-test between mCherry control and ChR2 animals for each frequency. Figure adapted from [53]

is responsible for the arousal-promoting effect with high frequency optical stimulation. This study established a causal link between Hcrt neuron activation and sleep-to-wake transitions, as suggested by previous single-unit recording studies described above [10, 11].

More recently, we examined whether Hcrt-mediated sleep-to-wake transitions are affected by circadian timing and sleep homeostasis. We found that stimulation of Hcrt neurons increased the probability of an awakening event throughout the entire light/dark period with an equal efficiency (*data not shown*) [53]. However, this effect was inversely correlated with sleep pressure induced by sleep deprivation. We found a significant decrease in the latencies to wakening for both sleep deprivation conditions at 20 Hz between *ChR2-mCherry* and *mCherry* control animals that received no sleep deprivation (Fig. 2). Four hours of sleep deprivation resulted in latencies that were not significantly different between animals ($p > 0.05$). These results suggest that the Hcrt system promotes wakefulness throughout the light/dark period by activating multiple downstream targets, which themselves are inhibited with increased sleep pressure. Finally, stimulation of Hcrt neurons was still sufficient to increase the probability of an awakening event in histidine decarboxylase-deficient knockout animals, suggesting that histamine neurons are not the main downstream target of Hcrt-mediated increase of arousal, as suggested previously [54].

In conclusion, these optogenetic studies support a role for the Hcrt system in "setting the arousal threshold" [8], resulting in a facilitation of wakefulness and possibly increased sensitivity to environmental stimuli when animals are asleep. Regarding the implication of the Hcrt system in brain reward and stress [55, 56], it is possible that activation of Hcrt neurons facilitates hyperarousal (defined as a transient "hyper-alertness" state triggered by salient stress or reward) when an animal is actively engaged in goal-oriented behaviors such as food- or drug-seeking behavior (see the chapter "The Hypocretins/Orexins: Master Regulators of Arousal and Hyperarousal" by de Lecea).

Perspectives

In addition to the heterogeneous cell types of the hypothalamus, Hcrt neurons show heterogeneity of functions [7]. Whether Hcrt neurons modulate brain functions separately or whether they integrate them into a coherent behavioral output, as recently suggested for arousal and depression-like symptoms during negative energy balance [7, 57], requires further investigation. To study the multiple brain areas innervated by Hcrt-containing axons, new experimental approaches such as optogenetics may help define which topological projections mediate specific behavioral transitions when the animal is asleep (sleep-to-wake transition) or awake (transition to hyperarousal). This may help define the biological mechanism, circuit dynamics, and neurotransmission imbalance underlying the triggering of cataplexy by emotions.

Milestones in experimental sleep research include the identification of genes, cell populations, and neural circuits regulating the sleep–wake cycle. However, many questions remain unanswered about the neurobiological mechanisms underlying behavioral state onset, maintenance, and stability. In particular, challenging questions in the field of experimental sleep research include how the neural circuits of the brain integrate homeostatic and circadian signals into a stable sleep–wake cycle. What are the kinetics of sleep- and wake-promoting circuits in an unrestrained, behaving animal? How do sleep and wake circuits interact together to stabilize the sleep–wake cycle?

Combining existing techniques discussed in this chapter to genetic tagging [58] and genetic manipulation of neural circuits [46] will help address such questions by expanding the experimental strategies. It will allow in vivo deconstruction of previously inaccessible genetically defined sleep–wake neuronal circuits located in the hypothalamus, basal forebrain, and brainstem with temporal and spatial resolutions relevant to physiological dynamics. Eventually, this should shed light on the functions of sleep and identify new therapeutic targets to cure sleep disorders and sleep-associated neuropsychiatric disorders, including metabolic imbalance, mood-related pathology, and cognitive impairment.

Acknowledgments A.A. is supported by fellowships from the Fonds National de la Recherche Scientifique ("Charge de Recherche"), NIH (K99), and NARSAD. L.d.L. is supported by grants from the National Institute on Drug Abuse, Defense Advanced Research Projects Agency, and National Alliance for Research on Schizophrenia and Depression.

References

1. Siegel JM. Do all animals sleep? Trends Neurosci. 2008;31:208–13.
2. Zimmerman JE, Naidoo N, Raizen DM, Pack AI. Conservation of sleep: insights from non-mammalian model systems. Trends Neurosci. 2008;31:371–6.
3. Pace-Schott EF, Hobson JA. The neurobiology of sleep: genetics, cellular physiology and subcortical networks. Nat Rev Neurosci. 2002;3:591–605.
4. Jouvet M, Michel F. Electromyographic correlations of sleep in the chronic decorticate & mesencephalic cat. C R Seances Soc Biol Fil. 1959;153:422–5.
5. de Lecea L et al. The hypocretins: hypothalamus-specific peptides with neuroexcitatory activity. Proc Natl Acad Sci USA. 1998;95:322–7.
6. Sakurai T et al. Orexins and orexin receptors: a family of hypothalamic neuropeptides and G protein-coupled receptors that regulate feeding behavior. Cell. 1998;92:573–85.
7. Sakurai T. The neural circuit of orexin (hypocretin): maintaining sleep and wakefulness. Nat Rev Neurosci. 2007;8:171–81.
8. Sutcliffe JG, de Lecea L. The hypocretins: setting the arousal threshold. Nat Rev Neurosci. 2002;3: 339–49.
9. Fort P, Bassetti CL, Luppi PH. Alternating vigilance states: new insights regarding neuronal networks and mechanisms. Eur J Neurosci. 2009;29:1741–53.
10. Mileykovskiy BY, Kiyashchenko LI, Siegel JM. Behavioral correlates of activity in identified hypocretin/orexin neurons. Neuron. 2005;46:787–98.
11. Lee MG, Hassani OK, Jones BE. Discharge of identified orexin/hypocretin neurons across the sleep-waking cycle. J Neurosci. 2005;25:6716–20.
12. Takahashi K, Lin JS, Sakai K. Neuronal activity of orexin and non-orexin waking-active neurons during wake-sleep states in the mouse. Neuroscience. 2008;153:860–70.
13. Eggermann E et al. The wake-promoting hypocretin-orexin neurons are in an intrinsic state of membrane depolarization. J Neurosci. 2003;23:1557–62.
14. Suntsova NV, Dergacheva OY, Burikov AA. The role of the posterior hypothalamus in controlling the paradoxical phase of sleep. Neurosci Behav Physiol. 2000;30:161–7.
15. Dergacheva OY, Meyers IE, Burikov AA. Effects of electrical stimulation of the posterior part of the hypothalamus on the spike activity of neurons in the oral nucleus of the pons. Neurosci Behav Physiol. 2005;35:865–70.
16. Nunez A, Moreno-Balandran ME, Rodrigo-Angulo ML, Garzon M, De Andres I. Relationship between the perifornical hypothalamic area and oral pontine reticular nucleus in the rat. Possible implication of the hypocretinergic projection in the control of rapid eye movement sleep. Eur J Neurosci. 2006;24: 2834–42.
17. Goutagny R, Luppi PH, Salvert D, Gervasoni D, Fort P. GABAergic control of hypothalamic melanin-concentrating hormone-containing neurons across the sleep-waking cycle. NeuroReport. 2005;16:1069–73.
18. Lu JW et al. Disinhibition of perifornical hypothalamic neurones activates noradrenergic neurones and blocks pontine carbachol-induced REM sleep-like episodes in rats. J Physiol. 2007;582:553–67.
19. Watson CJ, Soto-Calderon H, Lydic R, Baghdoyan HA. Pontine reticular formation (PnO) administration of hypocretin-1 increases PnO GABA levels and wakefulness. Sleep. 2008;31:453–64.
20. Moreno-Balandran E, Garzon M, Bodalo C, Reinoso-Suarez F, de Andres I. Sleep-wakefulness effects after microinjections of hypocretin 1 (orexin A) in cholinoceptive areas of the cat oral pontine tegmentum. Eur J Neurosci. 2008;28:331–41.
21. Matsuki T et al. Selective loss of GABA(B) receptors in orexin-producing neurons results in disrupted sleep/wakefulness architecture. Proc Natl Acad Sci USA. 2009;106:4459–64.
22. Danguir J, Nicolaidis S. Cortical activity and sleep in the rat lateral hypothalamic syndrome. Brain Res. 1980;185:305–21.
23. Jurkowlaniec E, Pracki T, Trojniar W, Tokarski J. Effect of lateral hypothalamic lesion on sleep-waking pattern and EEG power spectra in the rat. Acta Neurobiol Exp (Wars). 1996;56:249–53.
24. Denoyer M, Sallanon M, Buda C, Kitahama K, Jouvet M. Neurotoxic lesion of the mesencephalic reticular formation and/or the posterior hypothalamus does not alter waking in the cat. Brain Res. 1991;539:287–303.
25. Jurkowlaniec E, Trojniar W, Ozorowska T, Tokarski J. Differential effect of the damage to the lateral hypothalamic area on hippocampal theta rhythm during waking and paradoxical sleep. Acta Neurobiol Exp (Wars). 1989;49:153–69.
26. Cohen RA, Albers HE. Disruption of human circadian and cognitive regulation following a discrete hypothalamic lesion: a case study. Neurology. 1991;41:726–9.
27. Eisensehr I et al. Hypersomnia associated with bilateral posterior hypothalamic lesion. A polysomnographic case study. Eur Neurol. 2003;49:169–72.
28. Schwartz WJ, Stakes JW, Hobson JA. Transient cataplexy after removal of a craniopharyngioma. Neurology. 1984;34:1372–5.

29. Gerashchenko D, Blanco-Centurion C, Greco MA, Shiromani PJ. Effects of lateral hypothalamic lesion with the neurotoxin hypocretin-2-saporin on sleep in Long-Evans rats. Neuroscience. 2003;116:223–35.
30. Murillo-Rodriguez E, Liu M, Blanco-Centurion C, Shiromani PJ. Effects of hypocretin (orexin) neuronal loss on sleep and extracellular adenosine levels in the rat basal forebrain. Eur J Neurosci. 2008;28:1191–8.
31. Gerashchenko D, Blanco-Centurion CA, Miller JD, Shiromani PJ. Insomnia following hypocretin2-saporin lesions of the substantia nigra. Neuroscience. 2006;137:29–36.
32. Waite JJ et al. 192 immunoglobulin G-saporin produces graded behavioral and biochemical changes accompanying the loss of cholinergic neurons of the basal forebrain and cerebellar Purkinje cells. Neuroscience. 1995;65:463–76.
33. Dugovic C et al. Blockade of orexin-1 receptors attenuates orexin-2 receptor antagonism-induced sleep promotion in the rat. J Pharmacol Exp Ther. 2009;330:142–51.
34. Whitman DB et al. Discovery of a potent, CNS-penetrant orexin receptor antagonist based on an *n*, *n*-disubstituted-1,4-diazepane scaffold that promotes sleep in rats. ChemMedChem. 2009;4:1069–74.
35. Brisbare-Roch C et al. Promotion of sleep by targeting the orexin system in rats, dogs and humans. Nat Med. 2007;13:150–5.
36. Kelz MB et al. An essential role for orexins in emergence from general anesthesia. Proc Natl Acad Sci USA. 2008;105:1309–14.
37. Willie JT et al. Distinct narcolepsy syndromes in orexin receptor-2 and orexin null mice: molecular genetic dissection of non-REM and REM sleep regulatory processes. Neuron. 2003;38:715–30.
38. Chen L et al. REM sleep changes in rats induced by siRNA-mediated orexin knockdown. Eur J Neurosci. 2006;24:2039–48.
39. Gerashchenko D, Shiromani PJ. Different neuronal phenotypes in the lateral hypothalamus and their role in sleep and wakefulness. Mol Neurobiol. 2004;29:41–59.
40. Verret L et al. A role of melanin-concentrating hormone producing neurons in the central regulation of paradoxical sleep. BMC Neurosci. 2003;4:19.
41. Adamantidis A et al. Sleep architecture of the melanin-concentrating hormone receptor 1-knockout mice. Eur J Neurosci. 2008;27:1793–800.
42. Hassani OK, Lee MG, Jones BE. Melanin-concentrating hormone neurons discharge in a reciprocal manner to orexin neurons across the sleep-wake cycle. Proc Natl Acad Sci USA. 2009;106:2418–22.
43. Hara J et al. Thyrotropin-releasing hormone increases behavioral arousal through modulation of hypocretin/orexin neurons. J Neurosci. 2009;29:3705–14.
44. Gonzalez JA, Horjales-Araujo E, Fugger L, Broberger C, Burdakov D. Stimulation of orexin/hypocretin neurones by thyrotropin-releasing hormone. J Physiol. 2009;587:1179–86.
45. Wolfer DP, Crusio WE, Lipp HP. Knockout mice: simple solutions to the problems of genetic background and flanking genes. Trends Neurosci. 2002;25:336–40.
46. Luo L, Callaway EM, Svoboda K. Genetic dissection of neural circuits. Neuron. 2008;57:634–60.
47. Boyden ES, Zhang F, Bamberg E, Nagel G, Deisseroth K. Millisecond-timescale, genetically targeted optical control of neural activity. Nat Neurosci. 2005;8:1263–8.
48. Nagel G et al. Light activation of channelrhodopsin-2 in excitable cells of Caenorhabditis elegans triggers rapid behavioral responses. Curr Biol. 2005;15:2279–84.
49. Adamantidis AR, Zhang F, Aravanis AM, Deisseroth K, de Lecea L. Neural substrates of awakening probed with optogenetic control of hypocretin neurons. Nature. 2007;450:420–4.
50. Zhang F et al. Multimodal fast optical interrogation of neural circuitry. Nature. 2007;446:633–9.
51. Zhang F et al. Red-shifted optogenetic excitation: a tool for fast neural control derived from Volvox carteri. Nat Neurosci. 2008;11:631–3.
52. Tsai HC et al. Phasic firing in dopaminergic neurons is sufficient for behavioral conditioning. Science. 2009;324:1080–4.
53. Carter ME, Adamantidis A, Ohtsu H, Deisseroth K, de Lecea L. Sleep homeostasis modulates hypocretin-mediated sleep-to-wake transitions. J Neurosci. 2009;29:10939–49.
54. Eriksson KS, Sergeeva O, Brown RE, Haas HL. Orexin/hypocretin excites the histaminergic neurons of the tuberomammillary nucleus. J Neurosci. 2001;21:9273–9.
55. Winsky-Sommerer R et al. Interaction between the corticotropin-releasing factor system and hypocretins (orexins): a novel circuit mediating stress response. J Neurosci. 2004;24:11439–48.
56. Boutrel B et al. Role for hypocretin in mediating stress-induced reinstatement of cocaine-seeking behavior. Proc Natl Acad Sci USA. 2005;102:19168–73.
57. Lutter M et al. Orexin signaling mediates the antidepressant-like effect of calorie restriction. J Neurosci. 2008;28:3071–5.
58. Livet J et al. Transgenic strategies for combinatorial expression of fluorescent proteins in the nervous system. Nature. 2007;450:56–62.

Hypocretin/Orexin Receptor Functions in Mesopontine Systems Regulating Sleep, Arousal, and Cataplexy

Christopher S. Leonard, Mike Kalogiannis, and Kristi A. Kohlmeier

Keywords
Orexin receptor • Hypocretin receptor • Mesopontine systems • Narcolepsy • Cataplexy • Monoamines

Introduction

It is eminently clear from numerous chapters in this volume that the orexin (hypocretin) neuropeptides are necessary for the normal expression of waking and sleep. However, it remains fundamentally unclear how the absence of signaling by these peptides results in the symptoms of narcolepsy. Which of the many neurons bearing orexin receptors are necessary to sustain normal waking and sleep, and which of the numerous orexin actions are required for these processes? Does the simple loss of orexin's excitatory actions produce narcolepsy, or are there more subtle aspects to the loss of orexin signaling that result in plastic or trophic changes that give rise to the symptoms of narcolepsy and cataplexy?

Clues about the relevant neuronal populations come from the projection pattern of orexin neurons and the distribution of orexin receptors [1, 2], although it is likely that key targets remain to be identified. Structures at the dorsomedial and dorsolateral pontomesencephalic junction receive substantial input and contain a high density of receptors, and were noted early as candidate substrates for mediating the arousal-related functions of orexin [3]. The locus coeruleus (LC) is a prime example as it receives a high density of orexin terminals and expresses a high density of orexin-1 receptor message (OX1R). Neighboring structures including the dorsal raphe (DR), the laterodorsal tegmental (LDT) nuclei, and pedunculopontine tegmental (PPT) nuclei also receive substantial numbers of orexin fibers and express both OX1R and OX2R receptors.

These mesopontine regions and especially the noradrenergic neurons in the LC, the cholinergic neurons in the LDT/PPT, and the serotonergic neurons in the DR have long been linked to regulating global aspects of behavior including arousal, motivation, and reinforcement, the ultradian alternation between non-REM and REM sleep and, in the canine model of narcolepsy, the expression of cataplexy. Moreover, the wake-promoting and REM sleep-inhibiting effects of orexins delivered ICV [4, 5] may in part be mediated by actions in these regions, since microinjections of orexins in the LC [6] and LDT [7] have wake-promoting and REM sleep-suppressing actions. Hence, there has been considerable

C.S. Leonard (✉)
Department of Physiology, New York Medical College, Basic Sciences Building, Valhalla, NY 10595, USA
e-mail: chris_leonard@nymc.edu

C.R. Baumann et al. (eds.), *Narcolepsy: Pathophysiology, Diagnosis, and Treatment*,
DOI 10.1007/978-1-4419-8390-9_13,

interest in elucidating the actions of orexin peptides on the neurons of this region and more recently, in identifying the receptors mediating these actions.

Orexin has Two Distinct, Direct Actions on LDT, DR, and LC Neurons

Orexins Activate a Slow Excitatory Current in LDT and PPT Neurons, Enhance the Ca^{2+} Influx Mediated by L-Type Ca^{2+} Channels in LDT Neurons, and Stimulate Glutamatergic Afferent Neurons to the LDT

Following the discovery that disruption of orexin signaling produces a narcolepsy phenotype with cataplexy [8, 9], we began to examine the actions of orexin peptides using extracellular recording and whole-cell patch clamp methods in LDT neurons using mouse brain slices [10]. We found that orexin peptides have a direct and rather prolonged excitatory action on both cholinergic and non-cholinergic neurons. Indeed, the orexins produced a "noisy" inward current that was accompanied by an increase in membrane conductance which drives a membrane depolarization from resting potential sufficient to stimulate action potentials in quiescent unclamped cells. In a subsequent study, we determined that this direct excitation in LDT cholinergic neurons was primarily mediated by a noisy cation current [11]. Recently, a similar strong excitatory action of orexin-A was identified for rat PPT cholinergic neurons. In PPT, however, orexin-A produces excitation by activating a cation current and by closing K^+ channels [12].

The pioneering studies which identified the orexin peptides and their receptors relied on the fact that expressed orexin receptors mobilize Ca^{2+} from intracellular stores, as is common for receptors that couple to Gq/11 [13]. Since then, evidence from expression systems indicate that in addition to mobilizing Ca^{2+}, orexins at very low concentrations will preferentially stimulate a Ca^{2+} influx pathway mediated by TRP channels [14]. In contrast, studies of native receptors in hypothalamic neurons [15] and LDT and DR neurons [16] indicate that orexin-A elevates intracellular Ca^{2+} by membrane depolarization and the enhancement of Ca^{2+} entry through voltage-gated Ca^{2+} channels. Since this enhancement was abolished by the L-type Ca^{2+} channel antagonist nifedipine and was further enhanced by the agonist Bay-K, native orexin receptors appear to enhance Ca^{2+} influx selectively via L-type Ca^{2+} channels, at least in LDT and DR neurons [11]. Surprisingly, the cation current, which is probably mediated by TRP channels, was not a detectable source of Ca^{2+} entry in these neurons. In fact, the Ca^{2+} influx was enhanced in some neurons that lacked an inward current response and it was more sensitive to PKC inhibition, suggesting that these two actions of orexin-A are independent.

In addition to direct postsynaptic actions, the orexins were originally found to have potent presynaptic effects stimulating both EPSPs and IPSPs [15]. We, therefore, determined the effect of orexins on EPSPs and IPSPs in mouse LDT neurons. We found that orexin-A had a relatively weak mixed effect on spontaneously occurring IPSCs (sIPSCs) recorded in both cholinergic and non-cholinergic neurons. In some cells, it stimulated sIPSCs; in other cases, it reduced sIPSCs; and in others, it had no effect [10, 17]. In contrast, we found that orexin-A strongly stimulated glutamatergic afferent neurons presynaptic to both cholinergic and non-cholinergic LDT neurons. This was manifest as a prolonged increase in the sEPSC frequency and amplitude distributions and it was completely blocked in TTX, indicating that it required action potentials in presynaptic neurons. This finding indicates that orexin receptor activation produces spiking of local (i.e., within the slice and possibly the LDT) glutamate-releasing neurons that contact the recorded LDT neurons. While orexins did not increase the miniature EPSC (mEPSC) frequency, a classical indicator of presynaptic terminal action, orexin-A did enhance the electrically evoked EPSC (eEPSC) consistent with an additional excitatory action at presynaptic, glutamate-releasing terminals.

Collectively, these studies clearly indicate that OXRs are expressed on LDT and PPT cholinergic and non-cholinergic neurons and that one function of these receptors is to produce excitation. This is supported by the weak actions of orexins on GABAergic afferents and the strong stimulatory effect on glutamatergic afferents to LDT neurons. Another orexin receptor function revealed by these studies is the enhancement of Ca^{2+} influx produced by L-channels. The function of this action is not yet clear, but it interesting to note that L-channel signaling is consistently linked to changes in gene expression [18] (see below).

In DR Neurons, Orexins Activate a Noisy Cation Current, Enhance Ca^{2+} Influx Through L-Type Ca^{2+} Channels, and Trigger Endocanabinoid Signaling to Suppress Glutamate Synaptic Inputs

It has been established by several laboratories that orexin peptides directly excite serotonergic and GABAergic DR neurons [11, 19, 20]. This effect is mainly mediated by a noisy cation current, similar to that observed in LDT neurons. In addition, orexin-A also elevates intracellular Ca^{2+} in DR neurons – an effect mediated by the activation of voltage-gated Ca^{2+} channels through depolarization and by the enhancement of Ca^{2+} influx through L-type Ca^{2+} channels [11]. Interestingly, in recordings conducted under identical conditions, the inward current elicited by 300 nM orexin-A was ~2–3 times larger in DR neurons than in LDT neurons, indicating that the same input would produce greater excitation of DR neurons.

In addition to producing a direct excitation of serotonergic DR neurons, orexins stimulate local GABAergic inputs to these neurons [19] and also stimulate retrograde endocanabinoid signaling to dampen excitatory glutamatergic inputs [21]. In this elegant series of experiments, orexin-B was shown to decrease the mEPSC frequency and decrease the evoked EPSCs in serotonergic DR neurons via pathways sensitive to phospholipase C and DAG lipase signaling, but insensitive to the elevation of intracellular Ca^{2+} in DR neurons. These findings suggest that along with producing direct excitation, orexin signaling is bidirectional and may activate feedback inhibitory processes to stabilize firing of DR neurons.

In LC Neurons, Orexins Produce an Inward Current, Enhance Ca^{2+} Influx Through L-Type Ca^{2+} Channels, Enhance Firing Synchrony, and Trigger Somatodendritic Release of Norepinephrine

Studies in brain slices and dissociated LC neurons have shown that orexins directly excite and increase firing of noradrenergic (NA) LC neurons by activation of a cation current and the closure of K^+ channels [22–24]. Orexin-A also increases intracellular Ca^{2+} in LC neurons by a plasma membrane Ca^{2+} influx rather than by releasing Ca^{2+} from intracellular stores (see below). Like in LDT and DR neurons, part of this influx results from the enhancement of Ca^{2+} influx through L-type Ca^{2+} channels. Thus, L-type Ca^{2+} channels appear to be a common effector for orexin signaling.

In very young rodents, orexin-B also promotes synchronization of firing via gap junctions that are prevalent early in development. This suggests that orexin actions in LC neurons might influence aspects of early development since the LC establishes an early innervation of the forebrain and NA has trophic influences on synapse formation and dendritic growth [24].

LC neurons have also been shown to secrete NA from somatodendritic membrane in addition to synaptic terminals. This secretion is Ca^{2+} dependent and can be triggered by high firing rates. High K^+, NMDA, and orexin-A application all stimulate this release and it appears that orexin-A enhances this secretion by potentiating Ca^{2+} influx through NMDA receptors [25]. Since the liberated NA can activate somatodendritic autoreceptors, it is possible that this provides another inhibitory feedback pathway to limit LC firing and possibly the activity of LC afferents. Collectively, these data indicate that like LDT and DR neurons, orexins both directly excite LC neurons and enhance Ca^{2+} influx via L-type Ca^{2+} channels. Moreover, like DR neurons, but unlike

LDT neurons, orexins activate a mechanism that can dampen excitability of LC neurons.

Which Receptors Are Necessary for the Postsynaptic Actions of Orexins?

While considerable evidence for the function of each orexin receptor is available from studies using expression systems, relatively little information has been available for native orexin receptors. This is hampered by the lack of soluble, specific antagonists and by response (tachyphylaxis), making dose–response curves unreliable, at least under whole-cell conditions. This is also complicated because the two known orexin receptors have overlapping distributions and may be coexpressed in some neurons [26, 27]. We have utilized a genetic approach in which we assayed orexin-A actions in brain slices from mice lacking one or both of the two known receptors. Importantly, this genetic approach does not reveal the normal function of the remaining receptors since compensatory mechanisms may result in gain or loss of function. Rather, this approach reveals the residual capacity of the entire system without the targeted receptor. In the case of orexin receptors, this is not necessarily bad, since behavioral phenotypes of interest exist even though compensation may have taken place.

In the Absence of OX2Rs, Orexin-A Mediates Direct Excitation and Enhances Ca^{2+} Transients in LDT, DR, and LC Neurons

Using whole-cell patch clamp recordings with Ca^{2+} imaging, we examined the ability of orexin-A to activate the two postsynaptic actions described in WT mice in LDT, DR, and LC neurons (Fig. 1) [28]. In the absence of OX2Rs, 300 nM orexin-A produced both inward current and enhanced Ca^{2+} transients evoked by voltage jumps from −60 to −30 mVs, as observed for neurons studied in slices from wild-type mice in each nucleus. In the LDT, orexin-A also produced an increase in the frequency of sEPSCs (Fig. 1a, left panel). Moreover, we found that the same effectors activated by orexin-A in WT mice were activated by OX1Rs: a noisy current in LDT and DR neurons, and enhanced Ca^{2+} influx via L-type Ca^{2+} channels in LDT, LC, and DR neurons. These data suggest that signaling through the remaining OX1Rs is sufficient to support these two basic actions of orexin-A without a significant decrement in amplitude.

In the Absence of OX1Rs, Orexin-A Excitation Is Absent in LDT and LC Neurons and Attenuated in DR Neurons, While Orexin-A Enhancement of Ca^{2+} Transients Remains Intact

In similar studies using slices from OX1R–/– mice, we found that orexin-A (300 nM) failed to activate inward currents in LDT or LC neurons, suggesting an essential role for OX1Rs in these neurons. However, OX2Rs alone were clearly able to activate a noisy cation current in DR neurons, although it was smaller on average than in slices from WT or OX2R–/– mice. Thus, either receptor *can* activate a noisy cation current but only in DR neurons do both receptors activate this current. One explanation might be that LDT neurons and LC neurons lack functional OX2Rs. However, we found neurons in all three nuclei where OX2R activation alone enhanced Ca^{2+} influx produced by depolarization to the same extent as found in WT and OX2R–/– slices. Thus, functional OX2Rs are present even in the LC, but they do not mediate the depolarizing effects of orexin-A in LDT and LC neurons.

In order to better estimate the fraction of cells in each nucleus that responded to orexin-A with Ca^{2+} elevation, we bulk-loaded neurons in slices with Fura2-AM to record many cells simultaneously. We found that in slices from WT mice, about 70% of the fura2-AM-labeled cells in the LDT and DR responded to 300 nM orexin-A, while about 60% of cells in the LC responded. These fractions were significantly reduced in slices from OX1R–/– mice. Orexin-A produced

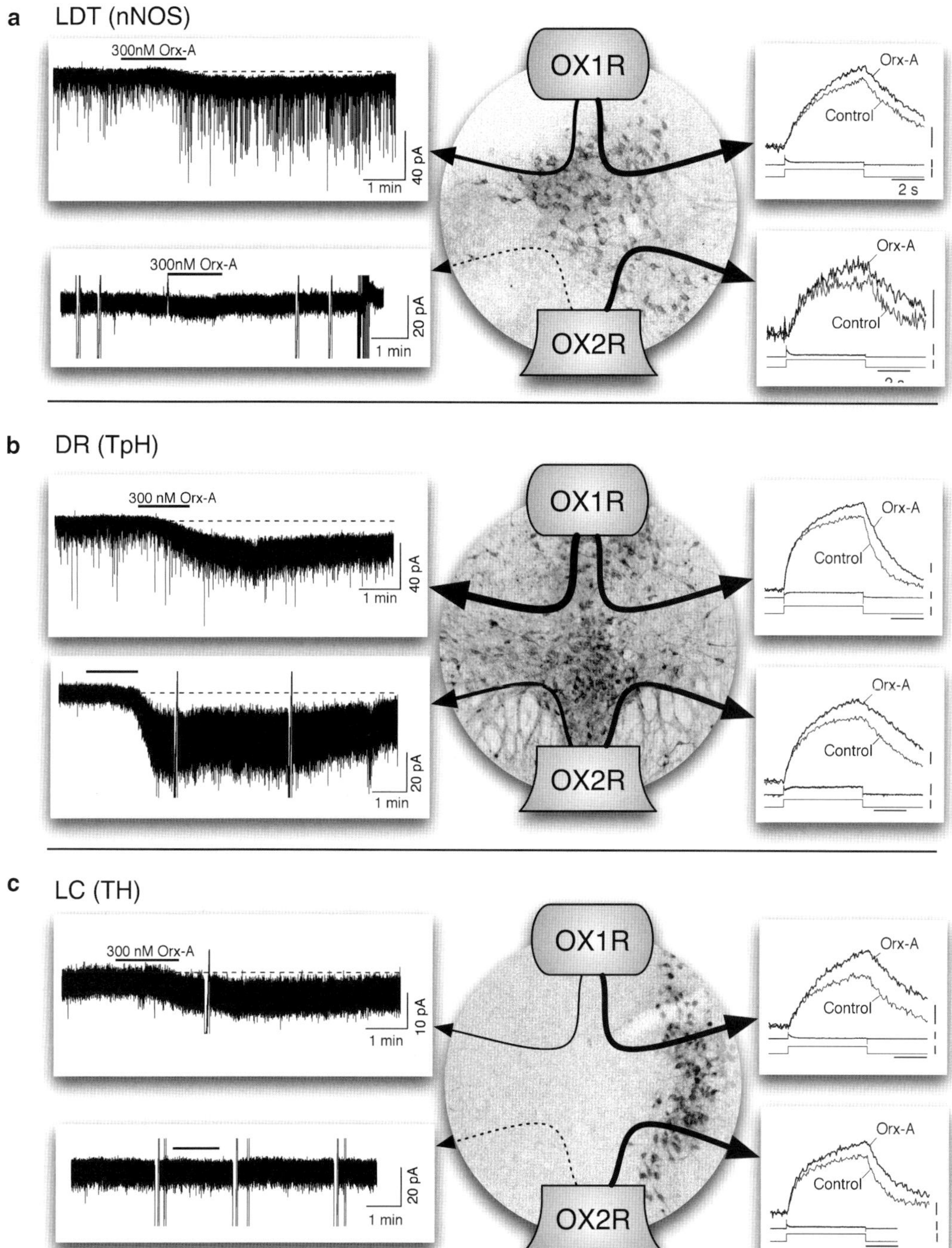

Fig. 1 Genetic dissection of orexin receptor actions in LDT, DR, and LC neurons. (**a**) In LDT neurons, activation of OX1Rs (in OX2R knockout mice) produces direct excitation by stimulating inward current and indirect excitation by stimulating glutamate sEPSCs. Activation of OX1Rs also enhances voltage-dependent Ca^{2+} influx. Activation of OX2Rs (in OX1R knockout mice) fails to produce either direct or indirect excitation, but enhances Ca^{2+} influx. (**b**) In DR neurons, activation of OX1Rs produces direct excitation by stimulating inward current and enhances voltage-dependent Ca^{2+} influx. Activation of OX2Rs stimulates a similar inward current and enhances voltage-dependent Ca^{2+} influx. (**c**) In LC neurons, activation of OX1Rs alone produces direct excitation by stimulating a small inward current and enhances voltage-dependent Ca^{2+} influx. Activation of OX2Rs alone fails to produce excitation, but enhances voltage-dependent Ca^{2+} influx

Ca^{2+} transients in only ~20% of cells in the LDT and only ~10% of cells in the LC, while in the DR, ~50% of labeled cells still responded. Thus, the loss of OX1Rs significantly attenuates the responsiveness of these structures.

In the Absence of Both Orexin Receptors, Orexin-A Excitation and Ca^{2+} Transient Enhancement Are Abolished

To interpret our results from single receptor knockouts, it is essential to know whether orexin-A, at the doses used, acts only through these receptors. We therefore tested the effect of orexin-A on slices from mice lacking both receptors. Consistent with our interpretation, orexin-A (300 nM) did not evoke membrane currents in whole-cell records or produce Ca^{2+} transients in fura-2 AM-loaded LDT, DR, and LC slices from double receptor knockout (DKO) mice. These cells were nonetheless viable since glutamate evoked large Ca^{2+} transients in each case.

Collectively, these data indicate that orexin signaling is essentially unimpaired in the LDT, DR, or LC in the absence of OX2Rs, but that in the absence of OX1Rs, orexin excitation is attenuated in DR neurons and abolished in LDT and LC neurons. In spite of this disruption, orexin modulation of Ca^{2+} influx persists via remaining OX2Rs.

Implications for Narcolepsy/Cataplexy

In addition to the implications for the cellular functions of native orexin receptors, these data have implications for the roles of these structures in narcolepsy. Orexin-deficient mice display cataplexy, sleep attacks, and the inability to maintain long bouts of waking and sleep [8, 29, 30] as do mice lacking both OXRs [31] (or see also below). Hence, the absence of orexin signaling in the LDT, DR, and LC in DKO mice is consistent with a role for this signaling in suppressing the symptoms of narcolepsy. However, in OX2R–/– mice, narcolepsy symptoms such as sleep/wake fragmentation and sleep attacks persist and appear similar to those in orexin peptide knockouts [29]. Since orexin-A actions are preserved in the LDT, DR, and LC of these mice, direct orexin signaling at these loci must contribute little to prolonging spontaneous bouts of waking and sleep or to suppressing sleep attacks, unless of course, orexin peptide release itself is deficient in OX2R–/– mice. This is unlikely since cataplexy, a hallmark of orexin peptide knockouts and DKOs, is only rarely seen in OX2R–/– mice. Available evidence from OXR1–/– also supports this view since these animals do not have sleep attacks or sleep/wake fragmentation [32], yet direct orexin-A excitation is absent in LDT and LC neurons and impaired in DR neurons. Thus, intact direct orexin peptide excitation in these structures is neither necessary nor sufficient to sustain normal spontaneous sleep/wake bouts duration or to suppress sleep attacks. Of course, direct orexin signaling at these structures may be important in other types of arousal processes (e.g., food-motivated arousal).

Our findings are consistent with a role for direct orexin excitation in the LDT, DR, and LC in suppressing cataplexy since OX2R–/– mice rarely express cataplexy. One possibility is that residual OXR1 signaling in the LDT, DR, or LC contributes to suppressing its expression. Nevertheless, the loss of orexin excitation in the LDT and LC is not sufficient to produce cataplexy since OX1R–/– mice do not express it. Perhaps residual OX2R excitation in the DR and/or preserved modulation of Ca^{2+} influx by orexin prevents the establishment of this behavior.

Orexin Receptor Knockout Mice Express Enhanced Cholinergic Properties in Laterodorsal Tegmental Neurons

A consistent finding from our Ca^{2+} imaging studies of LDT, DR, and LC neurons is that both OXRs enhance Ca^{2+} influx through L-type Ca2+ channels via a PKC-dependent mechanism. The consequences of this Ca^{2+} influx are unknown, although Ca^{2+} influx via L-type Ca^{2+} channels is consistently related to alterations of gene expression and plasticity [18]. This suggested to us that the loss of orexin signaling might drive neuronal adaptations that promote a narcolepsy phenotype. Since the neuropharmacology of narcolepsy points to an imbalance between central cholinergic and monoaminergic mechanisms [33], we examined whether the constitutive absence of orexin signaling is reflected by differences in the cholinergic properties of LDT cholinergic neurons – some of which innervate pontine regions linked with REM sleep induction and muscle atonia. Specifically, we compared between DKO and WT mice the number of cholinergic neurons in the LDT and the relative expression of cholinergic markers using RT-PCR [34]. We measured choline acetyltransferase (ChAT), which synthesizes ACh; the vesicular acetylcholine transporter (VAChT), which concentrates ACh into vesicles; the choline transporter (CHT1), which mediates high affinity choline uptake; and acetylcholinesterase (AChE), which hydrolyzes ACh.

For each gene, we estimated the initial template amount per unit sample RNA obtained from tissue punches centered on the LDT and motor nucleus of the fifth nerve (Mo5) from six independent experiments (Fig. 2). We found that ChAT, VAChT, and CHT1 message levels were two- to fourfold higher in the DKO samples. In contrast, no differences for AChE or cyclophilin levels were observed.

Interestingly, no significant differences were found when samples were isolated from whole brainstems. Since additional pools of cholinergic neurons were included in these samples, this suggests that the differences were localized to mesopontine regions. Indeed, levels for ChAT and VAChT were also elevated in DKO punch samples from the Mo5.

To test whether corresponding protein levels were also elevated, we compared Western blots prepared from whole brainstem, thalamus (which receives its major cholinergic input from the LDT/PPT), and cortex (which receives its major cholinergic input from the basal forebrain groups). While we were able to detect both ChAT and CHT1, we did not detect a systematic difference between genotypes, suggesting either that higher message levels were not translated into protein or that the differences were too localized to detect. We therefore compared ChAT immunoperoxidase and immunofluorescence labeling in LDT and Mo5 neurons from WT and DKO brains.

We first compared the numbers of LDT and Mo5 neurons between genotype by plotting and counting labeled somata in three matched sections (30-μm thick) chosen from the rostral, middle, and caudal portions of each nucleus. We found that the average number of ChAT+LDT and Mo5 neurons was not different between genotypes.

We next determined whether the intensity of ChAT staining was different between WT mice and DKO mice. Tissues from single WT and DKO mouse were processed together as pairs to limit staining variability ($n=7$ mice for each genotype). In the first series, we compared the optical density (OD) of ChAT+somata labeled by immunoperoxidase (Fig. 2b, c). DKO neurons were darker and there was little overlap in their distributions. In three of four cases, the average OD was greater in LDT neurons from the DKO mice. Identical measures of Mo5 neurons from the same sections showed no difference. Moreover, in the one LDT case where the average OD was lower in the DKO section, OD was also lower in the Mo5, suggesting that staining conditions might explain this variation. To determine if OD differences could be artifacts from the peroxidase reaction, we also conducted indirect ChAT immunofluorescence using sections from another three pairs of WT and DKO mice (Fig. 2d). In all three pairs, average somatic fluorescence was greater in neurons from the DKO mice, while there was no genotypic difference for Mo5 neurons.

From these data, we conclude that while the number of cholinergic neurons is not different in the absence of orexin signaling, there is an upregulation in the machinery necessary for acetylcholine synthesis and neurotransmission in a restricted population of mesopontine cholinergic neurons. Since a subpopulation of these neurons are associated with the production of REM sleep and cataplexy, we hypothesize that the capacity for cholinergic transmission is increased in these

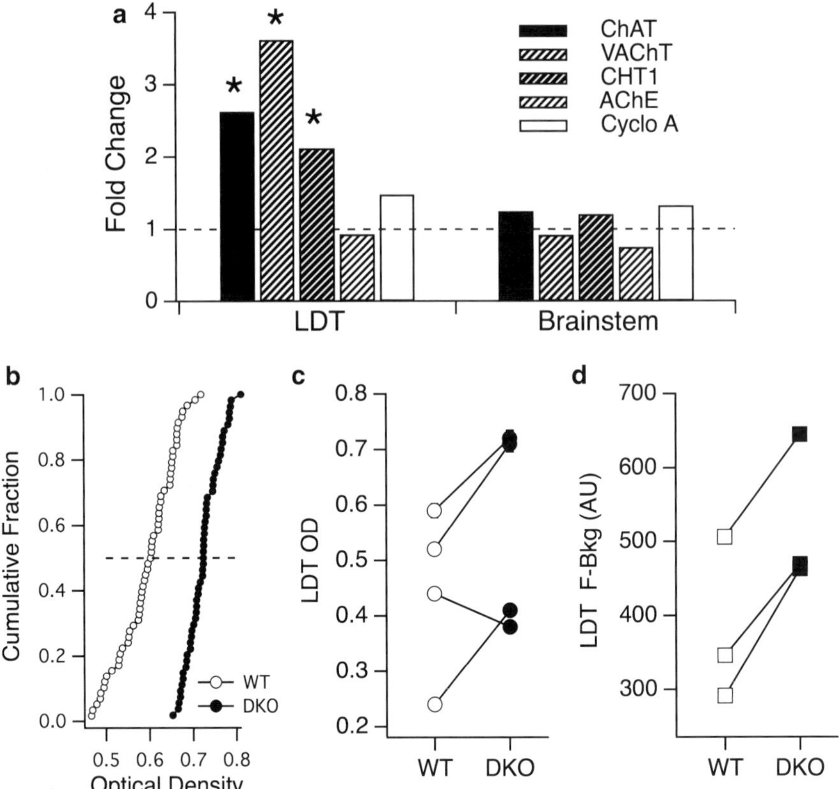

Fig. 2 Narcoleptic double orexin receptor knockout mice express enhanced cholinergic properties in LDT neurons. (**a**) LDT message levels for ChAT, VAChT, and CHT1 were approximately two- to fourfold higher in DKO samples than in WT samples. These differences have a restricted localization since message levels were not different in isolates from entire brainstems. Measurements are displayed as the ratio between the average DKO and WT values, and *asterisks* denote that the values were significantly different between genotype as determined by a *t*-test. (**b**) The optical density distribution of ChAT immunoperoxidase-labeled LDT neurons from DKO mice (*filled symbols*) was greater than that from WT mice. (**c**) Optical density (mean ± SEM) of LDT neurons from four experiments (1 DKO and 1 WT each). In three of the four experiments, ODs were higher in neurons from DKO mice. (**d**) Background-corrected ChAT fluorescence (mean ± SEM) from DKO and WT mice. Three of three pairs showed greater ChAT immunofluorescence in LDT neurons from DKO mice

neurons and that this promotes cataplexy by enhancing cholinergic outflow to REM atonia generating targets in the pontine reticular formation.

How these differences come about remains to be determined; however, it is intriguing to think that orexin receptor signaling may play a role in regulating neurotransmitter expression either developmentally or in an on-going manner. A consistent finding from developmental studies of the sympathetic nervous system and spinal cord is that decreasing Ca^{2+} influx or the frequency of Ca^{2+} spiking favors the cholinergic phenotype, while increasing Ca^{2+} influx suppresses or restricts this phenotype [35]. The importance of Ca^{2+} influx via NMDA receptors and L-type Ca^{2+} channels has also been recently illustrated in the developing hypothalamus, where persistent blockade of either of these Ca^{2+} influx pathways produces a dramatic increase in the number of cholinergic neurons and spontaneous cholinergic synaptic transmission [36, 37]. As indicated above, orexin receptors regulate both these pathways in LDT cholinergic neurons by stimulating glutamatergic synaptic inputs and increasing intracellular Ca^{2+} via direct depolarization and enhancement of Ca^{2+} entry through L-type Ca^{2+} channels. Attenuation of activity in these pathways is, therefore, a strong candidate mechanism for inducing these expression changes.

Implications for Narcolepsy/Cataplexy

Studies of narcoleptic canines have led to a consensus that an imbalance between cholinergic and monoaminergic transmission contributes importantly to the generation of narcolepsy/cataplexy [38]. Among the several changes that have been noted, a large increase in ACh release in the pons was observed during epochs of cataplexy provoked by the food elicited cataplexy test. Importantly, this elevation was not produced during epochs of locomotor activity or feeding without cataplexy, and neither was it produced in normal dogs under similar testing conditions [39]. Our findings suggest that this increase could be mediated by an increased capacity to synthesize, package, and release ACh by LDT neurons – some of which project to this region of the reticular formation (for example, see [40]). Our findings also indicate that mouse cholinergic LDT neurons are not increased in number and agree with a prior study [38] that was unable to replicate an earlier report of more cholinergic neurons in the LDT/PPT from narcoleptic canines [41]. Collectively, these data from narcoleptic DKO mice provide a plausible mechanism for key neurochemical events underlying cataplexy in the narcoleptic canines.

Can These Implied Changes in Brainstem Cholinergic Transmission Influence the Expression of Cataplexy in Mouse Models?

Based on studies of canine narcolepsy, it is a plausible but untested supposition that enhanced cholinergic transmission in the pontine reticular formation promotes cataplexy in mouse models of narcolepsy. We addressed this question [42] first with systemic injections of physostigmine and then by microinfusions of neostigmine targeted to the nucleus pontis oralis (nPO), a region where neostigmine induces REM sleep-like signs in several species including mouse [43], presumably by enhancing cholinergic output from cholinergic afferents from the LDT/PPT [44].

Under our home cage observation conditions (3 h starting 30 min before lights out), DKO mice displayed unambiguous behavioral arrests that were abrupt in onset and offset, were preceded and followed by purposeful behavior, and were variable in duration and frequency. In a group of 20 DKO mice, four had no arrests in this period, while three exhibited over 20 arrests. The mean number of arrests was 10 ± 2, with a mean arrest duration of 42 ± 4 s per mouse. Such arrests were never seen in WT mice ($n=6$). As expected, the number of arrests increased in the dark. The average time spent in arrests was ~20 s for the first 30 min and this increased to ~90 s per 30 min epoch in the last 90 min of observation.

To test whether globally enhancing cholinergic transmission altered arrest expression, we injected each animal with saline, or one of three doses of physostigmine IP (0.01, 0.03, and 0.08 mg/kg) 30 min prior to monitoring. In the first hour of monitoring, the two lowest doses produced a dramatic increase in time spent in arrests (Fig. 3a). The average time spent in arrests increased from 37 ± 14 s (saline) to 175 ± 74 s (0.01 mg/kg) and 174 ± 51 s (0.03 mg/kg:) following physostigmine injection ($p<0.05$, $n=6$). Consistent with the relatively short time course of central physostigmine action, there were no differences in the time spent in arrests at later times. Strikingly, this early increase was absent following 0.08 mg/kg physostigmine (28 ± 15 s), which appeared to suppress arrests even after the lights went out.

The increase in the average time spent in arrest was primarily due to a greater number of arrests, which increased from ~2 per mouse in the first recording hour following saline injection, to over 5 following physostigmine administration (0.01 and 0.03 mg/kg). Indeed, the distributions of arrest bout duration were not significantly different (K–S test, $p>0.05$), even though some longer arrests occurred following physostigmine (Fig. 3a right; squares vs. triangles). These long arrests were within the range of arrest durations observed over the total recording time following saline injections (thick line). This suggests that the longer arrests following physostigmine were simply a consequence of there being more arrests in the first hour. Collectively, these data indicate that arrest likelihood was increased by low doses of physostigmine, but that arrest dynamics were not altered.

We next examined the effect of atropine (0.5 mg/kg, IP). This strongly decreased the time spent in arrests over the last 2 h of recording (Fig. 3b). The average time decreased from 353 ± 77 s (saline) to 172 ± 74 s (atropine; $p < 0.05$, $n = 10$ DKO and 7 WT mice). This decrease was also attributable to fewer arrests rather than a change in bout duration, as indicated by the overlapping distributions (Fig. 3b right; K–S test, $p > 0.05$).

In a final series of experiments, we prepared DKO ($n = 8$) and WT mice ($n = 7$) for microinjection of cholinergic reagents into the pontine reticular formation. We then monitored behavior as in the previous experiments, starting 30 min after a single 50-nl microinjection of either artificial cerebrospinal fluid (aCSF), neostigmine (62.5 μM), or neostigmine + atropine (+100 μM atropine). We found that microinjections of neostigmine produced a clear increase in arrests, with the average time in arrests increasing from 68 ± 28 s (aCSF) to 231 ± 55 s (neostigmine; $p < 0.05$, $n = 8$ DKO and 7 WT mice; Fig. 3c). As indicated by the cumulative distributions (Fig. 3c), this was due to an increased number of arrests rather than a change in bout durations (K–S test, $p > 0.05$). Following identical injections with neostigmine + atropine, there was no increase in the time spent in arrests, indicating that muscarinic receptors are necessary for this effect (Fig. 3c). Importantly, identical injections in WT mice did not provoke behavioral arrests or increase the epochs of quiescence associated with sleep.

Collectively, these data indicate that augmenting cholinergic transmission systemically or locally at pontine targets of mesopontine cholinergic neurons promotes the expression of cataplexy-like behavioral arrests in DKO mice and that muscarinic receptor blockade prevents these actions.

Implications for Narcolepsy/Cataplexy

These findings have several implications for mechanisms underlying narcolepsy and cataplexy. First, they indicate that in the absence of orexin signaling, cholinergic transmission has an enhanced ability to activate muscle atonia circuits, since potentiating endogenous cholinergic transmission produces arrests in DKO but not WT mice. This is consistent with both a greater capacity for cholinergic outflow from mesopontine systems and a greater sensitivity at their targets. Second, these findings support the idea that brain cholinergic systems can both promote and suppress REM sleep signs such as postural atonia, since low doses of physostigmine increased behavioral arrests, while a higher dose suppressed them in DKO mice. Third, these findings indicate that brain cholinergic systems promote the switching into the behavioral arrests of narcolepsy but do not stabilize these arrests, since enhanced cholinergic transmission increased arrest number but not duration. Finally, these data suggest that the enhanced capacity for cholinergic transmission, implied by our expression study (see above), promotes the expression of cataplexy. This supports the idea that a functionally important adaptation to the loss of orexin signaling in mesopontine cholinergic systems is an increase in the capacity for cholinergic transmission, even though there is a loss of direct excitation.

Acknowledgments This study was supported by National Institutes of Health Grants HL64150 and NS27881. We would like to thank Drs. John Edwards, Iryna Gumenchuk, Masaru Ishibashi, Morten Kristensen, and Christopher Tyler along with Ms. Emily Hsu for their contributions to the experiments described in this chapter. We would also like to thank Dr. Masashi Yanagisawa and his colleagues for their contributions to this work, including engineering the knockout mouse lines.

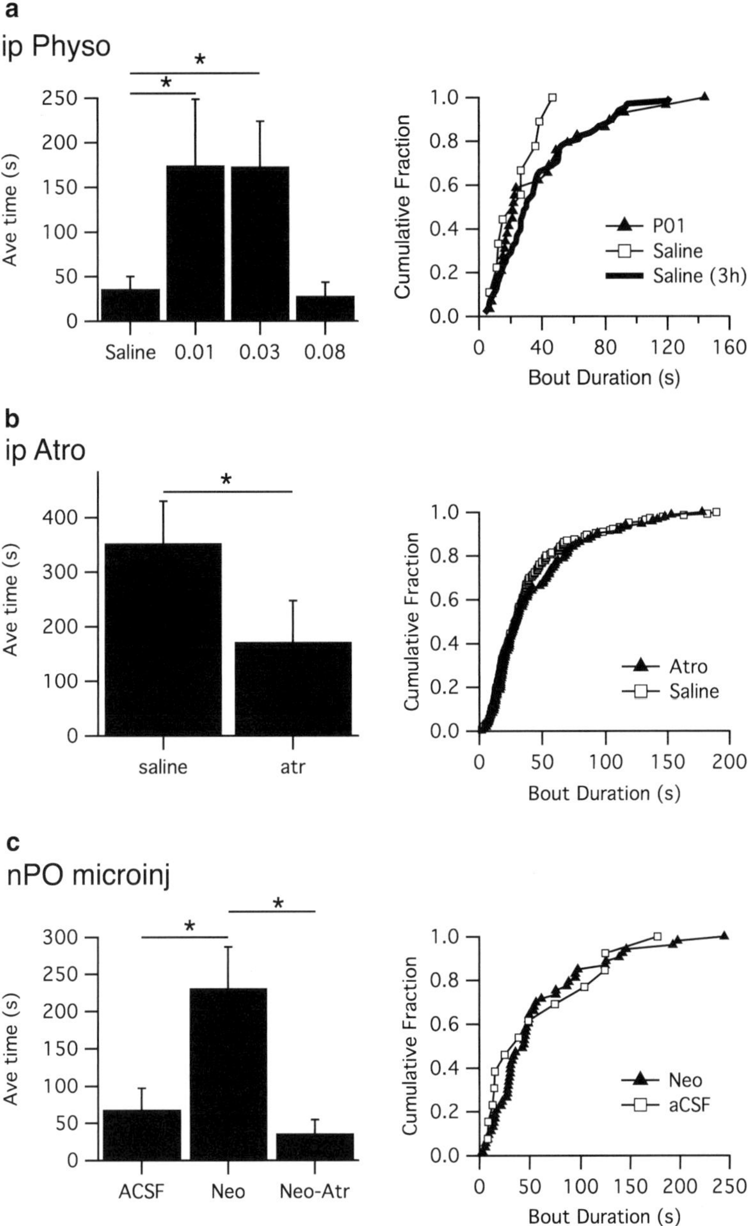

Fig. 3 Cholinergic modulation of behavioral arrests in DKO mice. (**a**) Low doses of physostigmine increased the time spent in behavioral arrests in the first hour of monitoring (*left*). Corresponding cumulative distribution of arrest durations (*right*). *Open symbols*: arrests after saline; *closed symbols*: arrests after physostigmine (0.01 mg/kg; both indicate arrests over the first recording hour); and *thick line*: arrests after saline from entire 3-h monitoring period. (**b**) Atropine decreased the time spent in arrests measured in the last 90 min of monitoring (*left*). Corresponding cumulative distribution of arrest durations (*right*). *Open symbols*: arrests after saline; *closed symbols*: arrests after atropine. (**c**) Microinjections of neostigmine increased the time in arrests, while microinjections of neostigmine + atropine did not (*left*). Corresponding cumulative distributions of arrest durations were not different (*right*)

References

1. Peyron C, Tighe DK, van den Pol AN, et al. Neurons containing hypocretin (orexin) project to multiple neuronal systems. J Neurosci. 1998;18(23): 9996–10015.
2. Marcus JN, Aschkenasi CJ, Lee CE, et al. Differential expression of orexin receptors 1 and 2 in the rat brain. J Comp Neurol. 2001;435(1):6–25.
3. Kilduff TS, Peyron C. The hypocretin/orexin ligand-receptor system: implications for sleep and sleep disorders. Trends Neurosci. 2000;23(8):359–65.
4. Hagan JJ, Leslie RA, Patel S, et al. Orexin A activates locus coeruleus cell firing and increases arousal in the rat. Proc Natl Acad Sci USA. 1999;96(19):10911–6.
5. Piper DC, Upton N, Smith MI, Hunter AJ. The novel brain neuropeptide, orexin-A, modulates the sleep-wake cycle of rats. Eur J Neurosci. 2000;12(2): 726–30.
6. Bourgin P, Huitron-Resendiz S, Spier AD, et al. Hypocretin-1 modulates rapid eye movement sleep through activation of locus coeruleus neurons. J Neurosci. 2000;20(20):7760–5.
7. Xi M, Morales FR, Chase MH. Effects on sleep and wakefulness of the injection of hypocretin-1 (orexin-A) into the laterodorsal tegmental nucleus of the cat. Brain Res. 2001;901(1–2):259–64.
8. Chemelli RM, Willie JT, Sinton CM, et al. Narcolepsy in orexin knockout mice: molecular genetics of sleep regulation. Cell. 1999;98(4):437–51.
9. Lin L, Faraco J, Li R, et al. The sleep disorder canine narcolepsy is caused by a mutation in the hypocretin (orexin) receptor 2 gene. Cell. 1999;98(3):365–76.
10. Burlet S, Tyler CJ, Leonard CS. Direct and indirect excitation of laterodorsal tegmental neurons by Hypocretin/Orexin peptides: implications for wakefulness and narcolepsy. J Neurosci. 2002;22(7): 2862–72.
11. Kohlmeier KA, Watanabe S, Tyler CJ, Burlet S, Leonard CS. Dual orexin actions on dorsal raphe and laterodorsal tegmentum neurons: noisy cation current activation and selective enhancement of Ca transients mediated by L-type calcium channels. J Neurophysiol. 2008;100:2265–81.
12. Kim J, Nakajima K, Oomura Y, Wayner MJ, Sasaki K. Electrophysiological effects of orexins/hypocretins on pedunculopontine tegmental neurons in rats: an in vitro study. Peptides. 2009;30(2):191–209.
13. Sakurai T, Amemiya A, Ishii M, et al. Orexins and orexin receptors: a family of hypothalamic neuropeptides and G protein-coupled receptors that regulate feeding behavior. Cell. 1998;92(4):573–85.
14. Larsson KP, Peltonen HM, Bart G, et al. Orexin-A-induced Ca^{2+} entry: evidence for involvement of TRPC channels and protein kinase C regulation. J Biol Chem. 2005;280(3):1771–81.
15. van den Pol AN, Gao XB, Obrietan K, Kilduff TS, Belousov AB. Presynaptic and postsynaptic actions and modulation of neuroendocrine neurons by a new hypothalamic peptide, hypocretin/orexin. J Neurosci. 1998;18(19):7962–71.
16. Kohlmeier KA, Inoue T, Leonard CS. Hypocretin/orexin peptide signaling in the ascending arousal system: elevation of intracellular calcium in the mouse dorsal raphe and laterodorsal tegmentum. J Neurophysiol. 2004;92(1):221–35.
17. Leonard CS, Tyler CJ, Burlet S, Watanabe S, Kohlmeier KA. Hypocretin/Orexin actions on mesopontine cholinergic systems controling behavioral state. In: de Lecea L, Sutcliffe JG, editors. Hypocretins: integrators of physiological functions. New York: Springer; 2005. p. 153–68.
18. West AE, Chen WG, Dalva MB, et al. Calcium regulation of neuronal gene expression. Proc Natl Acad Sci USA. 2001;98(20):11024–31.
19. Liu RJ, van den Pol AN, Aghajanian GK. Hypocretins (orexins) regulate serotonin neurons in the dorsal raphe nucleus by excitatory direct and inhibitory indirect actions. J Neurosci. 2002;22(21):9453–64.
20. Brown RE, Sergeeva OA, Eriksson KS, Haas HL. Convergent excitation of dorsal raphe serotonin neurons by multiple arousal systems (orexin/hypocretin, histamine and noradrenaline). J Neurosci. 2002; 22(20):8850–9.
21. Haj-Dahmane S, Shen RY. The wake-promoting peptide orexin-B inhibits glutamatergic transmission to dorsal raphe nucleus serotonin neurons through retrograde endocannabinoid signaling. J Neurosci. 2005; 25(4):896–905.
22. Horvath TL, Peyron C, Diano S, et al. Hypocretin (orexin) activation and synaptic innervation of the locus coeruleus noradrenergic system. J Comp Neurol. 1999;415(2):145–59.
23. Murai Y, Akaike T. Orexins cause depolarization via nonselective cationic and K+ channels in isolated locus coeruleus neurons. Neurosci Res. 2005;51(1): 55–65.
24. Van Den Pol AN, Ghosh PK, Liu RJ, Li Y, Aghajanian GK, Gao XB. Hypocretin (orexin) enhances neuron activity and cell synchrony in developing mouse GFP-expressing locus coeruleus. J Physiol. 2002; 541(Pt 1):169–85.
25. Chen XW, Mu Y, Huang HP, et al. Hypocretin-1 potentiates NMDA receptor-mediated somatodendritic secretion from locus ceruleus neurons. J Neurosci. 2008;28(12):3202–8.
26. Korotkova TM, Sergeeva OA, Eriksson KS, Haas HL, Brown RE. Excitation of ventral tegmental area dopaminergic and nondopaminergic neurons by orexins/hypocretins. J Neurosci. 2003;23(1):7–11.
27. Eriksson KS, Sergeeva O, Brown RE, Haas HL. Orexin/hypocretin excites the histaminergic neurons of the tuberomammillary nucleus. J Neurosci. 2001;21(23):9273–9.
28. Kohlmeier KA, Tyler CJ, Kalogiannis M, et al. Genetic dissection of orexin receptor functions in brainstem cholinergic and monoaminergic neurons: implications for orexinergic signaling in arousal and narcolepsy. (Submitted)

29. Willie JT, Chemelli RM, Sinton CM, et al. Distinct narcolepsy syndromes in orexin receptor-2 and orexin null Mice. Molecular genetic dissection of non-REM and REM sleep regulatory processes. Neuron. 2003;38(5):715–30.
30. Mochizuki T, Crocker A, McCormack S, Yanagisawa M, Sakurai T, Scammell TE. Behavioral state instability in orexin knock-out mice. J Neurosci. 2004; 24(28):6291–300.
31. Kisanuki YY, Chemelli RM, Tokita S, Willie JT, Sinton CM, Yanagisawa M. Behavioral and polysomnographic characterization of orexin-1 receptor and orexin-2 receptor double knockout mice. Sleep. 2001;24(Abstract Supplement):A22.
32. Kisanuki YY, Chemelli RM, Sinton CM, Williams SCR, Richardson JA, Hammer RE, et al. The role of orexin receptor type-1 (OX1R) in the regulation of sleep. Sleep. 2000;23 Suppl 2:A91.
33. Nishino S, Mignot E. Pharmacological aspects of human and canine narcolepsy. Prog Neurobiol. 1997;52(1):27–78.
34. Kalogiannis M, Grupke SL, Potter PE, et al. Narcoleptic orexin receptor knockout mice express enhanced cholinergic properties in laterodorsal tegmental neurons. Eur J Neurosci. 2010;32(1):130–42.
35. Spitzer NC, Root CM, Borodinsky LN. Orchestrating neuronal differentiation: patterns of Ca^{2+} spikes specify transmitter choice. Trends Neurosci. 2004; 27(7):415–21.
36. Belousov AB, O'Hara BF, Denisova JV. Acetylcholine becomes the major excitatory neurotransmitter in the hypothalamus in vitro in the absence of glutamate excitation. J Neurosci. 2001;21(6):2015–27.
37. Belousov AB, Hunt ND, Raju RP, Denisova JV. Calcium-dependent regulation of cholinergic cell phenotype in the hypothalamus in vitro. J Neurophysiol. 2002;88:1352–62.
38. Tafti M, Nishino S, Liao W, Dement WC, Mignot E. Mesopontine organization of cholinergic and catecholaminergic cell groups in the normal and narcoleptic dog. J Comp Neurol. 1997;379(2):185–97.
39. Reid MS, Siegel JM, Dement WC, Mignot E. Cholinergic mechanisms in canine narcolepsy-II. Acetylcholine release in the pontine reticular formation is enhanced during cataplexy. Neuroscience. 1994;59(3):523–30.
40. Semba K. Aminergic and cholinergic afferents to REM sleep induction regions of the pontine reticular formation in the rat. J Comp Neurol. 1993;330:543–56.
41. Nitz D, Andersen A, Fahringer H, Nienhuis R, Mignot E, Siegel J. Altered distribution of cholinergic cells in the narcoleptic dog. Neuroreport. 1995;6(11): 1521–4.
42. Kalogiannis M, Hsu E, Willie JT, et al. Cholinergic modulation of narcoleptic attacks in double orexin receptor knockout mice. PLoS One. 2011;6(4): e18697. doi:10.1371/journal.pone.0018697.
43. Lydic R, Douglas CL, Baghdoyan HA. Microinjection of neostigmine into the pontine reticular formation of C57BL/6J mouse enhances rapid eye movement sleep and depresses breathing. Sleep. 2002;25(8): 835–41.
44. Lydic R, Baghdoyan HA. Pedunculopontine stimulation alters respiration and increases ACh release in the pontine reticular formation. Am J Physiol. 1993; 264(3):R544–54.

Afferent Control of the Hypocretin/ Orexin Neurons

Thomas S. Kilduff, Junko Hara, Takeshi Sakurai, and Xinmin Xie

Keywords
Orexin • Hypocretin • Afferent control • Hypothalamus • GABA • Nociceptin

Since the discovery of the hypocretins (Hcrt) [1] or orexins [2] and the subsequent link to narcolepsy in animals [3, 4] and humans [5, 6] a decade ago, it has become apparent that the Hcrt system plays a central role in a number of physiological and behavioral functions. The well-established clinical symptomatology of narcolepsy including excessive daytime sleepiness and cataplexy gave rise to the hypothesis that the Hcrt system was important for the maintenance of wakefulness [7]. Although subsequent research has suggested that Hcrt activity may be related to wakefulness associated with motivated behaviors rather than wakefulness per se [8], other studies have demonstrated that the Hcrt system also has a role in other functions such as energy metabolism, reward, and addiction [9, 10].

Given the importance of the Hcrt system in such functions, understanding the control of the Hcrt neurons is of considerable interest. Afferent inputs to the Hcrt neurons have been mapped using conventional anterograde and retrograde neuroanatomical mapping techniques [11] as well as by a genetically encoded tracer [12]. These studies have revealed abundant projections from the allocortex, claustrum, lateral septum, bed nucleus of the stria terminalis, central nucleus of the amygdala (CeA), and hypothalamic regions such as the preoptic area, dorsomedial arcuate nucleus, lateral hypothalamus, and posterior hypothalamus to the Hcrt neurons. Projections from the brainstem appear to be relatively weaker, but significant inputs were found from the periaqueductal gray matter, dorsal and median raphe nuclei, and lateral parabrachial nucleus. Most of these projections were confirmed by anterograde tracing, with projections from the lateral septum, preoptic area, and posterior hypothalamus being particularly strong. Whereas hypothalamic regions preferentially innervate the medial and perifornical parts of the Hcrt neuronal field, projections from the brainstem generally target the lateral part of the field [11].

Although the inputs described above provide anatomical substrates whereby the Hcrt cells may be influenced, determination of the neurochemical identity, the postsynaptic effects, and the functional roles of such afferent inputs require the use of other techniques. Since the rodent brain contains fewer than 5,000 Hcrt neurons that are

T.S. Kilduff (✉)
Biosciences Division, SRI International, Menlo Park, CA 94025, USA
e-mail: thomas.kilduff@sri.com

C.R. Baumann et al. (eds.), *Narcolepsy: Pathophysiology, Diagnosis, and Treatment*,
DOI 10.1007/978-1-4419-8390-9_14, © Springer Science+Business Media, LLC 2011

not tightly clustered [7, 13], addressing such issues presents interesting challenges. Fortunately, research in this area has been facilitated by the introduction of a number of transgenic mouse strains, particularly those produced by the Sakurai laboratory. For example, mice expressing the enhanced green fluorescent protein under control of the *hcrt/orexin* promoter (*orexin/EGFP* mice) have facilitated in vitro studies of putative neurochemical inputs to the Hcrt neurons [14, 15]. *Orexin/YC2.1* mice, in which the calcium-sensing protein Yellow Chameleon 2.1 is under control of the *hcrt/orexin* promoter, have provided another tool to measure Hcrt neuron activation [16]. Finally, *orexin/ataxin-3* mice, a model of human narcolepsy in which the Hcrt neurons degenerate, have facilitated in vivo studies of Hcrt neuron function [17]. The following sections provide some examples that illustrate how such mouse strains have been used to advance our understanding of the Hcrt system.

Neuromodulatory Control of the Hcrt Neurons

As indicated above, study of the intrinsic properties of the Hcrt neurons and the neurotransmitters/neuromodulators to which they respond has been facilitated by the production of transgenic *orexin/EGFP* mice [14, 15, 18]. Table 1 provides a summary of the neurotransmitters/neuromodulators known to excite the Hcrt neurons based on whole-cell patch-clamp recordings of the Hcrt neurons in hypothalamic slices from these mice. Such cellular electrophysiological studies have revealed that Hcrt cells are excited by glutamate [14, 18], ghrelin [15], low glucose levels [15, 19, 20], ATP [21], corticotrophin releasing factor (CRF) [22], glucagon-like peptide 1 [23], cholecystokinin [16], neurotensin [16], vasopressin and oxytocin [16, 24].

Among these substances, the response of Hcrt neurons to CRF is of particular significance. In collaboration with the laboratories of Luis de Lecea and Tamas Horvath, we showed by both light and electron microscopy that CRF neurons contact Hcrt cells. Whole-cell patch-clamp recordings demonstrated that Hcrt neurons are excited by CRF through the CRF-R1. Moreover, both restraint and footshock stress activated Hcrt neurons in vivo, and this activation was absent in CRF-R1 knockout mice [22]. These results suggest that a CRF-to-Hcrt projection may underlie the alerting component of the anxiogenic response [22, 25]. Although the source of the CRF input is currently unknown, one possibility is the CRF-positive neurons in the CeA. Interestingly, the Hcrt1 (orexin-A) peptide excites a subset of CeA neurons that are characterized by low threshold burst firing [26]. Thus, a neural substrate for a CRF-mediated positive feedback circuit exists between these two brain regions that, when triggered by fear, may result in hyperactivation of the brain's arousal pathways. Neurobiological and clinical evidence implicates this amygdalo-hypothalamic axis in "fine tuning" of arousal states. We hypothesize that excessive activation of ascending monoaminergic systems as a consequence of mutually enforcing positive feedback in the amygdalo-hypothalamic circuit may result in hyperarousal and heightened startle and fear responses in conditions such as posttraumatic stress disorder. In this regard, a recent paper that implicates this system in panic anxiety is of interest [27].

Table 1 Excitatory modulators of hypocretin/orexin neurons

Substance	Receptor	Reference(s)
Glutamate	AMPA-Rs, NMDA-Rs, mGluRs	[14, 18]
Ghrelin	GHS-R	[15]
Low glucose		[15, 19, 20]
Cholecystokinin	CCK-A	[16]
Neurotensin	Unknown	[16]
Vasopressin	V1a	[16, 24]
Oxytocin	V1a	[16, 24]
Glucagon-like peptide	Unknown	[23]
CRF	CRF-R1	[22]
mACh (27% of cells)	M3	[12]
ATP	P2X	[21]
TRH	TRH-R1	[46, 47]

Table 2 indicates that Hcrt neurons are inhibited by GABA [14, 18, 28], norepinephrine, dopamine and epinephrine [14, 18, 29, 30], serotonin [14, 18, 31], neuropeptide Y [32], leptin [15], high glucose levels [15, 19, 20], adenosine [33], cannabinoids [34], dynorphin [35], Met-enkephalin [36], and nociceptin/ophanin FQ (N/OFQ) [37]. The first two studies that utilized *orexin/EGFP* mice determined that GABA inhibited Hcrt neurons through a $GABA_A$ receptor [14, 15, 18]. Using whole-cell patch-clamp recordings of Hcrt neurons, we subsequently showed that bath application of GABA or muscimol caused an early hyperpolarization mediated by Cl^- and a late depolarization mediated by the efflux of bicarbonate [28]. These $GABA_A$ receptor-mediated responses were blocked by picrotoxin and bicuculline. Under conditions of $GABA_A$ receptor blockade, GABA produced consistent hyperpolarization, decreased firing rate, and reduced input resistance in Hcrt neurons; the selective $GABA_B$ agonist (R)-baclofen caused a similar response. The (R)-baclofen effects were blocked by the $GABA_B$ antagonist CGP 52432 but persisted in the presence of tetrodotoxin, suggesting direct postsynaptic effects. The existence of $GABA_B$ modulation was supported by $GABA_{B(1)}$ subunit immunoreactivity on Hcrt cells colabeled with antisera to the Hcrt-2 peptide. Furthermore, $GABA_B$ receptor activation inhibited the presynaptic release of both glutamate and GABA: (R)-baclofen depressed the amplitude of evoked excitatory postsynaptic currents (EPSCs) and inhibitory synaptic currents (IPSCs), and also decreased the frequency of both spontaneous and miniature EPSCs and IPSCs with a modest effect on their amplitudes. These data suggested that, in addition to regulation through $GABA_A$ receptors, $GABA_B$ receptors modulate Hcrt neuronal activity via both pre- and postsynaptic mechanisms.

Table 2 Inhibitory modulators of hypocretin/orexin neurons

Substance	Receptor	Reference(s)
High glucose		[15]
GABA	$GABA_A$, $GABA_B$	[14, 18, 28]
Serotonin	5HT1A	[14, 18, 31]
Noradrenaline	α2	[14, 18, 29, 30]
Dopamine	α2	[14, 18, 29, 30]
Neuropeptide Y	Y1	[32]
Leptin	OB-R	[15]
mACh (6% of cells)	Unknown	[12]
Adenosine	A1R	[33]
Nociceptin (N/OFQ)	NOP	[37]
Cannabinoids	CB1R	[34]
Dynorphin	κ-opioid	[35]
Met-enkephalin	μ- and ∂-opioid	[36]
MCH	MCH-R1	[58]

The above results suggest that the promotion of nonrapid eye movement sleep through $GABA_B$ receptors may be through the Hcrt system. This hypothesis was recently tested in conditional knockout mice in which the $GABA_{B1}$ gene was specifically eliminated from Hcrt neurons ($GABA_{B1}^{flox/flox}$; *orexin-Cre* mice). In $GABA_{B1}^{flox/flox}$; *orexin-Cre* brain slices, the absence of $GABA_B$ receptors decreased the sensitivity of Hcrt neurons to both excitatory and inhibitory inputs [38]. This change in sensitivity was apparently due to augmented $GABA_A$-mediated inhibition that increases the membrane conductance and shunts postsynaptic currents (PSCs) in Hcrt neurons. Although neither the total amounts nor the distribution of sleep and wakefulness across the 24-h period were changed, $GABA_{B1}^{flox/flox}$; *orexin-Cre* mice exhibited severe fragmentation of sleep/wake states during both the light and dark periods (Fig. 1). The mean episode duration of both sleep and wake bouts was reduced in these mice, indicating a problem in the maintenance of sustained sleep or wake bouts. Thus, $GABA_B$ receptors on Hcrt neurons appear to be crucial for appropriate control of Hcrt tone throughout sleep and wakefulness and appear to play a role in consolidating sleep/wake bouts by stabilizing sleep/wake state transitions.

Nociceptin/Orphanin FQ: A Peptidergic Inhibitor of Hcrt Neuron Activity

Despite the extensive information on the response of Hcrt neurons to various neurotransmitters and neuromodulators summarized in Tables 1 and 2,

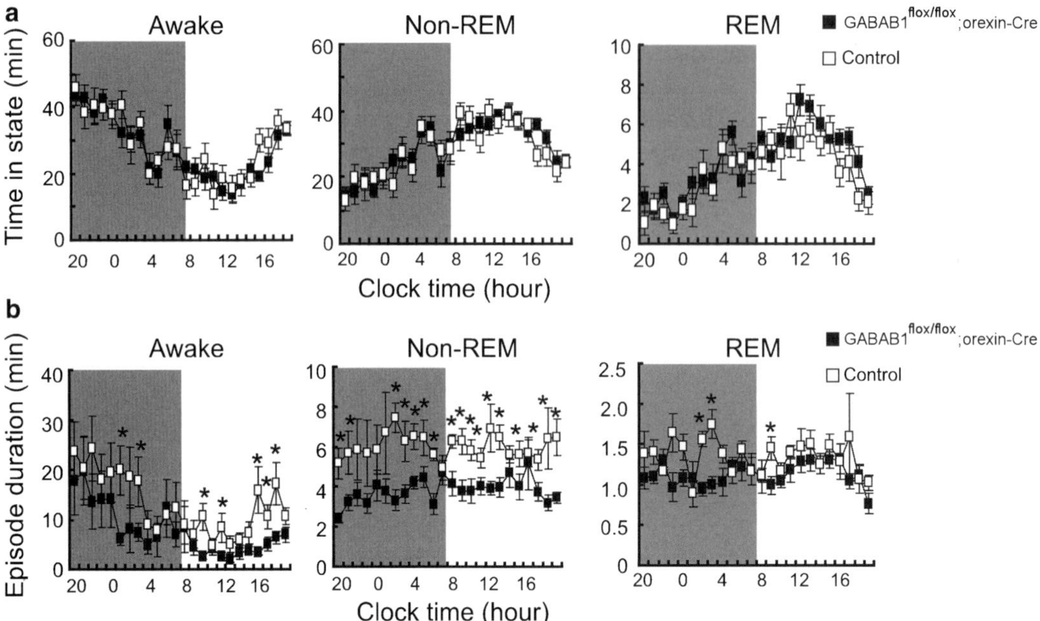

Fig. 1 $GABA_{B1}^{flox/flox}$; *orexin-Cre* mice have normal distribution of sleep and wakefulness but it is highly fragmented. Hourly amounts (**a**) and average episode duration (**b**) of awake, non-REM, and REM sleep states (mean ± SE) plotted over 24 h for control littermate ($n=11$) and $GABA_{B1}^{flox/flox}$; *orexin-Cre* mice ($n=10$). Data for the dark and light phases are displayed on *light gray* and *white backgrounds*, respectively. *$P<0.05$ by Student's *t*-test. Adapted from Matsuki et al. [37]

there have been few cases in which the functional significance of information obtained in vitro has been assessed in vivo [15, 22, 24, 31, 37]. Our experimental approach involves a combination of neuroanatomical, in vitro physiological, and in vivo behavioral studies using genetically modified mice, as exemplified by our studies on the N/OFQ and thyrotropin-releasing hormone (TRH) input to the Hcrt cells described in the following sections.

As indicated above, CRF provides excitatory input to the Hcrt neurons through the CRF-R1. One behavioral response in which CRF has been implicated is stress-induced analgesia (SIA) and SIA is blunted in Hcrt null mutant mice [39]. Since N/OFQ also blocks SIA [40, 41], we hypothesized that N/OFQ-mediated inhibition of Hcrt cells may have particular functional significance to counterbalance excitatory input from CRF neurons in SIA and other conditions [37]. N/OFQ cell bodies are widely distributed throughout the CNS and are abundant in the hypothalamus [42] as is its receptor, NOP [43]. In support of our hypothesized N/OFQ-Hcrt interaction, we showed at both the light and electron microscopic level that N/OFQ terminals innervate Hcrt neurons [37]. Application of N/OFQ to Hcrt cells in vitro in the presence of tetrodotoxin resulted in a dose-dependent hyperpolarization that was blocked by a NOP antagonist [37]. N/OFQ also indirectly modulated the activity of Hcrt cells by inhibition of presynaptic glutamate release, the inverse of CRF effects on Hcrt neurons [37]. Moreover, whereas CRF facilitates SIA, N/OFQ reduces SIA and the presence of Hcrt neurons is essential for SIA. Accordingly, we hypothesize that CRF and N/OFQ coordinately regulate the activity of the Hcrt neurons in SIA. The proposed interaction among these systems. Although currently supported only by data on SIA, the proposed interaction among these systems may have implications for other functions in which Hcrt neurons are involved.

Thyrotropin-Releasing Hormone: A Peptidergic Activator of Hcrt Neuron Activity

TRH is a tripeptide that has previously been shown to promote wakefulness [44] and to induce arousal from hibernation [45]. Expression of TRH receptor 1 (TRH-R1) is enriched in the tuberal and lateral hypothalamic area (LHA), where Hcrt neurons are located. Since the Hcrt system is implicated in sleep/wake control, we hypothesized that TRH provides excitatory input to the Hcrt cells [46].

Effect of TRH on Hcrt neurons. In vitro electrophysiological studies showed that bath application of TRH caused concentration-dependent membrane depolarization, decreased input resistance, and increased firing rate of identified Hcrt neurons (Fig. 2). In the presence of tetrodotoxin, TRH-induced inward currents that were associated with a decrease in frequency, but not amplitude, of miniature PSCs. Ion substitution experiments suggested that the TRH-induced inward current was mediated in part by Ca^{2+} influx. Although TRH did not significantly alter either the frequency or amplitude of spontaneous excitatory PSCs, TRH (100 nM) doubled the frequency of spontaneous inhibitory PSCs without affecting the amplitude of these events (Fig. 3), indicating increased presynaptic GABA release onto Hcrt neurons. In contrast, TRH significantly reduced the frequency, but not amplitude, of miniature

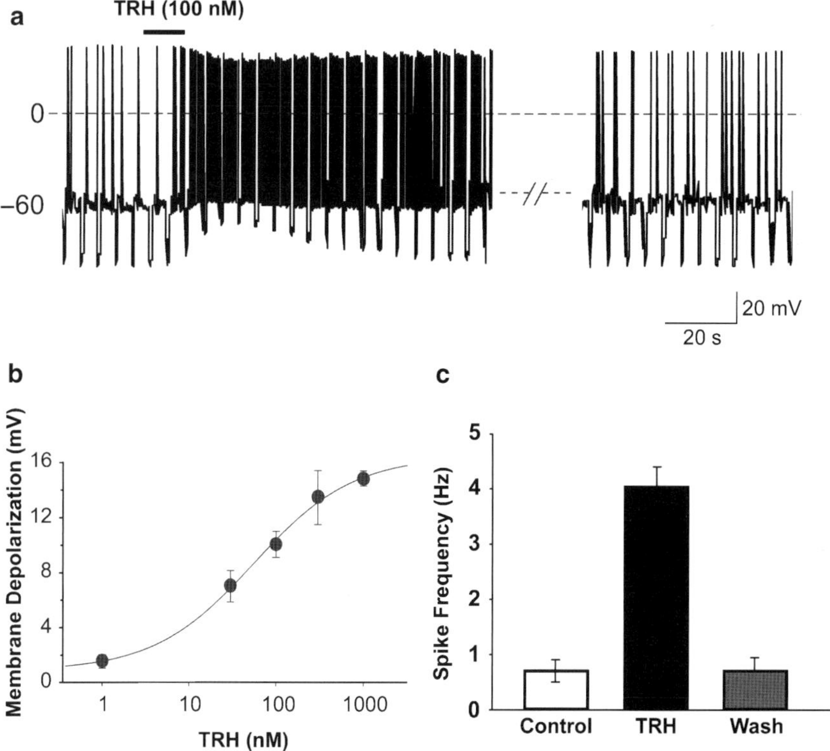

Fig. 2 TRH depolarizes hypocretin neurons in the LHA. (**a**) Current-clamp recording showing the excitatory effect of 100 nM TRH on the firing of action potentials of a hypocretin neuron. Membrane potential was adjusted to −60 mV by DC current injection; hyperpolarizing current pulses (−0.3 nA, 800 ms) were delivered every 5 s. (**b**) Concentration-dependence of the TRH effect on membrane potential. The IC_{50} value was estimated at 66 nM (Hill coefficient = 1.0). Data are mean ± SEM (n = 2–15 determinations per concentration). (**c**) The effects of TRH on spike frequency of Hcrt neurons are reversible after washout. Error bars indicate SEM (n = 5). Reproduced with permission from Hara et al. [46]

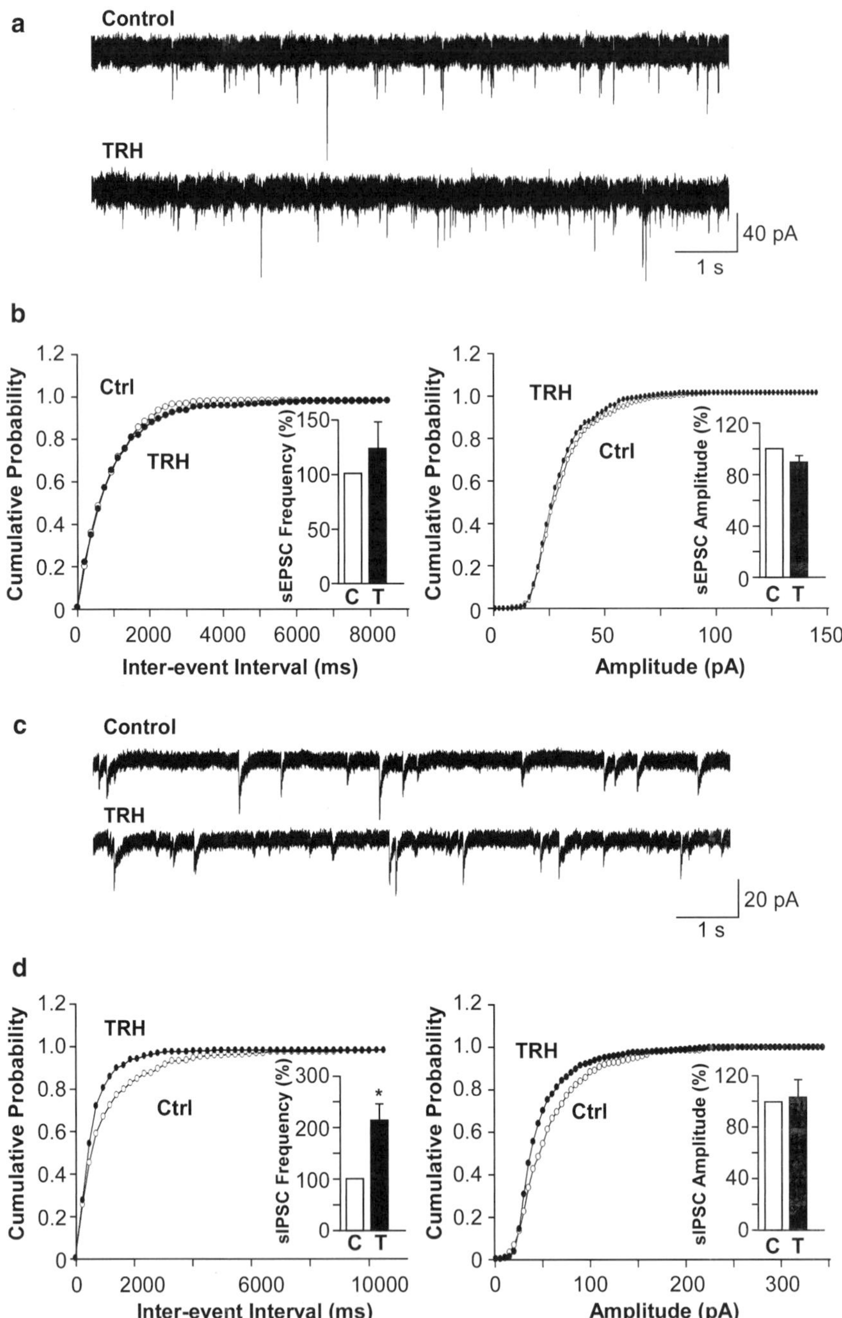

Fig. 3 TRH increases the frequency of sIPSCs, but not sEPSCs, recorded in hypocretin neurons. (**a**) Spontaneous EPSCs (downward deflections from baseline) were recorded by whole-cell voltage-clamp at a holding potential of −60 mV in the presence of bicuculline (40 μM). (**b**) TRH (100 nM) did not affect the frequency or amplitude of sEPSCs ($n=8$). (**c**) Spontaneous IPSCs (downward deflections) were recorded at a holding potential of −60 mV in the presence of AP-5 (50 μM) and DNQX (20 μM). (**d**) TRH doubled the frequency of sIPSCs ($p=0.027$) without affecting sIPSC amplitude ($n=14$). Reproduced with permission from Hara et al. [46]

excitatory PSCs without affecting miniature inhibitory PSC frequency or amplitude, indicating that TRH also reduces the probability of glutamate release onto Hcrt neurons. Similar results have also been obtained by the Burdakov laboratory [47].

Effect of local TRH injections into the Hcrt neuron region on locomotor activity (LMA) and body temperature (T_b). Local injections of TRH (0, 0.1, 1 μg) were made into the region of the hypothalamus containing the Hcrt neurons, as confirmed by visualizing co-administered fluorogold (Fig. 4a–d). When injected into this region in wild-type mice, TRH (1 μg) increased LMA, but LMA was not significantly affected in *Hcrt/ataxin-3* mice (Fig. 4e, f). In contrast, TRH local injection induced an increase in T_b in both wild-type and *Hcrt/ataxin-3* mice, suggesting that this effect is independent of Hcrt. Together, these

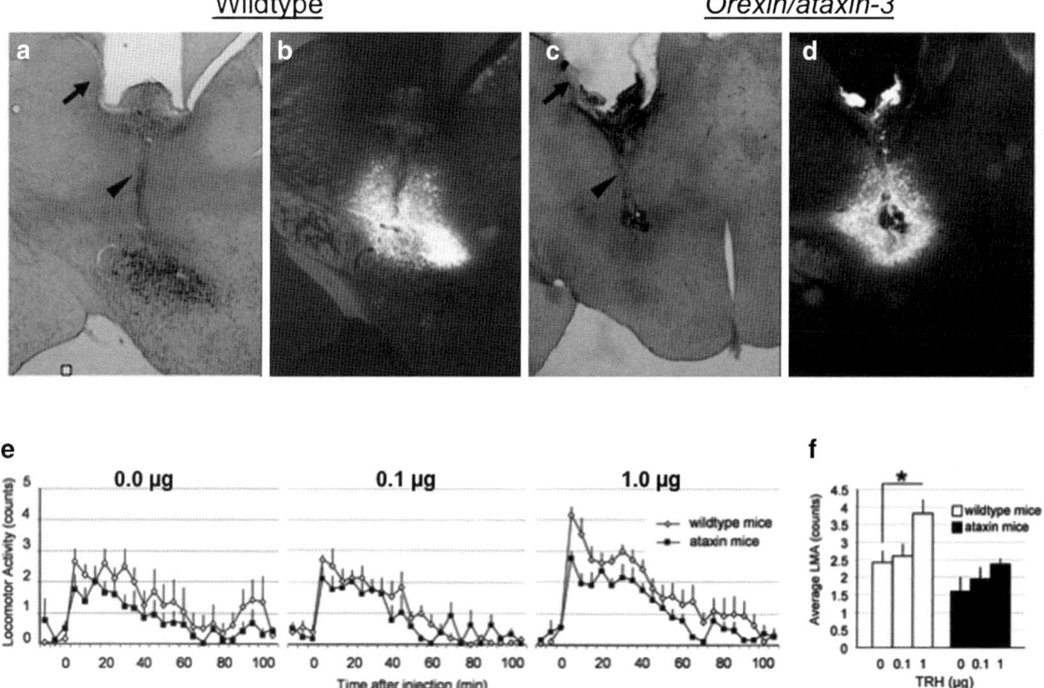

Fig. 4 Identification of injection sites and effect of unilateral TRH injections into the lateral hypothalamus on locomotor activity (LMA) in wild-type and *orexin/ataxin-3* mice. (**a**) Distribution of hypocretin-immunoreactive cells in a representative brain section of a wild-type mouse. *Arrows* show location of the guide cannula, and *triangles* show the tracks of the injection cannula. (**b**) Location of the injection site in the same brain section as (**a**) is indicated by fluorogold fluorescence. Note that the location of the injection site is dorsal to the distribution of hypocretin neurons. (**c**) In *orexin/ataxin-3* mice, there is a dramatic reduction in the number of hypocretin-immunoreactive cells (compare (**c**) vs. (**a**)). (**d**) The location of fluorogold injection site in the same brain section from the *orexin/ataxin-3* mouse immunostained in (**c**). (**e**) LMA counts recorded by the transmitters were averaged over 5 min intervals. The zero time point represents the LMA calculated during the first 5 min after the injection. Saline, 0.1 μg, or 1 μg of TRH were injected in a balanced order through a guide cannula implanted in the lateral hypothalamus in both wild-type ($n=6$) and *orexin/ataxin-3* mice ($n=6$). (**f**) Average LMA counts in the 10-min interval from 5 to 15 min following injection of TRH or saline. TRH at the dose of 1 μg produces significantly more LMA counts than saline injection in wild-type mice but not in *orexin/ataxin-3* mice. *$P<0.05$ compared to the saline group of the same genotype. Reproduced with permission from Hara et al. [46]

results are consistent with the hypothesis that TRH modulates behavioral arousal, in part, through the Hcrt system.

Activity and Activation of Hcrt Neurons In Vivo

The activity of the Hcrt neurons has also been measured in vivo [8, 48, 49]. Hcrt neurons discharge during active waking in association with movement when postural muscle tone is high. Hcrt cells decrease discharge during quiet waking in the absence of movement and virtually cease firing during sleep, when postural muscle tone is low or absent. During REM sleep, Hcrt neurons are relatively silent in association with muscle atonia with occasional burst discharge linked to phasic muscular twitches. Although based on a small sample of cells, Hcrt neurons appear to increase firing before the end of REM sleep and may anticipate the return of waking and muscle tone [49].

An exciting development to advance our understanding of Hcrt cells has resulted from the application of optogenetic technology [50–54] to activate these cells. Hcrt neurons were transfected in vivo with the blue light-sensitive channelrhodopsin protein (ChR2), allowing activation of these cells through implantation of a fiber optic bundle into the perifornical area [55]. Optogenetic photostimulation of Hcrt neurons increased the probability of transition to wakefulness from either SWS or REM sleep. Photostimulation using 5–30 Hz light pulse trains reduced the latency to wakefulness whereas 1 Hz trains did not. This study established a causal relationship between frequency-dependent activity of a genetically defined neural cell type and a specific mammalian behavior. Selective silencing of neurons using the yellow light-sensitive protein halorhodopsin (which is coupled to a Cl^- channel) is lagging behind ChR2-based photoactivation, but advances are being made [56, 57]. Use of the channelrhodopsin/halorhodopsin technologies will provide an additional tool with which to evaluate the relative importance of specific afferent inputs such as CRF, N/OFQ, and TRH to the Hcrt neurons and to determine in vivo the functional significance of physiological information obtained in vitro.

Acknowledgments This work was supported by NIH R01AG020584, R01MH061755, R01MH078194, R01 NS057464, and the Ministry of Education, Culture, Sports, Science and Technology of Japan.

References

1. de Lecea L, Kilduff TS, Peyron C, et al. The hypocretins: hypothalamus-specific peptides with neuroexcitatory activity. Proc Natl Acad Sci USA. 1998;95:322–7.
2. Sakurai T, Amemiya A, Ishii M, et al. Orexins and orexin receptors: a family of hypothalamic neuropeptides and G protein-coupled receptors that regulate feeding behavior. Cell. 1998;92:573–85.
3. Lin L, Faraco J, Li R, et al. The sleep disorder canine narcolepsy is caused by a mutation in the hypocretin (orexin) receptor 2 gene. Cell. 1999;98:365–76.
4. Chemelli RM, Willie JT, Sinton CM, et al. Narcolepsy in orexin knockout mice: molecular genetics of sleep regulation. Cell. 1999;98:437–51.
5. Peyron C, Faraco J, Rogers W, et al. A mutation in a case of early onset narcolepsy and a generalized absence of hypocretin peptides in human narcoleptic brains. Nat Med. 2000;6:991–7.
6. Thannickal T, Moore RY, Nienhuis R, et al. Reduced number of hypocretin neurons in human narcolepsy. Neuron. 2000;27:469–74.
7. Kilduff TS, Peyron C. The hypocretin/orexin ligand-receptor system: implications for sleep and sleep disorders. Trends Neurosci. 2000;23:359–65.
8. Mileykovskiy BY, Kiyashchenko LI, Siegel JM. Behavioral correlates of activity in identified hypocretin/orexin neurons. Neuron. 2005;46:787–98.
9. Sakurai T. The neural circuit of orexin (hypocretin): maintaining sleep and wakefulness. Nat Rev Neurosci. 2007;8:171–81.
10. Kilduff TS. Hypocretin/orexin: maintenance of wakefulness and a multiplicity of other roles. Sleep Med Rev. 2005;9:227–30.
11. Yoshida K, McCormack S, Espana RA, Crocker A, Scammell TE. Afferents to the orexin neurons of the rat brain. J Comp Neurol. 2006;494:845–61.
12. Sakurai T, Nagata R, Yamanaka A, et al. Input of orexin/hypocretin neurons revealed by a genetically encoded tracer in mice. Neuron. 2005;46:297–308.
13. Peyron C, Tighe DK, van den Pol AN, et al. Neurons containing hypocretin (orexin) project to multiple neuronal systems. J Neurosci. 1998;18:9996–10015.
14. Li Y, Gao XB, Sakurai T, van den Pol AN. Hypocretin/Orexin excites hypocretin neurons via a local glutamate neuron-A potential mechanism for orchestrating the hypothalamic arousal system. Neuron. 2002;36: 1169–81.

15. Yamanaka A, Beuckmann CT, Willie JT, et al. Hypothalamic orexin neurons regulate arousal according to energy balance in mice. Neuron. 2003;38:701–13.
16. Tsujino N, Yamanaka A, Ichiki K, et al. Cholecystokinin activates orexin/hypocretin neurons through the cholecystokinin A receptor. J Neurosci. 2005;25:7459–69.
17. Hara J, Beuckmann CT, Nambu T, et al. Genetic ablation of orexin neurons in mice results in narcolepsy, hypophagia, and obesity. Neuron. 2001;30:345–54.
18. Yamanaka A, Muraki Y, Tsujino N, Goto K, Sakurai T. Regulation of orexin neurons by the monoaminergic and cholinergic systems. Biochem Biophys Res Commun. 2003;303:120–9.
19. Burdakov D, Gerasimenko O, Verkhratsky A. Physiological changes in glucose differentially modulate the excitability of hypothalamic melanin-concentrating hormone and orexin neurons in situ. J Neurosci. 2005;25:2429–33.
20. Burdakov D, Jensen LT, Alexopoulos H, et al. Tandem-pore K+ channels mediate inhibition of orexin neurons by glucose. Neuron. 2006;50:711–22.
21. Wollmann G, Acuna-Goycolea C, van den Pol AN. Direct excitation of hypocretin/orexin cells by extracellular ATP at P2X receptors. J Neurophysiol. 2005;94:2195–206.
22. Winsky-Sommerer R, Yamanaka A, Diano S, et al. Interaction between the corticotropin-releasing factor system and hypocretins (orexins): a novel circuit mediating the stress response. J Neurosci. 2004;24: 11439–48.
23. Acuna-Goycolea C, van den Pol A. Glucagon-like peptide 1 excites hypocretin/orexin neurons by direct and indirect mechanisms: implications for viscera-mediated arousal. J Neurosci. 2004;24:8141–52.
24. Tsunematsu T, Fu LY, Yamanaka A, et al. Vasopressin increases locomotion through a V1a receptor in orexin/hypocretin neurons: implications for water homeostasis. J Neurosci. 2008;28:228–38.
25. Winsky-Sommerer R, Boutrel B, de Lecea L. Stress and arousal: the corticotrophin-releasing factor/hypocretin circuitry. Mol Neurobiol. 2005;32:285–94.
26. Bisetti A, Cvetkovic V, Serafin M, et al. Excitatory action of hypocretin/orexin on neurons of the central medial amygdala. Neuroscience. 2006;142:999–1004.
27. Johnson PL, Truitt W, Fitz SD, et al. A key role for orexin in panic anxiety. Nat Med. 2010;16:111–5.
28. Xie X, Crowder TL, Yamanaka A, et al. GABA(B) receptor-mediated modulation of hypocretin/orexin neurones in mouse hypothalamus. J Physiol. 2006;574:399–414.
29. Yamanaka A, Muraki Y, Ichiki K, et al. Orexin neurons are directly and indirectly regulated by catecholamines in a complex manner. J Neurophysiol. 2006;96:284–98.
30. Li Y, van den Pol AN. Direct and indirect inhibition by catecholamines of hypocretin/orexin neurons. J Neurosci. 2005;25:173–83.
31. Muraki Y, Yamanaka A, Tsujino N, Kilduff TS, Goto K, Sakurai T. Serotonergic regulation of the orexin/hypocretin neurons through the 5-HT1A receptor. J Neurosci. 2004;24:7159–66.
32. Fu LY, Acuna-Goycolea C, van den Pol AN. Neuropeptide Y inhibits hypocretin/orexin neurons by multiple presynaptic and postsynaptic mechanisms: tonic depression of the hypothalamic arousal system. J Neurosci. 2004;24:8741–51.
33. Liu ZW, Gao XB. Adenosine inhibits activity of hypocretin/orexin neurons by the A1 receptor in the lateral hypothalamus: a possible sleep-promoting effect. J Neurophysiol. 2007;97:837–48.
34. Huang H, Acuna-Goycolea C, Li Y, Cheng HM, Obrietan K, van den Pol AN. Cannabinoids excite hypothalamic melanin-concentrating hormone but inhibit hypocretin/orexin neurons: implications for cannabinoid actions on food intake and cognitive arousal. J Neurosci. 2007;27:4870–81.
35. Li Y, van den Pol AN. Differential target-dependent actions of coexpressed inhibitory dynorphin and excitatory hypocretin/orexin neuropeptides. J Neurosci. 2006;26:13037–47.
36. Li Y, van den Pol AN. Mu-opioid receptor-mediated depression of the hypothalamic hypocretin/orexin arousal system. J Neurosci. 2008;28:2814–9.
37. Xie X, Wisor JP, Hara J, et al. Hypocretin/orexin and nociceptin/orphanin FQ coordinately regulate analgesia in a mouse model of stress-induced analgesia. J Clin Invest. 2008;118:2471–81.
38. Matsuki T, Nomiyama M, Takahira H, et al. Selective loss of GABA(B) receptors in orexin-producing neurons results in disrupted sleep/wakefulness architecture. Proc Natl Acad Sci USA. 2009;106:4459–64.
39. Watanabe S, Kuwaki T, Yanagisawa M, Fukuda Y, Shimoyama M. Persistent pain and stress activate pain-inhibitory orexin pathways. Neuroreport. 2005;16:5–8.
40. Rizzi A, Marzola G, Bigoni R, et al. Endogenous nociceptin signaling and stress-induced analgesia. Neuroreport. 2001;12:3009–13.
41. Koster A, Montkowski A, Schulz S, et al. Targeted disruption of the orphanin FQ/nociceptin gene increases stress susceptibility and impairs stress adaptation in mice. Proc Natl Acad Sci USA. 1999;96: 10444–9.
42. Neal Jr CR, Mansour A, Reinscheid R, Nothacker HP, Civelli O, Watson Jr SJ. Localization of orphanin FQ (nociceptin) peptide and messenger RNA in the central nervous system of the rat. J Comp Neurol. 1999;406:503–47.
43. Neal Jr CR, Mansour A, Reinscheid R, et al. Opioid receptor-like (ORL1) receptor distribution in the rat central nervous system: comparison of ORL1 receptor mRNA expression with (125)I-[(14)Tyr]-orphanin FQ binding. J Comp Neurol. 1999;412:563–605.
44. Nishino S, Arrigoni J, Shelton J, Kanbayashi T, Dement WC, Mignot E. Effects of thyrotropin-releasing hormone and its analogs on daytime sleepiness and cataplexy in canine narcolepsy. J Neurosci. 1997;17:6401–8.
45. Stanton TL, Winokur A, Beckman AL. Reversal of natural CNS depression by TRH action in the hippocampus. Brain Res. 1980;181:470–5.

46. Hara J, Gerashchenko D, Wisor JP, Sakurai T, Xie X, Kilduff TS. Thyrotropin-releasing hormone increases behavioral arousal through modulation of hypocretin/orexin neurons. J Neurosci. 2009;29:3705–14.
47. Gonzalez JA, Horjales-Araujo E, Fugger L, Broberger C, Burdakov D. Stimulation of orexin/hypocretin neurones by thyrotropin-releasing hormone. J Physiol. 2009;587:1179–86.
48. Takahashi K, Lin JS, Sakai K. Neuronal activity of orexin and non-orexin waking-active neurons during wake-sleep states in the mouse. Neuroscience. 2008;153:860–70.
49. Lee MG, Hassani OK, Jones BE. Discharge of identified orexin/hypocretin neurons across the sleep-waking cycle. J Neurosci. 2005;25:6716–20.
50. Airan RD, Hu ES, Vijaykumar R, Roy M, Meltzer LA, Deisseroth K. Integration of light-controlled neuronal firing and fast circuit imaging. Curr Opin Neurobiol. 2007;17:587–92.
51. Deisseroth K, Feng G, Majewska AK, Miesenbock G, Ting A, Schnitzer MJ. Next-generation optical technologies for illuminating genetically targeted brain circuits. J Neurosci. 2006;26:10380–6.
52. Hwang RY, Zhong L, Xu Y, et al. Nociceptive neurons protect *Drosophila* larvae from parasitoid wasps. Curr Biol. 2007;17:2105–16.
53. Schneider MB, Gradinaru V, Zhang F, Deisseroth K. Controlling neuronal activity. Am J Psychiatry. 2008;165:562.
54. Boyden ES, Zhang F, Bamberg E, Nagel G, Deisseroth K. Millisecond-timescale, genetically targeted optical control of neural activity. Nat Neurosci. 2005;8:1263–8.
55. Adamantidis AR, Zhang F, Aravanis AM, Deisseroth K, de Lecea L. Neural substrates of awakening probed with optogenetic control of hypocretin neurons. Nature. 2007;450:420–4.
56. Gradinaru V, Thompson KR, Deisseroth K. eNpHR: a Natronomonas halorhodopsin enhanced for optogenetic applications. Brain Cell Biol. 2008;36:129–39.
57. Zhao S, Cunha C, Zhang F, et al. Improved expression of halorhodopsin for light-induced silencing of neuronal activity. Brain Cell Biol. 2008;36:141–54.
58. Rao Y, Lu M, Ge F, et al. Regulation of synaptic efficacy in hypocretin/orexin-containing neurons by melanin concentrating hormone in the lateral hypothalamus. J Neurosci. 2008;28:9101–10.

The Neural Basis of Sleepiness in Narcoleptic Mice

Thomas E. Scammell and Chloé Alexandre

Keywords
Narcolepsy • Mouse model • Behavioral state instability • Cataplexy • Sleep homeostasis

Everyone with narcolepsy has some degree of daytime sleepiness, and for most, sleepiness is the symptom of greatest concern. This chronic sleepiness is often severe and can substantially impair relationships, the ability to drive safely, and performance at school and work. Although it is now clear that narcolepsy with cataplexy is caused by an extensive and selective loss of the orexin/hypocretin-producing neurons, remarkably little is known about how this results in chronic sleepiness.

Several pathophysiological processes have been proposed to underlie the sleepiness of narcolepsy. These mechanisms include reduced activity of wake-promoting brain regions; inadequate inhibition of sleep-promoting systems; enhanced homeostatic sleep drive; impaired circadian promotion of wakefulness; and poor quality, nonrestorative sleep. All of these hypotheses have their strengths, but they have been difficult to test fully in clinical studies. In this chapter, we will discuss how mouse models of narcolepsy have provided novel insights into the pathophysiology of excessive daytime sleepiness in narcolepsy. The Appendix to this chapter contains a general discussion of using mice in sleep research.

Overview of Mouse Models of Narcolepsy

Researchers have produced several lines of mutant mice that shed light on narcolepsy and different aspects of orexin (hypocretin) signaling. Chemelli and colleagues produced knockout (KO) mice completely lacking the orexin-A and -B peptides [1]. These orexin KO mice have behavioral abnormalities quite similar to those in narcolepsy in humans such as very poor maintenance of wakefulness, moderately fragmented sleep, and frequent episodes of cataplexy (Fig. 1). To recapitulate the *acquired* loss of orexin neurons that occurs in people with narcolepsy, Hara and colleagues produced orexin/ataxin-3 mice [2]; in these mice, the prepro-orexin promoter drives the expression of ataxin-3, a toxic, polyglutamine protein that kills nearly all the orexin-producing neurons by about 12 weeks of age. These mice lack not only the orexin peptides but

T.E. Scammell (✉)
Department of Neurology, Beth Israel Deaconess Medical Center and Harvard Medical School, Center for Life Science, Room 705, 330 Brookline Ave., Boston, MA 02215, USA
e-mail: tscammel@bidmc.harvard.edu

C.R. Baumann et al. (eds.), *Narcolepsy: Pathophysiology, Diagnosis, and Treatment*,
DOI 10.1007/978-1-4419-8390-9_15,

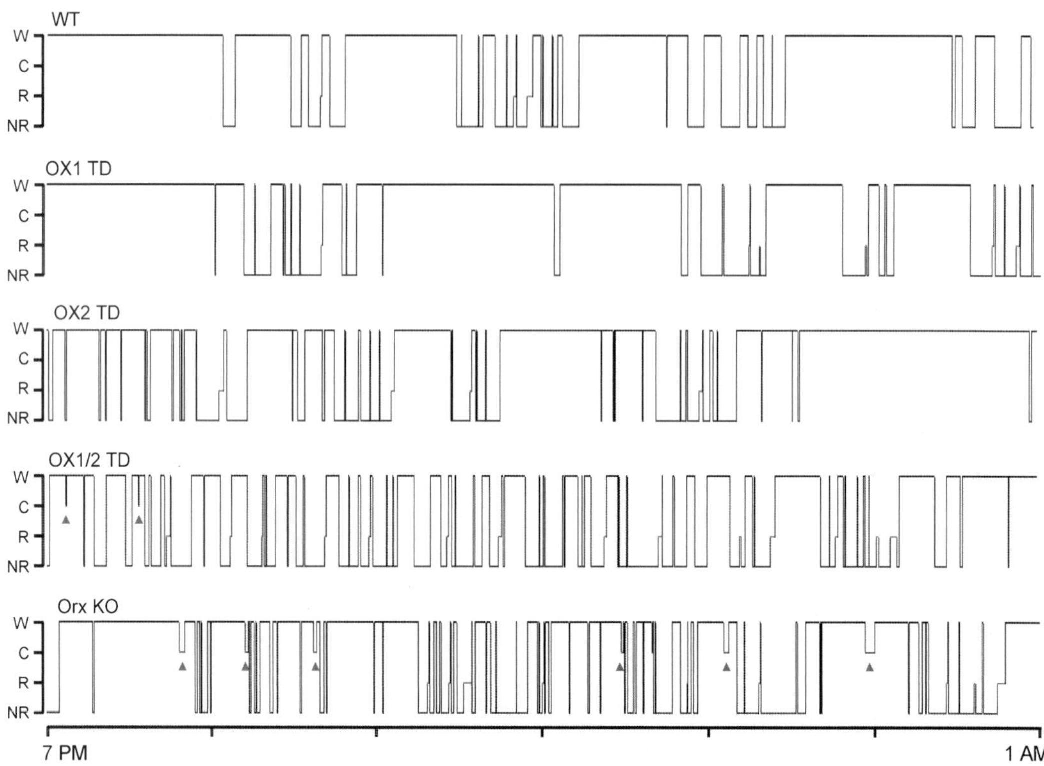

Fig. 1 Sleep–wake fragmentation in mice with disrupted orexin signaling. These hypnograms show wake (W), cataplexy (C), REM sleep (R), and NREM sleep (NR) over the first 6 h of the dark period in representative individual mice. WT (wild-type) mice have consolidated periods of wakefulness with interspersed periods of sleep. Mice lacking OX1 receptors (OX1 TD) have generally normal behavior. Mice lacking the OX2 receptor (OX2 TD) have moderately shorter wake bouts than normal. Mice lacking both OX1 and OX2 receptors (OX1/2 TD) exhibit severe sleep–wake fragmentation, almost as severe as that seen in orexin KO mice, but they have less cataplexy (*denoted by triangles*). These "TD" mice have transcriptional and translational disruption (TD) of the orexin receptors and no physiologic response to orexin-A in vitro [14]. (Unpublished data and figure courtesy of T. Mochizuki)

also other signaling molecules produced by these neurons, including dynorphin and glutamate [3, 4]. Interestingly, these mice have a slightly different phenotype than orexin KO mice, with a greater propensity for obesity and more REM sleep during the active period [5, 6]. A similar approach was used to produce rats lacking the orexin neurons [7], but later generations of this line appeared to have only a partial loss of orexin neurons and a milder phenotype, with little disruption of wakefulness and only rare cataplexy [8]. Possibly, the outbred genetics of these rats protected them against the toxic effects of ataxin-3.

In contrast to these mice completely lacking orexin signaling, mice lacking either OX1 or OX2 orexin receptors have less severe narcolepsy phenotypes. Current descriptions of the OX1R KO mice are preliminary, but these mice appear to have normal sleep/wake behavior and locomotion [9–11]. On the contrary, OX2 receptor KO mice have moderate sleepiness as inferred from their poor maintenance of wakefulness and frequent, abrupt transitions into NREM sleep. Although they have some sleep fragmentation, they have almost no cataplexy [12]. Preliminary reports indicate that double KO mice lacking both OX1 and OX2 receptors have severe sleepiness similar to that seen in orexin KO mice [10]. We have observed similar behavior in other, independently generated lines of OX1 and OX2 receptor null mice [13, 14]. Thus, just as in narcoleptic Dobermans [15], lack of signaling through the OX2 receptor appears to be a major contributor to the sleepiness in narcolepsy, but signaling through OX1 probably contributes as well.

Behavioral State Instability in Narcoleptic People and Mice

Sleepiness in narcolepsy is distinct from that in many other sleep disorders in that the total amount of sleep over 24 h is roughly normal, yet people and mice with narcolepsy have great difficulty maintaining long bouts of wakefulness. In fact, the sleepiness and fragmented sleep of narcolepsy may be a consequence of *unstable wakefulness and sleep*, with inappropriately low thresholds for crossing between states, resulting in frequent transitions into sleep and frequent arousals from sleep. Roger Broughton first raised this idea of behavioral state instability over 20 years ago, suggesting a loss of the "neurochemical glues" that help integrate neural activity [16]. Now, one can envision this as a loss of the coordinating activity of the orexin neurons on a variety of brain regions. For example, during wakefulness, orexins should help activate many wake-promoting neurons and inhibit sleep-promoting regions, but in the absence of orexins, some of these target neurons may have inappropriate patterns of activity that result in poor maintenance of wakefulness and sleep.

This lack of coordinated activity may also help explain some of the ambiguous states encountered in narcolepsy. People with narcolepsy often have hypnagogic hallucinations and sleep paralysis, suggestive of intermediate states between wakefulness and REM sleep [17]. In addition, one often encounters odd patterns during sleep recordings such as atonia or saccadic eye movements during stage 2 sleep. As proposed by Broughton [16], these phenomena may also reflect a lack of coordination across sleep- and wake-regulating brain regions.

This behavioral state instability is quite apparent in orexin KO mice that have considerably more transitions between sleep/wake states. In both the dark and light periods, orexin KO mice have about 50% more transitions between Wake and NREM sleep, and into and out of REM sleep [18]. Overall, there appears to be no bias for a particular state, and the mice show clear shifts toward shorter bouts of wakefulness and NREM sleep, especially in the dark period. (Interestingly, even though the number of transitions into and out of REM sleep is increased, the duration of REM sleep bouts is only slightly shorter, suggesting that the process for terminating REM sleep does not require orexins.) This same pattern of fragmentation is also apparent in OX2 receptor mutant dogs which have shorter and more frequent bouts of wakefulness, drowsiness, and deep sleep [19].

To gain a better understanding of how orexin deficiency impairs the maintenance of wakefulness, Diniz Behn and colleagues performed a survival analysis of wake bout durations in orexin KO and wild-type mice [20]. As shown in prior studies of mice [21, 22], about 90% of all wake bouts ended within 1 min in both groups of mice, presumably representing just short awakenings from sleep. After remaining awake for 1–2 min, wild-type mice generally remained awake, although after 17 min of wake, transitions into sleep became increasingly frequent, perhaps reflecting rising homeostatic sleep drive. In contrast, wake bouts in orexin KO mice were much less stable across the 2–17-min period, failing much earlier than normal. This pattern suggests that orexins play an essential role in maintaining wake bouts longer than 1–2 min (see Chapter "Mathematical Models of Narcolepsy" by Diniz Behn for more details).

Diniz Behn and colleagues also used a new state space analysis technique to visualize transitions between states and the variations within states [23, 24]. Using spectral analysis of the EEG, they mapped second-by-second EEG activity into a two-dimensional space in which one dimension is strongly influenced by EEG theta activity and the other by slow wave activity. In wild-type mice, this approach showed relatively discrete clusters of EEG activity corresponding to wake, NREM sleep, and REM sleep. In orexin KO mice, the wake and NREM clusters were closer together, with less time spent in deep NREM sleep and in active, theta-rich wakefulness and more time spent in the transitional region between wake and NREM sleep. In addition, in the midst of what should be stable, full wake, orexin KO mice exhibited rapid transitions into NREM sleep, suggesting that even high levels of arousal can be unstable. These findings suggest two explanations for the state instability of narcolepsy: transitions could be more frequent because narcoleptic animals spend more time in a "drowsy" state (in and near the transition region between wake and NREM sleep); and even apparently full alertness may be vulnerable to rapid transitions into sleep. This latter phenomena may be similar to the occasional "sleep attacks" seen in people with narcolepsy and needs to be investigated further as it suggests that in the absence of orexins, activity in wake-promoting systems can rapidly fail.

How Mouse Models Provide New Insights into the Sleepiness of Narcolepsy

Over the last two decades, many studies have sought to determine the neural mechanisms that underlie the sleepiness of narcolepsy, and mouse models of narcolepsy are helping advance these efforts.

Abnormal Sleep Homeostasis

Several researchers have hypothesized that people with narcolepsy may be especially sensitive to homeostatic sleep drive or they might accumulate sleep drive faster than normal. In support of this, people with narcolepsy often feel fairly alert soon after waking, but just a few hours later, they usually feel quite sleepy. Furthermore, some studies have shown that people with narcolepsy have enhanced responses to 16–24 h of sleep deprivation, with shorter sleep latencies and more deep NREM sleep than controls [25–27]. Building on these ideas, Zeitzer and colleagues hypothesized that orexin signaling may oppose homeostatic sleep drive because prolonged wakefulness activates the orexin neurons and increases extracellular levels of orexin in rats and monkeys [28–32].

However, several studies in narcoleptic rodents suggest that NREM and REM sleep homeostasis is unimpaired. First, orexin KO mice and orexin/ataxin-3 mice have essentially normal amounts of wake [2, 6, 18], and across 24 h, most people with narcolepsy have normal amounts of wake [33]. To examine homeostatic sleep drive, Mochizuki and colleagues deprived orexin KO and wild-type mice of sleep for 2, 4, or 8 h at the beginning of the light period [18]. Over the next several hours, orexin KO mice had normal amounts of recovery NREM sleep, normal sleep latencies, and normal increases in NREM delta power. They also recovered NREM and REM sleep at the same rate and to the same extent as wild-type mice. In addition, if homeostatic sleep drive were higher in narcolepsy, then levels of endogenous somnogens such as adenosine might be higher. Sleep deprivation increases levels of adenosine in the basal forebrain [34]; yet in rats with lesions of the lateral hypothalamus that kill most of the orexin neurons, adenosine levels in the basal forebrain were actually lower than expected after 6 h of sleep deprivation [35]. Considered along with several clinical studies showing essentially normal responses to sleep deprivation [36–38], these observations suggest that mice and people with narcolepsy accumulate and respond to homeostatic sleep drive normally.

Loss of the Circadian Promotion of Wakefulness

The suprachiasmatic nucleus generates a potent wake-promoting signal during the usual active period [39, 40], and the sleepiness of narcolepsy might be caused by a reduction in this arousing influence [37, 41, 42]. Anatomically, this hypothesis is well supported as the orexin neurons receive small direct and large indirect projections from the suprachiasmatic nucleus [43–47]. In the absence of the orexin neurons, wake-promoting brain regions may not respond fully to this circadian signal. Supporting this hypothesis, people and mice with narcolepsy can fall asleep at any time of day, and REM sleep which is normally under very strong circadian control can occur at any time [6, 37].

The orexin neurons are clearly under strong circadian control. In rats, activity of the orexin neurons is highest during the active period as indicated by the expression of fos, even in constant darkness [30]. In addition, extracellular levels of orexin exhibit a robust circadian rhythm, with high levels during the dark and subjective dark periods [31]. Furthermore, in rats with SCN lesions, the rhythm of orexin release is absent and mean orexin levels are lower, providing clear evidence for excitatory, circadian modulation of orexin neuron activity [31, 48].

Still, wake-promoting circadian signals may reach key arousal regions even in the absence of the orexin neurons. For example, many output signals of the SCN pass through the dorsomedial nucleus of the hypothalamus, and lesions of this nucleus disrupt the circadian rhythms of wakefulness and LC neuronal activity [45, 49]. The dorsomedial nucleus projects not only to the orexin neurons but also to the locus coeruleus and other wake-promoting brain regions. Most importantly, the circadian rhythms of wakefulness are normal in both orexin KO mice and orexin/ataxin-3 mice [6, 18], with robust increases in wakefulness at the expected times, even in constant darkness. Thus, it seems most likely that the orexin neurons are activated by circadian signals arising from the SCN, but they are not necessary for the circadian timing of wakefulness.

Poor Quality Sleep

In many sleep disorders such as obstructive sleep apnea or periodic limb movements of sleep, poor quality sleep results in daytime sleepiness, and similarly non-restorative sleep could contribute to the sleepiness of narcolepsy. Even in the absence of other sleep disorders such as sleep apnea, people with narcolepsy often have more spontaneous awakenings and more light NREM sleep than normal [50]. Based on our current understandings of orexins, this fragmented sleep is hard to explain as narcolepsy is caused by the loss of a wake-promoting peptide, yet it consistently occurs in narcoleptic people, dogs, and mice.

Khatami and colleagues have proposed that fragmented sleep may be a consequence of insufficiently intense NREM sleep. In unmedicated patients with narcolepsy, the usual exponential fall in EEG slow wave activity across NREM sleep episodes is disrupted, with a steeper decline in slow wave activity and more and longer awakenings in the middle of the night than usual [50]. Similar changes in slow wave activity have been observed in other, though not all, clinical studies [51].

Very little is known about the neural mechanisms that would reduce the intensity of NREM sleep, but two rodent studies provide some helpful insights. First, a fall in core body temperature is an essential element of normal sleep, and manipulations of skin and body temperature can improve sleep quality in people with narcolepsy [52]. In wild-type mice, body temperature usually falls 1–1.5°C over the first 20 min of sleep, but this fall in temperature is blunted in orexin KO mice [53]. This does not seem to be a consequence of fragmented sleep because orexin KO mice have blunted falls in temperature even during intense, sustained recovery sleep after 8 h of sleep deprivation. Possibly, a moderate fall in temperature is necessary for good quality sleep, and a smaller fall in temperature contributes to the poor sleep in narcolepsy.

Second, a reduction in sleep-promoting factors could give rise to lighter sleep. In rats with lesions of the lateral hypothalamus that kill most of the orexin neurons (and other nearby neurons), adenosine levels in the basal forebrain do not rise across 6 h of sleep deprivation [35]. As the orexin neurons play a key role in activating the basal forebrain [54], these lesions may reduce the activity-dependent rise in adenosine in this region, resulting in lower levels of adenosine and consequently lighter sleep. Still, animals with lateral hypothalamic lesions are much sleepier than expected from animals with orexin deficiency alone, and it would be helpful to repeat similar studies in animals lacking just the orexin neurons.

Currently, it is unclear whether these abnormalities in slow wave activity, thermoregulation, and adenosine signaling cause lighter sleep or are simply a consequence. In addition, most people with narcolepsy feel well rested in the morning and after taking a nap, and even narcoleptics with apparently good quality sleep can still have severe daytime sleepiness [55]. More work is needed to establish whether better sleep can improve the sleepiness in narcolepsy, but overall, it seems unlikely that poor quality sleep is a major cause of sleepiness in narcolepsy.

Weak Arousal Systems

One of the more compelling explanations for the sleepiness in narcolepsy is that in the absence of orexins, the activity of wake-promoting brain systems may be reduced or impersistent. Specifically, the orexin neurons provide a strong, excitatory drive to many arousal-promoting brain regions such as the tuberomammillary nucleus, locus coeruleus, raphe nuclei, and cholinergic neurons of the pons and basal forebrain (Fig. 2) [56–60]. In addition, selective activation of the orexin neurons can trigger awakenings from sleep, most likely through activation of these arousal regions [61, 62]. Under what conditions these systems are activated by orexins is unknown, as little is known about the precise pattern of orexin release. Preliminary observations indicate that the orexin neurons fire phasically when animals are engaged in motivated behaviors and locomotion [63, 64]. The orexin neurons may also have some tonic activity as van den Pol and colleagues showed that orexin-A excites the orexin neurons during wakefulness via a local glutamatergic neuron [65]. This unusual auto-excitation may promote sustained activity in the orexin neurons that then helps drive persistent activity in target regions. Possibly, in the absence of orexin signaling, activity in key wake-promoting regions lapses or is tonically reduced, especially when an individual is sedentary or bored.

Many pharmacologic and genetic studies have shown that reductions in monoamine or acetylcholine signaling promote sleepiness. Drugs that

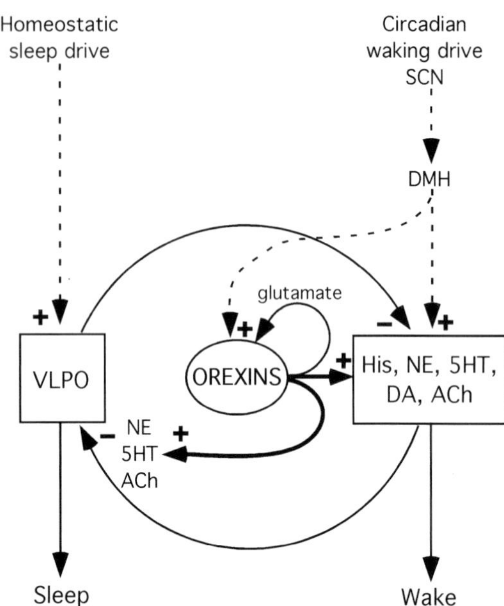

Fig. 2 Major pathways through which the orexin neurons help stabilize wakefulness. Neurons containing acetylcholine (ACh) and monoaminergic neurotransmitters such as histamine (His), norepinephrine (NE), serotonin (5HT), and dopamine (DA) play essential roles in promoting wakefulness through their projections to the cortex, thalamus, and hypothalamus [92]. On the contrary, neurons in the ventrolateral preoptic area (VLPO) and nearby regions promote sleep, by inhibiting the wake-promoting populations. During wakefulness, NE, 5HT, and ACh may inhibit VLPO neurons. These mutually inhibitory connections between the sleep- and wake-active populations ensure that one side is fully active, probably helping generate full arousal and sustained sleep. The orexin neurons regulate this system, sustaining wakefulness through their excitatory projections to arousal regions and perhaps also by enhancing the inhibition of VLPO neurons by increasing the release of NE, 5HT, and ACh. During wakefulness, orexin neurons may also maintain persistent activity via an autostimulatory loop mediated through local glutamate interneurons. The daily rhythm of wakefulness is driven by circadian signals from the SCN that are relayed through the dorsomedial nucleus of the hypothalamus (DMH) to various arousal regions. Homeostatic, sleep-promoting signals may indirectly excite VLPO neurons and inhibit wake-promoting neurons (not shown) in a manner independent of orexin signaling. Circadian and homeostatic influences are represented with dashed lines to indicate a lesser role in the sleepiness of narcolepsy

block signaling by dopamine, histamine, norepinephrine, or acetylcholine are sedating in both people and rodents [66]. Histamine-deficient mice have less wakefulness at the beginning of the active period, and mice lacking H1 receptors have fewer awakenings from sleep [67, 68]. Mice lacking norepinephrine return to sleep much more quickly than normal after a mild stress such as an injection of saline or transfer to a new cage [69].

Researchers are just beginning to examine whether orexin deficiency reduces the levels of wake-promoting neurotransmitters. Nishino, Kanbayashi, and colleagues recently showed that in patients with narcolepsy, histamine levels measured in lumbar CSF are lower than normal, especially in narcolepsy patients lacking orexins [70, 71]. Histamine levels are also lower in OX2 receptor null dogs [72]. The data on dopamine levels are less clear, and much less is known about the levels of other monoamines or acetylcholine [73–75]. In rats with lesions of the orexin neuron field, cortical levels of acetylcholine are reduced [76]. Whether the levels of wake-promoting neurotransmitters are reduced in mouse models of narcolepsy remains unknown, but it seems likely based on these studies.

In support of this hypothesis, the sleepiness in narcoleptic mice often improves with conditions that enhance monoamine signaling. For example, histamine H3 receptors inhibit the tuberomammillary and other monoaminergic neurons, and pharmacologic blockade of H3 receptors improves the sleepiness of orexin KO mice [77]. Similar effects are seen in narcoleptic mice treated with amphetamines that predominantly act by increasing dopamine tone [30]. The sleepiness of orexin KO mice also improves when they use a running wheel, which may promote arousal by increasing norepinephrine signaling [78, 79].

On the contrary, the sleepiness of narcoleptic mice differs importantly from that of mice lacking monoamines. For example, when a normal, wild-type mouse is placed in a clean cage in the middle of the sleep period, the mouse typically stays awake for about 45 min while becoming comfortable with its new environment. However, mice lacking histamine or norepinephrine fall asleep in only about 15 min, suggesting a major deficit in arousal [67, 69]. In contrast, after transfer to a new cage, orexin KO mice stay awake as long as wild-type mice [18]. Most likely, this mild stress successfully activates wake-promoting brain regions even in the absence of orexins.

Overall, much remains to be learned about the activity of arousal systems in orexin-deficient animals. The normal response to cage change suggests that at least under some conditions, arousal regions can be fully engaged, and certainly, untreated patients with narcolepsy can maintain full alertness for some time. Still, reduced signaling by monoaminergic and cholinergic wake-promoting systems may contribute to the sleepiness in narcolepsy, especially under sedentary conditions.

Increased Activity of Sleep-Promoting Systems

One could also hypothesize that in the absence of orexins, sleep-promoting neurons in the ventrolateral preoptic area (VLPO) or median preoptic nucleus are excessively active or not inhibited as they should be during wakefulness. These preoptic regions contain a moderate density of orexin fibers [56, 80], but VLPO neurons show no direct response to orexin-A, and orexin receptor mRNA is sparse in these regions [58, 81]. Still, the VLPO neurons are inhibited by norepinephrine, serotonin, and acetylcholine, and orexins may act presynaptically to enhance this inhibition as has been demonstrated with norepinephrine release in the cortex [82].

The main arguments for this perspective are that poor maintenance of wakefulness and rapid transitions into sleep may represent insufficient inhibition of sleep-promoting mechanisms. Orexin KO mice, orexin/ataxin-3 mice, and OX2 receptor null mice all have shorter than normal bouts of wake, and orexin KO mice can rapidly transition into NREM sleep, even from the midst of full wakefulness [23]. In addition, mathematical modeling of sleep architecture simulates sleep/wake behavior most accurately when one assumes that orexins enhance the inhibition of the VLPO by wake-promoting systems [20]. Specifically, in a neurobiologically realistic model, one can recapitulate key aspects of narcolepsy (fewer long bouts of wakefulness with no change in short awakenings from sleep) by assuming that loss of orexins reduces monoaminergic inhibition of the VLPO.

Conclusions

Experiments using mouse models of narcolepsy such as those discussed above have provided many new insights into the sleepiness in narcolepsy, but the fundamental causes remain unknown. The concept of behavioral state instability provides a helpful framework in which to view most symptoms of narcolepsy; low thresholds for transitioning between wakefulness and sleep, and inappropriate patterns of activity in sleep/wake regulatory regions could easily give rise to the sleepiness, fragmented behavior, and mixed states of narcolepsy. Mechanistically, it seems unlikely that the sleepiness is caused by a weak circadian promotion of wakefulness, altered sleep homoeostasis, or non-restorative sleep. Instead, much evidence suggests that sleepiness may result from reduced activity in a variety of wake-promoting regions or insufficient inhibition of sleep-promoting regions. In fact, redundancy is common in most neural systems, and a combination of these two factors may provide the best explanation.

Overall, we feel it is most likely that the orexin neurons play little role during short awakenings from sleep, but during sustained wakefulness, the orexin neurons increase and sustain activity in arousal regions while indirectly suppressing activity in sleep-promoting systems. In the absence of the orexin neurons, wake-promoting systems may have lapses in activity or sleep-promoting systems may become activated, thus initiating rapid and inappropriately timed transitions into sleep.

Much more work is needed to understand the neurobiologic causes of sleepiness and the broader questions of state instability. For example, do narcoleptic mice have reduced or less consistent activity in monoaminergic neurons during wakefulness? Are VLPO neurons sometimes active even during wakefulness? Are the patterns of activity across these systems uncoordinated, with simultaneous activity in both sleep and wake-promoting neurons or less coherence across wake-promoting systems? Would restoring orexin signaling to a subset of these pathways be sufficient to improve sleepiness? These questions will be challenging to address, but they should provide many helpful insights for improving the sleepiness of patients with narcolepsy.

Appendix

Advantages and Limitations of Using Mice to Understand Sleep and Sleepiness

Mice provide many opportunities to better understand the neurobiology of sleep that are not available through clinical research. Sleep in mice is much like that in humans, with reduced responsiveness to external stimuli, quick reversibility, and EEG patterns quite similar to those seen across human sleep/wake states [83]. The brains of the two species have remarkably similar anatomy and physiology, and 99% of mouse genes have homologs in the human genome [84]. In mice, consistent genetics and developmental conditions help minimize the variability that often confounds research with humans. Most importantly, numerous molecular biology tools allow one to test specific cellular mechanisms in mice by selective manipulation of genes. These and other factors make sleep research in mice a powerful approach for understanding the neural circuits and biochemical processes that underlie sleep and sleepiness.

Still, several challenges arise when using mice to study sleepiness, although most of these are quite surmountable. In the clinic, patients' description of their sleepiness and tools such as the Epworth Sleepiness Scale often drive treatment decisions, but of course, these subjective impressions cannot be studied in mice. Instead, researchers infer sleepiness from objective measures such as reductions in the amount of wakefulness, shorter wake bouts, and stronger responses to increased sleep pressure.

Compared to humans, healthy mice and other rodents have much more fragmented sleep/wake behavior. During the active period of mice, most wakefulness occurs in bouts lasting only 10–20 min, though sometimes mice will spontaneously remain awake for up to 2 or 3 h. (In part, these short wake bouts may be a consequence of the standard, stark recording environment with no social interaction, no opportunity for sustained running or exploration, and no need to forage for food. Sleep/wake behavior of mice in the wild with a much richer and more challenging environment may be quite different.) Despite the fact that normal mice have relatively short wake bouts, impaired maintenance of wakefulness is still very apparent in the especially short wake bouts of mice lacking wake-promoting neurotransmitters such as orexin, histamine, or norepinephrine [1, 18, 67, 69].

Measuring sleep in mice without disturbing their behavior is also a challenge. Most frequently, EEG and EMG signals are recorded using screw electrodes in the skull, thin wires in the neck extensor muscles, and long recording cables, but it is essential that the implant is tiny, the cable is lightweight and flexible, and the electrical slip ring atop the cage has very low torque. Recording equipment used in rats is generally not suitable for mice. Telemetry is appealing as mice are untethered, but some researchers feel the quality of the EEG signal is inadequate, and the implanted telemetry device can limit full movement. These factors must be considered when studying narcoleptic mice as easily disturbed behaviors such as cataplexy may be less frequent when mice are recorded with cables [85].

A great appeal of using mice for sleep research is that one can produce and study mice with a variety of genetic defects, but absence of a gene product can lead to developmental compensations that can alter or reduce the phenotype. For example, mice lacking the serotonin reuptake transporter have high extracellular levels of serotonin and surprisingly increased REM sleep, but when their serotonin signaling is reduced for just 2–4 weeks in the early postnatal period, the adult mice have normal amounts of REM sleep [86]. One might raise the same concern that some of the narcolepsy phenotype in orexin KO mice is caused by abnormal development, but this appears unlikely as their sleepiness and other narcolepsy-like behaviors can be reversed by restoring orexin signaling in adult mice using virus-mediated gene therapy [87, 88].

Mutant mice are generally maintained on specific inbred genetic backgrounds, and these genetic influences must be carefully considered as the amounts, timing, and maintenance of wakefulness differ considerably between different strains of mice. For example, during the active (dark) period, C57BL/6J mice are awake in 48% of the dark period, but AKR/J mice are awake only 36% of this time, and they spend less than half as much time in long wake bouts [89]. These strains also differ in their homeostatic responses to sleep deprivation, with much more sleep and EEG delta power in AKR/J mice than in C57BL/6J mice [89]. Thus, researchers must describe the background genetics of their mice and make strong efforts to backcross the mice for at least five generations to ensure that littermates are genetically quite homogeneous [90, 91].

Acknowledgments We appreciate the thoughtful comments on this text by A. Lim, and the generosity of T. Mochizuki in sharing preliminary data. Writing of this chapter was in part supported by the National Institutes of Health Grant NS055367 to T.E.S.

References

1. Chemelli RM, Willie JT, Sinton CM, et al. Narcolepsy in orexin knockout mice: molecular genetics of sleep regulation. Cell. 1999;98(4):437–51.
2. Hara J, Beuckmann CT, Nambu T, et al. Genetic ablation of orexin neurons in mice results in narcolepsy, hypophagia, and obesity. Neuron. 2001;30(2): 345–54.
3. Chou TC, Lee CE, Lu J, et al. Orexin (hypocretin) neurons contain dynorphin. J Neurosci. 2001;21(19): RC168–73.
4. Crocker A, Espana RA, Papadopoulou M, et al. Concomitant loss of dynorphin, NARP, and orexin in narcolepsy. Neurology. 2005;65(8):1184–8.
5. Hara J, Yanagisawa M, Sakurai T. Difference in obesity phenotype between orexin-knockout mice and orexin neuron-deficient mice with same genetic background and environmental conditions. Neurosci Lett. 2005;380(3):239–42.
6. Kantor S, Mochizuki T, Janisiewicz A, et al. Orexin neurons are necessary for the circadian control of REM sleep. Sleep. 2009;32:1127–34.

7. Beuckmann CT, Sinton CM, Williams SC, et al. Expression of a poly-glutamine-ataxin-3 transgene in orexin neurons induces narcolepsy-cataplexy in the rat. J Neurosci. 2004;24(18):4469–77.
8. Zhang S, Lin L, Kaur S, et al. The development of hypocretin (orexin) deficiency in hypocretin/ataxin-3 transgenic rats. Neuroscience. 2007;148(1):34–43.
9. Kisanuki Y, Chemelli R, Sinton C, et al. The role of orexin receptor type-1 (OX1R) in the regulation of sleep. Sleep. 2000;23:A91.
10. Sakurai T. The neural circuit of orexin (hypocretin): maintaining sleep and wakefulness. Nat Rev Neurosci. 2007;8(3):171–81.
11. Funato H, Tsai AL, Willie JT, et al. Enhanced orexin receptor-2 signaling prevents diet-induced obesity and improves leptin sensitivity. Cell Metab. 2009;9(1):64–76.
12. Willie JT, Chemelli RM, Sinton CM, et al. Distinct narcolepsy syndromes in orexin receptor-2 and orexin null mice: molecular genetic dissection of non-REM and REM sleep regulatory processes. Neuron. 2003;38(5):715–30.
13. Mochizuki T, Marcus J, Arrigoni E, et al. Rescue of fragmented sleep/wake behavior in orexin receptor-2 disrupted mice. Annual Meeting of the Society for Neuroscience. 2007.
14. Alexandre C, Mochizuki T, Arrigoni E, et al. Orexin acts in the basal forebrain to stabilize wakefulness. Paper presented at: Annual Meeting of the Society for Neuroscience, 2008; Washington, DC.
15. Lin L, Faraco J, Li R, et al. The sleep disorder canine narcolepsy is caused by a mutation in the hypocretin (orexin) receptor 2 gene. Cell. 1999;98(3):365–76.
16. Broughton R, Valley V, Aguirre M, Roberts J, Suwalski W, Dunham W. Excessive daytime sleepiness and the pathophysiology of narcolepsy-cataplexy: a laboratory perspective. Sleep. 1986;9(1 Pt 2): 205–15.
17. Scammell TE. The neurobiology, diagnosis, and treatment of narcolepsy. Ann Neurol. 2003;53(2):154–66.
18. Mochizuki T, Crocker A, McCormack S, Yanagisawa M, Sakurai T, Scammell TE. Behavioral state instability in orexin knock-out mice. J Neurosci. 2004;24(28):6291–300.
19. Nishino S, Riehl J, Hong J, Kwan M, Reid M, Mignot E. Is narcolepsy a REM sleep disorder? Analysis of sleep abnormalities in narcoleptic Dobermans. Neurosci Res. 2000;38(4):437–46.
20. Diniz Behn CG, Kopell N, Brown EN, Mochizuki T, Scammell TE. Delayed orexin signaling consolidates wakefulness and sleep: physiology and modeling. J Neurophysiol. 2008;99(6):3090–103.
21. Lo CC, Chou T, Penzel T, et al. Common scale-invariant patterns of sleep-wake transitions across mammalian species. Proc Natl Acad Sci USA. 2004; 101(50):17545–8.
22. Blumberg MS, Coleman CM, Johnson ED, Shaw C. Developmental divergence of sleep-wake patterns in orexin knockout and wild-type mice. Eur J Neurosci. 2007;25(2):512–8.
23. Diniz Behn CG, Klerman EB, Mochizuki T, Lin S-C, Scammell TE. Abnormal sleep/wake dynamics in orexin knockout mice. Sleep. 2010;33(3):297–306.
24. Gervasoni D, Lin SC, Ribeiro S, Soares ES, Pantoja J, Nicolelis MA. Global forebrain dynamics predict rat behavioral states and their transitions. J Neurosci. 2004;24(49):11137–47.
25. Tafti M, Rondouin G, Besset A, Billiard M. Sleep deprivation in narcoleptic subjects: effect on sleep stages and EEG power density. Electroencephalogr Clin Neurophysiol. 1992;83(6):339–49.
26. Tafti M, Villemin E, Carlander B, Besset A, Billiard M. Sleep in human narcolepsy revisited with special reference to prior wakefulness duration. Sleep. 1992;15(4):344–51.
27. Besset A, Tafti M, Nobile L, Billiard M. Homeostasis and narcolepsy. Sleep. 1994;17(8 Suppl):S29–34.
28. Zeitzer JM, Buckmaster CL, Lyons DM, Mignot E. Locomotor-dependent and -independent components to hypocretin-1 (orexin A) regulation in sleep-wake consolidating monkeys. J Physiol. 2004;557(Pt 3): 1045–53.
29. Zeitzer JM, Buckmaster CL, Parker KJ, Hauck CM, Lyons DM, Mignot E. Circadian and homeostatic regulation of hypocretin in a primate model: implications for the consolidation of wakefulness. J Neurosci. 2003;23(8):3555–60.
30. Estabrooke IV, McCarthy MT, Ko E, et al. Fos expression in orexin neurons varies with behavioral state. J Neurosci. 2001;21(5):1656–62.
31. Deboer T, Overeem S, Visser NA, et al. Convergence of circadian and sleep regulatory mechanisms on hypocretin-1. Neuroscience. 2004;129(3):727–32.
32. Desarnaud F, Murillo-Rodriguez E, Lin L, et al. The diurnal rhythm of hypocretin in young and old F344 rats. Sleep. 2004;27(5):851–6.
33. Vernet C, Arnulf I. Narcolepsy with long sleep time: a specific entity? Sleep. 2009;32(9):1229–35.
34. Porkka-Heiskanen T, Strecker RE, Thakkar M, Bjorkum AA, Greene RW, McCarley RW. Adenosine: a mediator of the sleep-inducing effects of prolonged wakefulness. Science. 1997;276(5316):1265–8.
35. Murillo-Rodriguez E, Liu M, Blanco-Centurion C, Shiromani PJ. Effects of hypocretin (orexin) neuronal loss on sleep and extracellular adenosine levels in the rat basal forebrain. Eur J Neurosci. 2008;28(6):1191–8.
36. Volk S, Schulz H, Yassouridis A, Wilde-Frenz J, Simon O. The influence of two behavioral regimens on the distribution of sleep and wakefulness in narcoleptic patients. Sleep. 1990;13(2):136–42.
37. Dantz B, Edgar DM, Dement WC. Circadian rhythms in narcolepsy: studies on a 90 minute day. Electroencephalogr Clin Neurophysiol. 1994;90(1): 24–35.
38. Khatami R, Landolt HP, Achermann P, et al. Challenging sleep homeostasis in narcolepsy-cataplexy: implications for non-REM and REM sleep regulation. Sleep. 2008;31(6):859–67.
39. Edgar DM, Dement WC, Fuller CA. Effect of SCN lesions on sleep in squirrel monkeys: evidence for

opponent processes in sleep-wake regulation. J Neurosci. 1993;13(3):1065–79.
40. Dijk DJ, Czeisler CA. Contribution of the circadian pacemaker and the sleep homeostat to sleep propensity, sleep structure, electroencephalographic slow waves, and sleep spindle activity in humans. J Neurosci. 1995;15(5 Pt 1):3526–38.
41. Kripke D. Biological rhythm disturbances might cause narcolepsy. In: Guilleminault C, Dement W, Passouant P, editors. Narcolepsy. New York: Spectrum; 1976. p. 475–83.
42. Broughton R, Krupa S, Boucher B, Rivers M, Mullington J. Impaired circadian waking arousal in narcolepsy-cataplexy. Sleep Res Online. 1998; 1(4):159–65.
43. Abrahamson EE, Leak RK, Moore RY. The suprachiasmatic nucleus projects to posterior hypothalamic arousal systems. Neuroreport. 2001;12(2):435–40.
44. Lu J, Zhang YH, Chou TC, et al. Contrasting effects of ibotenate lesions of the paraventricular nucleus and subparaventricular zone on sleep-wake cycle and temperature regulation. J Neurosci. 2001;21(13):4864–74.
45. Chou TC, Scammell TE, Gooley JJ, Gaus SE, Saper CB, Lu J. Critical role of dorsomedial hypothalamic nucleus in a wide range of behavioral circadian rhythms. J Neurosci. 2003;23(33):10691–702.
46. Yoshida K, McCormack S, Espana RA, Crocker A, Scammell TE. Afferents to the orexin neurons of the rat brain. J Comp Neurol. 2006;494(5):845–61.
47. Deurveilher S, Semba K. Indirect projections from the suprachiasmatic nucleus to major arousal-promoting cell groups in rat: implications for the circadian control of behavioural state. Neuroscience. 2005;130(1):165–83.
48. Zhang S, Zeitzer JM, Yoshida Y, et al. Lesions of the suprachiasmatic nucleus eliminate the daily rhythm of hypocretin-1 release. Sleep. 2004;27(4):619–27.
49. Aston-Jones G, Chen S, Zhu Y, Oshinsky M. A neural circuit for circadian regulation of arousal. Nat Neurosci. 2001;4(7):732–8.
50. Khatami R, Landolt HP, Achermann P, et al. Insufficient non-REM sleep intensity in narcolepsy-cataplexy. Sleep. 2007;30(8):980–9.
51. Nobili L, Besset A, Ferrillo F, Rosadini G, Schiavi G, Billiard M. Dynamics of slow wave activity in narcoleptic patients under bed rest conditions. Electroencephalogr Clin Neurophysiol. 1995;95(6): 414–25.
52. Fronczek R, Raymann RJ, Overeem S, et al. Manipulation of skin temperature improves nocturnal sleep in narcolepsy. J Neurol Neurosurg Psychiatry. 2008;79(12):1354–7.
53. Mochizuki T, Klerman EB, Sakurai T, Scammell TE. Elevated body temperature during sleep in orexin knockout mice. Am J Physiol Regul Integr Comp Physiol. 2006;291(3):R533–40.
54. Fadel J, Frederick-Duus D. Orexin/hypocretin modulation of the basal forebrain cholinergic system: insights from in vivo microdialysis studies. Pharmacol Biochem Behav. 2008;90(2):156–62.
55. Harsh J, Peszka J, Hartwig G, Mitler M. Night-time sleep and daytime sleepiness in narcolepsy. J Sleep Res. 2000;9(3):309–16.
56. Peyron C, Tighe DK, van den Pol AN, et al. Neurons containing hypocretin (orexin) project to multiple neuronal systems. J Neurosci. 1998;18(23): 9996–10015.
57. Burlet S, Tyler CJ, Leonard CS. Direct and indirect excitation of laterodorsal tegmental neurons by Hypocretin/Orexin peptides: implications for wakefulness and narcolepsy. J Neurosci. 2002;22(7): 2862–72.
58. Eggermann E, Serafin M, Bayer L, et al. Orexins/hypocretins excite basal forebrain cholinergic neurones. Neuroscience. 2001;108(2):177–81.
59. Eriksson KS, Sergeeva O, Brown RE, Haas HL. Orexin/hypocretin excites the histaminergic neurons of the tuberomammillary nucleus. J Neurosci. 2001;21(23):9273–9.
60. Horvath T, Peyron C, Diano S, et al. Hypocretin (orexin) activation and synaptic innervation of the locus coeruleus noradrenergic system. J Comp Neurol. 1999;415(2):145–59.
61. Adamantidis AR, Zhang F, Aravanis AM, Deisseroth K, de Lecea L. Neural substrates of awakening probed with optogenetic control of hypocretin neurons. Nature. 2007;450(7168):420–4.
62. Carter ME, Adamantidis A, Ohtsu H, Deisseroth K, de Lecea L. Sleep homeostasis modulates hypocretin-mediated sleep-to-wake transitions. J Neurosci. 2009;29(35):10939–49.
63. Mileykovskiy BY, Kiyashchenko LI, Siegel JM. Behavioral correlates of activity in identified hypocretin/orexin neurons. Neuron. 2005;46(5): 787–98.
64. Lee MG, Hassani OK, Jones BE. Discharge of identified orexin/hypocretin neurons across the sleep-waking cycle. J Neurosci. 2005;25(28):6716–20.
65. Li Y, Gao XB, Sakurai T, van den Pol AN. Hypocretin/Orexin excites hypocretin neurons via a local glutamate neuron-A potential mechanism for orchestrating the hypothalamic arousal system. Neuron. 2002;36(6): 1169–81.
66. España RA, Scammell TE. Sleep neurobiology for the clinician. Sleep. 2004;27(4):811–20.
67. Parmentier R, Ohtsu H, Djebbara-Hannas Z, Valatx JL, Watanabe T, Lin JS. Anatomical, physiological, and pharmacological characteristics of histidine decarboxylase knock-out mice: evidence for the role of brain histamine in behavioral and sleep-wake control. J Neurosci. 2002;22(17):7695–711.
68. Huang ZL, Mochizuki T, Qu WM, et al. Altered sleep-wake characteristics and lack of arousal response to H3 receptor antagonist in histamine H1 receptor knockout mice. Proc Natl Acad Sci USA. 2006;103(12):4687–92.
69. Hunsley MS, Palmiter RD. Norepinephrine-deficient mice exhibit normal sleep-wake states but have shorter sleep latency after mild stress and low doses of amphetamine. Sleep. 2003;26(5):521–6.

70. Nishino S, Sakurai E, Nevsimalova S, et al. Decreased CSF histamine in narcolepsy with and without low CSF hypocretin-1 in comparison to healthy controls. Sleep. 2009;32(2):175–80.
71. Kanbayashi T, Kodama T, Kondo H, et al. CSF histamine contents in narcolepsy, idiopathic hypersomnia and obstructive sleep apnea syndrome. Sleep. 2009;32(2):181–7.
72. Nishino S, Fujiki N, Ripley B, et al. Decreased brain histamine content in hypocretin/orexin receptor-2 mutated narcoleptic dogs. Neurosci Lett. 2001;313(3):125–8.
73. Montplaisir J, de Champlain J, Young SN, et al. Narcolepsy and idiopathic hypersomnia: biogenic amines and related compounds in CSF. Neurology. 1982;32(11):1299–302.
74. Faull KF, Guilleminault C, Berger PA, Barchas JD. Cerebrospinal fluid monoamine metabolites in narcolepsy and hypersomnia. Ann Neurol. 1983;13(3):258–63.
75. Faull KF, Guilleminault C, Berger PA, Barchas JD. Cerebrospinal fluid monoamine metabolites in narcolepsy: reanalysis. Ann Neurol. 1989;25(3):310–1.
76. Frederick-Duus D, Guyton MF, Fadel J. Food-elicited increases in cortical acetylcholine release require orexin transmission. Neuroscience. 2007;149(3):499–507.
77. Guo RX, Anaclet C, Roberts JC, et al. Differential effects of acute and repeat dosing with the H3 antagonist GSK189254 on the sleep-wake cycle and narcoleptic episodes in Ox–/– mice. Br J Pharmacol. 2009;157(1):104–17.
78. España RA, McCormack SL, Mochizuki T, Scammell TE. Running promotes wakefulness and increases cataplexy in orexin knockout mice. Sleep. 2007;30(11):1417–25.
79. Dishman RK. Brain monoamines, exercise, and behavioral stress: animal models. Med Sci Sports Exerc. 1997;29(1):63–74.
80. Chou TC, Bjorkum AA, Gaus SE, Lu J, Scammell TE, Saper CB. Afferents to the ventrolateral preoptic nucleus. J Neurosci. 2002;22(3):977–90.
81. Marcus JN, Aschkenasi CJ, Lee CE, et al. Differential expression of orexin receptors 1 and 2 in the rat brain. J Comp Neurol. 2001;435(1):6–25.
82. Hirota K, Kushikata T, Kudo M, Kudo T, Lambert DG, Matsuki A. Orexin A and B evoke noradrenaline release from rat cerebrocortical slices. Br J Pharmacol. 2001;134(7):1461–6.
83. Zepelin H, Siegel J, Tobler I. Mammalian Sleep. In: Kryger M, Roth T, Dement W, editors. Principles and practice of sleep medicine. 4th ed. Philadelphia: Elsevier; 2005.
84. Waterston RH, Lindblad-Toh K, Birney E, et al. Initial sequencing and comparative analysis of the mouse genome. Nature. 2002;420(6915):520–62.
85. Scammell TE, Willie JT, Guilleminault C, Siegel JM. A consensus definition of cataplexy in mouse models of narcolepsy. Sleep. 2009;32(1):111–6.
86. Alexandre C, Popa D, Fabre V, et al. Early life blockade of 5-hydroxytryptamine 1A receptors normalizes sleep and depression-like behavior in adult knock-out mice lacking the serotonin transporter. J Neurosci. 2006;26(20):5554–64.
87. Liu M, Thankachan S, Kaur S, et al. Orexin (hypocretin) gene transfer diminishes narcoleptic sleep behavior in mice. Eur J Neurosci. 2008;28(7):1382–93.
88. Ko B, Kantor S, Mochizuki T, Clark E, Clain E, Scammell T. Treatment of narcolepsy by gene therapy in mice lacking the orexin neurons. Paper presented at: Annual Meeting of the Society for Neuroscience, 2009; Chicago.
89. Franken P, Malafosse A, Tafti M. Genetic determinants of sleep regulation in inbred mice. Sleep. 1999;22(2):155–69.
90. Mutant mice and neuroscience: recommendations concerning genetic background. Banbury Conference on genetic background in mice. Neuron. 1997;19(4):755–9.
91. Crusio WE, Goldowitz D, Holmes A, Wolfer D. Standards for the publication of mouse mutant studies. Genes Brain Behav. 2009;8(1):1–4.
92. Saper CB, Scammell TE, Lu J. Hypothalamic regulation of sleep and circadian rhythms. Nature. 2005;437(7063):1257–63.

Mathematical Models of Narcolepsy

Cecilia Diniz Behn

Keywords
Mathematical models • Narcolepsy • Electroencephalogram • Spectral analysis • Prediction • Physiology

Introduction

Mathematical modeling offers a way to critically test experimentally derived theories, integrate experimental results across spatial and temporal scales, and generate predictions to drive bench science and influence clinical practice. Although, in general, mathematical modeling approaches have been applied to narcolepsy only recently, developments in modeling normal sleep/wake behavior have laid an excellent foundation for linking the experimental insights about the orexin (also known as hypocretin) system to the observed features of the narcolepsy phenotype. Thus, many recent models have addressed aspects of narcolepsy including fragmentation of sleep and wake behavior, altered cycling between rapid eye movement (REM) and non-REM (NREM) sleep, sleep onset REM periods (SOREMPs), and cataplexy.

In this chapter, I review both phenomenological and physiologically based models that capture elements of human narcolepsy and animal models of narcolepsy (Table 1). Phenomenological models synthesize theories about sleep/wake regulation and help to identify processes that may be altered in narcolepsy. Physiologically based models generate predictions about mechanisms underlying specific elements of the narcolepsy phenotype. The chapter concludes with a brief discussion of the strengths and limitations of existing models and some possible directions for future work.

Phenomenological Models of Narcolepsy in Humans

One of the most influential models in sleep research, the two-process model, describes the timing of human sleep/wake behavior resulting from the interaction of circadian (Process C) and homeostatic (Process S) processes [1]. This model has provided an intuitive framework for understanding the regulation of sleep/wake behavior and shaped the direction of the field. Although the two-process model was mainly phenomenological when it was postulated in the early 1980s, experimental work has identified biological correlates for each process: the molecular and electro physiological rhythms in the suprachiasmatic nucleus (SCN) underlie Process C, and the increase and decrease in electroen-

C. Diniz Behn (✉)
Department of Mathematics, University of Michigan, 2074 East Hall, 530 Church Street, Ann Arbor, MI 48109, USA
e-mail: cdbehn@umich.edu

C.R. Baumann et al. (eds.), *Narcolepsy: Pathophysiology, Diagnosis, and Treatment*, DOI 10.1007/978-1-4419-8390-9_16, © Springer Science+Business Media, LLC 2011

Table 1 Summary of current mathematical models of narcolepsy

	Human	*Animal*
Phenomenological	Electronic model [9]	
	Two-process-type model [1, 8]	
Physiological	Reciprocal interaction limit cycle model [34, 35]	Formal neuron quartet model [21]
	Combination reciprocal interaction and two-process-type model [43]	Brief wake bout relaxation oscillator model [15]
	Continuum neuronal population model [45]	Single-cell models [33, 50, 51]
	NREM–REM flip-flop relaxation oscillator model [46]	

cephalogram (EEG) slow-wave activity (SWA; power density in the 0.75–4 Hz band) during wake and sleep, respectively, provide a marker of Process S, though the biological essence of sleep drive remains unknown.

When measures of Process S or Process C are altered, as in adolescence or the polyphasic sleep of animals, appropriate modifications of the modeled processes capture the differences in sleep/wake behavior [2, 3]. Therefore, a necessary first step in applying the two-process model to narcolepsy was measuring Process C and Process S in patients with this disorder. Prior work suggested that while circadian rhythms were functional in narcolepsy, there may be abnormalities in homeostatic sleep regulation [4–6]. One indicator of abnormality was a steeper decline of SWA in patients with narcolepsy compared to controls [7]. This decline was related to an impaired accumulation of SWA during the second sleep cycle, but it was difficult to determine whether the impairment reflected the increased fragmentation of sleep present in these patients or represented different dynamics of Process S.

To investigate these possibilities, Khatami and colleagues conducted a 40-h sleep deprivation protocol to facilitate sleep consolidation [8]. This sleep deprivation protocol increased SWA and postponed the occurrence of sleep fragmentation, thereby allowing the authors to assess the time course of undisturbed SWA and, therefore, estimate the dynamics of Process S in narcolepsy. They found that the time course of Process S was normal suggesting that the alteration in Process S dynamics observed in narcolepsy results from increased sleep fragmentation rather than alterations in sleep homeostasis. Therefore, the two-process model alone may not be able to capture certain aspects of the narcolepsy phenotype, but the interactions between Process S and Process C will be important elements of more detailed models.

Another largely phenomenological model of narcolepsy from the 1980s was proposed by Lawder [9]. Lawder constructed a model for sleep/wake regulation from various analog and digital electronic computing elements. Although most model elements were not expressed in biological terms, several elements had explicit biological correlates. These included a homeostatic sleep drive, described as a "normal running toxin" (NRT), and an oscillation between REM-on and REM-off populations based on reciprocal interactions between locus coeruleus (LC) and the gigantocellular tegmental field (FTG) described by McCarley and Hobson [10]. In addition to sleep/wake state, model outputs included REM density (the number of eye movements per unit time) and motor inhibition. By including motor inhibition, typically associated with REM sleep, Lawder's modeling framework could simulate cataplexy and sleep paralysis.

Lawder attempted to model many aspects of the narcolepsy phenotype including sleep attacks, cataplexy, sleep paralysis, and hypnagogic hallucinations while maintaining normal durations of nocturnal sleep. By reducing the threshold for the transition from wake to sleep, he introduced sleep attacks when NRT was high and SOREMPs when NRT was low. Weakening the inhibition associated with an absence of sleep changed the balance between the wake phase and disinhibition of motor inhibition and REM density. Lawder interpreted this model behavior as simulated cataplexy, sleep paralysis, and hypnagogic hallucinations. He also referred to modeling instability of the LC/FTG oscillator with adjustable sensitivity to noise, though details were not provided.

Although phenomenological models are limited in their ability to predict mechanisms for the features they describe, they provide an important formalization of conceptual hypotheses. Furthermore, by linking conceptual elements such as Process S to clinical measures like SWA, phenomenological models provide general insights that can ultimately be investigated in an appropriate biological substrate.

Phenomenological Models of Narcolepsy in Animals

Phenomenological modeling approaches for nonhuman animals are less common, but process models, such as Markov models for sleep/wake behavior in the presence and absence of orexin, could provide insights into network theoretic properties of the physiological sleep/wake regulatory network. The development of such models requires rigorous quantification of sleep/wake architecture. One such approach, introduced by Lo and colleagues, involves the use of survival analysis for characterizing the distribution of wake and sleep bout durations [11]. This technique measures the probability that a given bout will "survive" long enough to reach a given duration, and the resulting survival curves can be fit to statistical distributions to evaluate the structure of the underlying data. Furthermore, survival analyses yield interesting insights into the theoretical properties of the neuronal networks under consideration. For example, power-law distributions often arise in the context of scale-free networks which tend to be more robust than random networks [12, 13].

Lo and colleagues showed that, across species, sleep bout durations followed an exponential distribution while wake bout durations followed a power-law distribution [11]. In orexin knockout (OXKO) mice, a rodent model of narcolepsy, these general distributions were maintained in both pups [14] and in adult animals [15]. However, in adult animals, Diniz Behn and colleagues identified several key differences in wake bout duration survival curves for wild type (WT) and OXKO mice (Fig. 1) [15]. Most strikingly, the survival curve for WT mice contained a plateau region between 100 and 1,000 s that was absent in the survival curve for OXKO mice. The relative flatness of the WT survival curve through this region indicates that wake bouts rarely terminate between 100 and 1,000 s. By contrast, the absence of the plateau in the OXKO survival

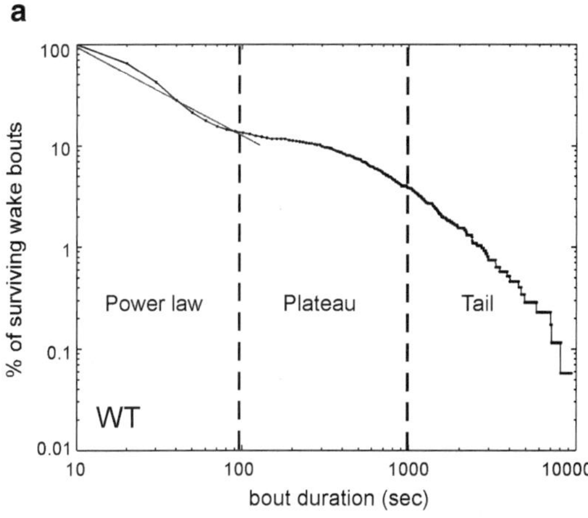

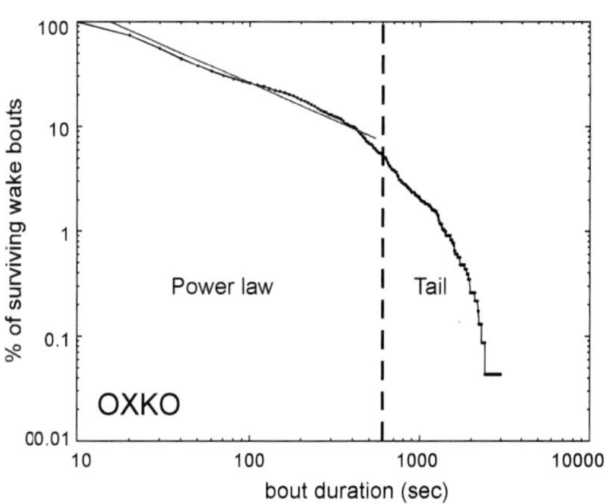

Fig. 1 Wake bout duration survival plots reveal a different structure in orexin knockout (OXKO) mice compared to wild type (WT) mice. (**a**) In WT mice, the survival curve shows three distinct regions: power law, plateau, and tail. (**b**) In OXKO mice, the plateau region is absent, and the power law behavior holds for wake bouts of durations up to 600 s. Note that the power law regions appear linear in these log–log plots

curve suggests that orexins are necessary for the maintenance of long bouts of wakefulness, but orexin deficiency had little impact on wake bouts less than 100 s in duration. The authors incorporated these findings into a physiologically based mathematical model of the sleep/wake regulatory network which is described in detail below. However, these data could also be used to develop process models for murine narcolepsy similar to those used to describe normal sleep/wake behavior [12, 16].

Physiologically Based Models of Narcolepsy in Animals

Extensive experimental work has identified many neuronal nuclei participating in sleep/wake regulation (Table 2). As the interactions among these neuronal populations have been established, different conceptual models for sleep/wake regulation have been proposed. For example, the mutual inhibition between wake-promoting monoaminergic neuronal populations and a sleep-promoting population in the ventrolateral preoptic nucleus (VLPO) has been postulated to act like an electrical flip-flop switch driving fast transitions between states of wake and NREM sleep [17], while reciprocal interactions between excitatory REM-promoting populations and inhibitory REM-off/wake-promoting populations is a long-standing proposed mechanism for the regulation of REM sleep [10]. Dynamic, physiologically based, mathematical frameworks provide formalisms for implementing these conceptual models to evaluate network structures and mechanisms in detail. For each physiologically based model, I summarize the assumed network structure of sleep/wake regulation (Fig. 2) and briefly discuss the modeling formalism applied. Although there are many similarities in the sleep/wake regulatory networks considered in these models, different model formalisms and inputs generate different behaviors and transition mechanisms. In each of these models, unless otherwise noted, the connectivity between different neuronal populations was specified by experimentally reported projections, and the sign of the coupling effects (excitatory or inhibitory) was chosen to be consistent with the action of the associated neurotransmitter on the postsynaptic population.

Because the experiments prescribing the anatomy and physiology of sleep/wake regulation focus on nonhuman animals, I begin the discussion of physiologically based models with models simulating animal sleep. Animals with disrupted orexin signaling exhibit sleepiness and cataplexy strikingly similar to that seen in patients with narcolepsy [18–20]. Therefore, the link between physiology and narcolepsy phenotype may be investigated directly in this context.

Modeling Approach Focusing on REM Sleep Abnormalities in Narcolepsy-Like Behavior in Rats

Tamakawa and colleagues introduced the most detailed of the physiologically based network structures that have been applied to model narcolepsy/narcolepsy-like behavior [21]. They employed a model structure including ten distinct neuronal populations to simulate polyphasic rat sleep. This is the only model in which the activity of the orexin population was explicitly modeled. The model exhibited emergent organization into a quartet of neuronal groups displaying similar state-dependent activity (these groups are denoted by the Wake, NREM, REM, and Wake/REM populations in Table 2 and Fig. 2c). Inputs to the orexin population included inhibition from sleep-promoting populations, circadian (sinusoidal) excitation, and auto-excitation. The orexin population projected to all of the wake-active populations in the model network.

Each neuronal population in the network was modeled by a formal neuron that represented the collective activity of the neurons in the population, and sleep/wake state was determined by the relative activity/inactivity of the relevant populations. The formal neurons tended to assume states of high or low activity with minimal intermediate activity. Although the simulated neuronal activities roughly followed experimentally determined profiles of neuronal activities at state transitions, the onset of activity in the orexin population

Table 2 Neuronal populations and modulatory inputs involved in sleep/wake regulation and their representation in physiologically based mathematical models of narcolepsy

Neuronal population/modulation	Massaquoi and McCarley [35]	Ferrillo et al. [43]	Tamakawa et al. [21]	Diniz Behn et al. [15]	Phillips and Robinson [45]	Rempe et al. [46]	Designation in Fig. 2
Orexin/hypocretin cells	○	○	•	○	○	○	OX
Locus coeruleus	•	•	•	•	•	•	Wake
Dorsal raphe			•	•	•	•	
Tuberomammillary nucleus			•	•	•	•	
Ventrolateral preoptic nucleus			•	•	•		NREM
Median preoptic nucleus			•				
Extended ventrolateral preoptic nucleus				•		•	eVLPO
Laterodorsal tegmental nucleus (REM active)	•	•	•	•	○		REM
Pedunculopontine tegmental nucleus (REM active)	•	•	•	•	○		
Mesopontine tegmental area (REM active)			•				
Sublaterodorsal nucleus						•	REM-on
Precoeruleus						•	
Laterodorsal tegmental nucleus (Wake or Wake/REM active)			•				Wake/REM
Pedunculopontine tegmental nucleus (Wake or Wake/REM active)			•				
Ventrolateral periaqueductal gray matter						•	REM-off
Lateral pontine tegmentum						•	
Basal forebrain (Wake/REM)			•				
Circadian input	○		•		•		C
Homeostatic input		•	•	•	•	•	H
Slow-wave activity		•					SWA

Solid dots indicate that population activity was modeled explicitly while *open dots* indicate that the population provided modulatory input to the model

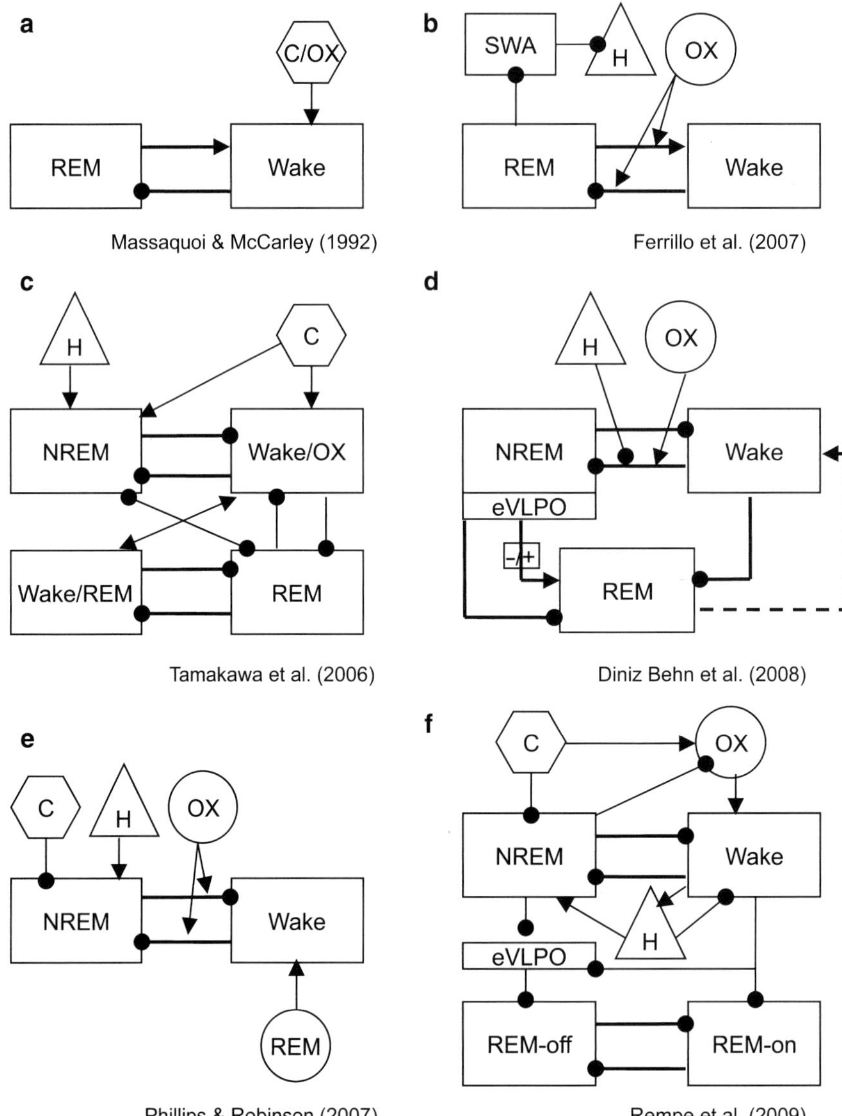

Fig. 2 Summary schematics of physiologically based models (inputs external to these networks and neuronal autoeffects are omitted for clarity). Uniform labeling has been used to facilitate comparison of model structures; the specific populations represented in each model are indicated in Table 2. Modeled neuronal populations are denoted with *rectangles*, and modulatory factors include circadian drive (**C**), homeostatic sleep drive (H), and orexin effects (OX). Note that the model structures contain many common elements. The model networks in panels **a–d** include reciprocal interactions between REM and Wake populations, and model networks in panels **c–f** include flip-flop interactions between NREM and Wake populations. (**a**) Altered circadian modulation of REM sleep observed in patients with narcolepsy may occur as a result of orexin-dependent mediation of circadian input to the Wake population. (**b**) Reducing the strength of interaction between REM and Wake populations produced SOREMPs and the longer REM–NREM cycles typical of narcolepsy. Slow-wave activity (SWA) was also simulated in this model. (**c**) When the orexin population was omitted from the full network structure of ten neuronal populations, simulated rat sleep/wake behavior showed an increase in REM sleep during the dark (active) period. The full model network is summarized in this schematic with a quartet of neuronal groups classified by similar activity profiles. (**d**) Modeling orexin effects as time-dependent modulation of inhibition from the Wake to the NREM population captured the stabilizing effect of orexin on mouse sleep/wake behavior. The indirect projection from eVLPO to the REM population is described with a net excitatory effect because the disinhibition of the intermediate population is not modeled explicitly. (**e**) Altered coupling strengths between NREM and Wake populations reproduced the fragmented sleep and wake behavior of patients with narcolepsy. (**f**) When orexin effects were modeled as a gating of excitatory circadian input to wake-promoting monoaminergic cell groups, SOREMPs were produced and a stabilizing effect on sleep/wake behavior was observed

following the transition from NREM sleep to wake was delayed for approximately 10 s. Although such a delay conflicts with experimental reports of anticipatory firing in these cells [22], the relevance of this delay in the model may be linked to the delayed onset of functional effects of orexin identified in the survival analysis.

Early reports of abnormal behavior in rodents with disrupted orexin signaling focused on the increase in REM sleep during the dark (active) period when REM sleep is normally suppressed [20, 23, 24]. To investigate this observation, Tamakawa and colleagues simulated the orexin knockout condition in their network model for rat sleep/wake regulation by removing the population of orexin neurons. The resulting simulated behavior showed an increase in the percentage of REM sleep during the dark period consistent with experimental results. This increase appeared to come mainly at the expense of NREM sleep, and the percent time spent in each state during the light period was not affected. Other summary statistics, such as mean bout durations and number of bouts, were not reported. However, the authors noted that the organization of sleep and wakefulness during the active period was unchanged suggesting that the loss of the orexinergic population did not result in fragmentation of sleep/wake behavior in their model.

Modeling Approach Focusing on Sleep/Wake Fragmentation in Murine Narcolepsy

Further experimental characterization of the OXKO mouse phenotype identified behavioral state instability as its primary feature [25–27]. Therefore, fragmented sleep/wake behavior was the main focus of the modeling approach employed by Diniz Behn and colleagues [15]. Based on the survival analysis of wake bout durations previously discussed, the authors concluded that the state instability in OXKO mice reflected a loss of the orexin-mediated maintenance of long bouts of wakefulness, but that brief wake bouts were minimally affected by orexin deficiency. Thus, orexin effects were modeled with a saturating sigmoidal function of time awake that minimized effects at wake onset, reflected time-dependent activation of orexin signaling, and eventually saturated in a manner consistent with biological signaling.

This model of the mouse sleep/wake regulatory network consisted of Wake-, NREM-, and REM-promoting populations, and orexin acted in the network to enhance the strength of inhibition from the Wake to the NREM population (Fig. 2d). The activity of each population was described with normalized Morris–Lecar equations [28, 29]. Like the formal neuron model, this modeling formalism primarily produces states of high and low activity with fast transitions through intermediate levels of activity. However, in appropriate parameter regimes, the normalized Morris–Lecar equations can also produce robust relaxation oscillations. Therefore, this modeling formalism included an intrinsic oscillatory state that reflected the inherent excitability of the modeled neuronal populations and provided a mechanism for brief awakenings. The resulting microarchitecture of simulated sleep/wake behavior highlighted the utility of this model for investigating mechanisms associated with behavioral state instability.

When simulated orexin effects were omitted, the model reproduced the fragmented sleep/wake behavior of OXKO mice. Analysis of the action of orexin in the model revealed that the key to this implementation was the differential effect of orexin on simulated brief and sustained wake bouts. The model structure used separate mechanisms to produce brief and sustained wake bouts: homeostatic drive governed sustained wakefulness while brief wake bouts were generated by intrinsic excitability in the wake-promoting populations [29]. Because the modeled orexin effect was minimized at wake onset, brief wake bouts were impacted minimally. By contrast, the modeled orexin effect was important for generating the sustained wake bouts typical of behavior in WT mice.

These modeling results lend support to the prediction that the functional effect of orexin is delayed by 1–2 min following wake onset. Although such a delay is difficult to reconcile with standard millisecond time courses associated with neuronal activity, the colocalization of orexin and dynorphin in orexin neurons may provide a potential mechanism for this delay [30]. In tuberomammillary neurons, direct excitatory orexin effects and disinhibition by dynorphin act together to increase excitability [31]. However, in orexin neurons, orexins exert indirect excitatory effects while dynorphin is directly inhibitory [32]. Furthermore, desensitization to dynorphin causes the net effect of the colocalized peptides to shift from inhibition to excitation [32]. Interestingly, desensitization to dynorphin takes approximately 90 s, consistent with the delay in functional orexin effects identified in the survival analysis. Therefore, the colocalization of peptides combined with the dynamics of desensitization may result in complex modulation of postsynaptic targets that is manifested as a delay in the onset of orexin effects. Preliminary investigation of this hypothesis in a network of Hodgkin–Huxley-type model orexin neurons supports this possibility [33, 51].

Physiologically Based Models of Narcolepsy in Humans

With the elucidation of sleep/wake regulatory circuitry in nonhuman animal models, several groups have applied animal neuronal network structures to model human sleep. Much of the relevant physiology is conserved across species, so the model structures obtained with this approach are probably reasonable approximations of human circuitry. However, modifications to model inputs and time scales were necessary to capture consolidated human sleep/wake behavior and ultradian REM–NREM cycling.

Modeling Approaches Focusing on REM Sleep Abnormalities in Narcolepsy

The McCarley–Hobson reciprocal interaction model for REM/NREM sleep regulation provided one of the earliest physiologically based approaches to sleep modeling [10]. In this model, reciprocal connectivity between an inhibitory Wake population and an excitatory REM sleep population (Fig. 2a) gave rise to oscillations that reproduced the ultradian structure of REM–NREM cycling in human sleep. Later versions of this model employed robust limit cycle mathematical structures and addressed circadian variation in the REM–NREM cycle [34, 35]. In both original and later versions of this model, population activity was represented by a single variable that changed continuously between states of REM and NREM sleep, and thresholding was used to determine the state of the model. Circadian modulation was explicitly included in the model by varying excitatory input to the Wake population. This variation resulted in the absence of REM sleep during the day and a progression from shorter REM cycles with smaller amplitude in the early part of the night to longer REM cycles at the end of the night, consistent with observed behavior.

Recently, McCarley has proposed that orexin is the factor or one of the factors providing circadian modulation of the modeled Wake population [36]. This hypothesis is consistent with evidence for direct and indirect projections from the SCN to the orexin neurons [37–39], and it suggests that the removal of circadian modulation from the limit cycle model would capture some of the abnormalities associated with REM sleep in patients with narcolepsy including the occurrence of REM sleep at abnormal circadian phases [40]. A recent study comparing REM sleep in ataxin-3 mice and mice lacking the orexin neuropeptides suggests that orexin neurons, but not the orexin peptides, are necessary for normal circadian suppression of REM sleep [41]. By contrast, circadian control of wake seems normal in both ataxin-3 mice and OXKO mice [25, 41]. A more complex modeling framework, including the action of colocalized orexin and dynorphin in orexin neurons, is needed to investigate the mechanisms underlying this observation.

In addition to changes in circadian variation in REM sleep with narcolepsy, sleep in narcolepsy patients is characterized by changes to the ultradian REM–NREM cycle including SOREMPs and longer REM–NREM cycle duration (120 min in patients with narcolepsy vs. 90 min in normals) [5, 8, 42]. To investigate the source of these observations, Ferrillo and colleagues merged the simple McCarley–Hobson reciprocal interaction model with the Borbely two-process model to study circadian and homeostatic regulation of REM–NREM cycling in patients with narcolepsy [43]. Their model consisted of four differential equations describing the dynamics of the homeostatic Process S, the SWA time course, and coupled REM sleep and Wake populations (Fig. 2b). The circadian process, Process C, was not included in this model. Activity in the REM sleep population influences SWA, but there was no feedback from Process S or SWA onto the REM–Wake population network. REM sleep was defined to occur when the relative values of the REM sleep population were greater than or equal to the momentary level of Process S. With this definition, the decrease in Process S over the course of the night caused an increase in the duration of REM sleep episodes despite the absence of explicit homeostatic or circadian variation in the activity of the REM sleep and Wake populations. This implementation contrasts with the direct circadian modulation of the Wake population in the limit cycle model of Massaquoi and McCarley [35].

Ferrillo and colleagues separately optimized model parameters for data from patients with narcolepsy and normals. The only significant difference in parameters occurred in the interaction parameters between the REM and Wake populations: the strength of coupling was reduced in the narcoleptic case. With the narcoleptic parameter set, the model produced an oscillating pattern with wider peaks in REM and Wake population activity and, consequently, captured the observed increase in REM–NREM cycle duration. Furthermore, the reduced strength of inhibition on the REM population was sufficient to allow SOREMPs. Thus, these findings suggest a mechanistic basis for both SOREMPs and the increase in REM–NREM cycle duration in narcolepsy.

Modeling Approaches Focusing on Sleep/Wake Fragmentation in Narcolepsy

Tamakawa and colleagues simulated human sleep/wake behavior with a model structure identical to the structure they used to simulate behavior in rats [21]. To achieve this, they increased the strength of circadian modulation, changed the rates of production and dissipation of the homeostatic sleep factors, and increased inhibition of the noncholinergic REM-on population by the GABAergic NREM sleep-promoting medial preoptic nucleus. The resulting simulated human sleep demonstrated a concentration of sleep in the appropriate circadian phase and reflected realistic human sleep architecture. In the model of human sleep, the wake-active orexin neurons played an important role in concentrating wakefulness at the appropriate circadian phase. The authors did not report simulations of the model of human sleep with the orexin population eliminated, but future work investigating this condition may provide useful insights into the role of orexin in circadian modulation of human sleep/wake behavior and the implications for disruption of this signaling in narcolepsy.

Although orexin neurons are primarily wake active [22, 44], fragmented sleep is common in patients with narcolepsy [40]. To investigate this conundrum, Phillips and Robinson used a network structure that focused on the mutual inhibition between Wake and NREM populations (Fig. 2e) [45]. REM–NREM cycling was not addressed in this model, although a constant excitatory input (representing cholinergic REM-promoting populations) was included. Each population was represented with a continuum neuronal population model that described mean cell body potential; mean firing rate was reported as a function of mean cell body potential. Like the formalisms described previously, the continuum neuronal population model describes average activity in a population of neurons rather than describing spiking of individual neurons, and most simulated behavior corresponded to high or low levels of activity with fast transitions through intermediate levels of activity.

The parameter space associated with the Phillips and Robinson model was partitioned into three key regions, and modulation by combined circadian and homeostatic inputs moved model trajectories among these regions (Fig. 3). For parameters associated with the Wake and Sleep regions, the model could express only wake or sleep, respectively. In the bistable region, either wake or sleep could be expressed. In general, when the trajectory entered the bistable region, it remained in its prior state; state transitions did not occur until the trajectory crossed into the other stable region. For example, if the model trajectory began in the stable wake region, it typically remained in the wake state through the bistable region and transitioned to sleep only when it reached the stable sleep region. However, within the bistable region a sufficiently strong noisy stimulus could induce a premature state transition.

Under baseline conditions, the model simulated human sleep/wake behavior. When parameters were varied to reduce the width of the bistable zone, noise-driven state transitions were more likely, and the stability of the sleep/wake flip-flop switch was decreased. Phillips and Robinson interpreted this as a mechanism for narcolepsy, and they inferred that the parameter governing the width of the bistable zone reflected activity of the orexin system. The role of this parameter in the model was consistent with orexin-mediated modulation of coupling strengths between Wake and NREM populations. Because the reduction of the bistable zone resulted in a symmetric reduction of stability of both the wake and sleep regions, this model suggested a basis for the fragmentation of sleep (as well as fragmentation of wake) observed in patients with narcolepsy.

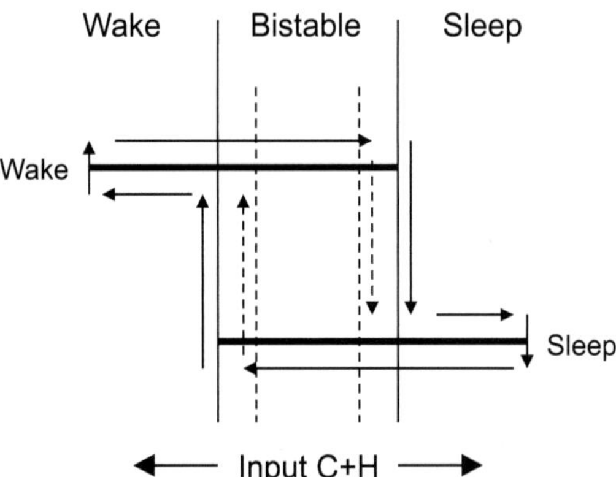

Fig. 3 The parameter space for the model proposed by Phillips and Robinson included three key regions: a stable wake region, a stable sleep region, and a bistable region in which both sleep and wake solutions, denoted by *thick horizontal lines*, exist. Combined circadian (C) and homeostatic (H) inputs moved the model among these regions (and, therefore, among behavioral states) in a prescribed trajectory loop denoted *by a series of arrows*. When the bistable region was reduced (*dashed lines*), the loop traced by the trajectory was decreased (*dashed arrows*). This resulted in a reduction of stability of the sleep/wake flip-flop switch, consistent with a disruption of orexin signaling

Rempe and colleagues also employed a model structure that included the sleep/wake flip-flop switch architecture (Fig. 2f) [46]. However, for REM sleep regulation, they incorporated the recently proposed NREM–REM flip-flop structure based on mutual inhibition between GABAergic REM-on and REM-off populations (Table 2) [47] and assumed an inhibitory projection from the VLPO core to the extended VLPO. They did not model the activity of the orexin population explicitly, but they included orexinergic modulation in their network. In the baseline model, orexin acted on the Wake population, and orexin effects were modulated by excitatory input from SCN and inhibitory input from the NREM population. The activity of each neuronal population was described by a Morris–Lecar system similar to the approach used by Diniz Behn and colleagues [15, 28]. As previously described, this formalism allowed the same fast transitions between high and low levels of activity that were exploited by the formal neuron and continuum neuronal population approaches [21, 45], but it also permitted intrinsic (relaxation) oscillations of population activity in appropriate parameter regimes.

In this model, removal of orexin input eliminated excitatory input to the system from the SCN. As a result, the model exhibited more frequent transitions between sleep and wake states, and the frequency of REM bouts was increased, consistent with observations in patients with narcolepsy. The increase in state transitions was related to the mechanism for the production of wake bouts. Under baseline conditions, the Wake population showed stable, sustained activity during the consolidated waking period; by contrast, in the absence of orexin input, the Wake population oscillated between high and low levels of activity every few hours (Fig. 4). This oscillation occurred because, when SCN inhibition to the NREM population was maximal, the NREM population could not activate even when it was released from Wake population-mediated inhibition. The authors interpreted the resulting pattern of short activation of the Wake population separated by periods of inactivation in both Wake and NREM populations as short bouts of wakefulness separated by bouts of NREM sleep. The possibility of NREM sleep without activation of sleep-promoting neuronal populations was a prediction of their model.

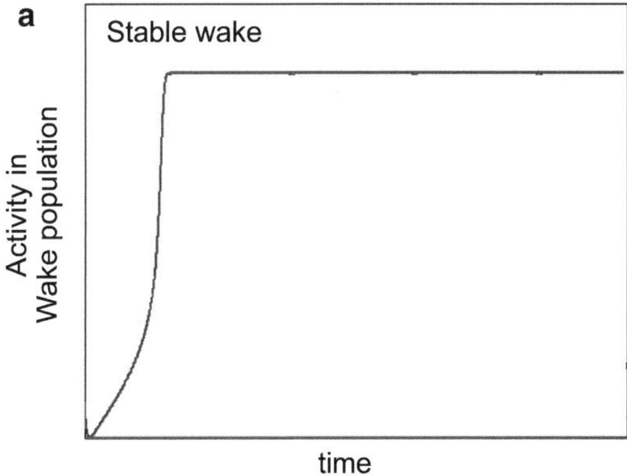

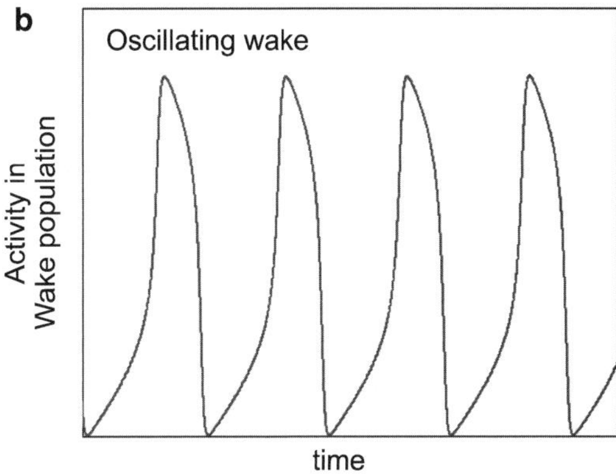

Fig. 4 In the model of Rempe and colleagues, the presence or absence of orexin results in different mechanisms for the production of wake bouts. (**a**) With orexin, the Wake population exhibits sustained activity and an extended wake bout is produced. The time course of this wake bout is governed by the homeostatic sleep drive. (**b**) Without orexin, activity in the Wake population oscillates between high and low levels resulting in wake bouts on the order of several hours

Rempe and colleagues also identified a mechanism for SOREMPs in their model. Without noise, transitions from wake to sleep were initiated by activation of the NREM population. However, in the presence of noise, sleep could be initiated by inactivation of the Wake population. When the Wake population inactivated, the REM population was released from inhibition before the NREM-promoting population, thereby resulting in a SOREMP. Because the mechanism for SOREMPs relies on inactivation of the Wake population, under normal conditions it occurred only at circadian phases when excitatory input to the Wake population was low. However, since the orexin population gates the excitatory circadian input to the system, the elimination of the orexin population permitted SOREMPs at all circadian phases.

Conclusions and Future Directions

In this chapter, I have discussed both phenomenological and physiologically based mathematical models of narcolepsy. I have focused on the concepts underlying each modeling approach and the characteristics of the narcolepsy phenotype they address rather than the mathematical details, but I have also tried to provide a sense of the analyses possible with these methods.

The reviewed models have addressed many aspects of narcolepsy (or narcolepsy-like behavior in nonhuman animals). The diversity of species represented in these models reflects the development of models based strictly on available animal data and allows for the extrapolation of these data to human physiology. Ideally, there exists a symbiotic relationship between experimental and theoretical approaches: mathematical models are based on existing experimental data, and model predictions suggest new ways of synthesizing data and drive future experiments. In the case of physiologically based models, analyses of the model mechanisms associated with specific behaviors have generated predictions about the role of orexin in sleep/wake regulation. Indeed, many of these models have predicted mechanisms and/or phenomena that await experimental investigation. These predictions include the role of orexin in circadian timing of REM sleep [36], delayed onset of functional orexin effects [15], and concurrent inactivity in Wake and NREM populations during NREM sleep in patients with narcolepsy [46]. Future work in which models are "reverse engineered" to probe mechanisms of robustness and state stability may help to identify novel targets for clinical treatment of narcolepsy.

Some aspects of the narcolepsy phenotype have not yet been addressed in a mathematical context or have been addressed incompletely. Many of the reviewed physiologically based models have grouped wake-promoting monoaminergic populations based on their coordinated behavior under normal conditions. However, since this coordination breaks down in pathological states such as cataplexy [48, 49], more detailed model structures are needed. Indeed, even the assumption in population modeling formalisms of coordinated activity within neuronal populations neglects the potentially significant role of single-cell behavior. Several recent mathematical modeling approaches have begun to focus on the properties of orexin neurons and the roles of these cells in modulation of sleep/wake behavior [33, 50], but additional work is necessary to fully exploit the power of single cell modeling to address orexin neuron function.

Clearly, there are many opportunities for future work in this field. Mathematical models of sleep have contributed to the understanding and treatment of jet lag, sleep problems in shift workers, and mood disorders. Mathematical models of narcolepsy are poised to synthesize experimental data and link spatial and temporal scales from the single neuron level to the level of behavior. As mathematical modeling of narcolepsy progresses, these theoretical approaches will continue to make important contributions to understanding the physiology of the disease and identifying novel treatment options.

References

1. Borbely AA. A two process model of sleep regulation. Hum Neurobiol. 1982;1(3):195–204.
2. Jenni OG, Achermann P, Carskadon MA. Homeostatic sleep regulation in adolescents. Sleep. 2005;28(11): 1446–54.
3. Tobler II, Franken P, Trachsel L, Borbely AA. Models of sleep regulation in mammals. J Sleep Res. 1992;1(2):125–7.
4. Dantz B, Edgar DM, Dement WC. Circadian rhythms in narcolepsy: studies on a 90 minute day. Electroencephalogr Clin Neurophysiol. 1994;90(1):24–35.
5. Nobili L, Besset A, Ferrillo F, Rosadini G, Schiavi G, Billiard M. Dynamics of slow wave activity in narcoleptic patients under bed rest conditions. Electroencephalogr Clin Neurophysiol. 1995;95(6):414–25.
6. Tafti M, Rondouin G, Besset A, Billiard M. Sleep deprivation in narcoleptic subjects: effect on sleep stages and EEG power density. Electroencephalogr Clin Neurophysiol. 1992;83(6):339–49.
7. Khatami R, Landolt HP, Achermann P, et al. Insufficient non-REM sleep intensity in narcolepsy-cataplexy. Sleep. 2007;30(8):980–9.
8. Khatami R, Landolt HP, Achermann P, et al. Challenging sleep homeostasis in narcolepsy-cataplexy: implications for non-REM and REM sleep regulation. Sleep. 2008;31(6):859–67.
9. Lawder RE. A proposed mathematical model for sleep patterning. J Biomed Eng. 1984;6(1):63–9.
10. McCarley RW, Hobson JA. Neuronal excitability modulation over the sleep cycle: a structural and mathematical model. Science. 1975;189(4196):58–60.
11. Lo CC, Chou T, Penzel T, et al. Common scale-invariant patterns of sleep-wake transitions across mammalian species. Proc Natl Acad Sci USA. 2004;101(50):17545–8.
12. Gall AJ, Joshi B, Best J, Florang VR, Doorn JA, Blumberg MS. Developmental emergence of power-law wake behavior depends upon the functional integrity of the locus coeruleus. Sleep. 2009;32(7): 920–6.
13. Albert R, Jeong H, Barabasi AL. Error and attack tolerance of complex networks. Nature. 2000;406(6794): 378–82.
14. Blumberg MS, Coleman CM, Johnson ED, Shaw C. Developmental divergence of sleep-wake patterns in orexin knockout and wild-type mice. Eur J Neurosci. 2006;25(2):512–8.
15. Diniz Behn CG, Kopell N, Brown EN, Mochizuki T, Scammell TE. Delayed orexin signaling consolidates wakefulness and sleep: physiology and modeling. J Neurophysiol. 2008;99(6):3090–103.
16. Blumberg MS, Seelke AM, Lowen SB, Karlsson KA. Dynamics of sleep-wake cyclicity in developing rats. Proc Natl Acad Sci USA. 2005;102(41):14860–4.
17. Saper CB, Chou TC, Scammell TE. The sleep switch: hypothalamic control of sleep and wakefulness. Trends Neurosci. 2001;24(12):726–31.

18. Chemelli RM, Willie JT, Sinton CM, et al. Narcolepsy in orexin knockout mice: molecular genetics of sleep regulation. Cell. 1999;98(4):437–51.
19. Hungs M, Mignot E. Hypocretin/orexin, sleep and narcolepsy. Bioessays. 2001;23(5):397–408.
20. Beuckmann CT, Sinton CM, Williams SC, et al. Expression of a poly-glutamine-ataxin-3 transgene in orexin neurons induces narcolepsy-cataplexy in the rat. J Neurosci. 2004;24(18):4469–77.
21. Tamakawa Y, Karashima A, Koyama Y, Katayama N, Nakao M. A quartet neural system model orchestrating sleep and wakefulness mechanisms. J Neurophysiol. 2006;95(4):2055–69.
22. Lee MG, Hassani OK, Jones BE. Discharge of identified orexin/hypocretin neurons across the sleep-waking cycle. J Neurosci. 2005;25(28):6716–20.
23. Hara J, Beuckmann CT, Nambu T, et al. Genetic ablation of orexin neurons in mice results in narcolepsy, hypophagia, and obesity. Neuron. 2001;30(2):345–54.
24. Willie JT, Chemelli RM, Sinton CM, et al. Distinct narcolepsy syndromes in Orexin receptor-2 and Orexin null mice: molecular genetic dissection of Non-REM and REM sleep regulatory processes. Neuron. 2003;38(5):715–30.
25. Mochizuki T, Crocker A, McCormack S, Yanagisawa M, Sakurai T, Scammell TE. Behavioral state instability in orexin knock-out mice. J Neurosci. 2004;24(28):6291–300.
26. Fujiki N, Cheng T, Yoshino F, Nishino S. Specificity of direct transitions from wake to REM sleep in orexin/ataxin-3 narcoleptic mice. Sleep. 2006;29:A225.
27. Scammell TE, Willie JT, Guilleminault C, Siegel JM. A consensus definition of cataplexy in mouse models of narcolepsy. Sleep. 2009;32(1):111–6.
28. Morris C, Lecar H. Voltage oscillations in the barnacle giant muscle fiber. Biophys J. 1981;35(1):193–213.
29. Diniz Behn CG, Brown EN, Scammell TE, Kopell NJ. A mathematical model of network dynamics governing mouse sleep-wake behavior. J Neurophysiol. 2007;97(6):3828–40.
30. Chou TC, Lee CE, Lu J, et al. Orexin (hypocretin) neurons contain dynorphin. J Neurosci. 2001;21(19): RC168.
31. Eriksson KS, Sergeeva OA, Selbach O, Haas HL. Orexin (hypocretin)/dynorphin neurons control GABAergic inputs to tuberomammillary neurons. Eur J Neurosci. 2004;19(5):1278–84.
32. Li Y, van den Pol AN. Differential target-dependent actions of coexpressed inhibitory dynorphin and excitatory hypocretin/orexin neuropeptides. J Neurosci. 2006;26(50):13037–47.
33. Williams K, Diniz Behn CG. A Hodgkin-Huxley-type model orexin neuron. Sleep. 2009;32:A25.
34. McCarley RW, Massaquoi SG. A limit cycle mathematical model of the REM sleep oscillator system. Am J Physiol. 1986;251(6 Pt 2):R1011–29.
35. Massaquoi SG, McCarley RW. Extension of the limit cycle reciprocal interaction model of REM cycle control. An integrated sleep control model. J Sleep Res. 1992;1(2):138–43.
36. McCarley RW. Neurobiology of REM and NREM sleep. Sleep Med. 2007;8(4):302–30.
37. Abrahamson EE, Leak RK, Moore RY. The suprachiasmatic nucleus projects to posterior hypothalamic arousal systems. Neuroreport. 2001;12(2):435–40.
38. Chou TC, Scammell TE, Gooley JJ, Gaus SE, Saper CB, Lu J. Critical role of dorsomedial hypothalamic nucleus in a wide range of behavioral circadian rhythms. J Neurosci. 2003;23(33):10691–702.
39. Deurveilher S, Semba K. Indirect projections from the suprachiasmatic nucleus to major arousal-promoting cell groups in rat: implications for the circadian control of behavioural state. Neuroscience. 2005;130(1): 165–83.
40. Nishino S. Clinical and neurobiological aspects of narcolepsy. Sleep Med. 2007;8(4):373–99.
41. Kantor S, Mochizuki T, Janisiewicz AM, Clark E, Nishino S, Scammell TE. Orexin neurons are necessary for the circadian control of REM sleep. Sleep. 2009;32(9):1127–34.
42. Nobili L, Ferrillo F, Besset A, Rosadini G, Schiavi G, Billiard M. Ultradian aspects of sleep in narcolepsy. Neurophysiol Clin. 1996;26(1):51–9.
43. Ferrillo F, Donadio S, De Carli F, Garbarino S, Nobili L. A model-based approach to homeostatic and ultradian aspects of nocturnal sleep structure in narcolepsy. Sleep. 2007;30(2):157–65.
44. Mileykovskiy BY, Kiyashchenko LI, Siegel JM. Behavioral correlates of activity in identified hypocretin/orexin neurons. Neuron. 2005;46(5): 787–98.
45. Phillips AJ, Robinson PA. A quantitative model of sleep-wake dynamics based on the physiology of the brainstem ascending arousal system. J Biol Rhythms. 2007;22(2):167–79.
46. Rempe MJ, Best J, Terman D. A mathematical model of the sleep/wake cycle. J Math Biol. 2009;60(5):615–44.
47. Lu J, Sherman D, Devor M, Saper CB. A putative flip-flop switch for control of REM sleep. Nature. 2006;441(7093):589–94.
48. John J, Wu MF, Boehmer LN, Siegel JM. Cataplexy-active neurons in the hypothalamus: implications for the role of histamine in sleep and waking behavior. Neuron. 2004;42(4):619–34.
49. Wu MF, John J, Boehmer LN, Yau D, Nguyen GB, Siegel JM. Activity of dorsal raphe cells across the sleep-waking cycle and during cataplexy in narcoleptic dogs. J Physiol. 2004;554(1):202–15.
50. Postnova S, Voigt K, Braun HA. A mathematical model of homeostatic regulation of sleep-wake cycles by hypocretin/orexin. J Biol Rhythms. 2009;24:523–35.
51. Williams K, Diniz Behn CG. Dynamic interactions between orexin and dynorphin may delay onset of functional orexin effects: a modeling study. J Bio Rhythms. 2011;26(2):171–81.

Part IV

The Key Role of the Hypothalamus

The Hypothalamus and Its Functions

Giovanna Zoccoli, Roberto Amici, and Alessandro Silvani

Keywords

Hypothalamus • Sleep-wake regulation • Reward • Emotion • Energy homeostasis • Cardiovascular regulation • Osmoregulation • Thermoregulation

This chapter provides a general overview of the hypothalamus with an emphasis on functional neuroanatomy. This lays a foundation for discussing the hypothalamic pathways that regulate wake–sleep behavior, body temperature, body fluid osmolarity, energy balance, and the cardiovascular system. In this picture, the role of hypothalamic hypocretin neurons in orchestrating behavioral and autonomic responses to environmental challenges will stand out. Knowledge of the integrative physiologic role of hypocretin neurons is a prerequisite for understanding the pathophysiology of patients with narcolepsy-cataplexy, in whom hypocretin neurons are lost.

Functional Neuroanatomy of the Hypothalamus

The hypothalamus is a highly integrative part of the diencephalon that is essential for reproduction, homeostasis, and survival. From rostral to caudal, the hypothalamus consists of the preoptic area (POA), the anterior hypothalamus, the tuberal hypothalamus, and the mammillary region. However, the hypothalamus can also be regarded as consisting of periventricular, medial, and lateral zones as these regions have different functions. Thompson and Swanson [1] recently proposed a more detailed functional–neuroanatomical partition, from medial to lateral: (1) *neuroendocrine motor zone*, which lies adjacent to the third ventricle and contains the endocrine motor neurons projecting to the posterior pituitary gland and to the median eminence; (2) *periventricular region*, which contains the *circadian timing network* and a *network generating visceromotor patterns*; (3) *behavioral control column*, which contains several well-defined nuclei involved in the control of motivated behavior and is mostly located in the medial zone; (4) *lateral zone* (or *lateral hypothalamic area, LHA*), which contains diffuse groups of neurons and is traversed by fibers of the medial forebrain bundle. The organization of the hypothalamus according to this functional–neuroanatomical partition and the principal connections between different hypothalamic zones and extrahypothalamic structures are outlined in Fig. 1.

G. Zoccoli (✉)
Department of Human and General Physiology, University of Bologna, Piazza di Porta S. Donato, 2, 40126 Bologna, Italy
e-mail: giovanna.zoccoli@unibo.it

C.R. Baumann et al. (eds.), *Narcolepsy: Pathophysiology, Diagnosis, and Treatment*,
DOI 10.1007/978-1-4419-8390-9_17,

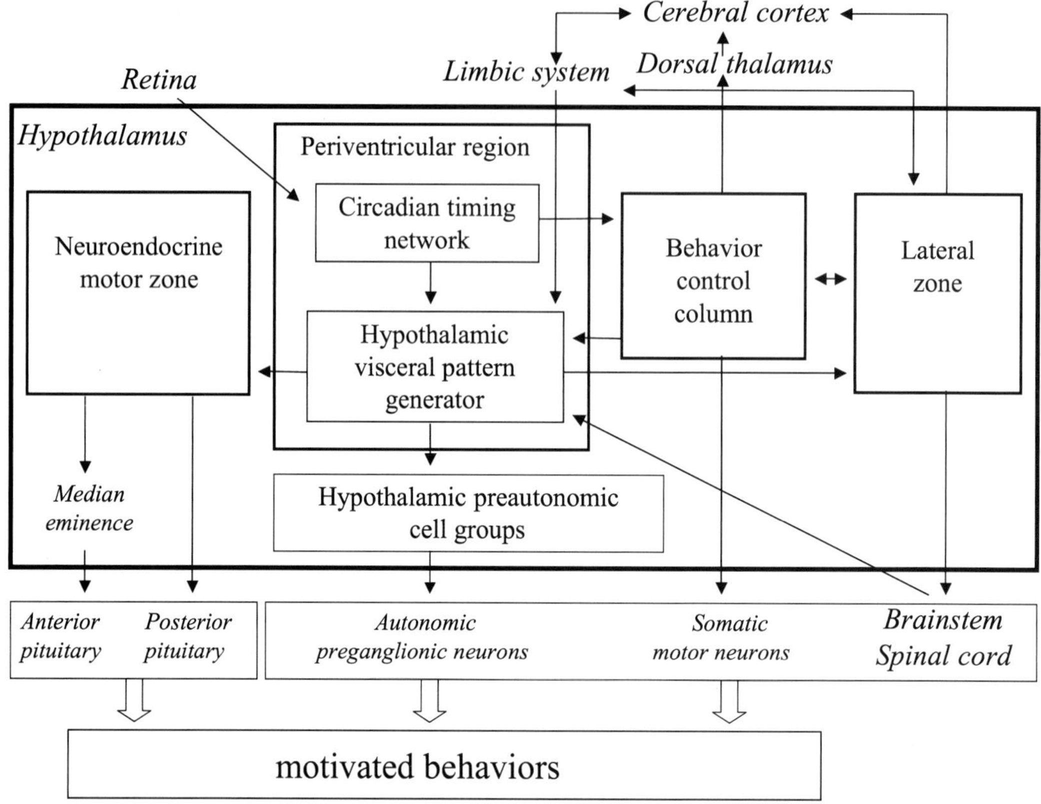

Fig. 1 Schematic drawing showing the functional–neuroanatomical partition of the hypothalamus according to Thompson and Swanson [1]. *Thin black arrows* indicate anatomical or functional connections between either different hypothalamic zones or between the hypothalamus and extrahypothalamic structures. *Thick white arrows* indicate the direct participation of the structures in the expression of motor-somatic, autonomic, or endocrine determinants of motivated behaviors

The *neuroendocrine motor zone* contains the magnocellular and parvicellular neurons. Magnocellular neurons produce oxytocin (OXY) or antidiuretic hormone (ADH) and are located in the paraventricular nucleus (PVN) and in the supraoptic nucleus (SO) (Table 1). Parvicellular neurons produce hormones regulating the anterior pituitary and are located in the anterior periventricular nucleus, the arcuate nucleus (ARC), and the PVN. Parvicellular neurons in the ARC release growth-hormone releasing hormone and dopamine, while those in the PVN mainly release corticotropin releasing hormone (CRH), somatostatin, and thyrotropin releasing hormone (TRH). Neurons that secrete gonadotropin releasing hormones are located in the POA.

The *behavioral control column* plays an essential role in regulating the expression of motivated behaviors. The cells in this column share a common pattern of neural outputs, controlling the somatomotor and autonomic centers of the brain stem and spinal cord and projecting to the LHA and the thalamocortical system. This region includes the medial preoptic nucleus, the anterior hypothalamic nucleus (AHN), part of the PVN, the ventromedial nucleus (VMH), part of the SO, the posterior hypothalamic area, and some small premammillary nuclei. Specific networks of neurons in this region determine the somatomotor programs in the brain stem and spinal cord that underlie motivated behaviors required for survival of the individual and the species

Table 1 Abbreviations

ADH	Antidiuretic hormone
AgRP	Agouti-related peptide
AHN	Anterior hypothalamic nucleus
ARC	Arcuate nucleus
BDNF	Brain-derived neurotrophic factor
CART	Cocaine–amphetamine-regulated transcript
CRH	Corticotropin releasing hormone
DMH	Dorsomedial nucleus
DRN	Dorsal raphe nuclei
GSN	Glucose sensitive neurons
HVPG	Hypothalamic visceral pattern generator
icv	Intracerebroventricular
LC	Locus coeruleus
LDT/PPT	Laterodorsal and peduncolopontine tegmental nuclei
LHA	Lateral hypothalamic area
MC3R and MC4R	Melanocortin receptors
MCH	Melanin-concentrating hormone
MePO	Median preoptic nucleus
NAc	Nucleus accumbens
NPY	Neuropeptide Y
NTS	Nucleus of the solitary tract
OXY	Oxytocin
PAG	Periaqueductal gray
POA	Preoptic area
POMC	Pro-opiomelanocortin
PVN	Paraventricular nucleus
RVLM	Rostral ventrolateral medulla
RVMM	Rostral ventromedial medulla
SCN	Suprachiasmatic nucleus
SO	Supraoptic nucleus
SPZ	Subparaventricular zone
TMN	Tuberomammillary nucleus
TRH	Thyrotropin releasing hormone
VLPO	Ventrolateral PO nucleus
VMH	Ventromedial nucleus
VTA	Ventral tegmental area
Y1R and Y5R	NPY receptors
α-MSH	α-Melanocyte-stimulating hormone

(e.g., defense, reproduction, ingestive behavior, and thermoregulatory behavior). Importantly, the same hypothalamic neurons also set the levels of autonomic and endocrine activity for each behavior. This integrated control on the cardiovascular, respiratory, and gastrointestinal systems is obtained through specific projections to the LHA and, mainly, to the hypothalamic nuclei that constitute the hypothalamic visceral pattern generator (HVPG).

The HVPG is located in the *periventricular region* of the hypothalamus, between the neuroendocrine motor zone and the behavioral control column. The HVGP is a neuronal network that includes the dorsomedial nucleus (DMH) and several small preoptic-anterior hypothalamic nuclei, such as the median preoptic nucleus (MePO). The HVPG coordinates the activity of the endocrine motor neurons with that of three groups of pre-autonomic hypothalamic neurons: (a) parvocellular neurons of the PVN, projecting to the rostral ventrolateral medulla (RVLM) and preganglionic cells in the brainstem and spinal cord [2]; (b) neurons in the VMH and ARC, constituting the retrochiasmatic pre-autonomic group; (c) neurons in the LHA and zona incerta, constituting the tuberal lateral pre-autonomic group. The HVPG also innervates autonomic circuits in the periaqueductal gray (PAG), pons, and medulla either directly or through the PVN and LHA. Sensory inputs reach the HVPG from the nucleus of the solitary tract (NTS), the lateral parabrachial nucleus, the spinal cord, and the subfornical organ. Besides synaptic input from the behavior control column, HVPG also receives inputs from the limbic system and, in particular, from the amygdala, which is involved in analyzing the emotional content of environmental conditions and promoting appropriate behavioral changes. The central nucleus of the amygdala also projects to the LHA and the PVN, producing the autonomic and neuroendocrine responses, respectively, that are elicited by fear conditioning.

Another pattern generator in the periventricular region is represented by the suprachiasmatic nucleus (SCN), which is the master circadian pacemaker in mammals, and by the adjacent subparaventricular zone (SPZ), which is the main target of the SCN. Synaptic inputs to the HVPG from the SCN allow autonomic responses to accompany motivated behaviors according to the circadian rhythm of the organism. The activity of SCN neurons is entrained to the light–dark cycle and determines the temporal organization of behavioral wake–sleep states. The SCN projects widely to hypothalamic (i.e., POA, AHN, SPZ, retrochiasmatic area, DMH, and LHA) and extra-hypothalamic (i.e., basal forebrain, midline thalamus, lateral geniculate complex, and PAG) regions [3]. The SPZ and DMH are crucial for driving circadian cycles and adapting them to environmental stimuli [4].

Another important component of the periventricular region is represented by nonendocrine neurons in the ARC. These neurons project to the PVN and other hypothalamic nuclei relevant to the control of eating, drinking, and reproductive behaviors, as well as to the reticular core and autonomic nuclei of the brainstem. The ARC and PVN, which are directly interconnected, thus contain both neuroendocrine and non-neuroendocrine neurons and send descending projections to brainstem reticular formation and autonomic centers. Neurons in the ARC have dendrites that lie outside the blood–brain barrier and are thus responsive to circulating hormones and markers of metabolic status.

The *lateral zone of the hypothalamus* represents a crossroad, connecting forebrain structures, the medial and periventricular zone of the hypothalamus, the brain stem, and the spinal cord. Recently, several neuropeptides have been described as being highly concentrated in the LHA. Hypocretin 1 and 2 (also known as orexin A and B) are neuropeptides selectively produced by a small number of neurons in the LHA and posterior hypothalamus [5, 6]. Hypocretins are a critical link between the wake–sleep cycle and the regulation of energy homeostasis and motivated behaviors such as food seeking, especially in the physiologic state of fasting stress. More recently, a role for hypocretins has also emerged regarding emotions, reward, and drug addiction [7]. Accordingly, hypocretin neurons receive inputs from the limbic system, which may mediate an increased state of arousal during emotional stimuli. Figure 2 illustrates the key role played by hypocretin neurons in the complex interplay among hypothalamic and extrahypothalamic structures involved in the control of different body functions.

Hypothalamic Control of Body Functions

The hypothalamus promotes specific behaviors coordinating somatic, autonomic, and endocrine motor activity on the basis of external and internal sensory information. The key role of the hypothalamus in this complex integrative activity, which is critical for the maintenance of body homeostasis and for reproduction, was evidenced in cats by Hess in the first half of the last century [8]. Different physiologic functions, such as thermoregulation, osmoregulation, regulation of energy balance and metabolism, and autonomic cardiovascular control are tightly entangled and share common hypothalamic substrates with the regulation of the wake–sleep states. The appropriate set of activity for physiologic functions is therefore organized at a hypothalamic level according to the different wake–sleep states leading to the concept that the wake–sleep state must be taken into account when physiologic functions are studied [9].

Wake–Sleep Regulation

In the early 1900s, von Economo noted that in some patients with "encephalitis lethargica," insomnia was the prominent symptom, and that in such cases, lesions were centered in the anterior hypothalamus, including the POA. This observation led von Economo to hypothesize the existence of a sleep center in the anterior hypothalamus [10]. Subsequently, Nauta showed that the disconnection of the anterior hypothalamus and POA from the diencephalon led rats to apparently irreversible insomnia [11]. The role of the POA in sleep induction was then confirmed by neurophysiologic studies in cats by Sterman and Clemente [12]. More recent immunohistochemical studies of c-Fos expression have allowed researchers to localize putative sleep regulatory neurons in both the ventrolateral preoptic nucleus (VLPO) [13] and the MePO [14]. Subsequent studies have highlighted a central role for VLPO neurons in sleep maintenance and for MePO

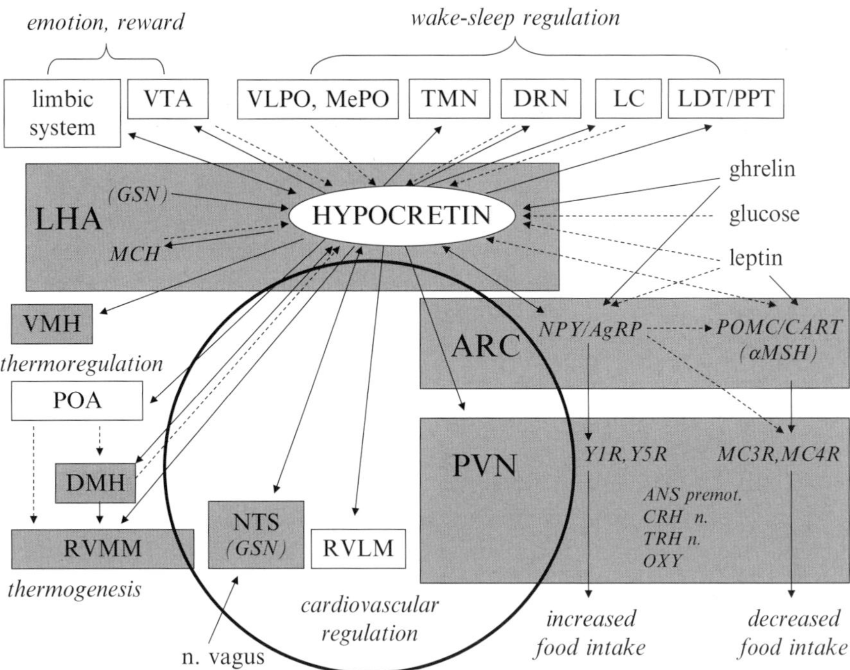

Fig. 2 Schematic drawing showing the connections of the hypocretin system with different hypothalamic or extrahypothalamic structures involved in physiologic processes, including wake–sleep regulation, thermoregulation, regulation of energy balance and metabolism, cardiovascular regulation, and processes related to emotion and reward. Structures directly involved in energy homeostasis are shown in *gray*. Structures involved in cardiovascular regulation are shown fully or partially within the *circle*. *Solid arrows* indicate excitatory projections, *dashed arrows* indicate inhibitory projections. *ANS* autonomic nervous system. All other abbreviations are listed in Table 1

neurons in wake–sleep transition and sleep homeostasis [15, 16].

The VLPO and MePO are reciprocally connected to wake-active neuronal groups, such as histaminergic neurons in the tuberomammillary nucleus (TMN), serotonergic neurons in the dorsal raphe nuclei (DRN), noradrenergic neurons in the locus coeruleus (LC), hypocretin neurons in the perifornical area of the LHA, and cholinergic neurons in the laterodorsal and pedunculopontine tegmental (LDT/PPT) nuclei and the nucleus basalis of the forebrain [17]. The VLPO and SCN are interconnected, so that circadian information may modulate VLPO activity. During wakefulness, VLPO neurons are inhibited by noradrenergic, serotonergic, and cholinergic inputs, while, conversely, histamine and hypocretins do not apparently modulate VLPO neuron activity [4]. It should be noted, however, that TMN neurons might inhibit VLPO neurons because they also release GABA. At sleep onset, GABAergic and galaninergic VLPO neurons and GABAergic MePO neurons start inhibiting wake-active monoaminergic and hypocretin neurons [15]. The reciprocal inhibitory interaction of these systems may give rise to a mechanism for the maintenance of stable vigilance states akin to a flip-flop switch [18].

Preoptic sleep-promoting neurons are activated by hypnogenic factors [15] such as prostaglandin D2, cytokine interleukin 1-β, and adenosine, which may act directly on postsynaptic A_{2A} receptors in VLPO neurons [19]. In addition, adenosine can directly inhibit wake-promoting neurons including those producing acetylcholine or hypocretin. Thermoregulatory processes may also increase excitatory drives on VLPO neurons [20].

Von Economo's original observation that in some patients with "encephalitis lethargica," a lesion of the posterior hypothalamus was associated with a decreased level of vigilance represented the first clear indication that this hypothalamic region contains neurons important for the maintenance of wakefulness. Anatomically, hypocretin neurons are good candidates for this role as they send excitatory inputs to the LC and TMN, which are relevant for the maintenance of wakefulness, as well as to the DRN and the mesencephalic ventral tegmental area (VTA), which contains dopaminergic neurons. Hypocretin neurons also excite a REM-off neuronal population in the ventrolateral PAG [21] and activate cholinergic neurons in the basal forebrain and LDT/PPT nuclei. In cats, hypocretin 1 injection in the LDT increases wakefulness and decreases REM sleep [22], while local hypocretin injection in the PPT strongly inhibits atonia in REM sleep [23]. Hypocretin neurons fire during active waking, decrease discharge during quiet waking, and cease firing during REM sleep. These neurons increase their activity just before the end of REM sleep announcing the return of wakefulness [24]. On the other hand, neurons in the TMN, LC, and DRN fire tonically in wakefulness, decrease their firing rate in NREM sleep, and cease firing in REM sleep. This suggests that the wake-promoting action of hypocretins results from the activation of these wake-active monoaminergic neurons. The importance of hypocretins in promoting wakefulness is highlighted by the demonstration of phenotypes bearing a remarkable similarity to human narcolepsy in mice lacking either the hypocretin gene (prepro-orexin knockout mice) or hypocretin neurons (orexin/ataxin-3 transgenic mice), as well as in mice and dogs with null mutations in the gene for the type 2 hypocretin receptor [25–28].

The posterior hypothalamus and the LHA also contain neurons releasing melanin-concentrating hormone (MCH). MCH neurons may have a role in the hypothalamic regulation and homeostasis of REM sleep [29], being strongly activated during REM sleep but not during NREM sleep or wakefulness [30]. The finding that intracerebroventricular (icv) administration of MCH produces a dose-dependent increase in REM sleep [29] also supports a role of MCH in REM sleep regulation. However, results obtained in transgenic mice which lack the R1 receptor for MCH are not totally in agreement with this hypothesis [31].

MCH and hypocretin neurons are interconnected, and MCH neurons may inhibit hypocretin neurons [32]. MCH neurons may thus contribute to the cessation of activity of hypocretin neurons during NREM and REM sleep probably through co-release of GABA [33]. MCH neurons may also inhibit GABAergic pontine neurons that gate REM sleep onset, thus increasing REM sleep amount. Conversely, the excitatory hypocretin input to these pontine neurons may prevent REM sleep occurrence [34].

Thermoregulation

The hypothalamus represents the most important integrative center for thermoregulation. Following the pioneering work of Nakayama and colleagues in 1961 [35], both cold- and warm-sensitive neurons have been described in the hypothalamus. However, both cold-defense and heat-defense responses are considered to be triggered by central warm-sensitive neurons, the activity of which is apparently modulated with a feed-forward mechanism by peripheral sensors of shell temperature and with a negative feedback mechanism by central sensors of core temperature [36]. Warm-sensitive neurons in the POA receive an inhibitory drive from the MePO, which is in turn activated by cutaneous cold thermoreceptors via an extrathalamic pathway that relays in the lateral parabrachial nucleus [37]. GABAergic projections from the POA may originate directly from warm-sensitive neurons and inhibit, either directly or via the DMH, the sympathetic premotor neurons in the rostral ventromedial medulla (RVMM). In turn, these neurons in the RVMM promote brown adipose tissue thermogenesis and cardiovascular adaptation to thermogenic needs (i.e., cutaneous vasoconstriction and increase in heart rate) [37]. Although the evidence of a direct action of hypocretin neurons on RVMM sympathetic premotor neurons is still lacking, it is worth noting that nonshivering thermogenesis is enhanced by direct stimulation of the LHA [38] (Fig. 2).

Thermoregulation and wake–sleep regulation strongly interact. In particular, thermoregulation is impaired during REM sleep, possibly as a consequence of a state-dependent modification of hypothalamic integrative activity during this sleep state [9]. During REM sleep, direct preoptic cooling or warming is ineffective in eliciting normal thermoregulatory responses [9], and the thermosensitivity of both cold-sensitive [9, 39] and warm-sensitive [9] neurons is significantly depressed. Moreover, a significant subpopulation of preoptic sleep-active neurons is warm-sensitive [40]. These neurons would account for the capacity of whole-body warming or increases in skin temperature to promote sleep in humans [41]. The sleep pattern is also apparently related to cellular activity at preoptic-hypothalamic level during both cold exposure and the following recovery period [42].

Osmoregulation

In 1947, Verney coined the term osmoreceptor to indicate specialized sensory elements in the brain that are critically involved in the reflex release of ADH [43]. At hypothalamic level, information from peripheral and central osmoreceptors is integrated with other visceral sensory modalities (e.g., blood volume, blood pressure, extracellular Na^+ concentration, and body temperature) for the maintenance of body fluid homeostasis [44]. Hyperosmolarity induces water intake and ADH release from the SO and PVN, which results in renal water reabsorption. At least in the rat, hyperosmolarity also induces OXY release from the SO and PVN, which results in natriuresis. Central osmoreceptors are located in the magnocellular neuroendocrine cells in the SO and PVN as well as in regions that lack the blood–brain barrier, including the organum vasculosum laminae terminalis and the subfornical organ, which are positioned at the rostral end of the third ventricle. Peripheral osmoreceptors are located in the stomach and the portal vein and send vagal afferents to the NTS. Information from the NTS and central osmoreceptors is integrated at the level of the PVN, SO, and MePO, which project to both the magnocellular and parvocellular components of the PVN and SO [44].

Osmoregulation apparently takes priority over other hypothalamic regulations. In fact, water-consuming thermoregulatory responses such as panting or sweating are suppressed in conditions of hyperosmolality [44] and anorexia develops during dehydration [45]. The interaction between osmoregulation and wake–sleep regulation is apparently weak or absent at hypothalamic level. The presence of two separate sets of sleep-related (mainly GABAergic) cells and osmosensitivity-related cells has been observed in the MePO by immunohistochemistry [46]. Furthermore, recent findings indicate that ADH is normally released during REM sleep after a central osmotic challenge [47] pointing to a thermoregulatory specificity of the impairment in the hypothalamic integrative activity during REM sleep.

Energy Homeostasis

The understanding of the role of the hypothalamus as a major center controlling food intake and body weight dates back to the nineteenth century, when Mohr described "hypothalamic obesity" produced by a tumor compressing the hypothalamus [48]. Studies performed in the first half of the last century identified the VMH as the "satiety center" and the LHA as the "hunger center" [49]. The notion of specific hypothalamic centers that control food intake and body weight has now been replaced by that of discrete neuronal pathways that generate integrated responses to afferent information on fuel stores. The ARC represents an important station on this pathway. ARC neurons are sensitive to different peripheral indicators of metabolic status. These indicators include leptin, which is secreted from adipose tissue and hence signals the abundance of fat stores; ghrelin, which is an appetite-stimulating hormone produced by the stomach prior to meals and during fasting; and insulin [50].

Most ARC neurons express leptin receptors and are regulated by leptin. ARC neurons producing the anorexigenic neuropeptides pro-opiomelanocortin (POMC), which is precursor of α-melanocyte-stimulating hormone (α-MSH), and cocaine–amphetamine-regulated transcript (CART) are activated by leptin. ARC neurons expressing orexigenic neuropeptide Y (NPY) and Agouti-related peptide (AgRP) are inhibited by leptin and activated by ghrelin. Central administration of NPY increases food intake, inhibits the thyroid axis, and decreases sympathetic nervous system outflow to brown adipose tissue, thus lowering energy expenditure. Conversely, stimulation of α-MSH receptors suppresses food intake, activates the thyroid axis, and increases energy expenditure [51]. Insulin receptors are also highly concentrated in ARC neurons. Neurons in the ARC send projections within the hypothalamus to the PVN, LHA, DMH, and VMH, which are second-order stations on the pathway controlling food intake.

The PVN integrates signals from the ARC and NTS and is sensitive to administration of many different peptides implicated in food intake regulation, such as cholecystokinin, ghrelin, hypocretin-1, leptin, and glucagon-like peptide [51]. Approximately 20% of ARC NPY neurons innervate the PVN. Stimulation of this pathway leads to increased food intake through direct stimulation of NPY receptors Y1R and Y5R and through AgRP antagonism of melanocortin receptors MC3R and MC4R in the PVN [52, 53]. Moreover, the administration of α-MSH into the PVN inhibits food intake and the orexigenic effect of NPY administration. Accordingly, neuropeptides synthesized in the PVN, such as CRH, TRH, and OXY, have an anorexigenic effect [53].

The LHA participates in the regulation of food intake, in part through the projections of MCH and hypocretin neurons to the cortex. Repeated icv administration of MCH produces hyperphagia in rats [54], while transgenic mice over-expressing the precursor of MCH are hyperphagic and develop obesity [55]. The activity of hypocretin neurons is also modulated by peripheral metabolic cues. Hypocretin neurons are activated by hypoglycemia, an effect that may be mediated by the NTS or by glucose sensitive neurons (GSN, representing about 40% of all LHA neurons) [56], and are directly inhibited by glucose [57]. Moreover, they are also stimulated by ghrelin and inhibited by leptin [51]. Hypocretin neurons exert an excitatory action on hypothalamic areas that are critical for feeding, drinking, and thermoregulatory behaviors. Moreover, reciprocal connections between hypocretin neurons and both subsets of ARC neurons (i.e., NPY/AgRP and POMC/CART) provide an indirect regulation of hypocretin neurons by leptin and ghrelin. In mice and rats, central hypocretin administration induces feeding behavior and generalized arousal [5] and enhances oxygen consumption and body temperature [58]. The hypocretin system thus acts as a sensor of the nutritional status of the body and responds with behavioral and physiologic solutions to maintain energy homeostasis, such as an increase of wakefulness to increase the probability of finding food [59] (Fig. 2). Hypocretin neurons may play a role in the phenomenon of hypoglycemia awareness, in which low glucose levels trigger autonomic and behavioral activation and may cause awakening from sleep [60]. During fasting, narcoleptic mice with genetic ablation of hypocretin neurons do not exhibit the expected increase in exploratory activity and spend less time awake than wild-type mice [60]. Lesions of the DMH inhibit the increase in wakefulness, locomotor activity, and body temperature with food restriction [7]. The DMH may act as a central food-entrainable oscillator, showing a robust circadian oscillation of gene expression only under restricted feeding. The DMH has widespread connections to the ARC, from which it receives NPY/AgRP projections [61], as well as to hypocretin neurons. Hypocretin neurons may thus play a role in the anticipatory locomotor activity induced by daily restricted feeding.

The VMH receives projections from NPY/AgRP and POMC/CART neurons and projects to the DMH and NTS. In VMH neurons, expression of brain-derived neurotrophic factor (BDNF) is regulated by the nutritional status and by melanocortin receptor signaling [62]. Mice with reduced BDNF signaling show a significant increase in food intake and body weight. On the other hand, BDNF infusion into the brain suppresses hyperphagic behavior in mice lacking melanocortin receptor signaling [62]. These results support the hypothesis that BDNF is another important effector through which the melanocortin system regulates food intake and body weight.

Cardiovascular Regulation

Behavioral activities such as exercise, fighting, and food seeking must be faced by the organism with an integrated response. Such a response is based on the coordinated hypothalamic activation of somatomotor, hormonal, and autonomic pathways. Accordingly, cardiovascular adaptive responses are mediated by different hypothalamic neuropeptides that are also involved in behavioral control. Hypocretin neurons project to the NTS and RVLM, which are part of a neuronal network critical for cardiovascular and respiratory regulation, as well as to the intermediolateral column of the spinal cord [63]. Hypocretin neurons also send strong efferents to the RVMM [64], which is critically involved in the autonomic cold-defense response and may mediate autonomic effects evoked by LHA activation [38]. Through these pathways, hypocretins may modulate cardiovascular adjustments to different motivated behaviors (Fig. 2). For example, a physiologic role of hypocretins in different features of the defense response has been suggested [65]. The defense response is characterized by increases in arterial pressure, heart rate, respiratory frequency, and resistance in visceral vascular beds, and by decreases in resistance in the airways and the skeletal muscle vascular bed. The defense response elicited by stimulation of the perifornical area is attenuated in genetically engineered mouse models of hypocretin deficiency [65]. These animals also show an attenuated defense response to emotional stressors with blunted increases in arterial pressure, heart rate, and locomotor activity [65]. In line with these results, icv administration of hypocretin 1 increases body temperature, arterial pressure, heart rate, renal sympathetic nerve activity, and plasma catecholamines in awake rats [63]. Hypocretin microinjection in the RVLM also increases arterial pressure and heart rate in anesthetized [66] and awake rats [67], and activates neuronal circuits that control vagal and sympathetic activity to the heart [68]. It is still unknown whether similar effects are physiologically exerted by endogenous hypocretins during wakefulness and/or during sleep. If confirmed, the finding that basal arterial pressure is lower by about 15 mmHg in hypocretin-deficient mice than in wild-type controls, would support this interpretation [65].

Other peptides involved in the hypothalamic control of motivated behaviors also modulate cardiovascular responses according to behavioral needs. ARC α-MSH neurons project to sympathetic preganglionic neurons in the spinal cord [69], to the PVN and LHA in the hypothalamus, and to the NTS, lateral parabrachial nucleus, and RVLM in the brainstem [70]. Acute icv administration of α-MSH increases sympathetic nerve activity, arterial pressure, and heart rate via the MC4R receptor [71]. Recent findings indicate that MC4R activation leads to an acute release of BDNF in the hypothalamus, and that this release is a prerequisite for MC4R-induced effects on cardiovascular function [72]. However, transgenic mice over-expressing melanocortins have increased arterial pressure but similar heart rate compared to wild-type controls [73]. In interpreting these findings, it must be kept in mind that compensatory mechanisms may lead to adaptive changes that oppose those caused by the lack of endogenous peptides, representing a potential limitation of the gene-targeting approach. Nonetheless, studies on the effect of long-term activation of the melanocortin system on arterial pressure and heart rate have also produced inconsistent results [74, 75]. On the other hand, it has been suggested that the melanocortin system mediates the increase in sympathetic activity and arterial pressure produced by leptin [76]. Hyperleptinemia, which characterizes diet-induced obesity, may contribute through this pathway to obesity-related hypertension. Other mechanisms are likely to be involved in the increase in arterial pressure associated with obesity, however, as arterial pressure is still significantly higher in leptin-deficient obese mice than in their lean wild-type littermates [77]. Intriguingly, the hypertensive derangement observed in obese leptin-deficient mice is modulated by the wake–sleep cycle, with sleep-dependent changes in arterial pressure buffering hypertension during REM sleep and the dark period in obese mice [77]. Once again, these data emphasize that the behavioral state must be taken into account when hypothalamic regulation is studied. In this respect, it is worth remarking that the hypothalamic neural substrates of the regulation of wake/sleep states and body functions such as energy homeostasis, thermoregulation, and circulation are extensively overlapped. This point is also stressed by growing laboratory and epidemiological evidence suggesting that sleep loss and poor quality sleep may produce metabolic derangements, including impaired glucose metabolism, increased appetite, and decreased energy expenditure, which may promote the development of obesity and diabetes mellitus [78, 79].

Hypothalamus, Emotion, and Reward

Although the LHA has long been implicated in reward and motivation, hypocretins have only recently been identified as key neurotransmitters involved in this function (Fig. 2). Hypocretin neurons have anatomical connections with many reward-associated brain regions, including the nucleus accumbens (NAc). Moreover, the largest source of inputs to A10 dopaminergic neurons in the VTA, which encode expectations regarding external rewards, originates in the hypothalamus, with a large and excitatory contribution from hypocretin neurons [80]. In turn, the mesocortical and mesolimbic dopaminergic pathways arising from the A10 group of the VTA innervate the prefrontal and temporal cortices and the limbic structures of the basal forebrain, including the NAc.

The first demonstration of a role of hypocretins in addiction was produced in 2003 by Georgescu and coworkers, who showed that morphine-induced physical dependence and withdrawal are regulated, at least in part, by hypocretin neurons [81]. To date, there is growing evidence that hypocretins play a role in multiple aspects of reward processing and addictive behaviors [82]. Such evidence is extensively discussed in other chapters of this book.

Hypocretin neurons project to the amygdala, and hypocretin receptors are abundant within this structure. In turn, the amygdala projects to the hypothalamus and brainstem, promoting physiologic and behavioral responses associated with emotion. In particular, axons from the central amygdaloid nucleus innervate MCH and hypocretin neurons in LHA. Inputs from the amygdala may thus regulate the activity of neurons expressing feeding-related peptides in the LHA in response to emotional stimuli [83]. Interestingly, narcoleptic patients display an abnormally high activation of the amygdala to positive emotions [84]. It is still unclear, however, whether this reflects a dysfunction of the pathway connecting the prefrontal and anterior cingulate cortices to the amygdala or a lack of direct modulation of amygdala activity by hypocretin neurons [84]. Finally, projections from the limbic system to hypocretin neurons may be implicated in the pathophysiology of cataplexy in narcolepsy-cataplexy patients. In particular, emotional stimuli may increase hypocretin release in the PPT, preventing atonia in normal subjects, while a depression of this system might explain the occurrence of cataplexy in narcoleptic patients [7].

In conclusion, the hypothalamus clearly represents a key neural structure for integrated control of different body functions, coordinating somatic, autonomic and endocrine activities which underlie different wake–sleep behaviors. Although these largely entangled regulatory processes involve a plethora of different hypothalamic neurotransmitters and neuropeptides, hypocretins stand out as key molecules. Thus, many of the apparently diverse signs and symptoms of narcoleptic patients may be rooted in the impairment of hypothalamic function that results from the lack of hypocretin signaling.

References

1. Thompson RH, Swanson LW. Structural characterization of a hypothalamic visceromotor pattern generator network. Brain Res Brain Res Rev. 2003;41:153–202.
2. Swanson LW, Sawchenko PE. Hypothalamic integration: organization of the paraventricular and supraoptic nuclei. Annu Rev Neurosci. 1983;6:269–324.
3. Moore RY, Speh JC, Leak RK. Suprachiasmatic nucleus organization. Cell Tissue Res. 2002;309: 89–98.
4. Saper CB, Scammell TE, Lu J. Hypothalamic regulation of sleep and circadian rhythms. Nature. 2005;437687: 1257–63.

5. Sakurai T, Amemiya A, Ishii M, et al. Orexins and orexin receptors: a family of hypothalamic neuropeptides and G protein-coupled receptors that regulate feeding behavior. Cell. 1998;92:573–85.
6. de Lecea L, Kilduff TS, Peyron C, et al. The hypocretins: hypothalamus-specific peptides with neuroexcitatory activity. Proc Natl Acad Sci USA. 1998;95:322–7.
7. Sakurai T. The neural circuit of orexin (hypocretin): maintaining sleep and wakefulness. Nat Rev Neurosci. 2007;8:171–81.
8. Hess W. Das Zwischenhirn: Syndrome Lokalisationen Funktionen (The diencephalon: syndrome localization function). Basel, Switzerland: Benno Schwabe; 1954.
9. Parmeggiani PL. Physiologic regulation in sleep. In: Kryger MH, Roth T, Dement WE, editors. Principles and practice of sleep medicine. 4th ed. Philadelphia: WB Saunders; 2005. p. 85–191.
10. von Economo C. Schlaftheorie. Ergebn Physiol. 1929;28:312–39.
11. Nauta W. Hypothalamic regulation of sleep in the rat. An experimental study. J Neurophysiol. 1946;9: 285–316.
12. Sterman MB, Clemente CD. Forebrain inhibitory mechanisms: sleep patterns induced by basal forebrain stimulation in the behaving cat. Exp Neurol. 1962;6:103–17.
13. Sherin JE, Shiromani PJ, McCarley RW, Saper CB. Activation of ventrolateral preoptic neurons during sleep. Science. 1996;271:216–9.
14. Gong H, McGinty D, Guzman-Marin R, Chew KT, Stewart D, Szymusiak R. Activation of c-fos in GABAergic neurones in the preoptic area during sleep and in response to sleep deprivation. J Physiol. 2004;556:935–46.
15. Szymusiak R, McGinty D. Hypothalamic regulation of sleep and arousal. Ann N Y Acad Sci. 2008;1129:275–86.
16. Dentico D, Amici R, Baracchi F, et al. c-Fos expression in preoptic nuclei as a marker of sleep rebound in the rat. Eur J Neurosci. 2009;30:651–61.
17. Fort P, Bassetti CL, Luppi PH. Alternating vigilance states: new insights regarding neuronal networks and mechanisms. Eur J Neurosci. 2009;29:1741–53.
18. Saper CB, Chou TC, Scammell TE. The sleep switch: hypothalamic control of sleep and wakefulness. Trends Neurosci. 2001;24:726–31.
19. Scammell TE, Gerashchenko DY, Mochizuki T, et al. An adenosine A2a agonist increases sleep and induces Fos in ventrolateral preoptic neurons. Neuroscience. 2001;107:653–63.
20. McGinty D, Alam MN, Szymusiak R, Nakao M, Yamamoto M. Hypothalamic sleep-promoting mechanisms: coupling to thermoregulation. Arch Ital Biol. 2001;139:63–75.
21. Lu J, Sherman D, Devor M, Saper CB. A putative flip-flop switch for control of REM sleep. Nature. 2006;441:589–94.
22. Xi MC, Morales FR, Chase MH. Effects on sleep and wakefulness of the injection of hypocretin-1 (orexin-A) into the laterodorsal tegmental nucleus of the cat. Brain Res. 2001;901:259–64.
23. Takakusaki K, Takahashi K, Saitoh K, et al. Orexinergic projections to the cat midbrain mediate alternation of emotional behavioural states from locomotion to cataplexy. J Physiol. 2005;568:1003–20.
24. Lee MG, Hassani OK, Jones BE. Discharge of identified orexin/hypocretin neurons across the sleep-waking cycle. J Neurosci. 2005;25:6716–20.
25. Chemelli RM, Willie JT, Sinton CM, et al. Narcolepsy in orexin knockout mice: molecular genetics of sleep regulation. Cell. 1999;98:437–51.
26. Hara J, Beuckmann CT, Nambu T, et al. Genetic ablation of orexin neurons in mice results in narcolepsy, hypophagia, and obesity. Neuron. 2001;30: 345–54.
27. Lin L, Faraco J, Li R, et al. The sleep disorder canine narcolepsy is caused by a mutation in the hypocretin (orexin) receptor 2 gene. Cell. 1999;98:365–76.
28. Thannickal TC, Moore RY, Nienhuis R, et al. Reduced number of hypocretin neurons in human narcolepsy. Neuron. 2000;27:469–74.
29. Verret L, Goutagny R, Fort P, et al. A role of melanin-concentrating hormone producing neurons in the central regulation of paradoxical sleep. BMC Neurosci. 2003;4:19.
30. Hanriot L, Camargo N, Courau AC, Leger L, Luppi PH, Peyron C. Characterization of the melanin-concentrating hormone neurons activated during paradoxical sleep hypersomnia in rats. J Comp Neurol. 2007;505:147–57.
31. Adamantidis A, Salvert D, Goutagny R, et al. Sleep architecture of the melanin-concentrating hormone receptor 1-knockout mice. Eur J Neurosci. 2008;27: 1793–800.
32. Rao Y, Lu M, Ge F, et al. Regulation of synaptic efficacy in hypocretin/orexin-containing neurons by melanin concentrating hormone in the lateral hypothalamus. J Neurosci. 2008;28:9101–10.
33. Alam MN, Kumar S, Bashir T, et al. GABA-mediated control of hypocretin-but not melanin-concentrating hormone-immunoreactive neurones during sleep in rats. J Physiol. 2005;563:569–82.
34. Luppi PH, Gervasoni D, Verret L, et al. Paradoxical (REM) sleep genesis: the switch from an aminergic-cholinergic to a GABAergic-glutamatergic hypothesis. J Physiol Paris. 2006;100:271–83.
35. Nakayama T, Eisenman JS, Hardy JD. Single unit activity of anterior hypothalamus during local heating. Science. 1961;134:560–1.
36. Romanovsky AA. Thermoregulation: some concepts have changed. Functional architecture of the thermoregulatory system. Am J Physiol Regul Integr Comp Physiol. 2007;292:R37–46.
37. Morrison SF, Nakamura K, Madden CJ. Central control of thermogenesis in mammals. Exp Physiol. 2008;93:773–97.
38. Cerri M, Morrison SF. Activation of lateral hypothalamic neurons stimulates brown adipose tissue thermogenesis. Neuroscience. 2005;135:627–38.

39. Alam MN, McGinty D, Szymusiak R. Preoptic/anterior hypothalamic neurons: thermosensitivity in rapid eye movement sleep. Am J Physiol. 1995;269: R1250–7.
40. Szymusiak R, Gvilia I, McGinty D. Hypothalamic control of sleep. Sleep Med. 2007;8:291–301.
41. Krauchi K, Cajochen C, Werth E, Wirz-Justice A. Functional link between distal vasodilation and sleep-onset latency? Am J Physiol Regul Integr Comp Physiol. 2000;278:R741–8.
42. Zamboni G, Jones CA, Domeniconi R, et al. Specific changes in cerebral second messenger accumulation underline REM sleep inhibition induced by the exposure to low ambient temperature. Brain Res. 2004;1022:62–70.
43. Verney E. The antidiuretic hormone and the factors which determine its release. Proc R Soc Lond B Biol Sci. 1947;135:25–106.
44. Bourque CW. Central mechanisms of osmosensation and systemic osmoregulation. Nat Rev Neurosci. 2008;9:519–31.
45. Watts AG. Dehydration-associated anorexia: development and rapid reversal. Physiol Behav. 1999;65: 871–8.
46. Gvilia I, Angara C, McGinty D, Szymusiak R. Different neuronal populations of the rat median preoptic nucleus express c-fos during sleep and in response to hypertonic saline or angiotensin-II. J Physiol. 2005;569:587–99.
47. Luppi M, Martelli D, Amici R, et al. Hypothalamic osmoregulation is maintained across the wake-sleep cycle in the rat. J Sleep Res. 2010;19:1–6.
48. Mohr H. Hypertrophie der Hypophysis cerebri und dadurch bedingter Druck auf die Hirngrundfläche, insbesondere auf die Sehnerven, das Chiasma derselben und linkseitigen Hirnschenkel. In: Hirschwald A, ed. Mittheilungen für neuropathologische Studien. Berlin: Wschr. ges. Heilk; 1840:565–571.
49. Stellar E. The physiology of motivation. Psychol Rev. 1954;61:5–22.
50. Schwartz MW, Woods SC, Porte Jr D, Seeley RJ, Baskin DG. Central nervous system control of food intake. Nature. 2000;404:661–71.
51. Wynne K, Stanley S, McGowan B, Bloom S. Appetite control. J Endocrinol. 2005;184:291–318.
52. Simpson KA, Martin NM, Bloom SR. Hypothalamic regulation of food intake and clinical therapeutic applications. Arq Bras Endocrinol Metabol. 2009; 53:120–8.
53. Woods SC, Seeley RJ, Porte Jr D, Schwartz MW. Signals that regulate food intake and energy homeostasis. Science. 1998;280:1378–83.
54. Qu D, Ludwig DS, Gammeltoft S, et al. A role for melanin-concentrating hormone in the central regulation of feeding behaviour. Nature. 1996;380:243–7.
55. Marsh DJ, Weingarth DT, Novi DE, et al. Melanin-concentrating hormone 1 receptor-deficient mice are lean, hyperactive, and hyperphagic and have altered metabolism. Proc Natl Acad Sci USA. 2002;99: 3240–5.
56. Cai XJ, Liu XH, Evans M, et al. Orexins and feeding: special occasions or everyday occurrence? Regul Pept. 2002;104:1–9.
57. Burdakov D, Jensen LT, Alexopoulos H, et al. Tandem-pore K+ channels mediate inhibition of orexin neurons by glucose. Neuron. 2006;50:711–22.
58. Wang J, Osaka T, Inoue S. Energy expenditure by intracerebroventricular administration of orexin to anesthetized rats. Neurosci Lett. 2001;315:49–52.
59. Zhang S, Zeiiter M, Sakurai T, Nishino S, Mignot E. Sleep/wake fragmentation disrupts metabolism in a mouse model of narcolepsy. J Physiol. 2007;581: 649–63.
60. Yamanaka A, Beuckmann CT, Willie JT, et al. Hypothalamic orexin neurons regulate arousal according to energy balance in mice. Neuron. 2003;38: 701–13.
61. Kalra SP, Dube MG, Pu S, Xu B, Horvath TL, Kalra PS. Interacting appetite-regulating pathways in the hypothalamic regulation of body weight. Endocr Rev. 1999;20:68–100.
62. Xu B, Goulding EH, Zang K, et al. Brain-derived neurotrophic factor regulates energy balance downstream of melanocortin-4 receptor. Nat Neurosci. 2003;6:736–42.
63. Shirasaka T, Takasaki M, Kannan H. Cardiovascular effects of leptin and orexins. Am J Physiol Regul Integr Comp Physiol. 2003;284:R639–51.
64. Berthoud HR, Patterson LM, Sutton GM, Morrison C, Zheng H. Orexin inputs to caudal raphe neurons involved in thermal, cardiovascular, and gastrointestinal regulation. Histochem Cell Biol. 2005;123:147–56.
65. Kayaba Y, Nakamura A, Kasuya Y, et al. Attenuated defense response and low basal blood pressure in orexin knockout mice. Am J Physiol Regul Integr Comp Physiol. 2003;285:R581–93.
66. Chen CT, Hwang LL, Chang JK, Dun NJ. Pressor effects of orexins injected intracisternally and to rostral ventrolateral medulla of anesthetized rats. Am J Physiol Regul Integr Comp Physiol. 2000;278:R692–7.
67. Machado BH, Bonagamba LGH, Dun SL, Kwok EH, Dun NJ. Pressor response to microinjection of orexin/hypocretin into rostral ventrolateral medulla of awake rats. Regul Pept. 2002;104:75–81.
68. Ciriello J, Li Z, de Oliveira CV. Cardioacceleratory responses to hypocretin-1 injections into rostral ventromedial medulla. Brain Res. 2003;991:84–95.
69. Koylu EO, Couceyro PR, Lambert PD, Kuhar MJ. Cocaine- and amphetamine-regulated transcript peptide immunohistochemical localization in the rat brain. J Comp Neurol. 1998;391:115–32.
70. Cone RD. Anatomy and regulation of the central melanocortin system. Nat Neurosci. 2005;8:571–8.
71. Ni XP, Butler AA, Cone RD, Humphreys MH. Central receptors mediating the cardiovascular actions of melanocyte stimulating hormones. J Hypertens. 2006;24:2239–46.
72. Nicholson JR, Peter JC, Lecourt AC, Barde YA, Hofbauer KG. Melanocortin-4 receptor activation stimulates hypothalamic brain-derived neurotrophic factor release to regulate food intake, body temperature

and cardiovascular function. J Neuroendocrinol. 2007;19:974–82.
73. Rinne P, Harjunpaa J, Scheinin M, Savontaus E. Blood pressure regulation and cardiac autonomic control in mice overexpressing alpha- and gamma-melanocyte stimulating hormone. Peptides. 2008;29:1943–52.
74. Hill C, Dunbar JC. The effects of acute and chronic alpha melanocyte stimulating hormone (alphaMSH) on cardiovascular dynamics in conscious rats. Peptides. 2002;23:1625–30.
75. Kuo JJ, DaSilva AA, Tallam LS, Hall JE. Role of adrenergic activity in pressor responses to chronic melanocortin receptor activation. Hypertension. 2004;43:370–5.
76. Haynes WG, Morgan DA, Djalali A, Sivitz WI, Mark AL. Interactions between the melanocortin system and leptin in control of sympathetic nerve traffic. Hypertension. 1999;33:542–7.
77. Silvani A, Bastianini S, Berteotti C, et al. Sleep modulates hypertension in leptin-deficient obese mice. Hypertension. 2009;53:251–5.
78. Spiegel K, Tasali E, Leproult R, Van Cauter E. Effects of poor and short sleep on glucose metabolism and obesity risk. Nat Rev Endocrinol. 2009;5:253–61.
79. Knutson KL, Spiegel K, Penev P, Van Cauter E. The metabolic consequences of sleep deprivation. Sleep Med Rev. 2007;11:163–78.
80. Fadel J, Deutch AY. Anatomical substrates of orexin-dopamine interactions: lateral hypothalamic projections to the ventral tegmental area. Neuroscience. 2002;111:379–87.
81. Georgescu D, Zachariou V, Barrot M, et al. Involvement of the lateral hypothalamic peptide orexin in morphine dependence and withdrawal. J Neurosci. 2003;23:3106–11.
82. Aston-Jones G, Smith RJ, Moorman DE, Richardson KA. Role of lateral hypothalamic orexin neurons in reward processing and addiction. Neuropharmacology. 2009;56 Suppl 1:112–21.
83. Nakamura S, Tsumori T, Yokota S, Oka T, Yasui Y. Amygdaloid axons innervate melanin-concentrating hormone- and orexin-containing neurons in the mouse lateral hypothalamus. Brain Res. 2009;1278:66–74.
84. Schwartz S, Ponz A, Poryazova R, et al. Abnormal activity in hypothalamus and amygdala during humour processing in human narcolepsy with cataplexy. Brain. 2008;131:514–22.

The Prehistory of Orexin/Hypocretin and Melanin-Concentrating Hormone Neurons of the Lateral Hypothalamus

Clifford B. Saper

Keywords
Hypothalamus • Melanin concentrating hormone • Feeding • Sleep-wake regulation • Orexin • Hypocretin

The discovery of the orexins (or hypocretins) and their receptors in 1998 marked a dramatic and rapid acceleration in our understanding of the neurons that contain these neurotransmitters, and their role in sleep and in narcolepsy [1, 2]. However, by 1998 the neurons that contained these neurotransmitters were already well known, and many aspects of their connections as well as their potential role in the regulation of wakefulness, had already been known for more than two decades. In this chapter, we consider the early history of studies that addressed the populations of neurons that we now know to contain the orexins, as well as those that have been shown to contain melanin-concentrating hormone (MCH), as the latter neurons in many ways play a role as a doppelganger for the orexin-containing neurons.

C.B. Saper (✉)
Department of Neurology, Program in Neuroscience, and Division of Sleep Medicine, Harvard Medical School and Beth Israel Deaconess Medical Center, Boston, MA 02215, USA
e-mail: csaper@bidmc.harvard.edu

Early Physiological Studies

The earliest studies implicating the lateral hypothalamus as containing wake-promoting neurons date back to the work of von Economo [3], during the epidemic of encephalitis lethargica that begin in 1915. He noted that patients who had lesions involving the junction of the midbrain and diencephalon generally presented with hypersomnolence. However, in patients who survived this syndrome, there often developed a mysterious illness that had been described by Gelineau and named narcolepsy [4]. In fact, in the wake of the world-wide epidemic of encephalitis lethargica, neurologists began diagnosing a wave of narcolepsy. Kinnier Wilson in London and Spiller in Philadelphia recognized this relationship in the early 1920s, when Kinnier Wilson reported that the incidence of new cases of narcolepsy was several times that which had been seen previously [5]. von Economo pointed out that in several such patients who came to autopsy, there were lesions found in the posterior hypothalamus [3].

C.R. Baumann et al. (eds.), *Narcolepsy: Pathophysiology, Diagnosis, and Treatment*,
DOI 10.1007/978-1-4419-8390-9_18, © Springer Science+Business Media, LLC 2011

These observations focused attention on the role of the lateral hypothalamus, particularly its posterior portions, in arousal. During this same period, Hess stimulated the hypothalamus in cats, reporting that he obtained arousal responses, including sympathetic activation, from sites in the posterior lateral hypothalamus [6]. By the mid-1930s Stephen Ranson and his colleagues at Northwestern University had reintroduced the stereotaxic instrument for making lesions in deep parts of the brain. In 1939, he reported that lesions of the posterior lateral hypothalamus in monkeys produced a long-lasting hypersomnolent state [7]. Thus, the concept that the posterior half of the lateral hypothalamus contained neurons that played an important role in both arousal and in autonomic activation was well established prior to World War II. Similar observations were later made by Nauta in rats [8], and by Swett and Hobson in cats [9], but none of these studies included EEG or EMG recordings or measurements.

Early Anatomical Studies

Following World War II, interest in the ascending reticular activating system superceded interest in hypothalamic arousal systems. However, when Nauta and Kuypers placed lesions in the paramedian midbrain reticular formation, they found that they could trace the ascending degenerating axons in a pathway that bifurcated, with one branch innervating the thalamus, while the other ran into the lateral hypothalamus [10]. They could not trace this pathway further, but Kuyper's interest in the lateral hypothalamus was ignited.

In the early 1970s with the advent of horseradish peroxidase (HRP) as the first retrograde axonal tracer, Kuypers undertook a pair of classic experiments with far-reaching results. In two brief papers, he showed in monkeys the patterns of retrograde labeling in the brain after injections of HRP into the cerebral cortex [11] and the spinal cord [12]. In both papers, he made a remarkable and unsettling observation: he noted the presence of retrogradely labeled neurons in the lateral hypothalamus at the tuberal level (roughly coextensive with the dorsomedial and ventromedial nuclei). Although Kuypers barely commented on these neurons in the original, brief publications, his observations drove substantial investigation in the succeeding years.

The pattern of retrograde labeling from the spinal cord was examined more closely by Saper and colleagues [13] and by Hancock [14]. Although attention at the time tended to focus on retrograde labeling of the paraventricular nucleus in rats, both papers identified neurons in the tuberal lateral hypothalamus that project to the spinal cord in rats, and Saper and colleagues also showed similar projections in cats and monkeys. Saper and coworkers then showed that the hypothalamic neurons projected primarily to the superficial laminae of the dorsal horn, and to both sympathetic and parasympathetic preganglionic neurons, in both the medulla and the spinal cord [13]. Thus, by the late 1970s it was well established that there was a population of tuberal lateral hypothalamic neurons with extensive autonomic connections, and these were posited to mediate the many autonomic responses that had been elicited from the lateral hypothalamus by Hess as well as later workers [6, 15].

The second major observation by Kuypers and colleagues was of the hypothalamic-cortical projection [11]. Prior to this discovery, such a projection would have been considered heretical, as the general principle had been established in the 1950s that the hypothalamus had no projections rostral to the septum. The hypothalamic-cortical projection system, however, was subsequently studied in detail by Saper in rats, and was shown to contain four separate components [16]. The most rostral of these was the tuberal lateral hypothalamic neurons that projected to the cerebral cortex. These neurons were shown to provide a roughly topographical projection to the cerebral cortex, with the most medial neurons in the perifornical region projecting to more medial cortical regions and neurons in the more lateral hypothalamus projecting to more lateral cortical areas.

Attempts to Characterize the Neurochemical Nature of the Lateral Hypothalamic Cells that Project to the Cerebral Cortex and the Spinal Cord

The first attempts to characterize the lateral hypothalamic neurons with these long projections were by Swanson and Kohler, who showed in 1984 that at least 95% of the neurons in the lateral hypothalamic area that project to either the spinal cord or the hippocampus are immunoreactive with some antibodies (but not all) against alpha melanocyte stimulating hormone (alpha-MSH) [17]. Subsequently, it was shown that at least 80% of the cortically projecting neurons and nearly 100% of the spinally projecting neurons in the lateral hypothalamic area also stained with certain alpha-MSH antisera [18, 19]. However, these neurons did not stain with other antisera against peptides derived from the precursor for alpha-MSH (pro-opiomelanocortin), such as beta-endorphin, indicating that the staining was probably a cross-reactivity. Swanson and Kohler then showed that at least 90% of the alpha-MSH-immunoreactive neurons in the lateral hypothalamus also contained acetylcholinesterase [20]. Thus, for all practical purposes, these two markers were found in nearly all of the lateral hypothalamic neurons with cortical and spinal projections, but the actual neurotransmitters contained in these neurons were still not known.

The mystery surrounding the actual identity of the neurotransmitters in the alpha-MSH immunoreactive neurons in the lateral hypothalamus was partially resolved in 1989, when Vale and his coworkers discovered MCH in neurons in this area [21, 22]. Shortly afterward, Lin and coworkers reported that at least some of the lateral hypothalamic neurons that project to the cerebral cortex stained with antisera against glutamic acid decarboxylase and gamma-aminobutyric acid (GABA) [23]. The relationship of these two observations was clarified by Elias and coworkers, who showed that many MCH neurons also coexpressed mRNA for both glutamic acid decarboxylase and the neuropeptide cocaine and amphetamine regulated transcript (CART). The MCH neurons were found to have wide-ranging projections, from the cerebral cortex to the spinal cord [24], similar to those previously shown from the lateral hypothalamus. Interestingly, one of the peptides contained in its precursor, NEI, cross-reacted with some alpha-MSH antisera [22], suggesting that the MCH neurons accounted for at least some of the alpha-MSH neurons with cortical and spinal projections that had previously been identified [17, 18]. However, about half of the neurons in the lateral hypothalamus that were retrogradely labeled from either the cerebral cortex or spinal cord stubbornly refused to stain with MCH antisera, suggesting the presence of a second population of lateral hypothalamic neurons with these projections.

The discovery of orexins (hypocretins) in the lateral hypothalamus in 1998 finally solved this puzzle [1, 25, 26]. Immunohistochemical studies soon showed that orexin/hypocretin neurons project widely in the CNS, from the cerebral cortex to the spinal cord. In fact, their projection patterns are remarkably similar to those of the MCH neurons [24–26]. Chou et al. then showed that the orexin and MCH neurons, taken together, accounted for all of the acetylcholinesterase-containing neurons in the lateral hypothalamic area [27], which had previously been shown to account for all of the cortically and spinally projecting neurons in this region [17, 18, 20].

At the same time, immunohistochemical studies of the descending orexin projection demonstrated that the orexin neurons innervate the spinal cord, with especially heavy projections to sympathetic preganglionic neurons, especially at the T1-2 level, and the superficial dorsal horn, although there are also more diffuse projections to the ventral horn [28]. This pattern is similar to, although more restricted than the spinal projections from the hypothalamus that had been traced using anterograde tracers [13], suggesting that the different hypothalamic-spinal projections may map out different territories. The projection pattern of the MCH neurons has not been studied in detail.

Conclusions

Although the discovery of the orexin/hypocretin peptides in 1998 [1, 2] and the demonstration of their role in narcolepsy in 1999 [25, 29] provided a landmark in understanding narcolepsy, this work really put a capstone on a series of discoveries that date back for over 70 years prior to that time. The role of posterior lateral hypothalamic neurons in arousal in general and in narcolepsy in particular was proposed as early as the 1920s [3, 5]. The first discovery of the cortical and spinal projections from the lateral hypothalamic neurons that were later found to contain orexins/hypocretins came in 1975 [11, 12], and by the 1980s the peculiar relationship of the orexin/hypocretin neurons with those containing MCH had been identified [17, 24, 27]. These studies showed that the orexin/hypocretin neurons and the MCH neurons share common projection patterns, and other cellular markers (such as acetylcholinesterase). Only with the recent studies recording from the two cell types has the almost polar opposite activity patterns of the orexin/hypocretin and MCH neurons come to light [30–32]. Yet both are implicated in different aspects of control of wake–sleep, feeding, and metabolism [33–37], and their receptors have opposite effects on their targets, suggesting that targets that orexin/hypocretin neurons excite during the wake period are inhibited by the MCH neurons during the sleep period. This yin–yang relationship may allow the two sets of neurons to cooperate in regulating the same processes. The work done before 1998 set the stage for our current attempts to unravel these relationships further, and to understand better the roles of both cell populations in health and in a wide variety of disease states.

Several questions that were first raised by these early studies now are much more approachable. A key problem that requires attention is the relative distribution of orexin and MCH axons in the central nervous system. Surprisingly, despite earlier studies that provided an overview of the connections, neither has been mapped in detail, and the relative contributions of each peptide in each terminal field have yet to be examined. Another issue that deserves attention is the topographic nature of the orexin and MCH projections within the cerebral cortex and elsewhere in the nervous system. Earlier tracing studies suggested that different parts of the field project to different cortical targets. Similarly, the presence of c-Fos protein in perifornical orexin neurons during wakefulness, but in lateral hypothalamic orexin neurons during drug-induced place preference suggests functional heterogeneity within these neurons. Understanding the roles of orexin and MCH cells in a range of physiological processes will require a more detailed understanding of their differential connectivity.

References

1. Sakurai T, Amemiya A, Ishii M, Matsuzaki I, Chemelli RM, Tanaka H, et al. Orexins and orexin receptors: a family of hypothalamic neuropeptides and G protein-coupled receptors that regulate feeding behavior. Cell. 1998;92:573–85.
2. de Lecea L, Kilduff TS, Peyron C, Gao X, Foye PE, Danielson PE, et al. The hypocretins: hypothalamus-specific peptides with neuroexcitatory activity. Proc Natl Acad Sci USA. 1998;95:322–7.
3. Von Economo C. Sleep as a problem of localization. J Nerv Ment Dis. 1930;71:249–59.
4. Gelineau JB. De la narcolepsie. Gaz Hop (Paris). 1880;53:626–8.
5. Wilson SAK. The narcolepsies. Modern problems in neurology. London: Edward Arnold & Co; 1928. p. 76–119.
6. Hess WR, Akert K. Experimental data on role of the hypothalamus in mechanism of emotional behavior. Arch Neurol Psychiatr. 1955;73:127–9.
7. Ranson SW. Somnolence caused by hypothalamic lesions in monkeys. Arch Neurol Psychiatr. 1939;41: 1–23.
8. Nauta WJH. Hypothalamic regulation of sleep in rats. An experimental study. J Neurophysiol. 1946;9:285–314.
9. Swett CP, Hobson JA. The effects of posterior hypothalamic lesions on behavioral and electrographic manifestations of sleep and waking in cat. Arch Ital Biol. 1968;106:270–82.
10. Nauta WJH, Kuypers HGJM. Some ascending pathways in the brain stem reticular formation. Reticular

Formation of the Brain: proceedings of the Henry Ford Hospital Symposium. Boston: Little-Brown; 1958. p. 3–30.
11. Kievet J, Kuypers HGJM. Basal forebrain and hypothalamic connections to the frontal and parietal cortex of the rhesus monkey. Science. 1975;187:660–2.
12. Kuypers HGJM, Maisky VA. Retrograde axonal transport of horseradish peroxidase from spinal cord to brain stem cell groups in the cat. Neurosci Lett. 1975;1:9–14.
13. Saper CB, Loewy AD, Swanson LW, Cowan WM. Direct hypothalamo-autonomic connections. Brain Res. 1976;117:305–12.
14. Hancock MB. Cells of origin of hypothalamo-spinal projections in the rat. Neurosci Lett. 1976;3:179–84.
15. Enoch EM, Kerr FWL. Hypothalamic vasopressor and vesicopressor pathways. I. Functional studies. Arch Neurol. 1967;16:290–306.
16. Saper CB. Organization of cerebral cortical afferent systems in the rat. II. Hypothalamocortical projections. J Comp Neurol. 1985;237:21–46.
17. Kohler C, Haglund L, Swanson LW. A diffuse of a-MSH-immunoreactive projection of the hippocampus and spinal cord form individual neurons in the lateral hypothalamic area and zona incerta. J Comp Neurol. 1984;223:501–14.
18. Saper CB, Akil H, Watson SJ. Lateral hypothalamic innervation of the cerebral cortex: immunoreactive staining for a peptide resembling but immunochemically distinct from pituitary/arcuate a-melanocyte stimulating hormone. Brain Res Bull. 1986;16:107–20.
19. Cechetto DF, Saper CB. Neurochemical organization of the hypothalamic projection to the spinal cord in the rat. J Comp Neurol. 1988;272:579–604..
20. Kohler C, Swanson LW. Acetylcholinesterase-containing cells in the lateral hypothalamic area are immunoreactive for alpha-melanocyte stimulating hormone (alpha-MSH) and have cortical projections in the rat. Neurosci Lett. 1984;49:39–43.
21. Vaughan JM, Fischer WH, Hoeger C, Rivier J, Vale W. Characterization of melanin-concentrating hormone from rat hypothalamus. Endocrinology. 1989;125:1660–5.
22. Nahon JL, Presse F, Bittencourt JC, Sawchenko PE, Vale W. The rat melanin-concentrating hormone messenger ribonucleic acid encodes multiple putative neuropeptides expressed in the lateral hypothalamus. Endocrinology. 1989;125:2056–65.
23. Lin C-S, Nicolelis MAL, Schneider JS, Chapin JK. A major direct GABAergic pathway from zona incerta to neocortex. Science. 1990;248:1553–6.
24. Bittencourt JC, Presse F, Arias C, Peto C, Vaughan J, Nahon JL, et al. The melanin-concentrating hormone system of the rat brain – an immunization and hybridization histochemical characterization. J Comp Neurol. 1992;319:218–45.
25. Chemelli RM, Willie JT, Sinton CM, Elmquist JK, Scammell T, Lee C, et al. Narcolepsy in orexin knockout mice: molecular genetics of sleep regulation. Cell. 1999;98:437–51.
26. Peyron C, Tighe DK, van den Pol AN, de Lecea L, Heller HC, Sutcliffe JG, et al. Neurons containing hypocretin (orexin) project to multiple neuronal systems. J Neurosci. 1998;18:9996–10015.
27. Chou TC, Rotman SR, Saper CB. Lateral hypothalamic acetylcholinesterase-immunoreactive neurons co-express either orexin or melanin concentrating hormone. Neurosci Lett. 2004;370:123–6.
28. Llewellyn-Smith IJ, Martin CL, Marcus JN, Yanagisawa M, Minson JB, Scammell TE. Orexin-immunoreactive inputs to rat sympathetic preganglionic neurons. Neurosci Lett. 2003;351:115–9.
29. Lin L, Faraco J, Li R, Kadotani H, Rogers W, Lin X, et al. The sleep disorder canine narcolepsy is caused by a mutation in the hypocretin (orexin) receptor 2 gene. Cell. 1999;98:365–76.
30. Lee MG, Hassani OK, Jones BE. Discharge of identified orexin/hypocretin neurons across the sleep-waking cycle. J Neurosci. 2005;25:6716–20.
31. Hassani OK, Lee MG, Jones BE. Melanin-concentrating hormone neurons discharge in a reciprocal manner to orexin neurons across the sleep-wake cycle. Proc Natl Acad Sci USA. 2009;106:2418–22.
32. Mileykovskiy BY, Kiyashchenko LI, Siegel JM. Behavioral correlates of activity in identified hypocretin/orexin neurons. Neuron. 2005;46:787–98.
33. Yamanaka A, Beuckmann CT, Willie JT, Hara J, Tsujino N, Mieda M, et al. Hypothalamic orexin neurons regulate arousal according to energy balance in mice. Neuron. 2003;38:701–13.
34. Willie JT, Chemelli RM, Sinton CM, Yanagisawa M. To eat or to sleep? Orexin in the regulation of feeding and wakefulness. Annu Rev Neurosci. 2001;24: 429–58.
35. Shimada M, Tritos NA, Lowell BB, Flier JS, Maratos-Flier E. Mice lacking melanin-concentrating hormone are hypophagic and lean. Nature. 1998;396:670–4.
36. Willie JT, Sinton CM, Maratos-Flier E, Yanagisawa M. Abnormal response of melanin-concentrating hormone deficient mice to fasting: hyperactivity and rapid eye movement sleep suppression. Neuroscience. 2008;156:819–29.
37. Verret L, Goutagny R, Fort P, Cagnon L, Salvert D, Leger L, et al. A role of melanin-concentrating hormone producing neurons in the central regulation of paradoxical sleep. BMC Neurosci. 2003;4:19.

Metabolic Influence on the Hypocretin/Orexin Neurons

Denis Burdakov

Keywords
Glucose • Feeding • Metabolism • Orexin • Hypocretin

Hypocretin Cells and Body Energy Levels: Reciprocal Regulation

The reciprocal connection between hypocretin neurons and body energy balance was recognized essentially at the moment of their discovery, when Sakurai et al. showed that (1) intracerebroventricular injection of orexin-A/hypocretin-1 dose dependently increases food intake and (2) prepro-orexin/hypocretin mRNA levels in the lateral hypothalamus are upregulated by fasting [1]. Thus, very early on, the hypocretin system was viewed as a feeding–promoting system activated by falling body energy levels, and this picture is still accurate. However, it is important to note that the net impact of hypocretin cell activity on body weight is probably either insignificant or negative, and not positive as would be expected from a purely feeding-promoting system. This is because hypocretin cell activity also stimulates metabolism, and this "energy-burning" action of hypocretins is presumably greater then their "energy-obtaining" role, since destruction of hypocretin cells leads to late-onset obesity thought to be due to a reduction of energy expenditure [2]. Thus it is more accurate to view the hypocretin system as an orchestrator network that engages multiple processes – wakefulness, hunger, reward-seeking behavior, increased locomotor activity – that help to facilitate successful food seeking. This orchestration is thought to be carried out by diverse projections of hypocretin neurons to wakefulness, reward, breathing, and autonomic centers [3, 4]. This chapter will focus on hypocretin neurons themselves and review evidence supporting the idea that unlike most other neurons in the brain, hypocretin cells act as specialized electrical sensors of ambient nutrient levels, and hence their impact on brain state may critically depend on body energy balance.

Hypocretin Cells as Intrinsic Electrical Detectors of Glucose

Several in vivo experiments indicate that body energy status is a key regulator of hypocretin neurons: prepro-hypocretin mRNA and c-fos expression in hypocretin neurons was increased upon fasting [1, 5–8]. Subsequent in vitro examination of the electrical activity of hypocretin cells indicated that this regulation is likely to be at least in part due to unusual intrinsic sensitivity of hypocretin neurons to glucose. Yamanaka et al. showed that the electrical properties of anatomically or

D. Burdakov (✉)
Department of Pharmacology, University of Cambridge, Cambridge CB2 1PD, UK
e-mail: dib22@cam.ac.uk

C.R. Baumann et al. (eds.), *Narcolepsy: Pathophysiology, Diagnosis, and Treatment*,
DOI 10.1007/978-1-4419-8390-9_19,

functionally (pharmacologically) isolated hypocretin neurons are directly sensitive to changes in the extracellular concentration of glucose and also to the appetite-regulating hormones leptin and ghrelin [9]. Leptin and glucose inhibited the electrical activity of hypocretin neurons, while ghrelin was stimulatory [9]. Increased circulating concentrations of leptin and glucose signal high energy levels and inhibit appetite, whereas increased ghrelin is thought to signal falling energy levels and stimulate feeding; these findings provide a cellular explanation for how hypocretin neurons may become activated during body energy depletion. The finding that mice whose hypocretin neurons are specifically destroyed failed to respond to fasting with increased wakefulness and activity [9] emphasized the potential importance of these signaling pathways in behavioral adaptation to physiological changes in body energy resources.

The inhibitory responses of hypocretin neurons to high glucose are unusual since most neurons in the brain are relatively insensitive to physiological elevations in ambient glucose levels, probably because their glucose-metabolizing enzymes are saturated already at relatively low glucose concentrations [10]. These responses indicate that hypocretin cells belong to a group of specialized "glucose-inhibited" neurons, a sparse group of cells also found in other hypothalamic regions and the brainstem [11]. Are the responses of hypocretin neurons to glucose likely to be physiologically relevant? This sensitivity and specificity of hypocretin cell glucose responses suggest that this is likely to be the case. Changes in glucose concentration in the brain rapidly follow those in the blood [12], and the extracellular glucose concentration range in the brain that corresponds to normal meal-to-meal fluctuation of glucose in the plasma (about 5–8 mM) is expected to be between 0.7 and 2.5 mM [13]. Under physiological conditions in rodents, brain glucose levels are unlikely to become lower than about 0.2 mM or much higher than about 5 mM [12, 13]. Within this concentration range (0.2–5 mM), glucose hyperpolarized and suppressed electrical activity in virtually all of the hypocretin neurons identified by biophysical and neurochemical fingerprinting or by targeted GFP expression in mouse brain slices [14, 15] (Fig. 1). Importantly, within the mouse lateral hypothalamus, these inhibitory responses were relatively specific to hypocretin neurons, since they were not seen in the neighboring cells containing melanin-concentrating hormone [14]. The glucose sensitivity of orexin neurons is also likely to be physiologically relevant as even small (5 Hz) changes in orexin cell firing can lead to behavioral responses [16], and physiological changes in glucose levels can easily evoke such alterations of firing, at least in vitro [15].

How Do Hypocretin Neurons Sense Glucose?

The cellular mechanisms responsible for glucose-induced inhibition of hypocretin neurons (as well as for other glucose-inhibited cells) are currently unclear and controversial. The final effectors appear to be plasma-membrane background K^+ channels

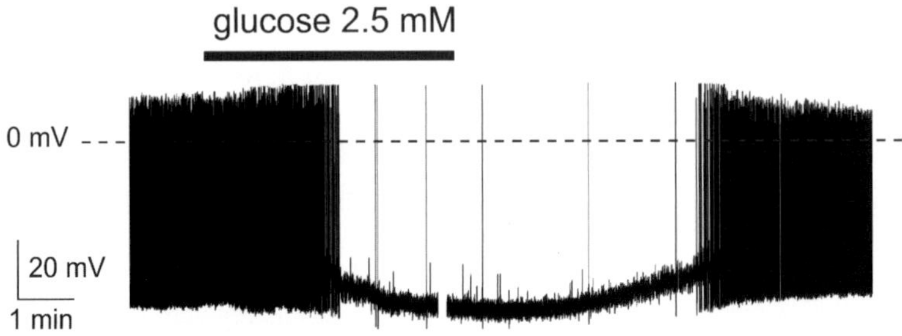

Fig. 1 Example of a glucose response in a single hypocretin neuron recorded using the whole-cell patch-clamp technique in an acute mouse brain slice preparation. Elevating extracellular glucose to 2.5 mM (from a baseline of 1 mM) caused membrane hyperpolarization and cessation of action potential firing

whose activity is profoundly increased by rising glucose levels [15]. This part of the mechanism appears to be shared by some glucose-inhibited neurons of the ventromedial hypothalamus [17], and by some glucose-inhibited neurons of invertebrates [18]. The glucose-stimulated K^+ currents in hypocretin cells exhibit leak-like biophysical properties consistent with channels such as KCNK; however, the exact molecular identity of glucose-stimulated channels have not been unequivocally elucidated, and it also remains possible that more than one type of K^+ channel contributes [19–21].

How glucose activates these inhibitory currents in hypocretin neurons is also unresolved. Traditional models of glucose sensing, inspired by studies of glucose-excited cells such as the beta cells of the pancreas, envisage a key role for intracellular glucose metabolism, with metabolic products such as ATP acting as second messengers modulating membrane channel activity [22–24]. The involvement of conventional glucose metabolism in the specialized effects of glucose on hypocretin neurons is, however, questioned by recent experiments. Blocking glucose metabolism using high concentrations of glucokinase inhibitors did not prevent glucose sensing in hypocretin cells [25]. In turn, similar to observations from neurons of the ventromedial hypothalamus [26], stimulating ATP production with lactate did not mimic glucose-induced hyperpolarization in hypocretin neurons, but was instead excitatory, as expected from the ubiquitous facilitation of neuronal activity by high energy levels [27]. Furthermore, 2-deoxyglucose, a non-metabolizable glucose analog, had effects similar to those of glucose on the membrane potential of hypocretin cells [25]. Such experiments suggest that the inhibitory effects of glucose on hypocretin cells do not require conventional intracellular metabolism of glucose in these neurons. Based on current evidence, it seems likely that glucose entry is not required at all, since intracellular infusion of glucose via the pipette tip does not initiate the inhibitory response while the same cell still responds to glucose applied extracellularly [25].The pharmacology of sugar sensing in hypocretin cells also hints at dissociation between sensing and uptake, since the sugar transporters GLUT1 and SGLT1 transport fructose and galactose, respectively, yet these sugars do not inhibit hypocretin cells [25].

Hypocretin Neurons as "Adaptive" Glucose Sensors

The combination of "sensor" and "effector" tasks is presumably what makes the hypocretin system such a prominent link between body energy status and behavior [9]. However, this multitasking may involve a paradox. If hypocretin cells are shut down by even a small rise in glucose, and loss of their activity causes narcolepsy, how is narcolepsy-free consciousness maintained after a meal or during diabetic hyperglycemia? One theoretical solution could be to delegate different functions to different sets of hypocretin neurons, whereby some cells measure energy status while others maintain cognitive arousal. Another theoretical solution, used by classical sensory systems such as the eye, is to encode changes in stimulus levels but not unchanging (baseline) stimulus levels, a phenomenon called adaptation [28]. This would allow hypocretin cells to maintain adequate tone across a wide range of glucose levels.

Recent experiments examining the intrinsic electrical properties of hypocretin neurons and correlating them with the time course of their glucose responses suggest that both these general mechanisms may be operative in the hypocretin system. Specifically, an estimated two-thirds of orexin neurons exhibit "adapting" firing responses to glucose; i.e. following glucose-induced hyperpolarization, in spite of continuing presence of high glucose levels, these cells depolarize back to firing threshold and resume their typical electrical activity [29]. This allows a large proportion of the orexin system to maintain electrical excitability irrespective of ambient glucose levels. The adapting and nonadapting hypocretin cells are also significantly different in their innate membrane potential properties, indicating that they are different cell types expressing different complements of ion channels [29]. The adaptation to continued presence of high glucose levels involves closure of leak-like K^+ channels, and acts as an automatic "sliding scale" that allows orexin cells to shift their firing sensitivity to match different ranges in glucose levels. Thus, it is possible that only part of the hypocretin system measures absolute glucose levels (nonadapting neurons), whereas another part (adapting neurons) is only sensitive to rapid changes in glucose levels and if the levels remain unchanged for a couple of minutes they self-erase their glucose response. These data point to a way of encoding of metabolic information in the hypocretin system that potentially allows it to track a wide range of sugar fluctuations without destabilizing consciousness.

Interactions Between Hypocretin Neurons and Other Hypothalamic Glucosensors

Apart from hypocretin neurons, the hypothalamus contains at least two other populations of glucose-inhibited cells, the appetite-promoting neuropeptide Y (NPY) neurons of the arcuate nucleus (ARC) and a group of cells in the ventromedial nucleus, whose role and neurochemical identity are not well understood [11]. Recent evidence suggests that the different groups of glucose-inhibited neurons of the hypothalamus are anatomically connected. Both cell bodies and dendritic processes of LH hypocretin neurons are in close apposition with NPY-containing nerve terminals that appear to come from the ARC since they coexpress agouti-related peptide (AgRP) [30]. In turn, ARC NPY/AgRP cells are surrounded by nerve terminals containing hypocretin [31]. Exogenously applied orexin stimulates ARC NPY/AgRP neurons in both Ca^{2+} imaging and patch-clamp electrophysiology assays [31]; [32]. Positive signals from hypocretin to NPY cells would seem to have a simple behavioral rationale of ensuring that states of high alertness and energy expenditure are associated with feelings of hunger. Much more difficult to rationalize is the observation that exogenously applied NPY robustly inhibits hypocretin neurons [33], which would imply a decrease in orexin cell activity when ARC NPY neurons are active. The functional importance of this negative feedback loop is likely to remain unclear until quantitative knowledge becomes available about how much hypocretin and NPY are released in the ARC and LH, respectively, under different behavioral circumstances.

Early models of hypothalamic function saw the mediobasal and lateral hypothalamic areas as behaviorally opposing centers of perhaps equal importance, interacting through reciprocal inhibitory connections [34]. Later, as the ARC became increasingly well characterized, it started to be called a "first-order" center that collects peripheral metabolic information and communicates it to "second-order" centers such as the LH [35]. However, subsequent experiments on the LH made it clear that, with regard to metabolic sensing at least, the first-order/second-order model is inaccurate. Similar to neurons of the ARC, LH hypocretin cells directly sense circulating indicators of body energy status, i.e., they are also "first-order" with regard to capturing metabolic information. This is well exemplified by glucose-inhibited neurons of the ARC and the LH. Apart from glucose, both types of neurons directly sense the "satiety" hormone leptin and the "hunger" hormone ghrelin. ARC NPY/AgRP neurons are directly depolarized and excited by ghrelin, and are hyperpolarized by leptin [32, 36]. As noted above, LH hypocretin neurons exhibit similar direct sensing responses; they are inhibited by leptin and excited by ghrelin [9]. Thus, the NPY neurons of the ARC and the hypocretin neurons of the LH appear to process several key inputs in a parallel rather than a sequential manner. The logic of this arrangement will probably only become clear when functional interactions with other neighboring circuits, such as those controlling temporal organization of behavior [37], become fully mapped and characterized.

Overview

The discovery of specialized nutrient-sensing capabilities of hypocretin neurons may have important implication for understanding neural links between body energy homeostasis and sleep/wake cycles, in both physiological conditions in experimental animals and pathological conditions in humans [38]. However, we still do not know the exact behavioral impact of nutrient sensing in these cells on brain state and behavior; this remains to be elucidated by examining the effects of selective disruption of nutrient sensing in hypocretin neurons on transitions between states of consciousness. Understanding this may be relevant for designing possible "diet therapies" for narcolepsy [39], for example, restricting carbohydrate intake [40], which could potentially increase the activity of surviving hypocretin neurons.

References

1. Sakurai T, Amemiya A, Ishii M, et al. Orexins and orexin receptors: a family of hypothalamic neuropeptides and G protein-coupled receptors that regulate feeding behavior. Cell. 1998;92(4):573–85.
2. Hara J, Beuckmann CT, Nambu T, et al. Genetic ablation of orexin neurons in mice results in narcolepsy, hypophagia, and obesity. Neuron. 2001;30(2): 345–54.
3. Peyron C, Tighe DK, van den Pol AN, et al. Neurons containing hypocretin (orexin) project to multiple neuronal systems. J Neurosci. 1998;18(23):9996–10015.
4. Sakurai T. The neural circuit of orexin (hypocretin): maintaining sleep and wakefulness. Nat Rev Neurosci. 2007;8(3):171–81.
5. Willie JT, Chemelli RM, Sinton CM, Yanagisawa M. To eat or to sleep? Orexin in the regulation of feeding and wakefulness. Annu Rev Neurosci. 2001;24:429–58.
6. Moriguchi T, Sakurai T, Nambu T, Yanagisawa M, Goto K. Neurons containing orexin in the lateral hypothalamic area of the adult rat brain are activated by insulin-induced acute hypoglycemia. Neurosci Lett. 1999;264(1–3):101–4.
7. Griffond B, Risold PY, Jacquemard C, Colard C, Fellmann D. Insulin-induced hypoglycemia increases preprohypocretin (orexin) mRNA in the rat lateral hypothalamic area. Neurosci Lett. 1999;262(2): 77–80.
8. Cai XJ, Widdowson PS, Harrold J, et al. Hypothalamic orexin expression: modulation by blood glucose and feeding. Diabetes. 1999;48(11):2132–7.
9. Yamanaka A, Beuckmann CT, Willie JT, et al. Hypothalamic orexin neurons regulate arousal according to energy balance in mice. Neuron. 2003;38(5): 701–13.
10. Burdakov D, Luckman SM, Verkhratsky A. Glucose-sensing neurons of the hypothalamus. Philos Trans R Soc Lond B Biol Sci. 2005;360(1464):2227–35.
11. Burdakov D, Gonzalez JA. Physiological functions of glucose-inhibited neurones. Acta Physiol (Oxf). 2009;195(1):71–8.
12. Silver IA, Erecinska M. Extracellular glucose concentration in mammalian brain: continuous monitoring of changes during increased neuronal activity and upon limitation in oxygen supply in normo-, hypo-, and hyperglycemic animals. J Neurosci. 1994;14(8):5068–76.
13. Routh VH. Glucose-sensing neurons: are they physiologically relevant? Physiol Behav. 2002;76(3):403–13.
14. Burdakov D, Gerasimenko O, Verkhratsky A. Physiological changes in glucose differentially modulate the excitability of hypothalamic melanin-concentrating hormone and orexin neurons in situ. J Neurosci. 2005;25(9):2429–33.
15. Burdakov D, Jensen LT, Alexopoulos H, et al. Tandem-pore K+ channels mediate inhibition of orexin neurons by glucose. Neuron. 2006;50(5):711–22.
16. Adamantidis AR, Zhang F, Aravanis AM, Deisseroth K, de Lecea L. Neural substrates of awakening probed with optogenetic control of hypocretin neurons. Nature. 2007;450(7168):420–4.
17. Williams RH, Burdakov D. Silencing of ventromedial hypothalamic neurons by glucose-stimulated K(+) currents. Pflugers Arch. 2009;458(4):777–83.
18. Glowik RM, Golowasch J, Keller R, Marder E. D-glucose-sensitive neurosecretory cells of the crab Cancer borealis and negative feedback regulation of blood glucose level. J Exp Biol. 1997;200(Pt 10): 1421–31.
19. Gonzalez JA, Jensen LT, Doyle SE, et al. Deletion of TASK1 and TASK3 channels disrupts intrinsic excitability but does not abolish glucose or pH responses of orexin/hypocretin neurons. Eur J Neurosci. 2009; 30(1):57–64.
20. Guyon A, Tardy MP, Rovere C, Nahon JL, Barhanin J, Lesage F. Glucose inhibition persists in hypothalamic neurons lacking tandem-pore K+ channels. J Neurosci. 2009;29(8):2528–33.
21. Burdakov D, Lesage F. Glucose-induced inhibition: how many ionic mechanisms? Acta Physiol (Oxf). 2009;198(3):295–301.
22. Ashcroft FM, Rorsman P. Electrophysiology of the pancreatic beta-cell. Prog Biophys Mol Biol. 1989;54(2):87–143.
23. Ashford ML, Boden PR, Treherne JM. Glucose-induced excitation of hypothalamic neurones is mediated by ATP-sensitive K+ channels. Pflugers Arch. 1990;415(4):479–83.
24. Levin BE, Routh VH, Kang L, Sanders NM, Dunn-Meynell AA. Neuronal glucosensing: what do we know after 50 years? Diabetes. 2004;53(10):2521–8.
25. Gonzalez JA, Jensen LT, Fugger L, Burdakov D. Metabolism-independent sugar sensing in central orexin neurons. Diabetes. 2008;57(10):2569–76.
26. Song Z, Routh VH. Differential effects of glucose and lactate on glucosensing neurons in the ventromedial hypothalamic nucleus. Diabetes. 2005;54(1):15–22.
27. Mobbs CV, Kow LM, Yang XJ. Brain glucose-sensing mechanisms: ubiquitous silencing by aglycemia vs.

hypothalamic neuroendocrine responses. Am J Physiol Endocrinol Metab. 2001;281(4):E649–54.
28. Carpenter R. Neurophysiology. 4th ed. London: Arnold; 2003.
29. Williams RH, Alexopoulos H, Jensen LT, Fugger L, Burdakov D. Adaptive sugar sensors in hypothalamic feeding circuits. Proc Natl Acad Sci USA. 2008; 105(33):11975–80.
30. Broberger C, De Lecea L, Sutcliffe JG, Hokfelt T. Hypocretin/orexin- and melanin-concentrating hormone-expressing cells form distinct populations in the rodent lateral hypothalamus: relationship to the neuropeptide Y and agouti gene-related protein systems. J Comp Neurol. 1998;402(4):460–74.
31. Muroya S, Funahashi H, Yamanaka A, et al. Orexins (hypocretins) directly interact with neuropeptide Y, POMC and glucose-responsive neurons to regulate Ca 2+ signaling in a reciprocal manner to leptin: orexigenic neuronal pathways in the mediobasal hypothalamus. Eur J Neurosci. 2004;19(6):1524–34.
32. van den Top M, Lee K, Whyment AD, Blanks AM, Spanswick D. Orexigen-sensitive NPY/AgRP pacemaker neurons in the hypothalamic arcuate nucleus. Nat Neurosci. 2004;7(5):493–4.
33. Fu LY, Acuna-Goycolea C, van den Pol AN. Neuropeptide Y inhibits hypocretin/orexin neurons by multiple presynaptic and postsynaptic mechanisms: tonic depression of the hypothalamic arousal system. J Neurosci. 2004;24(40):8741–51.
34. Oomura Y, Kimura K, Ooyama H, Maeno T, Iki M, Kuniyoshi M. Reciprocal activities of the ventromedial and lateral hypothalamic areas of cats. Science. 1964;143:484–5.
35. Schwartz MW, Woods SC, Porte Jr D, Seeley RJ, Baskin DG. Central nervous system control of food intake. Nature. 2000;404(6778):661–71.
36. Cowley MA, Smith RG, Diano S, et al. The distribution and mechanism of action of ghrelin in the CNS demonstrates a novel hypothalamic circuit regulating energy homeostasis. Neuron. 2003;37(4): 649–61.
37. Saper CB, Scammell TE, Lu J. Hypothalamic regulation of sleep and circadian rhythms. Nature. 2005;437(7063):1257–63.
38. Laposky AD, Bass J, Kohsaka A, Turek FW. Sleep and circadian rhythms: key components in the regulation of energy metabolism. FEBS Lett. 2008;582(1): 142–51.
39. Husain AM, Yancy Jr WS, Carwile ST, Miller PP, Westman EC. Diet therapy for narcolepsy. Neurology. 2004;62(12):2300–2.
40. Garma L, Marchand F. Non-pharmacological approaches to the treatment of narcolepsy. Sleep. 1994; 17(8 Suppl):S97–102.

Endocrine Abnormalities in Narcolepsy

Thomas Pollmächer, Marietta Keckeis, and Andreas Schuld

Keywords
Narcolepsy • Hypothalamus • Metabolism • Energy homeostasis • Endocrine systems • Orexin • Hypocretin • Obesity

Introduction

The profound influences of sleep and circadian rhythms on endocrine systems have already been discovered in the late 1960s. Growth hormone secretion, for example, is closely linked to the first hours of sleep, when slow-wave sleep (SWS) dominates [1] and sleep deprivation almost completely suppresses its release [2]. In contrast, circulating levels of cortisol, the effector hormone of the hypothalamo–pituitary–adrenal (HPA) axis, are the highest in the second half of the night and the early morning, but this time course is not affected by sleep deprivation, indicating circadian rather than sleep-dependent regulation [3].

Hormonal systems (see Table 1) individually interact with sleep regulation and circadian systems in a complex way. As a consequence, the hormonal secretion pattern during sleep profoundly differs from that in the waking state. Still, we know relatively little about the functional significance of sleep-related endocrine secretion patterns. One area which attracted increasing scientific attention during the last years is the possible role of sleep in the regulation of body weight and energy homeostasis. Epidemiological and experimental evidence indicates that sleep curtailment or disturbance contributes to the emergence of overweight and diabetes [4].

In this context, narcolepsy is a disorder of particular interest. In addition to the classical symptoms such as excessive daytime sleepiness and cataplexy, we know very well that narcoleptic patients are at increased risk for obesity and diabetes (for an overview, see [5]). These findings were considered as an epiphenomenon of minor importance until it was discovered that narcolepsy is closely linked to a deficient hypothalamic neuronal system producing the neurotransmitter orexins (hypocretins). These neuropeptides are involved in appetite regulation, energy homeostasis, regulation of endocrine systems, and sleep [6].

This chapter summarizes the present knowledge on the importance of the orexins for endocrine regulation and of endocrine abnormalities in narcolepsy.

T. Pollmächer (✉)
Center of Mental Health, Klinikum Ingolstadt, Krumenauer Str. 25, Ingolstadt 85049, Germany
and
Max Planck Institute of Psychiatry, Munich, Germany
e-mail: thomas.pollmaecher@klinikum-ingolstadt.de

C.R. Baumann et al. (eds.), *Narcolepsy: Pathophysiology, Diagnosis, and Treatment*,
DOI 10.1007/978-1-4419-8390-9_20, © Springer Science+Business Media, LLC 2011

Table 1 Overview of endocrine systems and organs

Neuroendocrine
Hypothalamo–pituitary–adrenal (HPA) system
Hypothalamo–pituitary–thyroid (HPT) system
Hypothalamo–pituitary–gonadal (HPG) system
Growth hormone system (GHRH/somatostatin)
Prolactin (dopamine regulated)
Melatonin (pineal gland)
Other
Parathyroid
Thymus
Pancreas
Adipose tissue

Basics of Endocrine Systems

The major principle of all endocrine systems is a complex long-distance regulatory network controlling the release of one or several hormones from a specific endocrine organ such as the pituitary, thyroid, or adrenal gland. Hormones enter the blood, which discriminates them from autocrine and paracrine mediators acting within specific tissue, and act at distant sites through specific receptors. Table 1 gives an overview on the major systems. In neuroendocrine systems, the endocrine gland is directly targeted by releasing factors from the central nervous system (CNS), which may be neurotransmitters or neuropeptides. In some of these systems, there are even hormonal cascades involving releasing factors and two endocrine glands. One of these, the hypothalamo–pituitary–adrenal (HPA) system, is schematically depicted in Fig. 1 to exemplify some regulatory principles. Upon appropriate stimulation, which can be rather diverse (e.g., among others, inflammatory cytokines and psychosocial stress for the HPA axis), the hypothalamic releasing factors corticotropin-releasing hormone (CRH) and arginine vasopressin (AVP) reach the anterior pituitary through the hypophyseal portal system and induce the release of the adrenocorticotropic hormone (ACTH). The latter is released in the blood stream, reaching the adrenals within minutes, where it induces the release of cortisol. The effector hormone cortisol exerts its effects on physiological functions (metabolism, hematopoiesis, lymphocyte function, and others) and, in parallel, sends negative feed back signals to the pituitary and the hypothalamus. Other neuroendocrine systems such as the hypothalamo–pituitary–thyroid (HPT) axis and the hypothalamo–pituitary–gonadal (HPG) axis are functionally similar.

Other endocrine systems – such as those involved in the regulation of growth hormone and prolactin secretion – depend exclusively on hypothalamic factors which stimulate, or in the case of prolactin inhibit, hormonal release from the pituitary. In addition, there are endocrine glands which are not under direct hypothalamic control, including the pineal gland, the parathyroid glands, and the endocrine pancreas. Finally, adipose tissue is not an endocrine gland per se, but it releases numerous hormonal substances with endocrine function, such as leptin [7]. Of course, these endocrine systems closely interact with the brain endocrine centers through various hormonal receptors in the hypothalamus, although lacking direct hypothalamic control.

From a CNS perspective, endocrine systems control complex physiological functions (metabolic rate, glucose metabolism, energy homeostasis, and size of the adipose tissue) which cannot be directly monitored or regulated by neurons.

Orexins and Endocrine Systems

Orexin-A and -B (also known as hypocretin-1 and hypocretin-2) are hypothalamic peptides that are produced in the posterior and lateral hypothalamus and cleaved from a common precursor protein (prepro-orexin). Orexins act through two specific receptors (OX-1 and OX-2). Although the number of orexin neurons is small, they project to wide areas of the brain, within and outside the hypothalamus, but not to the cerebellum. Orexin neurons are modulated by many neurotransmitters and neuromodulators. Orexin receptors are expressed in a variety of non-CNS structures,

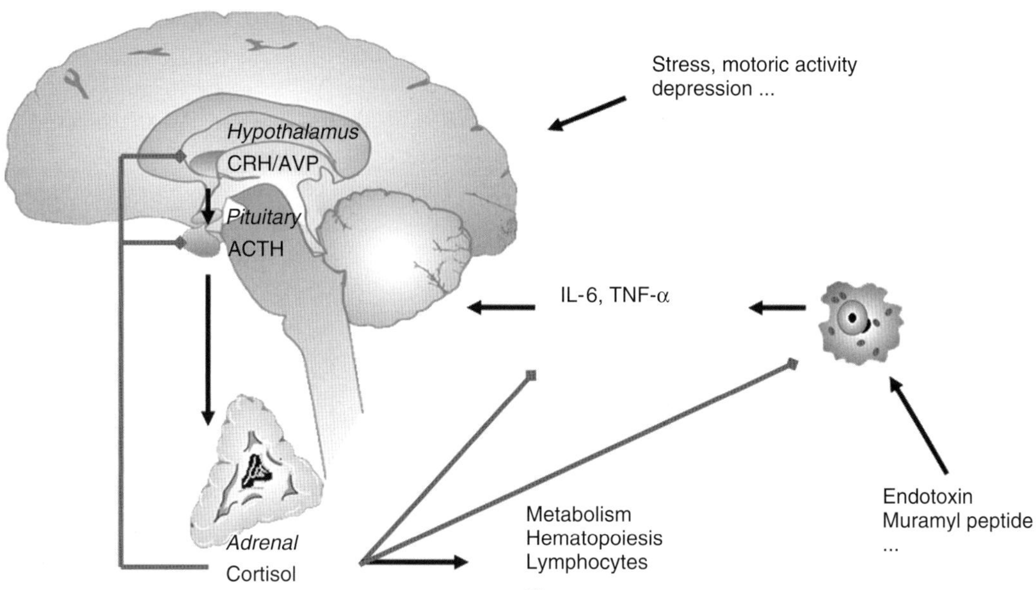

Fig. 1 Regulation of the major endocrine stress system, the hypothalamo–pituitary–adrenal (HPA) system. *CRH* corticotropin-releasing hormone, *AVP* arginine vasopressin, *ACTH* corticotropin, *IL-6* interleukin-6, *TNF-α* tumor necrosis factor-α

such as the pituitary, the adrenals, testis, ovary, pancreas, and others, but the functional significance of this expression is not yet established. Although present evidence suggests that orexin neurons do not directly project to the hypophyseal portal system, they clearly affect many endocrine systems. For a more general overview on orexins and their endocrine functions, refer to Tsujino and Sakurai [6], and for details on their interaction with neuroendocrine systems, see López et al. [8].

Orexins and the HPA, HPG, and HPS Systems (See Fig. 2)

Central injection of orexin-A leads to the expression of CRH mRNA in the periventicular nucleus (PVN) and induces a subsequent release of ACTH and cortisol [9]. This action of orexin may in part involve neuropeptide Y (NPY) [10]. Although the final role of peripheral actions of orexins remains to be elucidated, orexin induces cortisol release directly from the adrenals in the rat [11]. Possible actions of glucocorticoids on orexin neurons are not yet finally established. Hence, orexins seem to be potent activators of the HPA system. As of now, the functional relevance of HPA system activation is not fully understood. Behavioral activating effects of orexins may be meditated via this pathway, as well as by the modulation of stress and cardiovascular responses.

The interactions of orexins with the HPG system depend on gender and on the female menstrual cycle. Hence, it is not surprising that orexin exerts both stimulatory and inhibitory actions in this system, depending on the situation. For a detailed outline on the present knowledge, the reader is referred to the overview of López and colleagues [12].

Again at the hypothalamic and not at the pituitary level, orexins inhibit the HPT system by reducing the expression of thyrotropin-releasing hormone (TRH) [13], without any feedback actions of thyroid-stimulating hormone (TSH) or thyroid hormones on the orexin system. TRH–orexin interactions appear to be important for metabolic control, thermoregulation, and locomotor activity.

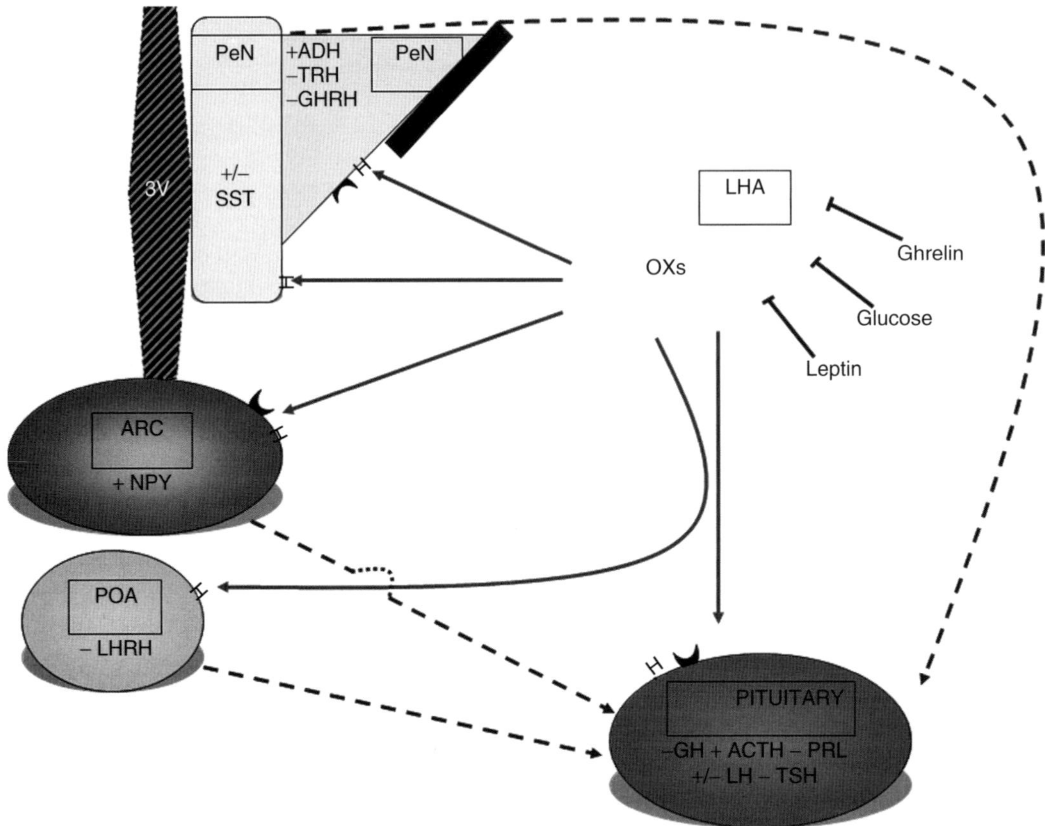

Fig. 2 Summary of endocrine actions and relationships of orexins. Modified from [8]

Orexins and the Growth Hormone Axis

Orexin neurons project to hypothalamic areas expressing growth hormone-releasing hormone (GHRH) and somatostatin (SST), and the same areas express orexin receptors. Orexin-A reduces pulsatile growth hormone (GH) release [14] probably by blunting GHRH expression, because there are no direct effects on GH-releasing cells in the pituitary. In addition, there is a stimulating effect of orexins on SST, which might involve NPY [15]. So far, no feedback actions of orexins on GH or GHRH have been demonstrated. Physiologically, the interactions between orexins and the growth hormone axis are not yet fully understood. Suppression of GHRH expression by orexins might mediate some effects on sleep, and also be important for the adaptation to fasting [14].

Orexins and Prolactin Release

The release of prolactin from the pituitary is controlled by a negative influence of dopamine release from the tuberoinfindubular dopamine neurons in the nucleus arcuatus. In most but not all studies, orexins suppressed prolactin release by an NPY-mediated blocking effect on dopamine release [16]. Prolactin, in turn, under certain conditions, suppresses orexin expression [17].

Orexins, the Pancreas, and Adipose Tissue

Orexins are not directly linked to the pancreas and the adipose tissue, but are involved in related regulatory systems: orexin neurons modulate multiple factors regulating energy balance and food intake. In particular, orexin neurons are glucose sensitive, with increasing glucose concentrations suppressing and decreasing concentrations stimulating neuronal activity [6]. Similarly, the fat cell hormone leptin – which communicates the extent of adipose tissue to the hypothalamus – also suppresses the activity of orexin neurons [18]. Ghrelin, an appetite-stimulating peptide released from the stomach, in turn, has potent activating effect on orexin neurons [19]. These findings indicate that orexin neurons act as sensors of the nutritional status of the body.

Endocrine Abnormalities in Narcolepsy

Studies on Endocrine Systems in Narcolepsy

Until the discovery of the orexins and their tight involvement in the pathophysiology of narcolepsy in 1998, the disorder was considered a paradigmatic pure disorder of sleep. Nonetheless, obesity [20] and an increased incidence of diabetes [21] have been described quite early. Moreover, shortly after the discovery of the interaction of sleep with growth hormone secretion, first endocrine studies were performed in narcoleptic patients [22, 23]. Until the discovery of orexins, however, scientific and clinical interest in this topic was sparse and motivated mainly by the basic interest in endocrine–sleep interactions. Table 2 lists all the studies done so far. It also summarizes the major results related to the different endocrine systems. Although the number of studies has increased since the discovery of orexins, the fundamental of data remains weak and, moreover, five studies done after the year 2000 rely on the very same seven subjects and seven matched controls [24–28]. These studies, on the contrary, are very well controlled in terms of medication, BMI matching, study design, and orexin deficiency, as opposed to many previous studies. At this stage, the results that are shown in Table 2 must still be considered rough and preliminary.

Studies on the HPA system indicate that cortisol secretion is not altered in narcolepsy. One study reported that ACTH secretion across 24 h is blunted [24]. Regarding the HPT system, reduced TSH and normal thyroid hormones have been reported [28]. One earlier case report supports TSH blunting [29], but this was not confirmed in a larger sample of patients [30]. Very few data are available for the HPG system, and only the data of Kok and colleagues are reliable, reporting a failure of pulsatile LH release [26]. The majority of available studies dealt with GH levels, but the results are very inconsistent. In narcolepsy, not only unchanged GH secretion, but also increased cross-sectional levels [31], irregular secretion, and changes in the 24-h secretion pattern [25] have been reported, and finally, there is evidence of decreased GH secretion [32]. Also, studies on prolactin levels are not consistent: of four available studies, two report decreased levels [23, 33], one an increase [30], and one no change as compared to non-narcoleptic controls [32]. The few studies on melatonin did not observe any abnormalities [34–36].

There is still a lively debate on leptin secretion in narcolepsy, because one frequent finding in narcolepsy is obesity [37]. Leptin levels are proportional to the extent of adipose tissue, and, moreover, orexin neurons are leptin sensitive. Two earlier well-controlled studies [27, 38] reported significant decrease in circulating leptin levels in narcoleptic patients. Later, however, two less controlled, but much bigger studies could not replicate this finding [39, 40].

Finally, glucose metabolism has been addressed in some studies so far. In narcolepsy, increased incidence of diabetes type II has been reported by two groups [21, 41], and we have preliminary data [42] suggesting increased fasting glucose levels in narcoleptic patients compared to that in population controls, but similar responses to an oral glucose load compared to that in BMI-matched controls. This may indicate that glucose metabolism is disturbed in relation to increased obesity. These data, however, need to be confirmed by larger studies.

Taken together, the present but often inconsistent data suggest some endocrine abnormalities in narcolepsy.

Table 2 Overview of endocrine studies in narcolepsy

System or hormone studied:	HPA	HPT	HPG	GH	Prolactin	Melatonin	Glucose	Leptin
Roberts et al. [21]							X	
Sjaastad et al. [44]			X					
Galagher [22]	X							
Besset et al. [45]	X			X				
Clark et al. [33]				X	X			
Higuchi et al. [23]	X			X	X			
Pavel et al. [34]						X		
Nygren and Röjdmark [29]		X						
Quabbe [46]				X				
Honda et al. [41]							X	
Sandyk [47]	X							
Schulz et al. [32]	X			X	X			
Blazejova et al. [35]	X					X		
Schuld et al. [38]								X
Kok et al. [24]	X							
Overeem et al. [25]				X				
Krahn et al. [36]						X		
Okun et al. [31]				X				
Kok et al. [26]			X					
Kok et al. [27]								X
Kok et al. [28]		X						
Arnulf et al. [39]								X
Chabas et al. [30]	X	X			X		X	
Dahmen et al. [40]								X
Pollmächer et al. [5]							X	

Possible Causes and Consequences of Endocrine Abnormalities in Narcolepsy

It is tempting to speculate that endocrine disturbances in narcolepsy are caused by orexin deficiency, because we believe that orexin cell loss is the key pathophysiology of the disorder. However, present data are not sufficient to support this assumption, and there are still many controversies. Moreover, most data are from patients who have been suffering from narcolepsy for years or decades, but in a chronic disorder, adaptive or compensatory mechanisms for the orexin deficiency may occur. For instance, we often see narcolepsy patients with a decrease of sleepiness and cataplexy over the long-term course of the disorder. In addition, there are confounding factors difficult to be controlled during endocrine studies: medication, even after withdrawal, may have long-term influence on the metabolism, and the altered sleep pattern in narcoleptic patients itself might influence endocrine systems [32]. Last but not least, altered immune parameters, e.g., altered cytokine levels that have been found in narcolepsy [43], are likely to affect endocrine systems.

The most obvious potential consequence of endocrine abnormalities in narcolepsy is obesity, but even this topic remains controversial (see Fig. 3). It is also possible that obesity is linked to decreased locomotion in narcolepsy patients, or to altered feeding behavior. Thus, the details of this association are unknown and deserve further studies.

Summary and Perspectives

Orexins (hypocretin) are potent neuroendocrine modulators that interact with almost every endocrine system, mainly on the hypothalamic level. These interactions are important for energy homeostasis and sleep regulation. In addition, the activity of orexin neurons is modulated by glucose,

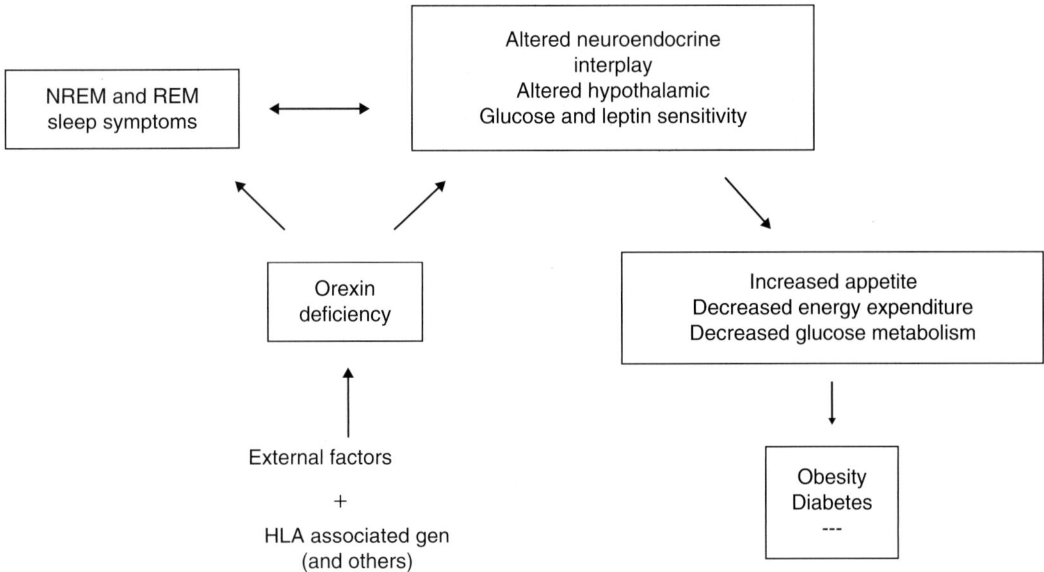

Fig. 3 Present pathophysiological model of narcolepsy as a complex neuroendocrine disorder

leptin, and ghrelin, suggesting that orexin neurons act as sensors of the nutritional status of the body. Therefore, narcolepsy – characterized by a complete acquired absence of orexin signaling – may nowadays be conceptualized as a complex neuroendocrine sleep–wake disorder. However, present knowledge on endocrine functions of the orexins in narcolepsy is still too limited to support a detailed understanding. On a clinical level, obesity is established as one aspect of the disease, but the underlying mechanisms remain to be elucidated. Better and more studies are needed in larger samples of well-defined patients, which warrants collaborative approaches. Furthermore, clinical studies may establish the epidemiology of other metabolic disturbances beyond obesity and diabetes in narcolepsy, which should allow us to get a better picture of the disease, enabling the estimation of additional health risks for patients.

References

1. Sassin JF, Parker DC, Mace JW, et al. Human growth hormone release: relation to slow-wave sleep and sleep-walking cycles. Science. 1969;165(892): 513–5.
2. Mullington J, Hermann D, Holsboer F, Pollmächer T. Age-dependent suppression of nocturnal growth hormone levels during sleep deprivation. Neuroendocrinology. 1996;64(3):233–41.
3. Steiger A. Neurochemical regulation of sleep. J Psychiatr Res. 2007;41(7):537–52.
4. Knutson KL, Van Cauter E. Associations between sleep loss and increased risk of obesity and diabetes. Ann N Y Acad Sci. 2009;1129:287–304.
5. Pollmächer T, Dalal MA, Schuld A. Immunoendocrine abnormalities in narcolepsy. Sleep Med Clin. 2007;2:293–302.
6. Tsujino N, Sakurai T. Orexin/hypocretin: a neuropeptide at the interface of sleep, energy homeostasis, and reward system. Pharmacol Rev. 2009;61(2):162–76.
7. Trayhurn P. Endocrine and signalling role of adipose tissue: new perspectives on fat. Acta Physiol Scand. 2005;184(4):285–93.
8. López M, Lage R, Tung YC, et al. Orexin expression is regulated by alpha-melanocyte-stimulating hormone. J Neuroendocrinol. 2007;19(9):703–7.
9. Jászberényi M, Bujdosó E, Pataki I, Telegdy G. Effects of orexins on the hypothalamic-pituitary-adrenal system. J Neuroendocrinol. 2000;12(12):1174–8.
10. Jászberényi M, Bujdosó E, Telegdy G. The role of neuropeptide Y in orexin-induced hypothalamic-pituitary-adrenal activation. J Neuroendocrinol. 2001;13(5):438–41.
11. Malendowicz LK, Tortorella C, Nussdorfer GG. Orexins stimulate corticosterone secretion of rat adrenocortical cells, through the activation of the adenylate cyclase-dependent signaling cascade. J Steroid Biochem Mol Biol. 1999;70(4–6):185–8.
12. López M, Nogueiras R, Tena-Sempere M, Diéguez C. Orexins (hypocretins) actions on the GHRH/somatostatin-GH axis. Acta Physiol (Oxf). 2010;198(3): 325–34.

13. Mitsuma T, Hirooka Y, Mori Y, et al. Effects of orexin A on thyrotropin-releasing hormone and thyrotropin secretion in rats. Horm Metab Res. 1999;31(11):606–9.
14. Seoane LM, Tovar SA, Perez D, et al. Orexin A suppresses in vivo GH secretion. Eur J Endocrinol. 2004;150(5):731–6.
15. López M, Seoane LM, García Mdel C, Diéguez C, Señarís R. Neuropeptidebut not agouti-related peptide or melanin-concentrating hormone, is a target peptide for orexin-A feeding actions in the rat hypothalamus. Neuroendocrinology. 2002;75(1):34–44.
16. Hsueh YC, Cheng SM, Pan JT. Fasting stimulates tuberoinfundibular dopaminergic neuronal activity and inhibits prolactin secretion in oestrogen-primed ovariectomized rats: involvement of orexin A and neuropeptide Y. J Neuroendocrinol. 2002;14(9):745–52.
17. García MC, López M, Gualillo O, et al. Hypothalamic levels of NPY, MCH, and prepro-orexin mRNA during pregnancy and lactation in the rat: role of prolactin. FASEB J. 2003;17(11):1392–400.
18. Kohno D, Suyama S, Yada T. Leptin transiently antagonizes ghrelin and long-lastingly orexin in regulation of Ca2+ signaling in neuropeptide Y neurons of the arcuate nucleus. World J Gastroenterol. 2008;14(41):6347–54.
19. Toshinai K, Date Y, Murakami N, et al. Ghrelin-induced food intake is mediated via the orexin pathway. Endocrinology. 2003;144(4):1506–12.
20. Daniels LE. Narcolepsy. Medicine. 1943;13(1):1–122.
21. Roberts HJ. The syndrome of narcolepsy and diabetogenic hyperinsulinism in the american negro: important clinical, social and public health aspects. J Am Geriatr Soc. 1965;13:852–85.
22. Gallagher BB. Regulation of cortisol secretion in Parkinson's syndrome and narcolepsy. J Clin Endocrinol Metab. 1971;32(6):796–800.
23. Higuchi T, Takahashi Y, Takahashi K, Niimi Y, Miyasita A. Twenty-four-hour secretory patterns of growth hormone, prolactin, and cortisol in narcolepsy. J Clin Endocrinol Metab. 1979;49(2):197–204.
24. Kok SW, Roelfsema F, Overeem S, et al. Dynamics of the pituitary-adrenal ensemble in hypocretin-deficient narcoleptic humans: blunted basal adrenocorticotropin release and evidence for normal time-keeping by the master pacemaker. J Clin Endocrinol Metab. 2002;87:5085–91.
25. Overeem S, Kok SW, Lammers GJ, et al. Somatotropic axis in hypocretin-deficient narcoleptic humans: altered circadian distribution of GH-secretory events. Am J Physiol Endocrinol Metab. 2003;284(3):641–7.
26. Kok SW, Roelfsema F, Overeem S, et al. Pulsatile LH release is diminished, whereas FSH secretion is normal, in hypocretin-deficient narcoleptic men. Am J Physiol Endocrinol Metab. 2004;287(4):630–6.
27. Kok SW, Meinders AE, Overeem S, et al. Reduction of plasma leptin levels and loss of its circadian rhythmicity in hypocretin (orexin)-deficient narcoleptic humans. J Clin Endocrinol Metab. 2002;87(2):805–9.
28. Kok SW, Roelfsema F, Overeem S, et al. Altered setting of the pituitary-thyroid ensemble in hypocretin-deficient narcoleptic men. Am J Physiol Endocrinol Metab. 2005;288(5):892–9.
29. Nygren A, Röjdmark S. Isolated thyrotropin deficiency in a man with narcoleptic attacks. Acta Med Scand. 1982;212(3):175–7.
30. Chabas D, Foulon C, Gonzalez J, et al. Eating disorder and metabolism in narcoleptic patients. Sleep. 2007;30(10):1267–73.
31. Okun ML, Giese S, Lin L, Einen M, Mignot E, Coussons-Read ME. Exploring the cytokine and endocrine involvement in narcolepsy. Brain Behav Immun. 2004;18(4):326–32.
32. Schulz H, Kiss E, Hasse D, et al. The 24-h hormonal secretion pattern of narcoleptic patients and control subjects. In: Horn J, editor. Sleep 1990. Bochum, FRG: Pontenagel; 1990. p. 375–8.
33. Clark RW, Schmidt HS, Malarkey WB. Disordered growth hormone and prolactin secretion in primary disorders of sleep. Neurology. 1979;29(6):855–61.
34. Pavel S, Goldstein R, Petrescu M. Vasotocin, melatonin and narcolepsy: possible involvement of the pineal gland in its patho-physiological mechanism. Peptides. 1980;1(4):281–4.
35. Blazejová K, Nevsímalová S, Illnerová H, Hájek I, Sonka K. Sleep disorders and the 24-hour profile of melatonin and cortisol. Sb Lek. 2000;101(4):347–51.
36. Krahn LE, Boeve BF, Oliver L, Silber MH. Hypocretin (orexin) and melatonin values in a narcoleptic-like sleep disorder after pinealectomy. Sleep Med. 2002;3(6):521–3.
37. Schuld A, Hebebrand J, Geller F, Pollmächer T. Increased body-mass index in patients with narcolepsy. Lancet. 2000;355(9211):1274–5.
38. Schuld A, Blum WF, Uhr M, et al. Reduced leptin levels in human narcolepsy. Neuroendocrinology. 2000;72(4):195–8.
39. Arnulf I, Lin L, Zhang J, et al. CSF versus serum leptin in narcolepsy: is there an effect of hypocretin deficiency? Sleep. 2006;29(8):1017–24.
40. Dahmen N, Engel A, Helfrich J, et al. Peripheral leptin levels in narcoleptic patients. Diabetes Technol Ther. 2007;9(4):348–53.
41. Honda Y, Doi Y, Ninomiya R, Ninomiya C. Increased frequency of non-insulin-dependent diabetes mellitus among narcoleptic patients. Sleep. 1986;9:254–9.
42. Beitinger P, Wehrle R, Fulda S, et al. Metabolic parameters in patients with narcolepsy. J Sleep Res. 2006;15 Suppl 1:79–80.
43. Himmerich H, Beitinger PA, Fulda S, et al. Plasma levels of tumor necrosis factor alpha and soluble

tumor necrosis factor receptors in patients with narcolepsy. Arch Intern Med. 2006;166(16): 1739–43.
44. Sjaastad O, Hultin E, Norman N. Narcolepsy: increased urinary secretion of estriol. Acta Neurol Scand. 1970;46(1):111–8.
45. Besset A, Bonardet A, Billiard M, Descomps B, de Paulet AC, Passouant P. Circadian patterns of growth hormone and cortisol secretions in narcoleptic patients. Chronobiologia. 1979;6(1):19–31.
46. Quabbe HJ. Hypothalamic control of GH secretion: pathophysiology and clinical implications. Acta Neurochir (Wien). 1985;75(1–4):60–71.
47. Sandyk R. ACTH and its relationship to the cataplectic syndrome. Int J Neurosci. 1989;47(3–4): 317–9.

Appetite and Obesity

Alice Engel and Norbert Dahmen

Keywords

Appetite • Obesity • Body mass index • Orexin • Hypocretin • Obesity • Leptin • Ghrelin

Introduction

Orexin (hypocretin) plays an important role in promoting wakefulness [1]. It stimulates wakefulness when injected in the cerebral ventricles, in the periventricular nucleus, dorsomedial hypothalamus, or lateral hypothalamus [2, 3]. Orexinergic neurons originate from the lateral hypothalamus and have projections to most parts of the central nervous system including the brain stem. The lack of orexin and/or orexin receptors is linked to narcolepsy [4–6].

Apart from its role in sleep–wake regulation, orexin stimulates feeding by activating the key feeding regulatory sites in the hypothalamus, and limbic and midbrain areas [7]. Fasting leads to an increase, and food intake to a decrease of the activity of orexinergic neurons. The role of the orexins in the initiation of food intake and energy metabolism suggests that orexin deficiency in narcoleptic patients should cause a reduction in body weight [8]. However, in stark contrast to this straightforward reasoning, one striking symptom of narcolepsy is obesity. The body mass index (BMI) of narcoleptic patients is about 15% higher than that in the general population. For many narcoleptic patients, it is painful to be considered not only unproductive and slow because of their excessive daytime sleepiness, but also as "fat and lazy" because of their overweight. The connection between narcolepsy and overweight has long been recognized by clinicians, but only during the last years have studies been aimed at the systematic exploration of possible explanations.

This chapter will cover some of the data of our and of other groups that substantiate the clinical observation that narcoleptic patients tend to be obese. We will then move forward to review some of the actual hypotheses that may underlie the phenomenon. In particular, we will address the following topics: (1) the leptin hypothesis of narcolepsy-associated obesity, (2) altered feeding behavior and eating disorders in narcoleptics, (3) reduced basal metabolic rate as a proxy for reduced energy expenditure, and (4) changes in body temperature and general motor activity.

N. Dahmen (✉)
Department of Psychiatry, University of Mainz, Untere Zahlbacher Straße 8, 55131 Mainz, Germany
e-mail: ndahmen@mail.psychiatrie.klinik.uni-mainz.de

C.R. Baumann et al. (eds.), *Narcolepsy: Pathophysiology, Diagnosis, and Treatment*,
DOI 10.1007/978-1-4419-8390-9_21, © Springer Science+Business Media, LLC 2011

Narcolepsy and Obesity

In 2000 and 2001, Schuld et al. and Dahmen et al. substantiated the clinical finding that patients with narcolepsy had a higher BMI than population-based controls [9, 10]. Schuld et al. recorded the BMI of 35 HLA-DR2-positive narcoleptic patients [9]. Dahmen et al. examined data from 132 narcoleptic patients, including four monozygotic discordant twin pairs. In both studies, there were no differences between medicated and drug-naïve patients. Thus, it was concluded that the higher BMI was linked to the pathophysiology of the disease (see Table 1 and Fig. 1).

Kok and colleagues measured BMI and waist circumference in 138 narcoleptic patients, and compared the results with anthropometric parameters measured in two large surveys of the general Dutch population [11]. They showed that obesity, defined as a BMI ≥ 30 kg/m^2, occurs more than twice as often among patients with narcolepsy than among the general (Dutch) population (see Fig. 2). Moreover, in 39% of narcoleptic patients, the waist circumference required medical

Table 1 Comparison of mean body mass index ± standard deviation [10]

	All	Women	Men
Narcoleptic patients (74 females, 58 males; mean age: 47.7 years)	28.2 ± 5.5	27.2 ± 5.8	29.5 ± 4.8
General population (1,386 females, 1,365 males; age group: 35–44 years)	23.5 ± 3.5 (p = 0.001)	22.4 ± 3.7 (p = 0.0003)	24.6 ± 3.2 (p = 0.16)
General population (1,294 females, 1,191 males; age group: 45–54 years)	24.4 ± 3.7 (p < 0.001)	23.2 ± 3.9 (p < 0.001)	25.5 ± 3.4 (p < 0.001)

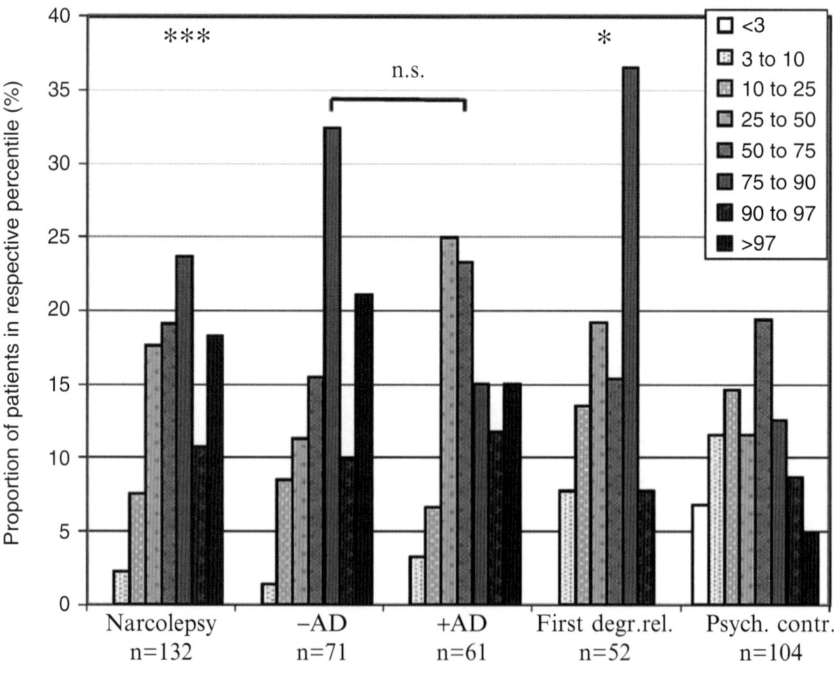

Fig. 1 Percent of patients within the percentiles of the German National Nutrition Survey. The proportion of patients above the 75th percentile is statistically increased in the narcoleptic patients and in the first-degree relatives of the narcoleptic patients but not in the psychiatric controls in comparison with the controls of the German National Nutrition survey (Chi-square test, two sided). There was no significant difference between patients with a lifetime tricyclic medication of less than 1 month (–AD) and more than 1 month (+AD). ***p < 0.05; *p < 0.001

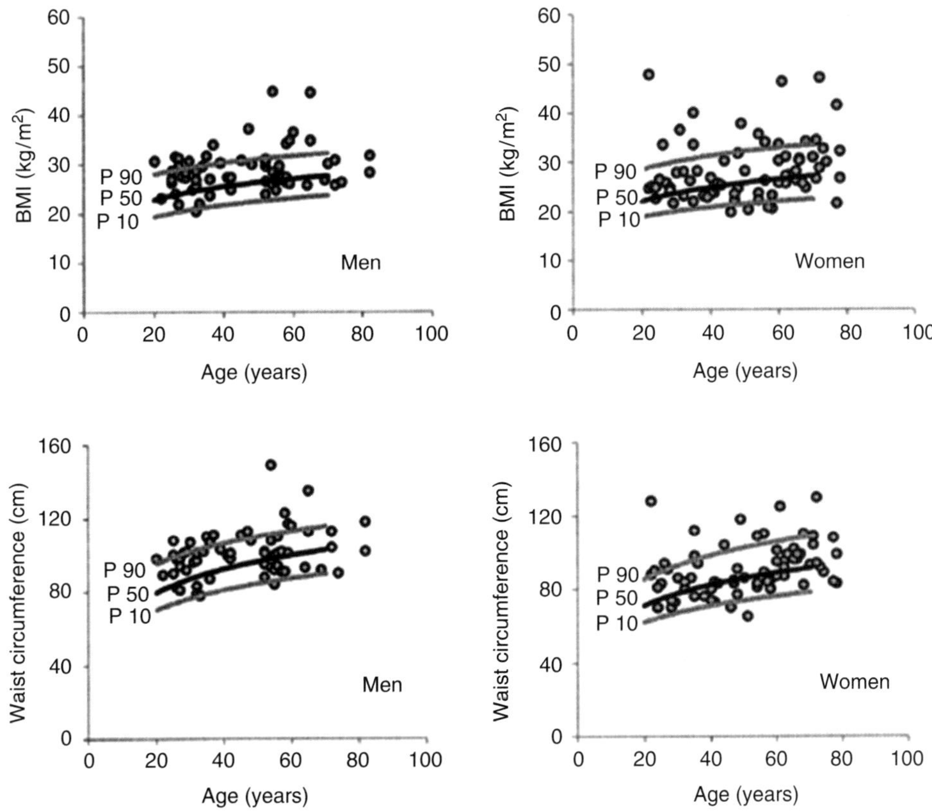

Fig. 2 BMI (*upper figures*) and waist circumference (*lower figures*) of narcoleptic patients (*dots*) plotted in the 10th, 50th, and 90th percentiles of the accompanying normal Dutch population [11]

interventions to prevent long-term complications of excess body fat.

Kotagal and colleagues compared the BMI of narcoleptic children to that of age- and gender-matched controls [12]. They studied the BMI measured when narcolepsy was diagnosed (age range 9–18 years). The median BMI in narcoleptic children was 22.93 compared with a BMI of 20.36 ($p=0.001$) of controls.

All authors (as well as other publications not mentioned here) concluded that increased BMI in narcolepsy might be a consequence of the pathophysiology of the disease itself, or secondary to narcoleptic symptoms.

The Leptin Hypothesis

Leptin is the product of the ob gene, is released by adipocytes, and is a proteohormone with a molecular weight of 16 kDa. It plays a role in the regulation of body fat mass [13]. The leptin-responsive hypothalamic network interacts with the orexinergic system. Leptin-responsive cells are stimulated by orexin and inhibited by leptin [14]. In the periphery, orexin and leptin change inversely in fasting subjects [15]. Schuld et al. compared 15 narcoleptic patients with 15 patients suffering from major depression, and 15 further patients suffering from various neurological disorders [16]. Serum leptin levels were lower in narcoleptic patients, whereas CSF leptin levels were only slightly and nonsignificantly decreased in narcoleptics. Kok et al. studied six male narcoleptic patients and six BMI- and age-matched healthy controls [17]. In this study, narcoleptic patients had an approximate 50% reduction of serum leptin levels compared to controls. Arnulf and her coworkers conducted a very large study, comprising 162 narcoleptic patients, 89 patients with other sleep–wake disorders, and 111 controls [18]. They did not find abnormal serum and CSF leptin levels in narcoleptic patients. Our own study was in agreement with the study of Arnulf: in 42 patients with narcolepsy, when compared to 31 BMI-matched controls, we found no significant differences between peripheral leptin levels [19].

The reason for the discrepancy between the latter data and the studies from Kok and Schuld remains elusive. In particular, the (replication) study of Kok and coworkers was carefully designed and included a measurement of 24-h serum leptin profiles in narcoleptic patients and matched controls. However, given the fact that the power in the two latter studies was higher, it is conceivable that the anomalies of leptin signaling in narcoleptic patients have been overestimated in the two initial reports. After all, there is currently active research going on in this field, and new study designs and the implementation of novel methods might yield new results. In the meantime, we assume that leptin levels may be considered as normal in most narcoleptic patients.

Eating Disorders

Changes in feeding behavior could be an explanation for the observed obesity in narcoleptics. Even small deviations from the normal balance of energy homeostasis could cause significant changes in body weight. In particular, studies into binge eating have shown a positive correlation between feeding behavior and obesity, and thus have attracted the interest of many narcolepsy researchers. Binge eating without subsequent purging leads to weight gain. Although obesity is not a diagnostic criterion for binge eating disorder (BED), the disorder is usually associated with obesity and unstable weight [20, 21]. Also, BED appears to be more prevalent with higher degrees of obesity [22–24]. More than one-third of obese individuals in weight-loss treatment programs report difficulties with binge eating [25]. In contrast, people with bulimia nervosa eat and then purge by vomiting, using laxatives, or other means.

There have been a few reports on symptoms of eating disorders in the past. In 1976, Bell first proposed the hypothesis that narcolepsy was associated with an eating disorder that includes increased intake, snacking, overweight, and postprandial drowsiness [26].

Pollak and Green analyzed eating patterns in six narcoleptic patients and seven controls while they were deprived of time cues [27]. For 18–22 consecutive days, the subjects lived in temporal isolation apartments of an Institute of Chronobiology with no clocks, windows, telephone, radio, etc. During the first 4 (controls) or 6 (narcoleptics) days, the subjects were given a 24-h sleep schedule that resembled the pattern at home, which was recorded for 2 weeks before entering the laboratory. Room lights and meals were scheduled during this period. For the remaining days in the "free-running period," the probands were allowed to retire to bed, arise, and eat whenever they chose. The room lights were lowered whenever a subject retired to bed and were increased again when a subject arose from bed. The probands were allowed to order a meal of their choice at any time. All meals were recorded, including the total caloric content of the meals and the ingested quantity. There were no significant differences in the number, frequency, size, or macronutrient composition of meals consumed by the narcoleptic subjects and controls, but narcoleptics were reported to eat snacks after dinner more often. During the initial scheduled period, the narcoleptic patients ate more often than controls. In the free-running period, the number of meals per biological day remained high in narcoleptic patients, but the meals were spread over a longer period spent out of bed. In summary, the data of Pollak and Green did not provide an evidence for altered eating behavior in narcolepsy.

Lammers collected data on spontaneous food choice by means of a cross-check dietary history of 12 patients and 12 age-, sex-, and social class-matched healthy controls [28]. Spontaneous food intake is considered to be partially governed by cerebral serotonin (5-HT) availability. Both this role of 5-HT in the regulation of food intake, on the one hand, and the fact that 5-HT reuptake inhibitors improve cataplectic symptoms in narcolepsy patients, on the other hand, suggest that 5-HT-mediated "carbohydrate craving" in narcoleptic patients could ameliorate their cataplexy. However, Lammers et al. found that narcoleptic patients consumed fewer kiloJoules of food per day (8,756 ± 2,312) than controls (10,640 ± 3,129) ($p = 0.001$) – mainly because of decreased consumption of carbohydrates.

Kotagal and coworkers found a tendency for increased weight gain in childhood narcolepsy (see above) [12]. The authors observed that 5 of 31 narcoleptic children showed signs of binge eating, whereas none of the 31 control children did. It was hypothesized that increased appetite could result from decreased leptin binding in the hypothalamus as a result of degeneration of the hypocretin neurons.

Bruck compared food intake patterns of 22 narcoleptic patients to 20 age- and socioeconomic status-matched controls over 3 days [29]. Patients suffering from narcolepsy initiated the consumption of sweets and snacking behavior more frequently than controls, and they consumed more total calories or carbohydrates through snacks. Furthermore, participants with narcolepsy consumed more food than controls, with the main difference occurring at mealtime. The authors also found that narcoleptic patients under stimulant medication consumed significantly more lunch carbohydrates and kilojoules than unmedicated subjects. Interestingly, a significant correlation was found in the control group between prelunch hunger/satiety ratings and both lunch carbohydrates and kilojoules consumed, but no such correlation was found in the narcolepsy group.

Chabas and colleagues performed a case–control study with 13 sleepy patients (seven of whom had typical narcolepsy symptoms) and 9 controls [30]. Eating behavior was evaluated using three psychometric tests: the Composite International Diagnostic Interview (CIDI-2), a face-to-face diagnostic test; the Eating Attitude Test (EAT-40), an auto-questionnaire with scores ranging from 0 to 120 that measures the propensity to anorectic behavior; and the Eating Disorder Inventory (EDI-2), an auto-questionnaire containing 11 primary symptom dimensions, including drive for thinness, bulimia, and body dissatisfaction.

No patient met all diagnostic criteria of anorexia or bulimia nervosa, but narcoleptics met more symptoms of these eating disorders than controls. Narcoleptic patients displayed elevated EAT-40 scores. Interestingly, EAT-40 scores were not linked to the age of the patients. Also, narcoleptic patients scored higher for bulimia ($p = 0.14$) and drive for thinness ($p = 0.15$), but not for body dissatisfaction ($p = 0.29$).

The narcoleptic patients of the Chabas study reported that eating interfered with their alertness and that they modified food intake accordingly. This included (1) avoiding food at lunch to be more alert in the afternoon versus consuming high amounts of food just before going to bed to reduce fragmentation of nocturnal sleep, (2) eating light snacks when feeling an upcoming sleep attack, (3) skipping lunch to nap during lunch time, and (4) ending up in irregular, unpredictable eating and sleeping schedules.

We examined 116 narcoleptic patients and 80 controls with the structured interview for anorectic and bulimic eating disorders (SIAB-S) [31]. The SIAB-S questionnaire consists of 81 items and is specifically designed to assess pathological eating patterns and related symptoms such as depressed mood, impaired social integration, and dysfunctional self-images and body perceptions. In contrast to most other eating questionnaires, the SIAB-S allows also for a diagnostic evaluation according to ICD10 and DSM IV, since 20 of the items are formulated specifically to assess diagnostic criteria. In addition, the SIAB allows defining subsyndromal categories such as bulimia nervosa binge (binge eating disorder without more than two eating attacks per week), bulimic syndrome (bulimia nervosa with eating attacks and either recurrent inappropriate compensatory behavior or a frequency of at least twice weekly but without undue influence of self-evaluation), and atypical bulimia (steady consumption of sweet or otherwise high caloric food throughout the day), as well as regular (inappropriate) compensatory behavior (disproportionate and undue effort to counteract food intake either through laxative medication, diuretics, or excessive sport), spitting (chewing and spitting of foodstuff without swallowing), and regurgitation (self-induced vomiting after eating). All symptoms are being asked for the present state (last 3 months) and for the worst condition/maximal symptom expression in the past.

The analysis showed no increased prevalence of bulimia, binge eating, and anorexia nervosa in narcoleptic patients when compared to controls. Similarly, none of the subsyndromal categories was more prevalent in narcoleptics than in controls. No significant differences in the frequency of eating attacks in narcoleptic patients and controls were detected. In line with the findings of Chabas et al., people with (present) hyperphagic traits (any hyperphagic diagnosis) showed a higher BMI (narcoleptics: 30.3±6.3; controls: 27.7±5.8) than persons without such traits (narcoleptics: 28.0±5.4; controls: 25.4±4.7), regardless of the presence or absence of the diagnosis of narcolepsy.

In a case–control study, Droogleever-Fortuyn and coworkers analyzed the eating behavior of 60 narcoleptic patients and 120 age- and sex-matched control subjects [32]. Furthermore, 32 narcoleptic patients were compared with 32 BMI-matched controls. Both studies were conducted using Chaps. 8 and 9 of the Schedules for Clinical Assessment in Neuropsychiatry (SCAN) version 2.1. From the SCAN data, symptoms of eating disorders were assessed on the item level. In addition, it was possible to classify symptoms into the DSM-IV diagnostic criteria for eating disorders such as anorexia nervosa, bulimia nervosa, and eating disorder not otherwise specified (EDNOS). Section 8 of the SCAN assesses physical functions such as changes in BMI over time. Section 9 assesses various symptoms of eating disorders, and the interference with daily activities due to these symptoms. The group found a total of 23% narcoleptics fulfilling the criteria for a clinical eating disorder, against none of the control subjects. Only one narcoleptic patient was classified as having anorexia nervosa. Four patients (7%) fulfilled the criteria of bulimia nervosa, and nine patients (15%) were classified with EDNOS. Within this category, two patients were subthreshold anorexia nervosa cases (having all criteria except amenorrhea), and one patient had subthreshold bulimia nervosa (having all criteria, but binge frequency was less than twice a week). Six patients met the criteria for a BED, five of whom were subthreshold because of missing the DSM-IV criterion of "marked distress regarding binge eating." In the study comparing narcoleptic patients with BMI-matched controls, the results regarding DSM-IV eating disorder diagnoses were essentially the same as in the other part of the study. Six of 32 patients in the second study spontaneously mentioned binging at night. Other patients reported that with increasing sleep pressure over the day, the resistance against binging diminished over the day. Droogleever-Fortuyn et al. concluded that eating disorders might be considered an integral part of the narcolepsy phenotype and proposed that the eating disorder could be a consequence of the hypocretin deficiency. Interestingly, no specific eating disorder was overrepresented in the narcoleptics: bulimia nervosa (4/60), anorexia nervosa (1/60), and EDNOS (9/60) were all present.

Night Eating Syndrome

The night eating syndrome is described as an eating disorder with a circadian delay of eating leading to disrupted sleep, while timing and duration of sleep do not seem to differ from other subjects [33]. In previous studies, a high prevalence of the night eating syndrome has been observed in obese patients [34]. Researchers in both the fields of eating and sleep have speculated that night eating syndrome may be associated with overweight in narcoleptics. For example, in the aforementioned study of Droogleever-Fortuyn et al., six patients spontaneously mentioned binging at night [32].

Therefore, a case-control was conducted [35]. The eating behavior at night was analyzed in 44 narcoleptic patients and 36 age-, BMI-, and sex-matched controls using the night eating questionnaire (NEQ). Participants were diagnosed with night eating syndrome if one of two criteria were met: (1) evening hyperphagia (more than one-third of total calories consumed after the evening meal) or (2) nocturnal awakenings with ingestions of food occurring three or more times per week. The diagnostic criteria of night eating syndrome were fulfilled in four narcoleptics and two controls ($p=0.438$). However, in the analysis of single questions used in the NEQ and several additional questions, narcoleptic patients showed significant differences in their eating habits: narcoleptics tended to wake up more often than controls (not sleeping more than 3 h without waking up: 39% of the narcoleptics and 6% of the controls, $p=0.007$). Therefore, narcoleptic patients ate more often during the night time more than once a month (eating at night: 43% of the narcoleptics and 11% of the controls, $p=0.002$). The frequency of night eating was more than one time per night in 46% of the narcoleptics and only 17% of the controls ($p=0.008$). The analysis of the consumed food at night showed that 48% of narcoleptics ate whole meals in combination with sweet or salty snacks, whereas the food consumed by the controls consisted mostly of snacks. Most narcoleptics (56%) slept between their night meals, whereas controls (0%) did not ($p=0.013$).

We also focused on the sleep-related eating disorder, which is characterized by partial arousals from sleep followed by rapid ingestion of food, commonly with at least partial amnesia for the episode. The diagnostic criteria are recurrent episodes of involuntary eating and drinking during the main sleep period, consumption of peculiar forms or combinations of food or inedible or toxic substances, and insomnia related to sleep disruption from repeated episodes of eating, with a complaint of non-restorative sleep, daytime fatigue or somnolence, sleep-related injuries, dangerous behaviors performed while in pursuit of food or while cooking food, morning anorexia, or adverse health consequences from recurrent binge eating of high caloric food. The disturbance must be independent from other sleep disorders, medical or neurologic disorders, mental disorders, and medication use or substance use disorder (hypoglycemic states, peptic ulcer disease, reflux esophagitis, Kleine–Levin syndrome, Kluver–Bucy syndrome, and nighttime extension of daytime anorexia nervosa, bulimia nervosa, and BED) (Diagnostic and Coding Manual) (Table 2). By applying a questionnaire, we detected six

Table 2 Analysis of eating disorders in the current state and past, Dahmen et al. [31] (with unpublished data)

Eating disorder	Narcoleptics	[Male/female]	Controls	[Male/female]	*p*-Values
Hyperphag					
Any hyperphag diagn. (present)	15 (12.5%)	[6/9]	14 (17.5%)	[0/0]	n.s.
Any hyperphag diagn. (past)	25 (21.6%)	[8/17]	18 (22.5%)	[11/7]	n.s.
Bulimia nervosa (present)	2 (1.7%)	[0/2]	0 (0%)	[0/0]	n.s.
Bulimia nervosa (past)	2 (1.7%)	[0/2]	0 (0%)	[0/0]	n.s.
Binge eating disorder (present)	3 (2.6%)	[1/2]	1 (0%)	[0/1]	n.s.
Binge eating disorder (past)	2 (1.7%)	[0/2]	1 (1.3%)	[0/1]	n.s.
Bulimia nervosa binge (present)	6 (5.2%)	[3/3]	1 (1.3%)	[0/1]	n.s.
Bulimia nervosa binge (past)	6 (5.2%)	[2/4]	5 (6.3%)	[1/4]	n.s.
Bulimic syndrome (present)	10 (8.6%)	[5/5]	11 (13.8%)	[4/7]	n.s.
Bulimic syndrome (past)	19 (16.4%)	[7/12]	15 (18.8%)	[8/7]	n.s.
Atypical bulimic (present)	8 (6.9%)	[3/5]	7 (8.8%)	[5/2]	n.s.
Atypical bulimia (past)	13 (11.2%)	[3/10]	7 (8.8%)	[6/1]	n.s.
Hypophag					
Anorexia nervosa (present)	0 (0%)	[0/0]	0 (0%)	[0/0]	n.s.
Anorexia nervosa (past)	0 (0%)	[0/0]	0 (0%)	[0/0]	n.s.
Anorectic syndrome (present)	0 (0%)	[0/0]	0 (0%)	[0/0]	n.s.
Anorectic syndrome (past)	0 (0%)	[0/0]	0 (0%)	[0/0]	n.s.
Other					
Regular compensatory (present)	9 (7.8%)	[1/8]	2 (2.5%)	[0/2]	n.s.
Regular compensatory (past)	14 (12.1%)	[2/12]	5 (6.3%)	[1/4]	n.s.
Spitting (present)	0 (0%)	[0/0]	0 (0%)	[0/0]	n.s.
Spitting (past)	0 (0%)	[0/0]	0 (0%)	[0/0]	n.s.
Regurgitation (present)	0 (0%)	[0/0]	0 (0%)	[0/0]	n.s.
Regurgitation (past)	2 (1.7%)	[0/2]	0 (0%)	[0/0]	n.s.

narcoleptics (14%) and no controls (0%) who could not or only partially remember their nightly eating ($p=0.021$). These patients realized their nightly eating often by seeing the kitchen changed or untidy in the morning, or by recognizing missing food or by finding food in their bed.

Energy Expenditure

Among other hypotheses, lower resting energy expenditure has been suspected to be the cause of the narcolepsy-associated obesity. In three studies, indirect calorimetry was used to measure the energy expenditure (EE) and to calculate the resting (basal) metabolic rate (BMR) (Table 3).

Chabas et al. published an evaluation of energy balance of seven typical narcoleptic patients and nine controls [30]. Narcoleptic patients tended to have lower rest EE than controls ($n=13$, $p=0.07$). Thus, it was concluded that narcolepsy-associated obesity could be the result of the difference in EE. However, cases and controls were not BMI matched and the results were only of weak significance.

In an attempt to replicate these findings, others determined the BMR of 15 male narcoleptic patients and 15 age- and BMI-matched controls using the Weir formula (Weir, 1949), and they found no difference ($p=0.77$) [36, 37]. To clarify this controversy, we determined EE with the same method that Chabas had utilized and calculated the BMR using the formula (Weir) employed in the Fronczek Study in 13 narcoleptics and 30 healthy BMI- and age-matched controls [38]. In our analysis, the BMR (EE) of nonobese narcoleptics was lower than that of BMI-matched controls. Only nonobese narcoleptics showed a BMR difference, which may indicate that narcolepsy induces a

Table 3 Results of BMR and EE measurements – a synopsis of the results of different studies [30, 36, 38]

Dahmen et al. (Deltatrac)	Patients	Controls	*p*-Values
N (male/female)	13 (2/11)	30 (8/22)	
Age (years)	36.54 ± 13.5	36.37 ± 14.4	0.971
BMI (kg/m²)	27.6 ± 5.8	27.5 ± 5.9	0.953
BMR (kcal/24 h)	1,573.5 ± 337.3	1,689.2 ± 318.4	0.289
BMR (kcal//kg/24 h)	20.2 ± 2.6	21.4 ± 3.3	0.242
EE (kcal/24 h)	1,438.31 ± 295.6	1,572.50 ± 302.7	0.186
EE (kcal/kg/day)	18.38 ± 1.6	19.8 ± 2.8	0.090
EE (kcal/kg/24 h) median (quartiles)	17.9 (17.2–19.6)	19.8 (17.6–21.4)	0.228
BMI < 30			
Age (years)	34.3 ± 10.8	30.2 ± 6.6	0.094
BMI	23.5 ± 2.0	23.8 ± 2.7	0.917
BMR (kcal/24 h)	1,409.5 ± 260.8	1,619.2 ± 308.3	0.139
BMR (kcal/kg/24 h)	20.8 ± 3.0	23.6 ± 2.4	0.022
EE (kcal/24)	1,282.5 ± 216.5	1,488.4 ± 298.5	0.122
EE (kcal/kg/24 h)	18.9 ± 1.8	21.7 ± 2.1	0.007
Chabas et al. (Deltatrac)	Patients	Controls	*p*-Values
N (male/female)	7 (2/5)	9 (3/6)	
Age (years)	22 (20–33)	29 (25–35)	Not matched
BMI (kg/m²)	28.6 (21.9–29.1)	22.9 (20.5–23.6)	Not matched
EE (kcal/kg/24 h)	21.3 (19.7–23.1)	23.6 (20.9–25.5)	0.07
Fronczek et al. (Oxycon B)	Patients	Controls	*p*-Values
N (male/female)	15 (15/0)	15 (15/0)	
Age (years)	29.2 ± 4.1	32.6 ± 16.2	0.55
BMI (kg/m²)	24.9 ± 2.6	26.2 ± 2.1	0.31
BMR (kcal/24 h)	1,767.1 ± 226.5	1,766.5 ± 226.5	0.99
BMR (kcal/kg/24 h)	19.9 ± 2.0	20.1 ± 2.2	0.77

The name of the used calorimeter is given in brackets

change in the individual BMI set point. Once the new (higher) set point is reached, no more BMR abnormalities can be detected. Given the small sample sizes of all current studies, an independent larger replication study would certainly be more than welcome. However, our findings fit well with recent pathophysiological models on narcolepsy. The leading hypothesis is that narcolepsy is an autoimmune disease that leads to a specific degeneration of hypothalamic hypocretin-containing cells. It is well known that hypothalamic damage of other origin – in particular by tumors – may lead to hypothalamic obesity (HO). Recently, it has been shown that energy expenditure rather than energy intake plays an important role in HO-associated weight gain. In addition, orexin-deficient narcoleptic mice eat less than wild-type littermates, whereas the situation is less clear in men [39].

Actigraphic Studies in Narcoleptic Patients

Only few studies have addressed the use of activity monitors in narcolepsy [40, 41]. The study of Durrer did not examine controls, but was intended to demonstrate the feasibility of the measurements. Bruck et al. reported that daytime waking in narcoleptics was characterized by significantly less active wakefulness than that in controls.

In 1995, Middelkoop and colleagues compared 14 unmedicated narcoleptic patients (8 women and 6 men; mean age 43 years) with a matched group of 14 healthy subjects [42]. Subjective nocturnal sleep was assessed by daily sleep logs in which the subjects indicated the time they went to bed at night, the "lights out" time, self-estimated sleep latency, the number of awakenings after sleep onset, time of definitive awakening, and rise time. Each 24-h period was divided into a diurnal and a nocturnal period. On the basis of the sleep log data, the nocturnal periods were defined as the periods from lights out until getting up in the morning, and the diurnal periods were the periods from getting up in the morning until lights out. For the diurnal period, the subjects had to record the number and times of naps and activity monitor off times. Parallel to the assessment of sleep by sleep logs, motor activity was continuously recorded by a solid state activity monitor worn on the wrist of the nondominant arm. The mean frequency of daytime napping and the mean number of nighttime awakenings were significantly higher in the narcoleptics. All narcoleptic patients reported at least one (range: 1–6) daily nap during the recording period. The amount and distribution of time in bed across the 24-h period differed significantly between narcoleptics and controls, with significantly higher mean time-in-bed values for the narcoleptics during the diurnal period ($p=0.001$). The analysis of the activity revealed a significantly smaller amplitude of the circadian activity and immobility rhythm in narcoleptics than in control subjects ($p=0.003$). Because comparisons were made only for the diurnal period or the nocturnal period, from the perspective of energy expenditure, it is unclear whether the higher activity of narcoleptics during the night compensates for the lower activity during the day time.

Mayer and colleagues compared core body temperature and motor activity recordings of 15 unmedicated narcoleptic patients with and without a nocturnal sleep-onset REM period (SOREMP) with those of 16 unmedicated, age- and sex-matched control subjects [43]. Although mean nocturnal motor activity did not differ between narcoleptic patients and controls, the analysis of wake stages showed significantly longer stages of quiet wakefulness in narcoleptic patients with SOREMPs than in controls. Compared with that of controls, temperature of both narcoleptic groups showed less rise of temperature curve in the morning, a dampening of temperature amplitude, phase advance of acrophase, and advance of temperature minimum after sleep onset. Maximal temperature decline occurred earlier in patients with SOREMPs during naps and sleep than in the other groups. Parallels between temperature and motor activity could be confirmed with controls and but not in narcoleptic patients.

Bruck et al. evaluated wrist actigraphy as an instrument to assess the daytime effects of stimulant medication in the treatment of narcolepsy [41]. They reported that narcoleptic patients differed significantly from matched control subjects in the frequency of naps, independent from medication (2.43 naps when medicated, and 3.94 naps when unmedicated).

Discussion

In this chapter, we discussed some of the current hypotheses on narcolepsy-related obesity.

The leptin hypothesis has been weakened by recent findings of normal leptin levels in large samples of narcoleptic patients. However, future investigations in further hormones and of the interplay of different hormone systems are likely to yield new insights.

The eating studies have not shown higher prevalences of specific eating disorders in narcolepsy. Some anomalies of eating behavior such as night snacking, however, appear to be frequent in narcoleptics. Currently, there are no convincing studies that address the question of the amount of food intake or exact food composition. This is surprising because adequate food diaries have been developed.

Three pilot studies have measured BMRs in narcoleptics and controls. The studies point to the intriguing possibility that narcolepsy is associated with a lowered BMR, or a change in the individual BMI set point.

The actigraphy studies so far have mainly proven the feasibility of the approach. Given the fact that actigraphical devices have become both cheaper and easier to handle, it might well be worthwhile to go on with similar studies.

References

1. Sakurai T. Roles of orexins in regulation of feeding and wakefulness. Neuroreport. 2002;13(8):987–95.
2. Hagan JJ, Leslie RA, Patel S, Evans ML, Wattam TA, Holmes S, et al. Orexin A activates locus coeruleus cell firing and increases arousal in the rat. Proc Natl Acad Sci U S A. 1999;96(19):10911–6.
3. Dube MG, Kalra SP, Kalra PS. Food intake elicited by central administration of orexins/hypocretins: identification of hypothalamic sites of action. Brain Res. 1999;842(2):473–7.
4. Lin L, Faraco J, Li R, Kadotani H, Rogers W, Lin X, et al. The sleep disorder canine narcolepsy is caused by a mutation in the hypocretin (orexin) receptor 2 gene. Cell. 1999;98:365–76.
5. Nishino S, Ripley B, Overeem S, Lammers GJ, Mignot E. Hypocretin (orexin) deficiency in human narcolepsy. Lancet. 2000;355:39–40.
6. Nishino S, Ripley B, Overeem S, Nevsimalova S, Lammers GJ, Vankova J, et al. Low cerebrospinal fluid hypocretin (orexin) and altered energy homeostasis in human narcolepsy. Ann Neurol. 2001; 50:381–8.
7. Mullett MA, Billington CJ, Levine AS, Kotz CM. Hypocretin I in the lateral hypothalamus activates key feeding-regulatory brain sites. Neuroreport. 2000;11(1): 103–8.
8. Siegel JM. Narcolepsy: a key role for hypocretins (orexins). Cell. 1999;98(4):409–12.
9. Schuld A, Hebebrand J, Geller F, Pollmächer T. Increased body-mass index in patients with narcolepsy. Lancet. 2000;355(9211):1274–5.
10. Dahmen N, Bierbrauer J, Kasten M. Increased prevalence of obesity in narcoleptic patients and relatives. Eur Arch Psychiatry Clin Neurosci. 2001;251: 85–9.
11. Kok SW, Overeem S, Visscher TL, Lammers GJ, Seidell JC, Pijl H, et al. Hypocretin deficiency in narcoleptic humans is associated with abdominal obesity. Obes Res. 2003;11(9):1147–54.
12. Kotagal S, Krahn LE, Slocumb N. A putative link between childhood narcolepsy and obesity. Sleep Med. 2004;5(2):147–50.
13. Zhang Y, Proenca R, Maffei M, Barone M, Leopold L, Friedman JM. Positional cloning of the mouse obese gene and its human momologue. Nature. 1994;372: 425–32.
14. Rauch M, Riediger T, Schmid HA, Simon E. Orexin A activates leptin-responsive neurons in the arcuate nucleus. Pflügers Arch Eur J Physiol. 2000;440:699–703.
15. Komaki G, Matsumoto Y, Nishikata H, Kawai K, Nozaki T, Takii M, et al. Orexin-A and leptin change inversely in fasting non-obese subjects. Eur J Endocrinol. 2001;144:645–51.
16. Schuld A, Blum WF, Uhr M, Haack M, Kraus T, Holsboer F, et al. Reduced leptin levels in human narcolepsy. Neuroendocrinology. 2000;72:195–8.
17. Kok SW, Meinders AE, Overeem S, Lammers GJ, Roelfsema F, Frolich M, et al. Reduction of plasma leptin levels and loss of its circadian rhythmicity in hypocretin (orexin)-deficient narcoleptic humans. J Clin Endocrinol Metab. 2002;87:805–9.
18. Arnulf I, Lin L, Zhang J, Russell IJ, Ripley B, Einen M, et al. CSF versus serum leptin in narcolepsy: is there an effect of hypocretin deficiency? Sleep. 2006;29(8):1017–24.
19. Dahmen N, Engel A, Helfrich J, Manderscheid N, Löbig M, Forst T, et al. Peripheral leptin levels in narcoleptic patients. Diabetes Technol Ther. 2007;9(4):348–53.

20. Fairburn CG, Doll HA, Welch SL, Hay PJ, Davies BA, O'Connor ME. Risk factors for binge eating disorder. Arch Gen Psychiat. 1998;55:425–32.
21. Spitzer RL, Yanovski SZ, Wadden T, Wing R. Binge eating disorder: its further validation in a multisite study. Int J Eating Disord. 1993;13:137–53.
22. Telch CF, Agras WS, Rossiter EM. Binge eating increases with increasing adiposity. Int J Eating Disord. 1988;7:115–9.
23. Hay P. The epidemiology of eating disorder behaviors: an Australian community-based survey. Int J Eating Disord. 1998;23:371–82.
24. Hay P, Fairburn C. The validity of the DSM-IV scheme for classifying bulimic eating disorders. Int J Eating Disord. 1998;23:7–15.
25. Yanovski SZ. Binge eating in obese persons. In: Fairburn CG, Brownell KD, editors. Eating disorders and obesity. 2nd ed. New York: Guilford Press; 2002. p. 403–7.
26. Bell IR. Diet histories in narcolepsy. In: Guilleminault C, Dement WC, Passouant P, editors. Narcolepsy. New York: Spectrum publications, Inc.; 1976. p. 221–7.
27. Pollak CP, Green J. Eating and its relationships with subjective alertness and sleep in narcoleptic subjects living without temporal cues. Sleep. 1990;13(6):467–78.
28. Lammers GJ, Pijl H, Iestra J, Langius JA, Buunk G, Meinders AE. Spontaneous food choice in narcolepsy. Sleep. 1996;19:75–6.
29. Bruck D. Food consumption patterns in narcolepsy. Sleep. 2003;26(Suppl):A272–3.
30. Chabas D, Foulon C, Gonzalez J, Nasr M, Lyon-Caen O, Willer JC, et al. Eating disorder and metabolism in narcoleptic patients. Sleep. 2007;30(10):1267–73.
31. Dahmen N, Becht J, Engel A, Thommes M, Tonn P. Prevalence of eating disorders and eating attacks in narcolepsy. Neuropsychiatr Dis Treat. 2008;4(1):257–61.
32. Droogleever-Fortuyn HA, Swinkels S, Buitelaar J, Renier WO, Furer JW, Rijnders CA, et al. High prevalence of eating disorders in narcolepsy with cataplexy: a case-control study. Sleep. 2008;31(3):335–41.
33. Stunkard AJ, Allison KC, O'Reardon JP. The night eating syndrome: a progress report. Appetite. 2005;45(2):182–6.
34. Rand CS, Macgregor AM, Stunkard AJ. The night eating syndrome in the general population and among postoperative obesity surgery patients. Int J Eat Disord. 1997;22(1):65–9.
35. Allison KC, Lundgren JD, O'Reardon JP, Martino NS, Sarwer DB, Wadden TA, et al. The Night Eating Questionnaire (NEQ): psychometric properties of a measure of severity of the Night Eating Syndrome. Eat Behav. 2008;9(1):62–72.
36. Fronczek R, Overeem S, Reijntjes R, Lammers GJ, van Dijk JG, Pijl H. Increased heart rate variability but normal resting metabolic rate in hypocretin/orexin-deficient human narcolepsy. J Clin Sleep Med. 2008;4:248–54.
37. Weir JB. New methods for calculating metabolic rate with special reference to protein metabolism. 1949. Nutrition. 1990;6:213–21.
38. Dahmen N, Tonn P, Messroghli L, Ghezel-Ahmadi D, Engel A. Basal metabolic rate in narcoleptic patients. Sleep. 2009;32(7):962–4.
39. Hara J, Yanagisawa M, Sakurai T. Difference in obesity phenotype between orexin-knockout mice and orexin neuron-deficient mice with same genetic background and environmental conditions. Neurosci Lett. 2005;380(3):239–42.
40. Bruck D, Kennedy GA, Cooper A, Apel S. Diurnal actigraphy and stimulant efficacy in narcolepsy. Hum Psychopharmacol. 2005;20(2):105–13.
41. Durrer M, Hess K, Dürsteler M. Narkolepsie und Aktivitätsmonitor. Schweizer Archiv für Neurologie und Psychiatrie. 1991;142:313–8.
42. Middelkoop HA, Lammers GJ, Van Hilten BJ, Ruwhof C, Pijl H, Kamphuisen HA. Circadian distribution of motor activity and immobility in narcolepsy: assessment with continuous motor activity monitoring. Psychophysiology. 1995;32(3):286–91.
43. Mayer G, Hellmann F, Leonhard E, Meier-Ewert K. Circadian temperature and activity rhythms in unmedicated narcoleptic patients. Pharmacol Biochem Behav. 1997;58(2):395–402.

Part V

Reward, Addiction, Emotions and the Hypocretin System

Effects of Orexin/Hypocretin on Ventral Tegmental Area Dopamine Neurons: An Emerging Role in Addiction

Stephanie L. Borgland

Keywords
Addiction • Reward • Ventral tegmental area • Orexin • Hypocretin • Motivation

Introduction

Addiction poses a significant threat to the health, social, and economic fabric of families, communities, and nations. The extent of worldwide psychoactive drug use is estimated at 2 billion alcohol users, 1.3 billion smokers, and 185 million drug users (UNDCP Statistics, 2002). Currently, there is no effective treatment for craving related to substance abuse. Several medications used clinically to counteract the effects of craving and relapse targeting different receptors have so far shown little efficacy. For example, drug-free retention rates with the use of opioid receptor antagonist, naltrexone, is often less that 20% in heroin addicts [1] and 30–40% in alcoholics [2]. Tricyclic antidepressants and serotonin reuptake inhibitors have been widely used in the treatment of cocaine addicts, but have not been efficacious [1]. Finally, trials using the $GABA_B$ receptor agonist, baclofen, have shown reductions in cocaine craving, but long-term outcome studies looking at craving and continued cocaine use are needed [3]. Taken together, novel targets are needed for the development of new medications for chronically relapsing forms of addiction. Interestingly, patients with narcolepsy are often treated with highly addictive amphetamine-like drugs such as methylphenidate, amphetamine, and γ-hydroxybutyrate [4], but they rarely become addicted to these drugs [5, 6]. Because narcolepsy results from a deficient orexin/hypocretin (ox/hcrt) system, the absence of addiction in narcoleptic patients treated with psychostimulants suggests the possibility of this system's involvement in the reinforcing aspect of addictive drugs. This chapter will discuss evidence for the interaction of these peptides in the neural circuits mediating drug-seeking behaviors.

Mesolimbic Dopamine Circuits Underlying Drug-Seeking Behaviors

Mesolimbic dopamine neurons of the ventral tegmental area (VTA) are critical mediators of drug-seeking behaviors. The VTA neurons and their widespread projections to the mesocortical limbic system are involved in both short- and long-term changes produced by most drugs of abuse (Fig. 1)

S.L. Borgland (✉)
Department of Anesthesiology, Pharmacology and Therapeutics, The University of British Columbia, 212-2176 Health Sciences Mall, Vancouver, BC, Canada V6T 1Z3
e-mail: Borgland@interchange.ubc.ca

C.R. Baumann et al. (eds.), *Narcolepsy: Pathophysiology, Diagnosis, and Treatment*,
DOI 10.1007/978-1-4419-8390-9_22, © Springer Science+Business Media, LLC 2011

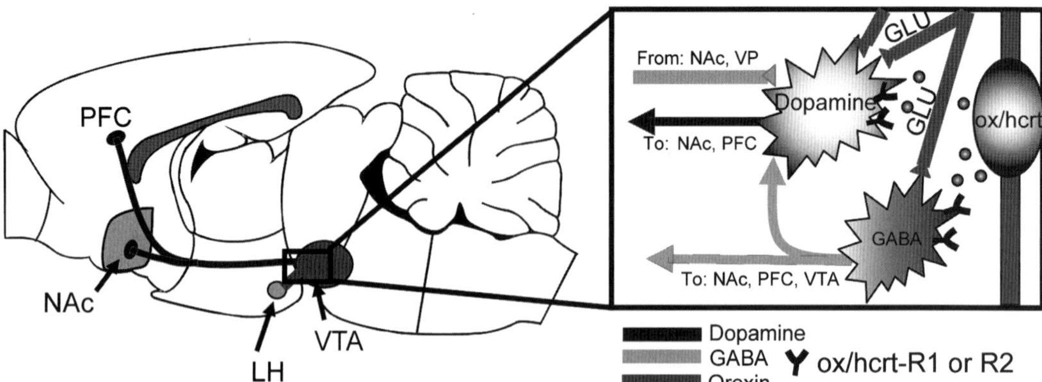

Fig. 1 Simplified schematic of the neural circuitry underlying drug-seeking behaviours in the rat brain. Dopamine neurons of the central tegmental area (VTA) project to the nucleus accumbens (NAc) and prefrontal cortex (PFC) comprising the mesocorticolimbic circuit. Dopamine neurons of the VTA receive glutamatergic input from the PFC, lateral hypothalamus (LH), lateral dorsal tegmentum and, bed nucleus stria terminalis. Dopamine neurons receive GABA input from local projections and that from the NAc and ventral pallidum (VP). VTA dopaminergic and GABAergic neurons also receive input from ox/hcrt-containing neurons of the LH. Glutamatergic synapses excite postsynaptic neurons and GABAergic synapses inhibit postsynaptic neurons. Dopamine release exerts more complex modulatory effects. The release of dopamine from VTA neurons increases in response to administration of all drugs of abuse

[7–9]. Increases in dopamine levels in the nucleus accumbens (NAc) have been observed after acute exposure to virtually every drug of abuse [7]. Furthermore, while acute withdrawal from chronic cocaine, morphine, and ethanol is associated with low dopamine levels in the NAc [10], an increase in dopaminergic neurotransmission has been observed after chronic cocaine exposures that induce behavioral sensitization [11, 12], a phenomenon associated with the progressive increase in behavioral responses to drugs elicited by repeated exposure to a drug of abuse observed in animals [13] and humans [14]. In general, drug-seeking behavior, relapse to compulsive drug use, and behavioral sensitization all produce transient or long-lasting modifications of dopamine release in the output regions [7–9].

Although there is an ongoing debate about the precise causal contribution made by the mesolimbic dopamine system to reward, there seems to be consensus that dopamine release in its target regions increases the probability for the animal to select actions that would enable the animal to obtain a reward [15, 16]. Because dopamine modulates synaptic plasticity in target neurons [9] and adjusts synaptic efficacy in appropriate neuronal circuits in learning networks, it may be responsible for "stamping in" and associatively reinforcing new links between the stimulus and response [17]. Consistent with this idea, dopamine levels in the NAc, a target region of the VTA, increase when trained animals respond to learned cues for natural [18] or drug rewards [19].

Burst Firing of VTA DA Neuron Encodes Motivationally Relevant Information

In vivo, VTA neurons exhibit two main patterns of firing activity: single spike firing, often irregular, and burst firing [20–23]. Several studies in rodents and primates have suggested that burst firing of VTA dopamine neurons encodes the occurrence of salient stimuli with a positive valence [24]. Furthermore, when spikes of dopamine neurons are clustered into bursts, the increase in extracellular dopamine in the projection areas is much larger than that observed for regularly spaced trains of action potential at the same average frequency due to nonlinear summation. It has been proposed that this is due to the fact that release outpaces dopamine uptake [25]. Thus, highly processed information is being transmitted to the dopamine neurons, and bursts of activity affect the forebrain to facilitate the expression of behaviors related to motivation.

Activity at excitatory synapses appears to be required for the transition of spike firing to burst firing. For example, intra-VTA microinjection of glutamate or stimulation of glutamatergic PFC afferents changed firing properties of in vivo-recorded VTA dopamine neurons from spike firing to bursts [26, 27]. Furthermore, studies have shown that burst firing patterns similar to that observed in vivo can be reproduced in vitro by bath application of NMDA, an agonist for *N*-methyl-d-aspartate receptors (NMDARs) [28, 29]. Therefore, an enhancement of excitatory synaptic transmission in the VTA likely promotes burst firing and consequently dopamine release.

Excitatory Synaptic Transmission on VTA DAergic Neurons

Enhancement of synaptic transmission in the VTA may not only play a central role in modulating the activity of dopamine neurons, but can also lead to the development and expression of drug-dependent behaviors [9]. When glutamate is released from presynaptic terminals in the VTA, it activates both ionotropic [α-amino-3-hydroxy-5-methyl-4-isoxazolepropionic acid receptors (AMPARs), kainate receptors, and NMDARs] and metabotropic glutamate receptors (mGluRs). Excitatory synapses onto VTA dopamine neurons can undergo long-term potentiation (LTP) and long-term depression (LTD), whereby an alteration in the postsynaptic response to glutamate can be reflected as the change in number or function of synaptic AMPARs. Unlike classical synaptic events that last from a few milliseconds to a few hundred milliseconds at most, LTP and LTD of AMPARs can last for days and may represent a key substrate of normal learning and memory processes [30]. Because of the general temporal properties of LTP and LTD, these processes may be contenders for mediating rapid onset and persistent neural adaptations in the reward circuitry.

LTP in dopamine neurons requires activation of NMDARs for its expression [31, 32]. In contrast, low frequency stimulation leads to an NMDAR-independent LTD through voltage-gated calcium channels [33] and activation of somatodendritic dopamine D2 receptors present on VTA dopamine neurons [34]. Furthermore, LTD in the VTA can be induced by activation of mGluRs and leads to a switch in AMPAR subunit composition [35] (Fig. 2).

Strength of excitatory synaptic input onto VTA dopamine neurons, as well as their activity and output, suggests that each contributes to reward-related behavior as LTP of VTA neurons appears to facilitate the transformation of neutral environmental stimuli to salient reward predictive cues [36]. The most compelling evidence about a role for these forms of plasticity in underlying drug-dependent behaviors stems from a study showing that in vivo exposure to cocaine promotes LTP-like potentiation of AMPAR-mediated synaptic responses in VTA dopaminergic cells [37]. Thus, the consequence of even a single exposure to an addictive drug has a long-lasting increase in the efficacy of glutamatergic synaptic transmission in VTA dopaminergic cells via potentiation of the number or function of AMPARs [37]. The effect was observed for at least five, but less than 10 days following a single exposure and, as with behavioral sensitization, the cocaine-induced potentiation of AMPAR-mediated synaptic transmission can be blocked by NMDAR antagonists. Although the number of experimenter-delivered cocaine injections does not alter the duration or magnitude of plasticity, after a single exposure there is a significant correlation between the augmentation of excitatory synaptic strength in the VTA and drug-induced locomotor activity, which is lost after repeated cocaine injections [38]. Drug-induced plasticity in the VTA is not unique to cocaine, as morphine, ethanol, and stress also produce a similar strengthening of excitatory synapses in the VTA [39]. These studies have used noncontingent experimenter-administered cocaine to assess drug effects on synaptic transmission of VTA neurons. However, because drug taking is a learned process and its actions can be reinforced in a context-dependent manner, it is important to

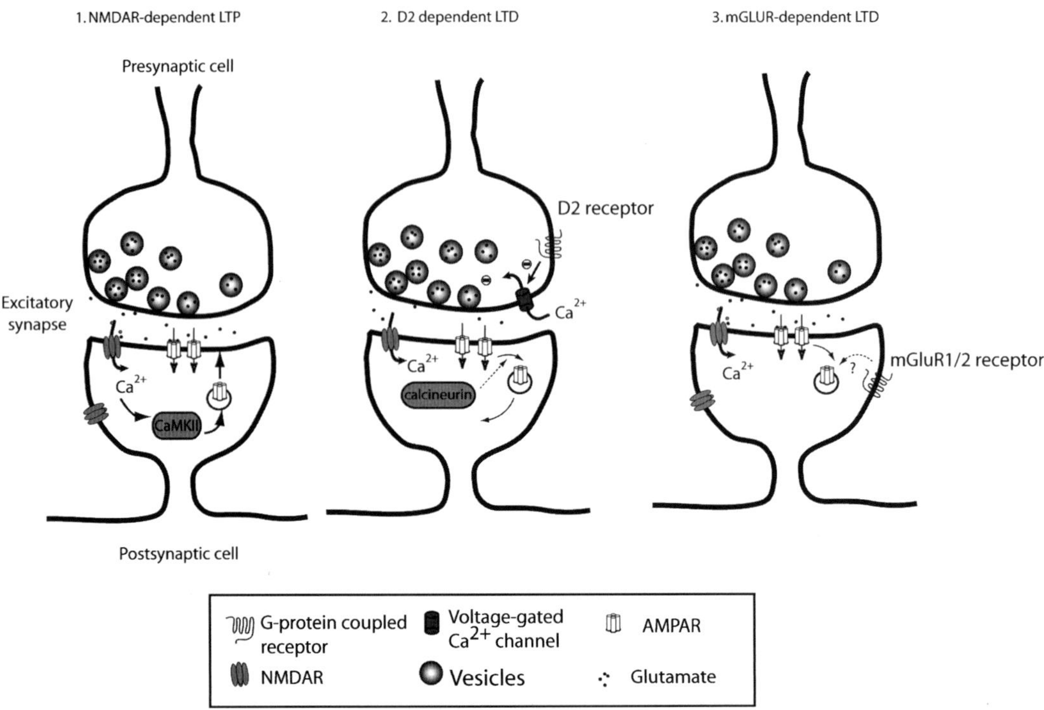

Fig. 2 Synaptic plasticity of VTA dopamine neurons. (1) NMDAR-dependent LTP is dependent on postsynaptic NMDAR activation and calcium/calmodulin-dependent protein kinase II (CaMKII) for its initiation. The voltage-dependent relief of the magnesium block of the NMDAR channel allows the synapse to detect coincident presynaptic release of glutamate and postsynaptic depolarization. AMPAR insertion into the postsynaptic membrane is a major mechanism underlying LTP expression. (2) NMDAR-dependent LTD is triggered by Ca2+ entry through postsynaptic NMDAR channels, leading to increases in the activity of the protein phosphatases calcineurin and protein phosphatase 1. The primary expression mechanism involves internalization of postsynaptic AMPARs and a downregulation of NMDAR5 by an unknown mechanism. In VTA DA neurons, activation of presynaptic D2 receptors inhibits Ca2+ entry through voltage-dependent calcium channels and reduces glutamate release. (3) Metabotropic glutamate receptor (mGluR)-dependent LTD occurs in VTA neurons. Activation of postsynaptic mGluRl/5 triggers the internalization of postsynaptic AMPARs, a switch in subunit composition of AMPARs to low-conducting, GluR2-containing AMPARs

understand whether voluntary drug intake alters synaptic efficacy in VTA neurons in a similar manner as passive drug exposure. In rats self-administering cocaine, AMPAR-mediated synaptic transmission was potentiated in dopamine neurons for up to 3 months after the last cocaine experience [40]. In contrast, rats passively receiving cocaine did not exhibit this persistent plasticity observed after 3 months. These results show that although the performance of an operant task to obtain cocaine produced a similar short-term potentiation of excitatory transmission, only voluntary responding for cocaine induced a persistent synaptic enhancement in the VTA. Self-administration of food also produced an increase in AMPAR-mediated synaptic transmission; however, this plasticity was only transiently expressed [36, 40]. Thus, these studies support the hypothesis that NMDARs and AMPARs play a major role in the development and expression of drug-dependent behaviors by virtually all major drugs of abuse. The abnormal learning process associated with drug addiction may be and sustained by activation of NMDARs and AMPARs.

Effects of Orexin/Hypocretin on the VTA

While it is clear that drugs of abuse can modulate synaptic transmission in VTA dopamine neurons, it is less known how endogenous substances can alter the synaptic efficacy of these neurons. Orexin/hypocretin (ox/hcrt) neuropeptides are ideal candidates for altering dopaminergic activity as ox/hcrt-containing neurons make close appositions with VTA dopamine neurons [41] and these neurons express both orexin/hypocretin receptor-1 (ox/hcrt-R1) and orexin/hypocretin receptor-2 (ox/hcrt-R2) [42, 43]. Furthermore, in addition to these peptides being implicated in the maintenance of arousal states and narcolepsy with cataplexy [44, 45], the ox/hcrt system plays a role in the regulation of reinforcement processes [46–51].

Although orexin-A/hypocretin-1 (oxA/hcrt-1)- and orexin-B/hypocretin-2 (oxB/hcrt-2)-synthesizing neurons represent a relatively small population of all lateral hypothalamic neurons (approximately 6,700 in rat; 52), afferents project widely through the brain including a significant projection to the VTA [52, 53]. The VTA receives input from these ox/hcrt-containing neurons composed of both synaptic terminals and mostly en passant fibers [54]. The relatively high presence of axonally located ox/hcrt-containing dense core vesicles suggests that oxA/hcrt-1 and oxB/hcrt-2 are likely primarily released into the VTA via extrasynaptic mechanisms [54].

The actions of oxA/hcrt-1 and oxB/hcrt-2 are mediated by two G-protein-coupled receptors: ox/hcrt-R1 and ox/hcrt-R2. Ox/hcrt-1 shows a tenfold selectivity for oxA/hcrt-R1, while oxA/hcrt-2 shows an equal affinity for both receptors [53]. Ox/hcrt-R1 is coupled exclusively to Gq-type G-proteins, whereas ox/hcrt-R2 is coupled to both Gi/o and Gq proteins [55]. Both receptors are highly expressed on both dopaminergic and GABAergic neurons of VTA [42, 43].

Ox/hcrt peptides have functional actions in the VTA. OxA/hcrt-1 has been demonstrated to increase the firing rate and, in some cases, cause burst firing of VTA dopamine neurons in rat brain slices [43, 56]. OxA/hcrt-1 has also been demonstrated to increase the firing rate of GABAergic neurons of the VTA [43]. Intra-VTA infusions of oxA/hcrt-1 in vivo have been found to increase extracellular dopamine levels in the prefrontal cortex (PFC) and in the shell region of the NAc, but not in the NAc core region [42, 57, 58]. The selective circuitry underlying oxA/hcrt-1-mediated increases in dopamine may be due to the effects of oxA/hcrt-1 on a subpopulation of VTA dopamine neurons. Accordingly, oxA/hcrt-1 significantly increased Fos immunoreactivity preferentially in small- to medium-sized dopamine neurons in the caudomedial VTA that project primarily to the PFC and NAc shell region [58].

Orexin/Hypocretin Neuropeptides Modulate Synaptic Transmission of VTA Neurons

OxA/hcrt-1-mediated increases in dopamine may be a result of its actions at glutamatergic synapses in the VTA. Brief exposure of either oxA or hcrt-1 to dopamine neurons in brain slices rapidly and transiently enhanced NMDAR-mediated excitatory postsynaptic currents (EPSCs) by promoting the insertion of NMDARs to the synapse (Fig. 3a). This effect was mediated by ox/hcrt-R1 activation of phospholipase C (PLC) and protein kinase C (PKC) [48]. Although oxA/hcrt-1 does not have acute effects on evoked AMPAR EPSCs, an increase in AMPAR miniature EPSC (mEPSC) amplitude and the AMPAR/NMDAR ratio was observed 3–4 h later, suggesting a delayed increase in the number or function of postsynaptic AMPARs. This delayed potentiation of AMPAR-mediated synaptic transmission was dependent on prior activation of NMDARs because oxA/hcrt-1-dependent potentiation of AMPAR mEPSCs was inhibited with APV, an NMDAR antagonist. These results suggest that oxA/hcrt-1 promotes an LTP-like induction at VTA dopaminergic synapses within a few hours.

OxB/hcrt-2 can also increase synaptic efficacy in VTA dopaminergic neurons. In a time course similar to that of oxA/hcrt-1, oxB/hcrt-2 can potentiate NMDAR-mediated currents via activation of ox/hcrt-R2 and PKC. However, when both peptides are applied together, these effects are additive, suggesting that oxB/hcrt-2 and oxA/hcrt-1 may mediate their effects through separate mechanisms [59]. Indeed, oxB/hcrt-2 increased AMPAR mEPSC frequency and caused a paired-pulse depression, suggestive

a Effects of oxA/hcrt-1 on naive or food self-administering rats.

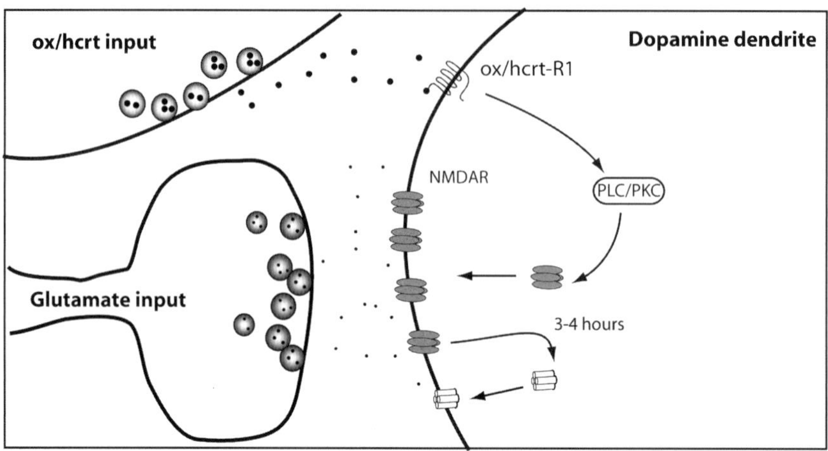

b Effects of oxA/hcrt-1 on cocaine or high fat chocolate food self-administering rats.

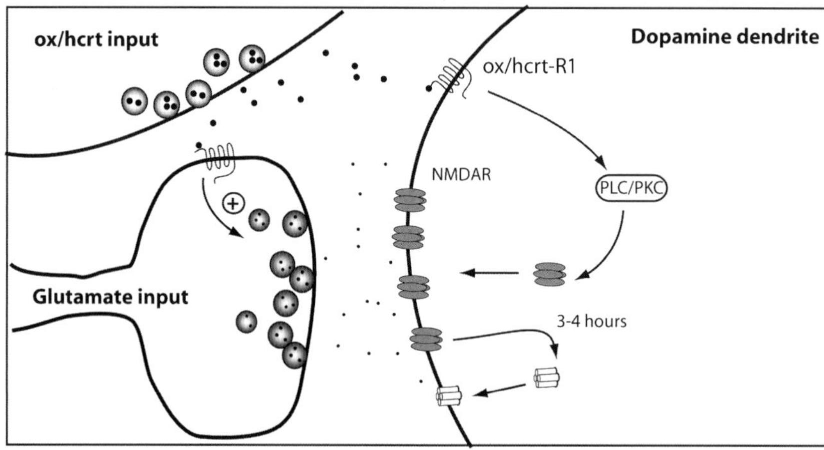

Fig. 3 OxA/Hcrt-1 modulates synaptic transmission in VTA dopamine neurons. (**a**) OxA/Hcrt-1 trafficks NMDARs to the synapse via activation of ox/hcrt-R1 and PLC/PKC. This is followed within 3–4 h by an NMDAR-dependent increase in AMPAR number or function at the synapse. (**b**) In animals self-administering cocaine or high-fat chocolate food, there is an additional increase in presynaptic glutamate release enhancing synaptic efficacy of VTA dopamine neurons

of an additional presynaptic increase in glutamate release. Surprisingly, this effect was not mediated by ox/hcrt-R1 or R2. Taken together, both oxA/hcrt-1 and oxB/hcrt-2 increase excitatory synaptic transmission in VTA DA neurons. Because NMDAR activation is required for the shift in tonic to burst firing, and burst firing promotes increased dopamine release, oxA-hcrt-1 or oxB-hcrt-2 potentiation of NMDARs may promote increased dopamine in target regions of the VTA.

Orexin/Hypocretin Receptor Signaling and Drug-Induced Plasticity

Some behavioral effects of oxA/hcrt-1 may be due to its neuroplastic effects at glutamatergic synapses in the VTA. Interestingly, the involvement of the VTA in the neuronal and behavioral changes caused by cocaine requires input from oxA/hcrt-1. The increase in AMPAR/NMDAR observed with in vivo cocaine exposure was blocked when the ox/hcrt-R1 antagonist was administered systemically before daily treatment with cocaine [48]. Furthermore, pretreatment with the ox/hcrt-R1 antagonist, either into the VTA or systemically, also abolishes the development of locomotor sensitization to repeated cocaine administration. However, administration of the inhibitor at the end of the cocaine treatment did not have any effect on the increase in locomotor activity, which indicates that while oxA/hcrt-1 is required for the acquisition of behavioral sensitization, it is not required for its expression. Taken together, these data suggest that oxA/hcrt-1 can enable neuroplasticity at excitatory synapses in the VTA and this action of oxA/hcrt-1 in the VTA is necessary for cocaine-induced plasticity of dopamine neurons and behavioral sensitization. It will be interesting to determine if the enhanced synaptic efficacy in dopamine neurons associated with other drugs of abuse, such as morphine or ethanol, also requires activation of ox/hcrt-R1.

Synaptic transmission of ox/hcrt neurons themselves can be altered by drugs of abuse. Ox/hcrt neurons of the lateral hypothalamus express μ-opioid receptors [60]. Furthermore, mice will self-administer morphine directly into the LH [61], suggesting that opioids may directly modulate ox/hcrt neurons to influence drug-seeking behaviors. Indeed, morphine inhibits ox/hcrt neurons by inhibiting calcium currents and opening a G-protein, inwardly rectifying potassium conductance. Furthermore, morphine suppresses glutamate release onto ox/hcrt neurons [62]. A reduction in presynaptic glutamate release is also observed with the application of cannabinoids [63]. Thus, it has been proposed that opioids and cannabinoids mediate their soporific effects by inhibiting synaptic activity in ox/hcrt neurons [62]. Other drugs of abuse might also directly modulate ox/hcrt neurons in the lateral hypothalamus. For example, Fos expression is increased in ox/hcrt neurons following acute administration of methamphetamine [64], amphetamine [41], and nicotine [65]; however, it is not known whether these drugs can modulate the synaptic efficacy of ox/hcrt neurons. In summary, in addition to ox/hcrt modulation of drug-induced plasticity in the VTA, drugs of abuse may also have direct effects on ox/hcrt neurons.

Orexin/Hypocretin Signaling in Addiction-Associated Behaviors

Ox/hcrt neurons are implicated in a variety of other addiction-associated behaviors [46–51, 66, 67]. The amount of c-Fos activation in ox/hcrt neurons was strongly correlated to preferences for cues associated with morphine and food reward [47]. Furthermore, both activation of ox/hcrt neurons and intra-VTA administration of oxA/hcrt-1 lead to a reinstatement of an extinguished conditioned place preference to morphine, food, or cocaine [47]. Ox/hcrt-R1 signaling in the VTA may be important for the association of the morphine with the environment where the drug was administered. Unilateral lesions of the lateral hypothalamus and intra-VTA administration of the ox/hcrt-R1 antagonist, SB334867, on the contralateral side block the acquisition of morphine conditioned place preference [50].

The ox/hcrt system has also been implicated in the reinstatement to self-administration of alcohol, abused drugs, or natural rewards, although its effects vary with the substance being self-administered. For example, stress- or cue-induced reinstatement, but not cocaine-primed reinstatement, of extinguished cocaine seeking is blocked by systemic administration of the ox/hcrt-R1 antagonist [46, 67]. Ox/hcrt-R1 signaling also has mixed effects on the maintenance of self-administration. The ox/hcrt-R1 antagonist, SB 334867, attenuates high-fat food pellet [68], ethanol [49, 66], and nicotine [51] self-administration, but not self-administration of cocaine [46, 67], water [49], or food [46]. These data suggest that the effects of ox/hcrt signaling on self-administration of natural rewards or drugs of abuse may be specific to the qualities of the reinforcer.

Effects of Orexin/Hypocretin on Motivation

Because ox/hcrt peptides are released during times of metabolic need, it has been hypothesized that they may serve to arouse and/or to motivate the animal to engage in foraging behavior, possibly through activation of VTA DA neurons [69, 70]. Therefore, ox/hcrt action in the VTA may drive motivated behavior for both natural and drug rewards. To establish whether the work required obtaining cocaine, regular food pellets or high-fat chocolate food pellets were altered by blockade of ox/hcrt-R1 signaling, and rats were tested in two paradigms measuring motivation. First, rats self-administered reinforcers under a progressive ratio schedule whereby the response requirement to earn a drug infusion or food pellet escalates after the delivery of each reinforcer. Second, rats were given a choice of climbing a barrier to obtain high-fat chocolate food or free access to the regular food arm. In both tests, systemic administration of SB 334867 significantly reduced the amount of work rats made to obtain high-fat chocolate food. Moreover, work for cocaine infusions in the progressive ratio test was reduced in the presence of SB 334867. Interestingly, the effort necessary for obtaining regular food, similar in composition to the rat's home cage chow, was not altered by SB 334867 [71], suggesting that ox/hcrt-R1 activation may produce effort selectively for highly salient rewards. These results are consistent with the effects of SB 334867 on motivation for another highly salient reinforcer, nicotine [51]. The effects of SB 334867 did not disrupt simple Pavlovian approach learning, suggesting that ox/hrct-R1 activation may contribute selectively to effort-related functions and does not appear to be required for learning an approach response induced by conditioned cues [71]. Consistent with these findings, the oxA/hcrt-1-mediated plasticity in the VTA was enhanced in animals self-administering cocaine or high-fat chocolate food, but not in those self-administering regular food (Fig. 3b). The increase in oxA/hcrt-1-mediated excitatory synaptic transmission observed in cocaine and high-fat chocolate food self-administering rats was likely due to increased presynaptic signaling of ox/hcrt-1Rs and consequently enhanced glutamate release. These results suggest that enhanced oxA/hcrt-1-mediated neural plasticity might underlie the motivational drive to obtain salient reinforcers.

Orexin/Hypocretin and Stressful Stimuli

Ox/hcrt neurons are activated by corticotropin-releasing factor (CRF) and other stressful stimuli [46, 72]. Therefore, oxA/hcrt-1-mediated plasticity in the VTA may also be enhanced by highly salient aversive stimuli. After 7 days of footshock, oxA/hcrt-1 could potentiate excitatory synaptic transmission; however, there was no additional presynaptic effect as was observed with cocaine or high-fat chocolate food self-administering rats. These results implicate ox/hcrt-R1-mediated neural plasticity in the VTA specifically for highly salient positive reinforcers, rather than simply arousing salient reinforcement.

A recent study has suggested that CRF and oxA/hcrt-1 may act on separate neural inputs into the VTA as an ox/hcrt-R1 antagonist did not block CRF-dependent footshock-induced reinstatement of cocaine seeking or associated glutamate or dopamine release in the VTA [73]. Additionally, reward and arousal functions of the oxA/hcrt-1 peptide have been proposed to be represented in different neural circuits elsewhere [74]. Evidence for separate circuits mediating arousal and rewarding functions of ox/hcrt signaling comes from studies that show differential regional activation of ox/hcrt neurons in the lateral hypothalamus with reinforcement or stress. Ox/hcrt neurons that are activated by reward-paired stimuli are located in the lateral hypothalamus and project to the VTA, whereas orexin/hypocretin neurons of the dorsal medial hypothalamus and perifornical area mediate arousal and response to footshock stress [47]. Furthermore, Fos activation in animals undergoing a stressful morphine withdrawal is specific to the dorsal medial and perifornical hypothalamus [75]. In another study, antipsychotic drugs that are associated with excessive weight gain preferentially activate ox/hcrt neurons in the lateral hypothalamus rather than the dorsal medial hypothalamus, and the amount of activation in the lateral hypothalamus correlated with the degree of weight gain [76]. Thus, distinct projections of ox/hcrt neurons may mediate the role of orexin in either stress or reward.

Future Treatments for Addiction

Because ox/hcrt-R1 signaling in the VTA appears to be important for the reinforcing effects of drugs of abuse, and oxA/hcrt-1-induced neural plasticity in the VTA might underlie the motivation to obtain highly salient reinforcers, compounds targeting ox/hcrt-R1 are likely to be excellent candidates for drug development in the treatment of addiction. Interestingly, an orally available mixed ox/hcrt-R1/R2 compound was not only found to be clinically useful for promoting sleep in humans, but also promoted some weight loss [77]. This drug (ACT-07857) blocked both ox/hcrt-R1 and R2 at nanomolar concentrations, but had little affinity for other G-protein-coupled receptors. Another possible use of this compound may be to reduce cocaine-, morphine-, or palatable food-triggered dopamine release, and potentially disrupt craving. However, use of this compound in clinical trials for weight loss or drug abstinence has not been reported. Antagonists targeting the ox/hcrt-R1 may be more selective for suppressing craving compared to inducing sleep. Cue-induced reinstatement to cocaine self-administration in rats was blocked by ox/hcrt-R1 but not ox/hcrt-R2 antagonists [67]. Furthermore, intra-VTA administration of oxA/hcrt-1, but not oxB/hcrt-2 which activates only ox/hcrt-R2, reinstates cocaine seeking [73]. Conversely, ox/hcrt-R2 knockout mice exhibit more overt behavioral abnormalities in sleep/wake transitions compared to ox/hcrt-R1 knockout mice [78]. Therefore, antagonists to ox/hcrt-R1 are promising candidates for the development of pharmacotherapies for the treatment of drug abuse.

Conclusions

In conclusion, the studies presented herein identify the ox/hcrt system as a new player in regulating the reinforcing properties of natural and drug rewards. However, while ox/hcrt signaling is involved in the reinstatement of extinguished drug-seeking behavior, its effects on the maintenance of drug seeking appear to be reinforcer specific. Thus, the orexin/hypocretin system may be integrating the animal's internal state (arousal and metabolic drive state) with other physiological (circadian) and environmental factors. (circadian and environmental) into goal-directed behaviors (motivation). OxA/hcrt-1-induced neural plasticity in the VTA may increase the likelihood of phasic burst firing of dopamine neurons and consequently enhance dopamine concentrations in target regions of the VTA. This may, in turn, drive motivated behaviors such as foraging or drug seeking.

References

1. Kreek MJ, Bart G, Lilly C, Laforge S, Nielsen DA. Pharmacogenetics and human molecular genetics of opiate and cocaine addictions and their treatments. Pharmacol Rev. 2005;57:1–26.
2. Johnson BA. Update on neuropharmacological treatments for alcoholism: scientific basis and clinical findings. Biochem Pharmacol. 2008;75:34–56.
3. Shoptaw S, Yang X, Rotheram-Fuller EJ, Hsieh YC, Kintaudi PC, Charuvastra VC, et al. Randomized placebo-controlled trial of baclofen for cocaine dependence: preliminary effects for individuals with chronic patterns of cocaine use. J Clin Psychiatry. 2003;64: 1440–8.
4. Nishino S, Mignot E. Pharmacological aspects of human and canine narcolepsy. Prog Neurobiol. 1997;52:27–78.
5. Akimoto H, Honda Y, Takahashi Y. Pharmacotherapy in narcolepsy. Dis Nerv Syst. 1960;21:704–6.
6. Guilleminault C, Carskadon M, Dement WC. On the treatment of rapid eye movement narcolepsy. Arch Neurol. 1974;30:90–3.
7. Robinson TE, Berridge KC. The neural basis of drug craving; an incentive-sensitization theory of addiction. Brain Res Rev. 1993;18:247–91.
8. Nestler EJ. Molecular mechanisms of drug addiction. Annu Rev Med. 2004;55:113–32.
9. Thomas MJ, Kalivas PW, Shaham Y. Neuroplasticity in the mesolimbic dopamine system and cocaine addiction. Br J Pharmacol. 2008;154:327–42.
10. Weiss F, Parsons LH, Schulteis G, Hyytiä P, Lorang MT, Bloom FE, et al. Ethanol self-administration restores withdrawal-associated deficiencies in accumbal dopamine and 5-hydroxytryptamine release in dependent rats. J Neurosci. 1996;16:3474–85.
11. Kalivas PW, Duffy P. Time course of extracellular dopamine and behavioral sensitization to cocaine. I. Dopamine axon terminals. J Neurosci. 1993;13:266–75.
12. Kalivas PW, Duffy P. Time course of extracellular dopamine and behavioral sensitization to cocaine. II. Dopamine perikarya. J Neurosci. 1993;13:276–84.
13. Post RM, Rose H. Increasing effects of repetitive cocaine administration in the rat. Nature. 1976;260:731–2.

14. Boileau I, Dagher A, Leyton M, Gunn RN, Baker GB, Diksic M, et al. Modeling sensitization to stimulants in humans: an [11C]raclopride/positron emission tomography study in healthy men. Arch Gen Psychiatry. 2006;63:1386–95.
15. Redgrave P, Gurney K. The short-latency dopamine signal: a role in discovering novel actions? Nat Rev Neurosci. 2006;7:967–75.
16. Berridge KC. The debate over dopamine's role in reward: the case for incentive salience. Psychopharmacol. 2007;191:391–431.
17. Wise RA. Dopamine, learning and motivation. Nat Rev Neurosci. 2004;5:483–94.
18. Roitman MF, Stuber GD, Phillips PE, Wightman RM, Carelli RM. Dopamine operates as a subsecond modulator of food seeking. J Neurosci. 2004;24:1265–71.
19. Phillips PE, Stuber GD, Heien ML, Wightman RM, Carelli RM. Subsecond dopamine release promotes cocaine seeking. Nature. 2003;422:614–8.
20. Grace AA, Bunney BS. The control of firing pattern in nigral dopamine neurons: burst firing. J Neurosci. 1984;4:2877–90.
21. Grace AA, Bunney BS. The control of firing pattern in nigral dopamine neurons: single spike firing. J Neurosci. 1984;4:2866–76.
22. Temper JM, Martin LM, Anderson DR. GABAA receptor-mediated inhibition of rat substantia nigra dopaminergic neurons by pars reticulata projection neurons. J Neurosci. 1995;15:3092–103.
23. Overton PG, Clark D. Burst firing in midbrain dopaminergic neurons. Brain Res Brain Res Rev. 1997;25:312–34.
24. Schultz W. Getting formal with dopamine reward. Neuron. 2002;36:241–63.
25. Wightman RM, Zimmerman JB. Control of dopamine extracellular concentration in rat striatum by impulse flow and uptake. Brain Res Rev. 1990;15:135–44.
26. Suaud-Chagny MF, Chergui K, Chouvet G, Gonon F. Relationship between dopamine release in the rat nucleus accumbens and the discharge activity of dopaminergic neurons during local in vivo application of amino acids in the ventral tegmental area. Neuroscience. 1992;49:63–72.
27. Murase S, Grenhoff J, Chouvet G, Gonon FG, Svensson TH. Prefrontal cortex regulates burst firing and transmitter release in rat mesolimbic dopamine neurons studied in vivo. Neurosci Lett. 1993;157:53–6.
28. Johnson SW, Seutin V, North RA. Burst firing in dopamine neurons induced by N-methyl-D-aspartate: role of electrogenic sodium pump. Science. 1992;258: 665–7.
29. Overton PG, Clark D. Iontophroetically administered drugs acting at the N-methyl-D-aspartate receptor modulate burst firing in A9 dopamine neurons in the rat. Synapse. 1992;10:131–40.
30. Malenka RC, Bear MF. LTP and LTD: an embarrassment of riches. Neuron. 2004;44:5–21.
31. Bonci A, Malenka RC. Properties and plasticity of excitatory synapses on dopaminergic and GABAergic cells in the ventral tegmental area. J Neurosci. 1999;19:3723–30.
32. Overton PG, Richards CD, Berry MS, Clark D. Long-term potentiation at excitatory amino acid synapses on midbrain dopamine neurons. Neuroreport. 1999;10: 221–6.
33. Jones S, Kornblum JL, Kauer JA. Amphetamine blocks long term synaptic depression in the ventral tegmental area. J Neurosci. 2000;20:5575–80.
34. Thomas MT, Malenka RC, Bonci A. Modulation of long-term depression by dopamine in the mesolimbic system. J Neurosci. 2000;20:5581–6.
35. Bellone C, Lüscher C. mGluRs induce a long-term depression in the ventral tegmental area that involves a switch of the subunit composition of AMPA receptors. Eur J Neurosci. 2005;21:1280–8.
36. Stuber GD, Klanker M, de Ridder B, Bowers MS, Joosten RN, Feenstra MG, et al. Reward predictive cues enhance excitatory synaptic strength onto midbrain dopamine neurons. Science. 2008;321:1690–2.
37. Ungless MA, Whistler JL, Malenka RC, Bonci A. A single cocaine exposure in vivo induces long-term potentiation in dopamine neurons. Nature. 2001;411:583–7.
38. Borgland SL, Malenka R, Bonci A. Acute and chronic cocaine-induced potentiation of synaptic strength in the VTA: electrophysiological and behavioral correlates in individual rats. J Neurosci. 2004;24:7482–90.
39. Saal D, Dong Y, Bonci A, Malenka RC. Drugs of abuse and stress trigger a common synaptic adaptation in dopamine neurons. Neuron. 2003;37:577–82.
40. Chen BT, Bowers MS, Martin M, Hopf FW, Guillory AM, Carelli RM, et al. Cocaine but not natural reward self-administration nor passive cocaine infusion produces persistent LTP in the VTA. Neuron. 2008;59:288–97.
41. Fadel J, Deutch AY. Anatomical substrates of orexin-dopamine interactions: lateral hypothalamic projections to the ventral tegmental area. Neuroscience. 2002;111:379–87.
42. Narita M, Nagumo Y, Hashimoto S, Narita M, Khotib J, Miyatake M, et al. Direct involvement of orexinergic systems in the activation of the mesolimbic dopamine pathway and related behaviors induced by morphine. J Neurosci. 2006;26:398–405.
43. Korotkova TM, Sergeeva OA, Eriksson KS, Haas HL, Brown RE. Excitation of ventral tegmental area dopaminergic and non-dopaminergic neurons by orexin/hypocretins. J Neurosci. 2003;23:7–11.
44. Adamantidis A, de Lecea L. The hypocretins as sensors for metabolism and arousal. J Physiol. 2009;587: 33–40.
45. Siegel JM. Hypocretin (orexin): role in normal behavior and neuropathology. Annu Rev Psychol. 2004;55: 125–48.
46. Boutrel B, Kenny PJ, Specio SE, Martin-Fardon R, Markou A, Koob GF, et al. Role for hypocretin in mediating stress-induced reinstatement of cocaine-seeking behavior. Proc Natl Acad Sci U S A. 2005;102: 19168–91.
47. Harris GC, Wimmer W, Aston Jones G. A role for lateral hypothalamic orexin neurons in reward seeking. Nature. 2005;437:556–9.
48. Borgland SL, Taha S, Sarti F, Fields HL, Bonci A. Orexin A signaling in dopamine neurons is critical for

cocaine induced synaptic plasticity and behavioral sensitization. Neuron. 2006;49:589–601.
49. Lawrence AJ, Cowen MS, Yang H-J, Chen F, Oldfield B. The orexin system regulates alcohol-seeking in rats. Br J Pharmacol. 2006;148:752–9.
50. Harris GC, Wimmer M, Randall-Thompson JF, Aston-Jones G. Lateral hypothalamic orexin neurons are critically involved in learning to associate an environment with morphine reward. Behav Brain Res. 2007;183: 43–51.
51. Hollander JA, Lu Q, Cameron MD, Kamenecka TM, Kenny PJ. Insular hypocretin transmission regulates nicotine reward. Proc Natl Acad Sci U S A. 2008;105: 19480–5.
52. Peyron C, Tighe DK, van den Pol A, de Lecea L, Heller HC, Sutcliffe JG, et al. Neurons containing hypocretin (orexin) project to multiple neuronal systems. J Neurosci. 1998;18:9996–10015.
53. Sakurai T, Amemiya A, Ishii M, Matsuzaki I, Chemelli RM, Tanaka H, et al. Orexin and orexin receptors: a family of hypothalamic neuropeptides and G protein-coupled receptors that regulate feeding behavior. Cell. 1998;92:573–85.
54. Balcita-Pedicino JJ, Sesack SR. Orexin axons in the rat ventral tegmental area synapse infrequently onto dopamine and gamma-aminobutyric acid neurons. J Comp Neurol. 2007;503:668–85.
55. Zhu Y, Miwa Y, Yamanaka A, Yada T, Shibahara M, Abe Y, et al. Orexin receptor type-1 couples exclusively to pertussis toxin-insensitive G-proteins, while orexin receptor type-2 couples to both pertussis toxin-sensitive and -insensitive G-proteins. J Pharmacol Sci. 2003;92:259–66.
56. Muschamp JW, Dominguez JM, Sato SM, Shen RY, Hull EM. A role for hypocretin (orexin) in male sexual behavior. J Neurosci. 2007;27:2837–45.
57. Vittoz NM, Berridge CW. Hypocretin/orexin selectively increases dopamine efflux within the prefrontal cortex: involvement of the ventral tegmental area. Neuropsychopharmacology. 2006;31:384–95.
58. Vittoz NM, Schmeichel B, Berridge CW. Hypocretin /orexin preferentially activates caudomedial ventral tegmental area dopamine neurons. Eur J Neurosci. 2008;28:1629–40.
59. Borgland SL, Storm E, Bonci A. Orexin B/hypocretin-2 increases glutamatergic synaptic transmission to ventral tegmental area neurons. Eur J Neurosci. 2008;29:1–12.
60. Georgescu D, Zachariou V, Barrot M, Mieda M, Willie JT, Eisch AJ, et al. Involvement of lateral hypothalamic peptide orexin in morphine dependence and withdrawal. J Neurosci. 2003;23:3106–11.
61. Cazala P, Darracq C, Saint-Marc M. Self-administration of morphine into the lateral hypothalamus in the mouse. Brain Res. 1987;416:283–8.
62. Li Y, van den Pol AN. Mu-opioid receptor-mediated depression of the hypothalamic hypocretin/orexin arousal system. J Neurosci. 2008;28:2814–9.
63. Huang H, Acuna-Goycolea C, Li Y, Cheng HM, Obrietan K, van den Pol AN. Cannabinoids excite hypothalamic melanin-concentrating hormone but inhibit hypocretin/orexin neurons: implications for cannabinoid actions on food intake and cognitive arousal. J Neurosci. 2007;27:4870–81.
64. Estabrooke IV, McCarthy MT, Ko E, Chou TC, Chemelli RM, Yanagisawa M, et al. Fos expression in orexin neurons varies with behavioral state. J Neurosci. 2001;21:1656–62.
65. Pasumarthi RK, Reznikov LR, Fadel J. Activation of orexin neurons by acute nicotine. Eur J Pharmacol. 2006;535:172–6.
66. Richards J, Simms J, Steesland P, Taha SA, Borgland SL, Bonci A, et al. Inhibition of orexin-1/hypocretin-1 receptors inhibits yohimbine-induced reinstatement of ethanol and sucrose seeking in long-evans rats. Psychopharmacol. 2008;199:109–17.
67. Smith R, See RE, Aston-Jones G. Orexin/hypocretin signalling at the OX1 receptor regulates cue-elicited cocaine seeking. Eur J Neurosci. 2009;30:493–503.
68. Nair SG, Golden SA, Shaham Y. Differential effects of the hypocretin 1 receptor antagonist SB334867 on high-fat food self-administration and reinstatement of food seeking in rats. Br J Pharmacol. 2008;154:1–11.
69. Yamanaka A, Beuckmann CT, Willie JT, Hara J, Tsujino N, Mieda M, et al. Hypothalamic orexin neurons regulate arousal according to energy balance in mice. Neuron. 2003;38:701–13.
70. Wisor JP, Nishino S, Sora I, Uhl GH, Mignot E, Edgar DM. Dopaminergic role in stimulant-induced wakefulness. J Neurosci. 2001;21:1787–94.
71. Borgland SL, Chang SJ, Bowers MS, Thompson JL, Vittoz NM, Floresco S, et al. Orexin A/Hypocretin1 promotes motivation selectively for highly salient positive reinforcers. J Neurosci. 2009;29:11215–25.
72. Winsky-Sommerer R, Yamanaka A, Diano S, Borok E, Roberts AJ, Sakurai T, et al. Interaction between the corticotropin-releasing factor system and hypocretins (orexins): a novel circuit mediating stress response. J Neurosci. 2004;24:11439–48.
73. Wang B, You Z, Wise RA. Reinstatement of cocaine-seeking by hypocretin (orexin) in the ventral tegmental area: independence from the local CRF network. Biol Psychiatry. 2009;65:857–62.
74. Harris GC, Aston-Jones G. Arousal and reward: a dichotomy in orexin function. Trends Neurosci. 2006;29:571–7.
75. Sharf R, Guarnieri DJ, Taylor JR, DiLeone RJ. Orexin mediates the expression of precipitated morphine withdrawal and concurrent activation of the nucleus accumbens shell. Biol Psych. 2008;64:175–83.
76. Fadel J, Bubser M, Deutch AY. Differential activation of orexin neurons by antipsychotic drugs associated with weight gain. J Neurosci. 2002;22:6742–6.
77. Brisbare-Roch C, Dingemanse J, Koberstein R, Hoever P, Aissaoui H, Flores S, et al. Promotion of sleep by targeting the orexin system in rats, dogs and humans. Nat Med. 2007;13:150–5.
78. Willie JT, Chemelli RM, Sinton CM, Tokita S, Williams SC, Kisanuki YY, et al. Distinct narcolepsy syndromes in Orexin receptor-2 and Orexin null mice: molecular genetic dissection of Non-REM and REM sleep regulatory processes. Neuron. 2003;8:715–30.

Orexin/Hypocretin, Drug Addiction, and Narcolepsy

Ralph J. DiLeone, Maysa Sarhan, and Ruth Sharf

Keywords

Orexin • Hypocretin • Dopamine • Ventral tegmental area • Abuse • Addition • Conditioned place preference

The hypothalamic neuropeptide orexin (hypocretin) influences a range of behaviors ranging from feeding to sleep and arousal. While genetic studies in rodents first demonstrated a role of orexin in narcolepsy with cataplexy, more recent work has brought attention to the role of orexin in drug addiction. Orexin influences many aspects of drug addiction, including drug dependence, reward, and reinstatement. Data suggest that orexin acts through its receptors to influence the mesocorticolimbic dopamine pathways that are known to underlie responses to drugs of abuse and the development of addiction. Importantly, orexin may also influence positive and negative (stress) states that can in turn influence addiction. The orexin neurons project broadly and the receptors are widely expressed, making it challenging to determine the circuits that are central to these effects. It may be that overlapping or distinct circuits are mediating orexin's influence on reward/addiction or arousal/narcolepsy.

R.J. DiLeone (✉)
Department of Psychiatry, Ribicoff Research Facilities, Connecticut Mental Health Center, Yale University School of Medicine, 34 Park Street, New Haven, CT 06508, USA
e-mail: ralph.dileone@yale.edu

Orexin and Its Receptors

The orexin peptides, orexin A and orexin B, are derived from proteolytic processing of the precursor prepro-orexin. The orexins are produced in hypothalamic neurons in the perifornical area (PFA), dorsomedial hypothalamus (DMH), and lateral hypothalamus (LH). Orexin neurons represent a distinct population from melanin-concentrating hormone (MCH)-expressing neurons that are also located in the PFA and LH [1]. The G-protein-coupled orexin receptors have different selectivity to orexin ligands. Orexin 1 receptor (Ox1r) is selective for orexin A, whereas the orexin 2 receptor (Ox2r) is less selective. The Ox1r couples to G_q while Ox2r is primarily $G_{i/o}$ coupled [2], and both Ox1r and Ox2r have been found on soma and pre- and postsynaptic processes in the LH [3].

Anatomical Basis for Interactions with Brain Dopamine Circuits

Most abused drugs increase extracellular DA levels at the level of the nucleus accumbens (NAc), and neuroadaptations in this system is thought to underlie many elements of addiction [4]. As reviewed below, the orexin system has a number

C.R. Baumann et al. (eds.), *Narcolepsy: Pathophysiology, Diagnosis, and Treatment*, DOI 10.1007/978-1-4419-8390-9_23, © Springer Science+Business Media, LLC 2011

of potential connections with this well-studied dopamine circuitry.

The orexin receptors are broadly expressed. Moreover, orexin neurons innervate many regions [5, 6], including prominent axonal projections to the ventral tegmental area (VTA) which expresses orexin receptors [7–9]. As with many neuropeptides, it is not clear if orexin is released synaptically or extra-synaptically. While low power analysis has shown that orexin terminals contact tyrosine hydroxylase (TH)-positive cells [8, 10], electron microscopic studies indicated that orexin-positive axons infrequently synapse onto dopamine (DA) and GABA neurons in the VTA [11].

Orexin's Effects on Neuronal Excitation and Plasticity in the VTA

Orexin exerts an excitatory action in the VTA on both DA and non-DA cells [10, 12]. Intra-VTA application of orexin results in increased Fos expression in DA neurons located specifically in the caudomedial portion of the VTA [13] and increases DA at the level of the nucleus accumbens shell and the medial prefrontal cortex [14, 15]. Intra-VTA orexin also results in potentiation of NMDA receptor-mediated excitatory postsynaptic currents [16], indicating a role for orexin in long-term neural plasticity. This is reviewed elsewhere in this book.

In addition to effects on the midbrain dopamine neurons of the VTA, orexin may also directly influence the target regions of dopamine neurons, such as the nucleus accumbens and prefrontal cortex. Orexin is excitatory in the mPFC [17] but some reports suggest inhibition [18], and others excitation [19] in the nucleus accumbens (Fig. 1).

Drugs of Abuse Regulate Orexin Neuron Excitation and Gene Expression

Fos expression is increased in orexin neurons following acute administration of methamphetamine [20] and amphetamine [8]. Amphetamine increases Fos expression in neurons specifically in the DMH, but not in PFA or LH [8]. Acute administration of cocaine does not affect either of the orexin receptors [21, 22] or peptide levels [21, 22], and acute morphine fails to alter orexin mRNA levels [23] (Table 1).

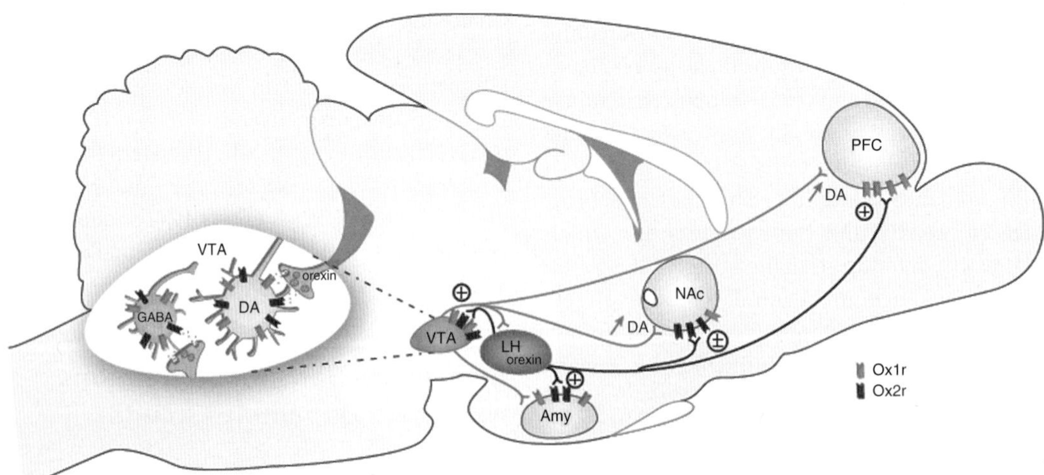

Fig. 1 Schematic representation of the orexin system and its interactions with the mesocorticolimbic reward pathway. Orexin projections activate the ventral tegmental area (VTA), amygdala (Amy), and prefrontal cortex (PFC). Conflicting data suggest both activation and inhibition of the nucleus accumbens (NAc) by orexin. In the VTA, orexin activates both dopamine and GABA neurons. Orexin's activation of the VTA results in increased dopamine (DA) via projections to the NAc and PFC. Orexin's action in the PFC is due to stimulation of Ox1rs. Orexin's action in the NAc and amygdala is due to stimulation of Ox2rs

Table 1 Summary of behavioral data on the role of orexin in response to drugs of abuse

Acute locomotion		
Cocaine	SB-334867 has no effect	Borgland et al. [16]
Morphine	OKO shows reduction	Narita et al. [15]
	SB-332867 has no effect	Sharf et al. [29]
	OKO shows normal behavior	Sharf et al. [29]
Sensitization		
Cocaine	SB-334867 blocks development not expression	Borgland et al. [16]
Morphine	SB-334867 has no effect	Sharf et al. [29]
	OKO shows normal behavior	Sharf et al. [29]
Withdrawal		
Cocaine	No activation of orexin neurons	Zhou et al. [22]
Morphine	Increases CRE in orexin cells	Georgescu et al. [26]
	Increases Fos in orexin cells	Sharf et al. [27]
	Increases orexin mRNA in LH	Zhou et al. [23]
	OKO shows reduced withdrawal symptoms	Georgescu et al. [26]
	SB-334867 attenuates withdrawal symptoms	Sharf et al. [27]
Self-administration		
Cocaine	SB-334867 or orexin shows no effect	Boutrel et al. [37]; Aston-Jones et al. [38]
Nicotine	SB-334867 reduces	Hollander et al. [35]
Ethanol	SB-338467 reduces	Lawrence et al. [36]
	Orexin A increases	Schneider et al. (2007)
Self-administration extinction		
Cocaine	SB-334867 reduces extinction responding	Aston-Jones et al. [38]
Self-administration reinstatement		
Cocaine	Orexin A reinstates	Boutrel et al. [37]; Wang et al. [39]
	SB-334867 attenuates cue-primed reinstatement	Aston-Jones et al. [38]
Ethanol	SB-334867 attenuates cue-primed reinstatement	Lawrence et al. [36]
	Fos increases in orexin cells following CS presentations	Dayas et al. [48]
Conditioned place preference		
Cocaine	Increases Fos in orexin neurons	Harris et al. [33]
	Decreases orexin mRNA in LLH	Zhou et al. [22]
	SB-334867 has no effect	Sharf et al. [29]
Morphine	Increases Fos in orexin neurons	Harris et al. [33]
	SB-334867 reduces CPP	Harris et al. [33]; Sharf et al. [29]; Narita et al. [15]
	OKO shows no CPP	Narita et al. [15]
	OKO shows normal behavior	Sharf et al. [29]

In addition to effects of acute drugs, chronic drugs of abuse also influence the orexin system via changes in orexin neural activation and mRNA levels of orexin or its receptors. Chronic administration of nicotine is associated with increased expression of prepro-orexin mRNA in the LH [24]. Orexin A expression is increased primarily in the DMH, whereas orexin B is increased in DMH and paraventricular hypothalamus (PVN) [24]. Furthermore, chronic administration of nicotine is associated with increases of Ox2r mRNA and, to a lesser degree, increases of Ox1r mRNA [24]. Chronic amphetamine exposure results in increased Fos expression in orexin neurons [25]. In contrast, chronic morphine treatment and steady-state chronic cocaine [22] fail to affect orexin mRNA levels or Fos expression in orexin neurons in the LH. Nonetheless, Ox2r levels are increased in the NAc following chronic cocaine exposure; a change that persists even months following cessation of cocaine treatment [21]. Although orexin mRNA levels are unchanged in

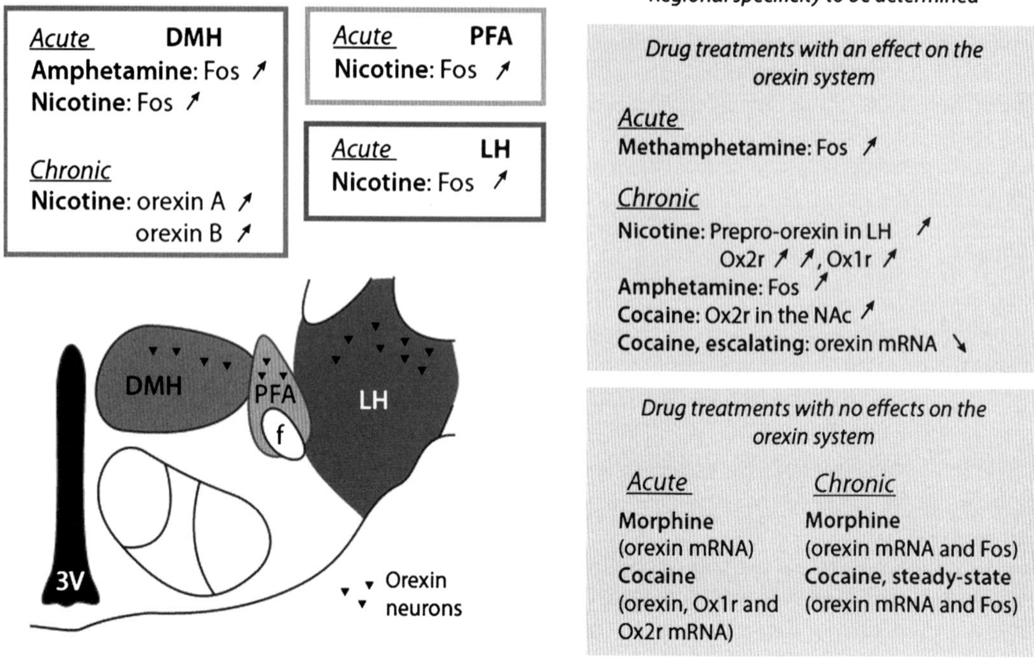

Fig. 2 Effects of acute or chronic drug administration on orexin cells. *DMH* dorsal medial hypothalamus, *PFA* perifornical area, *LH* lateral hypothalamus, *3V*, third ventricle, *f*, fornix. Data derived from [8, 20–27, 47, 49]

response to steady-state cocaine, chronic escalating dose cocaine treatment results in decreased orexin mRNA levels in the LH [22]. Also, it should be recognized that negative results with gene expression analysis could be a result of the timepoint chosen for the mRNA analysis (Fig. 2).

The Role of Orexin in Withdrawal from Drugs of Abuse

The first study establishing a functional role of orexin in addiction demonstrated an involvement in morphine withdrawal [26]. Precipitated morphine withdrawal leads to the induction of CRE activity in orexin neurons [26] and Fos expression in orexin cells [26, 27]. The increases in Fos expression were restricted to the DMH and PFA, and were not seen in the LH [27]. Similarly, spontaneous morphine withdrawal increases orexin mRNA in the LH [23]. Activation of orexin neurons in response to withdrawal appears to be opiate specific and is not seen following spontaneous withdrawal following chronic cocaine exposure [22].

The orexin system appears to be necessary for the expression of withdrawal. Treatment with the selective OxR1 antagonist SB-334867 prior to naloxone-precipitated withdrawal attenuates somatic morphine withdrawal symptoms [27]. The somatic withdrawal intensity correlates with cFos expression in the NAc shell, and blockade of Ox1r indirectly attenuates NAc cellular activation [27]. Interestingly, the VTA does not show changes in cFos expression in the same animals, suggesting a possible non-VTA mechanism by which orexin alters NAc neuronal activity.

Effects of Orexin on Locomotor Response and Sensitization

Drug-induced hyperlocomotion and behavioral sensitization are argued to represent a form of drug-induced neural plasticity [28] that may contribute to dependence and addictive processes.

Thus far, investigation of orexin's role in drug hyperlocomotion and sensitization has been limited. Orexin knockout mice show reduced hyperlocomotion in response to acute morphine exposure [15]. Systemic administration of the selective OxR1 antagonist SB-334867 does not affect acute cocaine-induced hyperlocomotion and activity in previously sensitized animals [16]. However, systemic and intra-VTA blockade of Ox1r blocks the development of cocaine sensitization [16]. However, recent data suggest that SB-334867 treated mice and OKO mice respond normally following an acute morphine treatment, and no changes in behavioral sensitization are seen following chronic morphine administration [29].

Orexin in Conditioned Place Preference for Drugs of Abuse

The conditioned place preference (CPP) is a paradigm for assessing reward-related behavior. An animal is placed in an enclosed environment, which is divided into two or more distinct compartments, each containing distinctive environmental cues that differentiate it from the other(s). After the animal is allowed to move freely among the compartments, it is then confined to one of the compartments while being presented with a rewarding stimulus. The animal forms an association between the rewarding stimulus and the environmental cues during this conditioning session. After the session, when allowed to roam freely among the compartments, the animal will spend more time in the reward-paired chamber if the drug was reinforcing. In the CPP test, animals are evaluated in a drug-free state and preference is indicative of the rewarding effects of the context rather than of the drug directly, as reviewed in [30]. It should also be noted that some discrepancy has been reported between CPP data and self-administration data [31, 32]. Nonetheless, CPP remains a useful experimental protocol in the investigation of neural mechanisms underlying drug addiction.

Animals that exhibit a preference for an environment previously paired with food, morphine, or cocaine show increased Fos expression in LH orexin neurons [33]. Interestingly, administration of SB-334867 reduces the expression of morphine place preference [29, 33], but not of cocaine place preference [29]. Previous reports also suggest a role for orexin in the formation of a place preference to a morphine-paired environment. Unilateral lesions of the LH and intra-VTA administration of SB-334867 on the contralateral side also block the development of morphine CPP [34], suggesting that orexin acting in the VTA is essential for the development of morphine CPP. A role for the VTA in the expression of morphine CPP was also suggested from intra-VTA SB-334867 suppression of morphine place preference [15].

Orexin in Self-Administration, Extinction, and Reinstatement

Self-administration studies are powerful measures that best mimic conditions of addiction whereby the subject is controlling the drug delivery and drug-seeking behavior. SB-334867 administration results in decreased self-administration of nicotine [35] and alcohol [36]. In contrast, intracerebroventricular (icv) administration of orexin A [37] or treatment with SB-334867 [38] fails to affect responding during active cocaine self-administration. Orexin plays a role in "drug relapse" as modeled by extinction and reinstatement paradigms. Here, following self-administration training, animals continue to lever press or nose poke, but no drug is delivered following previously correct responses until eventually animals cease to respond at high levels ("extinction"). Animals can then be tested for reinstatement, or relapse, by assessing lever pressing after extinction. Reinstatement of the extinguished response can be induced by presentation of various stimuli, including the drug (drug-primed reinstatement) or cues paired with the previously active response (cue-primed reinstatement). While icv administration of orexin A reinstates extinguished responses [37], blockade experiments suggest a more specific role for drug reinstatement. With both alcohol [36] and cocaine [38], treatment with

SB-334867 fails to affect drug-primed reinstatement, but attenuates cue-primed reinstatement. It is possible that orexin plays a more general role in mediating cue- or context-drug associations.

The involvement of orexin in reinstatement of extinguished drug seeking may reflect a role in mediating stress response. Stress-induced reinstatement of cocaine self-administration is abolished by SB-334867 [37]. These data suggest that drug seeking is induced by activation of the stress pathway. Interestingly, intra-VTA CRF antagonists do not block orexin-A-induced reinstatement of cocaine seeking, and foot shock stress-induced cocaine reinstatement is not blocked by intra-VTA SB-334867 [39], suggesting that orexin and CRF have independent actions and that blockade of stress-induced reinstatement by SB-334867 is not mediated via the VTA. It has been proposed that distinct populations of neurons mediate the role of the orexins in stress and reward [33, 40].

Orexin and Reward

The self-administration data summarized thus far suggest a role for orexin in the rewarding properties of some drugs of abuse. Additional evidence for the role of orexin in reward processing comes from CPP and cue-induced reinstatement data, suggesting that orexin functioning contributes to drug seeking when the rewarding stimulus is no longer present. These data are suggestive of a reward component, but the CPP data more directly suggest that processing of environmental cues and drug-associated contexts, and drug-seeking induced by the presence of such cues are mediated, at least in part, by the orexin system. However, this fails to explain data demonstrating that activation of the orexin system results in spontaneous reinstatement of an extinguished response.

In brain stimulation reward (BSR) studies, animals will lever press for the delivery of intracranial electrical stimulation and animals will robustly respond for stimulation of the LH [41]. Interestingly, orexin A administration has been shown to elevate stimulation thresholds, suggesting a decrease in reward sensitivity [37]. In contrast, SB-334867 abolishes nicotine-mediated reductions in BSR thresholds, suggesting that blockade of Ox1r results in decreased sensitivity to reward [35]. Based on these findings, it remains inconclusive whether orexin-mediated alterations in drug self-administration are directly related to reward processing.

Orexin, Neurocircuits, and Addiction: Implications for Narcolepsy

It is clear that the orexin system is engaged by drugs of abuse and that orexin plays a role in drug responses and behaviors that are relevant to addiction. Orexin can alter drug self-administration, drug-associated cue processing and reward, and stress responses during drug abstinence in drug-dependent animals (somatic withdrawal and reinstatement of extinguished drug seeking). Although orexin can modulate all of these effects, it is possible that they are mediated by specific anatomical substrates and pathways. Distinct orexin neuronal subpopulations may serve disparate functions. Reward and cue processing appears to be mediated by LH orexin neurons [33, 40] acting in the VTA [34], whereas drug-related stress or aversive responses may be mediated by DMH and PFA orexin neurons acting directly or indirectly through other brain regions. Although much evidence supports a role for orexin functioning in VTA in reward processing [34, 40], other brain regions are likely to be involved, such as the basolateral amygdala and infralimbic prefrontal cortex [42].

Orexin has also been shown to mediate sleep and arousal processes. Canine narcolepsy has been attributed to a genetic disruption of the Ox2r gene [43] and OKO mice demonstrate behavioral arrests similar to those seen in narcoleptic episodes [44]. Are the same neurocircuits relevant for drugs of abuse and for narcolepsy? One possibility is that the neural circuits mediating orexin's effects with drugs of abuse are independent of those mediated in narcolepsy. Since orexin axonal projections and receptor expression are so broad, this assumption seems not only possible, but also likely. On the contrary, some of the pathways implicated in addiction (e.g., VTA) seem likely to play a role in orexin's ability to activate behaviors and thus contribute to a state of arousal. In fact, orexin itself can modulate locomotor activity [10, 45], suggesting a general role in this process, independent of drugs of abuse.

However, it is important to note that comparisons between an evolved role for the molecule and pharmacological insults (drugs of abuse) are difficult to make. That is, the orexin neuropeptide system evolved to balance a complex set of animal behaviors. As with all neurochemicals and pathways that mediate addiction, it may be that drugs of abuse act via a specific component of this neuropeptide system that otherwise serves to integrate behaviors, such as arousal, motivation, and the formation of cue associations. In addition, orexin has been implicated in other biological functions such as regulation of body temperature and heart rate [46], supporting a broader role for the molecule in animal physiology and behavior. More research is needed to elucidate specific neuroanatomical substrates and receptor mechanisms of the role of orexin in drug dependence and addiction. Moreover, more analysis of its role in nondrug rewards (e.g., food, gambling, and sex) will help to describe more completely the influence of orexin on a diverse set of motivated and reward-related behaviors.

References

1. Broberger C, De Lecea L, Sutcliffe JG, Hokfelt T. Hypocretin/orexin- and melanin-concentrating hormone-expressing cells form distinct populations in the rodent lateral hypothalamus: relationship to the neuropeptide Y and agouti gene-related protein systems. J Comp Neurol. 1998;402(4):460–74.
2. Sakurai T, Amemiya A, Ishii M, et al. Orexins and orexin receptors: a family of hypothalamic neuropeptides and G protein-coupled receptors that regulate feeding behavior. Cell. 1998;92(4):573–85.
3. van den Pol AN, Geo XB, Obrietan K, Kilduff TS, Belousov AB. Presynaptic and postsynaptic actions and modulation of neuroendocrine neurons by a new hypothalamic peptide, hypocretin/orexin. J Neursci. 1998;18(19):7962–71.
4. Koob GF, Sanna PP, Bloom FE. Neuroscience of addiction. Neuron. 1998;21(3):467–76.
5. Mondal MS, Nakazato M, Date Y, Murakami N, Yanagisawa M, Matsukura S. Widespread distribution of orexin in rat brain and its regulation upon fasting. Biochem Biophys Res Commun. 1999;256(3): 495–9.
6. Nambu T, Sakurai T, Mizukami K, Hosoya Y, Yanagisawa M, Goto K. Distribution of orexin neurons in the adult rat brain. Brain Res. 1999;827(1–2): 243–60.
7. Baldo BA, Daniel RA, Berridge CW, Kelley AE. Overlapping distributions of orexin/hypocretin- and dopamine-beta-hydroxylase immunoreactive fibers in rat brain regions mediating arousal, motivation, and stress. J Comp Neurol. 2003;464(2):220–37.
8. Fadel J, Deutch AY. Anatomical substrates of orexin-dopamine interactions: lateral hypothalamic projections to the ventral tegmental area. Neuroscience. 2002;111(2):379–87.
9. Peyron C, Tighe DK, van den Pol AN, et al. Neurons containing hypocretin (orexin) project to multiple neuronal systems. J Neurosci. 1998;18(23):9996–10015.
10. Nakamura T, Uramura K, Nambu T, et al. Orexin-induced hyperlocomotion and stereotypy are mediated by the dopaminergic system. Brain Res. 2000; 873(1):181–7.
11. Balcita-Pedicino JJ, Sesack SR. Orexin axons in the rat ventral tegmental area synapse infrequently onto dopamine and gamma-aminobutyric acid neurons. J Comp Neurol. 2007;503(5):668–84.
12. Korotkova TM, Sergeeva OA, Eriksson KS, Haas HL, Brown RE. Excitation of ventral tegmental area dopaminergic and nondopaminergic neurons by orexins/hypocretins. J Neurosci. 2003;23(1):7–11.
13. Vittoz N, Schmeichel B, Berridge C. Hypocretin/orexin preferentially activates caudomedial ventral tegmental area dopamine neurons. Eur J Neurosci. 2008;28(8):1629–40.
14. Vittoz NM, Berridge CW. Hypocretin/orexin selectively increases dopamine efflux within the prefrontal cortex: involvement of the ventral tegmental area. Neuropsychopharmacology. 2006;31(2):384–95.
15. Narita M, Nagumo Y, Hashimoto S, et al. Direct involvement of orexinergic systems in the activation of the mesolimbic dopamine pathway and related behaviors induced by morphine. J Neurosci. 2006;26(2):398–495.
16. Borgland SL, Taha SA, Sarti F, Fields HL, Bonci A. Orexin A in the VTA is critical for the induction of synaptic plasticity and behavioral sensitization to cocaine. Neuron. 2006;49(4):589–601.
17. Xia J, Chen X, Song C, Ye J, Yu Z, Hu Z. Postsynaptic excitation of prefrontal cortical pyramidal neurons by hypocretin-1/orexin A through the inhibition of potassium currents. J Neurosci Res. 2005;82(5):729–36.
18. Martin G, Fabre V, Siggins GR, De Lecea L. Interaction of the hypocretins with neurotransmitters in the nucleus accumbens. Regul Pept. 2002;104(1–3):111–7.
19. Mukai K, Kim J, Nakajima K, Oomura Y, Wayner MJ, Sasaki K. Electrophysiological effects of orexin/hypocretin on nucleus accumbens shell neurons in rats: an in vitro study. Peptides. 2009;30(8): 1487–96.

20. Estabrooke IV, McCarthy MT, Ko E, et al. Fos expression in orexin neurons varies with behavioral state. J Neurosci. 2001;21(5):1656–62.
21. Zhang GC, Mao LM, Liu XY, Wang JQ. Long-lasting up-regulation of orexin receptor type 2 protein levels in the rat nucleus accumbens after chronic cocaine administration. J Neurochem. 2007;103(1):400–7.
22. Zhou Y, Cui CL, Schlussman SD, et al. Effects of cocaine place conditioning, chronic escalating-dose "binge" pattern cocaine administration and acute withdrawal on orexin/hypocretin and preprodynorphin gene expressions in lateral hypothalamus of Fischer and Sprague-Dawley rats. Neuroscience. 2008; 153(4):1225–34.
23. Zhou Y, Bendor J, Hofmann L, Randesi M, Ho A, Kreek MJ. Mu opioid receptor and orexin/hypocretin mRNA levels in the lateral hypothalamus and striatum are enhanced by morphine withdrawal. J Endocrinol. 2006;191(1):137–45.
24. Kane JK, Parker SL, Matta SG, Fu Y, Sharp BM, Li MD. Nicotine up-regulates expression of orexin and its receptors in rat brain. Endocrinology. 2000;141(10):3623–9.
25. Morshedi MM, Meredith GE. Repeated amphetamine administration induces Fos in prefrontal cortical neurons that project to the lateral hypothalamus but not the nucleus accumbens or basolateral amygdala. Psychopharmacology. 2008;197(2):179–89.
26. Georgescu D, Zachariou V, Barrot M, et al. Involvement of the lateral hypothalamic peptide orexin in morphine dependence and withdrawal. J Neurosci. 2003;23(8):3106–11.
27. Sharf R, Sarhan M, DiLeone RJ. Orexin mediates the expression of precipitated morphine withdrawal and concurrent activation of the nucleus accumbens shell. Biol Psychiatry. 2008;64(3):175–83.
28. Robinson TE, Becker JB, Priesty SK. Long-term facilitation of amphetamine-induced rotational behavior and striatal dopamine release produced by a single exposure to amphetamine: sex differences. Brain Res. 1982;253:231–41.
29. Sharf R, Guarnieri DJ, Taylor JR, DiLeone RJ. Orexin mediates morphine place preference, but not morphine-induced hyperactivity or sensitization. Brain Res. 2010;1317:24–32.
30. Bardo MT, Bevins RA. Conditioned place preference: what does it add to our preclinical understanding of drug reward? Psychopharmacology (Berl). 2000;153(1):31–43.
31. Bardo MT, Valone JM, Bevins RA. Locomotion and conditioned place preference produced by acute intravenous amphetamine: role of dopamine receptors and individual differences in amphetamine self-administration. Psychopharmacology (Berl). 1999;143(1): 39–46.
32. Deroche V, Le Moal M, Piazza PV. Cocaine self-administration increases the incentive motivational properties of the drug in rats. Eur J Neurosci. 1999;11(8):2731–6.
33. Harris GC, Wimmer M, Aston-Jones G. A role for lateral hypothalamic orexin neurons in reward seeking. Nature. 2005;437(7058):556–9.
34. Harris GC, Wimmer M, Randall-Thompson JF, Aston-Jones G. Lateral hypothalamic orexin neurons are critically involved in learning to associate an environment with morphine reward. Behavioural Brain Research. 2007;183(1):43–51.
35. Hollander JA, Lu Q, Cameron MD, Kamenecka TM, Kenny PJ. Insular hypocretin transmission regulates nicotine reward. Proc Natl Acad Sci U S A. 2008; 105(49):19480–5.
36. Lawrence AJ, Cowen MS, Yang HJ, Chen F, Oldfield B. The orexin system regulates alcohol-seeking in rats. Br J Pharmacol. 2006;148(6):752–9.
37. Boutrel B, Kenny PJ, Specio SE, et al. Role for hypocretin in mediating stress-induced reinstatement of cocaine-seeking behavior. Proc Natl Acad Sci U S A. 2005;102(52):19168–73.
38. Aston-Jones G, Smith RJ, Moorman DE, Richardson KA. Role of lateral hypothalamic orexin neurons in reward processing and addiction. Neuropharmacology. 2009;56:112–21.
39. Wang B, You ZB, Wise RA. Reinstatement of cocaine seeking by hypocretin (orexin) in the ventral tegmental area: independence from the local corticotropin-releasing factor network. Biol Psychiatry. 2009; 65(10):857–62.
40. Harris GC, Aston-Jones G. Arousal and reward: a dichotomy in orexin function. Trends Neurosci. 2006;29(10):571–7.
41. Olds J, Milner P. Positive reinforcement produced by electrical stimulation of septal area and other regions of rat brain. J Comp Physiol Psychol. 1954;47(6): 419–27.
42. Hamlin AS, Clemens KJ, McNally GP. Renewal of extinguished cocaine-seeking. Neuroscience. 2008;151: 659–70.
43. Lin L, Faraco J, Li R, et al. The sleep disorder canine narcolepsy is caused by a mutation in the hypocretin (orexin) receptor 2 gene. Cell. 1999;98(3):365–76.
44. Chemelli RM, Willie JT, Sinton CM, et al. Narcolepsy in orexin knockout mice: molecular genetics of sleep regulation. Cell. 1999;98(4):437–51.
45. Hagan JJ, Leslie RA, Patel S, et al. Orexin A activates locus coeruleus cell firing and increases arousal in the rat. Proc Natl Acad Sci U S A. 1999;96(19): 10911–6.
46. Zheng H, Patterson LM, Berthoud HR. Orexin-A projections to the caudal medulla and orexin-induced c-Fos expression, food intake, and autonomic function. J Comp Neurol. 2005;485(2):127–42.
47. Pasumarthi R, Reznikov LR, Fadel J. Activation of orexin neurons by acute nicotine. Eur J Pharmacol. 2006;535(1–3):172–6.
48. Dayas CV, McGranahan TM, Martin-Fardon R, Weiss F. Stimuli linked to ethanol availability activate hypothalamic CART and orexin neurons in a reinstatement model of relape. Biol Psychiatry. 2008;63(2):152–7.
49. Pasumarthi RK, Fadel J. Activation of orexin/hypocretin projections to basal forebrain and paraventricular thalamus by acute nicotine. Brain Res Bull. 2008;77(6):367–73.

Emotional Processing in Narcolepsy

Sophie Schwartz

Keywords
Narcolepsy with cataplexy • Functional MRI • Hypocretin/orexin • Amygdala • Hypothalamus • Ventral tegmental area • Ventral medial prefrontal cortex • Sleep • Emotion • Conditioning • Reward • Addiction

Introduction

Human narcolepsy with cataplexy (NC) is a major sleep–wake disorder caused by a deficiency in hypocretins (Hcrt or orexins) [1–4]. Recent data suggest that the pathophysiology of NC may also involve emotional dysfunctions: emotions trigger cataplexy in NC patients; most anticataplectic medications are antidepressants; NC patients seem less likely to become addicted to stimulant medications; and depression and psychosocial dysfunctions are common in NC [5]. In this chapter, I review data from animal physiology and from human clinical and neuroimaging studies that converge to show a role of the Hcrt system in emotional functions. Specifically, recent data indicate that the Hcrt system may modulate activity in emotion- and reward-related brain regions (e.g., amygdala, medial prefrontal cortex, ventral midbrain, striatum), and that these functions of the Hcrt system may be impaired in NC patients. Human NC could thus be associated with abnormal brain responses to emotions caused by Hcrt deficiency.

This hypothesis is supported by three fMRI studies reported in the present chapter, in which we measured changes of regional brain activity in NC patients compared to healthy matched controls. The first study investigated brain responses to positive emotions using humorous pictures; the second study assessed changes in amygdala activity during aversive conditioning; the third study used a game-like paradigm to activate reward brain circuits. The results from these studies revealed functional abnormalities in the hypothalamus, amygdala, ventral tegmental area (VTA), striatum, and medial prefrontal cortex (mPFC) during affective processing in NC patients. These findings suggest that Hcrt activity modulates emotional states in addition to consolidating sleep–wake states [6]. These fMRI data also help bridge the gap between molecular/cellular studies in animals and clinical symptoms of NC patients.

S. Schwartz (✉)
Department of Neuroscience, University of Geneva, Michel-Servet 1, 1211 Geneva 4, Switzerland
and
Geneva Center for Affective Sciences, Geneva, Switzerland
and
Geneva Center for Neuroscience, Geneva, Switzerland
e-mail: sophie.schwartz@unige.ch

C.R. Baumann et al. (eds.), *Narcolepsy: Pathophysiology, Diagnosis, and Treatment*, DOI 10.1007/978-1-4419-8390-9_24, © Springer Science+Business Media, LLC 2011

Emotional Peculiarities in Narcolepsy

Cataplexy is a diagnostic feature of NC and is strongly associated with hypothalamic Hcrt depletion [7, 8]. Cataplexy attacks are most usually triggered by emotional experiences, mainly positive emotions such as joking, laughing, or feeling elated, as well as by the mere anticipation of emotions, for example, when playing games [5, 9–14]. Khatami et al. [15] also recently demonstrated that NC patients fail to exhibit amygdala-dependent startle potentiation during the presentation of unpleasant stimuli. These observations document abnormal responses to both positive and aversive signals in NC. Other emotional disturbances in NC include frequent depression and psychosocial dysfunction [5, 16]. Depression might be caused by Hcrt deficiency (e.g., via its impact on emotional regulation systems, see below), as well as by the major psychosocial handicap related to daytime sleepiness and cataplexy. Interestingly, most anti-cataplectic medications are antidepressants. Moreover, behavioral evidence also exists for reduced drug addiction in NC patients, as the patients are often treated with addictive amphetamine-like drugs but addiction seems uncommon [5, 14, 17].

Does cataplexy observed in animals also somehow relate to emotional or rewarding conditions? Like humans, dogs with narcolepsy due to a mutation of the Hcrt-2 receptor gene have cataplexy, fragmented sleep, and excessive daytime sleepiness [18]. In these animals, positive excitation, such as the presentation of food or playing with congeners may elicit cataplexy attacks. The presentation of appetitive food can be used as an experimental tool to trigger cataplexy episodes in narcoleptic dogs ("Food Elicited Cataplexy Test") [18]. Cataplexy-like episodes can also be observed in Hcrt knockout mice [19]. It was recently demonstrated that positive emotions or reward (e.g., wheel running, highly palatable food) trigger cataplexy in these knockout mice [20, 21]. Together, these observations in human NC, as well as in rodent and dog models of narcolepsy, suggest possible interactions between the hypothalamic Hcrt system and emotion and/or reward brain circuits.

Involvement of the Hypocretin System in Motivated Behaviors

Recent animal work suggests that Hcrt signaling may contribute to the regulation of emotional processes including affective responses to stress and neural plasticity related to addiction [22–24]. Studies in rodents provide evidence for both anatomical and functional links between the Hcrt system and the dopamine system, the latter being critically involved in reward processes and motivated behaviors [25]. First of all, hypothalamic Hcrt neurons project densely to reward-associated brain regions, including the nucleus accumbens (NAcc) and dopaminergic VTA [26]. Hcrt receptors are expressed at the surface of VTA dopamine neurons [27, 28] and HCRT administration increases the firing rate of VTA neurons [29]. Finally, Hcrt was found to be involved in drug-seeking behaviors and associated mesolimbic dopaminergic activity [22, 23, 28, 30–33]. These recent findings in rodents suggest that the Hcrt system influences dopamine activity in brain reward circuits and impacts the expression of motivated behaviors and addiction.

Below, I report three brain imaging studies aiming at a better understanding of the consequences of Hcrt deficiency on regional brain activity in NC patients (1) during the processing of positive emotions elicited by humorous stimuli, (2) during emotional learning assessed by aversive conditioning, as well as (3) during the anticipation and experience of rewards in a game-like task.

Neuroimaging of Emotional Processing in Narcolepsy with Cataplexy

General Experimental Procedures

In all three studies reported below, we scanned 14 drug-free NC patients with clear-cut cataplexy (based on clinical examination and standard questionnaires) and 14 healthy volunteers matched for age, gender, and body-mass index.

Hypocretin-1 level (<120 pg/ml) in the cerebrospinal fluid was determined in 10 of the 14 patients and was confirmed to be low or undetectable in all 10. Depression levels measured by the Beck Depression Inventory (BDI) did not differ between patients and matched controls. For more details about the patient population, see [34–36]. In each study, two to five subjects had to be excluded because of poor behavioral performance and/or because they fell asleep in the scanner; for each excluded subject (patient or control), we also excluded the corresponding matching subject.

Whole-brain event-related fMRI data were acquired on a Philips Intera 3.0-T whole-body system (Philips Medical Systems, Best, NL), with scanning parameters optimized to avoid signal loss in hypothalamic, amygdala, and ventral prefrontal regions. The functional MRI data were first analyzed at the individual level using the standard general linear approach in SPM (http://www.fil.ion.ucl.ac.uk/spm) to reveal brain regions in which fMRI activity was significantly modulated by the experimental conditions (e.g., visual pictures). Linear contrasts between conditions of interest from each subject were then submitted to group level analyses using ANOVAs. These analyses allow the identification of regions showing main effects of conditions in both groups for specific experimental conditions, as well as differences in activation between the groups.

Increased Amygdala Response to Humor

Because NC patients often have cataplexy attacks when they experience positive emotions, we hypothesized that the processing of emotional signal within limbic-affective and/or mesolimbic reward circuits may be altered in the patients. To test this hypothesis, we compared neural activity elicited by humorous and neutral pictures in NC patients and healthy volunteers. Our prediction was that NC patients would show abnormal brain responses in regions previously reported to be activated by humorous stimuli in normal controls, including the hypothalamus, amygdala, and ventral striatum [37–40]. Below, we briefly describe the experimental paradigm and the results from this first fMRI study (more details can be found in [34]).

NC patients and healthy volunteers were scanned while they watched humorous or neutral picture-sequences. These mini-sequences comprised a first picture that was always neutral followed by a second picture that revealed either a funny or a neutral (nonfunny) element (Fig. 1). On each trial, the participants judged whether they found the sequence funny or not by pressing a button.

For each participant, we first computed linear contrasts between the humor and neutral pictures, as classified by the participant during scanning, and then performed a group analysis (see above, General experimental procedures). This analysis allowed us to identify regions showing group differences in humor-specific activation.

Behavioral responses recorded during scanning showed that NC patients and their matched controls did not differ in the proportion of images judged as humorous. This suggests that there was no general alteration of humor appreciation in the patients. At the brain level, during humor processing, NC patients showed increased amygdala activity compared to controls, together with reduced medial prefrontal activity and a lack of hypothalamic response (compared to controls; Fig. 1).

The absence of hypothalamic activity in NC patients is consistent with hypothalamic dysfunction in these patients [7, 8]. Humor-selective increase of activity in regions that integrate emotion- and reward-related functions (amygdala, insula, and ventral striatum, including NAcc) suggests exaggerated brain reactivity to positive emotions. These results provide a neural basis for the patients' subjective reports and well-documented clinical observation that positive emotions often trigger cataplexy attacks in NC patients. Our results have lately been replicated by Reiss et al. [41] in another fMRI study of humor processing in NC patients. Note, however, that in this latter study, the hypothalamic response to humor was increased (rather than reduced) in the patients, while decreased hypothalamic activity was suggested in one patient during cataplexy.

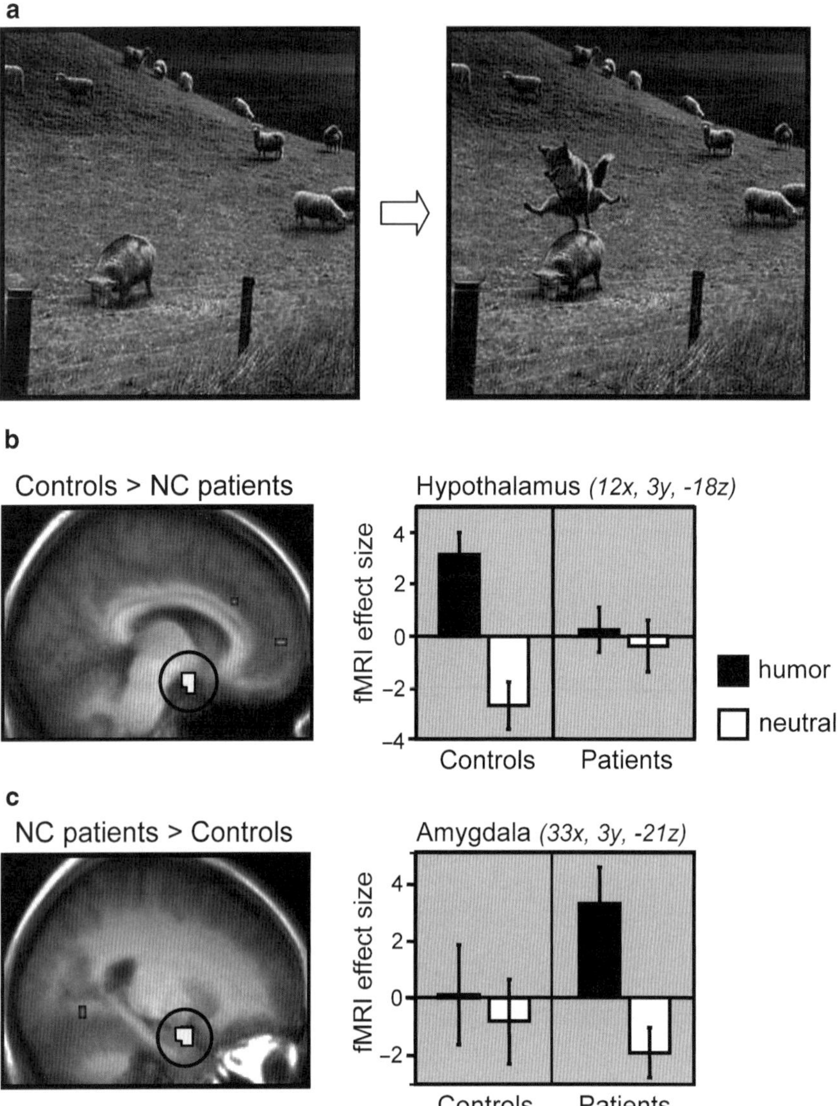

Fig. 1 (**a**) Mini-sequence of pictures judged as funny. On each trial, a neutral picture was presented (3 s), followed by a brief blank (300 ms) and by a second presentation of the picture with a new element that could be either neutral or humorous. Subjects made a humor judgment response after the offset of the second picture. (**b**) Right hypothalamus response to humor for controls but not for NC patients; signal change measured in the hypothalamus illustrates selective activation for humorous stimuli in the controls. (**c**) Increased amygdala response to humor in NC patients compared to controls; parameter estimates show increased fMRI signal to humorous sequences in the patients but not in the controls. Statistical maps are overlaid on the mean-normalized T1-structural scan; threshold at $P<0.001$

Moreover, we did not find any modification in humor judgment with our stimuli whereas NC patients in the Reiss et al. study rated significantly fewer humorous cartoons as funny compared to controls. Differences in the stimuli used may explain these partly diverging results: we used pictures whose funny quality could be comprehended almost instantly, unlike Reiss et al. who used cartoons with text that may load more on cognitive resources.

While connections from the amygdala to the hypothalamus are known to modulate reflex responses to emotional stimuli [42, 43], our fMRI results in NC patients suggest that the hypothalamus might also influence amygdala activity during positive emotions, possibly via direct projections from hypothalamic Hcrt neurons to the amygdaloid complex [27, 44–46]. Because the Hcrt system also sends projections to the VTA [26], these projections may also mediate increased prefrontal dopamine (DA) efflux [47], which in turn could suppress amygdala response [48, 49]. Thus, reduced hypothalamic activation and exaggerated amygdala response to humor could result from the loss or dysfunction of hypothalamic Hcrt neurons and/or from reduced inhibition of the amygdala by the mPFC (whose activity was also reduced in NC) leading to an abnormal release of amygdala response to positive emotions in narcolepsy.

Reduced Amygdala Activity During Aversive Conditioning

In a second fMRI experiment, we used an aversive conditioning paradigm to test whether NC patients show an abnormal amygdala response not only to positive stimuli, as in the humor experiment, but also to unpleasant stimuli [36]. Our second main goal was to assess the time-course of amygdala activity during conditioning.

We scanned unmedicated NC patients with clear-cut cataplexy and matched healthy controls (see General experimental procedures) while they underwent three successive sessions of conditioning (acquisition) followed by two sessions of extinction. The subjects were presented with triangles that could be colored either in blue or in yellow. One color (conditioned stimulus, CS+) signaled a possible upcoming aversive unconditioned stimulus (US), which was a brief painful electrical stimulation delivered on one finger on half of the CS+ trials during acquisition (i.e., partial reinforcement conditioning paradigm). The other color was never associated with an electrical stimulation (nonconditioned stimulus, CS–). During extinction, no CS+ was paired with the US anymore.

Analysis of functional MRI data revealed that activity of the right amygdala was strongly enhanced during the painful stimulation (US) in both controls and NC patients, while an increase of amygdala activity during the CS+ (not paired with the US) compared to the CS– was observed in the controls only (Fig. 2). This pattern of results in the patients demonstrates that the neural circuit associated with fear response could be activated by the US, but that there was no learning of the aversive properties of the visual CS+ [43, 50, 51].

A subsequent connectivity analysis revealed a negative functional coupling between the right amygdala and the medial prefrontal cortex in the controls during the acquisition of the conditioned response (CS+; Fig. 2c), in line with previous reports of an inverse functional relationship between mPFC and amygdala activity [48, 49]. The same analysis on the NC patients did not show any change in functional coupling between the amygdala and any other brain region. Therefore, narcolepsy with cataplexy may also be associated with altered regulatory processes involving connections between the mPFC and the amygdala. Together, our data suggest that both decreased activation of the medial prefrontal cortex (mPFC) and hypocretin deficiency may be involved in reduced amygdala activation during aversive conditioning in NC.

These results have important clinical and neurobiological implications. The amygdala dysfunction found in this study during aversive learning fits with the previous report of a blunted startle eyeblink response to unpleasant (aversive) stimuli, a finding that can be seen not only in human NC but also in animals and patients with amygdala lesions or dysfunctions [15]. Together with the observation of an exaggerated response during processing of humorous pictures [34] (see above), the current data suggest a central role of the amygdala in the pathophysiology of human NC. Importantly, because the patients had normal amygdala response to the CS+ when paired with an actual aversive stimulation, it is unlikely that reduced amygdala activity during the presentation of the CS+ alone

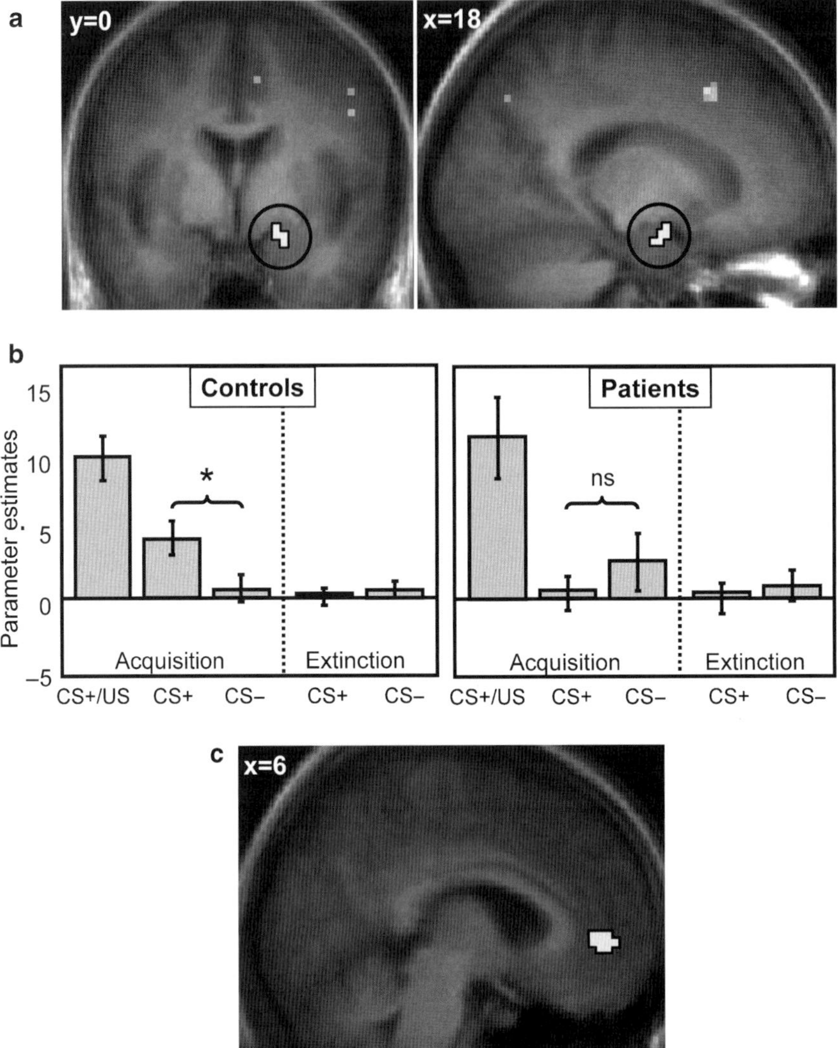

Fig. 2 (**a**) Increased amygdala activity in controls compared to NC patients, selectively during the presentation of the conditioned stimulus [contrast: CS+>CS–]. The statistical map is overlaid on coronal and sagittal sections of the mean-normalized T1-structural scan; threshold at $P<0.001$. (**b**) Mean parameter estimates of fMRI signal extracted from the amygdala peak ($x, y, z = 18, 0, -18$) for the main conditions during acquisition and extinction showing the expected increased activation level for the conditioned stimulus CS+ compared to the nonconditioned stimulus CS– during the acquisition in controls. By contrast, while the patients' amygdala showed a robust response to the US, there was no indication of a conditioned response to the CS+. (**c**) Increased functional coupling between the amygdala and the medial prefrontal cortex for the CS+ during the acquisition in controls

would be due to more efficient downregulation of negative emotions in the patients [52]. Because dopaminergic blockade can reduce the inhibitory influence of the PFC on the amygdala [53], changes in amygdala activity in the patients may also be due to an effect of Hcrt on the dopaminergic system [28, 29], which could also explain reduced medial PFC – amygdala connectivity. Together with recent animal data supporting a role of the Hcrt system in neural plasticity related to stressful or rewarding conditions [22, 23, 28–30], the lack of learning of a conditioned response observed here in NC patients strongly supports the involvement of the Hcrt system in emotional learning and regulation.

Abnormal Activity in Reward Brain Circuits

NC patients may be less prone to become addicted to amphetamine-like treatments (used against daytime sleepiness) [5, 14, 17]. Moreover, recent studies in rodents provide evidence for anatomical and functional links between the HCRT system and the dopaminergic reward system [26–29], and Hcrt was found to be involved in drug-seeking behaviors [22, 23, 28, 30–32]. The goal of our third fMRI study was to test whether Hcrt deficiency in NC patients may impact mesolimbic reward function, using a game-like task [35].

During scanning, NC patients and healthy matched controls performed an adapted version of a monetary incentive delay task, which is known to powerfully activate the mesolimbic and midbrain reward system [54]. In this task, the subjects could win (or lose) points if they rapidly pressed a key while a visual target was briefly shown (Fig. 3a). Each trial started with a preparation period, during which the subjects saw a cue indicating the potential gain (+1 or +5 points) or potential loss (−1 or −5 points) associated with that particular trial. After a variable delay, the visual target that required a rapid key press was briefly presented on the screen. The trial ended with a feedback display, telling the subjects whether they just won or lost the trial. To ensure a fixed proportion of successful trials (50–60%), a tracking algorithm adjusted the duration of each target presentation (i.e., task difficulty) based on each subject's current performance.

Behaviorally, NC patients had similar reaction times as controls during the game-like task. Moreover, both populations obtained similar final outcomes, ranging from 21.5 to 24.8 points (Mann–Whitney, $U=61$, $p=0.52$). In both NC patients and controls, high motivational cues (+5/−5 vs. +1/−1 points) activated a network of brain regions involved in the expectation of a reward, including the ventral striatum [25, 55]. Controls had increased fMRI signal in a region of the ventral midbrain, compatible with the VTA ($-3x$, $-24y$, $-21z$; Fig. 3b) [56, 57], while NC patients did not show such a modulation of activity by high cues. During winning on both positively and negatively cued trials (i.e., at the onset of the feedback display; Fig. 3a), activity in ventromedial prefrontal cortex (vmPFC) and NAcc increased during successful trials in the controls only. Both these regions receive dense dopaminergic projections from the VTA [47] and are known to be involved in the regulation of emotional processes and reward [56, 58]. As suggested by recent animal work [22, 23, 28, 30], Hcrt depletion may affect dopaminergic modulation within brain reward networks, which would also be consistent with the observation that the midbrain/VTA remained unresponsive to highly motivating cues in NC patients. Critically, response to large cues in the NAcc and the vmPFC correlated positively with disease duration, suggesting that some adaptive mechanisms in patients who had long been suffering from narcolepsy could restore neural activity in these regions (in the absence of associated VTA activation). This result is consistent with the patients' reports of increased control over cataplexy attacks with time [59]. Finally, on trials associated with the most positive emotions on this task, namely when winning on rewarded trials (i.e., actual gains on positively cued trials), NC patients showed a strong increase in amygdala/SLEA activity, which is in line with our first fMRI study with humorous stimuli described above [34].

These fMRI results provide the first evidence for abnormal brain responses to reward in NC patients, and thus suggest an implication of Hcrt activity in the regulation of brain reward function in humans. By providing a whole-brain description of reward-related activity in humans, these findings significantly extend existing results from animal cellular neurophysiology and suggest that the Hcrt system could be a useful target for future treatments of addiction.

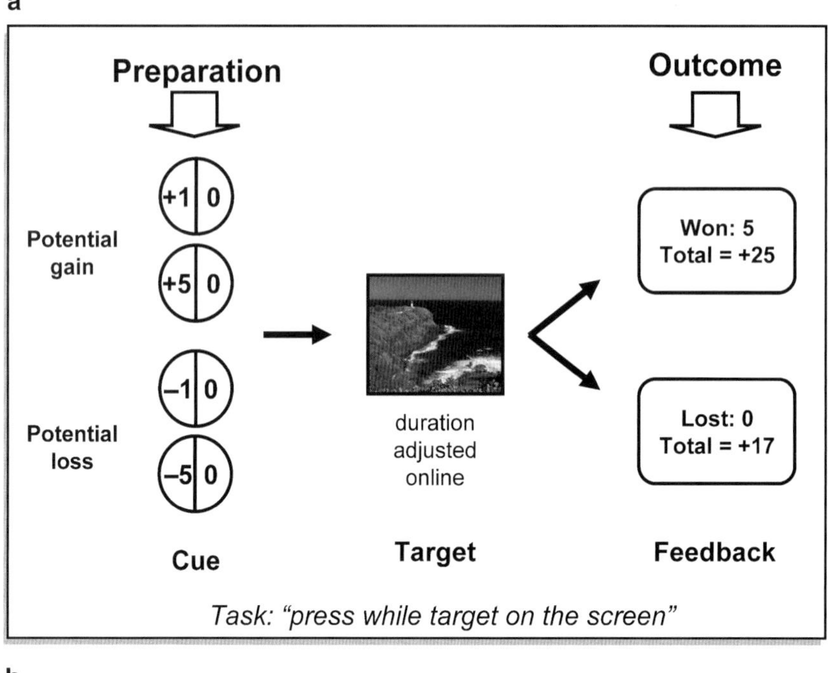

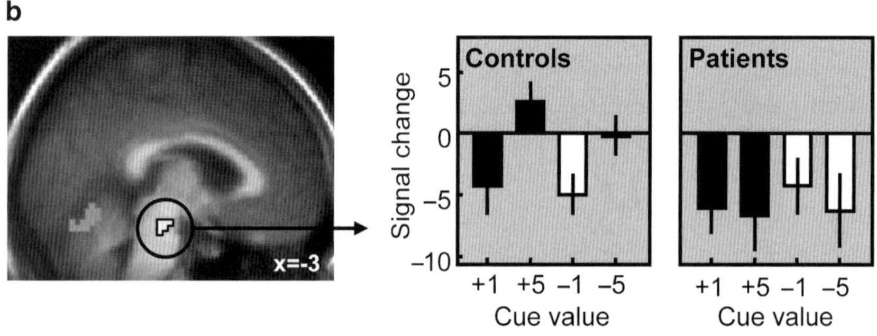

Fig. 3 (**a**) The game-like task. Each trial started with a cue indicating how many points were to be played on that particular trial. Participants had to press a button during the brief presentation of a target picture. To obtain a balanced proportion of won and lost trials in each condition, the duration of the target was automatically adjusted online. The trial ended with a panel showing the obtained reward. (**b**) Increased response to large vs. small cue values in ventral midbrain (including VTA) in controls compared to NC patients. The statistical map is overlaid on the mean-normalized T1-structural scan; threshold $p<0.001$. Parameter estimates extracted from the peak of this region (–3*x*, –24*y*, –21*z*) illustrate increased activation for highest incentives (large cue values: +5 and –5) in controls but not in NC patients

Conclusions

Narcolepsy with cataplexy is a major sleep–wake disorder caused by a deficiency in hypothalamic Hcrt. The recent animal physiology and human neuroimaging data reviewed in this chapter indicate that emotional disturbances emerge as another main facet of NC. Across three functional MRI studies conducted on NC patients and healthy controls, our research shows that NC is associated with a disruption of the physiological modulation of affective and reward processing. A plausible neurobiological interpretation of our findings is that the Hcrt system, through direct or indirect (via the VTA and the mPFC) effects, influences key components of emotion networks (such as the amygdala or the VTA) to afford rapid and efficient adaptation to emotional challenges, and enhances neural plasticity related to emotional learning, reward processing, and addiction. Clinically, these data indicate that human NC (and possibly also other disorders associated with hypocretin deficiency [60]) should be considered not only as a sleep–wake disorder but also as a condition which can impair emotion regulation. Finally, our fMRI studies on narcolepsy with cataplexy illustrate the usefulness of modern brain imaging methods to investigate brain functions in patients with complex neurological disorders.

Acknowledgments The author thanks her collaborators on the fMRI studies reported here for their inestimable scientific, clinical, and technical contributions: Claudio L. Bassetti, Peter Boesiger, Ramin Khatami, Aurélie Ponz, Rositsa Poryazova, and Esther Werth. This work was supported the Swiss National Science Foundation (grants # 3200B0-104100, #320030-118272, #3100A0-102133), by the National Centre of Competence in Research (NCCR) Affective sciences financed by the Swiss National Science Foundation (# 51NF40-104897), and the Geneva Center for Neurosciences.

References

1. de Lecea L, Kilduff TS, Peyron C, et al. The hypocretins: hypothalamus-specific peptides with neuroexcitatory activity. Proc Natl Acad Sci USA. 1998;95(1): 322–7.
2. Peyron C, Faraco J, Rogers W, et al. A mutation in a case of early onset narcolepsy and a generalized absence of hypocretin peptides in human narcoleptic brains. Nat Med. 2000;6(9):991–7.
3. Sakurai T, Amemiya A, Ishii M, et al. Orexins and orexin receptors: a family of hypothalamic neuropeptides and G protein-coupled receptors that regulate feeding behavior. Cell. 1998;92(4):573–85.
4. Thannickal TC, Moore RY, Nienhuis R, et al. Reduced number of hypocretin neurons in human narcolepsy. Neuron. 2000;27(3):469–74.
5. Bassetti C, Aldrich MS. Narcolepsy. Neurol Clin. 1996;14(3):545–71.
6. Saper CB, Scammell TE, Lu J. Hypothalamic regulation of sleep and circadian rhythms. Nature. 2005; 437(7063):1257–63.
7. Mignot E, Lammers GJ, Ripley B, et al. The role of cerebrospinal fluid hypocretin measurement in the diagnosis of narcolepsy and other hypersomnias. Arch Neurol. 2002;59(10):1553–62.
8. Baumann CR, Bassetti CL. Hypocretins (orexins) and sleep-wake disorders. Lancet Neurol. 2005;4(10): 673–82.
9. Mattarozzi K, Bellucci C, Campi C, et al. Clinical, behavioural and polysomnographic correlates of cataplexy in patients with narcolepsy/cataplexy. Sleep Med. 2008;9(4):425–33.
10. Sturzenegger C, Bassetti CL. The clinical spectrum of narcolepsy with cataplexy: a reappraisal. J Sleep Res. 2004;13(4):395–406.
11. Anic-Labat S, Guilleminault C, Kraemer HC, Meehan J, Arrigoni J, Mignot E. Validation of a cataplexy questionnaire in 983 sleep-disorders patients. Sleep. 1999;22(1):77–87.
12. Mignot E, Hayduk R, Black J, Grumet FC, Guilleminault C. HLA DQB1*0602 is associated with cataplexy in 509 narcoleptic patients. Sleep. 1997; 20(11):1012–20.
13. Overeem S, Mignot E, van Dijk JG, Lammers GJ. Narcolepsy: clinical features, new pathophysiologic insights, and future perspectives. J Clin Neurophysiol. 2001;18(2):78–105.
14. Guilleminault C, Carskadon M, Dement WC. On the treatment of rapid eye movement narcolepsy. Arch Neurol. 1974;30(1):90–3.
15. Khatami R, Birkmann S, Bassetti CL. Amygdala dysfunction in narcolepsy-cataplexy. J Sleep Res. 2007; 16(2):226–9.
16. Bassetti C. The spectrum of narcolepsy. In: Bassetti C, Billiard M, Mignot E, editors. Narcolepsy and hypersomnia. New York: Informa Healthcare; 2007. p. 97–108.
17. Nishino S, Mignot E. Pharmacological aspects of human and canine narcolepsy. Prog Neurobiol. 1997; 52(1):27–78.
18. Nishino S, Tafti M, Sampathkumaran R, Dement WC, Mignot E. Circadian distribution of rest/activity in narcoleptic and control dogs: assessment with ambulatory activity monitoring. J Sleep Res. 1997;6(2):120–7.
19. Chemelli RM, Willie JT, Sinton CM, et al. Narcolepsy in orexin knockout mice: molecular genetics of sleep regulation. Cell. 1999;98(4):437–51.
20. Clark EL, Baumann CR, Cano G, Scammell TE, Mochizuki T. Feeding-elicited cataplexy in orexin knockout mice. Neuroscience. 2009;161(4):970–7.
21. Espana RA, McCormack SL, Mochizuki T, Scammell TE. Running promotes wakefulness and increases cataplexy in orexin knockout mice. Sleep. 2007;30(11):1417–25.
22. Harris GC, Wimmer M, Aston-Jones G. A role for lateral hypothalamic orexin neurons in reward seeking. Nature. 2005;437(7058):556–9.
23. Boutrel B, Kenny PJ, Specio SE, et al. Role for hypocretin in mediating stress-induced reinstatement of cocaine-seeking behavior. Proc Natl Acad Sci USA. 2005;102(52):19168–73.
24. Lutter M, Sakata I, Osborne-Lawrence S, et al. The orexigenic hormone ghrelin defends against depressive symptoms of chronic stress. Nat Neurosci. 2008; 11(7):752–3.
25. Schultz W. Multiple reward signals in the brain. Nat Rev Neurosci. 2000;1(3):199–207.
26. Fadel J, Deutch AY. Anatomical substrates of orexin-dopamine interactions: lateral hypothalamic projections to the ventral tegmental area. Neuroscience. 2002;111(2):379–87.

27. Marcus JN, Aschkenasi CJ, Lee CE, et al. Differential expression of orexin receptors 1 and 2 in the rat brain. J Comp Neurol. 2001;435(1):6–25.
28. Narita M, Nagumo Y, Hashimoto S, et al. Direct involvement of orexinergic systems in the activation of the mesolimbic dopamine pathway and related behaviors induced by morphine. J Neurosci. 2006; 26(2):398–405.
29. Korotkova TM, Sergeeva OA, Eriksson KS, Haas HL, Brown RE. Excitation of ventral tegmental area dopaminergic and nondopaminergic neurons by orexins/hypocretins. J Neurosci. 2003;23(1):7–11.
30. Borgland SL, Taha SA, Sarti F, Fields HL, Bonci A. Orexin A in the VTA is critical for the induction of synaptic plasticity and behavioral sensitization to cocaine. Neuron. 2006;49(4):589–601.
31. Georgescu D, Zachariou V, Barrot M, et al. Involvement of the lateral hypothalamic peptide orexin in morphine dependence and withdrawal. J Neurosci. 2003;23(8):3106–11.
32. Borgland SL, Chang SJ, Bowers MS, et al. Orexin A/hypocretin-1 selectively promotes motivation for positive reinforcers. J Neurosci. 2009;29(36):11215–25.
33. Borgland SL, Storm E, Bonci A. Orexin B/hypocretin 2 increases glutamatergic transmission to ventral tegmental area neurons. Eur J Neurosci. 2008;28(8): 1545–56.
34. Schwartz S, Ponz A, Poryazova R, et al. Abnormal activity in hypothalamus and amygdala during humour processing in human narcolepsy with cataplexy. Brain. 2008;131(Pt 2):514–22.
35. Ponz A, Khatami R, Poryazova R, et al. Abnormal activity in reward brain circuits in human narcolepsy with cataplexy. Ann Neurol. 2010;67(2):190–200.
36. Ponz A, Khatami R, Poryazova R, et al. Reduced amygdala activity during aversive conditioning in human narcolepsy. Ann Neurol. 2010;67(3):394–8.
37. Wild B, Rodden FA, Grodd W, Ruch W. Neural correlates of laughter and humour. Brain. 2003;126 (Pt 10):2121–38.
38. Mobbs D, Greicius MD, Abdel-Azim E, Menon V, Reiss AL. Humor modulates the mesolimbic reward centers. Neuron. 2003;40(5):1041–8.
39. Moran JM, Wig GS, Adams Jr RB, Janata P, Kelley WM. Neural correlates of humor detection and appreciation. Neuroimage. 2004;21(3):1055–60.
40. Watson KK, Matthews BJ, Allman JM. Brain activation during sight gags and language-dependent humor. Cereb Cortex. 2007;17(2):314–24.
41. Reiss AL, Hoeft F, Tenforde AS, Chen W, Mobbs D, Mignot EJ. Anomalous hypothalamic responses to humor in cataplexy. PLoS One. 2008;3(5):e2225.
42. Price JL. Free will versus survival: brain systems that underlie intrinsic constraints on behavior. J Comp Neurol. 2005;493(1):132–9.
43. LeDoux JE. Emotion circuits in the brain. Annu Rev Neurosci. 2000;23:155–84.
44. Bisetti A, Cvetkovic V, Serafin M, et al. Excitatory action of hypocretin/orexin on neurons of the central medial amygdala. Neuroscience. 2006;142(4):999–1004.
45. Date Y, Ueta Y, Yamashita H, et al. Orexins, orexigenic hypothalamic peptides, interact with autonomic, neuroendocrine and neuroregulatory systems. Proc Natl Acad Sci USA. 1999;96(2):748–53.
46. Peyron C, Tighe DK, van den Pol AN, et al. Neurons containing hypocretin (orexin) project to multiple neuronal systems. J Neurosci. 1998;18(23):9996–10015.
47. Vittoz NM, Berridge CW. Hypocretin/orexin selectively increases dopamine efflux within the prefrontal cortex: involvement of the ventral tegmental area. Neuropsychopharmacology. 2006;31(2):384–95.
48. Maren S, Quirk GJ. Neuronal signalling of fear memory. Nat Rev Neurosci. 2004;5(11):844–52.
49. Phelps EA, Delgado MR, Nearing KI, LeDoux JE. Extinction learning in humans: role of the amygdala and vmPFC. Neuron. 2004;43(6):897–905.
50. LaBar KS, Gatenby JC, Gore JC, LeDoux JE, Phelps EA. Human amygdala activation during conditioned fear acquisition and extinction: a mixed-trial fMRI study. Neuron. 1998;20(5):937–45.
51. Williams LM, Phillips ML, Brammer MJ, et al. Arousal dissociates amygdala and hippocampal fear responses: evidence from simultaneous fMRI and skin conductance recording. Neuroimage. 2001;14(5): 1070–9.
52. Ochsner KN, Ray RD, Cooper JC, et al. For better or for worse: neural systems supporting the cognitive down- and up-regulation of negative emotion. Neuroimage. 2004;23(2):483–99.
53. Marowsky A, Yanagawa Y, Obata K, Vogt KE. A specialized subclass of interneurons mediates dopaminergic facilitation of amygdala function. Neuron. 2005;48(6):1025–37.
54. Knutson B, Adams CM, Fong GW, Hommer D. Anticipation of increasing monetary reward selectively recruits nucleus accumbens. J Neurosci. 2001;21(16): RC159.
55. Knutson B, Taylor J, Kaufman M, Peterson R, Glover G. Distributed neural representation of expected value. J Neurosci. 2005;25(19):4806–12.
56. O'Doherty JP, Buchanan TW, Seymour B, Dolan RJ. Predictive neural coding of reward preference involves dissociable responses in human ventral midbrain and ventral striatum. Neuron. 2006;49(1):157–66.
57. D'Ardenne K, McClure SM, Nystrom LE, Cohen JD. BOLD responses reflecting dopaminergic signals in the human ventral tegmental area. Science. 2008; 319(5867):1264–7.
58. Wager TD, Davidson ML, Hughes BL, Lindquist MA, Ochsner KN. Prefrontal-subcortical pathways mediating successful emotion regulation. Neuron. 2008; 59(6):1037–50.
59. Passouant P, Billard M. The evolution of narcolepsy with age. In: Guilleminault C, Dement WC, Passouant P, editors. Narcolepsy. New York: Spectrum; 1976. p. 179–96.
60. Fronczek R, Baumann CR, Lammers GJ, Bassetti CL, Overeem S. Hypocretin/orexin disturbances in neurological disorders. Sleep Med Rev. 2009;13(1): 9–22.

Depression in Narcolepsy

Michael Lutter

Keywords
Depression • Narcolepsy • Cognition • Orexin • Hypocretin • Affective regulation

Introduction

Major Depressive Disorder (MDD) is thought to develop from a complex combination of environmental stressors superimposed on to a vulnerable neural substrate [1]. Many diseases increase the risk of developing MDD, either from the psychosocial stress of the illness or because the illness causes direct dysfunction of the nervous system. Narcolepsy is one such illness in which the disease can cause both significant life stress and direct neuronal dysregulation. In the past decade significant advances have been made in understanding the neuronal basis of narcolepsy since the recognition that loss of orexin neurons (also known as hypocretin neurons) plays an important role in the development of certain forms of the disorder [2, 3]. Mouse models of narcolepsy now allow us to specifically test the role of orexin signaling in the regulation of mood and affect. Reduced levels of orexin peptide have been reported in human patients with MDD and in rodent models of chronic stress further strengthening the connection between orexin and depression. In this chapter, we will review the clinical findings on depression in narcolepsy, discuss the neurobiology of orexin signaling, and speculate on the mechanisms that mediate this association.

M. Lutter (✉)
Department of Psychiatry, University of Texas Southwestern Medical Center, Dallas, TX 75390-9127, USA
e-mail: Michael.Lutter@UTSouthwestern.edu

Clinical Findings

Common Psychiatric Comorbidities of Narcolepsy

Clinicians have long noted the association between narcolepsy and psychiatric illness. In a case review of male veterans with a clinical diagnosis of narcolepsy, Krishnan and colleagues reported that 16 out of 24 patients interviewed had a psychiatric diagnosis by DSM-III criteria. The most common diagnoses in this group were Adjustment Disorder, Major Depression, and Alcohol Dependence. The authors also noted that narcolepsy with cataplexy was associated with more severe psychiatric disturbances [4]. Kales and colleagues conducted clinical interviews with 50 patients with narcolepsy and cataplexy as well as controls. They reported increased rates of psychopathology in patients with narcolepsy that

C.R. Baumann et al. (eds.), *Narcolepsy: Pathophysiology, Diagnosis, and Treatment*,
DOI 10.1007/978-1-4419-8390-9_25,

they attributed to the severe psychosocial impairment associated with the illness [5]. Overall rates of comorbid depression in narcolepsy have been reported to occur in between 20 and 37% of patients [4–8].

It should be noted that some of these studies were conducted before modern diagnostic methods for detecting narcolepsy and depression were routinely employed in research studies. In a 2002 study, Vourdas and colleagues found no significant differences in rates of depression using DSM-IV criteria in 45 patients diagnosed with narcolepsy by International Classification of Sleep Disorders Criteria [9]. This discrepancy may be due to differences in methodology or may reflect better treatment outcomes related to improved understanding of the illness and use of nonstimulant medications.

Interestingly, only a handful of Case Studies have reported the occurrence of narcolepsy with other psychiatric diagnoses related to depressed mood including Bipolar Disorder, Posttraumatic Stress Disorder, and Substance Abuse. This finding is particularly surprising given the fact that two medications with high abuse potential, psychostimulants and gamma-hydroxybutyrate, are commonly used in the treatment of narcolepsy.

Psychosocial Consequences of Narcolepsy and Quality of Life

Multiple studies have consistently found that narcolepsy is a chronic and severely debilitating condition that dramatically affects quality of life in multiple domains of functioning. In the following section, we will discuss functional impairments in patients with narcolepsy including work, social, and marital difficulties. Unique aspects affecting the quality of life in pediatric population will also be addressed. Finally, though successful treatment with pharmacotherapy can improve the symptoms of the illness, we will discuss some of the functional consequences of treatment.

Patients with narcolepsy often face difficulties in the work place due to their symptoms. Several symptoms, including excessive daytime sleepiness and recurrent naps, impair concentration and productivity. In the absence of appropriate diagnosis and psychoeducation, these workers may be labeled as lazy or unmotivated. Furthermore, the sudden onset of sleep and loss of muscle tone creates safety hazards for workers in dangerous situations. Impairments in the ability to operate a car may also limit the range of jobs available to them [10–12].

Narcolepsy impacts interpersonal relationships in numerous ways. Similar to workplace considerations, the symptoms of narcolepsy may be wrongly attributed to a partner being accident prone or lazy. In addition to the personal embarrassment of the patient from specific behaviors, narcolepsy is also associated with a number of physical problems, such as the likelihood of being overweight and male patients can suffer from impotence, thus further impacting self-esteem. Finally, social isolation and avoidance of public places are frequent coping strategies employed by patients with narcolepsy, and though these methods may limit the embarrassment associated with the symptoms, they can further worsen depressive symptoms [10–12].

Several special considerations must be made for the treatment of pediatric populations. In a sample of 42 pediatric patients, Stores and colleagues found significantly higher rates of depressive symptoms, difficulties in school, and behavioral problems in children with narcolepsy compared to control subjects [13]. In addition to these issues, children may be more sensitive to self-image and peer group concerns given their developmental stage. In this context, treatment planning for children with narcolepsy should include a comprehensive plan encompassing all aspects of the child's life, including family, peer group, school, and extra-curricular activities.

Finally, treatment providers should not discontinue evaluation and management of psychosocial issues after effective pharmacotherapy has been achieved. Medications used for the treatment of narcolepsy will often have side effects that affect the quality of life including irritability and loss of appetite. Additionally, a period of psychosocial adjustment typically follows successful treatment of any chronic illness. Wilson and colleagues conducted an excellent psychosocial evaluation of 33 patients with narcolepsy following successful treatment of symptoms. They found several common problems in these patients including concerns about increased expectations at home and work, the need to make up for lost time, and frustration over delays in diagnosis and treatment [14].

Studies of Cerebrospinal Fluid

Measuring levels of neurotransmitters in human patients has always been difficult given the invasive nature of sampling methods. Despite the inability to collect localized measurements, some information can still be obtained from analysis of cerebrospinal fluid (CSF). Several studies have measured levels of orexin peptide in the CSF of patients with various psychiatric disorders. While the information these studies provide is somewhat nonspecific, the findings are nonetheless very compelling at times.

Salomon and colleagues reported the first study to measure CSF levels of orexin in depressed and control subjects. Using indwelling catheters, this group measured CSF levels continuously for 24 h in 15 depressed and 14 control subjects. Additionally, CSF levels were determined a second time in the depressed group after treatment with antidepressant medication. In control subjects, orexin-A levels displayed a circadian rhythm with a maximum peak in the middle of the night between 0100 and 0200 h [15]. Overall, the mean concentration of orexin-A did not differ significantly between the control and depressed subjects; however, depressed patients did demonstrate a significant blunting in the diurnal variation of orexin-A CSF samples with an overall reduction in the difference between the amplitude of the peak and trough values. Patients who responded to antidepressant medication did show some recovery in the diurnal cycle of orexin-A levels but did not return back to the level of control subjects.

Brundin and colleagues reported orexin levels in 66 patients with mood disorders shortly after a suicide attempt [16]. Compared to dysthymia and adjustment disorder, those patients with MDD who had recently attempted suicide had significantly lower levels of orexin-A peptide in their CSF. Furthermore a significant correlation existed between orexin-A and levels of other neuropeptides, including delta sleep inducing peptide-like immunoreactivity (DSIP-LI), somatostatin and corticotrophin-releasing factor (CRF). This finding suggests that disruption in orexin signaling along with deficiency of other hypothalamic neuropeptides may contribute to impairment in sleep and appetite observed in patients with MDD.

In a separate study, Brundin and colleagues also conducted a subanalysis of specific psychiatric symptoms in a group of 101 patients who had recently attempted suicide. Using the Comprehensive Psychopathology Rating Scale, they reported a significant negative correlation between orexin-A levels and three psychiatric symptoms including lassitude (difficulty initiating movement), slowness of movement, and overall severity of illness [17]. These results suggest that lower levels of orexin-A may directly contribute to the psychomotor retardation often observed in patients with MDD. Importantly, a follow-up study in ten patients who attempted suicide measured CSF levels of orexin-A at 6 and 12 months after the attempt, finding that orexin-A levels were significantly elevated at both time points [18]. Furthermore, there was a significant negative correlation between orexin-A levels in the CSF and scores on the Suicide Assessment Scale. These findings indicate a correlation between improvement in suicidality and levels of orexin-A in the CSF.

Finally, Pedrazzoli and colleagues measured levels of orexin-A in rats subjected to REM sleep deprivation. REM sleep deprivation is an interesting state characterized by hyperphagia as well as an antidepressant-like improvement in mood. Following 96 h of REM sleep deprivation, rats displayed significantly elevated levels of orexin-A in the CSF [19]. Orexin-A levels returned to normal following a period of REM rebound sleep when the antidepressant-like response is lost. These findings are consistent with research in human patients in which overnight sleep deprivation is used as a short-term treatment for depression [20].

Preclinical Findings

While substantial clinical evidence demonstrates higher rates of depression in patients with narcolepsy, the mechanism of this association is not clear. As we covered earlier, patients with narcolepsy have tremendous psychosocial difficulties that increase the stress of daily living. However, in accord with the stress-diathesis model of MDD, the possibility also exists that the loss of orexin neurons in the lateral hypothalamus may increase the underlying vulnerability to the development of depressive symptoms. Unfortunately, it is very difficult to distinguish the relative contributions of psychosocial stress and neuronal susceptibility in human populations. In this regard, animal models of narcolepsy and depression have provided great insight into the role of orexin signaling in depressive illness. As noted earlier, MDD affects multiple domains of brain function including affective, cognitive, and homeostatic systems. In this next section, we will review the current research on orexin signaling in these areas.

Affective Regulation

Animal models of depression have provided some insight into the role of orexin signaling in the regulation of mood. Most of the animal studies conducted so far ask two distinct questions: (1) Is orexin signaling altered in rodent models of depression? (2) Are orexin-null mice, an animal model of narcolepsy, at increased risk of developing depressive symptoms? In this section, we will review the research in these two areas and address these questions.

Allard and colleagues conducted a series of experiments using the Wistar-Kyoto (WKY) line of rats. WKY rats are a genetically bred line that displays many features consistent the diagnosis of MDD, including increased behavioral despair, increased anxiety and disrupted sleep similar to patterns observed in patients with MDD including sleep fragmentation and decrease latency to REM [21]. Because of the known role of orexin in sleep regulation, the authors speculated that orexin signaling might be disrupted in the WKY rat line. Consistent with their hypothesis, they found fewer orexin-positive neurons in the lateral hypothalamus of WKY rats. Furthermore, the average size of the neurons present was ~15% smaller than the size of neurons in nondepressed control rats. In a second study, the authors studied the effect of REM sleep deprivation on the WKY line and reported that while REM sleep deprivation did increase the number of orexin-positive neurons in the lateral hypothalamus, a control intervention also increased the quantity of orexin neurons [22]. Therefore, it is unclear if orexin signaling is required for the antidepressant effect of REM sleep deprivation in WKY rats.

In a separate series of studies, Lutter and colleagues investigated orexin signaling in the chronic social defeat model of stress. This model utilizes repeated episodes of social subordination to induce behavioral deficits in mice similar to symptoms observed in patients with MDD including social isolation and decreased preference for a sucrose-containing solution. Chronic social defeat stress is considered a model of MDD because chronic, but not acute treatment with antidepressants will reverse the behavioral deficits. Following exposure to social stress, defeated mice have lower expression of prepro-orexin mRNA and fewer orexin-positive neurons in the lateral hypothalamus than control mice [23]. This reduction in orexin expression was associated with increased levels of dimethyl-lysine 9 modification of histone 3 on the orexin promoter and, because the dimethyl-lysine 9 modification is associated with transcriptional repression, this finding suggests that chronic stress may suppress orexin expression via an epigenetic mechanism.

Recently, Davis and colleagues analyzed dominance behavior in the visible burrow system. The visible burrow system is a model of chronic social stress in which male rats establish dominance hierarchies in a social environment. They reported that rats that went on to become dominant displayed significantly lower levels of anxiety in the elevated plus maze prior to entry in the visible burrow system [24]. Furthermore, dominant rats had higher levels of orexin receptor 1 expression in the medial prefrontal cortex, an area of the brain involved in executive functioning, than control or subordinate rats. These data suggest that increased orexin 1 receptor signaling in the medial prefrontal cortex may be protective against the effects of social hierarchy stress.

Taken together, these studies suggest that decreased orexin signaling may contribute to depressive symptoms induced by chronic stress. A natural extension of this hypothesis is that orexin-null mice, a model of narcolepsy, would be more sensitive to the effects of chronic stress. Two studies by Lutter and colleagues attempted to address this question. In the first study, they compared the effect of calorie restriction on depressive behaviors in wild-type and orexin-null mice. Both wild-type and orexin-null mice given free access to chow displayed no difference in the forced swim test, a rodent model of behavioral despair. However, when placed on a calorie-restricted diet representing 60% of ad lib fed calories, a significant reduction in behavioral despair was observed in wild-type but not orexin-null mice [23]. Additionally, wild-type, but not orexin-null, mice subjected to social defeat also show improvement in social isolation after calorie restriction.

In a second study, Lutter and colleagues identified a novel antidepressant role for the stomach-derived hormone ghrelin. Ghrelin is released from the stomach in response to a state of negative energy balance and coordinates several behavioral adaptations to reduced calorie availability including increased appetite and decreased metabolic rate [25]. The authors demonstrated that ghrelin signaling reduces behavioral despair and social isolation in the forced swim test and chronic social defeat stress models of depression, respectively [26]. Interestingly, the effect of ghrelin in the forced swim test was dependent upon the presence of orexin neurons, consistent with previous reports that demonstrate a direct effect of ghrelin on orexin neurons [27]. These findings suggest that orexin neurons may integrate several signals of peripheral energy balance and coordinate behavioral adaptations to reduced calorie availability including arousal [27], locomotion [28], and affective [23] regulation.

Ito and colleagues confirmed the importance of orexin signaling in the affective regulation. They reported that intracerebroventricular (icv) infusion of orexin-A peptide reduced behavioral despair in the forced swim test [29]. Furthermore, orexin-A administration was associated with an increase in proliferation of neurons in the dentate gyrus, an effect thought to contribute to its anti-depressant-like actions.

Finally, in a study looking at the role of orexin signaling in the reinstatement of cocaine seeking behaviors, Boutrel and colleagues reported that icv administration of orexin-A increased the intracranial self stimulation (ICSS) threshold [30]. ICSS threshold evaluates the function of the reward system circuitry and is considered a measure of anhedonia. An increased threshold would be consistent with negative regulation of the reward circuit and induction of anhedonia. This study suggests that icv administration of orexin-A may produce a stress-like state. Given the complexity of the orexin system with two ligands and two receptors with different neuroanatomical expression, it is likely that distinct signaling pathways may regulate both positive and negative affective states.

Regulation of Cognition

Cognitive disruptions including impairment in concentration and feelings of guilt and worthlessness are among the most impairing features of MDD. Animal models of cognition include learning and memory tasks such as the Morris water maze (MWM) and passive avoidance tasks, which measure the ability of an animal to acquire, consolidate and retrieve memories. A growing body of literature has recognized the importance of the hippocampus in learning and memory. Given the expression of orexin receptor 1 in the hippocampus, several groups have studied the effect of orexin signaling on learning and memory.

Aou and colleagues studied the effect of orexin-A peptide in young adult male rats in the MWM. They reported that icv administration of orexin-A impaired both spatial learning and spatial memory after training in the water maze [31]. Furthermore, electrophysiological studies demonstrated that orexin-A signaling suppressed long-term potentiation in the Schaffer collateral-CA1 neurons of the hippocampus. In a separate series of studies, Akbari and colleagues studied similar questions using the orexin receptor 1-selective antagonist SB-334867 in adult rats. They found that infusion of SB-334867 into the CA1 region of the hippocampus impaired acquisition, consolidation, and retrieval of memories in the MWM [32]. In a second study, this group also reported that SB-334867 infusion into the dentate gyrus inhibited acquisition and consolidation of spatial memories in the MWM [33]. Perhaps one explanation for this apparent discrepancy in results is the age of the animals used. Selbach and colleagues analyzed the effect of orexin-A signaling on Schaffer collateral-CA1 synapses in hippocampal slices from mice of different ages. Orexin-A application caused long-term potentiation in synapses from adult mice, while it caused long-term depression in hippocampal slices from 3- to 4-week-old juvenile mice [34].

In addition to regulating spatial memory, several studies have also looked at orexin signaling in passive avoidance learning. Telegdy and Adamik determined the effect of icv orexin-A injection on a one-way passive avoidance task [35]. They found that orexin-A administration increased learning, consolidation, and retrieval of passive avoidance memories in a dose-dependent manner. Jaeger and colleagues analyzed orexin-A signaling on memory retention in the T-maze foot shock avoidance and one trial step down passive avoidance tasks [36]. Orexin-A signaling improved memory retention in both tasks consistent with previous findings. Finally, Akbari and colleagues determined the effect of orexin receptor-1 blockade in a foot shock passive avoidance task [37]. They reported that infusion of SB-334867 into the CA1 region of the hippocampus impaired retrieval of passive avoidance memories while infusion into the dentate gyrus inhibited acquisition and consolidation but not retrieval of passive avoidance memories. Together these studies suggest a role for orexin signaling in the regulation of learning and memory, supporting the theory that disruption of orexin signaling in either narcolepsy or MDD may have deleterious effects on cognitive function.

Regulation of Homeostatic Function

Homeostatic functions help an animal maintain consistency within the internal milieu by regulating processes as diverse as arousal, feeding, metabolism, water balance, and body temperature. Homeostatic regulation of the body is classically assigned to the hypothalamus because of its known role in integrating multiple peripheral signals of the internal milieu and coordinating adaptive responses through neuroendocrine outputs. MDD is associated with disruption of multiple homeostatic processes including sleep (insomnia or hypersomnia), appetite (anorexia or hyperphagia) and locomotor activity (psychomotor agitation or retardation). Given the complexity of homeostatic regulation, as well as the fact that most symptoms can be disrupted in opposite directions and still qualify as diagnostic criteria for MDD, it is not surprising that no single model has produced a satisfying explanation of how MDD affects multiple homeostatic functions. Given the central role of the orexin system within the lateral hypothalamus, one plausible hypothesis is that disruption of orexin signaling by stress may contribute to some of the neurovegetative symptoms observed in MDD. Several other chapters within this book detail the role of orexin signaling in neuroendocrine function. Therefore, in the following sections, we will briefly review research pertaining to orexin signaling in the regulation of appetite, sleep, and locomotor activity as it relates to MDD.

Change in appetite is one of the core features of MDD. While substantial progress has been made in understanding the neurobiological control of feeding and body weight regulation, relatively little is known about how MDD disrupts these circuits [38]. Orexin signaling has long been thought to regulate feeding behaviors in mice [39] and stress-induced disruptions in orexin signaling could contribute to the symptoms of MDD. In particular, Baldo and colleagues found that inhibition of neurons in the nucleus accumbens by a GABA receptor agonist produces a potent feeding response that is accompanied by activation of orexin neurons in the lateral hypothalamus [40]. In a later study, Zheng and colleagues confirmed the importance of orexin signaling in this pathway when they were able to block this feeding response by infusing an orexin receptor antagonist into the ventral tegmental area [41]. These findings indicate that a striatal–hypothalamic-ventral tegmental area circuit may be important for the regulation of food intake and support a role for orexin disruption in MDD-induced appetite changes.

The regulation of sleep by orexin is covered extensively in other sections of this book, so we will briefly touch upon topics relevant to MDD. Descriptive studies have long noted similarities in the sleep disruptions observed in narcolepsy and MDD. In particular, shortened REM latency and increased REM density are features common to both diagnoses [42, 43]. This observation led some to speculate that MDD and narcolepsy share a common pathophysiology [42], because regulation of REM–nonREM transitions are thought to be regulated by a balance between inhibitory monoaminergic and excitatory cholinergic signaling [43]. Consistent with this theory, antidepressant medications such as fluoxetine that increase levels of the aminergic neurotransmitter serotonin also increase REM latency [44]. Given the central role of orexin signaling in stabilizing sleep–wake transitions through the regulation of monoaminergic neurotransmitters [45], it is tempting to speculate that the reduction in orexin levels observed in MDD and narcolepsy may contribute to the similarities in sleep disturbances. It should be noted, however, that when Pollmacher and colleagues conducted a detailed analysis of REM latencies in patients with either MDD or narcolepsy, they found significant differences in the distribution pattern [46]. In particular, they found that patients with narcolepsy have a bimodal distribution of REM latency, whereas depressive patients showed a more continuous pattern. This difference suggests that while orexin signaling may contribute to symptoms in each illness, a single shared mechanism is not likely.

Finally, several studies have demonstrated that orexin signaling increases locomotor activity. Orexin-A signaling increases firing of noradrenergic neurons in the locus coeruleus to increase arousal and locomotor activity [47, 48]. Additionally, infusion of orexin-A into the nucleus accumbens stimulates a potent locomotor response in mice [49] indicating that orexin may act at multiple sites to stimulate activity. Together these findings support a role for orexin signaling in the regulation of arousal and activity.

Conclusions

Patients with narcolepsy appear to be at higher risk for the development of depressive symptoms. While the exact cause of this risk is not known, narcolepsy is a complex disease that affects not only neuronal control of sleep–wake transitions, but also affects psychological and social functioning of the individual. The increased stress from this impairment in daily functioning likely acts in concert with neuronal impairments in affective, cognitive, and neuroendocrine regulation to generate depressive symptoms. It will be interesting to see if pharmacologic agonists for the orexin system currently in development for the treatment of narcolepsy also regulate depressive symptoms.

References

1. Krishnan V, Nestler EJ. The molecular neurobiology of depression. Nature. 2008;455(7215):894–902.
2. Chemelli RM, Willie JT, Sinton CM, et al. Narcolepsy in orexin knockout mice: molecular genetics of sleep regulation. Cell. 1999;98(4):437–51.
3. Lin L, Faraco J, Li R, et al. The sleep disorder canine narcolepsy is caused by a mutation in the hypocretin (orexin) receptor 2 gene. Cell. 1999;98(3):365–76.
4. Krishnan RR, Volow MR, Miller PP, Carwile ST. Narcolepsy: preliminary retrospective study of psychiatric and psychosocial aspects. Am J Psychiatry. 1984;141(3):428–31.
5. Kales A, Soldatos CR, Bixler EO, et al. Narcolepsy-cataplexy. II. Psychosocial consequences and associated psychopathology. Arch Neurol. 1982;39(3):169–71.
6. Sours JA. Narcolepsy and other disturbances in the sleep-waking rhythm: a study of 115 cases with review of the literature. J Nerv Ment Dis. 1963;137:525–42.
7. Mosko S, Zetin M, Glen S, et al. Self-reported depressive symptomatology, mood ratings, and treatment

outcome in sleep disorders patients. J Clin Psychol. 1989;45(1):51–60.
8. Vandeputte M, de Weerd A. Sleep disorders and depressive feelings: a global survey with the Beck depression scale. Sleep Med. 2003;4(4):343–5.
9. Vourdas A, Shneerson JM, Gregory CA, et al. Narcolepsy and psychopathology: is there an association? Sleep Med. 2002;3(4):353–60.
10. Douglas NJ. The psychosocial aspects of narcolepsy. Neurology. 1998;50(2 Suppl 1):S27–30.
11. Broughton R, Ghanem Q, Hishikawa Y, Sugita Y, Nevsimalova S, Roth B. Life effects of narcolepsy in 180 patients from North America, Asia and Europe compared to matched controls. Can J Neurol Sci. 1981;8(4):299–304.
12. Broughton RJ, Guberman A, Roberts J. Comparison of the psychosocial effects of epilepsy and narcolepsy/cataplexy: a controlled study. Epilepsia. 1984;25(4):423–33.
13. Stores G, Montgomery P, Wiggs L. The psychosocial problems of children with narcolepsy and those with excessive daytime sleepiness of uncertain origin. Pediatrics. 2006;118(4):e1116–23.
14. Wilson SJ, Frazer DW, Lawrence JA, Bladin PF. Psychosocial adjustment following relief of chronic narcolepsy. Sleep Med. 2007;8(3):252–9.
15. Salomon RM, Ripley B, Kennedy JS, et al. Diurnal variation of cerebrospinal fluid hypocretin-1 (Orexin-A) levels in control and depressed subjects. Biol Psychiatry. 2003;54(2):96–104.
16. Brundin L, Bjorkqvist M, Petersen A, Traskman-Bendz L. Reduced orexin levels in the cerebrospinal fluid of suicidal patients with major depressive disorder. Eur Neuropsychopharmacol. 2007;17(9):573–9.
17. Brundin L, Petersen A, Bjorkqvist M, Traskman-Bendz L. Orexin and psychiatric symptoms in suicide attempters. J Affect Disord. 2007;100(1–3):259–63.
18. Brundin L, Bjorkqvist M, Traskman-Bendz L, Petersen A. Increased orexin levels in the cerebrospinal fluid the first year after a suicide attempt. J Affect Disord. 2009;113(1–2):179–82.
19. Pedrazzoli M, D'Almeida V, Martins PJ, et al. Increased hypocretin-1 levels in cerebrospinal fluid after REM sleep deprivation. Brain Res. 2004;995(1):1–6.
20. Morgan AJ, Jorm AF. Self-help interventions for depressive disorders and depressive symptoms: a systematic review. Ann Gen Psychiatry. 2008;7:13.
21. Allard JS, Tizabi Y, Shaffery JP, Trouth CO, Manaye K. Stereological analysis of the hypothalamic hypocretin/orexin neurons in an animal model of depression. Neuropeptides. 2004;38(5):311–5.
22. Allard JS, Tizabi Y, Shaffery JP, Manaye K. Effects of rapid eye movement sleep deprivation on hypocretin neurons in the hypothalamus of a rat model of depression. Neuropeptides. 2007;41(5):329–37.
23. Lutter M, Krishnan V, Russo SJ, Jung S, McClung CA, Nestler EJ. Orexin signaling mediates the antidepressant-like effect of calorie restriction. J Neurosci. 2008;28(12):3071–5.
24. Davis JF, Krause EG, Melhorn SJ, Sakai RR, Benoit SC. Dominant rats are natural risk takers and display increased motivation for food reward. Neuroscience. 2009;162(1):23–30.
25. Zigman JM, Nakano Y, Coppari R, et al. Mice lacking ghrelin receptors resist the development of diet-induced obesity. J Clin Invest. 2005;115(12):3564–72.
26. Lutter M, Sakata I, Osborne-Lawrence S, et al. The orexigenic hormone ghrelin defends against depressive symptoms of chronic stress. Nat Neurosci. 2008;11(7):752–3.
27. Yamanaka A, Beuckmann CT, Willie JT, et al. Hypothalamic orexin neurons regulate arousal according to energy balance in mice. Neuron. 2003;38(5):701–13.
28. Mieda M, Williams SC, Sinton CM, Richardson JA, Sakurai T, Yanagisawa M. Orexin neurons function in an efferent pathway of a food-entrainable circadian oscillator in eliciting food-anticipatory activity and wakefulness. J Neurosci. 2004;24(46):10493–501.
29. Ito N, Yabe T, Gamo Y, et al. I.c.v. administration of orexin-A induces an antidepressive-like effect through hippocampal cell proliferation. Neuroscience. 2008;157(4):720–32.
30. Boutrel B, Kenny PJ, Specio SE, et al. Role for hypocretin in mediating stress-induced reinstatement of cocaine-seeking behavior. Proc Natl Acad Sci USA. 2005;102(52):19168–73.
31. Aou S, Li XL, Li AJ, et al. Orexin-A (hypocretin-1) impairs Morris water maze performance and CA1-Schaffer collateral long-term potentiation in rats. Neuroscience. 2003;119(4):1221–8.
32. Akbari E, Naghdi N, Motamedi F. Functional inactivation of orexin 1 receptors in CA1 region impairs acquisition, consolidation and retrieval in Morris water maze task. Behav Brain Res. 2006;173(1):47–52.
33. Akbari E, Naghdi N, Motamedi F. The selective orexin 1 receptor antagonist SB-334867-A impairs acquisition and consolidation but not retrieval of spatial memory in Morris water maze. Peptides. 2007;28(3):650–6.
34. Selbach O, Bohla C, Barbara A, et al. Orexins/hypocretins control bistability of hippocampal long-term synaptic plasticity through co-activation of multiple kinases. Acta Physiol (Oxf). 2010;198(3):277–85.
35. Telegdy G, Adamik A. The action of orexin A on passive avoidance learning. Involvement of transmitters. Regul Pept. 2002;104(1–3):105–10.
36. Jaeger LB, Farr SA, Banks WA, Morley JE. Effects of orexin-A on memory processing. Peptides. 2002;23(9):1683–8.
37. Akbari E, Motamedi F, Naghdi N, Noorbakhshnia M. The effect of antagonization of orexin 1 receptors in CA1 and dentate gyrus regions on memory processing in passive avoidance task. Behav Brain Res. 2008;187(1):172–7.
38. Kishi T, Elmquist JK. Body weight is regulated by the brain: a link between feeding and emotion. Mol Psychiatry. 2005;10(2):132–46.
39. Sakurai T, Amemiya A, Ishii M, et al. Orexins and orexin receptors: a family of hypothalamic neuropeptides

and G protein-coupled receptors that regulate feeding behavior. *Cell.* 1998;92(5):1 page following 696.

40. Baldo BA, Gual-Bonilla L, Sijapati K, Daniel RA, Landry CF, Kelley AE. Activation of a subpopulation of orexin/hypocretin-containing hypothalamic neurons by GABAA receptor-mediated inhibition of the nucleus accumbens shell, but not by exposure to a novel environment. Eur J Neurosci. 2004;19(2): 376–86.
41. Zheng H, Patterson LM, Berthoud HR. Orexin signaling in the ventral tegmental area is required for high-fat appetite induced by opioid stimulation of the nucleus accumbens. J Neurosci. 2007;27(41):11075–82.
42. Hudson JI, Pope HG, Sullivan LE, Waternaux CM, Keck PE, Broughton RJ. Good sleep, bad sleep: a meta-analysis of polysomnographic measures in insomnia, depression, and narcolepsy. Biol Psychiatry. 1992;32(11):958–75.
43. Riemann D. Insomnia and comorbid psychiatric disorders. Sleep Med. 2007;8 Suppl 4:S15–20.
44. Wilson S, Argyropoulos S. Antidepressants and sleep: a qualitative review of the literature. Drugs. 2005;65(7):927–47.
45. Saper CB, Scammell TE, Lu J. Hypothalamic regulation of sleep and circadian rhythms. Nature. 2005;437(7063):1257–63.
46. Pollmacher T, Mullington J, Lauer CJ. REM sleep disinhibition at sleep onset: a comparison between narcolepsy and depression. Biol Psychiatry. 1997;42(8):713–20.
47. Hagan JJ, Leslie RA, Patel S, et al. Orexin A activates locus coeruleus cell firing and increases arousal in the rat. Proc Natl Acad Sci USA. 1999;96(19):10911–6.
48. Horvath TL, Peyron C, Diano S, et al. Hypocretin (orexin) activation and synaptic innervation of the locus coeruleus noradrenergic system. J Comp Neurol. 1999;415(2):145–59.
49. Thorpe AJ, Kotz CM. Orexin A in the nucleus accumbens stimulates feeding and locomotor activity. Brain Res. 2005;1050(1–2):156–62.

Part VI

REM Sleep Dysregulation and Motor Abnormalities in Narcolepsy

The Clinical Features of Cataplexy

Sebastiaan Overeem

Keywords
Cataplexy • Motor regulation • REM sleep • Narcolepsy • Orexin • Hypocretin • Atonia • Trigger

Introduction

Cataplexy: A Key Feature of Narcolepsy

Cataplexy is usually defined as "an episodic, bilateral loss of muscle tone, with preserved consciousness and triggered by emotions" [1, 2]. Early in the twentieth century, authors started to realize that cataplexy was a defining symptom of narcolepsy. In 1902, Loewenfeld was the first who firmly advocated reserving the term narcolepsy for a combination of hypersomnia and attacks of muscle weakness [3]. He recognized that isolated sleep attacks could be a nonspecific phenomenon, indicative of many other diseases. Henneberg introduced the term "cataplexy" in 1916, deriving it from the Greek verb καταπλησσω, meaning "to strike down" [4]. Around the same time, Redlich introduced the name "affektive Tonusverlust," stressing the role of emotional triggers [5].

S. Overeem (✉)
Department of Neurology, Donders Institute for Neuroscience, Radboud University Nijmegen Medical Centre, 9101, Nijmegen 6500, HB, The Netherlands
and
Sleep Medicine Center "Kempenhaeghe", Heeze, The Netherlands
e-mail: s.overeem@neuro.umcn.nl

Clinical Definition

For patients suffering from narcolepsy with cataplexy, the average time between symptom onset and final diagnosis is more than 10 years [6]. The reasons why narcolepsy is often not recognized or accurately diagnosed are likely related to the fact that excessive daytime sleepiness – the central symptom – is not only usually easy to identify but also the least specific, and is often regarded as secondary to insufficient sleep. In contrast, cataplexy – the other core symptom and virtually 100% specific – is much more difficult to diagnose.

In the current International Classification of Sleep Disorders (ICSD-2), cataplexy is descriptively defined as "sudden and transient episodes of loss of muscle tone triggered by emotions" [1]. In addition, "episodes must be triggered by strong emotions – most reliably laughing or joking – and must be generally bilateral and brief (less than 2 min)." In practice, this definition is often difficult to apply. First, this is due to the fact that the cataplexy phenotype differs widely in narcoleptic patients. It can range from rare occurrences

C.R. Baumann et al. (eds.), *Narcolepsy: Pathophysiology, Diagnosis, and Treatment*,
DOI 10.1007/978-1-4419-8390-9_26,

of partial attacks triggered by hearty laughter, to frequent complete attacks triggered by a range of emotions. Second, feelings of muscle weakness when laughing out loud are regularly reported in the general, healthy population [7].

Epidemiology

Cataplexy is the single most specific symptom of narcolepsy; its occurrence in other disorders is very rare. Using clinical criteria to diagnose narcolepsy, it was found to be present in about 70% of patients. However, in patients with a proven hypocretin-1 deficiency, the presence of cataplexy was much higher (around 93%) [8]. While excessive daytime sleepiness is most often the presenting symptom of narcolepsy, cataplexy comes first in approximately 5–8% of patients. Although most symptoms of narcolepsy remain relatively stable once established, cataplexy may diminish and sometimes almost disappear over several decades.

Clinical Features

Cataplexy is an intriguing phenomenon that has to be diagnosed based on patient history alone. Therefore, it is important to be aware of the variations in clinical presentation and to have a good knowledge about key features such as the patterns of muscle weakness and triggering factors. Nothing can replace clinical experience gained by talking extensively to narcoleptic patients, but here some of the most important features are described. In various parts of the text, reference will be made to a questionnaire study on the clinical features of cataplexy, which we recently performed in 109 patients with unequivocal cataplexy and a proven hypocretin-1 deficiency [9].

Pattern of Muscle Weakness

Involved muscle groups

Cataplexy is caused by a loss of muscle tone, so a flaccid paresis or paralysis ensues. Although cataplexy can affect virtually all skeletal muscles, there are various muscle groups that are preferentially involved. Often, there is muscle weakness in the neck, leading to a forward drop of the head. In addition, facial weakness may lead to sagging of the jaw and difficulty in speaking. In children, facial cataplexy may be particularly prominent, even leading to a so-called cataplectic facies (Fig. 1) [10]. When sleepiness is not readily recognized, this may even lead to confusion with neuromuscular disorders such as myasthenia gravis [11]. The lower limbs are often involved in cataplexy as well, leading to buckling of the knees. When cataplexy is very brief or subtle in the legs, it is sometimes difficult to differentiate from normal sensations of "weak with laughter" [7]. Respiratory muscles are not involved, although patients sometimes have some feeling of shortness of breath. Eye movements are not compromised. In the large majority of attacks, cataplexy is bilateral. However, some asymmetry may be present, and patients sometimes report one side of the body to be more distinctly affected.

Partial versus complete attacks

Most often, cataplectic attacks are partial, preferentially involving the muscle groups described above. However, complete attacks involving virtually the whole body are rather common. In our series, nearly half of patients experienced both partial and complete attacks. A small percentage reported experiencing only complete attacks. In any case, it is important to always inquire separately about the presence of both types. Most patients have rather consistent individual patterns of involved muscles.

Associated features

Partial attacks can be very subtle and are sometimes recognized only by experienced observers such as the patient's partner. Sometimes, cataplexy is limited to a sort of "fleeting sensations of weakness in the body" [12]. It is not uncommon for patients to interpret such phenomena as normal.

Cataplexy always starts abruptly, but this does not mean that the muscle weakness is immediately at its maximum. In complete attacks, it usually takes several seconds until complete muscle weakness develops (Fig. 2). Importantly, muscle

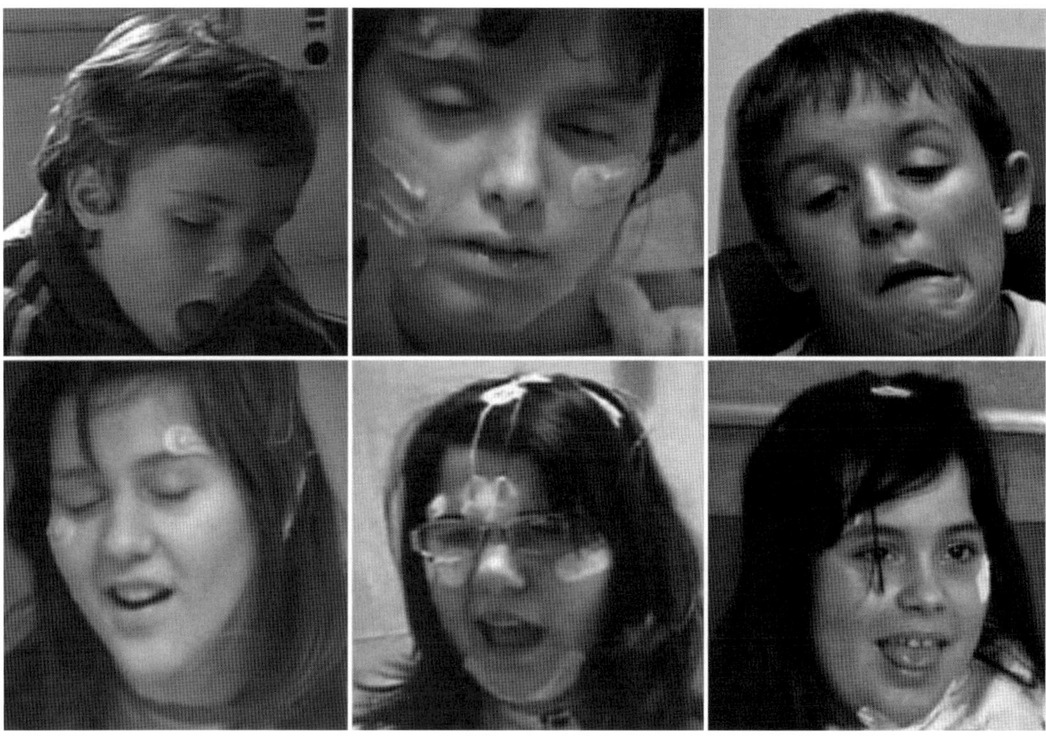

Fig. 1 Examples of a "cataplectic facies" in children, as described by Serra et al. Figure reprinted from [10]

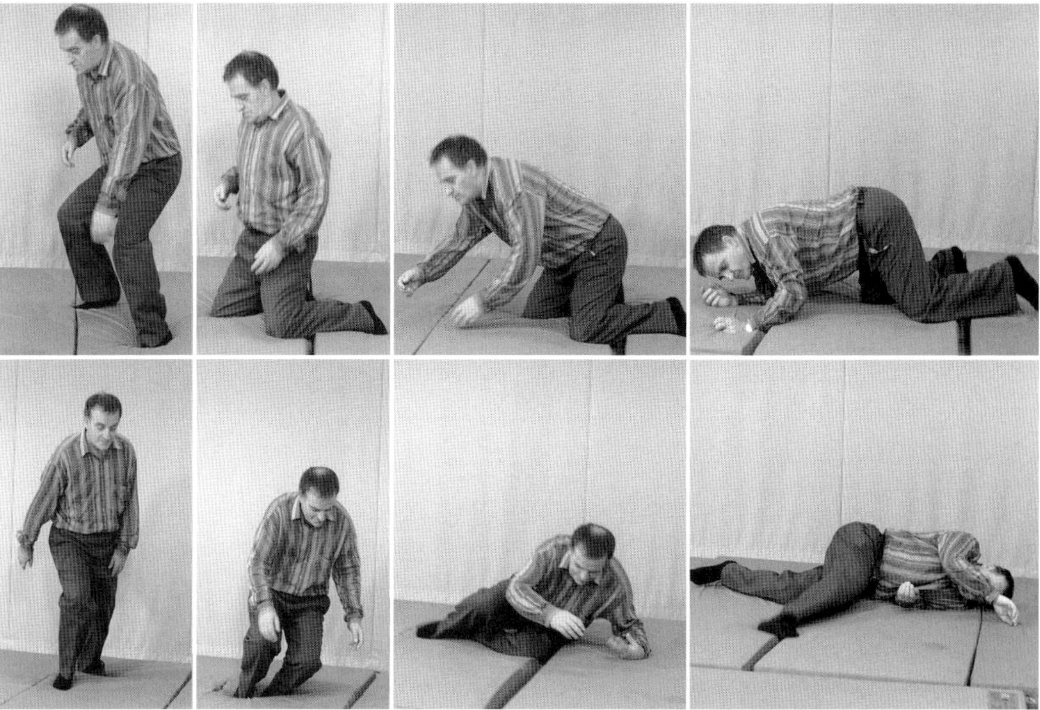

Fig. 2 Photographs showing two complete cataplectic attacks. Note that the muscle paralysis is not complete immediately and the patient can break his fall. Figure reprinted from [2]

function returns very quickly after an attack, with muscle tone restored within seconds.

Although the central characteristic of cataplexy is atonia, positive motor phenomena are quite often observed. Especially in the face, grimacing, jerks, and small muscle twitches occur frequently. Limbs may tremble or shake. It is unclear whether these positive motor symptoms are part of the cataplectic attack, or result from temporal interruptions of muscle atonia in combination with the patient "fighting" the attack.

Neurological examination during cataplexy reveals muscle atonia, in combination with a complete loss of stretch reflexes [12–14].

Triggers

Emotions as triggers

One of the most defining features of cataplexy is the triggering by emotions [15, 16]. In Fig. 3, various situations that can trigger cataplexy are shown, sorted in descending order based on the frequency of occurrence. Although virtually the whole gamut of emotions has been reported as potential triggers, mirth is the most frequent and usually the most potent [17]. Several situations are associated with this emotion: laughing out loud, telling a joke, making a witty remark, and so forth. Anger is an important "non-humorous" trigger. Many patients have the experience that cataplexy can be triggered by unexpectedly "bumping into" a friend or acquaintance. Note that the intensity and probably also the "unexpectedness" of an emotion are important factors determining how likely cataplexy is evoked.

Other triggers and spontaneous attacks

Not only the actual emotion may elicit cataplexy, the "anticipation" of something special or funny may do so as well: just before reaching the punch line of a joke, or visualizing the perfect smash

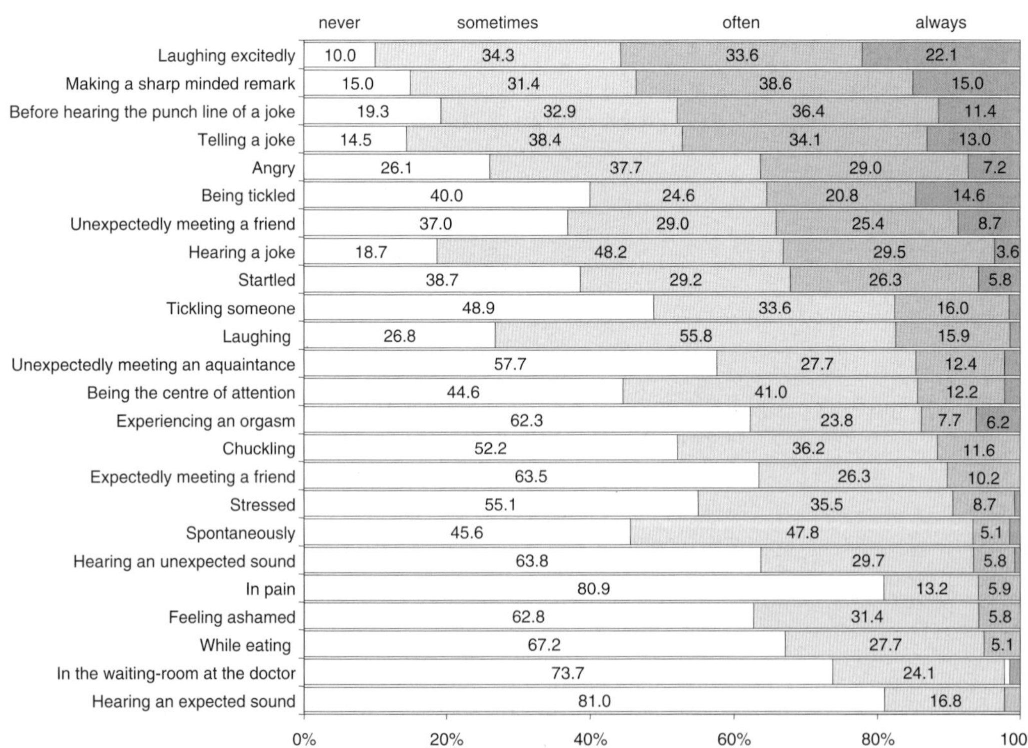

Fig. 3 Cataplexy triggers in patients with hypocretin-1 deficiency ($n = 109$). The triggers are sorted in descending order, based on the frequency of answers scored as "often" or "always"

when playing tennis. Although not happening frequently, many patients report having spontaneous cataplectic attacks, without a clear identifiable trigger.

Frequency and Duration

Frequency

The frequency of cataplexy varies widely between patients. It may range from less than once per month, to more than ten attacks per day. In an individual patient, the attack frequency is typically rather stable [18]. However, various circumstances can influence the likelihood of getting cataplexy. For example, sleep deprivation, severe sleepiness, or tiredness can lower the threshold for cataplexy [9]. Many patients recognize the presence of "bad days," in which the frequency of cataplexy is much higher, often without a specific cause. A certain "intimacy" is often required to have attacks in the company of others. Vice versa, cataplexy is typically very difficult to evoke during medical consultation or in the laboratory setting.

Duration

Brevity is an important feature of cataplexy: typical attacks do not last long. In our series, more than 90% of partial attacks and about 85% of complete attacks lasted less than 2 min. The majority of partial attacks were particularly short, lasting less than 10 s. Sometimes, when a trigger continues (e.g., because of friends who keep joking), consecutive short attacks may merge into what seems an unusually long bout of cataplexy. A so-called status cataplecticus has been described: long-lasting attacks happening in such a high frequency that patients are severely disabled. In these circumstances, attacks are often triggered by very minor situations. Sudden withdrawal of anti-cataplectic drugs, especially antidepressants, is the most important reason for inducing a status cataplecticus [2]. In addition, the antihypertensive drug prazosin can aggravate cataplexy, even up to a status.

Other Features

Warning signs, associated hallucinations

Many patients feel that a cataplectic attack is about to happen, enabling them to take countermeasures, such as sitting down. Examples of such warning signs are "strange" feelings in the head or sensations of warmth or nervousness.

Importantly, consciousness is preserved during cataplexy. This is an important differential diagnostic criterion. Long attacks can sometimes gradually shift into (REM) sleep. In our questionnaire study, almost a quarter of patients reported that hallucinatory experiences occurred during cataplexy. This has been reported before by Guilleminault et al. [12], who described a combination of retained environmental awareness together with vivid hallucinations in several patients.

Tricks to influence cataplexy

Many patients develop tricks to try and prevent a cataplectic attack from starting, or shorten it. Examples of such tricks are "trying to think of something else," "putting tension on muscles, for example, by clenching the fist," or "pressing against a firm support surface." In some patients, however, trying to resist an attack results in the prolongation of it. Only when they "let go," the attack becomes complete, but then subsides quickly.

Consequences of Cataplexy

When cataplexy is complete and occurs in a standing patient, it may lead to falls. Injury is uncommon, even in more severe attacks, because most people are able to find support or sit down at the onset of an attack. However, when assessed retrospectively, almost half of the patients have sustained some form of injury.

Cataplexy may result in severe limitations in daily life and thereby affect quality of life. Not only as a consequence of the attacks themselves, but also due to patients avoiding potentially emotional situations [19]. Over 25% of patients reported severe limitations not only in the domains of family life and recreation, but also in education and employment [9].

Diagnosis of Cataplexy

Diagnostic Classification

As stated above, in its "typical" form, cataplexy may be relatively easily diagnosed, even in the absence of an objective test. However, when the symptoms are subtle or not "text book like," a proper diagnosis may be much more difficult. To complicate matters, a distinction between "typical" and "atypical" cataplexy has been made in the literature, although "atypical cataplexy" has not been clearly defined and codified. In some patients, "atypical cataplexy" turns out to be real cataplexy, while in others, it refers to non-cataplectic events.

To circumvent the above-mentioned issues, a diagnostic grading system has been introduced for many diseases. In such systems, disorders are typically ranked as definite, probable, or possible. It seems that such a classification would be very appropriate for cataplexy. Classification could be based on the number of key features that are present, such as muscle atonia, triggered by a number of specifically defined emotions, short duration, preserved consciousness, and quick recovery. In addition, as proposed by Honda [20], a clear and prompt response to antidepressant medication could be added as well. The most important thing would be to rate such a system in a prospective study, in patients who are referred for the first time.

Differential Diagnosis

When cataplexy has typical features and occurs with a sufficiently high frequency, the diagnosis may be relatively straightforward. In these cases, important differential diagnostic options include the common normal feelings of weakness with emotions, particularly laughter [7, 17, 21]. Cataplexy can usually be differentiated from syncope or seizures because consciousness is preserved, at least at the beginning of an attack. Rare startle syndromes do not result in muscle weakness, but stiffness. Psychogenic reactions may occur.

Cataplexy may rarely occur in other disorders than narcolepsy, such as Niemann–Pick disease type C, Norrie disease, or diencephalic tumors. Obviously, there are other defining symptoms in these cases. Isolated cataplexy has been reported in – very rare – families [12].

Diagnostic Tools

The difficulty to detect and classify cataplexy accurately in the clinical setting is accentuated by the lack of a formal, objective diagnostic instrument. Cataplexy is notoriously difficult to evoke in the – unfamiliar and often uncomfortable – hospital environment. Several questionnaires have been developed that may have diagnostic value, but some are very long [17] and none have been prospectively evaluated in the specific target population (i.e., new patients with possible cataplexy) [22]. It is possible to try and evoke cataplexy in the hospital, for example, by showing humorous videos [14]. However, the yield of such attempts is low and false negatives occur with a high rate. Complicated "immersive" virtually reality-based systems aimed to deliver cataleptogenic emotional stimuli have been developed, but not clinically tested [23]. There have been attempts to use certain neurophysiological techniques to detect a propensity for cataplexy outside actual attacks, such as H-reflex recordings and repetitive transcranial magnetic stimulation, but these do not have diagnostic potential [24, 25].

This leaves a careful and detailed patient history, the primary diagnostic tool. In addition, it may be very helpful to ask a patient to try and record some attacks at home on video.

Cataplexy in Animal Models

For many years, the canine model of narcolepsy has been at the forefront of many pivotal discoveries on the pathophysiology of the disorder. The importance of the canine model was determined for a large part by the fact that their cataplexy very closely resembles the human phenotype. The pattern of muscle weakness shows combinations of partial (preferentially in the hind limbs) and complete attacks. Various circumstances can provoke attacks, such as playing with other dogs or caretakers, or eating favorable food items. The latter is such a potent trigger that a quantitative test for cataplexy has been based on it (the food-elicited cataplexy test) [26]. A recent consensus meeting yielded an informative overview of the features of cataplexy in various animal species (Table 1) [27].

When it was discovered that defects in hypocretin neurotransmission are the key pathophysiological mechanism leading to narcolepsy, several new animal models were developed very soon. Currently, there are a number of small rodent models, both in the mouse and in the rat. In most of these models, behavioral arrests have been observed that seem to be equivalent to cataplexy (Table 1) [27]. For obvious reasons, it remains much more difficult to classify events as cataplexy in rodents compared to dogs, for example, because "emotional" triggers are not easy to imagine. However, a useful working definition has been developed [27], which is especially important to distinguish cataplexy from REM sleep.

Conclusion

Cataplexy remains an intriguing symptom, which severely hampers quality of life of patients. It is also of important diagnostic value, being the only specific feature of narcolepsy. Future studies on the (effector) mechanisms of cataplexy may lead to diagnostic tools to diagnose cataplexy in patients. Until then, efforts should be made to formalize and refine the current diagnostic classification.

Table 1 Cataplexy across species. Reprinted from Scammell et al. [27]

	Human	Dog	Mouse
Behavioral features	Abrupt loss of postural muscle tone	Abrupt loss of postural muscle tone	Abrupt loss of postural muscle tone
Level of consciousness	Awake (memory of episodes intact)	Probably awake (visual tracking intact)	Uncertain, but possibly awake at onset (response to visual stimuli intact)
Triggers	Strong, generally positive emotions (e.g., laughter, joking, and playing) immediately before cataplexy	Probably positive emotions (e.g., playing and eating palatable food) immediately before cataplexy	Active behaviors with likely emotional content (e.g., running, climbing, vigorous grooming, and social interaction)
Duration	Brief (seconds to a few minutes)	Brief (seconds to a few minutes)	Brief (seconds to a few minutes)
EEG	Wake pattern and sometimes features of REM sleep	Wake pattern in cortex with theta activity in hippocampus	Theta activity similar to REM sleep
EMG	Atonia, sometimes with intermittent lapses in tone at onset	Atonia, sometimes with intermittent lapses in tone at onset	Atonia, sometimes with intermittent lapses in tone at onset
Response to therapy	Suppressed by monoamine reuptake blockers (e.g., antidepressants) and sodium oxybate	Suppressed by monoamine reuptake blockers	Suppressed by clomipramine

Cataplexy has many similarities in people, dogs, and mice with narcolepsy including abrupt postural atonia and improvement with clomipramine. During cataplexy in people and dogs, some authors report an EEG pattern similar to that of wake, while others describe characteristics of REM sleep, despite preservation of consciousness

References

1. ICSD. International classification of sleep disorders. 2nd ed. Westchester, IL: American Academy of Sleep Medicine; 2005.
2. Overeem S, Mignot E, van Dijk JG, Lammers GJ. Narcolepsy: clinical features, new pathophysiologic insights, and future perspectives. J Clin Neurophysiol. 2001;18:78–105.
3. Loewenfeld L. Über Narkolepsie. Münch Med Wochenschr. 1902;80:1041–5.
4. Henneberg R. Über genuine Narkolepsie. Neurol Centralbl. 1916;35:282–90.
5. Redlich E. ZurNarkolepsiefrage. Monatschr Psychiat Neurol. 1915;37:85.
6. Morrish E, King MA, Smith IE, Shneerson JM. Factors associated with a delay in the diagnosis of narcolepsy. Sleep Med. 2004;5:37–41.
7. Overeem S, Lammers GJ, van Dijk JG. Weak with laughter. Lancet. 1999;354:838.
8. Mignot E, Lammers GJ, Ripley B, et al. The role of cerebrospinal fluid hypocretin measurement in the diagnosis of narcolepsy and other hypersomnias. Arch Neurol. 2002;59:1553–62.
9. Overeem S, van Nues SJ, van der Zande WL, Donjacour CE, van Mierlo P, Lammers GJ. The clinical features of cataplexy: a questionnaire study in narcolepsy patients with and without hypocretin-1 deficiency. Sleep Med. 2011;12(1):12–8.
10. Serra L, Montagna P, Mignot E, Lugaresi E, Plazzi G. Cataplexy features in childhood narcolepsy. Mov Disord. 2008;23:858–65.
11. Dhondt K, Verhelst H, Pevernagie D, Slap F, Van CR. Childhood narcolepsy with partial facial cataplexy: a diagnostic dilemma. Sleep Med. 2009;10:797–8.
12. Guilleminault C, Gelb M. Clinical aspects and features of cataplexy. Adv Neurol. 1995;67:65–77.
13. Guilleminault C, Wilson RA, Dement WC. A study on cataplexy. Arch Neurol. 1974;31:255–61.
14. Krahn LE, Boeve BF, Olson EJ, Herold DL, Silber MH. A standardized test for cataplexy. Sleep Med. 2000;1:125–30.
15. Mattarozzi K, Bellucci C, Campi C, et al. Clinical, behavioural and polysomnographic correlates of cataplexy in patients with narcolepsy/cataplexy. Sleep Med. 2008;9:425–33.
16. Krahn LE, Lymp JF, Moore WR, Slocumb N, Silber MH. Characterizing the emotions that trigger cataplexy. J Neuropsychiatry Clin Neurosci. 2005;17:45–50.
17. Anic-Labat S, Guilleminault C, Kraemer HC, Meehan J, Arrigoni J, Mignot E. Validation of a cataplexy questionnaire in 983 sleep-disorders patients. Sleep. 1999;22:77–87.
18. Gelb M, Guilleminault C, Kraemer H, et al. Stability of cataplexy over several months – information for the design of therapeutic trials. Sleep. 1994;17:265–73.
19. Broughton WA, Broughton RJ. Psychosocial impact of narcolepsy. Sleep. 1994;17:S45–9.
20. Honda Y. Clinical features of narcolepsy: Japanese experiences. In: Honda Y, Juiji T, editors. HLA in Narcolepsy. Berlin: Springer-Verlag; 1988. p. 24.
21. Sturzenegger C, Bassetti CL. The clinical spectrum of narcolepsy with cataplexy: a reappraisal. J Sleep Res. 2004;13:395–406.
22. Moore WR, Silber MH, Decker PA, et al. Cataplexy Emotional Trigger Questionnaire (CETQ) – a brief patient screen to identify cataplexy in patients with narcolepsy. J Clin Sleep Med. 2007;3:37–40.
23. Augustine K, Cameron B, Camp J, Krahn L, Robb R. An immersive simulation system for provoking and analyzing cataplexy. Stud Health Technol Inform. 2002;85:31–7.
24. Lammers GJ, Overeem S, Tijssen MA, van Dijk JG. Effects of startle and laughter in cataplectic subjects: a neurophysiological study between attacks. Clin Neurophysiol. 2000;111:1276–81.
25. Overeem S, Afink J, Bakker M, et al. High frequency repetitive transcranial magnetic stimulation over the motor cortex: no diagnostic value for narcolepsy/cataplexy. J Neurol. 2007;254:1459–61.
26. Baker TL, Foutz AS, McNerney V, Mitler MM, Dement WC. Canine model of narcolepsy: genetic and developmental determinants. Exp Neurol. 1982;75:729–42.
27. Scammell TE, Willie JT, Guilleminault C, Siegel JM. A consensus definition of cataplexy in mouse models of narcolepsy. Sleep. 2009;32:111–6.

Parasomnias in Narcolepsy with Cataplexy

Yves Dauvilliers and Régis Lopez

Keywords
Parasomnia • REM sleep behavior disorder • Sleep paralysis • Hypnagogic • Hallucinations • Sleepwalking • Nocturnal terror • Narcolepsy

Narcolepsy with cataplexy is a disabling disorder characterized by excessive daytime sleepiness (EDS) and abnormal rapid eye movement (REM) sleep manifestations including cataplexy (sudden loss of muscle tone triggered by strong emotions), sleep paralysis, hypnagogic hallucinations, and sleep-onset REM periods [1].

Parasomnias are undesirable physical events or experiences that occur during entry into sleep, within sleep, or during arousals from sleep [2]. Bed partners may complain or note concern over frequent movements during the sleep period that may lead to severe injuries. Parasomnias are classified according to the type of sleep in which they occur, namely, non-rapid eye movement (NREM) sleep parasomnias including mainly sleep terrors and sleepwalking, and rapid eye movement (REM) sleep parasomnias including mainly REM sleep behavior disorder, recurrent isolated sleep paralysis, and nightmares [2]. Other parasomnias including sleep enuresis, sleep-related groaning, exploding head syndrome, sleep-related hallucinations, and sleep-related eating disorder (SRED) were without clear related stage. The natural history of age at onset, time of night of the events, characteristics of the behavior, memory for the events, and family history are important in distinguishing the parasomnia etiology.

Only few studies have focused on the frequency of NREM and REM parasomnias in narcolepsy. Even if clearly underestimated and not included in the tetrad narcolepsy symptoms, the parasomnia frequencies in both NREM and REM sleep were higher in narcolepsy compared to that in the general population [3–6]. Typically, nighttime sleep of narcoleptics was interrupted with several and long awakenings, and abnormal movements or behaviors during REM or NREM sleep, which may also be severe [3–6].

REM Behavior Disorder

REM behavior disorder (RBD) is currently classified as a parasomnia related to REM sleep, characterized by loss of REM sleep muscle atonia with a constant enactment of dream content [2, 7]. Sleep behaviors may produce injuries to the patient and/or the bed partner, being the main reason for consultation. Analyses of remembered

Y. Dauvilliers (✉)
Centre de Référence Nationale Maladie Rare – Narcolepsie et Hypersomnie Idiopathique, Service de Neurologie, Hôpital Gui-de-Chauliac, Inserm U1061, Montpellier, France
e-mail: ydauvilliers@yahoo.fr

C.R. Baumann et al. (eds.), *Narcolepsy: Pathophysiology, Diagnosis, and Treatment*,
DOI 10.1007/978-1-4419-8390-9_27, © Springer Science+Business Media, LLC 2011

dreams reveal an increased proportion of aggressive contents, without any problem of daytime aggressiveness [8]. Polysomnographically, the EMG activity revealed an excessive amount of sustained or intermittent elevation of tonic submental tone, or excessive phasic submental or (upper or lower) limb EMG twitching during REM sleep [2, 9, 10]. In cases of absence of clinical symptoms, this activity is called REM sleep without atonia (RWA) [2]. The importance of REM sleep without atonia was stressed in the latest international classification of sleep disorders (ICSD-2) [2]. Indeed, this classification only advised considering a diagnosis of RBD in the "presence of REM sleep without atonia," but no quantitative parameters were given [2].

The prevalence of clinical RBD remains largely unknown in the general population. A large telephone survey assessing violent behaviors during sleep suggested a prevalence of about 0.5% in the general population [11]. A male predominance (90%) in those aged over 50 years was always noted; however, milder forms of RBD with less aggressive behaviors and less clinical consultation requested are possible to occur in women [7]. The occurrence of RBD in narcolepsy was reported since the earlier description, initially called "ambiguous sleep" [12], and was further well described in 1986 by Schenck and coworkers [7]. The prevalence of RBD in narcolepsy with cataplexy seems to be fairly high. According to two recent studies, it is clinically evident in 45–61% of patients and is polysomnographically detectable in 36–43% of them [13, 14]. Patients with narcolepsy with cataplexy are more frequently affected by RBD than those without cataplexy [13], and in many patients with cataplexy, RBD can be induced or aggravated by anti-cataplectic treatment, i.e., antidepressants [15]. RBD may also be an early sign in childhood narcolepsy with cataplexy [16]. However, in the presence of associated parasomnias in narcolepsy–cataplexy, RBD was frequently the last parasomnia that occurred [6]. RBD is not an every night phenomenon in patients with narcolepsy, and episodes are usually less violent in narcoleptic patients than in other patients with secondary RBD [13]. Nightmares were also frequently reported in narcolepsy–cataplexy, especially in patients also presenting RBD [6].

An increased index of motor dyscontrol characterized by an increased electromyographic activity in REM sleep in the presence or absence of clinical RBD is an intrinsic finding of narcolepsy–cataplexy [6, 10, 17]. Hence, patients with narcolepsy–cataplexy often present dissociated sleep including REM sleep without atonia, frequent shift from REM to NREM sleep, and mixed features of REM and NREM sleep stages simultaneously such as the presence of atonia in sleep stage 2 and/or the presence of sleep spindles in REM sleep leading to ambiguous sleep [3, 6, 10, 18] (Fig. 1).

The prevalence of RWA, phasic EMG activity, and REM density is higher in narcolepsy compared to that in controls, while patients with idiopathic RBD have a higher prevalence of RWA and a lower REM density than narcoleptic patients and controls [10]. Recent findings showed that the altered REM sleep atonia index in patients with narcolepsy with cataplexy is mostly due to an increase in short-lasting EMG activity (approximately from 0 to 5 s) that might differentiate these patients from those with other forms of secondary RBD (i.e., Parkinson'disease and multiple system atrophy) [19, 20]. In addition, RBD in narcolepsy also differed from the idiopathic form because of the different sex ratio (in the idiopathic form, RBD mostly affects men) and its much earlier age at onset which precede narcolepsy in one-third of patients [13, 15]. Even if apparently a similar REM parasomnia, all these differences might indicate different neurochemical and neurophysiological mechanisms in RBD associated with narcolepsy—cataplexy, neurodegenerative disorders, or the idiopathic form [10, 19, 21] (Fig. 2).

There is no available report of prospective, double-blind, placebo-controlled trial of any specific drug for RBD in narcoleptic subjects, but only a few case reports of narcoleptic subjects with RBD. The use of clonazepam was frequently reported as successful [22–24]. Alternative treatments were needed in cases of nonresponders and/or intolerance to clonazepam, especially in the context of EDS. In one study involving 14

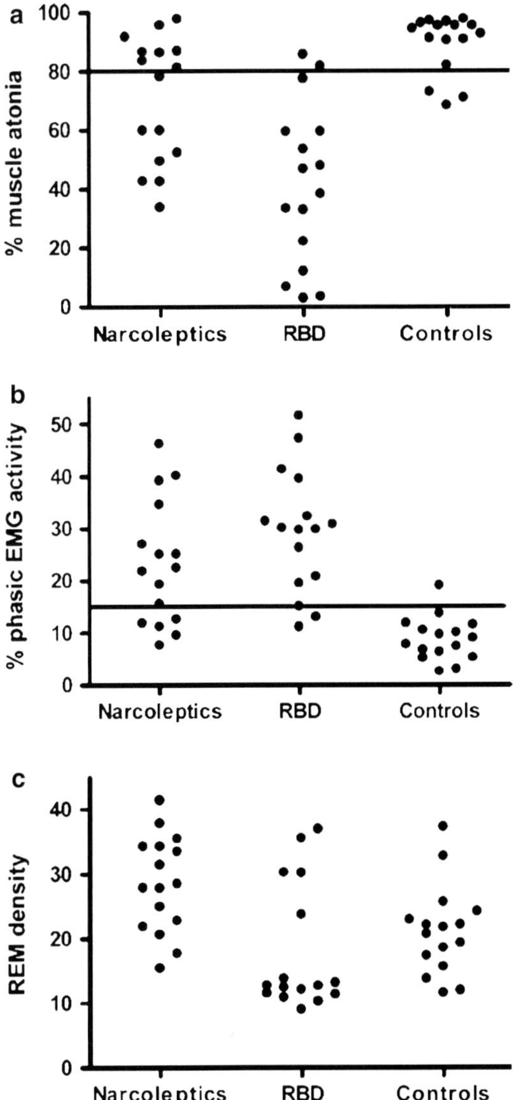

Fig. 1 Distribution of (**a**) REM sleep muscle atonia, (**b**) REM sleep phasic EMG activity, and (**c**) REM density in 16 patients with narcolepsy, 16 patients with "idiopathic" RBD, and 16 normal controls. Cutoffs at 80% and 15% were noted for the normal percentage of REM sleep muscle atonia and chin phasic EMG in REM sleep, respectively. From Fig. 1 page 847 of the reference: Dauvilliers Y, et al. (2007) REM sleep characteristics in narcolepsy and REM sleep behavior disorder. Sleep 30:844–9

patients, two of whom had narcolepsy, melatonin at a dose of 3–12 mg per night was used successfully in half of the cases [25]. Given the beneficial effects of sodium oxybate on disturbed nocturnal sleep, this medication might be of interest; however, no systematic study of sodium oxybate has ever been conducted on RBD of narcoleptics [26].

Sleep Paralysis

Sleep paralysis (SP) is a transient episode during which the patient upon falling asleep or awakening is unable to perform voluntary movements, speak, sometimes open the eyes, or breathe normally, while being mentally awakened [2]. The patient is fully aware during this state and able to recall the event. This feeling of being "blocked in an armor" is frightening, even terrifying, and can be associated with hypnagogic hallucinations in up to 75% of cases. Sleep paralysis is another example of state dissociation with elements of REM sleep (muscle atonia and dream imaging) existing into wakefulness. It occurs either at sleep onset or upon awakening from REM sleep at night, and less frequently during naps [27]. Recent findings confirm that SP is not intrinsically a sleep-onset or sleep-offset phenomenon, but rather is dependent on arousal during an REM period. Hence, SP will preferentially occur early or late in the sleep period whenever arousal exceeds waking thresholds during REM sleep [28]. Body position also seems to influence the occurrence of sleep paralysis, being mostly reported in the supine position [28]. Time of night seems to influence the intensity of the episodes, since fear was found to be significantly less intense at the end of sleep than at the beginning of sleep [29]. Most SP episodes last a few minutes and usually end spontaneously or, sometimes, after sensory stimulation (patient is touched or is alerted to a sound), or after an intense effort of the patient to "break" the paralysis episode. Some subjects establish a code with their bedpartner so that the latter can touch them or move their head or limbs and thereby put an end to the symptom. If the event still persists, the subject usually reenters sleep and awakens later.

Recurrent isolated SP is a common, generally benign parasomnia, experienced by normal

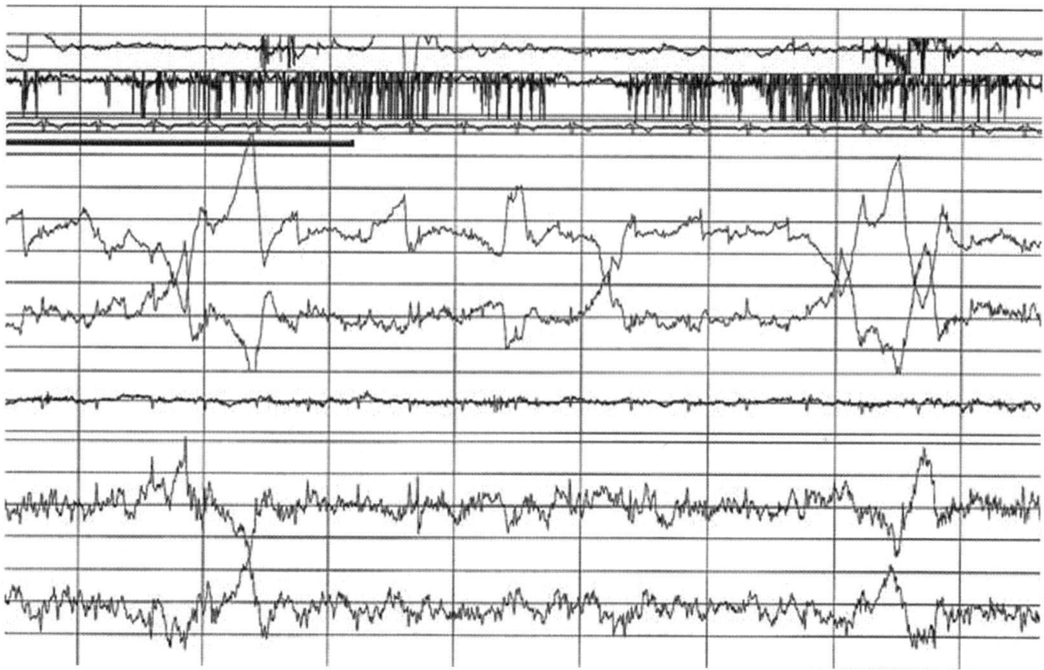

Fig. 2 REM sleep without atonia. Epoch of 30 s of persistence of tonic and phasic EMG activity during REM sleep on bilateral anterior tibial muscles. Montage from top to bottom: electromyogram of the right and left anterior tibial muscles, electrocardiogram, right and left electrooculogram, chin electromyogram, and electroencephalogram (C4-A1, C3-A2). Recording speed: 1 cm/s

subjects in up to 10% of the general population depending on definition, age of subjects, and sociocultural factors [2]. One study reported that the prevalence of subjects who experienced at least one SP episode in their lifetime is 6.2% in the general population, with severe SP (at least one episode per week) occurring in only 0.8% [29]. Earlier onset of SP episodes was associated with more frequent episodes. In addition, multiple episodes of SP occur, especially in anxiety and depression disorders, and after severe sleep deprivation, psychological stress, schedule disruption, or ingestion of alcohol and/or anxiolytic medication. However, based on the recent results of the Wisconsin Sleep Cohort Study population, depression of greater severity appeared to be strongly associated with the presence of SP (OR 5.0) that persists after several adjustments including antidepressant use, age, sex, and body mass index (BMI) [30].

Sleep paralysis is experienced by 40–60% of narcoleptic patients with cataplexy, being more frequent, more intense, and often associated with hallucinations in comparison to SP in the general population and in non-cataplectic central hypersomnias [31]. However, there is no specific SP characteristic that individualizes SP in narcolepsy–cataplexy and in the general population. The frequent association between SP and narcolepsy may be explained by the high number of awakenings with frequent sleep-wake state transitions, especially in REM sleep.

Hypnagogic and Hypnopompic Hallucinations

Hypnagogic (at sleep onset) or hypnopompic (upon awakening) hallucinations are mainly auditory (phone ringing or walking in the stairway), visual (threatening figure, passing shadows when driving, animals, or person), or somesthetic (out-of-body experience or brushing). Many patients were unwilling to talk about the symptoms. Even if patients criticize well these phenomena afterward, some of them reported difficulty to differentiate dream from reality, and can be rarely misdiagnosed as schizophrenic [3, 32]. These events may be repetitive but are usually not stereotypic. Sleep-related hallucinations can be found in normal subjects, but mainly only a few times in their life [33]. The prevalence in the general population is up to 20%, with the same risk factors as already reported for sleep paralysis, i.e., after sleep deprivation, a change in sleep schedule, ingestion of alcohol, anxiety, and depression disorders [33]. However, the Wisconsin Sleep Cohort Study population revealed that depression was less strongly related to hypnagogic hallucinations in comparison to SP, with anxiety (independently related to hypnagogic hallucinations) partly explaining this relationship [30].

Recurrent hypnagogic hallucinations are experienced by 40–80% of patients with narcolepsy–cataplexy, occurring more regularly and with more intensity in that condition compared to the general population and to non-cataplectic central hypersomnias [31]. Evoked awakening studies performed in patients with narcolepsy–cataplexy revealed that hypnagogic hallucinations occured when patients fall asleep directly into REM sleep [34]. Hypnagogic hallucinations in narcoleptics are sometimes so scary that the subject becomes fearful of going to bed. Hence the frequent sleep-wake state transitions in narcolepsy–cataplexy together with the early onset of REM sleep explained certainly the presence of frequent hallucinations in that condition [3].

Sleep paralysis and sleep-related hallucinations are generally more prevalent in young narcoleptics and tend to decrease with time, although some patients will experience these auxiliary symptoms throughout their life time [34].

Treatment of hallucinations and sleep paralysis is considered as a treatment of REM-associated phenomena as for cataplexy [26]. Most studies have focused much more on the treatment of cataplexy. Improvement of cataplexy through antidepressants and/or sodium oxybate is most often associated with reduction of hallucinations and sleep paralysis [26].

Sleepwalking and Nocturnal Terrors

Sleepwalking (SW) and nocturnal terrors are partial arousals, occurring in the first third of the sleep period, typically from slow wave sleep (SWS), patients being amnestic for the events [2]. Sleepwalking events can be minor behaviors such as sitting up in bed or can include complex and elaborate behaviors, including walking, cooking, eating, dressing, or rarely driving a car. Nocturnal terrors are other forms of disorders of arousals from SWS accompanied by a cry or piercing scream, simple behavioral manifestations of intense fear with a predominance of autonomic expression. The onset of the event is abrupt and patients exhibit significant tachycardia, sweating, rapid breathing, and mydriasis. A high degree of overlap exists between sleep terrors and somnambulism; however, autonomic features exist mainly in the former condition. Duration of episodes may vary from a few seconds to several minutes. During both sleepwalking and nocturnal terror events, patients appear confused and disoriented, and may become violent with possibilities of injuries to themselves and bed partners. The reports from witnesses of the behavior are essential for clinical evaluation. Episodes of sleepwalking and nocturnal terrors may be triggered not only by sleep deprivation but also by anxiety.

Despite widespread prevalence of these NREM parasomnias affecting up to 20% of children and 1–3% of adults [35], neither its pathophysiology nor its mechanisms responsible for these incomplete arousals are well understood [36, 37]. Only few sleep architectural differences have been reported between sleepwalkers and controls, with a higher number of arousals in SWS and a lower power in slow-wave activity during the first NREM cycle [36]. A strong genetic contribution is supported by frequent positive family history (from 50 to 80%) without any clear mode of transmission, and a higher concordance rate of SW is observed in monozygotic (MZ) twins (50%) than in DZ twins (10–15%) [38]. An important overlap exists in genetic predisposition to sleepwalking and night terrors. We have reported the first genetic association between the HLA DQB1*05 subtypes and sleepwalking, especially in the familial forms, the polymorphic amino acid Ser74 being the most closely associated DQB1 polymorphism [39]. In addition to a genetic susceptibility in SW, precipitating factors such as conditions and substances that increase SWS and/or make arousals from sleep more difficult may trigger episodes [37]. As an example, we may note that somnambulism episodes may be triggered or aggravated by the use of sodium oxybate (Fig. 3).

Only few reports on the frequency of sleepwalking and nocturnal terror episodes are available in the literature of narcolepsy. Hence, the frequency of these NREM parasomnias was higher in narcolepsy compared to the general population, with around three times higher episodes [40]. In addition, narcoleptic patients often display several associated parasomnias [41]. Hence, an association of multiple parasomnias including SW, night terrors, nightmares, enuresis, and bruxism has been reported in narcolepsy and especially in narcoleptic RBD patients [6, 41]. Most of these parasomnias occur in childhood with onset often preceding narcolepsy, and may switch to other phenotypes with the course of the disease and/or the aging process [6]. RBD is mostly the last parasomnia to occur in narcolepsy. In addition, family's narcolepsy study revealed that the relative risk to develop parasomnias in the first-degree relative is very high, higher than that in the general population and also higher than the risk to develop one of the major narcolepsy symptoms, i.e., excessive daytime sleepiness or cataplexy [42]. These findings support a common genetic predisposition of motor dyscontrol during sleep in narcolepsy–cataplexy.

Sleep-Related Eating Disorder

Sleep-related eating disorder is a recently recognized nocturnal disorder in which subjects are unaware of having eaten while asleep [2]. A continuum with the nocturnal eating syndrome (NES)

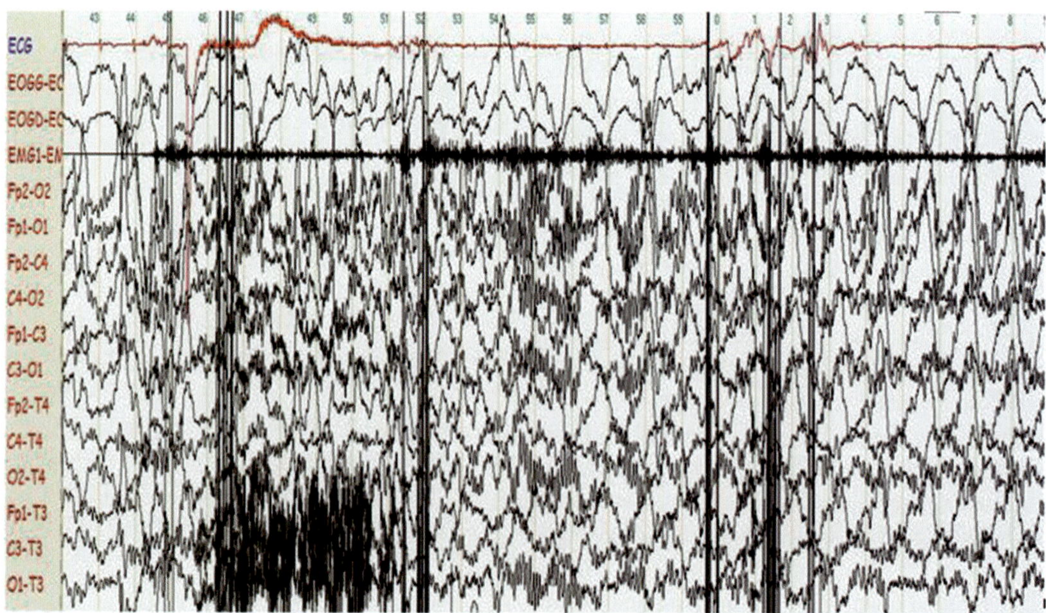

Fig. 3 Disorder of arousal. Epoch of 30 s of arousal from slow wave sleep (sleepwalking episode), with hypersynchronous delta waves persisting in the context of increased muscle activity. Montage from top to bottom: electrocardiogram, right and left electrooculogram, chin electromyogram, and a full electroencephalogram montage. Recording speed: 1 cm/s

characterized by a pattern of late-night binge eating is still subject to debate. It affects all ages and both sexes, although it is more common in young women with a prevalence estimate between 1 and 2% of the general population [2]. Narcolepsy has been found in some (small) cohorts of NES patients [43, 44], but to our better knowledge, no report exists of an association between narcolepsy and SRED.

Patients with narcolepsy–cataplexy are significantly more overweight, with BMIs approximately 10–20% higher than that of matched healthy control subjects [1]. This weight increase does not seem to result from a higher caloric intake, but a lower basal metabolism and changes in eating behavior [45]. A previous study measuring spontaneous food intake in a small cohort of narcolepsy–cataplexy reported that patients ingested fewer calories and especially fewer carbohydrates than controls [46]. However, snacking is common, with up to 80% of patients eating bedtime snacks. Recent findings reported that narcoleptic patients may suffer from a mild eating behavior disorder, with an emphasis on a craving for food and binge eating behavior, especially in overweight patients [43, 44]. However, this eating disorder is classified as atypical eating disorder, i.e., eating disorders not specified as proposed in ICSD-2 [2]. This increased appetite can be attributed not only to a secondary effect of medications (mainly tricyclic antidepressants), but also to ingestion of carbohydrates during the night, as it has been postulated that narcoleptics use carbohydrates as a sleep inducer to help themselves. Hypocretin is thought to regulate not only sleep but also metabolism and appetite, possibly explaining the association with nocturnal eating. Altogether, eating disorders seem to be an integral part of the narcolepsy phenotype. However, the prevalence of the night eating syndrome is still unknown in narcolepsy [44].

Conclusion

Despite few literatures in the field, NREM and REM parasomnias remain the frequently associated conditions in narcolepsy–cataplexy, with onset often preceding the narcolepsy condition. The coexistence of such diseases may suggest a common pathophysiology. This review pinpointed the frequent abnormalities in both NREM and REM sleep motor regulation that results in dissociated sleep/wake states. However, motor dyscontrol in narcolepsy is not restricted to sleep, involving also wakefulness with the presence of cataplexy and the increase in periodic leg movements index, suggesting a common neurobiological defect of motor inhibition. We may hypothesize that dysfunction in the hypocretin system leads to motor dyscontrol in sleep/wake states. A close relationship between HLA, motor activity, and sleep has also been highlighted, suggesting some immune-related regulation of motor control during sleep. Finally, the recognition of associated comorbidities in narcolepsy–cataplexy may help to diagnose the condition earlier and treat the patients. However, based on available information, it is difficult to provide guidance for prescribing in parasomnias associated with narcolepsy other than to recommend conventional medications.

References

1. Dauvilliers Y, Arnulf I, Mignot E. Narcolepsy with cataplexy. Lancet. 2007;369:499–511.
2. American Academy of Sleep Medicine. International Classification of Sleep Disorders. 2nd ed: Diagnostic and coding manual. Westchester, IL: American Academy of Sleep Medicine; 2005.
3. Dauvilliers Y, Billiard M, Montplaisir J. Clinical aspects and pathophysiology of narcolepsy. Clin Neurophysiol. 2003;114:2000–17.
4. Ferri R, Miano S, Bruni O, et al. NREM sleep alterations in narcolepsy/cataplexy. Clin Neurophysiol. 2005;116:2675–84.
5. Plazzi G, Serra L, Ferri R. Nocturnal aspects of narcolepsy with cataplexy. Sleep Med Rev. 2008;12:109–28.
6. Mayer G, Meier-Ewert K. Motor dyscontrol in sleep of narcoleptic patients (a lifelong development?). J Sleep Res. 1993;2:143–8.
7. Schenck CH, Bundlie SR, Ettinger MG, Mahowald MW. Chronic behavioral disorders of human REM sleep: a new cathegory of parasomnia. Sleep. 1986;9:293–308.
8. Fantini ML, Corona A, Clerici S, Ferini-Strambi L. Aggressive dream content without daytime aggressiveness in REM sleep behavior disorder. Neurology. 2005;11(65):1010–5.

9. Lapierre O, Montplasir J. Polysomnographic features of REM sleep behavior disorder: development of a scoring method. Neurology. 1992;42:1371–4.
10. Dauvilliers Y, Rompré S, Gagnon JF, Vendette M, Petit D, Montplaisir J. REM sleep characteristics in narcolepsy and REM sleep behavior disorder. Sleep. 2007;30:844–9.
11. Ohayon MM, Caulet M, Priest RG. Violent behavior during sleep. J Clin Psychiatry. 1997;58:369–76.
12. de Barros-Ferreira M, Lairy GC. Ambiguous sleep in narcolepsy. In: Guilleminault C, Dement WC, Passouant P, editors. Narcolepsy. Proceedings of the first International Symposium on Narcolepsy, July 1975, Montpellier, France, New York: Spectrum publications; 1976. p. 57–75.
13. Nightingale S, Orgill JC, Ebrahim IO, de Lacy SF, Agrawal S, Williams AJ. The association between narcolepsy and REM behavior disorder (RBD). Sleep Med. 2005;6:253–8.
14. Mattarozzi K, Bellucci C, Campi C, et al. Clinical, behavioural and polysomnographic correlates of cataplexy in patients with narcolepsy/cataplexy. Sleep Med. 2008;9:425–33.
15. Schenck CH, Mahowald MW. Motor dyscontrol in narcolepsy: rapid-eye-movement (REM) sleep without atonia and REM sleep behavior disorder. Ann Neurol. 1992;32:3–10.
16. Nevsimalova S, Prihodova I, Kemlink D, Lin L, Mignot E. REM behavior disorder (RBD) can be one of the first symptoms of childhood narcolepsy. Sleep Med. 2007;8:784–6.
17. Geisler P, Meier-Ewert K, Matsubayshi K. Rapid eye movements, muscle twitches and sawtooth waves in the sleep of narcoleptic patients and controls. Electroencephalogr Clin Neurophysiol. 1987;67:499–507.
18. Hishikawa Y, Wakamatsu H, Furuya E, et al. Sleep satiation in narcoleptic patients. Electroencephalogr Clin Neurophysiol. 1976;41:1–18.
19. Ferri R, Franceschini C, Zucconi M, et al. Searching for a marker of REM sleep behavior disorder: submentalis muscle EMG amplitude analysis during sleep in patients with narcolepsy/cataplexy. Sleep. 2008;31:1409–17.
20. Ferri R, Manconi M, Plazzi G, et al. A quantitative statistical analysis of the submentalis muscle EMG amplitude during sleep in normal controls and patients with REM sleep behavior disorder. J Sleep Res. 2008;17:89–100.
21. Plazzi G, Corsini R, Provini F, et al. REM sleep behavior disorders in multiple system atrophy. Neurology. 1997;48:1094–7.
22. Schuld A, Kraus T, Haack M, Hinze-Selch D, Pollmächer T. Obstructive sleep apnea syndrome induced by clonazepam in a narcoleptic patient with REM-sleep-behavior disorder. J Sleep Res. 1999;8:321–2.
23. Yeh SB, Schenck CH. A case of marital discord and secondary depression with attempted suicide resulting from REM sleep behavior disorder in a 35-year-old woman. Sleep Med. 2004;5:151–4.
24. Schenck CH, Mahowald MW. Long-term, nightly benzodiazepine treatment of injurious parasomnias and other disorders of disrupted nocturnal sleep in 170 adults. Am J Med. 1996;100:333–7.
25. Boeve BF, Silber MH, Ferman TJ. Melatonin for treatment of REM sleep behavior disorder in neurologic disorders: results in 14 patients. Sleep Med. 2003;4: 281–4.
26. Billiard M, Bassetti C, Dauvilliers Y, Dolenc-Groselj L, Lammers GJ, Mayer G, et al. EFNS Task Force EFNS guidelines on management of narcolepsy. Eur J Neurol. 2006;13:1035–48.
27. Hishikawa Y, Shimizu T. Physiology of REM sleep, cataplexy, and sleep paralysis. Adv Neurol. 1995;67: 245–71.
28. Girard T, Cheyne A. Timing of spontaneous sleep-paralysis episodes. J Sleep Res. 2006;15:222–9.
29. Ohayon MM, Zulley J, Guilleminault C, Smirne S. Prevalence and pathologic associations of sleep paralysis in the general population. Neurology. 1999;52: 1194–2000.
30. Szklo-Coxe M, Young T, Finn L, Mignot E. Depression: relationships to sleep paralysis and other sleep disturbances in a community sample. J Sleep Res. 2007;16:297–312.
31. Dauvilliers Y, Paquereau J, Bastuji H, Drouot X, Weil JS, Viot-Blanc V. Psychological health in central hypersomnias: the French Harmony study. J Neurol Neurosurg Psychiatry. 2009;80:636–41.
32. Fortuyn HA, Lappenschaar GA, Nienhuis FJ, et al. Psychotic symptoms in narcolepsy: phenomenology and a comparison with schizophrenia. Gen Hosp Psychiatry. 2009;31:146–54.
33. Ohayon MM. Prevalence of hallucinations and their pathological associations in the general population. Psychiatry Res. 2000;97:153–64.
34. Ribstein M. Hypnagogic hallucinations. In: Guilleminault C, Dement WC, Passouant P, editors. Narcolepsy. Proceedings of the first international symposium on narcolepsy, July 1975. Montpellier, France, New York: Spectrum Publications; 1976. p. 145–60.
35. Laberge L, Tremblay RE, Vitaro F, Montplaisir J. Development of parasomnias from childhood to early adolescence. Pediatrics. 2000;106:67–74.
36. Pilon M, Zadra A, Joncas S, Montplaisir J. Hypersynchronous delta waves and somnambulism: brain topography and effect of sleep deprivation. Sleep. 2006;29:77–84.
37. Zadra A, Pilon M, Montplaisir J. Polysomnographic diagnosis of sleepwalking: effects of sleep deprivation. Ann Neurol. 2008;63:513–9.
38. Dauvilliers Y, Maret S, Tafti M. Genetics of normal and pathological sleep in humans. Sleep Med Rev. 2005;9:91–100.
39. Lecendreux M, Bassetti C, Dauvilliers Y, Mayer G, Neidhart E, Tafti M. HLA and genetic susceptibility to sleepwalking. Mol Psychiatry. 2003;8:114–7.
40. Reynolds III CF, Christiansen CL, Taska LS, et al. Sleep in narcolepsy and depression. Does it all look alike? J Nerv Ment Dis. 1983;171:290–5.
41. Schenck CH, Boyd JL, Mahowald MW. A parasomnia overlap disorder involving sleepwalking, sleep terrors,

and REM sleep behavior disorder in 33 polysomnographically confirmed cases. Sleep. 1997;20:972–81.
42. Mayer G, Lattermann A, Mueller-Eckhardt G, Svanborg E, Meier-Ewert K. Segregation of HLA genes in multicase narcolepsy families. J Sleep Res. 1998;7:127–33.
43. Chabas D, Foulon C, Gonzalez J, Nasr M, Lyon-Caen O, Willer JC, et al. Eating disorder and metabolism in narcoleptic patients. Sleep. 2007;30:1267–73.
44. Fortuyn HA, Swinkels S, Buitelaar J, et al. High prevalence of eating disorders in narcolepsy with cataplexy: a case-control study. Sleep. 2008;31:335–41.
45. Kotagal S, Krahn LE, Slocumb N. A putative link between childhood narcolepsy and obesity. Sleep Med. 2004;5:147–50.
46. Lammers GJ, Pijl H, Iestra J, Langius JA, Buunk G, Meinders AE. Spontaneous food choice in narcolepsy. Sleep. 1996;19:75–6.

The Motor System and Narcolepsy: Periodic Leg Movements and Restless Legs Syndrome

Luigi Ferini-Strambi

Keywords
Restless legs syndrome • Periodic leg movements • Dopamine • Orexin • Hypocretin • Narcolepsy

Introduction

Restless legs syndrome (RLS) is a sensorimotor disorder with the cardinal symptoms consisting of an urge to move the legs because of unpleasant sensations, appearing during rest or inactivity, worsening at evening or during the night, which are partially or totally recovered by movement [1]. Depending on the severity and frequency of the symptoms, RLS is often associated with insomnia and an impairment of quality of life [2]. Periodic leg movements (PLM) during sleep may be observed in more than 80% of RLS patients [3]. The pathogenesis of RLS is still unknown, but there are several pieces of evidence that advert to a possible dysfunction of descending dopaminergic neurons from the hypothalamic A11 region to the intermediolateral and dorsal spinal gray matter [4]. It is well known that dopamine receptors consist of five different subtypes, classified within the two subfamilies of D1-like receptors (D1 and D5) and D2-like receptors (D2, D3, and D4) [5]. The A11 axons extend over the whole spinal cord to the D2 subfamily receptors, which have the highest densities in the superficial layers of the dorsal horn at cervical and lumbar levels. The connection between A11 and the intermediolateral spinal column is possibly implicated in the hypothalamic control of vegetative functions, while the one between A11 and the dorsal gray matter may play a role in the regulation of sensory input from the posterior root. There are three aspects that suggest a dysfunction of this pathway inducing RLS-like symptoms: dopamine production, correlation to a circadian endogenous pacemaker, and regulation of the spinal sensory input. Low doses of dopamine and the dopamine agonist pergolide reduce the monosynaptic spinal stretch reflex amplitude in wild-type mice but not in D3 knock-out (KO) mice [6]. These results suggest a limitation of the spinal reflex by dopamine and an excitation of the spinal reflex by D3 antagonists, acting in both cases on D3 receptors.

RLS worsens with antidopaminergic treatment [7], can be induced by hyposideremia (iron is the coenzyme of the tyrosine hydroxylase, the limiting enzyme in the dopamine synthesis) [8], and shows a circadian symptom and PLM distribution, which is inversely related to the levels of dopamine in both blood and cerebrospinal fluid [9]. Moreover,

L. Ferini-Strambi (✉)
Department of Neuroscience, Sleep Disorders Center, Università Vita-Salute San Raffaele, Milan, Italy
e-mail: ferinistrambi.luigi@hsr.it

C.R. Baumann et al. (eds.), *Narcolepsy: Pathophysiology, Diagnosis, and Treatment*,
DOI 10.1007/978-1-4419-8390-9_28, © Springer Science+Business Media, LLC 2011

the dopamine hypothesis in RLS is strongly supported by the fact that both RLS and PLM respond dramatically to medication with L-dopa and dopamine agonists, even at very low dosages [10].

The hypocretin (orexin) system, which is deficient in patients with narcolepsy, has widespread excitatory projections throughout the central nervous system including the dopaminergic system. Therefore, can we establish links between narcolepsy and RLS/PLM, in terms of etiopathogenetic, clinical, and polysomnographic patterns?

Ethiopathogenetic Links

Dopaminergic System

The role of dopamine in narcolepsy has been a matter of controversy for many years. Dopaminergic dysfunction is possibly involved in narcolepsy, because reduced levels of dopamine and its metabolite homovanillic acid are found in the cerebrospinal fluid (CSF) of narcolepsy patients [11]. In postmortem studies, D2 receptor binding was found to be increased in the basal ganglia of narcoleptic patients [12]. D1 receptor studies have shown either increased [12] (caudate and medial globus pallidus) or unchanged [13] (caudate nucleus and putamen) binding pattern. However, in vivo studies with positron emission tomography (PET) [14] or single photon emission computed tomography (SPECT) [15] have not found changes in D1 or D2 receptors in narcolepsy. Dopamine transporter is a protein situated in the presynaptic terminal and participating in the reuptake of dopamine: the activity of dopamine transporter regulates the synaptic content of dopamine. A study that evaluated unmedicated patients with narcolepsy and healthy controls with PET, using a dopamine transporter ligand, found no change in striatal dopamine transporter [16]. However, this finding does not exclude extrastriatal changes.

The systemic administration of dopaminergic D2/D3 agonists or their local perfusion into the midbrain dopaminergic nuclei [i.e., ventral tegmental area (VTA) and substantia nigra (SN) or diencephalic dopaminergic nuclei] significantly enhances cataplexy, while D2/D3 antagonists, such as sulpiride, reduce this symptom in the canine model [17, 18]. It is known that D2/D3 autoreceptors are enriched in the SN and the VTA, and that systemic or local administration of D2/D3 agonists into these structures significantly reduces the firing rate of dopaminergic neurons [19, 20]. Thus, changes in the activity of dopaminergic neurons in the midbrain may be critical for the regulation of cataplexy, which is often interpreted as being an REM sleep-associated phenomenon [21]. This result, however, contrasts with the well-documented lack of evidence of a role of midbrain dopaminergic systems in the regulation of REM sleep. D2/D3 receptor-mediated dopaminergic activity in the midbrain and diencephalon may thus be more specifically involved in the regulation of cataplexy than in REM sleep.

Hypocretin System

It is known that hypocretins increase arousal and motor activity and that they interact with the dopamine system [22]. Because all these associations of the hypocretin system are important aspects of RLS, it has been postulated that RLS may involve abnormal hypocretin transmission. Because high dosages of hypocretin lead to increased locomotion in animals [23, 24], it was postulated that motor restlessness in patients with RLS could be caused by increased hypocretin levels in the CSF. Furthermore, the excitatory and wake-promoting effects of hypocretin [23] could mediate the relatively high level of alertness during daytime and the nocturnal sleep disturbances in RLS. Alternatively, increased CSF hypocretin may be caused by increased motor activity and wakefulness in patients with RLS, because animal data suggest that hypocretin activity increases significantly with motor activity and during active wakefulness when compared to quiet wakefulness [25]. Furthermore, hypocretin neurons are activated during wakefulness when somatosensory activity is present [26].

Regarding CSF hypocretin levels in RLS, the literature reveals conflicting results. Some authors found increased CSF hypocretin levels in patients with early-onset RLS [27]. The finding was made at 11 p.m., several hours after the habitual onset of RLS symptoms in these patients. Hypocretin levels in age-matched control subjects and patients with late-onset RLS did not differ at the same circadian time. In this line, some authors consider that idiopathic RLS may consist of two different phenotypes [28]. In one subtype, "early-onset" RLS symptoms start before the age of 45 years and develop slowly, and patients have a high familial aggregation. In the second phenotype ("late-onset" RLS), RLS starts after the age of 45 years, and patients have a rapid progression of symptom severity and only weak familial aggregation. Thus, the above-mentioned finding of increased CSF hypocretin in early-onset RLS may support the view that this phenotype may be genetically determined and involve different pathomechanisms compared to late-onset RLS [25]. However, up to date, there is no established upper limit of CSF hypocretin levels, i.e., we still do not know whether pathologically increased hypocretin levels exist and may have any clinical significance.

Another group [29] did not find any significant increase in CSF hypocretin (samples collected at 6 p.m.) neither in early-onset nor in late-onset RLS. In addition, the authors found no significant correlation between CSF hypocretin levels and the subjective severity of RLS or objective polysomnographic measures. In another study with ten mostly treated patients with RLS, there was also no significant increase when CSF was collected in the morning or early afternoon [30]. One possible explanation for the lack of replication of the initial study [25] could be that abnormal hypocretin transmission may occur only in the late evening when RLS symptoms reach a maximum. The authors [29] concluded that continuous measurements of CSF hypocretin will be needed to clarify definitely the role, if any, of this neuropeptide in RLS. A recent preliminary study with a continuous monitoring of CSF monoamines and hypocretin in three patients with RLS compared to three controls found no evidence of significant 24-h rhythm abnormalities of CSF hypocretin [31]; however, these results need to be confirmed in larger populations.

Clinical and Polysomnographic Links

There is not much available information on the prevalence of RLS in narcoleptic patients. In a series of 47 consecutive newly diagnosed patients with narcolepsy, none of the included subjects had symptoms of RLS [32]. Another study evaluated 169 narcoleptic patients, and again, none of these had a diagnosis of RLS [33].

Concerning PLM, uncontrolled studies performed in a limited number of patients reported PLM in about 25–50% of narcoleptic patients [34]. Interestingly, it has been observed that narcoleptic hypocretin-deficient canines, like narcoleptic humans, also exhibit jerky, unilateral, or bilateral slow leg movements during sleep [35]. The movements in dogs are characterized by repetitive dorsiflexions of the ankle, lasting 0.5–1.5 s, and occur at regular intervals of 3–20 s, thus showing similarities to PLM in humans. More recently, one controlled study found that 66% of narcoleptics had a PLM index > 5 (49% with PLM index > 10) versus 10% of controls [32]. In another controlled study [33], more narcoleptics than controls had a PLM index > 5 (67% versus 37%) and a PLM index > 10 (53% versus 21%).

It is not known whether there is an impact of PLM on nocturnal sleep and daytime functioning in narcolepsy. A study conducted on a large cohort of 530 patients [36] found that narcoleptics with PLM index > 5 have poorer sleep, with greater arousal and awakening indices, increased stage 1 NREM, and reduced sleep efficiency. This finding of disturbed sleep architecture could, however, be related to the older age of the group with PLM index > 5, compared to the group with PLM index < 5 [36]. Thus, it seems to be important to control for age when studying the effects of PLM in narcolepsy. In one of the two controlled above-mentioned studies [32], age-matched narcoleptics with PLM had more arousals than those without PLM. Concerning the macrostructure of sleep, an increased percentage of stage 1 NREM, a decreased percentage of REM, and a decreased REM efficiency have been reported in narcoleptics with PLM [33]. These last findings are indicators of a partial REM sleep disruption, suggesting that PLM may be a part of the REM sleep dissociation phenomena characteristic of narcolepsy, with frequent shifts from REM to NREM sleep, and the presence of intermediate stages of sleep where features of REM and NREM sleep are present simultaneously [37]. The negative correlation between PLM index and REM efficiency suggests that PLM may be an indicator of the severity of REM disruption in narcolepsy, or alternatively may contribute to REM sleep disruption in this condition. The increase in stage 1 NREM sleep may be interpreted as a consequence of this process. Since a significant decrease in sleep latency has been observed on the Multiple Sleep Latency Test in narcoleptics with high PLM index [33], some authors suggested that REM disruption may play a role in diurnal symptoms of narcolepsy patients with PLM [33]. In the general population, on the contrary, it has been reported that even a high number of PLM is not usually accompanied by excessive daytime sleepiness [38]. Therefore, we may hypothesize that PLM is not a homogenous phenomenon, and thus different characteristics in narcoleptics compared to controls could determine a different functional impact of PLM. In narcoleptics, we find that PLM is more frequently associated with arousals and microarousals than in age-matched controls [32, 33]. Also, PLM indices in both NREM and REM sleep are higher in narcoleptics than in controls [33]. Moreover, it has been reported that 65% of patients with narcolepsy had a PLM index > 5 in REM sleep compared with only 27% of the controls [33]. This result may be due to impaired mechanism responsible for motor inhibition during REM sleep in narcolepsy. Indeed, REM sleep behavior disorder is often found in association with narcolepsy.

Some authors analyzed PLM in narcoleptics compared to normal controls and patients with RLS [39]. They found that the distribution of inter-PLM intervals was clearly bimodal in narcoleptics and RLS, with one peak at 2–4 s and another at around 22–26 s. However, all periodicity parameters were significantly lower in the narcolepsy group. Moreover, the distribution of the number of PLM per hour of sleep is bell shaped in normal controls and patients with narcolepsy, whereas patients with RLS show a progressive decrease throughout the night (Fig. 1, personal data). The reduction in periodicity of PLM in narcolepsy could be related to decreased arousal fluctuations during sleep [40]. Indeed, PLM are known to be modulated by a well-known periodic sleep phenomenon called cyclic alternating pattern (CAP), which is one of the most systematic and expressive constructs devised to describe arousals, their composition, and their timing during sleep. CAP has been reported to be reduced in patients with narcolepsy [40]. This finding could also explain the low index of PLM associated with microarousals in narcoleptics compared with that in RLS patients [41].

Therapeutic Links

Some authors evaluated the sleep organization of narcoleptic patients before and after suppression of PLM with L-dopa. L-dopa and placebo were administered in a double-blind crossover design to six narcoleptic patients [42]. Each treatment period lasted 2 weeks. A significant reduction of PLM was found after treatment with L-dopa, but this treatment did not improve sleep organization. On the contrary, L-dopa increased wake time after sleep onset [42]. This finding supports the hypothesis that PLM do not play a major role in the nocturnal sleep

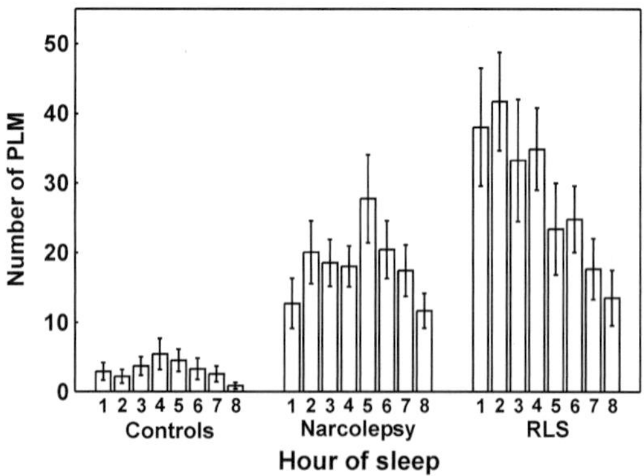

Fig. 1 Distribution of number of periodic leg movements (PLM) per hour of sleep (mean ± SD) in healthy controls (n=25), patients with narcolepsy (n=45), and patients with RLS (n=25)

disruption observed in narcolepsy. Similar results have been reported in narcoleptics with PLM treated with bromocriptine [43]. However, additional studies with a large number of patients are needed to assess the impact of PLM treatment on sleep architecture and daytime vigilance in narcolepsy.

In the last years, γ-hydroxybutyrate (GHB, also known as sodium oxybate) has re-emerged as a major treatment option for narcolepsy with cataplexy. About 20 years ago, one study evaluated 12 narcoleptic patients (six with PLM and six without) before and after treatment with GHB taken at bedtime for 1 month [44]. Treatment resulted in decreased REM sleep latency and increased REM efficiency without change in the total duration of REM sleep. Moreover, GHB was associated with the appearance of pathological levels of PLM in patients who were unaffected before treatment.

Recently, others described a case of a narcoleptic patient who presented with severe typical RLS under GHB treatment. The patient did not experience similar symptoms previous to treatment, and his complaints were reversible after withdrawal of GHB [45]. Thus, it is conceivable that GHB triggers RLS symptoms or PLM in narcolepsy through a decrease in dopamine neurotransmission. It is known that chronic treatment of GHB has major effects on dopaminergic transmission, with an upregulation of D1 and D2 mRNA expression in several brain regions, including the striatum [46]. The diencephalic A11 system could be another high-affinity GHB neuron receptor.

For treating cataplexy, other widely used medications include the antidepressants venlafaxine, imipramine, and protriptyline, which are usually used at lower dosages than that used in depression [47]. Interestingly, RLS is a potential side effect of several antidepressant drugs, and the majority of published papers focusing on this issue are case reports. In a recent prospective study, patients treated for the first time with an antidepressant were prospectively evaluated with regard to the question of whether RLS occurred or pre-existing RLS worsened as a result of the drug [48]. In 9% of patients, RLS was observed as a side effect related to the antidepressant treatment, and the side effect occurred after a median of 2.5 days. These findings should force treating physicians to ask the patients systematically about the presence of RLS during treatment with GHB or antidepressant drugs.

References

1. Allen RP, Picchietti D, Hening WA, Trenkwalder C, Walters AS, Montplaisi J. Restless legs syndrome: diagnostic criteria, special considerations, and epidemiology. A report from the restless legs syndrome diagnosis and epidemiology workshop at the National Institutes of Health. Sleep Med. 2003;4(2):101–19.
2. Allen RP. Restless legs syndrome effects on quality of life. In: Ondo WG, editor. Restless legs syndrome diagnosis and treatment. New York: Informa Healthcare; 2007. p. 199–203.
3. Montplaisir J, Boucher S, Poirier G, Lavigne G, Lapierre O, Lesperance P. Clinical, polysomnographic, and genetic characteristics of restless legs syndrome: a study of 133 patients diagnosed with new standard criteria. Mov Disord. 1997;12(1):61–5.
4. Rye DB. Parkinson's disease and RLS: the dopaminergic bridge. Sleep Med. 2004;5(3):317–28.
5. Kvernmo T, Hartter S, Burger E. A review of the receptor-binding and pharmacokinetic properties of dopamine agonists. Clin Ther. 2006;28(8):1065–78.
6. Clemens S, Hochman S. Conversion of the modulatory actions of dopamine on spinal reflexes from depression to facilitation in D3 receptor knock-out mice. J Neurosci. 2004;24(50):11337–45.
7. Inami Y, Horiguchi J, Nishimatsu O, Sasaki A, Sukegawa T, Katagiri H, et al. A polysomnographic study on periodic limb movements in patients with restless legs syndrome and neuroleptic-induced akathisia. Hiroshima J Med Sci. 1997;46(4):133–41.
8. Chaudhuri KR. Iron and restless legs syndrome: the story unfolds. Sleep Med. 2006;7(5):395–6.
9. Michaud M, Dumont M, Selmaoui B, Paquet J, Fantini ML, Montplaisir J. Circadian rhythm of restless legs syndrome: relationship with biological markers. Ann Neurol. 2004;55(3):372–80.
10. Manconi M, Ferri R, Zucconi M, Oldani A, Fantini ML, Castronovo V, et al. First night efficacy of pramipexole in restless legs syndrome and periodic leg movements. Sleep Med. 2007;8(5):491–7.
11. Montplaisir J, de Champlain J, Young SN, et al. Narcolepsy and idiopathic hypersomnia: biogenic amines and related compounds in CSF. Neurology. 1982;32:1299–302.
12. Aldrich MS, Hollingeswoth Z, Penney JB. Dopamine-receptor autoradiography of human narcoleptic brain. Neurology. 1992;42:410–5.
13. Kish SJ, Mamelak M, Slimovitch C, et al. Brain neurotransmitter changes in human narcolepsy. Neurology. 1992;42:229–34.
14. Rinne JO, Hublin C, Partinen M, et al. Positron emission tomography study of human narcolepsy: no

increase in striatal dopamine D2 receptors. Neurology. 1995;45:1735–8.
15. Hublin C, Launes J, Nikkinen P, Partinen M. Dopamine D2-receptors in human narcolepsy: a SPECT study with 125I-IBZM. Acta Neurol Scand. 1995;90:186–9.
16. Rinne JO, Hublin C, Nagren K, Helenius H, Partinen M. Unchanged striatal dopamine transporter availability in narcolepsy: a PET study with [11C]-CFT. Acta Neurol Scand. 2004;109:52–5.
17. Honda K, Riehl J, Mignot E, Nishino S. Dopamine D3 agonists into the substratia nigra aggravates cataplexy but does not modify sleep. Neuroreport. 1999;10:3111–8.
18. Okura M, Riehl J, Mignot E, Nishino S. Sulpiride, a D2/D3 blocker, reduces cataplexy but not REM sleep in canine narcolepsy. Neuropsychopharmacology. 2000;23:528–38.
19. Miller JD, Farber J, Gatz P, Roffwarg H, German DC. Activity of mesencephalic dopamine and non-dopamine neurons across stages of sleep and waking in the rat. Brain Res. 1983;273:133–41.
20. Piercey MF, Hoffmann WE, Smith MW, Hyslop DK. Inhibition of dopamine neuron firing by pramipexole, a dopamine D3 receptor-preferring agonist: comparison to other dopamine receptor agonists. Eur J Pharmacol. 1996;312:35–44.
21. Nishino S, Honda K, Riehl J, Mignot E. Extracellular single unit recordings of dopaminergic neurons in the ventral tegmental area in narcoleptic dobermans. Sleep. 2000;23(S2):A1a.
22. Taheri S, Zeitzer JM, Mignot E. The role of hypocretins (orexins) in sleep regulation and narcolepsy. Annu Rev Neurosci. 2002;25:283–313.
23. Hagan JJ, Leslie RA, Patel S, et al. Orexin A activates locus coeruleus cell firing and increases arousal in rat. Proc Natl Acad Sci USA. 1999;96:10911–6.
24. Nacamura T, Uramura K, Nambu T, et al. Orexin-induced hyperlocomotion and stereotypy are mediated by the dopaminergic system. Brain Res. 2000;873:181–7.
25. Wu MF, John J, Maidment N, Lam HA, Siegel JM. Hypocretin release in normal and narcoleptic dogs after food and sleep deprivation, eating, and movement. Am J Physiol Regul Integr Comp Physiol. 2002;283:R1079–86.
26. Torterolo P, Yamuy J, Sampogna S, Morales FR, Chase MH. Hypocretinergic neurons are primarily involved in activation of the somatomotor system. Sleep. 2003;26:25–30.
27. Allen R, Mignot E, Ripley B, Hishino S, Early C. Increased CSF hypocretin (orexin) in idiopathic restless legs syndrome. Neurology. 2002;59:639–41.
28. Allen RP, Earley CJ. Restless legs syndrome: a review of clinical and pathophysiologic features. J Clin Neurophysiol. 2001;18:128–47.
29. Stiasny-Kolster K, Mignot E, Ling L, Möller JC, Cassel W, Oertel WH. CSF hypocretin-1 levels in restless legs syndrome. Neurology. 2003;61(10):1426–9.
30. Mignot E, Lammers GJ, Ripley B, et al. The role of cerebrospinal fluid hypocretin measurement in the diagnosis of narcolepsy and other hypersomnias. Arch Neurol. 2002;59:1553–62.
31. Poceta JS, Parsons L, Engelland S, Kripke DF. Circadian rhythm of CSF monoamines and hypocretin-1 in restless legs syndrome and Parkinson's disease. Sleep Med. 2009;10(1):129–33.
32. Bahammam A. Periodic leg movements in narcolepsy patients: impact on sleep architecture. Acta Neurol Scand. 2007;115:351–5.
33. Dauvilliers Y, Pennestri MH, Petit D, Dang-Vu T, Lavigne G, Montplaisir J. Periodic leg movements during sleep and wakefulness in narcolepsy. J Sleep Res. 2007;16:333–9.
34. Baker TL, Guilleminault C, Nino-Murcia G, Dement WC. Comparative polysomnographic study of narcolepsy and idiopathic central nervous system hypersomnia. Sleep. 1986;9:232–42.
35. Okura M, Fujiki N, Ripley B, et al. Narcoleptic canines display periodic leg movements during sleep. Psychiat Clin Neurosci. 2001;55:243–4.
36. Harsh J, Peszka J, Hartwig G, Mitler M. Night-time sleep and daytime sleepiness in narcolepsy. J Sleep Res. 2000;9:309–16.
37. Dauvilliers Y, Billiard M, Montplaisir J. Clinical aspects and pathophysiology of narcolepsy. Clin Neurophysiol. 2003;114:2000–17.
38. Scofield H, Roth T, Drake C. Periodic limb movements during sleep: population prevalence, clinical correlates, and racial differences. Sleep. 2008;31(9):1221–7.
39. Ferri R, Zucconi M, Manconi M, Bruni O, Ferini-Strambi L, et al. Different periodicity and time structure of leg movements during sleep in narcolepsy/cataplexy and restless legs syndrome. Sleep. 2006;29:1587–94.
40. Ferri R, Miano S, Bruni O, et al. NREM sleep alterations in narcolepsy/cataplexy. Clin Neurophysiol. 2005;116:2675–84.
41. Sforza E, Nicolas A, Lavigne G, Gosselin A, Petit D, et al. EEG and cardiac activation during periodic leg movements in sleep. Neurology. 1999;52:786–91.
42. Boivin DB, Montplaisir J, Poirier G. The effects of L-dopa on periodic leg movements and sleep organization in narcolepsy. Clin Neuropharmacol. 1989;12(4):339–45.
43. Boivin DB, Lorrain D, Montplaisir J. Effects of bromocriptine on periodic limb movements in human narcolepsy. Neurology. 1993;43:2134–6.
44. Bédard MA, Montplaisir J, Godbout R, Lapierre O. Nocturnal gamma-hydroxybutyrate. Effect on periodic leg movements and sleep organization of narcoleptic patients. Clin Neuropharmacol. 1989;12(1):29–36.
45. Abril B, Carlander B, Touchon J, Dauvilliers Y. Restless legs syndrome in narcolepsy: a side effect of sodium oxybate? Sleep Med. 2007;8:181–3.
46. Maitre M. The gamma-hydroxybutyrate signalling system in brain: organization and functional implications. Prog Neurobiol. 1997;51:337–61.
47. Roth T. Narcolepsy: treatment issues. J Clin Psychiatry. 2007;68 Suppl 13:16–9.
48. Rottach K, Schaner B, Kirch M, et al. Restless legs syndrome as a side effect of second generation antidepressants. J Psychiat Res. 2009;43:70–5.

Part VII

The Borderlands of Narcolepsy

Spectrum of Narcolepsy

Claudio L. Bassetti

Keywords
Narcolepsy without cataplexy • Secondary narcolepsy • Monosymptomatic narcolepsy • Familial narcolepsy • Pseudocataplexy

Introduction

The debate about the borderland of narcolepsy and the existence of different forms of narcolepsy started early after the description of the disease and lasts until today.

In 1926, Adie made a review of the 25 cases reported in the literature (adding five personal observations), discussing narcolepsy as a specific disease ("disease sui generis") of neuroendocrinological (thalamo-hypophyseal) origin [1]. Two years later, in a second article in the journal "Brain," Wilson challenged this view suggesting, in analogy to epilepsy, the existence of primary and secondary types of narcolepsy, including in the latter group cases related to trauma, psychopathology, tumors, and circulatory problems [2].

In 1960, Yoss and Daly introduced the concept of the narcoleptic tetrade [3]. The existence of "incomplete" forms of narcolepsy, including narcolepsy without cataplexy and monosymptomatic narcolepsy, was pointed out in reviews of large patient series [4, 5].

The discovery of three biological markers of narcolepsy [sleep-onset REM periods (SOREMPs), HLA, and cerebrospinal fluid (CSF) hypocretin deficiency] was soon followed by the recognition of atypical cases without SOREMs ("NREM narcolepsy"), HLA markers ("HLA-negative" narcolepsy), or normal CSF hypocretin. Furthermore, these markers have also been found in non-narcoleptic patients and to some extent even in normal subjects.

This chapter will review our current knowledge about the borderland of narcolepsy, a better understanding of which is essential for the management of this disorder.

Narcolepsy Variants

Narcolepsy Without Cataplexy/ Monosymptomatic Narcolepsy

Narcolepsy without cataplexy (so-called monosymptomatic narcolepsy) [6] is uncommon and probably represents only 10–15% of patients with narcolepsy [7, 8]. In the literature, a higher frequency of monosymptomatic narcolepsy (with a ratio to classical narcolepsy as high as 0.4–1:1)

C.L. Bassetti (✉)
Department of Neurology, Neurocenter EOC of Southern Switzerland, Ospedale Civico, Lugano, Switzerland
e-mail: claudio.bassetti@eoc.ch

C.R. Baumann et al. (eds.), *Narcolepsy: Pathophysiology, Diagnosis, and Treatment*,
DOI 10.1007/978-1-4419-8390-9_29, © Springer Science+Business Media, LLC 2011

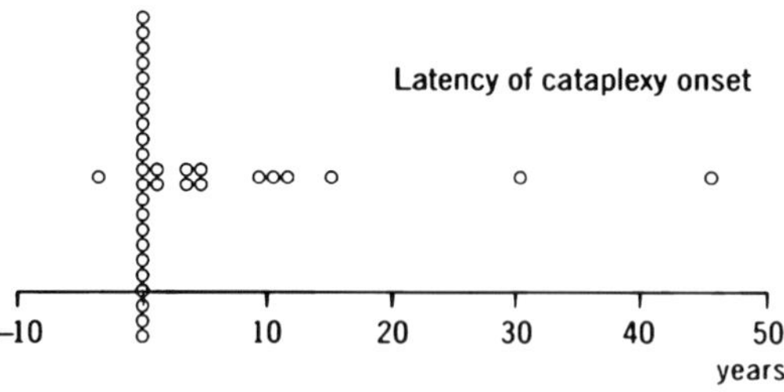

Fig. 1 Latency of cataplexy onset after appearance of first symptoms in 42 narcoleptics. Modified from Sturzenegger and Bassetti (J Sleep Res 2004)

probably reflects the inclusion of patients with EDS of other origin, or on the base of multiple sleep latency test (MSLT) results alone [9–11].

Although the existence of monosymptomatic narcolepsy has been questioned, its existence is proven by the following arguments:

1. EDS and cataplexy can appear separated by several years from each other in patients with narcolepsy with cataplexy (Fig. 1). Cataplexy usually appears concomitantly or within a few years from the onset of EDS; in rare cases, it follows EDS by an interval of up to 40–45 years [3, 12–14]. On the contrary, cataplexy can precede EDS by as much as 28 years [5].
2. In narcolepsy, the frequency and severity of symptoms can vary considerably. Hence, depending on the severity and frequency criteria chosen for diagnosing cataplexy, patients with EDS may be diagnosed as having narcolepsy with or without cataplexy. Cataplexy can in fact occur as often as hundreds of times daily or as rarely as a few times in a lifetime. Furthermore, its severity can range from a mild feeling of weakness of the extremities which is unnoticeable by others to severe spells with falls to the ground. Finally, cataplexy often decreases in severity several years after its appearance.
3. Patients with narcolepsy with cataplexy often have a positive family history of imperative ("narcolepsy-like") EDS but rarely of cataplexy, a phenomenon that also suggests the existence of monosymptomatic (milder, incomplete) forms of narcolepsy [15, 16]. This hypothesis is supported by the observation of longer mean sleep latencies on MSLT in patients without cataplexy when compared with narcoleptics with definite ("clear-cut", "true") cataplexy [14]. Monosymptomatic narcolepsy could indeed be conceptualized as a variant of narcolepsy with cataplexy due to insufficient genetic and/or environmental components.

Narcolepsy without cataplexy is a diagnosis of exclusion (Tables 1 and 2). The following criteria can be suggested for its recognition (slightly modified after [17]): (1) severe EDS typically starting in the second or third decade of life, with increased sleep propensity also during active situations (Fig. 2) and with daily or almost daily napping/"sleep attacks"; (2) sleep latency ≤8 min and ≥2 SOREMs on MSLT; (3) no evidence by history, clinical findings, or ancillary tests for other causes of EDS (such as sleep apnea–hypopnea syndromes, chronic sleep insufficiency/irregular sleep–wake habits (Fig. 3), EDS associated with psychiatric disease, substance abuse, neurological disorders, head trauma, etc.).

A history of frequent sleep paralysis/hallucinations, a history of narcolepsy in the family, a DQB1*0602 positivity, and the detection of low CSF hypocretin-1 levels give further support to the diagnosis of narcolepsy without cataplexy. A DQB1*0602 positivity was reported in 41% of patients with narcolepsy without cataplexy, and in 55–85% of those with atypical or mild cataplexy [18]. We found a DQB1*0602 positivity in 14 (89%) of 16 narcoleptics with non-definite (that is mild, rare, or atypical) cataplexy (Fig. 2) [14], and in eight of nine narcoleptics without cataplexy [19]. Low/absent CSF hypocretin-1 levels were found in 17% of the 113 patients with narcolepsy without cataplexy reported in the literature (review in [16]). In a series of nine narcoleptics without cataplexy, we found normal levels in eight patients and low (but detectable) levels in one patient [19]. In a series of nine patients with narcolepsy without cataplexy, Krahn et al. reported normal CSF hypocretin-1 levels in all patients [20]. Thannickal et al. reported a decrease in hypocretin neurons in the lateral hypothalamus of a single patient with narcolepsy without cataplexy [21].

Table 1 Narcolepsy without cataplexy: Diagnostic criteria and differential diagnosis

Diagnostic criteria (modified after [17])
1. Severe EDS present for at least 3 months, with increased sleep propensity also during active situations and with daily or almost daily napping/"sleep attacks"
2. Sleep latency ≤8 min and ≥2 SOREMs on multiple sleep latency test (MSLT)
3. No evidence by history, clinical findings, polysomnography, or other ancillary tests for other causes of EDS (see differential diagnosis)
4. *Supportive criteria*: frequent sleep paralysis/hallucinations (50% of cases), positive of narcolepsy in the family, DQB1*0602 positivity (>40–50%), low CSF hypocretin-1 levels (10–20%).
Differential diagnosis
1. Sleep-disordered breathing[a]
2. Chronic sleep deprivation/insufficiency[a]
3. Depression with EDS and/or hypersomnia[a]
4. Restless legs/periodic limb movement in sleep with EDS
5. Circadian disorders/irregular sleep–wake pattern[a]
6. Severe sleep fragmentation (insomnias, parasomnias)
7. Sedating medications, drug abuse
8. Abuse/withdrawal from stimulants
9. Neurological disorders (head trauma, stroke, hypothalamic lesions, etc.)[a]
10. Medical disorders (hypothyroidism, obesity, hepatic/uremic encephalopathy, hypercapnia, etc.)
11. Periodic hypersomnias
12. Multifactorial (age, medication, etc.)

[a]These conditions may present occasionally with a "narcolepsy-like" phenotype characterized by severe EDS, sleep paralysis, cataplexy-like episodes, and SOREM [7, 11, 14, 24, 47, 49]

The existence of an NREM variant of narcolepsy without cataplexy, that is, a narcolepsy form without SOREM on (repeated) MSLT, is controversial/very rare (<5% of all cases of narcolepsy). Follow-up studies in narcoleptics who develop or lose clinical or polygraphic REM phenomena over time give support to the existence of such an "NREM narcolepsy" [12, 22]. Many of these patients are probably today diagnosed as having a (narcolepsy-like or monosymptomatic) variant of idiopathic hypersomnia with short sleep duration [8].

In 10% of patients with narcolepsy, cataplexy precedes the onset of EDS usually by several months and rarely by a few years. *Isolated cataplectic attacks* are typical of childhood narcolepsy and, particularly when frequent, it should always evoke the possibility of a symptomatic form of the disease [23, 24]. Cataplexy without EDS may be accompanied by normal sleep latencies but SOREMPs on the MSLT (personal observation).

Rarely, narcolepsy first manifests with attacks of *isolated sleep paralysis* with or without *hallucinations*. EDS and cataplexy usually follow within months or a few years. Sleep paralysis, particularly when occurring frequently and at sleep onset and when associated with EDS, should evoke the possibility of (monosymptomatic) narcolepsy, although other sleep disorders must be considered in the differential diagnosis [7].

Familial Narcolepsy

Familial forms of narcolepsy with cataplexy, first reported by Westphal [25], are rare and represent only 1–10% of all cases of narcolepsy [15, 26, 27]. Multiplex families with more than two affected individuals are exceptional [16].

Clinical, neurophysiological and HLA features of familial narcolepsy have been reported by several authors, but usually not in detail [15, 16, 26–31]. No significant differences were found in clinical and neurophysiological findings between familial and sporadic forms of narcolepsy [26]. However, only 50–70% of patients are HLA-DQB1*0602 positive.

Families with both multiple sclerosis and narcolepsy affecting different generations were described by Yoss and Daly, and Nevsimalova et al. [15, 28].

Schrader et al. reported the occurrence of both narcolepsy with cataplexy and multiple sclerosis in a twin pair [32]. A family with narcolepsy associated with deafness, autosomal dominant ataxia, and hypocretin deficiency was reported in Scandinavia [33, 34].

Familial forms of sleep paralysis and cataplexy with and without other narcoleptic symptoms (EDS and sleep paralysis) are exceptional [35–38].

Table 2 The Swiss Narcolepsy Score (SNS)

SWISS NARCOLEPSY SCALE(Sturzenegger & Bassetti, J Sleep Res. 2004;13(4):395–406)					
Q1	How often are you unable to fall asleep?				
	[1] never	**[2]** rarely (less than once a month)	**[3]** sometimes (1–3 times a month)	**[4]** often (1–2 times a week)	**[5]** almost always
Q2	How often do you feel bad or not well rested in the morning?				
	[1] never	**[2]** rarely (less than once a month)	**[3]** sometimes (1–3 times a month)	**[4]** often (1–2 times a week)	**[5]** almost always
Q3	How often do you take a nap during the day?				
	[1] never	**[2]** I would like, but cannot	**[3]** 1–2 times a week	**[4]** 3–5 times a week	**[5]** almost daily
Q4	How often have you experienced weak knees/buckling of the knees during emotions like laughing, happiness, or anger?				
	[1] never	**[2]** rarely (less than once a month)	**[3]** sometimes (1–3 times a month)	**[4]** often (1–2 times a week)	**[5]** almost always
Q5	How often have you experienced sagging of the jaw during emotions like laughing, happiness, or anger?				
	[1] never	**[2]** rarely (less than once a month)	**[3]** sometimes (1–3 times a month)	**[4]** often (1–2 times a week)	**[5]** almost always
Calculation:					
Narcolepsy score = 6×Q1 + 9×+Q2 − 5×Q3 − 11×Q4 − 13×Q5 + 20					
Evaluation/Analysis:					
Narcolepsy score <0: narcoleptic					
Narcolepsy score >0: non-narcoleptic hypersomniac					

In a bicenter study (Zurich–Leyden), a SNS score <0 was found to have a sensitivity of 94% and a specificity of 89% for the diagnosis of hypocretin-deficient narcolepsy (in preparation)

	N-hd	*N*	*NpC*	P-*value*
Number	12	41	16	
Age	38.3 ± 15.9	43.3 ± 18.4	53.7 ± 14.2	0.01*; 0.047**
% males	50	61	50	
Epworth Sleepiness Score	16 ± 3.8	17.4 ± 5.1	17.9 ± 4.7	
SPAS score	8.6 ± 2.9	10.2 ± 3.5	11.8 ± 4.1	0.046
Ullanlinna Narcolepsy Score	21.4 ± 5.8	26.5 ± 6.8	30.2 ± 8.1	0.02
New narcolepsy score[†]	−33.42 ± 28.9	−49 ± 24.4	−47.1 ± 34.1	
HLA DQB1*0602 (%)	90	94	89	
≥2 SOREM in MSLT (%)	100	96	89	
Sleep latency in MSLT (min)	1.6 ± 1.8	2.4 ± 1.8	3.7 ± 2.1	0.045
Sleep latency of < 5 min (%)	100	94	75	

P-values are given as *statistical significance between N-hd and NpC; **statistical significance between N and NpC.

Fig. 2 Comparison of patients with narcolepsy with cataplexy with proven hypocretin deficiency (N-hd) and without hypocretin determination (N), and patients with possible/no cataplexy (NpC). With permission from Sturzenegger and Bassetti (J Sleep Res 2004). New narcolepsy score: Swiss narcolepsy score (Table 2). Sleep propensity during active situations (SPAS): this score is calculated based on four questions concerning sleepiness while driving, standing, eating, and active conversation (normal value for controls: 3.8±0.9) (Sturzenegger and Bassetti, J Sleep Res 2004)

In a study of 23 patients with narcolepsy with cataplexy from nine different families, Mignot et al. reported normal CSF hypocretin-1 levels in 12 (60%) of 20 patients. Nine (41%) of 22 patients were HLA-DQB1*0602 negative and had normal CSF hypocretin levels [39].

Narcolepsy in Twins

Twenty monozygotic narcoleptic twin pairs have been reported in the literature [16, 31, 40–43]. Thirteen (65%) of these 20 pairs are discordant for narcolepsy. Unfortunately, clinical,

a

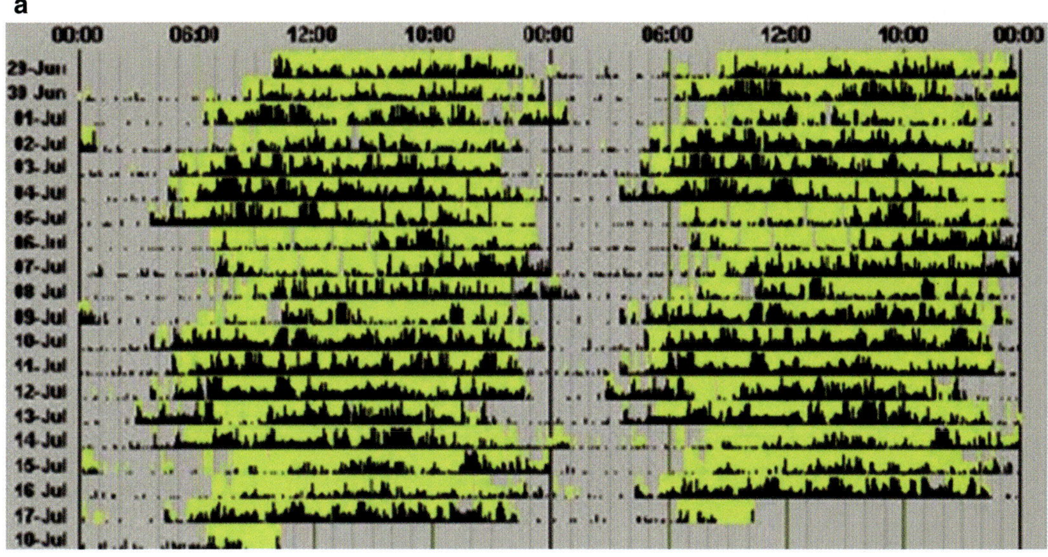

b

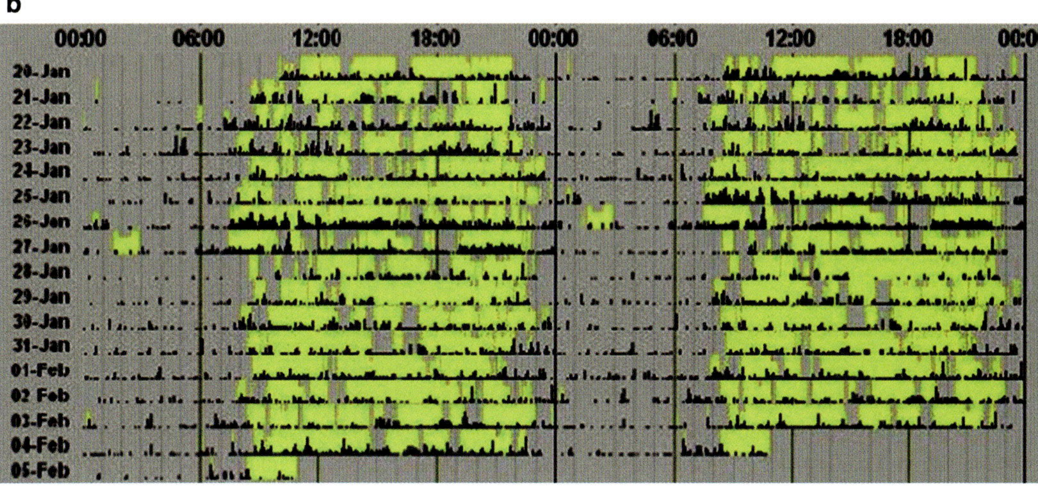

Fig. 3 Severe, narcolepsy-like EDS in a patient with chronic sleep deprivation. (**a**) *Baseline*. Epworth sleepiness score: 20/24. Mean sleep latency: 3.8 min (with 3 SOREMs) on MSLT, 12.9 min on MWT. During a 2-week actigraphy, the patients spent 26% of the time "sleeping." (**b**) *After sleep extension*. Epworth sleepiness score: 4/24. Mean sleep latency: 12.1 min (without SOREM) on MSLT, 31.5 min on MWT. During a 2-week actigraphy, the patients spent 41% of the time "sleeping"

neurophysiological, and HLA findings in most twin pairs are not know in details (the first 16 pairs are reviewed by [31]).

Sixteen (89%) of the 18 twin pairs tested were HLA DQB1*062 positive. Conversely, two of the five concordant twin pairs were HLA DQB1*062 negative. The only concordant twin pair had normal CSF hypocretin-1 levels [43]. A second twin pair discordant for narcolepsy and hypocretin-1 deficiency was reported by Dauvilliers et al. [44].

Non-hypocretin-Deficient Narcolepsy

In a review of the 175 patients with narcolepsy with cataplexy reported in the literature, low/absent CSF hypocretin-1 levels were found in 89 and 59% of patients, respectively [16]. A similar sensitivity was reported by Mignot et al. (87% of 101 patients [39]), Dauvilliers et al. (88% of 26 patients [19]), Krahn et al. (71% of 17 patients [20]), and Baumann and Bassetti (89% of 17 patients [16]). It should be noted, however, that the presence of a definite ("true," "clear-cut") cataplexy was specified in these patients only by Mignot et al. and Baumann and Bassetti, who used identical criteria [45].

Patients with definite cataplexy and normal CSF hypocretin-1 levels also exist. Most of them are HLA negative or have a positive family history for EDS or narcolepsy. In a series of 65 patients with narcolepsy and normal hypocretin-1 levels, Mignot et al. found that 25 (38%) had a typical cataplexy, 17 (26%) had a positive family history, and 41 (66%) were negative for HLA-DQB1*0602 [39]. Similarly, 6 of 11 non-hypocretin-deficient narcoleptics reported by other groups had a positive family history and were HLA negative [16, 19, 20].

Narcolepsy Mimics

Secondary Narcolepsy

In a review of the literature, Nishino and Kanbayashi reported in 2005 a total of 116 cases of symptomatic narcolepsy [24]. The most common causes were inherited disorders (34%, Prader–Willie syndrome, Norrie's disease, Möbious syndrome, Niemann–Pick disease type C, etc.), tumors (29%), head trauma (16%), and multiple sclerosis/demyelinating disorders (9%). Stroke and encephalitis were only rarely reported as a cause of secondary narcolepsy. In 70% of cases, the responsible lesion was found in the hypothalamus, and in other regions of the brain such as brainstem (9%), posterior fossa (6%), or multiple locations (9%), the lesion was less frequently seen.

Many of these reports are questionable/uncertain. Very often, the clinical description is insufficient. Furthermore, the majority of patients present with no or only atypical cataplexy or with cataplexy, but without EDS. Also, cataplexy-like episodes in the course of neurodegenerative disorders or brainstem lesions differ, for example, in terms of sensorimotor manifestations, triggering factors, duration, or accompanying features from cataplexy in patients with idiopathic narcolepsy. Finally, in some cases (e.g., narcolepsy following mild traumatic brain injury), it is possible that brain disease triggered an underlying condition rather than being per se the cause of narcolepsy [46].

Less than 50% of patients with symptomatic narcolepsy are HLA-DQB1*0602 positive and/or CSF hypocretin-1 deficient [24].

Narcolepsy-Like EDS with Sleep-Onset REM Episodes in Sleep-Deprived Subjects and Patients with Other Diseases

Excessive daytime sleepiness with MSLT results suggestive of narcolepsy is not infrequently observed and often leads to the (mis) diagnosis of monosymptomatic narcolepsy. Up to 40% of patients with proven *chronic sleep deprivation/insufficiency* are first diagnosed (in part due to suggestive MSLT findings) with narcolepsy [47].

In a large study of the community-based Wisconsin Sleep Cohort Study (289 males and 267 females, age 35–70 years), multiple SOREMPs were observed in 13% of males and 6% of females. An MSLT result suggesting narcolepsy (mean sleep latency ≤8 min and ≥2 SOREMPs) was observed in 6% of males and 1% of females. Increased sleepiness, sleep paralysis, cataplexy-like episodes, shift work, sleep restriction, decreased mean lowest oxygen saturation at night, and antidepressant drugs were associated (mainly in men) with these MSLT results.

This observation of high prevalence of a narcolepsy-like phenotype in the general population led the authors to suggest a higher than expected prevalence of narcolepsy without cataplexy. Alternatively, we may consider – until proven otherwise – that the presence of such a phenotype (with a "positive" MSLT) is suggestive yet still insufficient for the diagnosis of narcolepsy. This more "parsimonious" attitude is supported by the following considerations:

1. Abnormal MSLT findings can be observed in non-sleepy patients [48].

2. Severe "narcolepsy-like" EDS and cataplexy-like episodes can be observed in the absence of SOREMs.
3. Almost one-third of sleepy patients with "positive MSLT" findings do not have narcolepsy but sleep-disordered breathing, depression, sleep deprivation, or other cause of EDS [49].
4. In a consecutive series of 47 patients diagnosed at our institution as having chronic sleep deprivation (behaviorally induced insufficient sleep), 45% had a severe, "narcolepsy-like" subjective sleepiness (Epworth score >14/24), 52% a mean sleep latency of <5 min on MSLT, and 34% a mean sleep latency of <12 min on MWT (unpublished observations).
5. In patients with "narcolepsy-like" EDS and "positive MSLT," for example, in the course of chronic sleep deprivation, appropriate treatment can reverse both (Fig. 3).

Considering the association of HLA positivity with both SOREM and narcolepsy–cataplexy, and the absence of a biological marker of narcolepsy without cataplexy, the question whether the above-described narcolepsy-like phenotype belongs (or not) to the spectrum of narcolepsy remains open [11].

Patients with brain disorders and particularly those with hypothalamic damage can present with a narcolepsy-like phenotype, without cataplexy, characterized by severe EDS, SOREM, and sometimes hypocretin deficiency or HLA positivity (see below, *symptomatic narcolepsy*) [24].

Cataplexy-Like Symptoms in Healthy Subjects, Sleepy Patients, and Patients with Other Diseases

The diagnosis of "true"/"clear-cut"/"definite"/"classical" cataplexy is easy in patients with typical history, frequent or observed spells with documentation of transient muscle atonia, and tendon areflexia (Table 3) [50, 51]. Not uncommonly, however, history is insufficient or misleading and the differential diagnosis of cataplexy may become difficult (Table 3).

Table 3 Cataplexy: Diagnostic criteria and differential diagnosis

Diagnostic criteria
1. Loss of muscle tone, usually bilaterally and restricted to single/multiple muscle groups (non-generalized)
2. Sudden onset, usually triggered by emotions (most commonly laughing)
3. Short duration (typically <1–2 min)
4. Preserved state of consciousness (at least at the beginning of the spell)
5. *Supportive criteria*: improvement with anti-cataplectic drugs
Differential diagnosis
1. Cataplexy-like episodes in normal subjects/sleepy patients ("weak with laughter," "physiologic cataplexy")
2. Episodes of sudden weakness/falls in patients with psychiatric disorders and (rarely) with narcolepsy (pseudocataplexy)
3. Cataplexy-like episodes in the course of brain disorders (symptomatic narcolepsy)
4. Other neurological "spells":
– Falls (vertebrobasilar ischemia, myopathy/myasthenia, cryptogenic attacks of women, etc.)
– Syncope (including geloplegia = laugh syncope)
– Atonic-astatic seizures [including gelastic (laughing) seizures]
– Sleep attacks, "absence-like" episodes, automatic behaviors
– Startle syndromes (idiopathic, secondary, familial)

Cataplexy-like episodes ("physiological cataplexy") following strong emotions can also be observed in non-narcoleptic normal subjects [14, 52–54]. Such episodes that are reflected by a popular expression such as "weak with laughter," "cela me coupe les jambes," may occur in 3–30% of healthy adults. Healthy subjects (most often woman) may report a urine loss with laughing, a symptom that can also accompany cataplexy. Cataplexy-like episodes tend to be more common in subjects complaining of EDS. They usually involve only the lower limbs (rarely the jaw and facial muscles) and may be triggered by positive and negative emotions (e.g., stress, anxiety, and sorrow) [14].

Cataplexy-like episodes ("pseudocataplexy") can also occur in psychiatric patients [55–58]. Noteworthy, epidemiological studies have shown an association of sleep paralysis and hallucinations in the general population with the anxiety and psychiatric symptoms [59, 60]. Abrupt discontinuation of aminergic reuptake inhibitors may be involved in individual cases [61]. The following points should raise the possibility of "pseudocataplexy"/functional spells: prolonged duration, unusual triggers or circumstances, coexisting psychiatric diagnosis, multiple physician visits, signs or symptoms that resolve by distraction, fluctuating intensity of symptoms, expression of relief or indifference, crying or wimpering, and absence of stereotypy. Clinical examination (loss of muscle tendon reflexes in "true" cataplexy) and video-polysomnographic recordings may be helpful in the differential diagnosis, particularly in the rare cases in which cataplexy and pseudocataplexy may occur in the same patient [62].

As mentioned above, cataplexy-like episodes can also be observed in the course of *symptomatic cataplexy*, sometimes in isolation [24].

Other neurological spells including *gelastic seizures* and laugh syncope (*geloplegia*) are usually easy to be differentiated by history from cataplexy [63–65].

The phenomenological similarity between emotionally triggered cataplexy and cataplexy-like episodes of healthy/sleepy subjects and psychiatric patients suggests the possibility of an overlapping underlying pathophysiology. This hypothesis is further supported by similarities in neurophysiology (suppression of the H-reflex) and neuropharmacology (improvement with noradrenaline and serotonin reuptake inhibitors). In addition, recent data have shown an altered processing of emotional stimuli in narcoleptic patients, with a disturbed interaction between hypothalamus, amygdala, frontal cortex, and ventral tegmental area [66–69].

Taking into consideration all these data, it is tempting to speculate that cataplexy, cataplexy-like episodes, and similar phenomena (including symptomatic cataplexy and gelastic seizures) may represent abnormal variations of hypocretin-related appetite (reward) and aversive behaviors which lead to a more or less complete and enduring recruitment of ponto-medullo-spinal atonia (REM) mechanisms.

Hypocretin-Deficient Conditions Without Narcoleptic Phenotype

In addition to narcolepsy, hypocretin deficiency (decreased CSF hypocretin-1 levels and/or reduction in number of hypocretin neurons in the lateral hypothalamus) has been found in humans in a variety of conditions including the following four groups of disorders (Fig. 4):

1. Hypothalamic damage such as stroke, tumor, or infections (e.g., Morbus Whipple) [24]
2. Parkinson's disease (advanced stage) [70–72]
3. Autoimmune disorders such as Guillan–Barré Syndrome and paraneoplastic encephalopathies [39, 73, 74]
4. Traumatic brain injury (acute phase) [39, 75–77]

The presentation and discussion of these conditions goes beyond the scope of this chapter. The following comments are, however, relevant in the discussion concerning the spectrum of narcolepsy.

First, chronically undetectable levels of CSF hypocretin-1 have been found so far only in narcoleptic patients. This suggests the involvement of very unique (profound and irreversible) mechanisms of neuronal damage in narcolepsy. Second, hypocretin-1 deficiency disorders other than narcolepsy do not necessarily go along with EDS or other narcoleptic symptoms. This observation, together with the knowledge of the existence of non-hypocretin-deficient narcolepsy (see above), suggests the possibility that even in idiopathic narcolepsy, the clinical manifestations may be due at least in part to non-hypocretin-dependent mechanisms.

Conclusions

The spectrum of narcolepsy remains unclear. Current evidence favors the existence of the following four main forms of narcolepsy:

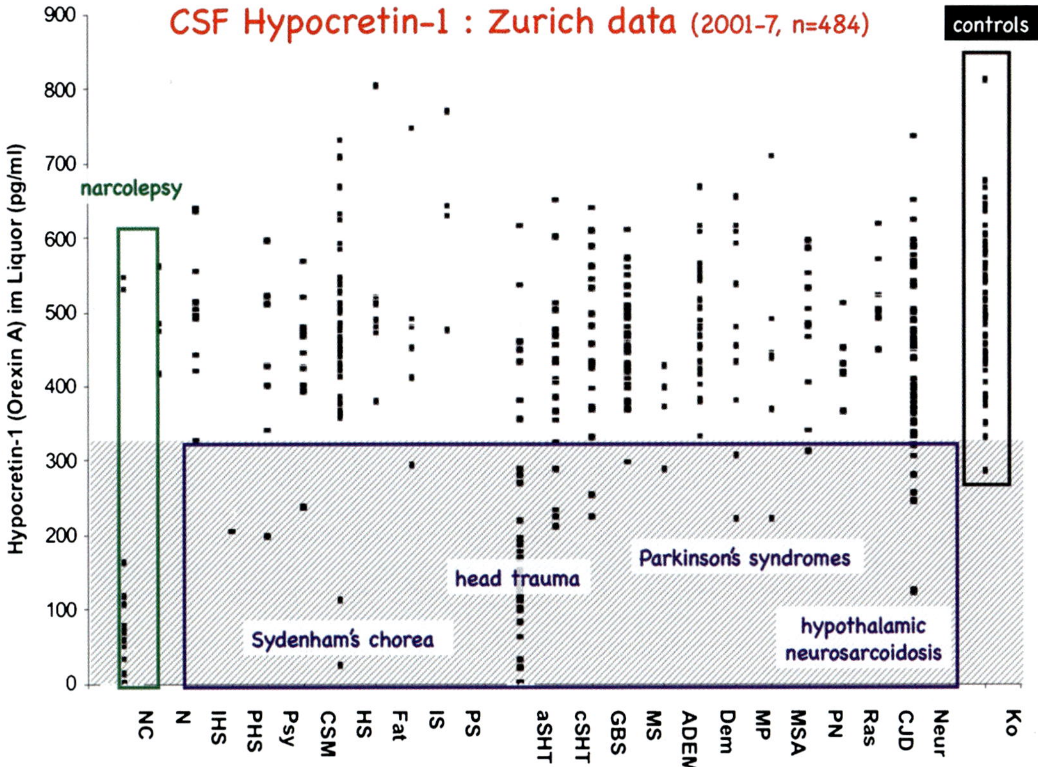

Fig. 4 CSF hypocretin-1 in a series of 484 neurological patients from Zurich, Switzerland. *NC* narcolepsy with cataplexy, *N* narcolepsy without cataplexy, *IHS* idiopathic hypersomnia, *PHS* periodic hypersomnia, *PSY* psychiatric hypersomnias, *IS* chronic sleep deprivation, *aSHT* acute traumatic brain injury, *cSHT* chronic traumatic brain injury, *GBS* Guillain–Barré syndrome, *MP* Parkinson's disease, *MSA* multisystem atrophy, *Neur* other neurological diseases, *KO* normal controls

1. Idiopathic, HLA-positive, hypocretin-1-deficient narcolepsy with cataplexy ("classical narcolepsy")
2. A milder variant of idiopathic narcolepsy without cataplexy and with normal hypocretin-1 levels
3. Familial narcolepsy (more commonly without HLA positivity and hypocretin deficiency)
4. Symptomatic narcolepsy (more commonly without HLA positivity and hypocretin deficiency)

This spectrum reflects the complex etiology of narcolepsy which arises from a variable combination of genetic and environmental factors. Future research will have to address the exact nature (and frequency) of narcolepsy without cataplexy and the relationship between narcolepsy-like symptoms in healthy subjects and patients with "classical narcolepsy," and neurological and psychiatric disorders.

References

1. Adie WJ. Idiopathic narcolepsy: a disease sui generis: with remarks on the mechanism of sleep. Brain. 1926;49:257–306.
2. Wilson SA. The narcolepsies. Brain. 1928;51: 63–109.
3. Yoss RE, Daly DD. Narcolepsy. Med Clin North Am. 1960;44:955–68.
4. Roth B. Narkolepsie. In: Narkolepsie und Hypersomnie. Berlin: VEB Verlag; 1962. p. 130–80.
5. Parkes JD, Baraitser M, Marsden CD, Asselman P. Natural history, symptoms and treatment of the

narcoleptic syndrome. Acta Neurol Scand. 1975;52(5): 337–53.
6. Roth B. Narcolepsy and hypersomnia. Basel: Karger Libri; 1980.
7. Aldrich M. The clinical spectrum of narcolepsy and idiopathic hypersomnia. Neurology. 1996;46:393–401.
8. Bassetti C, Aldrich M. Idiopathic hypersomnia. A study of 42 patients. Brain. 1997;120:1423–35.
9. Billiard M. Diagnosis of narcolepsy and idiopathic hypersomnia. An update based on the International Classification of Sleep Disorders, 2nd edition. Sleep Med Rev. 2007;11:377–88.
10. Silber MH, Krahn LE, Olson EJ, Pankratz VS. The epidemiology of narcolepsy in Olmsted County, Minnesota: a population-based study. Sleep. 2002;25: 197–202.
11. Mignot E, Lin L, Finn L, et al. Correlates of sleep-onset REM periods during the Multiple Sleep Latency Test in community adults. Brain. 2006;129:1609–23.
12. Billiard M, Besset A, Cadilhac J. The clinical and polygraphic development of narcolepsy. In: Guilleminault C, Lugaresi E, editors. Sleep/wake disorders: natural history, epidemiology, and long-term evolution. New York: Raven Press; 1983. p. 171–85.
13. Yoss RE, Daly DD. Criteria for the diagnosis of the narcoleptic syndrome. Mayo Clin Proc. 1957;32: 320–8.
14. Sturzenegger C, Bassetti C. The clinical spectrum of narcolepsy with cataplexy: a reappraisal. J Sleep Res. 2004;13:395–406.
15. Nevsimalova S, Mignot E, Sonka K, Arrigoni JL. Familial aspects of narcolepsy-cataplexy in the Czech republic. Sleep. 1997;20:1021–6.
16. Baumann CR, Dauvilliers Y, Mignot E, Bassetti CL. Normal CSF hypocretin-1 (orexin A) levels in dementia with Lewy bodies associated with excessive daytime sleepiness. Eur Neurol 2004;52:73–6.
17. American Academy of Sleep Medicine. The international classification of sleep disorders. 2nd ed. Westchester, IL: American Academy of Sleep Medicine; 2005.
18. Mignot E, Hajduk R, Black J, Grumet FC, Guilleminault C. HLA DQB1*0602 is associated with cataplexy in 509 narcoleptic patients. Sleep. 1997;20(11):1012–20.
19. Dauvilliers Y, Baumann CR, Maly FE, Billiard M, Bassetti C. CSF hypocretin-1 levels in narcolepsy, Kleine-Levin syndrome, other hypersomnias and neurological conditions. J Neurol Neurosurg Psychiatry. 2003;74:1667–73.
20. Krahn LE, Pankratz VS, Olivier L, Boeve BF, Silber M. Hypocretin (orexin) levels in cerebrospinal fluid of patients with narcolepsy: relationship to cataplexy and HLA DQB1*0602 status. Sleep. 2002;25(7): 733–6.
21. Thannickal TC, Nienhuis R, Siegel JM. Localized loss of hypocretin (orexin) cells in narcolepsy without cataplexy. Sleep. 2009;32(8):993–8.
22. Guilleminault C, Mignot E, Partinen M. Controversies in the diagnosis of narcolepsy. Sleep. 1994;17:S1–6.
23. Challamel MJ, Mazzola ME, Nevsimalova S, Cannard C, Louis J, Revol M. Narcolepsy in children. Sleep. 1994;17:17–20.
24. Nishino S, Kanbayashi T. Symptomatic narcolepsy, cataplexy, and hypersomnia, and their implications in the hypothalamic hypocretin/orexin system. Sleep Med Rev. 2005;9:269–310.
25. Westphal C. Eigentümliche mit Einschlafen verbundene Anfälle. Arch Psychiat Nervenkr. 1877;7: 631–5.
26. Billiard M, Pasquié-Magnetto V, Heckmann M, et al. Family studies in narcolepsy. Sleep. 1994;17:S54–9.
27. Guilleminault C, Mignot E, Grumet FC. Familial patterns of narcolepsy. Lancet. 1989;335:1376–9.
28. Daly DD, Yoss RE. A family with narcolepsy. Mayo Clin Proc. 1959;34:313–9.
29. Nevsimalova-Bruhova S, Roth B. Heredofamilial aspects of narcolepsy and hypersomnia. Schweiz Arch Neurol Neurochir Psychiatr. 1972;110:45–54.
30. Mayer G, Lattermann A, Mueller-Echardt G, Sbanborg E, Meier-Ewert K. Segregation of HLA genes in multicase narcolepsy families. J Sleep Res. 1998;7:127–33.
31. Mignot E. Genetic and familial aspects of narcolepsy. Neurology. 1998;50 Suppl 1:S16–22.
32. Schrader H, Gotlibsen OB, Skomedal GN. Multiple sclerosis and narcolepsy/cataplexy in a monozygotic twin. Neurology. 1980;30(1):105–8.
33. Melberg A, Dahl N, Hetta J, Valind S. Neuroimaging study in autosomal dominant cerebellar ataxia, deafness, and narcolepsy. Neurology. 1999;53:2190–2.
34. Melberg A, Ripley B, Lin L, Hetta J, Mignot E, Nishino H. Hypocretin deficiency in familial symptomatic narcolepsy. Ann Neurol. 2001;49(1):136–7.
35. Roth B, Bruhova S, Berkova L. Familial sleep paralysis. Schweiz Arch Neurol Neurochir Psychiatr. 1968;102:321–30.
36. Gelardi JAM, Brown JW. Hereditary cataplexy. J Neurol Neurosurg Psychiatr. 1967;30:455–7.
37. Dahlitz M, Parkes JD. Sleep paralysis. Lancet. 1993;341:406–7.
38. Hartse KM, Zorick F, Sicklesteel J, Roth T. Isolated cataplexy: a familial study. Henry Ford Hosp Med J. 1988;36:24–7.
39. Mignot E, Lammers GJ, Ripley B, et al. The role of cerebrospinal fluid hypocretin measurement in the diagnosis of narcolepsy and other hypersomnias. Arch Neurol. 2002;59:1553–62.
40. Partinen M, Hublin C, Kaprio J, Koskenvuo M, Guilleminault C. Twin studies in narcolepsy. Sleep. 1994;17:S13–6.
41. Kaprio J, Hublin C, Partinen M, Heikkilä K, Koskenvuo M. Narcolepsy-like symptoms among adult twins. J Sleep Res. 1996;5(1):55–60.
42. Honda M, Honda Y, Uchida S, Miyazaki S, Tokunaga K. Monozygotic twins incompletely concordant for narcolepsy. Biol Psychiatry. 2001;49:943–7.
43. Khatami R, Maret S, Werth E, et al. A monozygotic twin pair concordant for narcolepsy-cataplexy without any detectable abnormality in the hypocretin (orexin) pathway. Lancet. 2004;363:1199–200.

44. Dauvilliers Y, Maret S, Bassetti CL, et al. A monozygotic twin pair discordant for narcolepsy and CSF hypocretin-1. Neurology. 2004;62:2137–8.
45. Anic-Labat S, Guilleminault C, Kraemer HC, Meehan J, Arrigoni J, Mignot E. Validation of a cataplexy questionnaire in 983 sleep-disorders patients. Sleep. 1999;22:77–87.
46. Lankford DA, Wellmann JJ, O'Hara C. Posttraumatic narcolepsy in mild to moderate closed head injury. Sleep. 1994;17:25–8.
47. Roehrs TA, Roth A. Chronic insufficient sleep and its recovery. Sleep Med. 2003;4(1):5–6.
48. Bishop C, Rosenthal L, Helmus T, Roehrs T, Roth T. The frequency of multiple sleep onset REM periods among subjects with no excessive daytime sleepiness. Sleep. 1996;19(9):727–30.
49. Aldrich MS, Chervin RD, Malow BA. Value of the multiple sleep latency test (MSLT) for the diagnosis of narcolepsy. Sleep. 1997;20(8):620–9.
50. Dyken ME, Yamada M, Lin-Dyken DC, Seaba P, Yeh M. Diagnosing narcolepsy through the simultaneous clinical and electrophysiologic analysis of cataplexy. Arch Neurol. 1996;53:456–60.
51. Rubboli G, d'Orsi G, Zaniboni A, et al. A video-polygraphic analysis of the cataplectic attack. Clin Neurophysiol. 2000;111 Suppl 2:S120–8.
52. Paskind HA. Effect of laughter on muscle tone. Arch Neurol Psychiatry. 1932;28:623–8.
53. Partinen M. Sleeping habits and sleep disorders of Finnish men before, during and after military service. Ann med Milit Fenn. 1982;57:1–96.
54. Overeem S, Lammers GJ, van Dijk JG. Weak with laughter. Lancet. 1999;354:838.
55. Rosenthal C. Ueber das Auftreten von halluzinatorisch-kataplektischen Angstsyndrom, Wachanfällen und ähnlichen Störungen bei Schizophrenen. Monatsschr Psychiatr Neurologie. 1939;102:11–38.
56. Krahn LE, Hansen MR, Shepard JW. Pseudocataplexy. Psychosomatics. 2001;42:356–8.
57. Kishi Y, Konishi S, Koizumi S, et al. Schizophrenia and narcolepsy: a review with a case report. Psychiatry Clin Neurosci. 2004;58:117–24.
58. Krahn LE. Reevaluating spells initially identified as cataplexy. Sleep Med. 2005;6:537–42.
59. Ohayon MM, Zulley J, Guilleminault C, Smirne S. Prevalence and pathologic associations of sleep paralysis in the general population. Neurology. 1999;52(6):1194–200.
60. Ohayon M. Prevalence of hallucinations and their pathological associations in the general population. Psychiatry Res. 2000;97:153–64.
61. Nissen C, Feige B, Nofzinger EA, et al. Transient narcolepsy-cataplexy syndrome after discontinuation of the antidepressant venlafaxine. J Sleep Res. 2005;14(2):207–8.
62. Plazzi G, Khatami R, Serra L, Pizza F, Bassetti CL. Pseudocataplexy in narcolepsy with cataplexy. Sleep Med 2010;11:591–4.
63. Oppenheim H. Ueber Lachschlag. Monatsschr Psychiatr Neurologie. 1902;11:242–7.
64. Totah AR, Benbadis SR. Gelastic syncope mistaken for cataplexy. Sleep Med. 2002;3:77–8.
65. Bellman M, Pathan AH, Sinclair L. A girl who laughed and fell down. Lancet. 2000;355:2216.
66. Khatami R, Birkmann S, Bassetti CL. Absence of aversive startle reflex potentiation in human narcolepsy-cataplexy: implications for amygdala dysfunction. J Sleep Res. 2007;16:226–9.
67. Schwartz S, Ponz A, Poryazova R, Werth E, Bösiger P, Bassetti CL. Abnormal activity in hypothalamus and amygdala during humour processing in humany narcolepsy with cataplexy. Brain. 2008;131:514–22.
68. Ponz A, Khatami R, Poryazova R, et al. Abnormal activity in reward brain circuits in human narcolepsy with cataplexy. Ann Neurol. 2010;67:190–200.
69. Ponz A, Khatami R, Poryazova R, et al. Reduced amygdala activity during aversive conditioning in human narcolepsy. Ann Neurol. 2010;67:394–8.
70. Overeem N, Van Hilten JJ, Ripley B, Mignot E, Nishino H, Lammers C. Normal hypocretin-1 levels in Parkinson's disease patients with excessive daytime sleepiness. Neurology. 2002;58:498–9.
71. Baumann CR, Scammell TE, Bassetti CL. Parkinson's disease, sleepinesss and hypocretin/orexin. Brain. 2008;131:e91.
72. Fronczeck R, Overeem S, Lee SY, et al. Hypocretin (orexin) loss in Parkinson's disease. Brain. 2007;130:1577–85.
73. Nishino S, Kanbayashi T, Fujiki N, et al. CSF hypocretin levels in Guillain-Barré syndrome and other inflammatory neuropathies. Neurology. 2003;61:823–5.
74. Overeem N, Dalmau J, Bataller L, et al. Hypocretin-1 CSF levels in anti-Ma2 associated encephalitis. Neurology. 2004;62(1):138–40.
75. Baumann CR, Stocker R, Imhof HG, et al. Hypocretin-1 (orexin A) deficiency in acute traumatic brain injury. Neurology. 2005;65:147–9.
76. Baumann CR, Werth E, Stocker R, Luwid S, Bassetti CL. Sleep-wake disturbances 6 months after traumatic brain injury. Brain. 2007;130:1873–83.
77. Baumann CR, Bassetti CL, Valko PO, Haybaeck J, Keller M, Clark E, Stocker R, Tolnay M, Scammell TE. Loss of hypocretin (orexin) neurons with traumatic brain injury. Ann Neurol 2009;66:555–9.

Secondary Narcolepsy

Philipp O. Valko and Rositsa Poryazova

Keywords

Secondary narcolepsy • Tumor • Stroke • Inflammation • Neurodegeneration • Hypocretin • Orexin

Introduction

Idiopathic narcolepsy with cataplexy is caused by a deficiency of the hypothalamic hypocretin system [1]. Selective loss of hypocretin-producing neurons in the dorsolateral hypothalamus with sparing of adjacent melanin-concentrating hormone cells, a tight linkage to the HLA DQB1*0602 haplotype, and the recent discovery of specific antibodies (Trib2) suggest an autoimmune etiology [2–6]. The disorder is characterized by excessive daytime sleepiness (EDS) and cataplexy. In addition, fragmented nocturnal sleep, hypnagogic or hypnopompic hallucinations, sleep paralysis, periodic limb movements in sleep (PLMS), intense dreaming, and REM sleep behavior disorder (RBD) are frequently encountered [7]. According to the diagnostic criteria of the International Classification of Sleep Disorders, diagnosis of narcolepsy with cataplexy can be made in the presence of persisting EDS for at least 3 months, definite history of clear-cut cataplexy and specific findings in the multiple sleep latency test (MSLT) [mean sleep latency ≤8 min, ≥2 sleep onset rapid eye movement (SOREM) periods] [8]. If cataplexy is absent or atypical, narcolepsy without cataplexy is diagnosed [8].

If a medical condition leads to sleep–wake disturbances bearing resemblance to idiopathic narcolepsy or fulfilling the current diagnostic criteria of narcolepsy, the term "secondary" or "symptomatic" narcolepsy has generally been used. Although a coincidental occurrence of two unrelated disorders cannot always be ruled out, secondary narcolepsy must be suspected if its symptom onset shows a close time relationship with the underlying medical condition. However, the designation "secondary narcolepsy" is misleading, since the authentic presence of secondary forms of narcolepsy with cataplexy remains doubtful [9]. In most cases, it would be more appropriate to speak of a "narcolepsy-like phenotype," because the reported sleep–wake disturbances usually differ from that of idiopathic narcolepsy, and the presence and nature of cataplexy is often questionable. For instance, many patients with so-called secondary narcolepsy were reported to suffer from hypersomnia, whereas the amount of sleep per 24 h is typically not altered in idiopathic narcolepsy. Hence, if hypersomnia

P.O. Valko (✉)
Department of Neurology, University Hospital of Zurich, Frauenklinikstrasse 26, 8091 Zurich, Switzerland
e-mail: philipp.valko@usz.ch

C.R. Baumann et al. (eds.), *Narcolepsy: Pathophysiology, Diagnosis, and Treatment*,
DOI 10.1007/978-1-4419-8390-9_30,

emerges in the context of secondary narcolepsy, additional injury to sleep–wake regulating structures other than the hypocretin system must be suspected. The majority of secondary narcolepsy cases are caused by hypothalamic lesions. However, while narcolepsy with cataplexy is characterized by a selective damage to the hypothalamic hypocretin system, secondary narcolepsy forms are usually due to a more widespread destruction affecting various hypothalamic (and even other) nuclei. This fact explains why invasion of the hypothalamus by a neoplastic, neurodegenerative, or inflammatory process hardly ever produces the full and highly specific clinical picture of idiopathic narcolepsy with cataplexy.

Along with the ascending arousal system of the brainstem and cell groups in the basal forebrain, the hypothalamus is crucial for arousing the cerebral cortex and the thalamus [10]. Moreover, a complex network of hypothalamic nuclei is critically involved in both homeostatic and circadian regulation of sleep [10]. The hypocretins are believed to stabilize a mutually inhibitory system between sleep and wake inducing cell groups, and hypocretin deficiency may therefore lead to frequent and unwanted transitions between wakefulness and sleep [11, 12]. This peculiar and distinguishing symptom of idiopathic narcolepsy, however, rarely appears in the setting of secondary narcolepsy.

A second problem concerns the low specificity of the diagnostic criteria for narcolepsy without cataplexy. Multiple SOREM periods have been found to occur regularly also in other sleep–wake disorders with EDS, including sleep apnea syndrome and behaviorally induced insufficient sleep syndrome [13, 14]. Furthermore, a MSLT study in normal adults showed that 4.2% of males and 0.4% of women met the criteria for narcolepsy without cataplexy [15]. Finally, many neurological disorders can manifest with symptoms that are also characteristic for narcolepsy, for instance EDS, hallucinations, or RBD, and will be discussed here despite the additional presence of other symptoms, which entirely distinguish them from idiopathic narcolepsy.

In 1880, Gélineau recognized narcolepsy with cataplexy as a primary sleep disorder; in his original paper, he analyzed 14 cases and already differentiated between idiopathic and secondary forms [16]. Later on, cases of secondary narcolepsy were reported during the encephalitis lethargica pandemic (1916–1923) [17], and von Economo, based on the observation of inflammatory lesions in the posterior hypothalamus, was first to suggest that this region played a crucial role in the promotion of wakefulness and that narcolepsy reflects a disorder originating in this brain area [18, 19]. The experience with secondary forms led to a better recognition and characterization of idiopathic narcolepsy with cataplexy, and it was Adie in 1926 to confirm that narcolepsy was "a disease sui generis" [20].

While these historical cases of secondary narcolepsy helped in recognizing idiopathic narcolepsy with cataplexy as a distinct sleep–wake disorder, and identifying the role of the hypothalamus in sleep–wake regulation, they nevertheless remain poorly documented, as sleep studies were not yet performed, and neuroimaging or CSF measures were not available. More recently, reduced CSF hypocretin-1 levels have been found in several patients with secondary narcolepsy, indicating that impairment of hypocretin signaling may contribute to the emergence of narcolepsy-like symptoms in these patients. The discovery of the hypothalamic hypocretins promoted a large body of studies to explore their possible involvement in different neurological disorders presenting with narcolepsy-like symptoms. The disorders that will be revisited in this chapter can be divided in two categories. The first group consists of sporadic disorders of the central nervous system (CNS) leading to injury of the hypothalamus or – more rarely – of the upper brainstem tegmentum. This group includes neoplastic, inflammatory or infectious diseases, and traumatic brain injury. The second group comprises hereditary diseases of infancy on one hand, and neurodegenerative disorders affecting mainly the elderly patient on the other hand.

Sporadic Disorders of the CNS with Narcolepsy-Like Symptoms

Inflammatory and Immune-Mediated Neurological Disorders

Multiple Sclerosis

Multiple sclerosis (MS) is an inflammatory disorder characterized by pleomorphic clinical symptoms as a result of disseminated demyelination within the brain and spinal cord. MS is much more prevalent in Northern countries and uncommon in equatorial climates. Similarly, the reported prevalence of narcolepsy with cataplexy in Great Britain, France, Czech Republic, or Finland is higher than in Israel [21]. Co-occurrence of both MS and narcolepsy with cataplexy in the same individual seems to be more common than expected by chance, which is of particular interest, as it may improve our understanding of the putative autoimmune etiology of the latter. In 1928, Guillain and Alajouanine already reported cases of narcolepsy in patients with MS [22]. Later on, new reports appeared on sporadic and familial narcolepsy with cataplexy and coexistent MS [23–25]. The association with the same HLA-DR2 haplotype led to the assumption of a common genetic susceptibility to MS and narcolepsy [26, 27]. Furthermore, a questionnaire-based study including 70 patients with definite MS suggested that the prevalence of sleep attacks, sleep paralysis, and cataplexy was higher than in normal subjects [28]. However, all 47 patients with relapsing-remitting MS or optic neuritis had normal CSF hypocretin-1 levels, irrespective whether the lumbar puncture was performed in attack or in remission [29]. In addition, in almost all recent cases of co-occurrent narcolepsy and MS, inflammatory lesions in the hypothalamus (mostly bilaterally) were found in brain magnetic resonance imaging (MRI), thus indicating a secondary origin of narcolepsy and arguing against a common immunogenetic etiology [30–34]. For instance, Oka et al. described a 22-year-old woman with acute hypersomnia [33]. This patient did not present cataplexy, sleep paralysis, or hypnagogic hallucinations, but a first MSLT performed 11 days following symptom onset showed a mean sleep latency of 2.8 min and 5/5 sleep onset REM (SOREM) periods, and CSF hypocretin-1 level was undetectable [33]. Brain MRI revealed bilateral hypothalamic plaques. Hypersomnia resolved within 2 weeks following administration of glucocorticoids. Four months later, both MSLT and CSF hypocretin-1 concentration returned to normal. Similar cases of MS patients with acute hypothalamic plaques causing transient severe EDS, narcolepsy-like MSLT findings and undetectable CSF hypocretin-1 levels and remission under steroid treatment have been reported [32, 35, 36]. Some authors hypothesized that other narcoleptic symptoms, namely cataplexy, might have developed if the disease duration was longer, thereby giving way to a more profound and irreversible destruction of the hypothalamic hypocretin system [33]. In line with this assumption is the recent report of the first case of idiopathic narcolepsy with cataplexy, in which undetectable CSF hypocretin-1 levels completely recovered to normal concentrations after the administration of intravenous immunoglobulin perfusions [37].

Neuromyelitis Optica and Acute Disseminated Encephalomyelitis

Neuromyelitis optica (NMO) – also called Devic disease – is a variant of MS characterized by uni- or bilateral optic neuritis in combination with a severe transverse or ascending myelitis. Recent immunohistochemical studies identified a highly specific antibody directed against aquaporin 4 (AQP4), suggesting a humoral immunogenetic pathomechanism [38]. Brain MRI in NMO patients may show additional abnormalities with a certain predominance of the diencephalic region [39]. Interestingly, AQP4 is strongly expressed in the hypothalamus [40, 41]. In the last 5 years, five NMO patients with secondary narcolepsy or coma-like episodes in the context of bilateral hypothalamic lesions have been reported [42–45]. In one patient, narcolepsy was the presenting symptom, and optic neuritis and transverse myelitis developed only 2 months later [45]. Lumbar puncture was performed in three of them, and all showed reduced CSF hypocretin-1 levels [42, 44, 45]. Serum anti-AQP4 antibodies were positive in these patients. In two patients, repeated CSF hypocretin-1 measurements were performed and showed significant improvement after steroid treatment, accompanied by reduction of EDS and regression of the hypothalamic lesions [42, 45]. Similar to the above-mentioned MS patients with secondary narcolepsy, these NMO cases illustrate that inflammatory attacks to the hypothalamus induced by specific anti-AQP4 antibodies may secondarily produce transient narcoleptic symptoms, probably by means of an acute but reversible inhibition of hypocretin function. Alternatively, demyelination of hypocretin neurons projecting through the periaqueductal gray, where AQP4 expression is particularly high, may lead to an interruption of hypocretin signaling.

Acute disseminated encephalomyelitis (ADEM) is an acute monophasic inflammatory disorder with multifocal demyelination, usually associated with a recent infection or vaccination. Six ADEM cases with bilateral hypothalamic involvement and secondary symptoms of narcolepsy or hypersomnia have been reported [46–50]. Reversible alterations of CSF hypocretin-1 concentrations were documented in four patients [47–49]. Gledhill et al. described a 38-year-old woman with ADEM and inflammatory lesions in the hypothalamus and extension into the upper brainstem; she had 4/4 SOREM periods in the MSLT, low CSF hypocretin-1 level (87 pg/ml), and was DQB1*0602 positive [49]. Similar to the majority of other secondary narcolepsy cases in MS, NMO, or ADEM, cataplexy did not evolve in this woman, possibly due to successful treatment with steroids and subsequent improvement of hypocretin level.

Guillain–Barré Syndrome

Guillain–Barré syndrome (GBS) is an acute autoimmune polyradiculoneuritis preceded by an upper respiratory or gastrointestinal infection. Different antiganglioside antibodies are targeting peripheral nerves, but CNS symptoms are frequent as well. Almost one in three GBS patients may report hallucinations, illusions, and RBD with intense dream mentation [51–53]. The syndrome of inappropriate antidiuretic hormone secretion (SIADH) is also common in GBS and might be explained in the context of acute dysautonomia or by a direct hypothalamic involvement [54]. After narcolepsy, GBS was the second disorder that was identified to result sometimes in abnormally low or undetectable CSF hypocretin-1 levels [55, 56]. A consecutive study explored CSF hypocretin-1 levels in 28 patients with GBS, 12 with Miller-Fisher syndrome (MFS), 12 with chronic inflammatory demyelinating polyneuropathy (CIDP), and 48 control subjects [57]. CSF hypocretin-1 levels were undetectable in 7 GBS patients with severe autonomic disturbances, and abnormally low in 11 GBS and 5 MFS patients [57]. As all GBS patients with undetectable hypocretin-1 levels were Japanese, a follow-up study was carried out that included 11 Caucasian GBS patients and 1 MFS patient, and revealed normal CSF hypocretin-1 levels in 9/11 GBS patients and in the MFS patient but moderately decreased values in 2 GBS patients [58]. Pathophysiologic or immunogenetic differences in Japanese and Caucasian GBS forms may account for this discrepancy [57, 58]. The underlying pathomechanism of hypocretin deficiency in GBS remains unclear. The hypothesis of an immune-mediated injury to hypocretin-containing neurons in GBS entailed a screening study for antiganglioside antibodies in idiopathic narcolepsy, which failed, however, to detect increased titers of these antibodies [59].

Anti-Ma2-Associated Encephalitis and Other Immune-Mediated Limbic Encephalitis

Anti-Ma2-associated encephalitis is a paraneoplastic autoimmune disorder, most often associated with germ-cell tumors of the testis and non-small cell lung cancer [60–62]. Usual targets of anti-Ma2 antibodies are limbic, diencephalic, and upper brainstem structures. Patients present with memory loss, eye movement abnormalities, and endocrine dysfunction. EDS has also been reported as a presenting symptom [63]. In a large series of anti-Ma2-associated encephalitis, the prevalence of symptomatic narcolepsy was 32% [62]. Another study explored CSF hypocretin-1 levels in six patients with positive anti-Ma2 antibodies; four of them had EDS [64]. All patients with EDS had undetectable CSF hypocretin-1 levels; none had cataplexy [64]. Brain MRI revealed mesiotemporal, periventricular, hypothalamic, and mesencephalic abnormalities [64]. So far, sleep studies have been performed in only two patients with anti-Ma2 associated encephalitis, and both fulfilled the diagnostic criteria for narcolepsy [65, 66]. The patient described by Landolfi et al. is notable for the additional presence of cataplexy. Compta et al. reported a 69-year-old man with paraneoplastic anti-Ma2-associated encephalitis, evidenced on brain MRI by T2-weighted hyperintense lesions involving the dorsolateral midbrain, both hippocampi and amygdala, and characterized clinically by severe hypersomnia with multiple irresistible naps, memory loss, diplopia, and gait disturbance [66]. RBD with vigorous dream-enacting behaviors was detected on nocturnal polysomnography. MSLT showed a mean sleep latency of 7 min with 4/5 SOREM periods, and CSF hypocretin-1 level was low (49 pg/ml). This case is a very rare example of a neurological condition that simultaneously gives rise to both secondary narcolepsy and secondary RBD. In the same year, Mathis et al. reported another patient with secondary narcolepsy/RBD [67]. This 30-year-old man additionally presented sleep paralysis, hypnagogic hallucinations, and cataplexy-like episodes. Brain MRI showed an isolated lesion in the mediotegmental pons, probably caused by acute brainstem encephalitis. MSLT displayed a mean sleep latency of 5 min with two SOREM periods. In contrast to the case of Compta et al., however, this patient was found to have normal CSF hypocretin-1 levels, suggesting that hypocretin deficiency is not a prerequisite for the emergence of secondary narcolepsy/RBD.

Secondary RBD without narcolepsy have been reported in five patients with nonparaneoplastic limbic encephalitis associated with voltage-gated potassium channel (VGKC) antibodies [68], and in a patient with acute aseptic limbic encephalitis [69]. In contrast to the two cases with secondary narcolepsy/RBD, these six patients showed MRI abnormalities in the mesial temporal lobes without any evidence of brainstem involvement [68, 69]. CSF hypocretin-1 levels were measured in four patients with VGKC limbic encephalitis, resulting normal in three and intermediate in one patient [68].

Behçet's Disease

Very recently, two reports appeared on secondary narcolepsy in Behçet's disease [70, 71]. Behçet's disease is a chronic inflammatory multisystem disorder characterized by recurrent oral and genital ulcers, uveitis, and skin lesions. CNS involvement is observed in 10–20% of Behçet patients [72]. Hsieh et al. reported a 44-year-old woman with Behçet's disease and secondary narcolepsy, as evidenced clinically by EDS, sleep paralysis, hypnagogic hallucinations, and insomnia, and by typical MSLT findings including 4/4 SOREM periods [70]. Baumann et al. reported a 31-year-old man with recurrent oral and painful genital ulcers since childhood and a 2-month-history of headache, hypersomnia (12–14 h sleep per day), and EDS (ESS score 14/24) [71]. Flair and T2-weighted cranial MRI images revealed increased signal intensity in bilateral subthalamic diencephalic areas with a small left-sided subthalamic gadolinium enhancement (Fig. 1a–c). Additional lesions were found in the pons and the cerebral pedunculi (not shown). CSF hypocretin-1 level was decreased (215 pg/ml; normal: >320 pg/ml). Treatment with high-dose prednisone and azathioprine led to a regression of EDS (ESS score 7/24) and hypersomnia (7–8 h/day), the hyperintense lesions vanished on T2-weighted cranial MRI images (Fig. 1d), and CSF hypocretin-1 returned to normal (400 pg/ml).

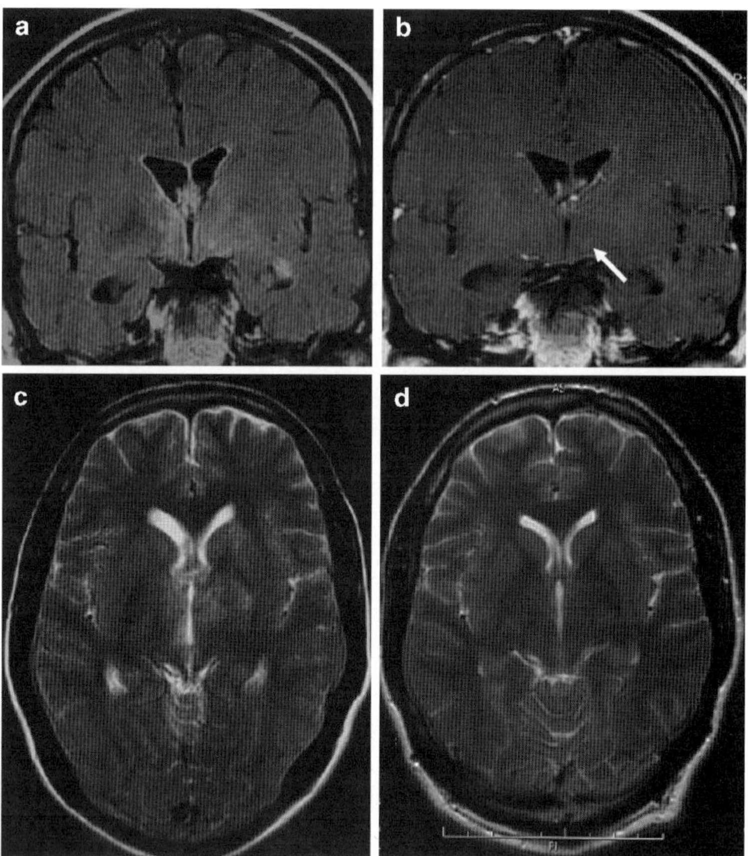

Fig. 1 (**a**) Flair imaging demonstrates extended areas of signal hyperintensity in bilateral thalamic and subthalamic areas as well as a small area of signal hyperintensity in the left hippocampus. (**b**) On gadolinium-enhanced T1-weighted imaging there is a small area of gadolinium enhancement in the left thalamus (*arrow*). (**c**) T2-weighted imaging shows left-dominant areas of signal hyperintensity in both thalami. (**d**) Follow-up examination 6 months later demonstrates remission of this finding

Infections of the CNS

A few cases of secondary narcolepsy occurring in the setting of CNS infections have been reported. Neurocysticercosis is the most common CNS parasitic disease endemic in Central and South America, characterized by small parenchymal, subarachnoidal or intraventricular cysts, and meningitis. Watson et al. described a patient with secondary narcolepsy without cataplexy, resulting from neurocysticercosis [73]. The patient had both subjective and objective EDS including SOREM periods on MSLT. Brain MRI revealed a hypothalamic cyst, but CSF hypocretin-1 level was normal [73].

Maia et al. reported a 45-year-old man with irresistible sleep attacks, insomnia, and new-onset sleepwalking [74]. In addition, he presented supranuclear ophthalmoplegia, oculomasticatory myorhythmia, hyperphagia, and subcortical dementia. The patient was HLA-DQB1*0602 positive, but MSLT was normal. CSF hypocretin-1 level was not available. Brain MRI revealed T2-weighted signal alterations in the caudate head, hippocampus, and right amygdala with extension to the diencephalic–hypothalamic region. Duodenal biopsy led to the diagnosis of Whipple disease (WD). Another patient with WD presented with transient total sleep loss [75]. Cerebral [^{18}F]deoxyglucose positron emission tomography (FDG-PET) showed hypermetabolism in diencephalic and brainstem areas, including the mesencephalic tectal plate and pontine tegmentum. Surprisingly, CSF hypocretin-1 levels were diminished in the acute phase and at 6-month follow-up (113 and 89 pg/ml, respectively) [75].

Several cases of sarcoidosis with involvement of hypothalamic–diencephalic structures have been reported [76–80]. Hypothalamic sarcoidosis was associated with secondary narcolepsy including four SOREM periods on MSLT [76–78], with secondary Kleine–Levine syndrome [79], and with hypersomnia and decreased CSF hypocretin-1 level (110 pg/ml) [80]. Cataplexy was reported in none of these patients.

While secondary narcolepsy due to CNS infections seems to be very rare, streptococcal infections have recently been recognized as a probable environmental trigger for idiopathic narcolepsy. Two hundred DQB1*0602 patients with narcolepsy and cataplexy and low CSF hypocretin-1 levels were found to display significantly higher antistreptococcal antibodies compared to 200 age-matched healthy controls [81].

Tumors of the CNS

CNS tumors account for a large proportion of cases with secondary narcolepsy. The majority of cases reported so far are associated with direct tumor invasion of the hypothalamus or adjacent structures [82–84]. In contrast, secondary narcolepsy or cataplexy due to brainstem tumors occurs very rarely. Two cases associated with a pontine pilocytic astrocytoma [85] and a fourth ventricular subependymoma [86] have been reported, the former being characterized by isolated cataplexy. Craniopharyngioma, typically located in the suprasellar region, seems to be the most frequent tumor associated with secondary narcolepsy. Since this tumor affects primarily children younger than 10 years, secondary narcolepsy is often diagnosed by pediatricians. Out of 115 pediatric patients with craniopharyngeoma, 35 patients (30%) presented with EDS, and 4 of them had MSLT findings fulfilling the diagnostic criteria of narcolepsy [87].

Differently from idiopathic narcolepsy, symptomatic narcolepsy due to brain tumors can be reversible after tumor resection, chemo- or radiotherapy, or embolization. Dauvilliers et al. recently reported a 46-year-old man with hypersomnia and irresistible sleep attacks while eating or speaking, but without cataplexy, sleep paralysis, or hallucinations [88]. A primary CNS B-cell lymphoma with infiltration in the hypothalamic region was diagnosed. CSF hypocretin-1 level was undetectable. After treatment with corticosteroids and chemotherapy, hypersomnia and sleep attacks resolved completely. Lumbar puncture was repeated and normal CSF hypocretin-1 level was found. There are previous reports with disappearance of narcoleptic symptoms after tumor treatment [89–92]. In the case of Nokura et al., CSF hypocretin-1 was measured once and was decreased [92], the other three cases were published prior to the discovery of the hypocretins [89–91]. In one case, secondary narcolepsy improved after embolization and radiosurgery of an arteriovenous malformation located around the third ventricle [90].

Tumor infiltration in the hypothalamus does not necessarily lead to narcoleptic symptoms. Conversely, it appears as a peculiarity that many cases of secondary narcolepsy evolved as a complication after removal of the tumor by surgery or radiotherapy [93]. Such iatrogenic cases include a patient with acromegaly and development of secondary narcolepsy 2 weeks after radiotherapy for his pituitary adenoma [94]; transient cataplexy after removal of a craniopharyngioma [95]; a 16-year-old girl with severe EDS after removal of a hypothalamic Grade 2 pilocystic astrocytoma, with mean sleep latency of 1.7 min on MSLT and decreased CSF hypocretin-1 level (102 pg/ml) [96]; a 28-year-old woman with a choroid plexus carcinoma of the pineal gland, who developed severe EDS, sleep paralysis, and hypnagogic hallucinations after pinealectomy followed by chemo-/radiotherapy, showing normal CSF hypocretin-1 level [97]; a 11-year-old girl with a hypothalamic craniopharyngioma, who developed narcolepsy characterized by hypersomnia, mean sleep latency of 1.4 min with 3/5 SOREM periods on MSLT, increased amount

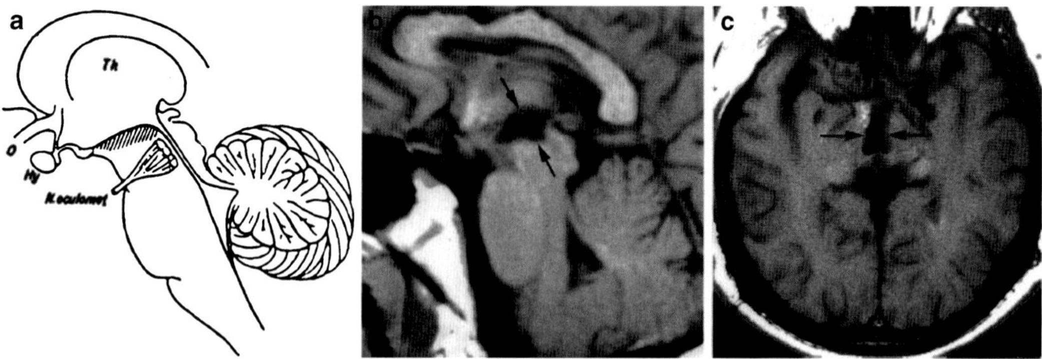

Fig. 2 (**a**) Sagittal section, showing brainstem, mesencephalon, and cerebellum. The hatching shows regions in which lesions produced hypersomnia (adapted from von Economo). (**b**, **c**) T1-weighted sagittal (**b**) and axial (**c**) MR images. Lesions are shown between the *arrows*, demonstrating ex vacuo changes, located in the posterior hypothalamus and in the rostral midbrain. *Th* thalamus, *o* optic nerve, *Hy* hypothalamus, *N. oculomot* occulomotor nerve

of REM sleep (41%) on polysomnography, and low CSF-hypocretin-1 level (93 pg/ml) [98]. Several mechanisms may account for iatrogenic narcolepsy, including direct removal or destruction of wakefulness promoting areas such as hypocretin-containing neurons or by injury to ascending arousal pathways or hypocretin projections. Moreover, Scammell et al. reported a patient with secondary narcolepsy caused by a diencephalic ischemic stroke after removal of a large craniopharyngioma (Fig. 2) [99]. This case is remarkable for the presence of the full narcolepsy tetrad, including clear-cut cataplexy, sleep paralysis, and hypnagogic hallucinations, with a mean sleep latency of 0.5 min with 4/4 SOREM periods on MSLT, and decreased CSF hypocretin-1 level (167 pg/ml) [99].

In adults, metastatic brain tumors are more frequent than primary CNS tumors. Nevertheless, to the best of our knowledge, there have not yet been reports on secondary narcolepsy in the setting of metastasis to the brain.

Hereditary Diseases with Narcolepsy-Like Symptoms

Prader–Willi Syndrome

Prader–Willi syndrome (PWS) is a complex genetic disorder, which is associated with a deletion at 15q11–q13. The disease occurs in up to 1 in 20,000 births, and affects both sexes equally. PWS is characterized by hypotonia (floppy infant), areflexia, dysmorphic facies, hypoplastic genitalia, arthrogryposis, and mental retardation [100]. Several symptoms suggest a hypothalamic dysfunction, including temperature instability, growth retardation due to deficiency of growth hormone releasing hormone, and hyperphagia with early childhood obesity ("adiposogenital dystrophy") [101]. Obesity is also a common feature in childhood narcolepsy, occurring in up to 25% of affected children [102]. In addition, sleep–wake disturbances are frequent in PWS and typically comprise EDS, hypersomnia, SOREM periods in MSLT and reduced REM sleep latency or SOREM on nocturnal polysomnography, which have been reported also in PWS patients without comorbid sleep apnea syndrome or after successful treatment with continuous positive air pressure (CPAP) [103–107]. Furthermore, cataplexy in PWS seems to be quite common, and the reported incidences varied among 16, 18, and 23% [108–110]. Tobias et al. reported three PWS patients with typical cataplectic attacks and EDS, despite excellent weight control [111]. The mother of an 18-year-old female PWS patient reported recurrent attacks with sudden loss of muscle tone, usually induced by laughter, leading to falls to the floor and persisting for only several seconds [111]. Only one of the three PWS patients possessed the HLA DR15 (DR2) DQB1*0602 haplotype [111]. One study with four PWS patients showed CSF hypocretin-1 levels that were significantly decreased compared to ten young healthy controls (164 ± 47 pg/ml vs. 265 ± 49 pg/ml, $p = 0.02$) [100]. The deficiency of CSF hypocretin-1 levels appeared to be correlated with the severity of EDS as measured by MSLT [100]. Intermediate CSF hypocretin-1 concentrations in PWS have been reported also by two other groups [112, 113]. However, the number of hypocretin-1 neurons in eight PWS adults and three PWS infants did not differ from that in 11 controls, matched for age, sex, and postmortem delay [114]. Therefore, the involvement of impaired hypocretin-1 signaling and its contribution to the narcolepsy-like symptoms in PWS remain uncertain. Hypersomnia and generalized hypoarousal in PWS, as hypothesized by several groups, might be explained by hypothalamic dysfunction [103, 105, 115]. This is supported by the autoptic detection of structural alterations to various hypothalamic nuclei including the suprachiasmatic nucleus (SCN) [116]. The origin of cataplexy in PWS, on the other hand, remains unclear, in particular, whether it is associated with the hypothalamic dysfunction or rather with a so far unknown brainstem pathology.

Niemann–Pick Disease Type C

Niemann–Pick disease type C (NPC) is a lysosomal storage disorder of autosomal recessive inheritance leading to accumulation of cholesterol and glycosphingolipids in the brain and in visceral organs [117]. The clinical picture of NPC is very heterogeneous, and symptom onset may range from birth to adults over 50 years. Neurological signs of NPC are ataxia, epilepsy, mental retardation, spastic tetraplegia, dysphagia, dysarthria, supranuclear vertical gaze palsy, auditory disturbance, and dystonia [117, 118]. The main narcolepsy-like symptoms that may occur in NPC are cataplexy and EDS. Three of the five juvenile NCP patients reported by Vankova et al. showed shortened mean sleep latencies in the MSLT [119]. One patient had also cataplexy and was found to be HLA DQB1*0602 positive; moreover, he had multiple SOREM periods in the MSLT [119]. In this study, CSF hypocretin-1 levels were measured in four patients, showing reduced values in two patients, while the other two patients had values in the lower normal range. Hypocretin deficiency in NPC was reported also by other groups [112, 120, 121]. NPC is among the most frequent medical conditions leading to secondary cataplexy, especially in childhood [122–125]. Almost one in four NPC patients presents cataplexy [117]. Treatment with anticataplectic antidepressants may be successful but atypical responses have also been reported [126]. The pathomechanism of cataplexy in NPC is not understood. A hypothalamic involvement with impaired hypocretin signaling has been suggested, although CSF hypocretin-1 levels are, if at all, only moderately reduced and brain MRI did not show any morphological changes in the hypothalamus of affected patients [120]. Alternatively, the neurodegeneration in NPC might affect the rostral pontine tegmentum, an area comprising the locus coeruleus, the pedunculopontine nuclei, and the dorsal raphe nuclei. These structures are known to play a cardinal role in regulating the transition from NREM to REM sleep, and receive major input from hypothalamic hypocretin-containing neurons [118]. Indeed, several neurological symptoms in NPC suggest a brainstem involvement, such as dysphagia, vertical gaze palsy and auditory disturbances, and neuropathological changes in different brainstem nuclei including also the locus coeruleus, which were recently observed in a mouse model of NPC [118]. Human neuropathological studies, however, are still warranted to elucidate the etiology of cataplexy in NPC.

Moebius Syndrome

Moebius syndrome is a rare disorder with congenital facial diplegia and abducens nerve palsies. Neuropathological studies demonstrated midline brainstem necrosis, nuclear agenesis, and subtle brainstem hypoplasia [127]. Using a standard questionnaire pertaining to sleep–wake disturbances in 19 patients with Moebius syndrome, five patients (26%) had EDS accompanied by an abnormal tendency to collapse with laughter in three of them [115]. One of these patients with cataplectic attacks described facial jerks on the incompletely paretic right side. Parkes gave an additional description of a 26-year-old female patient with Moebius syndrome, who started to respond to emotional stimuli with sudden paralysis when she was 6 months old. Later on, laughter but not other forms of excitement produced momentary muscle atonia with falls and mutism. At the age of 21 years, she had to interrupt her college education due to severe EDS (ESS score 16/24). Furthermore, she had one to two sleep paralysis per week, and reported vivid and frequent dream mentation at sleep onset. In conclusion, Moebius syndrome belongs to a small number of genetic disorders that can cause secondary childhood narcolepsy with cataplexy, conceivably because the midbrain lesions involved also hypocretin-responsive areas or REM-regulating pathways.

Myotonic Dystrophy Type 1

Myotonic dystrophy type 1 (DM1) is the most prevalent muscular dystrophy in adults. DM1 is a multisystem disorder caused by a CTG repeat expansion with an autosomal dominant pattern of inheritance. The most frequent symptoms are myotonia and weakness of facial and sternocleidomastoid muscles, frontal balding, ptosis, and dystrophic changes in nonmuscular tissues including lens of eye, testicle and other endocrine glands, heart, skin, and brain. EDS and hypersomnia are among the most frequent sleep–wake disturbances in DM1 [128–135]. Of note, EDS may precede the manifestation of other symptoms of DM1. For instance, Ciafaloni et al. reported a patient with severe EDS (ESS score 16/24), mean sleep latency of 1.5 min and four SOREM periods in the MSLT, but normal CSF hypocretin-1 level (305 pg/ml) [136]. Prior to the diagnosis of DM1, this patient had been considered for many years to suffer from narcolepsy. In addition, the patient displayed an increased amount of REM sleep (38%), instead of increased NREM1 sleep and REM sleep fragmentation characteristic for narcolepsy. In a recent polysomnographic study, 70% of DM1 patients had either subjective EDS (ESS score $\geq$ 11) or objective EDS (mean sleep latency in MSLT $\leq$ 8 min), and 50% were found to have both [137]. SOREM periods were found in 37%, and 26% had $\geq$2 SOREM periods [137]. Furthermore, narcolepsy-like features, including cataplexy, sleep paralysis and hypnagogic hallucinations, seem to be common in DM1. The overall prevalences of cataplexy, sleep paralysis, and hypnagogic hallucinations in DM1 are 23, 33, and 23%, respectively, but if the patients indicate also EDS, the respective prevalences are as high as 43, 52, and 33%, respectively [137]. Cataplexy was more frequently reported by DM1 patients with $\geq$2 SOREM periods than by patients with $\leq$1 SOREM period (55% vs. 13%). Other groups, however, failed to find any differences in frequencies of cataplexy, sleep paralysis, and hypnagogic hallucinations between DM1 patients and healthy controls [129, 138, 139]. Data on CSF hypocretin-1 levels in DM1 are inconsistent. A first report on seven DM1 patients found lower values compared to controls [113, 139], but a recent larger study failed to detect any differences in the CSF hypocretin-1 concentrations between 38 DM1 patients and 33 healthy controls, suggesting that the pathophysiology underlying EDS and frequent occurrence of SOREM periods in DM1 is different from narcolepsy [136]. On the other hand, several lines of evidence indicate that EDS and hypersomnia in DM1 are not merely due to frequently comorbid sleep-disordered breathing but rather a consequence of a disruption of sleep–wake regulatory circuits in the CNS. For instance, EDS may occur at early disease stages when respiratory muscles function is still preserved and EDS may persist despite adequate CPAP treatment [136, 140]. Furthermore, neuropathological studies showed that hypersomnia was associated with a neuronal loss in the dorsal raphe nuclei and in the reticular formation [141].

Neurodegenerative Disorders with Narcolepsy-Like Symptoms

Idiopathic Parkinson's Disease

Sleep–wake disorders are frequent in most neurodegenerative disorders presenting with parkinsonism. The frequent presence of narcolepsy-like features in idiopathic Parkinson's disease (PD), including severe EDS with sleep attacks, RBD with intense dreaming, hallucinations, and disturbed nocturnal sleep, led to the assumption that the underlying pathomechanism of two disorders might share common aspects [142–146]. Using MSLT, some patients with PD and EDS present ≥2 SOREM periods and meet the criteria for narcolepsy without cataplexy [147, 148]. In selected patients with PD and significant clinical sleepiness [145] and severe hallucinations [144], the frequency of ≥2 SOREM periods reached 39 and 60%, respectively. With the exception of two cases, these patients appeared all to be negative for HLA DQB1*0602 [144, 145]. Another group confirmed that HLA typing does not predict RBD and hallucinations in PD [148]. Conversely, Autret et al. reported several cases of secondary narcolepsy that were HLA-DR2 positive and suggested that the emergence, nature, and severity of secondary narcolepsy were not only induced by injury to certain brain areas but also by a specific genetic predisposition [149]. Clear-cut cataplexy does not occur in PD. Although sudden falls in patients with PD have been described, they clearly differ from cataplexy in narcoleptic patients, as they are not provoked by positive emotions but rather by freezing or stressful situations [144]. Also, the phenotype of hallucinations in PD is clearly distinct from the one seen in narcolepsy. Last not least, EDS in PD is often caused or exacerbated by dopaminergic treatment.

Despite the high frequency of narcolepsy-like features, normal lumbar CSF hypocretin-1 levels have been found in almost all patients with PD [55, 113, 150–152]. Maeda et al. reported a patient with a 15-year-history of mild idiopathic PD and EDS, in whom CSF hypocretin-1 was low (86 pg/ml) [153]. However, this patient had suffered from EDS already since adolescence, indicating that decreased hypocretin-1 levels might be due to comorbid narcolepsy. Two groups reported reduced hypocretin-1 levels in CSF samples drawn from the ventricular system of patients with PD [154, 155]. More recently, neuropathological studies of the hypothalamus of deceased patients with PD were performed [155, 156]. A significant hypocretin cell loss ranging from 23 to 62% was found. However, this cell loss was not specific, since the number of adjacent melanin-concentrating hormone cells was also reduced by 12–74% [156]. The presence of Lewy bodies in the hypothalamic hypocretin cell area indicates that widespread neurodenegeration in PD affects also the hypocretinergic system, and may contribute to the frequent sleep–wake disturbances [155]. While the association between hypocretin cell loss and occurrence and severity of sleep–wake disturbances during lifetime remains to be established, the degenerative changes in the hypocretin cells were shown to be better correlated with disease progression than that in neuromelanin cells of the substantia nigra [156].

Alzheimer's Disease

Insomnia, frequent nocturnal awakenings, EDS and disturbed sleep–wake cycles are among the most prevalent sleep–wake disturbances in patients with Alzheimer's disease (AD) [157–160]. Normal CSF hypocretin-1 levels were measured by different groups [55, 113, 161, 162], but wake fragmentation has been shown to correlate with lower CSF hypocretin-1 concentrations [163]. Studies in older rats suggested a slight reduction of hypocretin neurons and decreased CSF hypocretin-1 concentrations [164]. A 40% loss of hypocretin cells have been found in ten AD brains, including four patients with known sleep disturbances (Fronczek et al., unpublished data). Furthermore, significant impairment of hypocretin signaling in AD could be caused by disturbed circadian control of hypocretin release. There is strong evidence indicating that degeneration of the SCN plays a central role in AD-associated sleep–wake disturbances [165]. SCN lesions in rats eliminated the daily rhythm of hypocretin-1, indicating that hypocretin-1 release is under SCN control [166–168]. Furthermore, mean hypocretin-1 levels in SCN-lesioned animals were significantly lower than mean values in control animals, underscoring the importance of the SCN in promoting hypocretin-1 upregulation during wakefulness [167]. Studies on the circadian rhythm of hypocretin-1 release in patients with AD and their correlation with sleep–wake disturbances have not been done.

Lewy-Body Dementia, Progressive Supranuclear Palsy, Corticobasal Degeneration

Lewy-body dementia (LBD) is characterized by a parkinsonian syndrome, dementia, fluctuations in vigilance and attention, and hallucinations. Along with recurrent visual daytime hallucinations, this neurodegenerative disorder presents other narcolepsy-like features including EDS, insomnia, and prominent RBD. However, normal CSF hypocretin-1 levels were found by different groups [152, 162], suggesting a distinct pathophysiology of EDS, hallucinations and RBD in LBD and narcolepsy.

CSF hypocretin-1 levels were also determined in patients with progressive supranuclear palsy (PSP) and corticobasal degeneration (CBD), and were normal in all but two PSP and one CBD case [162]. In PSP, a disorder that may lead to RBD and REM sleep without atonia [169], CSF hypocretin-1 levels were inversely correlated with duration and severity of the disease [162]. The authors hypothesized that the moderately decreased CSF hypocretin-1 levels in three patients might indicate that the hypothalamic hypocretin system is also affected by these neurodegenerative disorders. Neuropathological studies have not been done.

Multiple System Atrophy

Sleep disorders are more common in patients with multiple system atrophy (MSA) than in patients with PD after the same duration of the disease [170]. In MSA, 70% of patients suffer from sleep disturbances including EDS, sleep fragmentation, and RBD [170]. In addition, severe autonomic failure in MSA suggests a hypothalamic involvement of the neurodegenerative process. Abdo et al. did not find any differences in CSF hypocretin-1 levels between 12 patients with MSA (six patients with predominant cerebellar, six patients with predominant parkinsonian phenotype) and 11 age-matched controls [171]. Another group confirmed this result [172]. The normal findings of CSF hypocretin-1 levels in MSA contrasts with a 70% reduction of hypocretin cells found in seven patients with MSA, including three patients with a history suggestive of EDS [173]. Furthermore, abundant glial cytoplasmic inclusions were observed in the dorsolateral but not in the periventricular hypothalamus, indicating a possible regional susceptibility of the hypocretin neurons to the neurodegenerative process [173]. The discrepancy of normal CSF hypocretin-1 values despite significant destruction of hypocretin neurons suggest that the remaining hypocretin neurons are able to compensate by upregulating hypocretin release. Conversely, a rat study demonstrated a 50% decrease in CSF hypocretin-1 levels in the context of a 73% loss of hypocretin-producing neurons [174].

Huntington's Disease

Huntington's disease (HD), distinguished by the triad of dominant inheritance, choreoathetosis, and dementia, is caused by an abnormal expansion of a trinucleotide repeat (CAG) sequence. Sleep–wake disturbances, such as insomnia, circadian rhythm disturbances, nocturnal awakenings, EDS, and RBD, are reported in up to 88% of patients with HD [175–178]. Alteration of energy expenditure is common in both HD and narcolepsy, suggesting a hypothalamic dysfunction in HD as well. Neuropathological studies described a neuronal loss, which primarily affects the striatum and the cortex. However, a neuronal loss of up to 90% has been demonstrated in the tuber nucleus of the lateral hypothalamus of deceased patients with HD [179, 180]. In end-stage R6/2 mice, the most widely used rodent model of HD, a dramatic atrophy and loss of hypocretin neurons with a decrease of CSF hypocretin-1 levels by 72% were found [181]. Interestingly, these mice showed recurrent episodes with sudden interruption of purposeful motor activity resembling cataplectic attacks of hypocretin-1 knockout mice [181]. Comparing the hypothalamus of five deceased patients with HD and four controls, a 32% reduction of hypocretin neurons were reported by the same group [181]. Aziz et al. confirmed this finding and reported a 30% reduction of the total number of hypocretin-1 neurons compared to controls [182]. In contrast to the above-mentioned neuropathological study in PD [156], Aziz et al. did not find a simultaneous reduction of adjacent MCH-expressing neurons, suggesting different vulnerability of hypocretin and MCH neurons to the neurodenegeration in HD and PD [182]. In this study, hypocretin-1 levels in postmortem ventricular CSF did not differ in patients with HD and controls [182]. Similarly, normal CSF hypocretin-1 levels in patients with HD were reported by four other groups [183–186]. In summary, despite the high frequency of sleep–wake disturbances and the reported hypocretin cell loss, secondary narcolepsy in HD does not occur. None of 25 patients with HD reported by Arnulf et al. exhibited multiple SOREM periods in the MSLT, and the frequency of narcolepsy-like symptoms such as cataplexy, sleep paralysis, and hypnagogic hallucinations were equal in patients with HD and healthy controls [178]. At the moment, the significance of impaired hypocretin signaling in HD and its contribution to clinical symptoms remain uncertain. On the other hand, the observation of imperfect REM sleep atonia and overt RBD in some patients with HD additionally points to brainstem pathology.

Amyotrophic Lateral Sclerosis

Cases with secondary narcolepsy have not been reported in amyotrophic lateral sclerosis (ALS), but the presence of autonomic disturbances and the occasional occurrence of sleep disorders led to the hypothesis of an additional hypothalamic dysfunction in this neurodegenerative disorder of the motor system. Furthermore, ALS may be associated with frontotemporal dementia (FTD), and the frontal cortex is known to be innervated by dense hypocretinergic projections [187]. In the only study exploring the hypocretin system in ALS, CSF hypocretin-1 levels were normal in all 20 patients, including 1 patient with ALS-FTD [188].

Conclusion

In conclusion, we know very well that narcolepsy-like symptoms may occur in a huge variety of disorders, most of them – but not all – with hypothalamic involvement, leading to so-called secondary or symptomatic narcolepsy. However, the specific narcoleptic phenotype with clear-cut pathognomonic cataplexy and low CSF hypocretin levels occurs only rarely, which corroborates the assumption that narcolepsy with cataplexy is a unique and specific disorder, whereas symptomatic narcolepsy would usually better be described as a symptomatic narcolepsy-like phenotype.

References

1. Bassetti CL. Narcolepsy. Selective hypocretin (orexin) neuronal loss and multiple signalling deficiencies. Neurology. 2005;65:1152–3.
2. Peyron C, Faraco J, Rogers W, Ripley B, Overeem S, Charnay Y, et al. A mutation in a case of early onset narcolepsy and a generalized absence of hypocretin peptides in human narcoleptic brains. Nat Med. 2000;6:991–7.
3. Thannickal TC, Moore RY, Nienhuis R, Ramanathan L, Gulyani S, Aldrich M, et al. Reduced number of hypocretin neurons in human narcolepsy. Neuron. 2000;27:469–74.
4. Honda Y, Doi Y, Juji T, Satake M. Narcolepsy and HLA: positive DR2 as a prerequisite for the development of narcolepsy. Folia Psychiatr Neurol Jpn. 1984;38:360.
5. Mignot E, Lin X, Arrigoni J, et al. DQB1*0602 and DQA1*0102 (DQ1) are better markers than DR2 for narcolepsy in Caucasian and black Americans. Sleep. 1994;17:S60–7.
6. Cvetkovic-Lopes V, Bayer L, Dorsaz S, Maret S, Pradervand S, Dauvilliers Y, et al. Elevated tribbles homolog 2-specific antibody levels in narcolepsy patients. J Clin Invest. 2010;120:713–9.

7. Dauvilliers Y, Arnulf I, Mignot E. Narcolepsy with cataplexy. Lancet. 2007;369:499–511.
8. American Academy of Sleep Medicine. International classification of sleep disorders, 2nd edition: diagnostic and coding manual. Westchester, IL: American Academy of Sleep Medicine; 2005.
9. Bonduelle M, Degos C. Symptomatic narcolepsies: a critical study. Adv Sleep Res. 1974;3:313–32.
10. Saper CB, Scammell TE, Lu J. Hypothalamic regulation of sleep and circadian rhythms. Nature. 2005;437:1257–63.
11. Adamantidis A, de Lecea L. Physiological arousal: a role for hypothalamic systems. Cell Mol Life Sci. 2008;65:1475–88.
12. Saper CB, Cano G, Scammell TE. Homeostatic, circadian, and emotional regulation of sleep. J Comp Neurol. 2005;493:92–8.
13. Aldrich MS, Chervin RD, Malow BA. Value of the multiple sleep latency test (MSLT) for the diagnosis of narcolepsy. Sleep. 1997;20:620–9.
14. Marti I, Valko PO, Khatami R, Bassetti CL, Baumann CR. Multiple sleep latency measures in narcolepsy and behaviourally induced insufficient sleep syndrome. Sleep Med. 2009;10:1146–50.
15. Mignot E, Lin L, Finn L, Lopes C, Pluff K, Sundstrom ML, et al. Correlates of sleep-onset REM periods during the Multiple Sleep Latency Test in community adults. Brain. 2006;129:1609–23.
16. Gélineau JB. De la narcolepsie. Gaz Hôp (Paris). 1880;53:626–8. 635–7.
17. Symonds CP. Narcolepsy as a symptom of encephalitis lethargica. Lancet. 1926;12:1214–5.
18. Von Economo C. Sleep as a problem of localization. J Nerv Ment Dis. 1930;71:249–59.
19. Von Economo C. Encephalitis lethargica: its sequelae and treatment. Oxford University Press: London; 1931.
20. Adie WJ. Idiopathic narcolepsy: a disease sui generis: with remarks on the mechanism of sleep. Brain. 1926;49:257–306.
21. Mignot E. Genetic and familial aspects of narcolepsy. Neurology. 1998;50:S16–22.
22. Guillain G, Alajouanine T. La somnolence dans la sclérose en plaques. Les episodes aigus ou subaigus de la sclérose en plaques pouvant simuler l'encéphalite léthargique. Ann Med. 1928;24: 111–8.
23. Berg O, Hanley J. Narcolepsy in two cases of multiple sclerosis. Acta Neurol Scand. 1963;39:252–7.
24. Ekbom K. Familial multiple sclerosis associated with narcolepsy. Arch Neurol. 1966;15:337–44.
25. Schrader H, Gotlibson O, Skomedal G. Multiple sclerosis and narcolepsy/cataplexy in a monozygotic twin. Neurology. 1980;30:105–8.
26. Younger DS, Pedley TA, Thorpy MJ. Multiple sclerosis and narcolepsy: possible similar genetic susceptibility. Neurology. 1991;41:447–8.
27. Celius EG, Harbo HF, Egeland T, Vartdal F, Vandvik B, Spurkiand A. Sex and age at diagnosis are correlated with the HLA-DR2, DQ6 haplotype in multiple sclerosis. J Neurol Sci. 2000;178:132–5.
28. Poirier G, Montplaisir J, Dumont M, et al. Clinical and sleep laboratory study of narcoleptic symptoms in multiple sclerosis. Neurology. 1987;37:693–5.
29. Knudsen S, Jennum PJ, Korsholm K, Sheikh SP, Gammeltoft S, Frederiksen JL. Normal levels of cerebrospinal fluid hypocretin-1 and daytime sleepiness during attacks of relapsing-remitting multiple sclerosis and monosymptomatic optic neuritis. Mult Scler. 2008;14:734–8.
30. Rao DG, Singhal BS. Secondary narcolepsy in a case of multiple sclerosis. J Assoc Physicians India. 1997;45:321–2.
31. Iseki K, Mezaki T, Oka Y, et al. Hypersomnia in MS. Neurology. 2002;59:2006–7.
32. Kato T, Kanbayashi T, Yamamoto K, Nakano T, Shimizu T, Hashimoto T, et al. Hypersomnia and low CSF hypocretin-1 (orexin-A) concentration in a patient with multiple sclerosis showing bilateral hypothalamic lesions. Intern Med. 2003;42:743–5.
33. Oka Y, Kanbayashi T, Mezaki T, Iseki K, Matsubayashi J, Murakami G, et al. Low CSF hypocretin-1/orexin-A associated with hypersomnia secondary to hypothalamic lesion in a case of multiple sclerosis. J Neurol. 2004;251:885–6.
34. Nozaki H, Shimohata T, Kanbayashi T, Sagawa Y, Katada S, Satoh M, et al. A patient with anti-aquaporin 4 antibody who presented with recurrent hypersomnia, reduced orexin (hypocretin) level, and symmetrical hypothalamic lesions. Sleep Med. 2009;10:253–5.
35. Kanbayashi T, Shimohata T, Nakashima I, Yaguchi H, Yabe I, Nishizawa M, et al. Symptomatic narcolepsy in patients with neuromyelitis optica and multiple sclerosis. New neurochemical and immunological implications. Arch Neurol. 2009;66:1563–6.
36. Vetrugno R, Stecchi S, Plazzi G, Lodi R, D'Angelo R, Alessandria M, et al. Narcolepsy-like syndrome in multiple sclerosis. Sleep Med. 2009;10:389–91.
37. Dauvilliers Y, Abril B, Mas E, Michel F, Tafti M. Normalization of hypocretin-1 in narcolepsy after intravenous immunoglobulin treatment. Neurology. 2009;73:1333–4.
38. Lennon VA, Wingerchuk DM, Kryzer TJ, Pittock SJ, Lucchinetti CF, Fujihara K, et al. A serum autoantibody marker of neuromyelitis optica: distinction from multiple sclerosis. Lancet. 2004;364:2106–12.
39. Pittock SJ, Lennon VA, Krecke K, Wingerchuk DM, Lucchinetti CF, Weinshenker BG. Brain abnormalities in neuromyelitis optica. Arch Neurol. 2006;63: 390–6.
40. Amiry-Moghaddam M, Ottersen OP. The molecular basis of water transport in the brain. Nat Rev Neurosci. 2003;4:991–1001.
41. Pittock SJ, Weinshenker BG, Lucchinetti CF, Wingerchuk DM, Corboy JR, Lennon VA. Neuromyelitis optica brain lesions localized at sites of high aquaporin 4 expression. Arch Neurol. 2006;63:964–8.
42. Nozaki H, Katada S, Sato M, Tanaka K, Nishizawa M. A case with hypersomnia and paresthesia due to

diffuse MS lesions from hypothalamus to spine. Rinsho Shinkeigaku. 2004;44:59.
43. Poppe AY, Lapierre Y, Melancon D, Lowden D, Wardell L, Fullerton LM, et al. Neuromyelitis optica with hypothalamic involvement. Mult Scler. 2005;11:617–21.
44. Carlander B, Vincent T, Le Floch A, Pageot N, Camu W, Dauvilliers Y. Hypocretinergic dysfunction in neuromyelitis optica with coma-like episodes. J Neurol Neurosurg Psychiatry. 2008;79:333–4.
45. Baba T, Nakashima I, Kanbayashi T, Konno M, Takahashi T, Fujihara K, et al. Narcolepsy as an initial manifestation of neuromyelitis optica with anti-aquaporin-4 antibody. J Neurol. 2009;256:287–8.
46. Kanbayashi T, Goto A, Hishikawa Y, et al. Hypersomnia due to acute disseminated encephalomyelitis in a 5-year-old girl. Sleep Med. 2001;2: 347–50.
47. Kubota H, Kanbayashi T, Tanabe Y, Takanashi J, Kohno Y. A case of acute disseminated encephalomyelitis presenting hypersomnia with decreased hypocretin level in cerebrospinal fluid. J Child Neurol. 2002;17:537–9.
48. Arii J, Kanyabashi T, Tanabe Y, Sawaishi S, Kimura A, Watanabe K, et al. CSF hypocretin-1 (orexin-A) levels in childhood narcolepsy and neurologic disorders. Neurology. 2004;63:2440–2.
49. Gledhill RF, Bartel PR, Yoshida Y, Nishino S, Scammell TE. Narcolepsy caused by acute disseminated encephalomyelitis. Arch Neurol. 2004;61:758–60.
50. Yoshikawa S, Suzuki S, Kanbayashi T, Nishino S, Tamai H. Hypersomnia and low cerebrospinal fluid hypocretin levels in acute disseminated encephalomyelitis. Pediatr Neurol. 2004;31:367–70.
51. Schmidt-Degenhard M. Oneiric perception in intensively treated panplegic polyradiculitis patients. Nervenarzt. 1986;57:712–8.
52. Wegener K, Tassan P, Josse M, Bolgert F. An oneiroid experience during severe acute polyradiculoneuritis. Ann Med Psychol. 1995;153:121–6.
53. Cochen V, Arnulf I, Demeret S, Neulat ML, Gourlet V, Drouot X, et al. Vivid dreams, hallucinations, psychosis and REM sleep in Guillain-Barre syndrome. Brain. 2005;128:2535–45.
54. Hochman M, Kobetz S, Handwerker J. Inappropriate secretion of antidiuretic hormone associated with Guillain-Barré syndrome. Ann Neurol. 1982;11: 322–3.
55. Ripley B, Overeem S, Fujiki N, et al. CSF hypocretin/orexin levels in narcolepsy and other neurological conditions. Neurology. 2001;57:2253–8.
56. Kanbayashi T, Ishiguro H, Aizawa R, et al. Hypocretin-1 (orexin-A) concentrations in cerebrospinal fluid are low in patients with Guillain-Barré syndrome. Psychiatry Clin Neurosci. 2002;56:273–4.
57. Nishino S, Kanbayashi T, Fujiki N, et al. CSF hypocretin levels in Guillain-Barré syndrome and other inflammatory neuropathies. Neurology. 2003;61: 823–5.
58. Baumann CR, Bassetti CR. CSF hypocretin levels in Guillain-Barré syndrome and other inflammatory neuropathies. Neurology. 2004;62:2337.
59. Overeem S, Geleijns K, Garssen MP, Jacobs BC, van Doorn PA, Lammers GJ. Screening for anti-ganglioside antibodies in hypocretin-deficient human narcolepsy. Neurosci Lett. 2003;341:13–6.
60. Voltz R, Gultekin SH, Rosenfeld MR, et al. A serologic marker of paraneoplastic limbic and brainstem encephalitis in patients with testicular cancer. N Engl J Med. 1999;340:1788–95.
61. Rosenfeld MR, Eichen JG, Wade DF, Posner JB, Dalmau J. Molecular and clinical diversity in paraneoplastic immunity to Ma proteins. Ann Neurol. 2001;50:339–48.
62. Dalmau J, Graus F, Villarejo A, Posner JB, Blumenthal D, Thiessen B, et al. Clinical analysis of anti-Ma2-associated encephalitis. Brain. 2004;127: 1831–44.
63. Rojas-Marcos I, Graus F, Sanz G, Robledo A, Diaz-Espejo C. Hypersomnia as presenting symptom of anti-Ma2-associated encephalitis: case study. Neuro Oncol. 2007;9:75–7.
64. Overeem S, Dalmau J, Bataller L, Nishino S, Mignot E, Verschuuren J, et al. Hypocretin-1 CSF levels in anti-Ma2 associated encephalitis. Neurology. 2004; 62:138–40.
65. Landolfi JC, Nadkarni M. Paraneoplastic limbic encephalitis and possible narcolepsy in a patient with testicular cancer: case study. Neuro Oncol. 2003;5:214–6.
66. Compta Y, Iranzo A, Santamaría J, Casamitjana R, Graus F. REM sleep behavior disorder and narcoleptic features in anti-Ma2-associated encephalitis. Sleep. 2007;30:767–9.
67. Mathis J, Hess CW, Bassetti C. Isolated mediotegmental lesion causing narcolepsy and rapid eye movement sleep behaviour disorder: a case evidencing a common pathway in narcolepsy and rapid eye movement sleep behaviour disorder. J Neurol Neurosurg Psychiatry. 2007;78:427–9.
68. Iranzo A, Graus F, Clover L, Morera J, Bruna J, Vilar C, et al. Rapid eye movement sleep behavior disorder and potassium channel antibody-associated limbic encephalitis. Ann Neurol. 2006;59:178–82.
69. Lin FC, Liu CK, Hsu CY. Rapid-eye-movement sleep behavior disorder secondary to acute aseptic limbic encephalitis. J Neurol. 2009;256:1174–6.
70. Hsieh CF, Lai CL, Liu CK, Lan SH, Hsieh SW, Hsu CY. Narcolepsy and Behçet's disease: report of a Chinese-Taiwanese case. Sleep Med. 2010;11:426–8.
71. Baumann CR, Bassetti CL, Hersberger M, Jung HH. Excessive daytime sleepiness in Behçet's disease with diencephalic lesions and hypocretin dysfunction. Eur Neurol. 2010;63:190.
72. Serdaroglu P. Behçet's disease and the nervous system. J Neurol. 1998;245:197–205.
73. Watson NF, Doherty MJ, Zunt JR. Secondary narcolepsy following neurocysticercosis infection. J Clin Sleep Med. 2005;1:41–2.

74. Maia LF, Marta M, Lopes V, Rocha N, Lopes C, Martins-da-Silva A, et al. Hypersomnia in Whipple disease. Case report. Arq Neuropsiquiatr. 2006; 64:865–8.
75. Vorderholzer U, Riemann D, Gann H, et al. Transient total sleep loss in cerebral Whipple's disease: a longitudinal study. J Sleep Res. 2002;11:321–9.
76. Rubinstein I, Gray TA, Moldofsky H, Hoffstein V. Neurosarcoidosis associated with hypersomnolence treated with corticosteroids and brain irradiation. Chest. 1988;94:205–6.
77. Servan J, Marchand F, Garma L, Seilhean D, Hauw J, Delattre J. Narcolepsy disclosing neurosarcoidosis. Rev Neurol (Paris). 1995;151:281–3.
78. Malik S, Boeve BF, Krahn LE, Silber MH. Narcolepsy associated with other central nervous system disorders. Neurology. 2001;57:539–41.
79. Afshar K, Engelfried K, Sharma OP. Sarcoidosis: a rare cause of Kleine-Levine-Critchley syndrome. Sarcoidosis Vasc Diffuse Lung Dis. 2008;25:60–3.
80. Nakazato Y, Kondo S, Okhuma A, Ito Y, Tamura N, Araki N. Neurosarcoidosis presenting as spontaneously remitting hypersomnia. J Neurol. 2009;256: 1929–31.
81. Aran A, Lin L, Nevsimalova S, Plazzi G, Hong SC, Weiner K, et al. Elevated anti-streptococcal antibodies in patients with recent narcolepsy onset. Sleep. 2009;32:979–83.
82. Aldrich MS, Naylor MW. Narcolepsy associated with lesions of the diencephalon. Neurology. 1989; 39:1505–8.
83. Marcus C, Trescher W, Halbower A, Lutz J. Secondary narcolepsy in children with brain tumors. Sleep. 2002;25:435–9.
84. Rosen G, Bendel A, Neglia J, Moertel C, Mahowald M. Sleep in children with neoplasms of the central nervous system: case review of 14 children. Pediatrics. 2003;112:e46–54.
85. D'Cruz OF, Vaughn BV, Gold SH, et al. Symptomatic cataplexy in pontomedullary lesions. Neurology. 1994;44:2189–91.
86. Ma TK, Ang LC, Mamelak M, Kish SJ, Young B, Lewis AJ. Narcolepsy secondary to fourth ventricular subependymoma. Can J Neurol Sci. 1996;23: 59–62.
87. Müller HL, Müller-Stöver S, Gebhardt U, Kolb R, Sörensen N, Handwerker G. Secondary narcolepsy may be a causative factor of increased daytime sleepiness in obese childhood craniopharyngioma patients. J Pediatr Endocrinol Metab. 2006;S19:423–9.
88. Dauvilliers Y, Abril B, Charif M, Quittet P, Bauchet L, Carlander B, et al. Reversal of symptomatic tumoral narcolepsy, with normalization of CSF hypocretin level. Neurology. 2007;69:1300–1.
89. Onofrj M, Curatola L, Ferracci F, Fulgente T. Narcolepsy associated with primary temporal lobe B-cells lymphoma in a HLA DR2 negative subject. J Neurol Neurosurg Psychiatry. 1992;55:852–3.
90. Clavelou P, Tournilhac M, Vidal C, Georget A, Picard L, Merienne L. Narcolepsy associated with arteriovenous malformation on the diencephalon. Sleep. 1995;18:202–5.
91. Ogata N, Yonekawa Y. Recurrent sleep attacks associated with a craniopharyngioma. J Clin Neurosci. 1997;4:499–500.
92. Nokura K, Kanbayashi T, Ozeki T, Koga H, Zettsu T, Yamamoto H, et al. Hypersomnia, asterixis and cataplexy in association with orexin A-reduced hypothalamic tumor. J Neurol. 2004;251:1534–5.
93. Snow A, Gozal E, Malhotra A, Tiosano D, Perlman R, Vega C, et al. Severe hypersomnolence after pituitary/hypothalamic surgery in adolescents: clinical characteristics and potential mechanisms. Pediatrics. 2002;110:e74.
94. Dempsey O, McGeoch P, de Silva R, Douglas N. Acquired narcolepsy in an acromegalic patient who underwent pituitary irradiation. Neurology. 2003;61: 537–40.
95. Schwartz WJ, Stakes JW, Hobson JA. Transient cataplexy after removal of a craniopharyngioma. Neurology. 1984;34:1372–5.
96. Arii J, Kanbayashi T, Tanabe Y, Ono J, Nishino S, Kohno Y. A hypersomnolent girl with decreased CSF hypocretin level after removal of a hypothalamic tumor. Neurology. 2001;56:1775–6.
97. Krahn LE, Boeve BF, Oliver L, Silber MH. Hypocretin (orexin) and melatonin values in a narcoleptic-like sleep disorder after pinealectomy. Sleep Med. 2002;3:521–32.
98. Tachibana N, Taniike M, Okinaga T, Ripley B, Mignot E, Nishino S. Hypersomnolence and increased REM sleep with low cerebrospinal fluid hypocretin level in a patient after removal of craniopharyngioma. Sleep Med. 2005;6:567–9.
99. Scammell T, Nishino S, Mignot E, Saper C. Narcolepsy and low CSF orexin (hypocretin) concentration after a diencephalic stroke. Neurology. 2001;56:1751–3.
100. Nevsimalova S, Vankova J, Stepanova I, Seemanova E, Mignot E, Nishino S. Hypocretin deficiency in Prader-Willi syndrome. Eur J Neurol. 2005;12: 70–2.
101. Swaab DF. Prader-Willi syndrome and the hypothalamus. Acta Pediatr Scand. 1997;423:50–4.
102. Nevsimalova S. Narcolepsy in childhood. Sleep Med Rev. 2009;13:169–80.
103. Vgontzas AN, Bixler EO, Kales A, Centurione A, Rogan PK, Mascari M, et al. Daytime sleepiness and REM abnormalities in Prader-Willi syndrome: evidence of generalized hypoarousal. Int J Neurosci. 1996;87:127–39.
104. Richdale AL, Cotton S, Hibbit K. Sleep and behaviour disturbance in Prader-Willi syndrome: a questionnaire study. J Intellect Disabil Res. 1999;43: 380–92.
105. Manni R, Politini L, Nobili L, Ferrillo F, Livieri C, Veneselli E, et al. Hypersomnia in the Prader-Willi syndrome: clinical-electrophysiological features and underlying factors. Clin Neurophysiol. 2001;112: 800–5.

106. Wagner MH, Berry RB. An obese female with Prader-Willi syndrome and daytime sleepiness. J Clin Sleep Med. 2007;3:645–7.
107. Camfferman D, McEnvoy RD, O'Donoghue F, Lushington K. Prader Willi syndrome and excessive daytime sleepiness. Sleep Med Rev. 2008;12:65–75.
108. Helbing-Zwanenburg B, Kamphuisen HA, Mourtazaev MS. The origin of excessive daytime sleepiness in the Prader-Willi syndrome. J Intellect Disabil Res. 1993;37:533–41.
109. Cassidy SB, McKillop JA, Morgan WJ. Sleep disorders in the Prader-Willi syndrome. 1989 David W Smith Workshop on Malformations and Morphogenesis, Madrid, Spain. Proc Greenwood Gent Centre. 1990;9:74–5.
110. Clift S, Dahlitz M, Parkes JD. Sleep apnoea in the Prader-Willi syndrome. J Sleep Res. 1994;3:121–6.
111. Tobias ES, Tolmie JL, Stephenson JBP. Cataplexy in the Prader-Willi syndrome. Arch Dis Child. 2002;87:170–1.
112. Mignot E, Lammers GJ, Ripley B, Okun M, Nevsimalova S, Overeem S, et al. The role of cerebrospinal fluid hypocretin measurement in the diagnosis of narcolepsy and other hypersomnias. Arch Neurol. 2002;59:1553–62.
113. Dauvilliers Y, Baumann CR, Carlander B, et al. CSF hypocretin-1 levels in narcolepsy, Kleine-Levin syndrome, and other hypersomnias and neurological conditions. J Neurol Neurosurg Psychiatry. 2003;74: 1667–73.
114. Fronczek R, Lammers GJ, Balesar R, Unmehopa UA, Swaab DF. The number of hypothalamic hypocretin (orexin) neurons is not affected in Prader-Willi syndrome. J Clin Endocrinol Metab. 2005;90:5466–70.
115. Parkes JD. Genetic factors in human sleep disorders with special reference to Norrie disease, Prader-Willi syndrome and Moebius syndrome. J Sleep Res. 1999;8:S14–22.
116. Swaab DF, Roozendaal B, Ravid R, Velis DN, Gooren L, Williams RS. Suprachiasmatic nucleus in aging, Alzheimer's disease, transsexuality and Prader-Willi syndrome. Prog Brain Res. 1987;72:301–10.
117. Sévin M, Lesca G, Baumann N, Millat G, Lyon-Caen O, Vanier MT, et al. The adult form of Niemann-Pick disease type C. Brain. 2007;130: 120–33.
118. Luan Z, Saito Y, Miyata H, Ohama E, Ninomiya H, Ohno K. Brainstem neuropathology in a mouse model of Niemann-Pick disease type C. J Neurol Sci. 2008;268:108–16.
119. Vankova J, Stepanova I, Jech R, Elleder M, Ling L, Mignot E, et al. Sleep disturbances and hypocretin deficiency in Niemann-Pick disease type C. Sleep. 2003;26:427–30.
120. Kanbayashi T, Abe M, Fujimoto S, Miyachi T, Takahashi T, Yano T, et al. Hypocretin deficiency in Niemann-Pick type C with cataplexy. Neuropediatrics. 2003;34:52–3.
121. Oyama K, Takahashi T, Shoji Y, Oyamada M, Noguchi A, Tamura H, et al. Niemann-Pick disease type C: cataplexy and hypocretin in cerebrospinal fluid. Tohoku J Exp Med. 2006;209:263–7.
122. Kandt RS, Emerson RG, Singer HS, Valle DL, Moser HW. Cataplexy in variant forms in Niemann-Pick C disease. Ann Neurol. 1982;12:284–8.
123. Denoix C, Rodriguez-Lafrasse C, Vanier MT, Navelet Y, Landrieu P. Cataplexie révélatrice d'une forme atypique de la maladie de Niemann-Pick type C. Arch Fr Pediatr. 1991;48:31–4.
124. Challamel MJ, Mazzola ME, Nevsimalova S, Cannard C, Louis J, Revol M. Narcolepsy in children. Sleep. 1994;17:S17–20.
125. Böor R, Reitter B. Kataplexie bei Morbus Niemann-Pick Typ C. Klin Pediatr. 1997;209:88–90.
126. Smit LS, Lammers GJ, Catsman-Berrevoets CE. Cataplexy leading to the diagnosis of Niemann-Pick disease type C. Pediatr Neurol. 2006;35:82–4.
127. Sudarshan A, Goldie WD. The spectrum of congenital facial diplegia (Moebius syndrome). Pediatr Neurol. 1985;1:180–4.
128. Phemister JC, Small JM. Hypersomnia in dystrophia myotonica. J Neurol Neurosurg Psychiatry. 1961;24:173–5.
129. Manni R, Zucca C, Martinetti M, Ottolini A, Lanzi G, Tartara A. Hypersomnia in dystrophia myotonica: a neuropsychological and immunogenetic study. Acta Neurol Scand. 1991;84:498–502.
130. Park J, Radtke R. Hypersomnolence in myotonic dystrophy: demonstration of sleep onset REM sleep. J Neurol Neurosurg Psychiatry. 1995;58:512–3.
131. Rubinsztein JS, Rubinsztein DC, Goodburn S, Holland AJ. Apathy and hypersomnia are common features of myotonic dystrophy. J Neurol Neurosurg Psychiatry. 1998;64:510–5.
132. Giubilei F, Antonini G, Bastianello S, Morino S, Paolillo A, Fiorelli M, et al. Excessive daytime sleepiness in myotonic dystrophy. J Neurol Sci. 1999;164:60–3.
133. Phillips MF, Steer HM, Soldan JR, Wiles CM, Harper PS. Daytime somnolence in myotonic dystrophy. J Neurol. 1999;246:275–82.
134. Khandelwal D, Bhatia M, Tripathi M, Sahota P, Jain S. Excessive daytime sleepiness: an unusual presentation of myotonic dystrophy. Sleep Med. 2002;3: 431–2.
135. Laberge L, Bégin P, Montplaisir J, Mathieu J. Sleep complaints in patients with myotonic dystrophy. J Sleep Res. 2004;13:95–100.
136. Ciafaloni E, Mignot E, Sansone V, Hilbert JE, Lin L, Lin X, et al. The hypocretin neurotransmission system in myotonic dystrophy type 1. Neurology. 2008;70:226–30.
137. Laberge L, Bégin P, Dauvilliers Y, Beaudry M, Laforte M, Jean S, et al. A polysomnographic study of daytime sleepiness in myotonic dystrophy type 1. J Neurol Neurosurg Psychiatry. 2009;80:642–6.
138. Gibbs JW, Ciafaloni E, Radtke RA. Excessive daytime somnolence and increased rapid eye movement pressure in myotonic dystrophy. Sleep. 2002;25: 672–5.

139. Martinez-Rodriguez JE, Lin L, Iranzo A, Genis D, Marti MJ, Santamaria J, et al. Decreased hypocretin-1 (orexin-A) levels in the cerebrospinal fluid of patients with myotonic dystrophy and excessive daytime sleepiness. Sleep. 2003;26:287–90.
140. Van Hilten JJ, Kerkhof GA, van Dijk JG, Dunnewold R, Wintzen AR. Disruption of sleep-wake rhythmicity and daytime sleepiness in myotonic dystrophy. J Neurol Sci. 1993;114:68–75.
141. Ono S, Takahashi K, Jinnai K, et al. Loss of serotonin-containing neurons in the raphe of patients with myotonic dystrophy: a quantitative immunohistochemical study and relation to hypersomnia. Neurology. 1998;50:535–8.
142. Frucht S, Rogers JD, Greene PE, Gordon MF, Fahn S. Falling asleep at the wheel: motor vehicle mishaps in persons taking pramipexole and ropinirole. Neurology. 1999;52:1908–10.
143. Ferreira JJ, Galitzky M, Montastruc JL, Rascol O. Sleep attacks and Parkinson's disease treatment. Lancet. 2000;355:1333–4.
144. Arnulf I, Bonnet AM, Damier P, et al. Hallucinations, REM sleep, and Parkinson's disease: a medical hypothesis. Neurology. 2000;55:281–8.
145. Arnulf I, Konofal E, Merino-Andreu M, et al. Parkinson's disease and sleepiness: an integral part of PD. Neurology. 2002;58:1019–24.
146. Hobson DE, Lang AE, Martin WR, Razmy A, Rivest J, Fleming J. Excessive daytime sleepiness and sudden-onset sleep in Parkinson disease: a survey by the Canadian Movement Disorders Group. JAMA. 2002; 287:455–63.
147. Rye DB, Johnston LH, Watts RL, Bliwise DL. Juvenile Parkinson's disease with REM sleep behaviour disorder, sleepiness, and daytime REM onset. Neurology. 1999;53:1868–70.
148. Onofrj M, Luciano AL, Iacono D, Thomas A, Stocchi F, Papola F, et al. HLA typing does not predict REM sleep behaviour disorder and hallucinations in Parkinson's disease. Mov Disord. 2003;18:337–40.
149. Autret A, Lucas B, Henry-Lebras F, et al. Symptomatic narcolepsies. Sleep. 1994;17:S21–4.
150. Overeem S, van Hilten JJ, Ripley B, Mignot E, Nishino S, Lammers GJ. Normal hypocretin-1 levels in Parkinson's disease patients with excessive daytime sleepiness. Neurology. 2002;58:498–9.
151. Baumann C, Ferini-Strambi L, Waldvogel D, Werth E, Bassetti CL. Parkinson's disease with excessive daytime sleepiness – a narcolepsy-like disorder? J Neurol. 2005;252:139–45.
152. Yasui K, Inoue Y, Kanbayashi T, Nomura T, Kusumi M, Nakashima K. CSF orexin levels in Parkinson's disease, dementia with Lewy bodies, progressive supranuclear palsy and corticobasal degeneration. J Neurol Sci. 2006;250:120–3.
153. Maeda T, Nagata K, Kondo H, Kanbayashi T. Parkinson's disease comorbid with narcolepsy presenting low CSF hypocretin/orexin level. Sleep Med. 2006;7:662.
154. Drouot X, Moutereau S, Nguyen JP, et al. Low levels of ventricular CSF orexin/hypocretin in advanced PD. Neurology. 2003;61:540–3.
155. Fronczek R, Overeem S, Lee SY, et al. Hypocretin (orexin) loss in Parkinson's disease. Brain. 2007;130:1577–85.
156. Thannickal TC, Lai YY, Siegel JM. Hypocretin (orexin) cell loss in Parkinson's disease. Brain. 2007;130:1586–95.
157. Van Someren EJ, Hagebeuk EE, Lijzenga C, Scheltens P, de Rooij SE, Jonker C, et al. Circadian rest-activity rhythm disturbances in Alzheimer's disease. Biol Psychiatry. 1996;40:259–70.
158. Ancoli-Israel S, Klauber MR, Jones DW, Kripke DF, Martin J, Mason W, et al. Variations in circadian rhythms of activity, sleep, and light exposure related to dementia in nursing-home patients. Sleep. 1997; 20:18–23.
159. Yesavage JA, Friedman L, Ancoli-Israel S, et al. Development of diagnostic criteria for defining sleep disturbances in Alzheimer's disease. J Geriatr Psychiatry Neurol. 2003;16:131–9.
160. Harper DG, Stopa EG, McKee AC, Satlin A, Fish D, Volicer L. Dementia severity and Lewy bodies affect circadian rhythms in Alzheimer disease. Neurobiol Aging. 2004;25:771–81.
161. Ebrahim IO, Semra YK, De Lacy S, et al. CSF hypocretin (orexin) in neurological and psychiatric conditions. J Sleep Res. 2003;12:83–4.
162. Baumann CR, Dauvilliers Y, Mignot E, Bassetti CL. Normal CSF hypocretin-1 (orexin A) levels in dementia with Lewy bodies associated with excessive daytime sleepiness. Eur Neurol. 2004;52: 73–6.
163. Friedman LF, Zeitzer JM, Lin L, et al. In Alzheimer disease, increased wake fragmentation found in those with lower hypocretin-1. Neurology. 2007;68: 793–4.
164. Desarnaud F, Murillo-Rodriguez E, Lin L, et al. The diurnal rhythm of hypocretin in young and old F344 rats. Sleep. 2004;27:851–6.
165. Harper DG, Stopa EG, Kuo-Leblanc V, McKee AC, Asayama K, Volicer L, et al. Dorsomedial SCN neuronal subpopulations subserve different functions in human dementia. Brain. 2008;131:1609–17.
166. Deboer T, Overeem S, Visser NAH, Duindam H, Frölich M, Lammers GJ, et al. Convergence of circadian and sleep regulatory mechanisms on hypocretin-1. Neuroscience. 2004;129:727–32.
167. Zhang S, Zeitzer JM, Yoshida Y, Wisor JP, Nishino S, Edgar DM, et al. Lesions of the suprachiasmatic nucleus eliminate the daily rhythm of hypocretin-1 release. Sleep. 2004;27:619–27.
168. Marston OJ, Williams RH, Canal MM, Samuels RE, Upton N, Piggins HD. Circadian and dark-pulse activation of orexin/hypocretin neurons. Molecular Brain. 2008;1:1–16.
169. Arnulf I, Merino-Andreu M, Bloch F, Konofal E, Vidailhet M, Cochen V, et al. REM sleep behaviour

disorder and REM sleep without atonia in patients with progressive supranuclear palsy. Sleep. 2005;28: 349–54.

170. Ghorayeb I, Yekhlef F, Chrysostome V, Balestre E, Bioulac B, Tison F. Sleep disorders and their determinants in multiple system atrophy. J Neurol Neurosurg Psychiatry. 2002;72:798–800.
171. Abdo WF, Bloem BR, Kremer HP, Lammers GJ, Verbeek MM, Overeem S. CSF hypocretin-1 levels are normal in multiple-system atrophy. Parkinsonism Relat Disord. 2008;14:342–4.
172. Martinez-Rodriguez JE, Seppi K, Cardozo A, Iranzo A, Stampfer-Kountchev M, Wenning G, et al. Cerebrospinal fluid hypocretin-1 levels in multiple system atrophy. Mov Disord. 2007;22:1822–4.
173. Benarroch EE, Schmeichel AM, Sandroni P, Low PA, Parisi JE. Involvement of hypocretin neurons in multiple system atrophy. Acta Neuropathol. 2007;113:75–80.
174. Gerashchenko D, Murillo-Rodriguez E, Lin L, et al. Relationship between CSF hypocretin levels and hypocretin neuronal loss. Exp Neurol. 2003;184: 1010–6.
175. Starr A. A disorder of rapid eye movements in Huntington's chorea. Brain. 1967;90:545–64.
176. Wiegand M, Moller AA, Lauer CJ, Stolz S, Schreiber W, Dose M, et al. Nocturnal sleep in Huntington's disease. J Neurol. 1991;238:203–8.
177. Taylor N, Bramble D. Sleep disturbance and Huntington's disease. Br J Psychiatry. 1997; 171:393.
178. Arnulf I, Nielsen J, Lohmann E, Schieffer J, Wild E, Jennum P, et al. Rapid eye movement sleep disturbances in Huntington Disease. Arch Neurol. 2008;65:482–8.
179. Kremer HP, Roos RA, Dingjan G, Marani E, Bots GT. Atrophy of the hypothalamic lateral tuberal nucleus in Huntington's disease. J Neuropathol Exp Neurol. 1990;49:371–82.
180. Kremer HP, Roos RA, Dingjan GM, Bots GT, Bruyn GW, Hofman MA. The hypothalamic lateral tuberal nucleus and the characteristics of neuronal loss in Huntington's disease. Neurosci Lett. 1991;132: 101–4.
181. Petersen A, Gil J, Maat-Schieman ML, et al. Orexin loss in Huntington's disease. Hum Mol Genet. 2005;14:39–47.
182. Aziz NA, Swaab DF, Pijl J, Roos RA. Hypothalamic dysfunction and neuroendocrine and metabolic alterations in Huntington's disease: clinical consequences and therapeutic implications. Rev Neurosci. 2007;18:223–51.
183. Gaus SE, Lin L, Mignot E. CSF hypocretin levels are normal in Huntington's disease patients. Sleep. 2005;28:1607–8.
184. Meier A, Mollenhauer B, Cohrs S, et al. Normal hypocretin-1 (orexin-A) levels in the cerebrospinal fluid of patients with Huntington's disease. Brain Res. 2005;1063:201–3.
185. Björkqvist M, Petersen A, Nielsen J, et al. Cerebrospinal fluid levels of orexin-A are not a clinically useful biomarker for Huntington disease. Clin Genet. 2006;70:78–9.
186. Baumann CR, Hersberger M, Bassetti CL. Hypocretin-1 (orexin A) levels are normal in Huntington's disease. J Neurol. 2006;253:1232–3.
187. Peyron C, Tighe DK, van Den Pool AN, de Lecea L, Heller HC, Sutcliffe JG, et al. Neurons containing hypocretin (orexin) project to multiple neuronal systems. J Neurosci. 1998;18:9996–10015.
188. Van Rooij FG, Schelhaas HJ, Lammers GJ, Verbeek MM, Overeem S. CSF hypocretin-1 levels are normal in patients with amyotrophic lateral sclerosis. Amyotroph Lateral Scler. 2009;10:487–9.

Posttraumatic Narcolepsy

Christian R. Baumann and Rositsa Poryazova

Keywords
Traumatic brain injury • Narcolepsy • Orexin • Hypocretin • Hypersomnia • Secondary injury

Does narcolepsy occur after traumatic brain injury (TBI)? Although there is a long-standing history of case reports on posttraumatic narcolepsy, this question remains essentially unanswered. This chapter reviews current clinical and pathophysiological evidence on whether or not posttraumatic narcolepsy may exist, how it might be caused, and how to treat it.

Traumatic Brain Injury and Sleep

TBI is caused by a physical trauma such as a blow to the head or a penetrating head injury, and is the leading cause of death and disability in young adults worldwide [1, 2]. Persisting deficits after TBI commonly include neuropsychological and psychiatric symptoms [3]. Also, sleep–wake disturbances (SWD) are common after TBI. Posttraumatic excessive daytime sleepiness (EDS), fatigue, and hypersomnia are the most prevalent SWD after TBI in a number of studies (Fig. 1) [4–11].

The reported prevalence of EDS and fatigue after TBI ranges from 19% up to 65% [4–11]. Methodological differences between the studies may account for this large range: first, EDS is not uniformly defined and assessed. Some studies used questionnaires with different scales such as the Epworth sleepiness scale, or the Stanford sleepiness scale, while others performed electrophysiological sleep laboratory tests. Second, time intervals between TBI and the assessment of EDS varied considerably between studies, ranging from 3 months to more than 4 years. In our own prospective study on SWD 6 months after trauma in 65 consecutive patients, we found EDS in 28% of patients, when assessed with the Epworth sleepiness scale [4]. The severity of EDS was not associated with severity of TBI. Fatigue was present in 17% of TBI patients. In most of these patients, we could not identify the causes of posttraumatic vigilance impairment other than the trauma itself.

On the other hand, it is a well-known fact that many TBI patients suffer from comorbid psychiatric or neurological disorders including SWD, which are risk factors for trauma [5, 11]. After TBI, the diagnosis of pretraumatic SWD relies solely on the information given by the patient and cannot be ruled out in all trauma patients. For example, EDS is a major cause for motor vehicle accidents, but due to sleep misperception, many patients may not recall a short sleep attack preceding their car crash. EDS and sleep attacks can

C.R. Baumann (✉)
Department of Neurology, University Hospital Zurich, 8091 Zurich, Switzerland
e-mail: christian.baumann@usz.ch

C.R. Baumann et al. (eds.), *Narcolepsy: Pathophysiology, Diagnosis, and Treatment*,
DOI 10.1007/978-1-4419-8390-9_31,

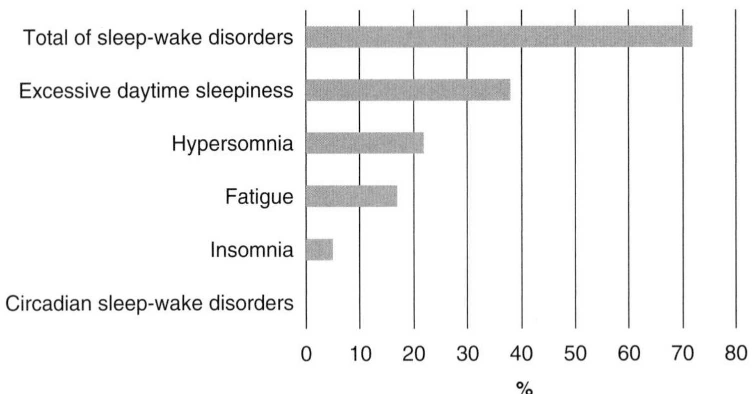

Fig. 1 Sleep–wake disorders 6 months after traumatic brain injury in 65 consecutive patients in the first systematic prospective trial on posttraumatic sleep–wake disorders

have many causes. For example, patients with obstructive sleep apnea or chronic sleep deprivation are more prone to motor vehicle accidents than the general population [12, 13]. Many patients, however, are not aware of being sleep-deprived despite the presence of severe EDS. Obstructive sleep apnea, on the other hand, has been suspected to occur de novo after TBI [4, 11]. Guilleminault et al. found obstructive sleep apnea in 59 of 184 patients with head and neck trauma, 6 of them were identified as snorers before TBI, but none of them exhibited EDS prior to the accident [11]. Periodic limb movement during sleep is another cause of posttraumatic EDS, and has been observed in 2–7% of TBI patients [4–9]. Again, whether these sleep-associated motor symptoms were present already before TBI or not remain elusive in many patients.

Last but not least, epilepsy and its treatments, depression, and treatment of chronic pain can contribute to posttraumatic EDS. Most neuroactive drugs, particularly many antidepressants, suppress REM sleep, which is associated with a lower frequency of sleep onset REM sleep periods (SOREMPs). On the other hand, the discontinuation of these drugs can lead to REM rebound with multiple SOREMPs [14].

Thus, the relationship between SWD and TBI is complex because of bidirectional influences: in some patients, SWD contribute to TBI, and in others, TBI causes de novo SWD. Finally, we should not forget that posttraumatic EDS often causes psychosocial problems. Guilleminault et al. reported that 103 of 184 patients with posttraumatic EDS had insurance, financial, and other medical/legal issues linked to their head trauma and their subsequent report of EDS [11].

Traumatic Brain Injury and Narcolepsy: A Difficult Relationship

Do some of the TBI patients who suffer from posttraumatic EDS have posttraumatic narcolepsy? In our 65 consecutive TBI patients, we identified two young male patients who met the diagnostic criteria of narcolepsy without cataplexy: they had low mean sleep latencies on multiple sleep latency test (MSLT) and multiple SOREMPs [4]. On the other hand, convincing reports of cataplexy were not given by any of our patients.

Symptomatic narcolepsy exists, and there are several disorders that can cause it, mainly by affecting the hypothalamus. This is reviewed in the chapter by Valko and colleagues. The first reports on posttraumatic narcolepsy are almost 100 years old. The French neurologist Jean Lhermitte, the Austrian neurologist Emil Redlich, and the British neurologist William Adie described patients with narcolepsy following TBI already in the early twentieth century [15–17]. Wilson Gill discussed a patient with traumatic narcolepsy in a Lancet article in 1941, and he described three other cases with SWD following TBI [18]. However, these individuals probably did not suffer from narcolepsy with cataplexy.

The first patient had attacks of irresistible sleep and depression, the second had drowsiness after laughter but without any loss of muscle tone, and the third had dreamless sleep attacks. This last patient reported also episodes in which he had "an inability to breathe and a feeling that someone had hit me behind the knees causing the legs to give way." These episodes occurred when the patient concentrated on mental problems, and based on current knowledge, it seems unlikely that these were true cataplexy attacks [19].

Later reports did not shed more light on this difficult topic. In 1989, Good and colleagues reported on a case with posttraumatic narcolepsy, and announced in their title a patient with the complete syndrome [20]. However, cataplexy in this case was not typical, and sleep laboratory tests to confirm the diagnosis were not performed. Five years later, the largest case series of patients with posttraumatic narcolepsy with cataplexy was reported by Lankford and colleagues [21]. Based on history, nocturnal polysomnography, and MSLT, nine patients were diagnosed with posttraumatic narcolepsy, five of them with cataplexy. Furthermore, five of seven patients tested were positive for HLA haplotypes associated with narcolepsy, some of them with only mild TBI. The authors concluded that even minor injury to the central nervous system can trigger narcolepsy in patients genetically at risk for the disorder. Still, sleep laboratory-based diagnosis of narcolepsy in this and other studies is limited by the low specificity of positive MSLT findings. This problem is discussed in the chapter on the "Diagnosis of Narcolepsy" by Iranzo. Briefly, low mean sleep latencies and multiple sleep onset REM periods on the MSLT, which are diagnostic for narcolepsy, can also be seen in other SWD, such as sleep apnea, or even in a significant portion of healthy subjects with chronic sleep deprivation [22–26]. Thus, mean sleep latency lower than 8 min and multiple SOREMPs have been reported in up to 6% of males and 1% of females in a community-based study [22]. Furthermore, since the publication of the abovementioned reports, the diagnostic criteria for narcolepsy have changed, which makes the comparisons of current trials with older studies difficult.

Guilleminault and colleagues found narcolepsy without cataplexy in 5 of 184 patients with head and neck trauma [11]. In 2005, Nishino and Kanbayashi reviewed patients with symptomatic EDS, and found 19 patients with posttraumatic narcolepsy [27]. Ebrahim and colleagues reviewed 21 literature cases with posttraumatic narcolepsy, and they added two cases [28]. Clear-cut cataplexy or decreased hypocretin levels or both were not reported in any case. In a recent study, Castriotta et al. found posttraumatic narcolepsy without cataplexy in 5 of 87 TBI patients [5], and Masel identified only one narcolepsy patient in a TBI cohort of 71 cases [6]. Bruck and Broughton summarized the puzzling question of posttraumatic narcolepsy with a case report titled "Diagnostic Ambiguities in a Case of Posttraumatic Narcolepsy with Cataplexy" [29]. In their case, they identified a TBI patient with posttraumatic EDS and cataplexy-like episodes, but they did not measure cerebrospinal fluid (CSF) levels of hypocretin, which would be the most specific confirmation of narcolepsy with cataplexy.

In summary, there have been many reports of posttraumatic narcolepsy over the last 100 years, but to date, there is still no convincing report on posttraumatic narcolepsy with typical cataplexy and proven hypocretin deficiency. The diagnosis of posttraumatic narcolepsy without cataplexy, on the other hand, lacks specificity and is especially problematic. Posttraumatic narcolepsy may have a heterogeneous etiology. In some cases, narcolepsy could be a preexisting condition, leading to accidents and TBI. Furthermore, we cannot rule out the possibility of a genetic predisposition toward posttraumatic narcolepsy, namely in association with narcolepsy-related HLA markers. Last not least, narcolepsy and narcolepsy-like symptoms could be due to the injury itself.

Pathophysiology of Posttraumatic EDS and Narcolepsy-Like Symptoms

Why do patients with TBI so often develop EDS and narcolepsy-like symptoms? Damage to wake-promoting structures and neurotransmitter systems might contribute to these symptoms. Crompton reported that almost half of the patients with moderate to severe TBI have histological damage to the hypothalamus and to the brainstem [30, 31].

Further-more, TBI-associated damage often includes the posterolateral hypothalamus, where wake-promoting hypocretin neurons are localized [32–34]. In accordance with these findings, we found low or undetectable lumbar or ventricular CSF hypocretin levels in 87% of patients within 4 days after TBI [35]. In patients with severe TBI, hypocretin levels were low or undetectable in 97% of the patients, even in wake patients. This acute hypocretin deficiency could be caused by damage to the hypocretin neurons and their axonal projections, or alternatively, production and release of hypocretin could be transiently downregulated. The second explanation would make sense because sleep plays a significant role in synaptic plasticity after brain damage [36–38]. Thus, low hypocretin levels could produce more sleep and facilitate recovery from TBI. In a follow-up study of these patients 6 months after TBI, we observed an increase in CSF hypocretin levels in all patients [4]. None of the patients had an undetectable hypocretin level 6 months after TBI, supporting the hypothesis that hypocretin release is transiently reduced after TBI. Still, four subjects had low levels, which suggest permanent damage to the hypocretin system. Indeed, in a pilot postmortem study in four patients with severe TBI and four controls without TBI, we found a 30% reduction in the number of hypocretin neurons (Fig. 2) [39].

An almost complete loss of hypocretin neurons causes narcolepsy with cataplexy [40, 41]. However, it is still unknown whether some hypocretin neurons die in narcolepsy without cataplexy. These patients usually have normal CSF hypocretin levels. A preliminary postmortem study of patients with narcolepsy without cataplexy revealed a 33% loss of hypocretin neurons, but this study analyzed only one brain and part of another, so it needs to be substantiated by larger case series [42]. Gerashchenko and colleagues showed in a rodent model that decreased CSF hypocretin levels require a loss of >70% of the hypocretin cells [43]. Taking all these pieces of evidence together, we might assume that partial loss of hypocretin neurons in TBI can contribute to EDS and narcolepsy-like findings in TBI patients. In accordance with this assumption, we found that TBI patients with EDS 6 months after trauma had significantly lower CSF hypocretin levels compared to nonsleepy patients [4].

Treatment of Posttraumatic Narcolepsy

There are currently no guidelines for the treatment of posttraumatic narcolepsy. Gerard and Ivanhoe reported successful treatment of posttraumatic narcolepsy with methylphenidate in one patient [44]. In a prospective, double-blind, randomized and placebo-controlled pilot study in 20 TBI patient with EDS and fatigue, we found a

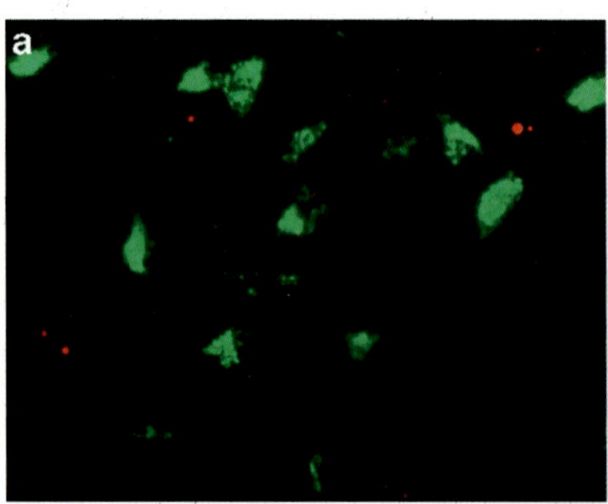

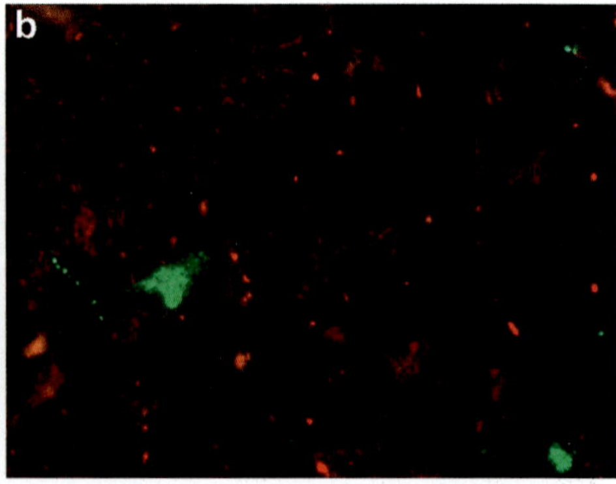

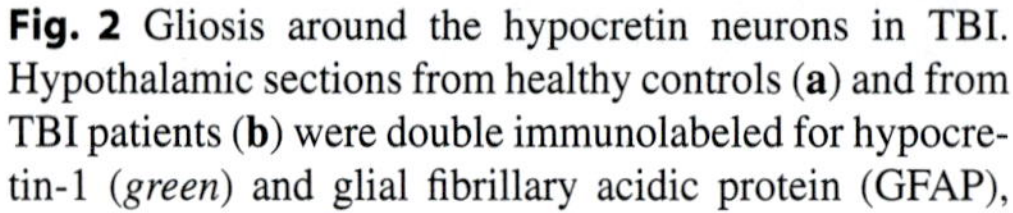

Fig. 2 Gliosis around the hypocretin neurons in TBI. Hypothalamic sections from healthy controls (**a**) and from TBI patients (**b**) were double immunolabeled for hypocretin-1 (*green*) and glial fibrillary acidic protein (GFAP), a marker of astrocytes (*red*). In TBI patients, there was a 30% loss of hypocretin cells, together with an increased number of astrocytes, indicating gliosis

benefit of modafinil on EDS but not on fatigue (unpublished findings). In general, we believe that treatment of posttraumatic narcolepsy should not differ from that of idiopathic narcolepsy. The same medications should be tried, including stimulants, antidepressants and perhaps with more caution, sodium oxybate.

References

1. Sosin DM, Sniezek JE, Thurman DJ. Incidence of mild and moderate brain injury in the United States, 1991. Brain Inj. 1996;10:47–54.
2. Bruns Jr J, Hauser WA. The epidemiology of traumatic brain injury: a review. Epilepsia. 2003;44:2–10.
3. Maegele M, Engel D, Bouillon B, Lefering R, Fach H, Raum M, et al. Incidence and outcome of traumatic brain injury in an urban area in Western Europe over 10 years. Eur Surg Res. 2007;39:372–9.
4. Baumann CR, Werth E, Stocker R, Ludwig S, Bassetti CL. Sleep-wake disturbances 6 months after traumatic brain injury: a prospective study. Brain. 2007;130:1873–83.
5. Castriotta RJ, Wilde MC, Lai JM, Atanasov S, Masel BE, Kuna ST. Prevalence and consequences of sleep disorders in traumatic brain injury. J Clin Sleep Med. 2007;3:349–56.
6. Masel BE, Scheibel RS, Kimbark T, Kuna ST. Excessive daytime sleepiness in adults with brain injuries. Arch Phys Med Rehabil. 2001;82:1526–32.
7. Verma A, Anand V, Verma NP. Sleep disorders in chronic traumatic brain injury. J Clin Sleep Med. 2007;3:357–62.
8. Watson NF, Dikmen S, Machamer J, Doherty M, Temkin N. Hypersomnia following traumatic brain injury. J Clin Sleep Med. 2007;3:363–8.
9. Parcell DL, Ponsford JL, Rajaratnam SM, Redman JR. Self-reported changes to nighttime sleep after traumatic brain injury. Arch Phys Med Rehabil. 2006;87:278–85.
10. Castriotta RJ, Lai JM. Sleep disorders associated with traumatic brain injury. Arch Phys Med Rehabil. 2001;82:1403–6.
11. Guilleminault C, Yuen KM, Gulevich MG, Karadeniz D, Leger D, Philip P. Hypersomnia after head-neck trauma: a medicolegal dilemma. Neurology. 2000;54:653–9.
12. Ellen RLB, Marshall SC, Palayew M, Molnar FJ, Wilson KG, Man-Son-King M. Systematic review of motor vehicle crash risk in persons with sleep apnea. J Clin Sleep Med. 2006;2:193–200.
13. Leger D. The cost of sleep-related accidents: a report for the National Commission on Sleep Disorders Research. Sleep. 1994;17:84–93.
14. Ristanovic RK, Liang H, Hornfeldt CS, Lai C. Exacerbation of cataplexy following gradual withdrawal of antidepressants: manifestation of probable protracted rebound cataplexy. Sleep Med. 2009;10:416–21.
15. Redlich E. Epilogomena zur Narkolepsiefrage. Zschr f ges Neurol und Psychiat. 1931;136:128–73.
16. Lhermitte J, Tourney T. Rapport sur le sommeil normal et pathologique. Rev neurol. 1927;34:1–752.
17. Adie WJ. Idiopathic narcolepsy: a disease sui generis; with remarks on the mechanisms of sleep. Brain. 1926;49:257–306.
18. Gill AW. Idiopathic and traumatic nacrolepsy. Lancet. 1941;1:474–6.
19. Anic-Labat S, Guilleminault C, Kraemer HC, Meehan J, Arrigoni J, Mignot E. Validation of a cataplexy questionnaire in 983 sleep-disorders patients. Sleep. 1999;22:77–87.
20. Good JL, Barry E, Fishman PS. Posttraumatic narcolepsy: the complete syndrome with tissue typing. Case report. J Neurosurg. 1989;71:765–7.
21. Lankford DA, Wellman JJ, O'Hara C. Posttraumatic narcolepsy in mild to moderate closed head injury. Sleep. 1994;17(8 Suppl):S25–8.
22. Mignot E, Lin L, Finn L, Lopes C, Pluff K, Sundstrom ML, et al. Correlates of sleep-onset REM periods during the Multiple Sleep Latency Test in community adults. Brain. 2006;129:1609–23.
23. Bishop C, Rosenthal L, Helmus T, Roehrs T, Roth T. The frequency of multiple sleep onset REM periods among subjects with no excessive daytime sleepiness. Sleep. 1996;19:727–30.
24. Singh M, Drake CL, Roth T. The prevalence of multiple sleep-onset REM periods in a population-based sample. Sleep. 2006;29:890–5.
25. Chervin RD, Aldrich MS. Sleep onset REM periods during multiple sleep latency tests in patients evaluated for sleep apnea. Am J Respir Crit Care Med. 2000;161:426–31.
26. Marti I, Valko PO, Khatami R, Bassetti CL, Baumann CR. Multiple sleep latency measures in narcolepsy and behaviourally induced insufficient sleep syndrome. Sleep Med. 2009;10:1146–50.
27. Nishino S, Kanbayashi T. Symptomatic narcolepsy, cataplexy and hypersomnia, and their implications in the hypothalamic hypocretin/orexin system. Sleep Med Rev. 2005;9:269–310.
28. Ebrahim IO, Peacock KW, Williams AJ. Posttraumatic narcolepsy – two case reports and a mini review. J Clin Sleep Med. 2005;1:153–6.
29. Bruck D, Broughton RJ. Diagnostic ambiguities in a case of post-traumatic narcolepsy with cataplexy. Brain Inj. 2004;18:321–6.
30. Crompton MR. Hypothalamic lesions following closed head injury. Brain. 1971;94:165–72.
31. Crompton MR. Brainstem lesions due to closed head injury. Lancet. 1971;1:669–73.
32. Thorley RR, Wertsch JJ, Klingbeil GE. Acute hypothalamic instability in traumatic brain injury: a case report. Arch Phys Med Rehabil. 2001;82:246–9.
33. Estabrooke IV, McCarthy MT, Ko E, et al. Fos expression in orexin neurons varies with behavioral state. J Neurosci. 2001;21:1656–62.

34. Mochizuki T, Scammell TE. Orexin/hypocretin: wired for wakefulness. Curr Biol. 2003;13:563–4.
35. Baumann CR, Stocker R, Imhof HG, Trentz O, Hersberger M, Mignot E, et al. Hypocretin-1 (orexin A) deficiency in acute traumatic brain injury. Neurology. 2005;65:147–9.
36. Huber R, Ghilardi MF, Massimini M, Tononi G. Local sleep and learning. Nature. 2004;430:78–81.
37. Stickgold R, Walker MP. Sleep-dependent memory consolidation and reconsolidation. Sleep Med. 2007;8:331–43.
38. Siengsukon C, Boyd LA. Sleep enhances off-line spatial and temporal motor learning after stroke. Neurorehabil Neural Repair. 2009;23:327–35.
39. Baumann CR, Bassetti CL, Valko PO, Haybaeck J, Keller M, Clark E, et al. Loss of hypocretin (orexin) neurons with traumatic brain injury. Ann Neurol. 2009;66:555–9.
40. Peyron C, Faraco J, Rogers W, et al. A mutation in a case of early onset narcolepsy and a generalized absence of hypocretin peptides in human narcoleptic brains. Nat Med. 2000;6:991–7.
41. Thannickal TC, Moore RY, Nienhuis R, Ramanathan L, Gulyani S, Aldrich M, et al. Reduced number of hypocretin neurons in human narcolepsy. Neuron. 2000;27:469–74.
42. Thannickal TC, Nienhuis R, Siegel JM. Localized loss of hypocretin (orexin) cells in narcolepsy without cataplexy. Sleep. 2009;32:993–8.
43. Gerashchenko D, Murillo-Rodriguez E, Lin L, Xu M, Hallett L, Nishino S, et al. Relationship between CSF hypocretin levels and hypocretin neuronal loss. Exp Neurol. 2003;184:1010–6.
44. Francisco GE, Ivanhoe CB. Successful treatment of post-traumatic narcolepsy with methylphenidate: a case report. Am J Phys Med Rehabil. 1996;75:63–5.

The Hypocretin System and Sleepiness in Parkinson's Disease

Techniques to Assess Hypocretin Functioning in a Neurodegenerative Disorder Other Than Narcolepsy

R. Fronczek

Keywords
Parkinson's disease • Excessive daytime sleepiness • Orexin • Hypocretin • Dopamine • Hallucinations

Introduction

Parkinson's disease (PD) is a neurodegenerative disorder with prominent motor symptoms, such as rigidity, tremor and hypokinesia. These symptoms are thought to be the result of decreased stimulation of the motor cortex by the basal ganglia. This decreased stimulation is due to a progressive and irreversible degeneration of dopaminergic neurons projecting from the substantia nigra to the striatum [1]. In addition, there are degenerative changes in many other parts of the brain, including the hypothalamus. Lewy bodies, the pathophysiological hallmark of PD, have been found in various brain regions, again including the hypothalamus [2]. These observations suggest hypothalamic changes in PD.

Although the motor symptoms are the most prominent features in many PD patients, there are many non-motor symptoms as well [3]. Autonomic dysfunction, sleep–wake disturbances, mood disturbances and cognitive decline can have a profound impact on the quality of daily life of patients suffering from PD. During the last decade, we observed growing scientific and clinical interest in these non-motor symptoms. Sleep-wake disturbances often belong to the most striking non-motor symptoms [4–6], and can precede motor symptoms for years [7].

Nighttime sleep disturbances in PD include insomnia with fragmented nocturnal sleep, rapid eye movement (REM) sleep behaviour disorder and periodic leg movements [5, 8, 9]. During the day, excessive daytime sleepiness (EDS) with frequent naps and so-called "sleep-attacks" have been reported in 15–50% of patients, as well as daytime sleep onset REM periods [4, 10, 11]. These symptoms share many characteristics with the symptoms of narcolepsy [12]. The latter is characterised by EDS and REM sleep dissociation phenomena such as cataplexy. Nighttime symptoms are fragmented nocturnal sleep and REM sleep behaviour disorder. Sleep onset REM periods on the multiple sleep latency test (MSLT) constitute the neurophysiological hallmarks of narcolepsy [13]. Human narcolepsy is caused by the loss of hypocretin-producing neurons. This is reflected by undetectable hypocretin-1 levels in the cerebrospinal fluid [14]. The hypocretins (hypocretin-1 and -2) – also known as orexins

R. Fronczek (✉)
Department of Neurology, Leiden University Medical Centre, Leiden, The Netherlands
e-mail: r.fronczek@lumc.nl

C.R. Baumann et al. (eds.), *Narcolepsy: Pathophysiology, Diagnosis, and Treatment*,
DOI 10.1007/978-1-4419-8390-9_32, © Springer Science+Business Media, LLC 2011

(orexin A and B) – are neuropeptides involved in sleep–wake regulation, metabolism, autonomic regulation and reward processing [15]. Hypocretin-producing neurons are exclusively located in the lateral hypothalamus, from where they project widely throughout the central nervous system [16].

Excessive Daytime Sleepiness in Parkinson's Disease

EDS is generally assessed with questionnaires that measure the ability to drift off or fall asleep during daytime activities such as driving a car, having a conversation or watching television. The most commonly used questionnaire is the Epworth sleepiness scale (ESS) [17]. PD patients consistently score higher on the ESS compared to healthy controls. Mean ESS values range from 4.9 ± 3.6 to 11.1 ± 5.9. When using the common cut-off point of 10, up to 50% of PD patients suffer from EDS [4, 10, 11, 18–22].

A "sleep attack" is defined as an event of sudden irresistible and overwhelming sleepiness that is not preceded by a feeling of sleepiness. In parts, EDS and particularly sleep attacks are possible side effects of the dopamine agonist used in the treatment of PD. Up to 27% of PD patients suffer from these sleep attacks [4, 10, 11, 19, 21, 23]. However, narcolepsy patients generally have even higher ESS scores (>12) [24].

Electrophysiological Studies in Parkinson's Disease

Even if they bear major limitations, electrophysiological tests are a more objective way to measure EDS. The MSLT is an important test in the diagnosis of narcolepsy [25]. During this 1-day test, subjects are requested to try to fall asleep at various testing times throughout the day, lying in a bed in a dark room, while an electroencephalographic recording is performed. The main outcome measures are the mean sleep latency and the occurrence of sleep onset REM periods. Non-sleep deprived healthy controls do not exhibit REM sleep during the MSLT, neither do patients with other hypersomnia disorders than narcolepsy. To fulfil the diagnostic criteria of narcolepsy, patients should have a mean sleep latency shorter than 8 min and a minimum of two sleep onset REM periods (see also chapter on "Current Diagnostic Criteria of adult Narcolepsy" by Iranzo) [26].

MSLT studies in PD patients yielded mean sleep latencies between 6.3 ± 0.6 min in patients with EDS and 8 ± 1 min in patients with hallucinations. More than two sleep onset REM periods have been observed in 70% of PD patients with hallucinations, and in 39% of patients with EDS [5, 12, 27]. Thus, a significant number of PD patients fulfil the MSLT criteria for narcolepsy, which supports the hypothesis that both disorders share some common pathophysiological pathway. Therefore, the question is whether or not the hypocretin system is damaged in PD as well. An overview on studies of EDS in PD is given in Fig. 1.

In Vivo Studies

Assessing the hypocretin system in humans is far from easy. Most information pertaining to hypocretin function comes from animal models. In humans, the only in vivo possibility is measuring hypocretin-1 concentration in cerebrospinal fluid as a reflection of the production in the posterolateral hypothalamus [28]. Using this technique, we can detect the subtotal depletion of hypocretin neurons in narcolepsy [29]. However, assessing hypocretin function in other disorders is difficult, because there is presumably only partial loss of hypocretin cells, or there are only changes downstream the hypocretin system. The latter are difficult to assess in patients, since it is extremely difficult to study receptor distribution in humans. Currently, there are no means to directly assess the complete hypocretin pathway. It is impossible to monitor downstream signalling and there is no indicator of synaptic activity.

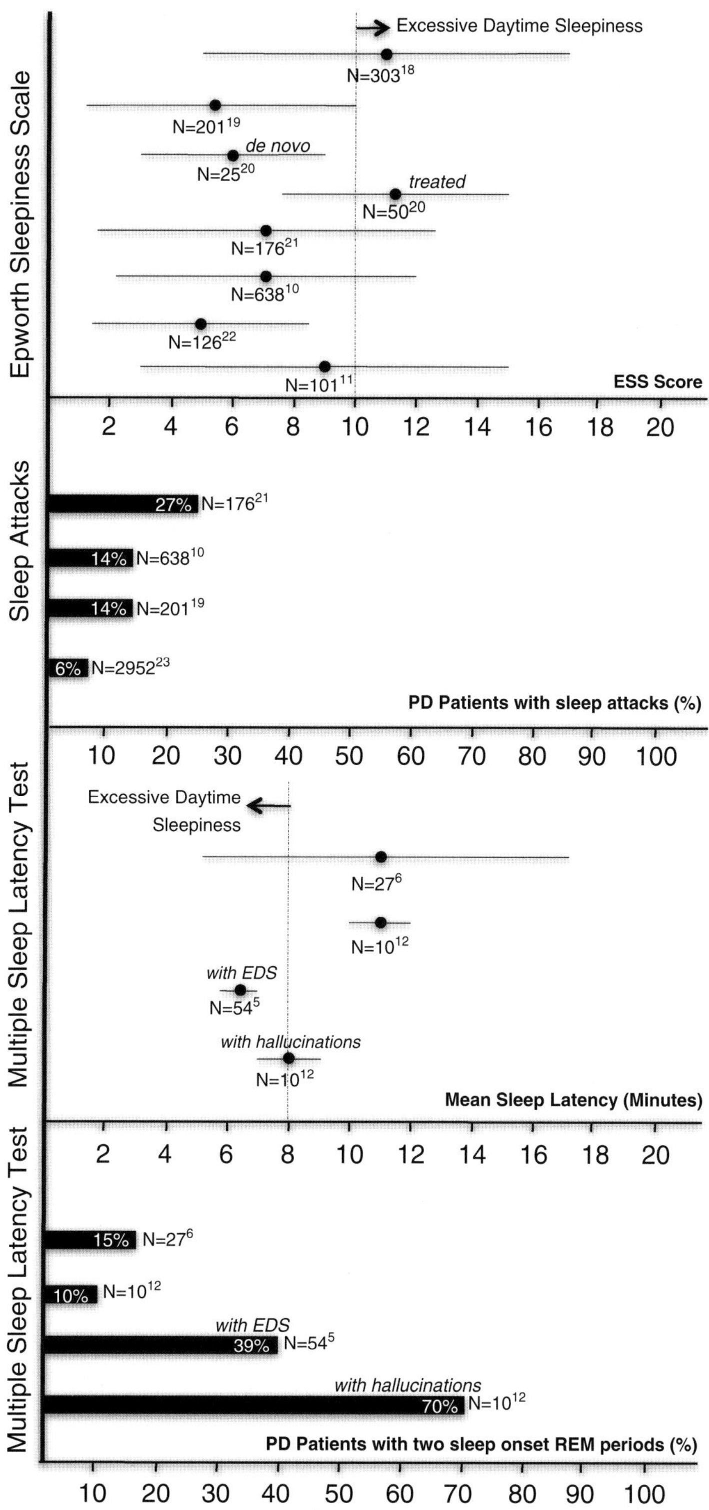

Fig. 1 Overview of post-mortem studies in which the hypocretin system was assessed in Parkinson's disease. Datapoints are medians, while lines represent the 25th and 75th percentiles. *CSF* cerebrospinal fluid

Cerebrospinal Fluid Measurements of Hypocretin-1

The currently available radioimmunoassay (RIA) kit to measure hypocretin-1 concentrations is neither sensitive nor specific enough to reliably detect the peptides in the blood or serum. Despite the pitfalls of this kit, however, it is possible to reliably measure concentrations of hypocretin-1 in cerebrospinal fluid [28]. Hypocretin-2 appears to be less stable in this compartment, and cannot be quantified. The commercially available hypocretin-1 RIA kit suffers from high inter-assay variability, making the use of standardised reference samples in different dilutions necessary [28]. Nevertheless, hypocretin-1 measurements in cerebrospinal fluid have been used as a reflection of central hypocretin signalling. Most often, spinal cerebrospinal fluid from lumber puncture is used, and ventricular cerebrospinal fluid is rarely tested [30]. Spinal cerebrospinal fluid hypocretin-1 concentrations change only when there is a relatively large reduction in hypocretin cell numbers. In a rodent study, lesioning 15% of hypocretin cells did not alter cerebrospinal fluid hypocretin-1 levels, but a loss of 70% of neurons resulted in a 50% decline of cerebrospinal fluid levels [31]. Therefore, it is possible to lose a substantial number of hypocretin cells without changes in cerebrospinal fluid levels. Note that this does not at all reflect the functional implications of a partial loss of hypocretin neurons. Therefore, the functional meaning of different hypocretin levels in the cerebrospinal fluid is unclear. It could be interpreted as a marker without an actual physiological function, or even be considered as "waste" that is leaving the brain by the spinal fluid after performing its function, showing only a drop when the number of hypocretin neurons is severely reduced [32].

Results in Parkinson's Disease

Several in vivo studies have been conducted to detect damage to the hypocretin system in PD. However, these studies only assessed ventricular and spinal cerebrospinal fluid hypocretin levels. Moreover, results have been conflicting. Studies by Overeem et al. ($n=3$), Ripley et al. ($n=7$), Baumann et al. ($n=10$) and Yasui et al. ($n=62$) – using spinal cerebrospinal fluid – have all found normal hypocretin-1 levels [28, 33–35]. Even PD patients who were selected because of clear sleep abnormalities did not show lowered hypocretin-1 concentrations in spinal cerebrospinal fluid [33]. In contrast, low levels and even absence of hypocretin-1 were found in ventricular cerebrospinal fluid in patients with late-stage PD [30]. In the latter study, an inverse correlation between hypocretin-1 levels and disease severity was reported.

The discrepancies between the study in ventricular cerebrospinal fluid and the studies in spinal cerebrospinal fluid could be due to the fact that hypocretin-1 concentrations in ventricular cerebrospinal fluid may more directly reflect hypocretin function. In the rat, it has been shown that hypocretin-1 is produced and released by fibres protruding into the lumen of the ventricles [36]. If ventricular concentrations are indeed much higher than spinal levels, this could be reflected by a ventriculolumber gradient. However, in one human study, hypocretin-1 was measured in six subsequent fractions of spinal cerebrospinal fluid, using up to 12 ml, and no clear gradient was found [28]. Obviously, ventricular cerebrospinal fluid is only available from patients with specific disorders, and therefore there are no normative data for hypocretin levels in this compartment.

Post-mortem Studies

Tissue Levels

Using the same RIA that is used for cerebrospinal fluid measurements, it is possible to assess the hypocretin-1 concentrations in brain tissue extract, from the densely innervated prefrontal cortex, the hypothalamus or the pons [16]. This technique has been used in post-mortem samples from narcolepsy patients, from patients with neurodegenerative disorders and from control subjects [14, 37, 38]. Hypocretin-1 concentrations in the brain parenchyma are higher than in cerebrospinal fluid, ranging between 500 and 1,000 pg/ml [14]. However, since no clear normative values are available and levels may vary considerably between specific brain regions, the technique is only suitable to directly compare patient groups with a carefully matched control group.

Neuronal Quantification

Probably, the best way of assessing the integrity of the hypocretin system is counting the number of hypocretin neurons in the lateral hypothalamus. This has been done in frozen hypothalamic sections using in situ hybridization, and in formalin-fixed material using immunocytochemistry [14, 37–40]. Although hypocretin-1 immunoreactive neurons are mainly restricted to the perifornical region of the lateral hypothalamus, it is essential to make sure that sections cover the complete hypocretin cell area when counting hypocretin neurons [38]. In this line, hypocretin neurons start to appear in the supraoptic area on the level where the fornix crosses the paraventricular nucleus. In subsequent levels, the fornix migrates to the corpora mammillaria while passing through an area with a high number of hypocretin cells. When the fornix abuts the corpora mammillaria, there are still many hypocretin-1 immunoreactive cell bodies visible [37]. A similar distribution pattern has been described using in situ hybridization [14]. Visually inspecting the distribution patterns of hypocretin neurons within individual hypothalami is helpful to verify the reliability of the counting procedure. Furthermore, a carefully matched control group is necessary, because of differences in antisera, exact counting techniques and observers [38]. In rats, estimates of the number of hypocretin containing neurons range from 1,000 to 4,000, depending on the antiserum and/or estimation method. In the human brain, this number was estimated to be 15,000–20,000 using in situ hybridization and 50,000–80,000 using immunocytochemistry [14, 39]. There is no known explanation for these large differences in hypocretin neuronal counts.

In heterogeneous populations of neurons, it is impossible to distinguish a loss of neurons from a mere loss of a cell marker, such as hypocretin-1. Strictly speaking, a decreased number of neurons that express hypocretin-1 does not mean that there is actually a loss of these neurons [41]. Neurons that are used to express hypocretin-1 could still be present and functionally active, but could just have stopped producing hypocretins for an unknown reason. To study this, the presence of markers that are co-expressed by hypocretin neurons, such as dynorphin and neuronal activity-regulated pentraxin (NARP) were examined [42]. In hypothalamic sections from narcolepsy patients with dramatic reduction in the number of hypocretin containing neurons, an equally dramatic reduction in the number of NARP and dynorphin containing neurons has been found [42, 43]. This provided evidence for the assumption that hypocretin neurons are really lost in narcolepsy. However, it might also be possible that hypocretin neurons still exist, but produce less peptides on a global level. On the other hand, the functional consequences of a "real" loss of neurons and a loss of "only" the marker are similar.

Results in Parkinson's Disease

Two studies assessed the hypocretin system in post-mortem brains of patients with Parkinson's disease (Fig. 2) [37, 40]. We (1) estimated the number of hypocretin neurons in post-mortem hypothalami using immunocytochemistry and an image analysis system (Fig. 3) and (2) quantified hypocretin levels in post-mortem ventricular cerebrospinal fluid and (3) prefrontal cortex using a RIA. Furthermore, presence of Lewy bodies was verified in the hypothalamic hypocretin cell area. We found (1) an almost 50% decrease of hypocretin neurons in PD patients when compared to controls (PD: 20,276; controls: 36,842; $p=0.016$), (2) a 40% decreased hypocretin-1 concentration in post-mortem ventricular cerebrospinal fluid (PD: 365.5 pg/ml; controls: 483.5; $p=0.012$) and (3) 25% reduced hypocretin-1 concentrations in prefrontal cortex (PD: 389.6 pg/g;

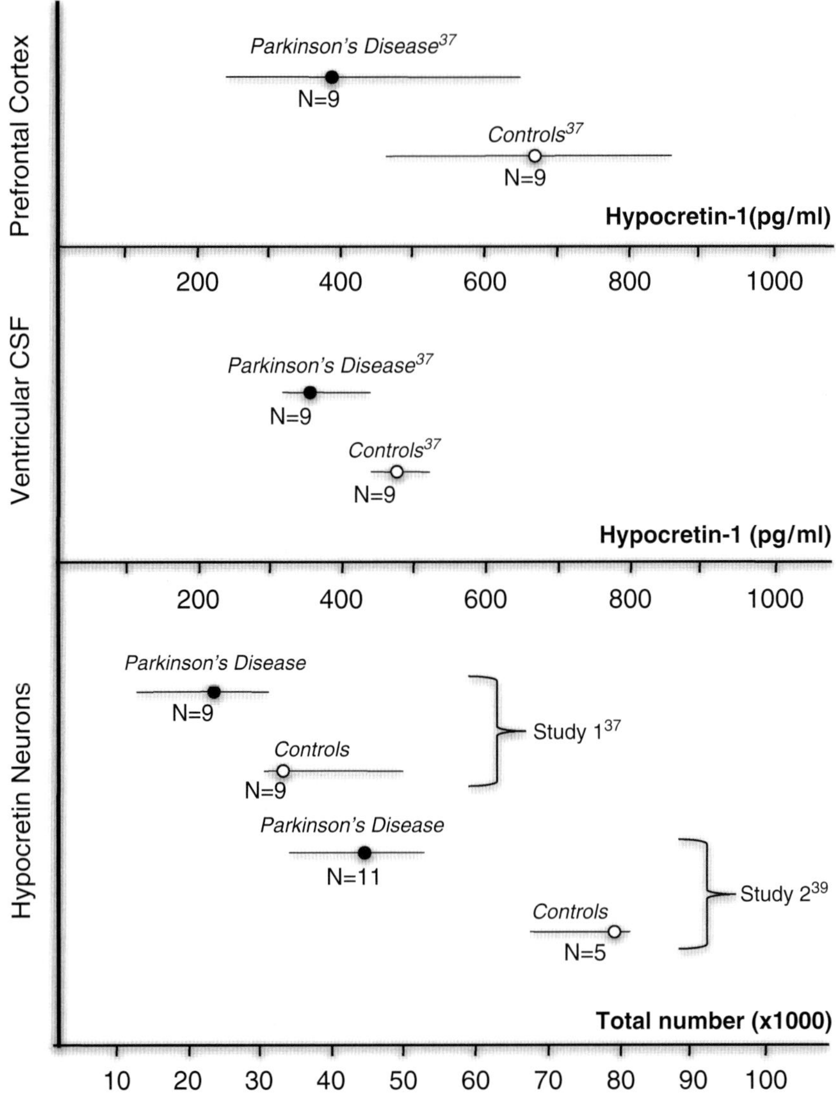

Fig. 2 Overview of studies looking into excessive daytime sleepiness in Parkinson's disease. Data are presented as means and standard deviations, or in percentages. Adapted from [4]. *ESS* Epworth sleepiness scale, *EDS* excessive daytime sleepiness, *PD* Parkinson's disease

controls: 676.6; $p=0.043$). Importantly, there was also a significant correlation between cell number and ventricular cerebrospinal fluid content in the combined group ($n=15$, $r=0.62$, $p=0.010$), suggesting that cerebrospinal fluid hypocretin levels do reflect to some extent the number of hypocretin neurons. In all PD patients, Lewy bodies were abundantly present in the perifornical region of the lateral hypothalamus, while only few Lewy bodies could be discerned in one control patient. However, hypocretin neurons that contained a Lewy body were rare and only one to two double-stained neurons could be discerned in sections that contained numerous hypocretin neurons [37].

Thannickal et al. found an increasing loss of hypocretin cells with PD disease progression. Similarly, they found an increased loss of melanin-concentrating hormone (MCH) neurons with disease severity. MCH neurons are situated

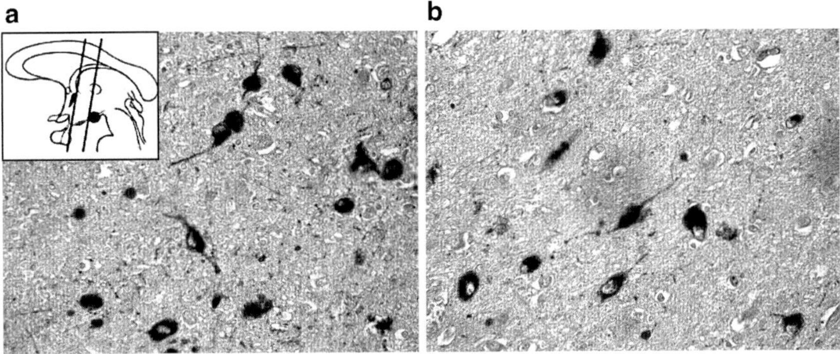

Fig. 3 Example of hypocretin-1 cell bodies in the lateral hypothalamus of a control patient (**a**) and a patient with Parkinson's disease (**b**). *Inset*: A schematic sagittal view of the hypothalamus; the paraventricular nucleus and the mammillary bodies are indicated in *dark*. The *lines* depict the first and the last section in which hypocretin cell bodies are found

in the same region as the hypocretin neurons in the lateral hypothalamus. Hypocretin and MCH cells were lost throughout the anterior to posterior extent of their hypothalamic distributions. The percentage loss of hypocretin cells was minimal in PD stage I (23%) and was maximal in PD stage V (62%). Similarly, the percentage loss of MCH cells was lowest in stage I (12%) and was highest in stage V (74%). There was no difference in the size of surviving hypocretin ($p=0.18$, $t=1.39$, df=14) and MCH ($p=0.28$, $t=1.39$, df=14) cells relative to controls [40].

These studies imply that hypocretin neurotransmission is affected in PD. The hypocretin-1 concentration in the prefrontal cortex was almost 40% lower in PD patients, while ventricular cerebrospinal fluid levels were almost 25% reduced. The total number of hypocretin neurons was 50–74% reduced compared to controls.

Relationship Neuronal Loss and Symptoms

There remains an intriguing question: whether or not the post-mortem findings of decreased hypocretin functioning in PD explain the sleep–wake symptoms that we see in PD [35]. Due to the retrospective character of the post-mortem studies, there were no clinical data on sleep–wake disturbances in the deceased patients. However, as described previously, many PD patients have been reported to experience EDS and a narcolepsy-like phenotype, including sleep onset REM periods and fragmented nocturnal sleep [5, 6]. This implicates that a significant proportion of the cases studied would have suffered from similar sleep disturbances. The link between the complete narcoleptic phenotype and a complete loss of hypocretin has been well established. Hypocretin is undetectable in the spinal cerebrospinal fluid of narcoleptic patients with cataplexy [29], and hypocretin knockout rodent models show the complete narcoleptic phenotype [44]. However, the exact relationship between loss of hypocretin containing neurons and the occurrence of clinical symptoms remains unknown. At this moment, it is impossible to quantify hypocretin neurons in vivo [38].

In brains from deceased PD patients who had been suffering from the disorder for many years, the degree of cell loss in the substantia nigra pars compacta turned out to be at least 75% [45]. However, the question is how many dopamine neurons must be lost to produce clinical symptoms [46]. The only disorders in which in vivo evidence for the relationship between neuronal loss and neurological function is available, are disorders of the motor neuron. In amyotrophic lateral sclerosis (ALS), the number of remaining motor units can be estimated using an electrophysiological calculation method (motor unit number estimation, MUNE). It has been shown that subjects can lose up to 25% of their motor units without a reduction of muscle strength, meaning that function can be maintained through compensatory processes [47].

Regarding the relation between decreased hypocretin neurotransmission and functional consequences, the only clues originate from animal studies. Microinjection of prepro-hypocretin short interfering RNAs (siRNA) in the perifornical hypothalamus resulted in a 60% reduction of prepro-orexin mRNA and a persistent increase in the amount of REM sleep [48]. In the aforementioned rodent study by Gerashchenko and colleagues, where 70% of hypocretin neurons were lesioned, an increase in REM sleep was observed as well [31]. Although these results were obtained in rodents, it is not improbable that the reduction in hypocretin neurotransmission found in post-mortem human studies contributes to the sleep problems commonly seen in PD. This would imply that the significant neuronal loss found in Parkinson's disease (reflected in lower ventricular cerebrospinal fluid levels) could at least partially explain sleep–wake disturbances (EDS, REM sleep behavior disorder) commonly seen in this disorder [32, 49]. Although these phenomena are often described in PD, however, there are no reports on cataplexy. Cataplexy is the essential feature of narcolepsy with cataplexy, linked to undetectable levels of hypocretin in the spinal cerebrospinal fluid. In contrast, hypocretin is usually detectable in the cerebrospinal fluid of narcolepsy without cataplexy, where REM sleep disturbances occur without cataplexy, comparable to the findings in PD [28]. It has been proposed that narcolepsy without cataplexy may be caused by a milder form of hypocretin deficiency compared with the almost complete loss of hypocretin in narcolepsy with cataplexy. Indeed, Thannickal et al. describe a high number of surviving hypocretin neurons, i.e. only partial hypocretin neuronal loss in the brain of a narcoleptic patient that did not suffer from cataplexy [39, 50]. The hypocretin findings in PD may support this hypothesis, since a partial reduction in the number of hypocretin neurons could be associated with REM sleep disturbances and sleep–wake abnormalities, but not to cataplexy.

The exact contribution of a loss of hypocretin neurons to sleep disturbances thus still needs to be evaluated. In this respect, it is likely that cell loss is not limited to only the hypocretin cell group in the hypothalamus. Many cell types are affected in PD throughout the brain, but vulnerability seems to be different. Indeed, deficiencies in other neurotransmitter systems or brain areas besides the hypocretin neurons have been proposed as an explanation for the sleepiness in PD. For example, Rye et al. mention the possible involvement of midbrain dopaminergic and noradrenergic neurons that influence sleep–wake states through thalamocortical pathways [51]. Mathis et al. describe a patient with an acute focal inflammatory lesion in the dorsomedial pontine tegmentum in the presence of normal cerebrospinal fluid hypocretin-1 levels [52]. This patient simultaneously developed narcolepsy and REM sleep behaviour disorder. This report underlines that other brain regions, e.g. the mediotegmental pontine area, can contribute to narcoleptic sleep–wake disturbances and REM sleep disorders, even in the absence of a clear hypocretin deficiency [52]. In fact, the combination of both, a loss of dopamine neurons and damage to other neuronal systems (such as the hypocretin system) could very well lead to the sleep disturbances in PD.

Conclusion

Current evidence proves that the disease process in PD also affects the hypothalamic hypocretin system [30, 37, 40]. It is now important to establish the correlation between hypocretin impairment and the occurrence of the various sleep disturbances. The functional relevance of a loss of hypocretin neurons in Parkinson's disease still needs to be studied. This could involve studying post-mortem hypothalami of Parkinson's disease patients, combined with a thorough documentation of their sleep–wake disturbances in the last few years of their lives. This will be a difficult and tedious task to accomplish. Future research in patients with Parkinson's disease may also focus on treatment, e.g. performing a treatment trial with narcolepsy medication, for instance, gamma hydroxybutyrate or hypocretin-agonists, when available for human use.

Concluding, sleep–wake disturbances are a core feature of Parkinson's disease [4–6]. Although the exact cause of these debilitating sleep symptoms is unknown, the use of dopamine agonists, disturbed nocturnal sleep due to motor symptoms, reduced levels of dopamine, loss of hypocretin neurons or damage to other neuronal systems could be contributing factors [4].

References

1. Jellinger KA. The pathology of Parkinson's disease. Adv Neurol. 2001;86:55–72.
2. Langston JW, Forno LS. The hypothalamus in Parkinson disease. Ann Neurol. 1978;3(2):129–33.
3. Chaudhuri KR, Healy DG, Schapira AH. Non-motor symptoms of Parkinson's disease: diagnosis and management. Lancet Neurol. 2006;5(3):235–45.
4. Arnulf I. Excessive daytime sleepiness in parkinsonism. Sleep Med Rev. 2005;9(3):185–200.
5. Arnulf I, Konofal E, Merino-Andreu M, et al. Parkinson's disease and sleepiness: an integral part of PD. Neurology. 2002;58(7):1019–24.
6. Rye DB. Excessive daytime sleepiness and unintended sleep in Parkinson's disease. Curr Neurol Neurosci Rep. 2006;6(2):169–76.
7. Schenck CH, Bundlie SR, Mahowald MW. Delayed emergence of a parkinsonian disorder in 38% of 29 older men initially diagnosed with idiopathic rapid eye movement sleep behaviour disorder. Neurology. 1996;46(2):388–93.
8. Rye DB. Parkinson's disease and RLS: the dopaminergic bridge. Sleep Med. 2004;5(3):317–28.
9. Gagnon JF, Bedard MA, Fantini ML, et al. REM sleep behavior disorder and REM sleep without atonia in Parkinson's disease. Neurology. 2002;59(4):585–9.
10. Hobson DE, Lang AE, Martin WR, Razmy A, Rivest J, Fleming J. Excessive daytime sleepiness and sudden-onset sleep in Parkinson disease: a survey by the Canadian Movement Disorders Group. JAMA. 2002;287(4):455–63.
11. Brodsky MA, Godbold J, Roth T, Olanow CW. Sleepiness in Parkinson's disease: a controlled study. Mov Disord. 2003;18(6):668–72.
12. Arnulf I, Bonnet AM, Damier P, et al. Hallucinations, REM sleep, and Parkinson's disease: a medical hypothesis. Neurology. 2000;55(2):281–8.
13. Overeem S, Mignot E, van Dijk JG, Lammers GJ. Narcolepsy: clinical features, new pathophysiologic insights, and future perspectives. J Clin Neurophysiol. 2001;18(2):78–105.
14. Peyron C, Faraco J, Rogers W, et al. A mutation in a case of early onset narcolepsy and a generalized absence of hypocretin peptides in human narcoleptic brains. Nat Med. 2000;6(9):991–7.
15. Siegel JM. Hypocretin (orexin): role in normal behavior and neuropathology. Annu Rev Psychol. 2004;55:125–48.
16. Peyron C, Tighe DK, van den Pol AN, et al. Neurons containing hypocretin (orexin) project to multiple neuronal systems. J Neurosci. 1998;18(23):9996–10015.
17. Johns MW. A new method for measuring daytime sleepiness: the Epworth sleepiness scale. Sleep. 1991; 14(6):540–5.
18. Ondo WG, Dat VK, Khan H, Atassi F, Kwak C, Jankovic J. Daytime sleepiness and other sleep disorders in Parkinson's disease. Neurology. 2001;57(8):1392–6.
19. Tan EK, Lum SY, Fook-Chong SM, et al. Evaluation of somnolence in Parkinson's disease: comparison with age- and sex-matched controls. Neurology. 2002;58(3):465–8.
20. Fabbrini G, Barbanti P, Aurilia C, Vanacore N, Pauletti C, Meco G. Excessive daytime sleepiness in de novo and treated Parkinson's disease. Mov Disord. 2002;17(5):1026–30.
21. Montastruc JL, Brefel-Courbon C, Senard JM, et al. Sleep attacks and antiparkinsonian drugs: a pilot prospective pharmacoepidemiologic study. Clin Neuropharmacol. 2001;24(3):181–3.
22. Kumar S, Bhatia M, Behari M. Excessive daytime sleepiness in Parkinson's disease as assessed by Epworth Sleepiness Scale (ESS). Sleep Med. 2003;4(4):339–42.
23. Korner Y, Meindorfner C, Moller JC, et al. Predictors of sudden onset of sleep in Parkinson's disease. Mov Disord. 2004;19(11):1298–305.
24. Black J, Houghton WC. Sodium oxybate improves excessive daytime sleepiness in narcolepsy. Sleep. 2006;29(7):939–46.
25. Littner MR, Kushida C, Wise M, et al. Practice parameters for clinical use of the multiple sleep latency test and the maintenance of wakefulness test. Sleep. 2005; 28(1):113–21.
26. American Academy of Sleep Medicine. International classification of sleep disorders – 2nd ed. Rochester, MN. 2005.
27. Rye DB, Bliwise DL, Dihenia B, Gurecki P. FAST TRACK: daytime sleepiness in Parkinson's disease. J Sleep Res. 2000;9(1):63–9.
28. Ripley B, Overeem S, Fujiki N, et al. CSF hypocretin/orexin levels in narcolepsy and other neurological conditions. Neurology. 2001;57(12):2253–8.
29. Nishino S, Ripley B, Overeem S, Lammers GJ, Mignot E. Hypocretin (orexin) deficiency in human narcolepsy. Lancet. 2000;355(9197):39–40.
30. Drouot X, Moutereau S, Nguyen JP, et al. Low levels of ventricular CSF orexin/hypocretin in advanced PD. Neurology. 2003;61(4):540–3.
31. Gerashchenko D, Murillo-Rodriguez E, Lin L, et al. Relationship between CSF hypocretin levels and

hypocretin neuronal loss. Exp Neurol. 2003;184(2): 1010–6.
32. Fronczek R, Overeem S, Lee SY, et al. Hypocretin (orexin) loss and sleep disturbances in Parkinson's disease. Brain. 2008;131(Pt 1):e88.
33. Overeem S, van Hilten JJ, Ripley B, Mignot E, Nishino S, Lammers GJ. Normal hypocretin-1 levels in Parkinson's disease patients with excessive daytime sleepiness. Neurology. 2002;58(3):498–9.
34. Yasui K, Inoue Y, Kanbayashi T, Nomura T, Kusumi M, Nakashima K. CSF orexin levels of Parkinson's disease, dementia with Lewy bodies, progressive supranuclear palsy and corticobasal degeneration. J Neurol Sci. 2006;250(1–2):120–3.
35. Baumann CR, Scammell TE, Bassetti CL. Parkinson's disease, sleepiness and hypocretin/orexin. Brain. 2008;131(Pt 3):e91.
36. Chen CT, Dun SL, Kwok EH, Dun NJ, Chang JK. Orexin A-like immunoreactivity in the rat brain. Neurosci Lett. 1999;260(3):161–4.
37. Fronczek R, Overeem S, Lee SY, et al. Hypocretin (orexin) loss in Parkinson's disease. Brain. 2007; 130(Pt 6):1577–85.
38. Fronczek R, Baumann CR, Lammers GJ, Bassetti CL, Overeem S. Hypocretin/orexin disturbances in neurological disorders. Sleep Med Rev. 2009;13(1):9–22.
39. Thannickal TC, Moore RY, Nienhuis R, et al. Reduced number of hypocretin neurons in human narcolepsy. Neuron. 2000;27(3):469–74.
40. Thannickal TC, Lai YY, Siegel JM. Hypocretin (orexin) cell loss in Parkinson's disease. Brain. 2007;130(Pt 6):1586–95.
41. Fronczek R. Hypocretin deficiency: neuronal loss and functional consequences. Leiden: 8 A.D.
42. Crocker A, Espana RA, Papadopoulou M, et al. Concomitant loss of dynorphin, NARP, and orexin in narcolepsy. Neurology. 2005;65(8):1184–8.
43. Blouin AM, Thannickal TC, Worley PF, Baraban JM, Reti IM, Siegel JM. Narp immunostaining of human hypocretin (orexin) neurons: loss in narcolepsy. Neurology. 2005;65(8):1189–92.
44. Hara J, Beuckmann CT, Nambu T, et al. Genetic ablation of orexin neurons in mice results in narcolepsy, hypophagia, and obesity. Neuron. 2001;30(2): 345–54.
45. Damier P, Hirsch EC, Agid Y, Graybiel AM. The substantia nigra of the human brain. II. Patterns of loss of dopamine-containing neurons in Parkinson's disease. Brain. 1999;122(Pt 8):1437–48.
46. Sulzer D. Multiple hit hypotheses for dopamine neuron loss in Parkinson's disease. Trends Neurosci. 2007;30(5):244–50.
47. Bromberg MB. Updating motor unit number estimation (MUNE). Clin Neurophysiol. 2007;118(1):1–8.
48. Chen L, Thakkar MM, Winston S, Bolortuya Y, Basheer R, McCarley RW. REM sleep changes in rats induced by siRNA-mediated orexin knockdown. Eur J Neurosci. 2006;24(7):2039–48.
49. Thannickal TC, Lai YY, Siegel JM. Hypocretin (orexin) and melanin concentrating hormone loss and the symptoms of Parkinson's disease. Brain. 2008;131(Pt 1):e87.
50. Thannickal TC, Nienhuis R, Siegel JM. Localized loss of hypocretin (orexin) cells in narcolepsy without cataplexy. Sleep. 2009;32(8):993–8.
51. Rye DB. The two faces of Eve: dopamine's modulation of wakefulness and sleep. Neurology. 2004;63(8 Suppl 3):S2–7.
52. Mathis J, Hess CW, Bassetti C. Isolated mediotegmental lesion causing narcolepsy and rapid eye movement sleep behaviour disorder: a case evidencing a common pathway in narcolepsy and rapid eye movement sleep behaviour disorder. J Neurol Neurosurg Psychiatry. 2007;78(4):427–9.

Idiopathic Hypersomnia

Ramin Khatami

Keywords
Idiopathic hypersomnia • Excessive daytime sleepiness • Sleep inertia • Narcolepsy

Introduction

Idiopathic hypersomnia (IHS) is a rare and poorly defined sleep disorder characterized by excessive daytime sleepiness (EDS) despite undisturbed nocturnal sleep. Historically, the concept of IHS was developed in the 1950s to separate patients with unexplained daytime sleepiness from those with classical narcolepsy. Two forms of IHS were distinguished solely on clinical grounds based on (1) typical features of EDS and sleep (continuous EDS, nonirresistible naps, difficulties upon awakening including sleep drunkenness, prolonged, and nonrefreshing naps), (2) a familial pattern, and (3) a poor response to treatment. Later, documentation of objective sleepiness together with the absence of more than two sleep onset REMs (SOREMs) in the multiple sleep latency test(MSLT) became important supportive criteria. Since the first description of IHS, the validity of its concept has been under constant debate and the diagnostic criteria have been changed several times. The distinction of two forms of IHS was abandoned in the sleep classification of 1990, but was reintroduced to the current classification of sleep disorders in 2005. Even recent studies have raised doubt on the specificity of any of the clinical features as well as on the value of MSLT criteria. Due to these diagnostic uncertainties, the pathophysiology of IHS is essentially unknown. Future work, in particular the development of specific biomarkers, is needed to avoid the risk that any disorder with EDS of uncertain origin will be classified as IHS.

R. Khatami (✉)
Center of Sleep Medicine, Klinik Barmelweid AG, CH 5017 Barmelweid, Switzerland
e-mail: Ramin.Khatami@barmelweid.ch

History and Nomenclature

The concept of IHS dates back to the 1950s, and the development of its conceptual framework is tightly linked to the history of narcolepsy. As soon as narcolepsy was clinically characterized by a specific combination of daytime sleepiness and cataplexy (and additional features such as sleep paralysis and hallucinations), it became increasingly clear that patients with chronic sleepiness but without cataplexy may represent a distinct form of hypersomnia. Several terms have been proposed, such as "essential narcolepsy" or "NREM-narcolepsy," to separate this entity from classical narcolepsy. Over

C.R. Baumann et al. (eds.), *Narcolepsy: Pathophysiology, Diagnosis, and Treatment*, DOI 10.1007/978-1-4419-8390-9_33, © Springer Science+Business Media, LLC 2011

Table 1 Nomenclature of idiopathic hypersomnia

Year	Nomenclature	Author/classification	Comment
1957	Independent sleep drunkenness	Roth [1]	First description of IHS
1967	Essential narcolepsy	Berti-Ceroni [2]	
1968	NREM narcolepsy	Passouant [3]	
1972	Hypersomnia with sleep drunkenness	Roth [4]	
1976	Idiopathic hypersomnia	Roth [5]	Two forms of IHS Monosymptomatic form Polysymptomatic form
1979	Idiopathic CNS hypersomnolence	Association of sleep diagnostic centers. Diagnostic classification of sleep and arousal disorders	One form of IHS
1990	Idiopathic hypersomnia	International Classification of Sleep Disorders, 1st edition (ICSD-1)	One form of IHS
2005	Idiopathic hypersomnia	International Classification of Sleep Disorders, 2nd edition (ICSD-2)	Two forms of IHS IHS with long sleep time IHS without long sleep time

the past decades, nomenclature of IHS has changed several times (Table 1).

Finally, Bedrich Roth introduced the term IHS to describe a series of patients with severe, non-imperative daytime sleepiness associated with long nonrefreshing naps, difficulties in awakening, and sleep drunkenness [4–6]. He distinguished two types, a monosymptomatic and a polysymptomatic form. While the monosymptomatic form was characterized by EDS alone, the polysymptomatic form presented with EDS, prolonged nocturnal sleep and difficulties upon awakening with sleep drunkenness, especially in the morning. In 1960, shortly after the discovery of REM sleep (in 1953), Vogel found periods of REM sleep at sleep onset as a typical polysomnographic feature of narcoleptic patients [7]. These periods of SOREM were absent in IHS and thus interpreted as a specific biological marker for narcolepsy. The emerging concept that narcolepsy and IHS were two distinct disorders that can be distinguished by clinical features and other markers, led to the introduction of "Idiopathic CNS hypersomnia" as a new category in the Diagnostic Classification of Sleep and Arousal Disorders (1979) [8]. The distinction of narcolepsy with cataplexy and IHS as certain entities also had some important implications for understanding the pathophysiology of daytime sleepiness. Sleepiness in narcolepsy, in particular the short irresistible "sleep attacks" were thought to be associated with a high REM sleep propensity. By contrast non-imperative sleepiness with long daytime naps in IHS was attributed to increased NREM sleep pressure. Accordingly, in the first edition of the International Classification of Sleep Disorders (1990) narcolepsy was defined as EDS that is typically associated with "... cataplexy and other REM sleep phenomena ...," whereas IHS was defined as a disorder with a normal or prolonged major sleep episode and "... excessive sleep episodes of non-REM sleep ...." Notably, at that time the two forms of IHS were merged into one "idiopathic hypersomnia" [9]. In 1996, Bassetti and Aldrich pointed to a substantial overlap of clinical features of narcolepsy and IHS. They proposed three clinical subtypes of IHS: a "classic" IHS roughly corresponding to Bedrich Roth's polysymptomatic form; a "narcoleptic"-like IHS with EDS resembling classical narcolepsy but without any signs of increased REM sleep propensity; and finally an intermediate "mixed" type of IHS [10]. At the same time Billiard et al. [11, 12] used continuous

Table 2 Diagnostic criteria of idiopathic hypersomnia with and without long sleep time

Complaints	
A.	The patient has a complaint of excessive sleepiness occurring almost daily for at least 3 months
B.-1.	The patient has prolonged nocturnal sleep time (more than 10 h) documented by interview, actigraphy, or sleep logs. Waking up in the morning or at the end of naps is almost always laborious
For IHS without long sleep time	
B.-2.	The patient has normal nocturnal sleep (greater than 6 h but less than 10 h), documented by interviews, actigraphy, or sleep logs
Polysomnography (PSG)	
C.	Nocturnal PSG has excluded other causes of daytime sleepiness
D.-1	The polysomnogram demonstrates a short sleep latency and a major sleep period that is prolonged to more than 10 h in duration
For IHS without long sleep time	
D.-2.	PSG demonstrates a major sleep period that is normal in duration (greater than 6 h but less than 10 h)
Multiple sleep latency test (MSLT)	
E.	If an MSLT is performed following overnight PSG, a mean sleep latency of less than 8 min is found and fewer than two SOREMPs are recorded. Mean sleep latency in idiopathic hypersomnia with long sleep time has been shown to be 6.2 ± 3.0 min
Diagnosis on exclusion	
F.	The hypersomnia is not better explained by another sleep disorder, medical or neurological disorder, mental disorder, medication use, or substance use disorder

Adapted from the ICSD-2 classification

24-h polysomnographic recordings and provided supportive data for the existence of Roth's polysymptomatic form. In the second edition of the International Classification of Sleep Disorders (2005) [13], the subdivision into two forms re-appeared, named as IHS with and without long sleep time (Table 2). Several studies published after the introduction of the revised ICSD-2 classification suggest that apart from the long sleep period no other features including sleep latencies on MSLT or HLA status help differentiate IHS patients with long or normal sleep periods [14–16].

Prevalence and Epidemiology of IHS

The exact prevalence of IHS remains unknown due to uncertainties of diagnostic criteria and in the absence of robust epidemiologic studies. Reported prevalence rates of 0.03–0.005% are based on surveys of patients seen in sleep referral centers. These surveys allowed an estimation of narcolepsy:IHS ratio between 10:1 and 3:2 [10, 11, 17–19]. An interesting observation is that the rate of narcolepsy:IHS appears to have decreased over time [20] probably due to better identification and diagnosis of other sleep disorders that mimic IHS (e.g., upper airway resistance syndrome or mood disorders). Only one recent retrospective clinical study reported a higher frequency of IHS, and found IHS approximately 60% as frequent as NC, even as high as 40% when stricter diagnostic criteria were applied [14]. Considering that narcolepsy is reported to affect 0.05–0.1% of European and American populations, IHS is a rare disorder.

As with narcolepsy, IHS usually begins during adolescence [10, 14, 15, 19]. Women are more often affected than men, and most cases run in families [14, 20]. In Roth's original population, 30–40% of patients had a familial background [21]. Difficulties awakening in the morning have a strong impact on quality of life, and often cause problems with school, work, and social activities [19, 22]. IHS is a lifelong disorder in most cases with stable EDS over years. However, spontaneous improvements and – in a few documented cases – disappearance of EDS can occur and clearly contrast with the natural history of persisting sleepiness in narcolepsy [10, 23].

Clinical Features

The essential feature of IHS is chronic EDS despite undisturbed nocturnal sleep. According to the recent ICSD-2 classification, nocturnal sleep may be prolonged (more than 10 h) or of normal duration (6–10 h). Additional features of IHS include sleep drunkenness, long but unrefreshing daytime naps, and difficulties upon awakening. Available data suggest that none of these additional features are more prevalent in one of the two IHS forms, and clinically, it is difficult to distinguish between the two forms [14]. For that reason clinical characteristics are summarized for both IHS forms.

Excessive Daytime Sleepiness

EDS refers to a subjective feeling of constant sleepiness, with an urge to sleep. The propensity to fall asleep during the daytime is lower in IHS than in narcolepsy patients [23, 24]. IHS patients are especially less prone to fall asleep during active situations [25]. Daytime naps are long and unrefreshing in 46–78% [4, 14, 15, 23]. In a recent study, daytime napping of greater than 1 h was the best clinical symptom to distinguish IHS from narcolepy with cataplexy with a sensitivity and specificity of 87% [14]. However, a subgroup of "narcoleptic-like" IHS patients [10] experiences short and refreshing naps [4, 23], indicating that a specific quality of EDS in IHS does not exist. Automatic behavior refers to a twilight state between wakefulness and sleep and reflects another dimension of sleepiness. It occurs in 40–50% of narcoleptic patients but is less prominent in IHS [10]. Automatic behavior may be provoked when patients fight against their sleepiness.

Nocturnal Sleep and Problems upon Awakening

Prolonged nocturnal sleep is by definition a cardinal feature in the long sleep time type of IHS. A recent systematic study [26] and observations in narcolepsy patients suggest that long sleep duration can also occur in narcolepsy with cataplexy [10, 16, 23] and raise doubts on the specificity of prolonged sleep. The quality of nocturnal sleep in both IHS forms is even more controversial. Deep and undisturbed sleep with few awakenings was originally described by Roth [4] and confirmed by others [14, 20]. This markedly contrasts with the complaints of narcolepsy patients of disrupted night sleep and early morning awakening. However, in one study up to 45% of IHS patients experienced a "narcolepsy-like" sleep with frequent awakening, sleep paralysis, and hypnagogic hallucinations [10]. Waking up from sleep at any time, either in the morning or when awakening from naps, is laborious for IHS patients and can present as sleep drunkenness, a prolonged state of disorientation, incoordination, irritation, and aggression upon awakening. It is described in 21–60% of IHS patients [4, 10, 14, 15, 27] and does not correlate with Epworth sleepiness scale (ESS) or sleep duration.

Associated Features

Autonomic symptoms (such as tension and migraine-like headache in up to 30%, orthostatic dysregulation, and Raynaund-like symptoms) and cognitive dysfunction (subjective complaints of memory and unspecific cognitive deficits due to sleepiness) are frequently described in IHS. None of these symptoms are diagnostic, specific, or discriminative for IHS. By contrast, weight gain, especially at the time of disease onset appears to be characteristic for narcolepsy [28], whereas IHS-patients are of normal weight [15, 19]. Obesity persists in NC while BMI remains normal in IHS or in narcoleptic patients without cataplexy in the course of disease [19]. Depression and anxiety are associated with IHS [10, 19, 29], and more with NC, with up to 30% [30] scoring high on standardized depressions scales [19].

Diagnosis of Idiopathic Hypersomnia

Diagnostic procedure has changed several times as definitions of IHS have changed, but it still remains a diagnosis of exclusion due to the lack of any specific clinical symptoms and diagnostic markers. Thus, both forms of IHS are diagnosed on a clinical basis and require polysomnography (PSG) and MSLT to distinguish IHS from other sleep disorders and to assess comorbidities.

An essential step is to take a detailed history of EDS and to document the duration of nocturnal sleep time. It is recommended to ask about sleep drunkenness, unrefreshing naps, and difficulties upon awakening and to exclude specific symptoms such as cataplexy or interrupted breathing while sleeping. There is no questionnaire validated for IHS, but the ESS is recommended for standardized assessment of daytime sleepiness. IHS patients usually score high on ESS, though slightly lower than narcoleptic patients with and without cataplexy [19, 31]. Actigraphy for at least 1–2 weeks in combination with sleep logs is extremely helpful for several reasons. First, documentation of long sleep time as required by ICSD-2 can be easily estimated from rest time during recordings. Second, recordings help exclude circadian disorders which may present as daytime hypersomnia. Third, behaviorally induced insufficient sleep syndrome (BIISS) can be confused with IHS and is usually distinguished by short rest times during the weekdays and long rest time during the weekend.

Polysomnographic recordings at night are usually unremarkable but may reveal unspecific signs of high sleep pressure, such as short sleep latency, few awakenings, high sleep efficiency, and high amounts of slow-wave sleep. Sleep architecture is generally normal, while SOREMs are rare they do not exclude IHS [10]. Sleep disordered breathing, including upper airway resistance syndrome and periodic limb movements in sleep should theoretically not be present in PSG. This ideal situation is rarely found in reality, and mild forms of these disorders should not interfere with the diagnosis of IHS. MSLT following overnight PSG is mandatory to exclude SOREMs during daytime sleep. Mean sleep latencies are short in IHS, mostly between 5 and 10 min, but longer than in narcolepsy patients. Standard protocols of PSG followed by MSLT are not ideal in the diagnostic work-up of IHS and face some theoretical and practical issues. The documentation of prolonged sleep time in PSG is impossible when patients have to be woken up for the MSLT in the morning. Thus diagnostic standard procedures interfere with the aim to document prolonged sleep and prevent distinguishing the two forms of IHS. Standard MSLT protocols with scheduled naps of 20 min every 2 h will fail to record the typical features of prolonged and unrefreshing naps. In addition, the value of proposed mean sleep latency of less than 8 min for diagnostic purposes is highly questionable for several reasons (see below). Alternatives for diagnostic work-up have been proposed [20]. A 24-h recording with sleep ad libitum may be more appropriate to document the amount of sleep (Fig. 1). Cognitive evoked potentials (P300) performed after awakening are an interesting tool to better understand the process of waking up in hypersomnias, but at the moment they are not helpful for diagnostic purpose.

Psychiatric evaluation is mandatory to identify hypersomnia associated with depression and anxiety. Comorbidity with psychiatric disorders is high and depressive symptoms may contribute to sleepiness, changes in sleep architecture, and the response to treatment. Brain imaging, HLA status, and CSF hypocretin are not helpful for diagnosing IHS.

Pathogenesis and Pathophysiology of IHS

The pathogenesis of IHS remains largely unknown. As with narcolepsy, genetic and environmental factors are discussed in IHS, but most information is from single cases with mild head trauma, diabetes, obesity, viral illness, and general anesthesia [10]. A familial background

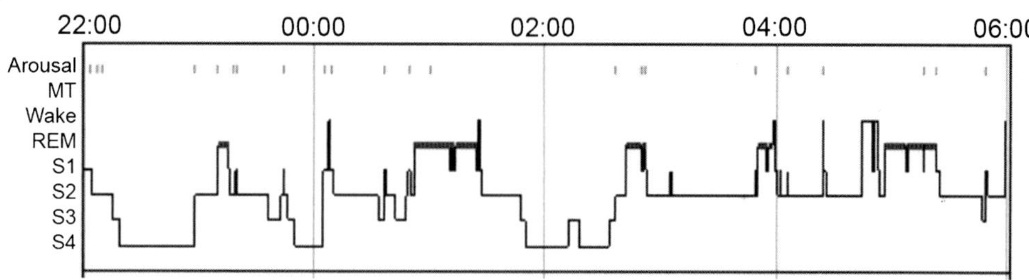

24h PSG with sleep ad libitum

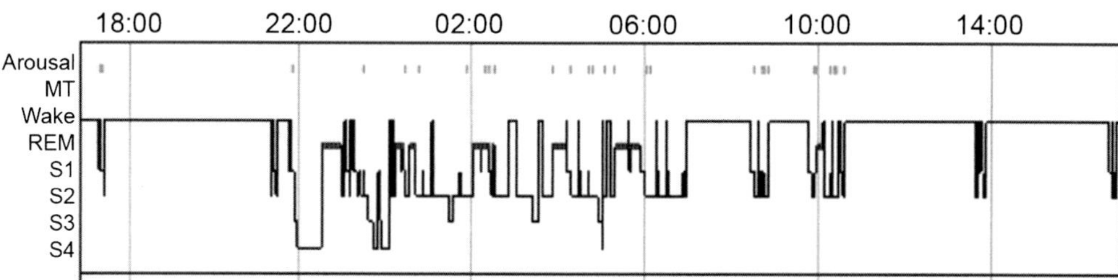

Fig. 1 A 21-year-old IHS patient with long sleep time. Overnight polysomnography (*upper panel*) and 24-h continuous recordings (*lower panel*). Sleep architecture is essentially normal (sleep latency 11 min to NREM2, amount of slow-wave sleep 24% of total sleep time, and sleep efficiency 94%). There were no signs of sleep disordered breathing or periodic limb movements in sleep. A 24-h recording with ad libitum sleep in the same patient shows a consolidated and prolonged nocturnal sleep period of 603 min plus 114 min daytime sleep

occurs in up to 50% of IHS. Rare cases of IHS and NC in the same family also support a genetic basis of IHS [10, 20]. Genetic analysis in IHS has mainly been limited to small studies that used low-resolution HLA typing. This may account for the negative, or only marginal (for Cw2 [32]) associations between HLA and IHS [10, 14, 15]. DRB1*15 (DR2) and DQB1*0602 occur at the same frequencies in IHS and control subjects [10, 14, 15, 19]. A recent genome-wide analysis in 135 Japanese patients and 569 controls found a single nucleotide polymorphism (SNP rs5770917) associated with "essential hypersomnia" [33].

The idea that sleepiness and cataplexy in narcolepsy are mediated by a high REM sleep propensity while sleepiness and difficulties upon awakening in IHS are related to a facilitation of non-REM sleep has some interesting implications for the pathophysiology of IHS. Facilitation of non-REM sleep may in theory be caused by dysfunctional monoaminergic systems. Consistent with this idea, a state of hypersomnia together with increased levels of monoamines could be induced by destroying parts of the noradrenergic locus coeruleus in cats [34]. Similarly, early human studies found disproportional levels of monoamines in the cerebrospinal fluid (CSF) in narcolepsy and IHS patients [35, 36]. These findings (which could not be confirmed by others [37]) were interpreted as an indicator for an increased monoaminergic turnover and a desynchronization of the noradrenergic system from the dopamine and the serotonin system.

Unlike in narcolepsy, hypocretin levels are normal in IHS [27, 38–40]. However, histamine signaling may be reduced in both disorders [41]. It is of particular interest that CSF histamine was normal in other non-CNS hypersomnias, such as obstructive sleep apnea.

Dysfunctional homeostatic or circadian regulation may contribute to the sleepiness of IHS [42, 43]. A quantitative marker of sleep homeostasis is the level and time course of slow-wave activity (SWA, power within 0.75–4.5 Hz in NREM sleep), which dissipates exponentially across successive NREM sleep episodes. NREM sleep homeostasis was essentially intact in IHS patients, as shown by an exponential decline of SWA across a baseline sleep episode [42]. However, the level of SWA in the first two NREM–REM sleep cycles was significantly lower than in controls. IHS patients may thus need a prolonged sleep time because of a lower intensity of their NREM sleep. However, as the level of SWA is a function of prior wake time, the decrease in SWA could simply be explained by the increased duration of sleep at night.

Challenging sleep homeostasis by sleep deprivation is more instructive to test the integrity of homeostatic sleep regulation. An ongoing study examines the effects of a sustained wake on NREM sleep regulation in IHS and healthy controls. Six drug-naïve IHS patients with long sleep time and six age- and sex-matched controls have completed this study consisting of an adaptation night, baseline night, 40-h of prolonged wakefulness, and a recovery night. Preliminary results suggest a similar time course of SWA in baseline sleep and an increase of SWA in both groups in recovery sleep after sleep deprivation (Hefti et al., unpublished observation). The increase of SWA appeared even higher in IHS patients than in controls, and SWA dissipated exponentially across subsequent NREM–REM sleep cycles. These observations suggest that NREM sleep regulation is virtually intact in IHS.

Only one study tested the circadian aspects of IHS and reported a nonsignificant phase delay of melatonin and cortisol rhythms compared to controls [43]. In a recent study, more evening types were found in IHS patients and Horne–Ostberg scores were lower in IHS than in controls, consistent with a delayed sleep phase in IHS [15].

Treatment

Treatment of IHS remains symptomatic and difficult in most cases. Nonpharmacological approaches and sleep hygiene are not as useful as in narcolepsy because of the unrefreshing nature of naps. Both extension of sleep time for several days to saturate the patients' sleep need [20] and sleep restriction [10] have been tried. Sleepiness improved with both regimes, but it remains unclear whether this effect was sustained.

Pharmacological recommendations are based on experiences with narcolepsy and are aimed at improving sleepiness. Specific treatment of other, more compromising features, such as difficulties in waking or sleep drunkenness are currently not available. A poor response to treatment was initially described by Roth, which is not confirmed by others. Medication is effective in up to 80% of IHS patients but seems to be less effective than in narcolepsy [10, 14, 19, 31]. Modafinil is the most commonly used drug in up to 80% of IHS cases [19]. Dosages between 200 and 400 mg/day (only 12% of cases above 400 mg/day) were effective in up to 62% in two uncontrolled observational studies [14, 19] with a follow-up time of 3.8 years in one of the study [14]. Similar to narcolepsy patient treatment, IHS may improve on the Epworth scale while mean sleep latencies on MSLT remain essentially unchanged [10]. Antidepressants may be more effective than stimulants in patients with prevailing depressive symptoms, but comparative studies are absent. Many other substances have been tried in single cases, including SSRIs, tricyclic antidepressants, dopaminergic substances, etc. with anecdotal improvements in some of these cases. In one study, slow release of melatonin effectively reduced sleepiness in 50% of the IHS patients [44].

Conclusions: Present Challenges and Future Perspectives

Diagnostic Challenges

Classical IHS with sleep drunkenness and unrefreshing naps is found in only 36% of IHS with long sleep time and in 54% of short sleep time type, but almost never in controls [15], indicating a poor sensitivity but a high specificity of the clinical features. Several single clinical features have been reported with a high specificity (up to 97%) and sensitivity (up to 87%), among them nap duration of more than 60 min, sleep drunkenness, and vivid dreams [14]. However, these factors have been evaluated in a study comparing IHS with classical narcolepsy with cataplexy, which is in daily practice not a diagnostic problem due to the high specificity of cataplexy. By contrast, the differentiation of IHS with normal sleep time and narcolepsy without cataplexy is often challenging. According to the current classification, the diagnosis hinges on the demonstration of two or more SOREMs during the MSLT, but the number of SOREMs can vary and is influenced by chronic sleep deprivation [45]. One future challenge will be to develop reliable markers for IHS that discriminate it from narcolepsy without cataplexy, hypersomnias with mood disorders, and BIISS.

The ICSD-2 recognizes IHS with long sleep time as a distinct entity and requires documentation of sleep duration. The problem of documentation of sleep duration by PSG is long-term monitoring, in particular 24-h continuous recording lacks standardization and normative values. In addition, the cut-off value of less than 10 h is based on experts' opinion rather than on systematic studies [46]. Very recently, data of 30 healthy controls have been published who underwent 24-h continuous recording with ad libitum sleep [15]. The 95% confidence interval for total sleep time was 493–558 min in controls versus 672–718 min in IHS patients with prolonged sleep time. Chronic sleep deprivation has been excluded by questionnaire in these subjects. Similarly, proposed documentation of a major sleep period of more than 10 h by actigraphy awaits standardization and normative values. The use of absolute sleep time values as diagnostic criteria somehow neglects that the individual increase of sleep need relative to the premorbid period may be more important and compromising for the patient.

The diagnostic value of MSLT for IHS is even more debatable. MSLT is useful to obtain quantitative parameters for the propensity to fall asleep and to measure REM sleep propensity as indexed by the number of SOREM sleep episodes. MSLT is with no doubt an appropriate diagnostic tool for narcolepsy, but the rationale why such a "narcolepsy-like approach" was adapted for the diagnosis of IHS is mostly historical. The sensitivity of mean sleep latencies during MSLT is low, considering that as many as 71% patients with long sleep time have normal values [15, 47]. In addition, IHS patients with normal mean latencies during MSLT (≥8 min) do not differ in any clinical or polysomnographic parameters compared to those with shorter mean sleep latencies (≤8 min) [14]. Efforts should be made to develop tests that reliably measure difficulties of waking up.

IHS: Distinct Subtypes or a Spectrum of the Disease?

An important goal in sleep medicine is to define sleep disorders as distinct clinical entities with specific diagnostic markers and explained biological basis. A first step in this direction is the description of homogenous phenotypes. Equally important is a better understanding of the underlying neurobiology. A possible approach is to look for spectrum of disorders that share the same symptom (e.g., EDS) and identify common contributors. IHS and narcolepsy may provide a unique possibility to pursue both goals.

Several subtypes of IHS have been proposed since the first description of IHS, but it remains unclear how useful these subdivisions are for clinical and scientific purposes. With regard to defining homogeneous phenotypes, Roth's polysymptomatic form corresponds to the "classical" type [10] and is now referred to as IHS with long sleep time in the current ICSD-2 classification. By contrast Roth's monosymptomatic form of IHS as defined by EDS alone is less robust. As with the monosymptomatic type doubts have been raised that IHS with normal sleep time is a distinct disorder. Due to its unspecificity it may be overdiagnosed and concerns have been raised that diagnosis could be used for any nonspecific disorder presenting with EDS [10, 47]. Indeed, clinical studies suggest that – except for sleep duration – no other clinical or paraclinical parameter is characteristic for IHS without long sleep time [14, 15]. The subdivision in a "classic," "narcoleptic-like," and "mixed" type based on clinical grounds raised some interesting issues

[10]. These different subtypes may reflect distinct etiologies that need specific treatment. The "classical" form shared many clinical features with psychiatric hypersomnias, in particular, with atypical depression and showed a good response to antidepressive agents [10]. Therefore, a diencephalic dysfunction that may involve monoamines has been hypothesized. Likewise, a "narcoleptic-like type" which improved after stimulants implicates a neurobiology that is closer to classical narcolepsy.

Alternatively, a substantial overlap between narcolepsy, IHS, and other CNS hypersomnias could be assumed. Sleep drunkenness is also found in hypersomnias associated with psychiatric or neurologic disorders. Similarly, long sleep time is not specific and is found in a subgroup of narcoleptic patients comparable to IHS [10, 26]. According to this idea, narcolepsy and IHS may represent heterogenous groups of multifactorial disorders, in which genetic and nongenetic factors determine the variable expression of symptoms. The HLA system may represent one factor that contributes to the presence and severity of symptoms. It has been shown that the HLA DQB1*0602 allele is associated with a shorter REM latency at night in normal controls [48]. Other genes adjacent to the HLA system may have similar effects. A single nucleotide polymorphism (SNP rs5770917) located between CPT1 and CHKH and the DRB1*0501–DQB1*0602 haplotype, which has previously been associated with narcolepsy [49] appears also to confer susceptibility for "essential hypersomnia" [33]. Since "essential hypersomnia" represent a group of heterogenous hypersomnias, this SNP may point to a common genetic background for CNS hypersomnias. Nongenetic factors, such as a partial loss of hypocretin neurons [50] or a decrease of histamine in the CSF [41] have been proposed to play a role in EDS. Environmental factors may be similarly important, but remain poorly defined. A minority of IHS patients experience a spontaneous resolution of the disease (10–25% demonstrated by several groups [10, 14, 23]). These patients are especially interesting because they clearly separate IHS from narcolepsy and may help to identify contributing environmental factors.

References

1. Roth B. EEG studies of a large series of cases of narcolepsy and hypersomnia. Cesk Neurol. 1957;20: 155–61.
2. Berti-Ceroni G, Coccagna G, Gambi D, Lugaresi E. Considerazioni clinico poligrafiche sull narcolessia essenziole "a somno lento". Sist Nerv. 1967;19:81–9.
3. Passouant P, Popoviciu L, Velok G, Baldy-Moulinier M. Etude polygraphique des narcolepsies au cours du nycthe´me`re. Rev Neurol. 1968;118:431–41.
4. Roth B, Nevsimalova S, Rechtschaffen A. Hypersomnia with "sleep drunkenness". Arch Gen Psychiatry. 1972;26:456–62.
5. Roth B. Narcolepsy and hypersomnia: review and classification of 642 personally observed cases. Schweiz Arch Neurol Neurochir Psychiatr. 1976;119:31–41.
6. Roth B, Bruhova S, Lehovsky M. REM sleep and NREM sleep in narcolepsy and hypersomnia. Electroencephalogr Clin Neurophysiol. 1969;26: 176–82.
7. Vogel G. Studies in psychophysiology of dreams. III. The dream of narcolepsy. Arch Gen Psychiatry. 1960;3:421–8.
8. Association of Sleep Disorders Centers. Diagnostic classification of sleep and arousal disorders. 1st ed. Prepared by the Sleep Disorders Classification Committee. Roffwarg HP, Chairman. Sleep. 1979;2:1–37.
9. Thorpy MJ, Diagnostic Classification Steering Committee. International classification of sleep disorders: diagnostic and coding manual. Rochester, MA: American Sleep Disorders Association; 1990.
10. Bassetti C, Aldrich MS. Idiopathic hypersomnia. A series of 42 patients. Brain. 1997;120:1423–35.
11. Billiard M. Idiopathic hypersomnia. Neurol Clin. 1996;14:573–82.
12. Billiard M, Merle C, Carlander B, Ondze B, Alvarez D, Besset A. Idiopathic hypersomnia. Psychiatr Clin Neurosci. 1998;52:125–9.
13. American Academy of Sleep Medicine. International classification of sleep disorders. 2nd ed. Diagnostic and coding manual. Westchester, IL. 2005.
14. Anderson KN, Pilsworth S, Sharples LD, Smith IE, Shneerson JM. Idiopathic hypersomnia: a study of 77 cases. Sleep. 2007;30:1274–81.
15. Vernet C, Arnulf I. Idiopathic hypersomnia with and without long sleep time: a controlled series of 75 patients. Sleep. 2009;32:753–9.
16. Heier MS, Evsiukova T, Vilming S, Gjerstad MD, Schrader H, Gautvik K. CSF hypocretin-1 levels and clinical profiles in narcolepsy and idiopathic CNS hypersomnia in Norway. Sleep. 2007;30:969–73.
17. Coleman RM, Roffwarg H, Kennedy SJ, Guilleminault C, Cinque J, Cohn A. Sleep wake disorders based on a polysomnographic diagnosis. A national cooperative study. JAMA. 1982;247:997–1003.
18. Matsunaga H. Clinical study on idiopathic CNS hypersomnolence. Jpn J Psychiatry Neurol. 1987;41: 637–44.

19. Dauvilliers Y, Paquereau J, Bastuji H, Drouot X, Weil JS, Viot-Blanc V. Psychological health in central hypersomnias: the French Harmony study. J Neurol Neurosurg Psychiatry. 2009;80:636–41.
20. Billiard M, Dauvilliers Y. Idiopathic hypersomnia. Sleep Med Rev. 2001;5:349–58.
21. Nevsimalova-Bruhova S, Roth B. Heredofamilial aspects of narcolepsy and hypersomnia. Schweiz Arch Neurol Neurochir Psychiatry. 1972;110:45–54.
22. Bayon V, Leger D, Philip P. Socio-professional handicap and accidental risk in patients with hypersomnias of central origin. Sleep Med Rev. 2009;13:421–6.
23. Bruck D, Parkes JD. A comparison of idiopathic hypersomnia and narcolepsy-cataplexy using self report measures and sleep diary data. J Neurol Neurosurg Psychiatry. 1996;60:576–8.
24. Komada Y, Inoue Y, Mukai J, Shirakawa S, Takahashi K, Honda Y. Difference in the characteristics of subjective and objective sleepiness between narcolepsy and essential hypersomnia. Psychiatry Clin Neurosci. 2005;59:194–9.
25. Sturzenegger C, Bassetti CL. The clinical spectrum of narcolepsy with cataplexy: a reappraisal. J Sleep Res. 2004;13:395–406.
26. Vernet C, Arnulf I. Narcolepsy with long sleep time: a specific entity? Sleep. 2009;32:1229–35.
27. Bassetti C, Gugger M, Bischof M, Mathis J, Sturzenegger C, Werth E, et al. The narcoleptic borderland: a multimodal diagnostic approach including cerebrospinal fluid levels of hypocretin-1 (orexin A). Sleep Med. 2003;4:7–12.
28. Schuld A, Hebebrand J, Geller F, Pollmacher T. Increased body-mass index in patients with narcolepsy. Lancet. 2000;355:1274–5.
29. Roth B, Nevsimalova S. Depression in narcolepsy and hypersommia. Schweiz Arch Neurol Neurochir Psychiatr. 1975;116:291–300.
30. Broughton R, Ghanem Q, Hishikawa Y, Sugita Y, Nevsimalova S, Roth B. Life effects of narcolepsy in 180 patients from North America, Asia and Europe compared to matched controls. Can J Neurol Sci. 1981;8:299–304.
31. Sasai T, Inoue Y, Komada Y, Sugiura T, Matsushima E. Comparison of clinical characteristics among narcolepsy with and without cataplexy and idiopathic hypersomnia without long sleep time, focusing on HLA-DRB1(*)1501/DQB1(*) 0602 finding. Sleep Med. 2009;10:961–6.
32. Poirier G, Montplaisir J, Decary F, Momege D, Lebrun A. HLA antigens in narcolepsy and idiopathic central nervous system hypersomnolence. Sleep. 1986;9: 153–8.
33. Miyagawa T, Honda M, Kawashima M, Shimada M, Tanaka S, Honda Y, et al. Polymorphism located between CPT1B and CHKB, and HLA-DRB1*1501-DQB1*0602 haplotype confer susceptibility to CNS hypersomnia (essential hypersomnia). PLOS One. 2009;4:1–3.
34. Petitjean F, Jouvet M. Hyersomnie et augmentation de`l acide 5-hydroxy-indolacetique cèrèbral par lèsion isthmique chez le chat. C R Seances Soc Biol Fil. 1970;164:2288–93.
35. Faull KF, Guilleminault C, Berger PA, Barchas JD. Cerebrospinal fluid monoamine metabolites in narcolepsy and hypersomnia. Ann Neurol. 1983;13:258–63.
36. Faull KF, Thiemann S, King RJ, Guilleminault C. Monoamine interactions in narcolepsy and hypersomnia: a preliminary report. Sleep. 1986;9:246–9.
37. Montplaisir J, de Champlain J, Young SN, Missala K, Sourkes TL, Walsh J, et al. Narcolepsy and idiopathic hypersomnia: biogenic amines and related compounds in CSF. Neurology. 1982;32:1299–302.
38. Mignot E, Lammers GJ, Ripley B, Okun M, Nevsimalova S, Overeem S, et al. The role of cerebrospinal fluid hypocretin measurement in the diagnosis of narcolepsy and other hypersomnias. Arch Neurol. 2002;59:1553–62.
39. Baumann CR, Bassetti CL. Hypocretins (orexins) and sleep-wake disorders. Lancet Neurol. 2005;4:673–82.
40. Kanbayashi T, Inoue Y, Chiba S, Aizawa R, Saito Y, Tsukamoto H, et al. CSF hypocretin-1 (orexin-A) concentrations in narcolepsy with and without cataplexy and idiopathic hypersomnia. J Sleep Res. 2002;11:91–3.
41. Kanbayashi T, Kodama T, Kondo H, Satoh S, Inoue Y, Chiba S, et al. CSF histamine contents in narcolepsy, idiopathic hypersomnia and obstructive sleep apnea syndrome. Sleep. 2009;32:181–7.
42. Sforza E, Gaudreau H, Petit D, Montplaisir J. Homeostatic sleep regulation in patients with idiopathic hypersomnia. Clin Neurophysiol. 2000;111: 277–82.
43. Nevsimalova S, Blazejova K, Illnerova H, Hajek I, Vankova J, Pretl M, et al. A contribution to pathophysiology of idiopathic hypersomnia. Suppl Clin Neurophysiol. 2000;53:366–70.
44. Montplaisir J, Fantini L. Idiopathic hypersomnia. Sleep Med Rev. 2001;5:361–2.
45. Marti I, Valko PO, Khatami R, Bassetti CL, Baumann CR. Multiple sleep latency measures in narcolepsy and behaviourally induced insufficient sleep syndrome. Sleep Med. 2009;10:1146–50.
46. Billiard M. Diagnosis of narcolepsy and idiopathic hypersomnia. An update based on the International Classification of Sleep Disorders, 2nd edition. Sleep Med Rev. 2007;11:377–88.
47. van den Hoed J, Kraemer H, Guilleminault C, Zarcone Jr VP, Miles LE, Dement WC, et al. Disorders of excessive daytime somnolence: polygraphic and clinical data for 100 patients. Sleep. 1981;4:23–37.
48. Mignot E, Young T, Lin L, Finn L. Nocturnal sleep and daytime sleepiness in normal subjects with HLA-DQB1*0602. Sleep. 1999;22:347–52.
49. Miyagawa T, Kawashima M, Nishida N, Ohashi J, Kimura R. Variant between CPT1B and CHKB associated with susceptibility to narcolepsy. Nat Genet. 2008;40:1324–8.
50. Thannickal TC, Nienhuis R, Siegel JM. Localized loss of hypocretin (orexin) cells in narcolepsy without cataplexy. Sleep. 2009;32:993–8.

Part VIII

The Diagnosis of Narcolepsy and the Assessment of Fitness to Drive

Current Diagnostic Criteria for Adult Narcolepsy

Alex Iranzo

Keywords
Narcolepsy • Diagnosis • Multiple sleep latency test • Polysomnography

Introduction

Narcolepsy is clinically characterized by excessive daytime sleepiness (EDS) and other symptoms including cataplexy, sleep paralysis, hypnagogic and hypnopompic hallucinations, automatic behaviors, disrupted nocturnal sleep, REM sleep behavior disorder, and moderate weight gain [1–5]. Vehicle accidents and social, familial, professional, and psychological problems are common and can dramatically impact quality of life. More than 95% of the cases are sporadic but the disease has been described in some families and in monozygotic twins. The risk of a first-degree relative developing narcolepsy with cataplexy (NC) is 10–40 times higher (1–2%) than that in the general population. The cause of NC is still unknown, although an autoimmune process occurring in the brain is suspected since the condition is linked to a loss of hypocretin- (orexin-) producing cells in the hypothalamus, low levels of cerebrospinal fluid (CSF) hypocretin [6–13], and the presence of the human leukocyte antigen (HLA) DQB1*0602 subtype [14].

A. Iranzo (✉)
Neurology Service and Multidisciplinary Sleep Unit, Hospital Clínic and Institut D'Investigació Biomèdiques August Pi i Sunyer (IDIBAPS), Barcelona, Spain
e-mail: airanzo@clinic.ub.es

Narcolepsy is not a common disorder, with a prevalence rate of 0.02–0.18% in North America and Europe [1]. Unfortunately, it is estimated that more than 80% of the narcoleptics remain undiagnosed and in those who are correctly diagnosed, the diagnosis is often established several years after symptom onset. In narcolepsy, underdiagnosis and misdiagnosis are frequent mainly due to (1) physicians' lack of awareness of the disease and (2) patients', relatives', and associates' inability to recognize the narcoleptic symptoms as pathological and thereby not seeking medical attention. Accurate diagnosis of the disease is crucial to (1) confirm the presence of a lifelong disorder linked to disability and socioeconomic difficulties, (2) start the available effective therapy (wake-promoting agents, sodium oxybate, and anticataplectics) to improve quality of life, (3) confirm the presence of a disease associated with a specific biological abnormality (deficient hypocretin transmission within the brain) that might be ideally treated with targeted treatments (hypocretin replacement, hypocretin agonists, cell transplantation, and gene therapy), (4) exclude other causes of EDS and falls that may mimic the symptoms of narcolepsy, and (5) identify malingering in people who feign symptoms to obtain psychoactive medications (e.g., stimulants and sodium oxybate) or social benefits (e.g., disability).

C.R. Baumann et al. (eds.), *Narcolepsy: Pathophysiology, Diagnosis, and Treatment*, DOI 10.1007/978-1-4419-8390-9_34, © Springer Science+Business Media, LLC 2011

The Current Diagnostic Criteria: ICSD-2

In 2005, the second edition of the International Classification of Sleep Disorders (ICSD-2) established the current diagnostic criteria for narcolepsy, and classified the disorder into three different entities: narcolepsy with cataplexy, narcolepsy without cataplexy (NwC), and narcolepsy caused by a medical condition [1]. The ICSD-2 distinguishes primary from symptomatic forms of narcolepsy, but neither familial from sporadic forms nor pediatric from adult forms – all of these have distinct clinical and biological features.

The ICSD-2 states that the diagnosis of NC can be made only by assessing a medical history: chronic EDS and clear-cut cataplexy, an almost pathognomonic symptom of the disease, are sufficient for the diagnosis [1]. The correct diagnosis of cataplexy only by history, however, can be very difficult. Unfortunately, there is no objective test to confirm cataplexy in patients reporting recurrent falls or episodes of weakness. Therefore, both over- and underdiagnosis of NC are frequent in clinical practice. In most cases, treating physicians need great knowledge of the disease for a correct diagnosis. In addition, the ability of the patients to give an accurate description of the symptoms is essential. Even in expert hands, apathy, tiredness, lack of energy, clinophobia, and hysteria can be confounded with EDS, and brief physiological episodes of weakness in the knees or psychogenic spells with cataplexy. For unexperienced physicians, true cataplectic episodes can easily be confounded with psychogenic symptoms, syncope, and atonic or reflex seizures. Misdiagnosis of narcolepsy in subjects with malingering and other conditions that may mimic the narcoleptic symptoms (e.g., some forms of epilepsy, fatigue, and conversion disorders) carries medical malpractice liability including misinformation and starting therapy with specific drugs such as stimulants and sodium oxybate. Conversely, false-negative diagnoses in narcoleptic patients often lead to inadequate treatment such as antiepileptics.

Even more, wrong diagnoses of narcolepsy cause serious problems in clinical trials. In most studies and in contradiction to the ICSD-2 criteria, clinicians do not only pertain to the history but include objective measures for the diagnosis of NC: low CSF hypocretin-1 levels and multiple sleep-onset REM periods (SOREMPs) in the MSLT are highly sensitive and specific for the diagnosis of NC. The ICSD-2 recommends "whenever possible" to confirm the diagnosis of NC by measuring the CSF hypocretin-1 level or performing the MSLT [1]. The practical reasons for this ambiguous recommendation are the lack of availability of these tests in many centers and the fact that these tests are expensive. These tests need experienced technicians and physicians scoring sleep and measuring hypocretin-1 concentration in the CSF, and they require spending one day in the sleep laboratory and performing a lumbar puncture.

In clinical practice, we and others perform the MSLT and HLA typing in all patients with suspected narcolepsy, despite they report clear-cut cataplexy, and save CSF hypocretin-1 measurement for difficult and controversial cases. Similarly, in 2002, Silber et al. [15] proposed the following diagnostic criteria for definite NC which included the use of clinical history plus the MSLT:

- History of EDS.
- History of cataplexy.
- Sleep latency of less than 8 min on MSLT.
- Two or more SOREMPs on MSLT, or 1 SOREMP on MSLT and 1 SOREMP on the preceding nocturnal polysomnogram.
- Apnea–hypopnea index less than 10 on nocturnal polysomnogram preceding the MSLT.

(The last three criteria can be replaced by cataplexy witnessed by a physician with documented recoverable areflexia, or cataplexy recorded by polysomnogram and video recording.)

It should be noted that these diagnostic criteria were stated in early 2002, before an abundant amount of data on CSF hypocretin-1 was available [15].

In contrast, the ICSD-2 states that for the diagnosis of NwC, an MSLT confirming short mean sleep latency and multiple SOREMPs in patients with EDS is mandatory [1]. This is explained by the fact that the symptoms of NwC are unspecific. The only known biological marker and prerequisite of the condition are the presence of enhanced REM sleep pressure, i.e., multiple SOREMPs on MSLT. Multiple SOREMPs can be observed in other conditions associated with EDS (e.g., chronic sleep deprivation, Parkinson's disease, and myotonic dystrophy), but these disorders – with the exception of chronic sleep deprivation – are often easy to diagnose and distinguish from NwC. Most importantly, the physical examination in narcolepsy is normal. The occurrence of multiple SOREMPs distinguishes NwC from other disorders associated with chronic EDS such as idiopathic hypersomnia. In NwC, low CSF hypocretin-1 levels occur in only about 10–15% of the patients, and HLA DQB1*0602 subtype is positive in only 40% [1]. In summary, the lack of MSLT in many centers may lead to underdiagnosis of NwC. On the contrary, since most sleep centers do not routinely perform actigraphy studies prior to polysomnography and MSLT, chronic sleep deprivation is often underrecognized. The ICSD-2 do not ask for actigraphy studies before these sleep tests; thus, NwC is probably often overdiagnosed in sleep-deprived patients.

The following paragraphs review the current diagnosis criteria for idiopathic narcolepsy in the adult, as proposed by the ICSD-2 [1].

Narcolepsy with Cataplexy

Comment on the Diagnostic Criteria of NC (Table 1)

Compared to the first diagnostic criteria established in 1990 (ICSD-1) [16], the following changes were made in the second edition (ICSD-2) in 2005 [1, 17]:

1. A better and more detailed definition of cataplexy is given.
2. Associated features such as sleep paralysis, hypnagogic hallucinations, automatic behaviors, and disrupted major sleep episodes were removed from the diagnostic criteria because they are nonspecific.
3. The mean sleep latency on MSLT was changed from less than 5 to less than 8 min because of a result of a recent meta-analysis.
4. Polysomnography and MSLT were recommended but not mandatory, given that cataplexy is a unique feature in NC.
5. The polysomnographic findings of sleep latency of less than 10 min and REM sleep latency of less than 20 min were removed because of low sensitivity: they are seen in less than 50% of the patients.
6. Determination of the CSF hypocretin-1 level was first introduced as a desirable but not a mandatory test.
7. HLA-DR2 positivity was removed due low specificity.

Table 1 ICSD-2 diagnostic criteria of narcolepsy with cataplexy

The patient has a complaint of EDS occurring almost daily for at least 3 months
A definite history of cataplexy, defined as sudden and transient episodes of loss of muscle tone triggered by emotion, is present
Note: To be labeled as cataplexy, these episodes must be triggered by strong emotions – most reliably laughing or joking – and must be generally bilateral and brief (less than 2 min). Consciousness is preserved, at least at the beginning of the episode. Observed cataplexy with transient reversible loss of deep tendon reflexes is a very strong, but rare, diagnostic finding
The diagnosis of NC should, whenever possible, be confirmed by nocturnal polysomnography followed by an MSLT; the mean sleep latency on the MSLT is less than or equal to 8 min, and two or more SOREMPs are observed following sufficient nocturnal sleep (minimum 6 h) CSF are less than or equal to 110 pg/mL or one-third of mean control values
Note: The presence of two or more SOREMPs during the MSLT is a very specific finding, whereas a mean sleep latency of less than 8 min can be found in 30% of the normal population. Low CSF hypocretin-1 levels (less than or equal to 110 pg/mL or one-third of mean normal control values) are found in more than 90% of patients with NC and almost never in controls or in other patients with other pathologies
The hypersomnia is not better explained by another sleep disorder, medical or neurological disorder, mental disorder, medication use, or substance use disorder

The Patient Has a Complaint of EDS Occurring Almost Daily for at Least 3 Months

The presence of EDS is a prerequisite for the diagnosis of narcolepsy. All patients with narcolepsy experience chronic EDS, and in the majority of cases, EDS is the earliest and most disabling complaint. However, there is a great variability in the clinical presentation of EDS in narcolepsy, ranging from mild to severe. In some patients with NC, the cause of consultation is difficulty maintaining sleep, but detailed questioning reveals experiencing EDS and cataplexy for years [1–6]. Duration of EDS was arbitrarily defined by the ICSD-2 as occurring for at least 3 months [1]. To distinguish narcolepsy from other disorders associated with recurrent EDS (e.g., Kleine–Levin syndrome), EDS in NC must be present almost daily.

There are two possible and often coexisting clinical presentations of EDS in narcolepsy [1]. Both types of EDS impair the patient's quality of life by causing social, professional, and familial problems, and by leading to automobile crashes. The first is a state of daytime sleep propensity which leads to regular napping. Patients experience a constant pressure for falling asleep and difficulty to stay awake. Patients fall asleep and take frequent short naps at inappropriate times, especially in non-stimulating situations such as watching television and reading. Some subjects are unable even to stay awake in active situations such as eating or talking. This type of EDS is nonspecific for narcolepsy, it is also common in other sleep–wake disturbances such as obstructive sleep apnea, idiopathic hypersomnia, depression, and nocturnal sleep deprivation. However, sleep episodes in patients with NC are typically short (5–15 min) and sometimes associated with vivid dreams. Naps are usually refreshing in NC.

On the contrary, EDS may present as sudden onset of sleep episodes [1]. These episodes are abrupt, brief, and unexpected, and have been reported to occur during active situations such as driving, eating, talking, having a shower, walking, being on the telephone, and writing. It is unclear whether these episodes, classically termed "sleep attacks," constitute a unique entity, or if they are merely an extreme manifestation of severe EDS. They have also been described in patients with Parkinson's disease, particularly in those under dopaminergic therapy. Sometimes narcoleptics with sleep attacks are not aware that they are sleepy because of associated amnesia or because patients have long habituated to the sensation of chronic and severe EDS. In this line, we observed that some narcoleptics do not recall having fallen asleep for a few minutes (in recorded stages N1 and N2) after being awakened by us (unpublished observations).

Clinical history is most important to evaluate the occurrence and characteristics of EDS. Clinically, it is crucial to distinguish EDS from fatigue, lack of energy, asthenia, and tiredness. Untreated true narcoleptics sometimes falsely report not experiencing EDS. This is because of a false perception of chronically habituated EDS or because patients are afraid of the consequences of EDS (e.g., losing the driving license and problems when applying for a job). In contrast, malingering also occurs; subjects falsely report EDS and other typical features of narcolepsy such as cataplexy that they have learnt from the Internet or books. In these situations, an objective test such as the MSLT is helpful to make a correct diagnosis [18].

A Definite History of Cataplexy, Defined as Sudden and Transient Episodes of Loss of Muscle Tone Triggered by Emotion, Is Present. Note: To Be Labeled as Cataplexy, These Episodes Must Be Triggered by Strong Emotions – Most Reliably Laughing or Joking – and Must Be Generally Bilateral and Brief (Less Than 2 Min). Consciousness Is Preserved, At Least at the Beginning of the Episode. Observed Cataplexy with Transient Reversible Loss of Deep Tendon Reflexes Is a Very Strong, But Rare, Diagnostic Finding

The ICSD-2 clearly states that the diagnosis of NC can be made by clinical history alone in patients reporting chronic EDS and clear-cut cataplexy. Although cataplexy may occasionally occur in subjects with structural brain lesions (e.g., hypothalamic tumors) and uncommon diseases

(Niemann–Pick disease type C, Prader–Willi syndrome, Coffin–Lowry syndrome, and Norrie disease), cataplexy in patients with both normal brain imaging and neurological examination is virtually pathognomonic for narcolepsy [1].

The ICSD-2 provides an excellent definition of a cataplectic event, indicating that it is an episode characterized by loss of muscle tone which is sudden, transient, brief, often bilateral, with conserved consciousness, and provoked by emotions, generally laughing and joking. Severity varies from partial and minimal symptoms (e.g., facial twitches, head dropping, and slurred speech) to falls and collapse. Duration varies from a few seconds to several minutes. During a cataplectic attack, physical examination reveals the loss of tendon reflexes. However, like in epilepsy, only a minority of patients experience their reported episodes in front of their physicians. Therefore, in most cases, the diagnosis of cataplexy is based on clinical history. Although it seems easy at first glance, the diagnosis of cataplexy by clinical history is often tricky and can be very difficult in some cases. It requires experienced clinicians and clear descriptions of the events by the patients. Overall, the maintenance of consciousness and emotional triggers allow for its distinction.

NC patients with undetectable CSF hypocretin-1 may experience only two or three episodes of cataplexy after experiencing EDS for many years. Others may not report triggering emotions prior to some cataplectic episodes. Validated cataplexy questionnaires may help in the diagnosis of NC [18]. However, even in expert hands, cases of reflex epilepsy, gelastic seizures, atonic seizures, hyperkaliemic periodic paralysis, syncope, orthostatic hypotension, drop attacks, and functional manifestations can occasionally be confounded with cataplexy. Again, malingering and conversion disorder may be confounded with cataplexy. In the absence of clear-cut cataplexy, the terms "atypical cataplexy" and "doubtful cataplexy" have been introduced. In many cases, these diagnostic ambiguities can be solved by performing an objective test such as the MSLT [19] or measuring hypocretin-1 in the CSF [10–12]. In the majority of patients with true NC, the MSLT shows more than one SOREMP, and CSF hypocretin-1 level is undetectable or low. This is why the ICSD-2 recommends performing MSLT and CSF hypocretin-1 level determination to corroborate the diagnosis of narcolepsy [1].

The Diagnosis of NC Should, Whenever Possible, Be Confirmed by Nocturnal Polysomnography Followed by an MSLT; The Mean Sleep Latency on the MSLT Is Less Than or Equal to 8 min and Two or More SOREMPs Are Observed Following Sufficient Nocturnal Sleep (Minimum 6 h) During the Night Prior to the Test. Note: The Presence of Two or More SOREMPs During the MSLT Is a Very Specific Finding, Whereas a Mean Sleep Latency of Less Than 8 Min Can Be Found in 30% of the Normal Population

MSLT is an important objective test for the diagnosis of narcolepsy [19]. The ICSD-2 recommends a regular sleep–wake schedule two weeks prior to the MSLT, which should be documented by sleep logs or – even better – by actigraphy [1]. Polysomnography and MSLT should be conducted after sufficient washout of drugs that influence sleep (e.g., central nervous stimulants, sedatives, sodium oxybate, and antidepressants). Central nervous system stimulants increase the sleep latency, sedatives and sodium oxybate decrease the sleep latency, and antidepressants suppress REM sleep. These compounds must be discontinued for at least five half-life times of the respective drug and long-lasting metabolite. Drugs that influence sleep (e.g., antidepressants, sodium oxybate, and stimulants) must be withdrawn long enough before the test. Furthermore, the MSLT must be preceded by nocturnal polysomnography showing sufficient total sleep time (i.e., sleep satiation) in the previous night. The minimum of 6 h of nocturnal sleep was arbitrarily defined by consensus. This is crucial since SOREMPs and short mean sleep latencies on the MSLT are common in shift workers, sleep-deprived people, and persons taking antidepressants [19, 20]. Polysomnography also provides information on possible other sleep–wake disturbances such as obstructive sleep apnea, periodic leg movements in sleep, and REM sleep behavior disorder that frequently coexist in narcolepsy [1]. Disruption of sleep architecture by brief awakenings of unknown origin is a frequent finding in NC. Compared to subjects with NwC and idiopathic hypersomnia, those with NC have almost 50% more stage N1, less stage N3, lower sleep efficiency, and 60% more awakenings. Compared with controls, patients with NC exhibit similar total sleep time per 24 h, but reduced sleep efficiency, short sleep latency (often below 10 min), short REM sleep latency, greater wake time after sleep onset, increased amount of N1, and similar amount of REM sleep.

Short sleep latency in the polysomnogram, however, is a nonspecific finding for narcolepsy since it can be frequently seen in healthy volunteers and in most of the cases with true EDS of any origin. A short REM sleep latency of less than 10 min in the polysomnogram is a more specific finding for NC, which occurs in about 50% of the cases. Short REM sleep latency can be seen in other conditions including major depression and situations linked to REM sleep deprivation (insufficient nocturnal sleep and very fragmented sleep due to obstructive sleep apnea).

In patients with obstructive sleep apnea already treated with continuous positive airway pressure (CPAP) in whom comorbid narcolepsy is suspected, we recommend performing both the nocturnal polysomnogram and MSLT using the CPAP mask that the patient habitually uses at home.

MSLT is considered the gold standard objective method for the assessment of EDS. It measures the ability to fall asleep when the subject is instructed to sleep in a soporific situation [19]. A SOREM period on the MSLT is defined by the occurrence of REM sleep within 15 min of sleep onset. The diagnostic sensitivity of the MSLT for the diagnosis of NC has been estimated around 60%, while the diagnostic specificity when two or more SOREMPs are present is up to 95%. If the presence of two or more SOREMPs is combined with a mean sleep latency of less than 5 min, then the diagnostic specificity rises to up to 97%.

The majority of the narcoleptics show a mean MSLT of less than 8 min. The mean sleep latency in healthy controls is 10.5±4.6, whereas it is 3.1±2.9 min in narcolepsy [1, 19]. However, about 10% of the patients with NC exhibit a sleep latency of more than 8 min or less than two SOREMPs in the MSLT or both [1, 19]. In our experience, there is an association between a higher number of SOREMPs and a shorter mean sleep latency (data not published). Multiple SOREMPs, however, may also be observed in a small proportion of normal young adults, particularly in those with chronic sleep deprivation, and also in major depression, alcohol withdrawal, treatment with REM suppressants (e.g., antidepressants), situations linked to fragmented sleep (e.g., severe obstructive sleep apnea and severe periodic limb movement disorder), and other diseases with associated EDS (e.g., multiple system atrophy, Parkinson's disease, Prader–Willi syndrome, Niemann–Pick disease type C, Norrie disease, Kleine–Levin syndrome, myotonic dystrophy, and hyperkaliemic periodic paralysis) [1–6] (Table 2).

Most of the studies have shown that the mean number of SOREMPs in NC is between 3 and 4.

Table 2 Conditions that are associated with (occasional) multiple SOREMPs on the MSLT

Narcolepsy with cataplexy
Narcolepsy without cataplexy
Very small percentage of the normal population
Major depression
Alcohol withdrawal
REM suppressant medications
Obstructive sleep apnea
Periodic limb movement disorder
Long sleepers
Sleep deprivation
Circadian rhythm disorders
Shift work
Multiple system atrophy
Parkinson's disease
Prader–Willi syndrome
Niemann–Pick disease type C
Norrie disease
Kleine–Levin syndrome
Myotonic dystrophy
Hyperkaliemic periodic paralysis

In our experience with 97 narcoleptics, 22.36% of SOREMPs occurred in the first nap, 19.80% in the second nap, 21.08% in the third nap, 15.65% in the fourth nap, and 21.08% in the fifth nap. The likelihood a SOREMP in the fourth nap was lowest. Shortening the MSLT to four naps decreases the capability of the test for the diagnosis of narcolepsy in patients who had one SOREMP [21].

The Diagnosis of NC Should, Whenever Possible, Be Confirmed by Measurement of Hypocretin-1 Levels in the CSF Which Has to Be Less Than or Equal to 110 pg/mL or One-Third of Mean Control Values. Note: Low CSF Hypocretin-1 Levels (Less Than or Equal to 110 pg/mL or One-Third of Mean Normal Control Values) Are Found in More Than 90% of Patients with NC and Almost Never in Controls or in Other Patients with Other Pathologies

Pathological studies in human brains of NC have demonstrated selective and prominent loss of the hypocretin-producing neurons in the hypothalamus [7, 8]. Hypocretin-1 is an excitatory neurotransmitter that promotes wakefulness and inhibits the presence of REM sleep. Hypocretin-1 is measurable in the CSF and levels can be assessed by radioimmunoassay. Many authors assume that CSF hypocretin-1 levels reflect the quantity of hypocretin-1-producing cells in the hypothalamus. In a rodent study, a decrease in 50% of the CSF hypocretin-1 level was obtained after losing 70% of the hypocretin-containing neurons. Fifteen percent loss of these cells did not alter the CSF hypocretin-1 level [9].

In humans, CSF hypocretin-1 level has been classified as undetectable (detection limit of assay of less than 40 pg/mL), low (<110 pg/mL), intermediate (110–200 pg/mL), and normal (>200 pg/mL, mean value about 350 pg/mL) [10–12]. These values are derived from the Stanford group – for other centers, other values may apply. The commercially available kits suffer from high interassay variability. It is recommended to measure all samples in duplicate and using internal controls due to assay variability. Therefore, hypocretin-1 measurements in CSF should be performed only by experienced centers, because false results are likely due to the pitfalls of the kit. In humans, the CSF hypocretin-1 level is not influenced by age, gender, narcolepsy duration, medications, and body mass index [2–4].

In patients with narcolepsy, levels depend on the presence of cataplexy and, particularly, the presence of HLA DQB1*0602 subtype [1–5]. CSF hypocretin-1 is undetectable or low in 90–95% of patients with NC, and in almost all narcoleptics with cataplexy carrying the HLA DQB1*0602 subtype. Hypocretin deficiency is present very early in the course of the disease, even 1–6 months after symptom onset. Undetectable CSF hypocretin-1 levels can be found in patients with NC who have only one SOREMP in the MSLT. In HLA-negative patients with cataplexy, CSF hypocretin-1 levels are usually normal or intermediate. In HLA-positive patients with cataplexy, the sensitivity and specificity of a low CSF hypocretin-1 value are 93% and 100%, respectively. CSF hypocretin-1 concentrations lower than 110 pg/mL have a predictive value of 94% for the diagnosis of NC. In HLA-negative patients with cataplexy, the sensitivity and specificity of low CSF hypocretin-1 value are 7% and 100%, respectively [1–4].

In subjects with familial NC, both low and normal CSF hypocretin-1 levels have been documented. Low levels have been found in familial cases with cataplexy when they were HLA positive.

In narcolepsy patients without cataplexy, CSF hypocretin-1 levels are low in only 10–15%, particularly if they are HLA positive. In patients with doubtful and atypical cataplexy, CSF hypocretin-1 levels can be either low or normal. In these doubtful and atypical cataplexy cases, normal CSF hypocretin-1 level argues against the diagnosis of NC, particularly if they are HLA positive [1].

Hypocretin-1 levels in the CSF are always normal in healthy controls without EDS and in most of the conditions associated with EDS (e.g., obstructive sleep apnea and idiopathic hypersomnia). Absent, low, or intermediate levels of CSF hypocretin-1 have been found in some subjects with hypothalamic tumors, anti-Ma2 paraneoplastic limbic encephalitis, Kleine–Levin syndrome,

Table 3 Conditions that are (occasionally) associated by absent, low, or intermediate levels of CSF hypocretin-1

Narcolepsy with cataplexy (90–95% of the cases)
Narcolepsy without cataplexy (10–15% of the cases)
Familial narcolepsy
Hypothalamic lesions (tumor, stroke, and demyelinating plaques)
Anti-Ma2 paraneoplastic limbic encephalitis
Kleine–Levin syndrome
Guillain–Barré syndrome
Prader–Willi syndrome
Niemann–Pick disease type C
Myotonic dystrophy
Traumatic brain injury
Central nervous system infections
Acute lymphocytic leukemia

Table 4 Secondary narcolepsy due to brain tumors

One-third of secondary narcolepsy is associated with the presence of primary tumors of the brain
Most frequent tumor location is the hypothalamus or adjacent structures
Craniopharyngioma in children is one the most frequent brain tumors associated with secondary narcolepsy
All patients present with various degrees of excessive daytime sleepiness
About 50% of the patients exhibit cataplexy
About 50% of the patients are HLA DR2/DQB1*0602 positive
Sleep-onset REM sleep episodes are frequent in the MSLT
CSF hypocretin-1 value can be undetectable, low, intermediate, or normal

Guillain–Barré syndrome, Prader–Willi syndrome, myotonic dystrophy, and head trauma (Table 3). These conditions cause symptomatic narcolepsy or narcolepsy due to a medical condition [22].

In patients with suspected NC, measuring CSF levels of hypocretin-1 [17] is indicated when there is a need to confirm or exclude the condition objectively when (1) MSLT is not available, (2) MSLT is not interpretable, equivocal, or feasible (e.g., in children), (3) MSLT is negative, (4) the patient is treated with REM suppressants such as antidepressants which cannot be discontinued before MSLT, (5) doubtful or atypical cataplexy is suspected, or (6) malingering or a psychogenic disorder is suspected.

The Hypersomnia Is Not Better Explained by Another Sleep Disorder, Medical or Neurological Disorder, Mental Disorder, Medication Use, or Substance Use Disorder

This point differentiates primary NC from symptomatic narcolepsy. Patients with symptomatic narcolepsy present with chronic EDS and at least one of the following: clear-cut cataplexy, short mean sleep latency and multiple SOREMPs in the MSLT, and low CSF hypocretin-1 concentration [1]. Thus, the criteria suggest that all causes of symptomatic narcolepsy (hypothalamic tumors, focal brain lesions, inherited diseases, paraneoplastic syndromes, neurodegenerative diseases, etc.) [22] need to be excluded for the diagnosis of idiopathic NC (Table 4). Therefore, detailed clinical history, physical examination, blood tests, and brain MRI should be performed in all cases in order to exclude secondary causes of narcolepsy.

Associated Features that May Help in the Diagnosis of NC But Are Not Included in the ICSD-2 Current Diagnostic Criteria

There are other clinical signs that may help to distinguish NC from mimicking conditions. They have not been included in the ICSD-2 diagnostic criteria because they are not specific for the diagnosis of narcolepsy. They are not highly useful in excluding or confirming the diagnosis of narcolepsy, but in some difficult cases, they may be of considerable help.

Clinical Onset and Clinical Course [1–6]

The majority of NC patients do not recall the exact onset of EDS and cataplexy. This is because of the insidious and gradual progression of symptoms in many patients. Others report an acute onset which may be linked to a viral infection, a head trauma, or a life event. In about 90% of NC patients, onset of EDS precedes or coincides with cataplexy onset [1]. Epidemiological studies revealed that the age of EDS onset follows a bimodal distribution with the highest peak at the age of 14 years and a smaller second peak at the age of 35 years [23]. However, EDS onset may occur in children before the age of 10 years (5%) and in adults above the age of 50 years. Cataplexy onset does not have a similar bimodal distribution. Late-life onset of cataplexy explains the late diagnosis (in many patients after the age of 40 years), together with the fact that many patients have been misdiagnosed for many years [24]. NC persists throughout lifetime, but in many patients, EDS and cataplexy improve with time. In contrast, nocturnal disturbances including sleep fragmentation, REM sleep behavior disorder, and obstructive sleep apnea worsen or first appear in patients older than 40 years [1].

Sleep Paralysis [1]

This is an abrupt intrusion of REM sleep muscle atonia into wake–sleep transitions, i.e., at falling asleep or waking up. Subjects experience a transient inability to move or speak, lasting from a few seconds to few minutes. The whole body is involved. The episodes may be associated with a sensation of suffocation and fear. Recurrent sleep paralysis occurs in approximately 50% of the patients with NC. They also occur in 3–5% of the general population (particularly in young sleep-deprived individuals) and in a few patients with other sleep disorders linked to EDS such as obstructive sleep apnea.

Hypnagogic and Hypnopompic Hallucinations [1]

These are dream-like REM sleep-related vivid perceptual experiences that occur while falling asleep (hypnagogic) and upon awakening (hypnopompic). They may be visual, auditory, vestibular, or tactile. They can be frightening and are often associated with sleep paralysis. Recurrent hallucinations are common in subjects with NC (30–60%), but they are also observed in 1–3% of the general population. Therefore, like sleep paralysis, hallucinations are neither highly sensitive nor highly specific for NC.

Coexistent Polysomnographic Abnormalities

Periodic leg movements in sleep [25], REM sleep behavior disorder [26], and obstructive sleep apnea [27] appear to be more common in NC than in the general population. In patients with narcolepsy, these polysomnographic abnormalities may occur at any age, but they are more frequent in the elderly population.

HLA Subtype [1, 14]

NC is tightly linked to the HLA DQB1*0602 subtype. About 95% of patients with NC are positive for the HLA DQB1*0602 subtype. This subtype is present in about 50% of narcolepsy patients with atypical cataplexy, in 40% of narcoleptics without cataplexy, and in 25% of the Caucasian general population. Therefore, the absence of the HLA DQB1*0602 subtype strongly argues against the diagnosis of NC, whereas its presence only supports the diagnosis. Given that the prevalence rate of NC is 0.02–0.18% in the general population, more than 99% of HLA DQB1*0602-positive individuals do not have narcolepsy. Interestingly, however, normal individuals who are HLA DQB1*0602 positive enter faster in REM sleep than those who are negative.

Narcolepsy Without Cataplexy

Narcolepsy without cataplexy represents about 20–40% of the all diagnosed narcoleptic cases. Available demographic, clinic, and biological data on NwC, however, are scarce due to lack of biological markers and heterogeneity of the disorder. Thus, it is not surprising that the current diagnostic criteria for NwC still bear significant limitations. Symptoms of NwC are unspecific, and there is no specific objective test. The current diagnostic criteria of NwC require clinical history and an MSLT. Clinical history must reveal chronic EDS in the absence of cataplexy, and MSLT must demonstrate a mean sleep latency of less than or equal to 8 min and two or more SOREMPs [1]. Thus, the diagnosis is not possible without MSLT. This test is useful because it (1) proves and quantifies EDS and (2) reveals a high REM sleep pressure by detecting multiple SOREMPS. Also, demonstration of multiple SOREMPs on the MSLT distinguishes NwC from idiopathic hypersomnia [1, 19]. Unfortunately, multiple SOREMPs on the MSLT, the only objective finding, lack specificity because it can be found in normal people and in other conditions associated with EDS [19, 20, 28]. If the ICSD-2 recommendations for performing an MSLT are strictly followed (e.g., stopping central nervous system medications and excluding subjects with sleep deprivation), however, many false-positive MSLT results can be excluded. Other objective tests (e.g., CSF hypocretin-1 concentration and HLA typing) and the presence of associated symptoms (e.g., sleep paralysis and hypnagogic hallucinations) are not included in the current diagnostic criteria because they are neither highly specific nor sensitive [1]. Low CSF hypocretin-1 levels are only observed in up 10–15% of the NwC patients, and HLA DQB1*0602 is positive in approximately 40% [1–5, 10–13].

Comment on the Diagnostic Criteria of NwC (Table 5)

The Patient Has a Complaint of EDS Occurring Almost Daily for at Least 3 Months

EDS in NwC resembles that of NC: patients report chronic EDS characterized by high sleep propensity and short, irresistible, and refreshing naps. Intensity of EDS varies, but overall it is less severe than in NC, particularly in subjects without the HLA DQB1*0602 subtype. Nocturnal sleep is less disrupted than in NC. Some subjects may report automatic behaviors during the daytime and others, difficulty in awakening in the morning, thereby arriving late at work or school [1].

Typical Cataplexy Is Not Present, Although Doubtful or Atypical Cataplexy-Like Episodes May Be Reported

The crucial difference to NC is the absence of cataplexy. Since cataplexy is tightly linked to low CSF hypocretin-1 and HLA positivity, it is not surprising to find low CSF hypocretin-1 levels and HLA positivity in only a minority of NwC patients. It is conceivable that some true narcoleptics with cataplexy are first diagnosed with NwC, because they have not yet developed cataplexy at the moment of examination. It is well known that in NC, EDS onset usually precedes cataplexy onset by several years [1, 23].

Interestingly, the ICSD-2 criteria include patients with doubtful cataplexy or atypical cataplexy-like episodes in the NwC group. Doubtful cataplexy often consists of atypical sensations of muscle weakness triggered by unusual emotions, long episodes of tiredness after emotions such as anger, or cataplexy-like episodes that occur only a few times during the whole patient life. Atypical cataplexy attacks are episodes that resemble cataplexy but are not triggered by emotions such as laughing and joking [1]. This understanding may reflect the limitations of clinical history in the diagnosis of this heterogeneous condition. Thus, patients with true cataplexy can be labeled as NwC when (1) they present mild or occasional cataplexy that can be easily overlooked by either the patient or the clinician, (2) patients are not capable of giving a good description of their typical cataplectic attacks, and (3) physicians are not able to recognize typical cataplexy as a pathological feature.

Table 5 ICSD-2 diagnostic criteria of narcolepsy without cataplexy

The patient has a complaint of EDS occurring almost daily for at least 3 months
Typical cataplexy is not present, although doubtful or atypical cataplexy-like episodes may be reported
The diagnosis of NwC must be confirmed by nocturnal polysomnography followed by an MSLT. In NwC, the mean sleep latency on the MSLT is less than or equal to 8 min, and two or more SOREMPs are observed following sufficient nocturnal sleep (minimum 6 h) during the night prior to the test
Note: The presence of two or more SOREMPs during the MSLT is a specific finding, whereas a mean sleep latency of less than 8 min can be found in up 30% of the normal population
The hypersomnia is not better explained by another sleep disorder, medical or neurological disorder, mental disorder, medication use, or substance use disorder

The Diagnosis of NwC Must Be Confirmed By Nocturnal Polysomnography Followed by an MSLT. In NwC, the Mean Sleep Latency on the MSLT Is Less Than or Equal to 8 min and Two or More SOREMPs Are Observed Following Sufficient Nocturnal Sleep (Minimum 6 h) During the Night Prior to the Test. Note: The Presence of Two or More SOREMPs During the MSLT Is a Specific Finding, Whereas a Mean Sleep Latency of Less Than 8 Min Can Be Found in up to 30% of the Normal Population

Similar to that in NC, polysomnography usually shows a short mean sleep latency and short mean REM sleep latency [1, 29]. However, nighttime sleep is much less fragmented and sleep architecture can be similar to that seen in idiopathic hypersomnia, with a sleep efficiency of 90% and normal amount of N3 [29].

The mean sleep latency on the MSLT is 3.1±2.9 min in NwC [19]. The utility of multiple SOREMPs on the MSLT as the prerequisite for the diagnosis of NwC was recently challenged by a population-based cohort study involving people not reporting EDS or cataplexy [20, 28]. The study evaluated 289 males and 267 females (age 35–70 years, 97% Caucasians) who completed a sleep diary for one week preceding a nocturnal polysomnogram which was followed by MSLT. At MSLT, 13.1% of men and 5.6% of women had two or more SOREMPs, and 5.9% of men and 1.1% of women met the full ICSD-2 MSLT criteria for the diagnosis of narcolepsy. In all, 4.2% of males and 0.4% of females met the diagnostic criteria for NwC (chronic subjective EDS, not experiencing cataplexy, a mean sleep latency of less than or equal to 8 min, and two or more SOREMPs on the MSLT). SOREMPs in males were associated with shift work, the use of antidepressants, HLA DQB1*0602 positivity, shorter REM sleep latency on the polysomnogram, shorter mean sleep latency on the MSLT, hypnagogic hallucinations, and cataplexy-like symptoms. Thus, the study suggested that NwC is much more common than expected or – what we believe – challenged the ICSD-2 statement that "the presence of multiple SOREMPs during the MSLT is a very specific finding for the diagnosis of NwC." This study reveals that the test is not valid neither for people taking sleep-active medications for sleep-deprived subjects such as shift workers [1, 28]. A recent MSLT study confirmed that multiple SOREMPs may occur in a significant proportion of sleep-deprived, otherwise healthy subjects [30].

The Hypersomnia Is Not Better Explained by Another Sleep Disorder, Medical or Neurological Disorder, Mental Disorder, Medication Use, or Substance Use Disorder

Again, the ICSD-2 warrants exclusion of symptomatic cases [1]. Most of the symptomatic cases of narcolepsy exhibit multiple SOREMPs or low CSF hypocretin-1 values or both, but lack cataplexy [22].

Associated Features that May Help in the Diagnosis of NwC But Are Not Included in the ICSD-2 Current Diagnostic Criteria

Clinical Onset and Clinical Course

Like in NC, onset in NwC typically occurs gradually during adolescence and persists through lifetime. In our experience, EDS in NwC responds better to stimulants than in NC [1].

Recurrent Sleep Paralysis

In NwC, the frequency of recurrent sleep paralysis is about 40%, which is slightly lower than that in patients with NC (about 50%) but much higher than that in the general population (3–5%) [1].

Recurrent Hypnagogic and Hypnopompic Hallucinations

The prevalence of recurrent hypnagogic and hypnopompic hallucinations in NwC is 15%, which is lower than that in patients with NC (about 30–60%) but again higher than that in the general population (1–3%) [1].

HLA Typing

The prevalence of HLA DQB1*0602 is about 40% in NwC, lower than in patients with NC (about 90%), but higher than in the general population (25%) [1, 14].

Considerations on the Current Diagnostic Criteria of NwC, Considering Its Uncertain Underlying Biological Basis

NC and NwC share the presence of EDS linked to multiple SOREMPs. This is why they have been labeled as narcolepsy, suggesting that both conditions share pathophysiological features. This assumption is supported by the fact that two other REM sleep-associated abnormalities such as sleep paralysis and hypnagogic and hyponopompic hallucinations are more frequent in both entities than in the general population [1]. Moreover, about 10–15% of the patients without cataplexy exhibit low CSF hypocretin-1 levels associated with HLA DQB1*0602 positivity, two highly specific and sensitive features of NC [1, 10–13]. Thus, the pathophysiological substrate in this small group of patients without cataplexy may be very similar, or the same, to the one occurring in NC. However, it seems that overall NwC is a less severe or a less complete condition because it differs from NC in several aspects such as absence of cataplexy, less severity of subjective EDS, lower prevalence of hallucinations and sleep paralysis, less disrupted nocturnal sleep, longer mean sleep latency on the MSLT, lesser mean number of SOREMPs on the MSLT, lower HLA DQB1*0602 positivity, and much lower percentage of subjects exhibiting low CSF hypocretin-1 levels [1–6].

A few studies have evaluated the CSF hypocretin-1 concentration in narcoleptics without cataplexy. These studies showed variable results describing that up to 10–15% of the cases may show decreased levels, particularly when they are DQB1*0602 or DR2 positive. Low values are more frequently observed in young individuals (presumably because they have not developed cataplexy yet) and in those with prominent REM sleep abnormalities such as sleep paralysis, hypnagogic and hypnopompic hallucinations, and high number of SOREMPs in the MSLT. CSF hypocretin-1 concentration is usually normal in patients with NwC who are negative for the DR2 or DQB1*0601 alleles [1–5, 10–13, 31]. In HLA-negative NwC patients, the sensitivity and specificity of low CSF hypocretin-1 value are 2% and 100%, respectively. In HLA-positive NwC patients, the number goes up to 31% and 100%, respectively [1–5, 10–13].

CSF hypocretin-1 levels may reflect the integrity of the hypocretin cells in the hypothalamus. However, only severe loss of hypocretin cells may be associated with decreased levels. Thus, a normal CSF hypocretin-1 level does not necessarily indicate that the hypocretin cell population in the hypothalamus is intact. This situation is seen in Parkinson's disease where neuropathological studies show some degree of hypocretin cell loss related to clinical severity [32, 33], but CSF hypocretin-1 concentration is normal in all clinical stages of the disease [34]. Preliminary evidence suggests that hypocretin-producing cells are partially lost in NwC [35].

The cause of NwC is unknown. The lack of specific cardinal symptoms on the one side and variable findings in terms of HLA typing and CSF hypocretin-1 levels on the other side suggest that NwC is a heterogeneous entity, whereas NC seems to be a more robust condition, tightly linked to hypocretin deficiency in the brain and the HLA DQB1*0602 subtype. Taken together, NwC is a heterogeneous entity that probably includes patients with (1) NC who have not yet expressed cataplexy, (2) NC genotypes with different phenotypes, particularly lacking cataplexy, (3) NwC genotypes and phenotypes, and (4) other sleep–wake disorders and even chronic sleep deprivation [29]. Therefore, the creation of the next ICSD-3 diagnostic criteria for NwC will be challenging.

References

1. American Academy of Sleep Medicine. ICSD-2. International Classification of Sleep disorders. 2nd ed: Diagnostic and coding manual. American Academy of Sleep Medicine, 2005.
2. Overeem S, Mignot E, van Dijk G, Lammers GJ. Narcolepsy: clinical features, new pathophysiological insights, and future perspectives. J Clin Neurophysiol. 2001;18:78–105.
3. Chakravorty SS, Rye DB. Narcolepsy in the older adult. Epidemiology, diagnosis and management. Drugs Aging. 2003;20:361–76.
4. Dauvilliers Y, Arnulf I, Mignot E. Narcolepsy with cataplexy. Lancet. 2007;369:499–511.
5. Plazzi G, Serra L, Ferri R. Nocturnal aspects of narcolepsy with cataplexy. Sleep Med Rev. 2008;12:109–28.

6. De Lecea L, Kilduff TS, Peyron C, et al. The hypocretins: hypothalamus-specific peptides with neuroexcitatory activity. Proc Natl Acad Sci USA. 1998;95:322–7.
7. Peyron C, Faraco J, Rogers W, et al. A mutation in a case of early onset narcolepsy and a generalized absence of hypocretin peptides in human narcoleptic brains. Nat Med. 2000;6:991–7.
8. Thannickal TC, Moore RY, Nienhuis R, et al. Reduced number of hypocretin cells in human narcolepsy. Neuron. 2000;27:469–74.
9. Gerashchencko D, Murillo-Rodriguez E, Lin L, et al. Relationship between CSF hypocretin levels and hypocretin neuronal loss. Exp Neurol. 2003;184: 1010–6.
10. Ripley B, Overeem S, Fujiki N, et al. CSF hypocretin/orexin levels in narcolepsy and other neurological conditions. Neurology. 2001;57:2253–8.
11. Mignot E, Lammers GJ, Ripley B, et al. The role of cerebrospinal fluid hypocretin measurement in the diagnosis of narcolepsy and other hypersomnias. Arch Neurol. 2002;59:1553–62.
12. Bourgin P, Zeitzer JM, Mignot E. CSF hypocretin-1 assessment in neurological disorders. Lancet Neurol. 2008;7:649–62.
13. Fronczek R, Baumann CR, Lammers GJ, Bassetti CL, Overeem S. Hypocretin/orexin disturbances in neurological disorders. Sleep Med Rev. 2009;13:9–22.
14. Mignot E, Hayduk R, Black J, Grumet FC, Guilleminault C. HLA DQB1*0602 is associated with cataplexy in 509 narcoleptics patients. Sleep. 1997;20:1012–20.
15. Silber MH, Krahan LE, Olson EJ. Diagnosing narcolepsy: validity and reliability of new diagnostic criteria. Sleep Med. 2002;3:109–13.
16. ICSD-International classification of sleep disorders. Diagnostic and coding manual. Diagnostic Classification Steering Committee, Thorpy MJ, Chairman. Rochester, MA. American Sleep Disorders Association, 1990.
17. Billiard M. Diagnosis of narcolepsy and idiopathic hypersomnia. An update based on the International Classification of Sleep disorders, 2nd edition. Sleep Med Rev. 2007;11:377–88.
18. Anic-Labat S, Guilleminault C, Kraemer HC, et al. Validation of a cataplexy questionnaire in 983 sleep-disorders patients. Sleep. 1999;22:77–87.
19. Aarand D, Bonnet M, Hurwitz M, Rosa R, Sangal B. The clinical use of the MSLT and MWT. Sleep. 2005;28:123–44.
20. Mignot E, Lin L, Finn L, et al. Correlates of sleep-onset REM periods during the multiple sleep latency test in community adults. Brain. 2006;129: 1609–23.
21. Santamaria J, Falup C, Iranzo A, Salamero M. Themporal distribution of sleep onset REM periods in narcolepsy. J Sleep Research. 2006;15 Suppl 1:76.
22. Nishino S, Kanbayashi T. Symptomatic narcolepsy, cataplexy and EDS, and their implications in the hypothalamic hypocretin/orexin system. Sleep Med Rev. 2005;9:269–310.
23. Dauvilliers Y, Montplaisir J, Molinari N, et al. Age at onset of narcolepsy in two large populations of patients in France and Quebec. Neurology. 2001;57:2029–33.
24. Rye DB, Dihenia B, Weissman JD, Epstein CM, Bliwise DL. Presentation of narcolepsy after 40. Neurology. 1998;50:459–65.
25. Bahammam A. Periodic leg movements in narcolepsy patients: impact on sleep architecture. Acta Neurol Scand. 2007;115:351–5.
26. Schenck CH, Mahowald M. Motor dyscontrol in narcolepsy: rapid-eye-movement (REM) sleep without atonia and REM sleep behaviour disorder. Ann Neurol. 1992;32:3–10.
27. Sansa G, Iranzo A, Santamaria J. Obstructive sleep apnea in narcolepsy. Sleep Med. 2009;11:93–5.
28. Allen RP. When, if ever, can we use REM-onset naps on the MSLT for the diagnosis of narcolepsy? Sleep Med. 2006;7:657–9.
29. Aldrich MS. The clinical spectrum of narcolepsy and idiopathic hypersomnia. Neurology. 1996;46: 393–401.
30. Marti I, Valko PO, Khatami R, Bassetti CL, Baumann CR. Multiple sleep latency measures in narcolepsy and behaviourally induced insufficient sleep syndrome. Sleep Med. 2009;10:1146–50.
31. Oka Y, Inoue Y, Kanbayashi T, et al. Narcolepsy without cataplexy: 2 subtypes based on CSF Hypocretin/orexin-A findings. Sleep. 2006;29:1439–43.
32. Thannickal TC, Lai YY, Siegel JM. Hypocretin (orexin) cell loss in Parkinson's disease. Brain. 2007;130:1586–95.
33. Fronczek R, Overeem S, Lee SY, et al. Hypocretin (orexin) loss in Parkinson's disease. Brain. 2007;130: 1577–85.
34. Compta Y, Santamaria J, Ratti L, et al. Excessive daytime sleepiness, cerebrospinal fluid hypocretin-1 and sleep architecture in Parkinson's disease with dementia. Brain. 2009;132:3308–17.
35. Thannickal TC, Nienhuis R, Siegel JM. Localized loss of hypocretin (orexin) cells in narcolepsy without cataplexy. Sleep. 2009;32:993–8.

The Arguments for Standardized Diagnostic Procedures

Geert Mayer

Keywords

Diagnosis • Differential diagnosis • Polysomnography • Actigraphy • Multiple sleep • Latency test • Narcolepsy • Orexin • Hypocretin

The diagnosis of full-blown narcolepsy with cataplexy can be established by medical history (see also the chapter "Current Diagnostic Criteria of Adult Narcolepsy" by Iranzo). There are very few diseases with the symptom cataplexy. If so – as in Coffin–Lowry syndrome – they can easily be identified by their course or the presence of other key symptoms. Mimics of cataplexy are only few. Isolated cataplexy (without narcolepsy) is very rare. However, many of our narcolepsy patients do not report the simple type of emotionally triggered cataplexy. They either grew up with the symptom and considered it as normal, or they thought is was physiological. If nobody in the patient's environment has seen the cataplexies, or if cataplexy attacks are rare, we run into trouble with our diagnosis. The same applies for patients who report excessive daytime sleepiness (EDS) and maybe one or two accessory symptoms (such as sleep paralysis or hypnagogic hallucination), or those who just report associated sleep disorders such as fragmented nocturnal sleep. Excessive daytime sleepiness is an unspecific symptom of all types of sleep–wake and other disorders (e.g., Parkinson's disease, Niemann-Pick disease, and hypothyreosis). Even experienced clinicians have diagnostic problems when the medical history is diffuse. Problematic are particularly patients who report symptoms which meet the clinical criteria for narcolepsy without cataplexy. And finally, what to do with patients who could have secondary forms of narcolepsy? How are we to understand the underlying pathology if we do not have standardized diagnostic procedures?

Moreover, in a world of easy access to special knowledge, we are frequently confronted with patients who report symptoms they have read about and that they attribute to their complaints. And, last but not least, there are also malingerers.

From the history of narcolepsy, we had to learn painfully that many of the symptoms that we thought were clearly defined had to be redefined, and many of the disorders that were named narcolepsy turned out to be distinct sleep–wake disorders. Because of the uncertainty and changes of definitions, it is difficult to compare earlier scientific results with present studies. Therefore, standard procedures would allow us to identify diseases and their phenotypes based on clear, defined criteria. The ICSD-2 gives the strongest

G. Mayer (✉)
Hephata Klinik, Schimmelpfengstrasse 6, 34613 Schwalmstadt, Germany
e-mail: geert.mayer@hephata.com

C.R. Baumann et al. (eds.), *Narcolepsy: Pathophysiology, Diagnosis, and Treatment*, DOI 10.1007/978-1-4419-8390-9_35, © Springer Science+Business Media, LLC 2011

argument in the definition of diagnostic criteria: "The hypersomnia is not better explained by another sleep disorder, medical or neurological disorder, mental disorder, medication use, or substance use disorder." However, there remain problems with these criteria, see also the chapter "Current Diagnostic Criteria of Adult Narcolepsy" by Iranzo.

Looking at the literature before the year 2000, there were only few studies in larger populations of narcolepsy patients that described detailed narcoleptic symptoms. Many studies were based on questionnaires, such as the Stanford Narcolepsy Questionnaire [1]. Some representative epidemiologic studies in the USA [2] may find similar prevalences as other well-designed studies from Europe [3], but it remains unclear what type of narcolepsy is reported. Also, because of the lack of specific sleep laboratory tests, it remains unclear what phenotype is reported in these studies. A study on the socioeconomic situation of narcoleptic patients in Denmark revealed that only a few narcoleptic patients had undergone standard diagnostic procedures (personal communication by P. Jennum/Copenhagen).

Hong et al. [4] tested the reliability of ICSD-2 criteria and the utility of the Stanford Sleep Inventory, polysomnography (PSG), multiple sleep latency test (MSLT), and cerebrospinal fluid (CSF) hypocretin-1 (hcrt-1) in a large number of patients with unexplained EDS. In this study, the diagnosis was easy in patients with narcolepsy with cataplexy. The authors found, however, that many of the other EDS patients including controls reported cataplexy-like events. The correct classification (i.e., narcolepsy or not?) of other patients with EDS is summarized by the authors: "Many patients without cataplexy were difficult to classify because of difficulties in interpreting the MSLT in the presence of sleep apnea or reduced sleep."

Recently, a large American cohort study [5] revealed MSLT features of narcolepsy in a significant proportion of subjects who never had a diagnosis of narcolepsy. This raises the question whether narcolepsy is underdiagnosed, or whether the methods and their interpretations are incorrect. There are only few studies that addressed the validity of the methods, especially of the MSLT. Geisler et al. [6] found one or more sleep onset REM periods (SOREMPs) in MSLT recordings of healthy subjects, patients with obstructive sleep apnea, and those with other disorders. Marti et al. [7] described more than two SOREMPs in patients with chronic sleep deprivation. These findings put our present diagnostic criteria to the test.

A practice parameter publication by Littner et al. [8] reviewed 13 studies (four studies with 39 narcolepsy patients and 40 controls, and nine studies with 255 narcolepsy patients without controls). The analysis showed significantly shortened sleep latencies ($p<0.001$) in polysomnographies of narcolepsy patients. Occurrence of two or more SOREMPs in the MSLT had a sensitivity of 0.78 and a specificity of 0.93. Mean sleep latencies of 8 or less minutes and two SOREMPs are the diagnostic criteria for narcolepsy as defined by ICSD-2 [9, 10]. However, there are narcolepsy patients without SOREMPs, and on the contrary, multiple SOREMPs sometimes occur in healthy sleep-deprived people. Therefore, the diagnosis of narcolepsy always has to take clinical symptoms into account.

HLA Association

In all, 85–95% of all narcolepsy patients are positive for the HLA allele DQB1*0602 [11]. The haplotype HLA DRB1*1501/DQB1*0602 and a pathologically low CSF hypocretin-1 level are present in 92% of narcolepsy patients with cataplexy [4]. However, HLA is neither specific nor sensitive for the diagnosis of narcolepsy.

HLA can be relevant for genetic counselling. In homocygeous narcolepsy patients, the risk of narcolepsy due to the allele increases two- to fourfold. The occurrence of the allele is independent from ethnicity. In heterocygeous narcolepsy patients, the risk increases with DQB1*0301, whereas it decreases with DQB1*01501 and DQB1*0601. HLA DQB1*0602 is positive in about 40–60% of a reference population of patients without narcolepsy and in 75% of familiar cases [12], making up for low specificity. Therefore, for routine diagnosis of narcolepsy, HLA testing is obsolete. However, it is important for genetic studies and the classification of different phenotypes of narcolepsy.

CSF Hypocretin-1

Low CSF hypocretin-1 (Stanford references: <110 pg/ml), defined as a third of mean control values, has a sensitivity of 86% and specificity of 95% [13] for the diagnosis of narcolepsy with cataplexy. Recent studies suggest a higher number of patients with narcolepsy with cataplexy who have CSF hypocretin-1 within the normal range [14]. Furthermore, narcolepsy patients with low CSF-hcrt-1 levels are younger at disease onset [15]. In general, narcolepsy patients with low CSF hypocretin-1 levels are HLA DQB1*0602 positive [16].

These HLA findings may us allow to discriminate several phenotypes of narcolepsy. (Costly) HLA typing, therefore, is important for classification purposes in scientific studies, for patients who do not have typical narcolepsy symptoms, or for those who display atypical sleep laboratory findings.

Vigilance Tests

Within the last years, these tests became more important for legal questions such as fitness to drive (see also the chapter "Fitness to Drive in Narcolepsy" by Mathis) and professional or educational performance. Tests such as the McWorth Clock have been used to assess the efficacy of stimulants [17]. Other neuropsychological tests have been developed and show good sensitivity and specificity.

Fronczek et al. used a new test for narcolepsy patients, the sustained attention to response task (SART) in combination with the Epworth Sleepiness Scale [18]. The authors applied the test prior to MSLT naps. Sensitivity and specificity of the MSLT for the diagnosis of narcolepsy were 80 and 100%, respectively, using a cut-off point of 5 min. For the SART, the respective values were 87 and 100%, using a five-error cut-off. The SART, measuring attention, was abnormal as often as the MSLT, measuring sleepiness. However, the SART and MSLT showed no correlation with each other or with the Epworth Sleepiness Scale.

There are many other tests that can be used to assess different forms of vigilance, attention, wakefulness, and sleepiness. The choice of method depends on the diagnostic question. Most questions cannot be answered with one test only. For many patients, it is best to use a combination of selected tests whose results have to be interpreted in the light of clinical symptoms.

Questionnaires

The Epworth Sleepiness Scale is the most frequently used self-administered questionnaire to evaluate daytime sleepiness in the past weeks [19]. A score above eleven indicates excessive daytime sleepiness [20, 21]. The scale has been validated for narcolepsy. In contrast to the Epworth Sleepiness Scale, the Stanford Sleepiness Scale asks for the actual feeling of being sleepy [22]. It is not specific for narcolepsy.

The Pittsburg Sleep Quality Index (PSQI) [23] consists of questions evaluating the quality of sleep within the past 4 weeks. It comprises 18 self-rating questions, and five questions that have to be filled in by a partner (if present). The components are subjective sleep quality, sleep latency, sleep duration, sleep efficiency, sleep disorders, consumption of hypnotics, and daytime sleepiness. An empirical cut-off score allows discriminating "bad" from "good" sleepers. However, the PSQI is not suited for differential diagnosis purposes.

The Ullanlinna Narcolepsy Scale is a validated instrument to distinguish between narcolepsy, sleep apnea, multiple sclerosis, and epilepsy [3]. The Swiss Narcolepsy Scale has a similar sensitivity (96 versus 98%), but a higher specificity (98 versus 56%) than the Ullanlinna Narcolepsy Scale [24]. The Stanford Narcolepsy Scale is a validated narcolepsy questionnaire (146 questions) focusing on cataplexy. A disadvantage of the questionnaire is the lack of severity scales for the major symptoms [25].

The Sleep-EVAL [26] is a knowledge-based expert system that would allow diagnosing virtually all sleep disorders. The system is based on the interviewee's response rather than on the interviewer's judgment. This instrument has been used to assess the prevalence of narcolepsy [2]. However, it must be assumed that there is a significant number of false-positive diagnoses. The SKID, finally, is a validated questionnaire based on DSM-III-R with a high sensitivity and specificity. It allows to follow a decision tree via systematic exploration of symptoms that ends up in a correct diagnosis [27].

Sleep Laboratory Tests

The MSLT measures the time required to fall asleep according to the instruction to try to fall asleep [9]. The test is carried out in a comfortable bed in a dark, quiet room, while standard EEG is recorded. The test is repeated four to five times during a day [28, 29]. The mean value of the sleep latencies of all naps is the mean sleep latency. Excessive daytime sleepiness is classified severe when the score is 5 min or less, moderate with a score of 5–10 min, and normal with a score of 10–20 min [9]. In a normal population, up to 6% may have sleep latencies of less than 5 min, and 34% of less than 10 min [6], which makes the specificity of low mean sleep latencies for narcolepsy questionable.

Today, one of the main indications of the MSLT is the assessment of sleep onset REM periods (SOREMPs) in suspected narcolepsy patients. In a few narcolepsy cases, multiple SOREMPs cannot be found on MSLT [30]. The presence of two or more SOREMPs has a sensitivity of 0.78 and a specificity of 0.93 for narcolepsy.

Maintenance of Wakefulness Test

When performing a maintenance of wakefulness test (MWT), the patient is instructed to stay awake as long as possible in a semi-recumbent position in a darkened room. There are four tests which last 20 or 40 min [29, 31]. A short MWT latency identifies abnormally low sleep resistance.

Reliability of ESS, MSLT, and MWT

A test–retest trial of the ESS in 104 healthy medical students was carried out with a 5-month interval [20]. The authors found a high Pearson correlation of $r=0.822$ between the two time points. In a second test–retest trial of the ESS in 56 healthy subjects, paired scores did not differ or differed only by one point in 54%, by two points in 20%, by three or four points in 23%, and by five points in 4% of the sample [32]. There are no correlations between ESS and MSLT [6].

The test–retest reliability of the MSLT has been examined in a study of 14 healthy men [33]. The MSLT was repeated 4–14 months after the first test. The Pearson correlation was as high as 0.97. The MWT has not been submitted to reliability testing as of yet.

Nocturnal polysomnography, preferably with video, is indicated for the diagnosis and differential diagnosis of primary sleep–wake disorders with excessive daytime sleepiness and parasomnias. Videometry is important for the precise characterization of movements or epileptic activity during sleep. As some comorbid disorders such as parasomnias and periodic limb movements during sleep are more prevalent in narcolepsy, it may be helpful, but not essential, to record electromyographic activity from more than one limb muscle [34].

Actigraphy

Actigraphy is used to evaluate rest–activity, i.e., sleep–wake, patterns over a prolonged period of time. A motion-sensitive device is attached to the wrist of the nondominant hand. Each movement is logged on a time dimension. The result is a time-based trend of movements from which sequences of relative activity versus rest can be inferred. These may be related to wakefulness and sleep, respectively. Algorithms to estimate sleep–wake patterns have been developed and confirmed against polysomnography. The sleep–wake rhythm under everyday life conditions may be monitored during several weeks. This method is helpful to verify sleep logs in insomnia or circadian rhythm disorders, and to rule out chronic sleep deprivation. Sleep deprived patients often present with low mean sleep latencies and occasionally with multiple SOREMPs on MSLT [7].

Imaging

MRI scans are necessary to rule out secondary forms of narcolepsy. To assess medication effects, measurement of regional blood flow can be useful [35]. Up to now, voxel-based morphometric investigations have revealed conflicting results [36–38]. Specialized functional imaging studies will be relevant to understand the pathophysiology of narcolepsy [39], but are not everyday clinical routine.

Summary

In summary, standardized diagnostic procedures are necessary for carrying out the following:

- Rule out all forms of symptomatic narcolepsy.
- Rule out all disorders and diseases that can cause single symptoms or an array of symptoms that also occur in narcolepsy.
- Identify different subtypes of narcolepsy.
- Understand the pathology and pathophysiology of narcolepsy.
- Assess legal requirements in education, profession, health care, insurance, drivers license, etc.

In general, it may make sense to combine several tests because single tests are specific for just one or two narcolepsy symptoms among many. There is ongoing need to improve our diagnostic tools, because many of them are unspecific and not sensitive enough to establish unequivocal diagnosis. Therefore, detailed medical history, clinical examinations, and regular follow-ups are still the most important diagnostic tools. Home videos of cataplexy and other narcolepsy-associated behavior may provide further important information.

References

1. Okun ML, Lin L, Pelin Z, Hong S, Mignot E. Clinical aspects of narcolepsy-cataplexy across ethnic groups. Sleep. 2002;25:27–35.
2. Ohayon MM, Priest RG, Zulley J, Smirne S, Paiva T. Prevalence of narcolepsy symptomatology and diagnosis in the European general population. Neurology. 2002;58:1826–33.
3. Hublin C, Kaprio J, Partinen M, et al. The Ullanlinna Narcolepsy Scale: validation of a measure of symptoms in the narcoleptic syndrome. J Sleep Res. 1994;3:52–9.
4. Hong S-C, Lin L, Jeong J-H, et al. A study of the diagnostic utility of HLA typing, CSF hypocretin-1 measurements, and MSLT testing for the diagnosis of narcolepsy in 163 Korean patients with unexplained excessive daytime sleepiness. Sleep. 2006;29(11):1429–38.
5. Mignot E, Lin L, Finn L, Lopes C, Pluff K, Sundstrom ML, et al. Correlates of sleep-onset REM periods during the Multiple Sleep Latency Test in community adults. Brain. 2006;129:1609–23.
6. Geisler P, Tracik F, Crönlein T, et al. The influence of age and sex on sleep latency in the MSLT-30 – a normative study. Sleep. 2006;29(5):687–92.
7. Marti I, Valko PO, Khatami R, Bassetti CL, Baumann CR. Multiple sleep latency measures in narcolepsy and behaviourally induced insufficient sleep syndrome. Sleep Med. 2009;10(10):1146–50.
8. Littner M, Kushida C, Wise M, Davila D, Morgenthaler T, et al. Practice parameters for clinical use of the multiple sleep latency test and the maintenance of wakefulness test. Sleep. 2005;28:113–21.
9. American Academy of Sleep Medicine. International classification of sleep disorders, Diagnostic and coding manual. 2nd ed. Westchester, IL: American Academy of Sleep Medicine; 2005.
10. Overeem S, Mignot E, van Dijk JG, et al. Narcolepsy: clinical features, new pathophysiologic insight and future perspectives. J Clin Neurophys. 2001;18:78–105.
11. Mignot E, Hayduk R, Black J, Grumet FC, Guilleminault C. HLA DQB1*0602 is associated with cataplexy in 509 narcoleptic patients. Sleep. 1997;20:1012–20.
12. Mignot E, Lin L, Rogers W, et al. Complex HLA-DR and -DQ interactions confer risk of narcolepsy-cataplexy in three ethnic groups. Am J Hum Genet. 2001;68:686–99.
13. Mignot E, Lammers GJ, Ripley B, et al. The role of cerebrospinal fluid hypocretin measurement in the diagnosis of narcolepsy and other hypersomnias. Arch Neurol. 2002;59:1553–62.
14. Heier MS, Evsiukova T, Vilming S, Gjerstad MD, Schrader H, Gautvik K. CSF hypocretin-1 levels and clinical profiles in narcolepsy and idiopathic CNS hypersomnia in Norway. Sleep. 2007;30:969–73.

15. Oka Y, Inoue Y, Kanbayashi T, et al. Narcolepsy without cataplexy: 2 subtypes based on CSF hypocretin-1/orexin-A findings. Sleep. 2006;29(11):1439–43.
16. Kanbayashi T, Inoue Y, Chiba S, et al. CSF hypocretin-1 (orexin-A) concentrations in narcolepsy with and without cataplexy and idiopathic hypersomnia. J Sleep Res. 2002;11:91–3.
17. Meier-Ewert K, Wismans L, Benter L. Narcolepsy patients with predominating cataplexy have lower vigilance. 1984, 7th Congress of the European Sleep Research Society, München (Abstrakt)
18. Fronczek R, Middelkoop HAM, Van Dijk JG, et al. Focusing on vigilance instead of sleepiness in the assessment of narcolepsy: high sensitivity of the sustained attention to response task (SART). Sleep. 2006;29(2):187–91.
19. Johns MW. A new method for measuring daytime sleepiness: the Epworth sleepiness scale. Sleep. 1991;14:540–5.
20. Johns MW. Reliability and factor analysis of the Epworth Sleepiness Scale. Sleep. 1992;15:376–81.
21. Johns M, Hocking B. Daytime sleepiness and sleep habits of Australian workers. Sleep. 1997;20:844–9.
22. Hoddes E, Zarcone V, Smythe H, et al. Quantification of sleepiness: a new approach. Psychophysiology. 1973;10:431–6.
23. Buysse DJ, Reynolds III CF, Monk TH, Berman SR, Kupfer DJ. The Pittsburgh Sleep Quality Index: a new instrument for psychiatric practice and research. Psychiatry Res. 1989;28:193–213.
24. Sturzenegger C, Bassetti C. The clinical spectrum of narcolepsy with cataplexy: a reappraisal. J Sleep Res. 2004;13:395–406.
25. Anic-Labat S, Guilleminault C, Kraemer HC, Meehan J, Arrigoni J, Mignot E. Validation of a cataplexy questionnaire in 983 sleep-disorders patients. Sleep. 1999;22:77–87.
26. Ohayon MM. Sleep-EVAL, Knowledge Based System for the Diagnosis of Sleep and Mental Disorders, Copyright Office, Canadian Intellectual Property Office [English Finnish, French, German, Italian, Portuguese, and Spanish versions]. Ottawa, Ontario: Industry Canada; 1994.
27. Schramm E, Hohagen F, Graßhoff U, Berger M. Strukturiertes interview für Schlafstörungen nach DSM-III-R (SIS-D). Weinheim: Beltz PVU; 1991.
28. Carskadon MA, Dement WC, Mitler MM, Roth T, Westbrook PR, Keenan S. Guidelines for the multiple sleep latency test (MSLT): a standard measure of sleepiness. Sleep. 1986;9:519–24.
29. Littner MR, Kushida C, Wise M, Davila DG, Morgenthaler T, Lee-Chiong T, et al. Standards of Practice Committee of the American Academy of Sleep Medicine. Practice parameters for clinical use of the multiple sleep latency test and the maintenance of wakefulness test. Sleep. 2005;28:113–21.
30. Dauvilliers Y, Gosselin A, Paquet J, Touchon J, Billiard M, Montplaisir J. Eff ect of age on MSLT results in patients with narcolepsy-cataplexy. Neurology. 2004;62:46–50.
31. Mitler MM, Gujavarty KS, Browman CP. Maintenance of wakefulness test: a polysomnographic technique for evaluation treatment efficacy in patients with excessive somnolence. Electroencephalogr Clin Neurophysiol. 1982;53:658–61.
32. Chung KF. Use of the Epworth Sleepiness Scale in Chinese patients with obstructive sleep apnea and normal hospital employees. J Psychosom Res. 2000;49:367–72.
33. Zwyghuizen-Doorenbos A, Roehrs T, Schaefer M, Roth T. Test-retest reliability of the MSLT. Sleep. 1988;11:562–5.
34. Frauscher B, Iranzo A, Högl B, Casanova-Molla J, et al. Quantification of electromyographic activity during REM sleep in multiple muscles in REM sleep behavior disorder. Sleep. 2008;31(5):724–31.
35. Joo EY, Seo DW, Tae WS, Hong SB. Effect of modafinil on cerebral blood flow in narcolepsy patients. Sleep. 2008;31(6):868–73.
36. Brenneis C, Brandauer E, Frauscher B, Schocke M, Trieb T, Poewe W, et al. Voxel-based morphometry in narcolepsy. Sleep Med. 2005;6(6):531–6.
37. Kaufmann C, Schuld A, Pollmacher T, Auer DP. Reduced cortical gray matter in narcolepsy: preliminary findings with voxel-based morphometry. Neurology. 2002;58(12):1852–5.
38. Draganski B, Geisler P, Hajak G, Schuierer G, Bogdahn U, Winkler J, et al. Hypothalamic gray matter changes in narcoleptic patients. Nat Med. 2002;8(11):1186–8.
39. Chabas D, Habert MO, Maksud P, et al. Functional imaging of cataplexy during status cataplecticus. Sleep. 2006;30(2):153–6.

Fitness to Drive in Narcolepsy

Johannes Mathis

Keywords

Driving regulations • Alertness • Vigilance • Narcolepsy • Maintenance of wakefulness test • Motor vehicle accidents

Abbreviations

ADHD	Attention-deficit hyperactivity disorder
DADT	Divided attention driving test
DMV	Department of Motor Vehicle
EDS	Excessive daytime sleepiness
IHS	Idiopathic hypersomnia without prolonged sleep
MVA	Motor vehicle accident
MSLT	Multiple sleep latency test
MWT	Maintenance of wakefulness test
OSAS	Obstructive sleep apnoea (hypopnea) syndrome

Accident Risk: Data from Official Statistics and Scientific Studies

According to projections by the World Health Organisation (WHO), road traffic injuries will be the third leading cause of global burden of disease and injury by the year 2020 [1] – in 50% of these accidents, the driver will be at fault. The main causes of death- or injury-producing accidents are excessive speed, alcohol consumption, inattention, and excessive daytime sleepiness (EDS).

Based on scientific data [2, 3], EDS causes 10–20% of all those motor vehicle accidents (MVAs) caused by the driver himself. This figure surpasses the percentage of alcohol-induced MVAs. Unfortunately, the awareness of this important cause of MVAs is low, even in official statistics. As long as official police statistics of many countries reveal much lower percentages of EDS-induced MVA (around 2–4%) [4–6], authorities will not adequately judge the significance of this socioeconomical impact [7]. This explains why there are almost no effective yet possibly expensive countermeasures. In this regard, France is an exception: Already in 1991, French authorities reported that 23% of accidents were caused

J. Mathis (✉)
Sleep Disorders Centre and Department of Neurology, Inselspital, Bern University Hospital, and University of Bern, Switzerland
e-mail: johannes.mathis@insel.ch

C.R. Baumann et al. (eds.), *Narcolepsy: Pathophysiology, Diagnosis, and Treatment*,
DOI 10.1007/978-1-4419-8390-9_36,

by "fatigue or inexplicable disorders" [8]. As acknowledged in the few systematic reviews on this topic [9, 10], well-designed studies are sparse, but the scientific evidence on a high prevalence of sleepiness-induced MVAs is convincing enough to propose concrete countermeasures [3].

The majority of EDS-induced accidents are caused by young healthy male subjects due to sleep deprivation, whereas MVAs due to disease-related EDS is less frequent. Overall, less than 5% of all accidents are due to a medical condition [11], and in a population-based study, there was no increased MVA risk for drivers with any medical condition [12]. The list of diagnoses obtained by a group in French Toulouse [8] in 110 drivers who were referred for medicolegal evaluation after EDS-induced MVA included 37 drivers with sleep deprivation, five with insomnia, 18 were under benzodiazepines, 34 had obstructive sleep apnea syndrome (OSAS), 11 suffered from narcolepsy, and eight were diagnosed as "idiopathic" including nonorganic hypersomnia.

Risk of Accidents in Narcolepsy

Among the most difficult responsibilities of treating physicians is not only to diagnose and treat disease-related EDS, but also to counsel the individual patient about the fitness to drive.

For treated narcolepsy patients, it is not at all clarified whether the risk for MVA surpasses the "generally accepted risk" within a given society, but the risk of untreated narcoleptics may be greater than the *mean* risk in the same society [13].

1. Comparing the odds ratio of accident risk of any particular minority with the *mean* risk of the general population is not entirely fair, because many societies accept higher risks for particular subgroups. For instance, many countries allow drivers to drink small quantities of alcohol prior to driving. Furthermore, males have a 1.15-fold risk of MVA compared to females, and young drivers below 24 years of age have a 5.2-fold risk compared to those aged 40–60 years [13]. The rate of MVA of any type is doubled in winter months compared to summer months, but the rate of *fatal* MVA is doubled in summer months compared to the winter months [14]. Therefore, patient groups should be compared to the range of risk in other subgroups.
2. The risk for MVAs in patients should be calculated for both untreated and treated subjects. Many treated and well-advised patients will use efficacious coping strategies to reduce the risk of MVAs. It is, however, still not known how to deal with these coping strategies when calculating average MVA risks. Sleepy subjects may reduce the number of driven kilometers or miles per year due to safety considerations, or they may abandon driving at all. Should we include these subjects when we calculate the average risk, and should the risk be calculated per driven distance or per year? Because coping strategies will influence the rate of any other type of MVA, we must analyze not only the rate of EDS-related MVAs, but also the rate of accidents of any cause.
3. It has been stated many times that otherwise healthy but sleep-deprived drivers deny having fallen asleep at the wheel after an MVA, and according to expert opinion, this is due to the low awareness and poor memory of the subjects regarding the feeling of sleepiness [15]. If we accept that sleepiness usually cannot be remembered shortly after an accident, how then should drivers be able to report EDS reliably as the cause of an accident that happened years ago? It is conceivable that narcoleptics who are certainly aware of their increased EDS may remember this particular cause of accidents more easily than healthy controls, which causes a bias derogatory to narcoleptics in retrospective questionnaire studies.
4. It is important to realize that strict driving regulations may be contraproductive, because many patients will not seek medical help and will remain untreated if they fear a great risk of losing their driver's license.

The question if narcoleptics may have an increased risk while driving has been addressed by only a few small studies in mostly untreated patients. However, all the cited studies below should be carefully analyzed with respect to the concerns mentioned above.

In a questionnaire study by Bartels and Kusakcioglu, 16% of 105 narcolepsy patients reported having caused an MVA due to falling asleep, compared to 1% among 105 healthy controls. Falling asleep at the wheel (not necessarily with an MVA) was reported by 40% of the narcoleptics and by 7% of the controls. Treatment in narcolepsy patients resulted in relief of undue drowsiness in 58%, but the authors gave no information on whether the MVA happened before or under treatment, and whether accidents were caused by other reasons than EDS. Furthermore, ten narcolepsy patients who were not driving were excluded from the analyses.

Roger Broughton performed a questionnaire survey on life effects in 180 patients with narcolepsy and in 180 controls in Asia, Europe, and Northern America. He included questions on accidents at work, at home, and while driving due to narcoleptic symptoms [16]. Patients (68.7%) or controls (37.2%) who did not drive were excluded. This study revealed more frequent EDS-induced MVAs (36.8%) in narcoleptics compared to matched controls (5.3%). The results did not allow analyzing the frequency of accidents before and under treatment. A surprisingly high proportion of narcoleptics (29%) reported cataplexy while driving, but it was not possible to assess the rate of accidents due to cataplexy. Data were not presented separately for different countries and, therefore, the influence of various public transportation systems and society effects could not be elucidated.

In a later Canadian study, Broughton et al. compared driving habits of 60 *treated* narcolepsy patients with that of 60 epileptic patients [17] and found that 73% of the narcoleptic patients still drove, compared to 32% among epileptic patients and 80% among normal controls. In this study, 34% of narcoleptics and 2% of controls reported lifetime MVAs, 64% reported frequent episodes with a concrete risk of an accident (controls 2%), and 72% reported falling asleep at the wheel (controls 13%). These lifetime events, however, included also MVAs prior to treatment.

Michael Aldrich reviewed retrospective clinical data of 25 narcoleptics [18] who were referred to a sleep disorder center over a 3-year period, and he excluded patients who did not drive. The age- and gender-matched control group of 279 men and 145 women were medical center employees. In a standardized questionnaire, male drivers with narcolepsy reported an average number of 2.0 lifetime MVAs from *any cause* per subject compared to 2.2 MVAs in the control group. Female drivers with narcolepsy reported 1.4 MVAs per subject lifetime compared to 1.5 in control subjects. However, the average number of *EDS-induced* lifetime MVAs per subject was 1.0 in male narcoleptics compared to 0.11 in male controls, and 0.42 in female narcoleptics compared to 0.06 in female controls. A correlation between the number of EDS-induced accidents and the frequency of subjective feelings of sleepiness was established only for the control group, and not for narcoleptics. Within each group, there was no significant difference in the mean sleep latency obtained in a multiple sleep latency test (MSLT) between subjects with EDS-related MVAs and those without. Unfortunately, it was not discussed whether the reported MVAs in patients happened before or during therapy, but since the questionnaire was filled out "at the initial visit," most probably data from periods without treatment were included as well.

In 1998, Michael Aldrich presented the impressive case report of a 65-year-old salesman who had been driving crash free 700,000 miles despite EDS since high school and a diagnosis of narcolepsy–cataplexy at age 35 years (oral communication). This case demonstrates that some narcoleptics are able to apply the most effective coping strategies despite severe EDS. Aldrich speculated that "awareness of sleepiness and good judgment concerning driving while drowsy are probably more important determinants of the ability to drive safely than the 'etiquette' of a narcolepsy diagnosis."

Findley et al. analyzed driving records obtained from the Department of Motor Vehicles (DMV) of the Commonwealth of Virginia (USA) from a group of 58 patients with untreated OSAS and from nine patients with untreated narcolepsy who were licensed as drivers [19]. The main focus of this study was to assess vigilance by using a simple driving simulator (Steer Clear). In fact, the authors were able to show that patients showing higher error rates in the Steer Clear had higher accident rates in the preceding years, while the mean latency in MSLT did not predict the accident risk. According to the performance in the Steer Clear, patients were arbitrarily divided into three groups. Patients with error rates above 4.5% had 0.38 accidents within the 5-year period, which was significantly more than that in the general population of Virginia with 0.18 accidents within 5 years. Since the nine narcoleptics had a mean error rate of 7%, they were most probably part of this latter group of 22 patients with "very poor performance." Thus, a higher accident risk in narcolepsy can be derived from this study.

In a questionnaire study on quality of life in 137 patients with narcolepsy and idiopathic hypersomnia without prolonged sleep (IHS), Ozaki and colleagues included a question on automobile accidents or near-miss incidents while driving [20]. In all, 42% of patients who had no driver's license and those without "usual driving habits" were excluded from this analysis. At least one event during the preceding 5 years was reported by 75% of the 28 patients with narcolepsy–cataplexy, 50% of the 27 patients with narcolepsy without cataplexy, and by 50% of patients with IHS. An Epworth sleepiness scale score of 16 or more was linked to an increased risk of MVA, with an odds ratio of 12.0 compared to those with an Epworth score of 10 or less. The relative accident risk of patients with an Epworth score of 11–15 was 4.25, which did not reach statistical significance.

More recently, Kotterba et al. analyzed questionnaires from 100 narcoleptic patients (57 females) from the German patient association. The authors compared the answers with those of 95 healthy controls recruited from the hospital staff and among medical students [21]. Accident rates per subject were calculated separately for both genders, in three age groups: 20–30, 30–40, and >40 years, and separately for the period before diagnosis and after diagnosis. In this study, the authors found no difference in the rate of any type of MVA between narcoleptics and controls. Most probably due to efficient coping strategies, narcoleptics who were older than 40 years reported fewer accidents than healthy controls. Diagnosis had a positive effect on accident rate only in female narcoleptics above the age of 30 years, but not in the young female group or in male narcoleptics. In the youngest group below 30 years, the accident rate per subject increased after diagnosis in males and females. However, this comparison of rates before and after diagnosis is problematic, because the examined time intervals are unknown and no statistical analysis was presented.

In summary, the risk of lifelong MVA in narcolepsy patients is probably increased compared to the *mean risk* of a given society. However, under optimal treatment and by using effective coping strategies, the mean risk related to a given time period (e.g., a year) is probably within the so-called *acceptable risk* range. To gain better insights, larger and specifically designed studies are warranted. The description of narcoleptics without any MVA during a whole life of driving underlines the importance of not judging the fitness to drive simply based on the diagnostic etiquette, but corroborates the need of sensitive and specific assessment batteries in order to identify patients with EDS of any cause with unacceptably high risk of MVA. Unfortunately, the optimal test battery to assess fitness to drive in sleepy patients is not yet defined.

Assessing Fitness to Drive

Some experts have tried to judge the fitness to drive in narcoleptics based on the performance in a *driving simulator*. It is acknowledged that *on-the-road studies* have greater face validity; however, they have other disadvantages such as the effect of the supervisor, the smaller range of road situations, and the significant safety considerations [22]. The aim of ongoing research is to show correlations between on-the-road tests and laboratory assessments [23], but not necessarily equality between the two. As already shown for elderly people, strategic and tactical compensation mechanisms may be more relevant in preventing accidents than any directly measured performance capacity during a test or on-the-road study [24]. However, the use of these very efficient coping strategies is not allowed in most test situations. Narcoleptic patients have adopted compensation strategies such as having a nap before driving, reducing driving time, not driving at night, or taking stimulants before driving or when feeling sleepy. It seems at best unfair to test their driving ability without allowing them to use their individual compensation strategies.

Findley et al. have used a simple computer-based reaction time test called "Steer Clear" which was designed as a two-lane street with a car [19]. The imposed task was to avoid hitting obstacles appearing in either lane corresponding to a "go or no-go" paradigm. They found a significant increase in the error rate in ten untreated narcoleptics and in 62 patients with OSAS, compared to that in age- and gender-matched controls in this 30-min reaction time test. Within the patient group, poorer performance on the test was associated with a higher rate of MVAs registered in the records of the DMV of the Commonwealth of Virginia of the previous 5 years. In particular, the group with a very poor performance in the Steer Clear test (>4.5% error rate) had a significantly greater rate of MVAs. In a second similar study [25], the investigators confirmed these results of the Steer Clear test and found a more pronounced time-on-task performance decrement in untreated narcoleptics compared to patients with OSAS, and in both patients groups compared to normal controls. No data were presented in either study on the accident rate in those under treatment.

George et al. have examined 16 untreated narcoleptics in possession of a driver's license with a 20-min divided attention driving test (DADT), and compared the results to 21 patients with OSAS and 21 normal controls [22]. Using a steering wheel, the subject must keep a cross on the PC monitor within a laterally moving rectangle while the distance between the cross and rectangle (= tracking error) is continuously measured, and its standard deviation is used to assess performance. The secondary task is to respond adequately to signals presented in the visual field periphery. On average, both patient groups performed worse than the control group, but half of the patients were within the range of the normal controls. During the course of the 20-min test, the tracking error increased steadily evidencing a fatigue effect, while normal controls had a constant performance throughout the test. There was no relationship between the apnea–hypopnea index and the DADT performance in OSAS, and only a weak relationship was found between the MSLT latency and performance ($r=0.34$) in narcoleptics.

Kotterba and colleagues have tested a small number of narcolepsy patients in a realistic driving simulator [26]. During 60 min, a condition of driving a monotonous highway was presented, interrupted by crossing deer, pedestrians, and other vehicles. Furthermore, the effects of speeding, slowing down, and aquaplaning were visually simulated. The mean accident rate (crashes with obstacles and driving off the road) was higher in the 13 mostly untreated (8 of 13) narcoleptics (3.2±1.8) compared to ten healthy controls (1.3±1.5; $p<0.01$). All participants had been active drivers for 2 years or more, and they used their cars at least 4 days per week. The worst performance in patients was observed under ideal daylight weather conditions, as it was described for sleepy healthy subjects after sleep deprivation. After treatment optimization in five narcolepsy patients, the mean accident number in this test decreased significantly from 4.4±1.5 to 2.6±1.1.

Driving Regulations for Narcolepsy Patients

In 1995, Pakola et al. reviewed regulations and guidelines for commercial and noncommercial drivers with sleep apnea and narcolepsy in USA, Canada, Australia, and several European countries [27]. Driving regulations in patients with EDS were reviewed in two publications focusing mainly on driving regulations for sleep apnea syndrome in various European countries [28, 29].

It is not at all known whether specific driving regulations for patient groups with EDS or specific diagnoses are effective in reducing road accidents. In many countries, nevertheless, driving regulations exist for private and professional drivers, with great differences among the countries. National differences in law may account for the differences in the choice between individual freedom and the protection of the society against a potential risk.

Irrespective of the legal situation, the clinician has a responsibility to inform his patient on the driving risks related to EDS and to discourage him from driving when sleepy [28]. In some countries (e.g., Finland, Hungary, Poland, Spain, and the United Kingdom), it is the physician's responsibility to inform the authorities. In other countries, reporting to the authorities is allowed without breaking normal confidentiality rules. Categorical reporting, i.e., reporting to the authorities solely based on a diagnosis – independent of the severity of EDS and of any coping strategies – would prevent many patients to seek medical help. Untreated patients, however, impose the greatest risk of MVAs and therefore, categorical reporting is not recommended by expert commissions [30].

Specific regulations for narcolepsy were available in Great Britain, the Netherlands, Sweden, Spain, Belgium, France, Australia, and in most US states as assessed in 1995 [27]. In other countries, the regulations are formulated less specifically for "Disorders with excessive sleepiness or tiredness" or more generally "The driver has no established medical history or clinical diagnosis likely to interfere with his ability to control and drive a motor vehicle safely." In a more recent review [29] among 25 European countries, only nine specifically mentioned EDS (Belgium, Finland, France, Germany, Hungary, Spain, Sweden, and the United Kingdom), and ten mentioned sleep apnoea syndrome (the same, including Poland) as a disorder requiring a medical assessment of the driving ability. Others only mention a state of "mental or physical incapacity to drive" where EDS belongs to. It is beyond the scope of this review to list the slightly different regulations of various countries, many undergoing continuous adaptations. Harmonization on the European level was repetitively claimed, but until recently with minor success [29].

Specific narcolepsy-related regulations usually forbid a narcoleptic patient to work as a *professional* driver. For *private* driving, a minimal time of 2–12 months with "well-controlled" EDS under treatment with regular assessments every 6–12 months is proposed [27]. The stricter regulations for professional drivers were usually explained by greater exposure time. A more important argument is the pressure from companies to continue driving and save time despite EDS. The request to drive a longer distance within a defined time may prevent the use of efficient countermeasures and coping strategies such as ceasing to drive and having a nap or drinking coffee in a rest area [31]. Instead, less efficient countermeasures such as opening the window or listening to loud music while continuing to drive [32] are used.

If they are not allowed to work any more as drivers, professional drivers should probably receive disability payments [27]. This may raise significant socioeconomic problems for society and the affected.

Assessment of Fitness to Drive in Narcolepsy

The alternative to categorical reporting is reporting based on the severity of EDS under optimal treatment, on the assumption that the risk of an MVA is related to residual EDS. Unfortunately, this approach is compromised by several uncertainties.

First, a direct relationship between accident risk and EDS has not been proven yet, and the ideal test to assess EDS is still unknown. In my personal view, such a test will never be introduced, because EDS is a multidimensional symptom that can hardly be measured by a single test. Therefore, we should search the optimal test battery to answer a particular question such as fitness to drive or to work for a treated but still sleepy patient. As far as available studies on sleepy healthy sleep-deprived subjects [33] and on sleepy patients [23] allow conclusions, real driving, driving simulator tests, and maintenance of wakefulness test (MWT) in the context of a clinical investigation by a sleep disorders expert might be the optimal assessments for the judgment of the fitness to drive in narcolepsy and other sleepy patients.

However, it is not feasible to apply extensive and costly assessment batteries with sophisticated techniques in every EDS patient. Moreover, some of these tests are not yet standardized for use in every sleep disorders center.

The Swiss Society of Sleep Research, Sleep Medicine, and Chronobiology has proposed recommendations for treating physicians, and sleep disorder centers regarding how fitness to drive in EDS patients could be assessed. Notably, in Switzerland, treating physicians have the right but not the duty to report patients who are unfit to drive [34]. These recommendations are based on the assumption that subjective EDS while driving is always realized by the driver before causing an MVA. We and others believe, however, that this assumption is not necessarily correct [15, 35].

The Swiss recommendations introduced three steps of assessment. The first is performed by the general practitioner. His most important task – besides an accurate diagnosis and treatment – is to counsel any sleepy patient on his individual risk and on the consequences of an EDS-induced MVA, and to discourage the patient from driving when feeling subjective sleepiness. The physician should also counsel his patient on effective coping strategies such as to stop and nap or drink coffee, and warn the patient from applying less effective countermeasures such as opening the window or listening to loud music while continuing to drive.

Patients who have already caused an EDS-induced MVA, and all professional drivers and patients with presumed reckless behavior should be referred to a sleep disorder center. This is the second step of assessment. In the Sleep Disorders Centres, clinical examination, the assessment of an accurate diagnosis, and the installation of an optimal treatment should be completed before assessing the individual's fitness to drive. The latter can be done by performing an MWT, a driving simulator test, or real driving. According to the most recent review [29], France is the only European country in which a normal MWT is officially required for professional drivers. Since MWT is probably the most widespread and also the most standardized of these time-consuming assessments, this test shall be reflected here. Originally, the MWT was introduced in analogy to the rules used for the MSLT [36]. This implies that patients were often left untreated during the test, and no measures to improve wakefulness – such as regular naps or drinking coffee – were allowed. This applies also for the best available normal values [37, 38]. The motivation to stay awake in this test for a healthy test participant is not at all comparable to the situation of a driver who is at risk of losing his driver's license. For sleepy patients, the treatment can obviously not be stopped and simple measures to improve wakefulness should be allowed, especially since these same measures are recommended to be used while driving. Until very recently, the judgment of the MWT result represented an expert opinion. Many experts used a limit of >20 min in the MWT for private driving and >30 min for professional driving. Philip et al. recently compared the MWT results with performance in a 90-min real-life driving session. He and his team showed that untreated sleep apnea patients were at an increased accident risk when having an MWT latency below 20 min. There was, however, a normal risk for the group with latencies >30 min [39]. Interestingly, the group of patients with MWT latencies between 20 and 30 min had an equally poor performance similar to those with MWT latencies below 20 min. A prospective study on the real-life accident frequency in patients under treatment and, therefore, under continuing medical counseling and while using coping strategies is not yet available.

Ambiguous cases and those suspected of reckless driving should be reported to the authorities, and these patients are required to undergo a real driving test. It is important to understand, however, that real driving tests often last only about 1 h and do not expose the driver to situations particularly sensitive to the individual's impairment due to the disease. Patients with a major complaint of EDS should probably be exposed to a long-lasting driving session on a highway during the night, as demonstrated by the group of Philip in healthy sleep-deprived subjects [40]. Obviously, the alerting effect of the expert person in the car prevents a completely normal situation in real driving tests. Therefore, a single real driving test in isolation should *not* be considered as *the* gold standard test to judge fitness to drive, but should be used in combination with results from other tests.

Driving Under the Influence of Stimulants

EDS in narcolepsy is best treated by a combination of sufficient nighttime sleep and regular daytime naps [41], in combination with stimulants such as modafinil, methylphenidate, or amphetamines.

Taking stimulants while driving may be a problem in many countries, since it has been shown in drug abusers and also in healthy subjects that high-dose stimulants induce impaired attention, causing accidents [42]. Amphetamine abusers may show aggressive risk-taking and antisocial behaviour, extending to confusion, hallucinations, and psychotic episodes. We have observed a narcolepsy patient who developed a psychotic episode after he increased the dosage of modafinil to 1,000 mg per day on his own. However, in therapeutic dosages, modafinil and amphetamines do not increase the accident risk [43, 44]. More importantly, it could be shown that d-amphetamine in low dosages improved reaction times [45] and reduced impulsivity in healthy subjects, as we know it from ADHD patients [46].

Tunnel vision is a particularly well-recognized cause of accidents in both conditions: in EDS after sleep deprivation, and under the influence of stimulants. However, treating EDS with stimulants in healthy sleep-deprived subjects did not worsen the tunnel effect, but on contrary improved it [44].

To my best knowledge, there are no data available to show that treatment with stimulants reduces the real-life risk of MVAs in EDS patients. However, mean latency in MWT is significantly prolonged in narcolepsy under treatment with stimulants [47], and a driving simulator study showed improved performance [48]. In summary, in the absence of pertinent studies, it can be assumed that treatment of EDS with low to moderate dosages of stimulants reduces the risk of accidents. Sufficient nighttime sleep and regular daytime naps reduce EDS [41] and improve performance in narcolepsy, although often not to a normal level [49].

For the treating physician, informing the patient about the risk of driving under drugs is an important task. He should advise patients to refrain from driving at the start of a drug therapy or when increasing the dose. This is not only valid for stimulants, but also for sedatives, opioids, and other psychoactive drugs. Narcolepsy patients must take stimulants according to medical prescription, and must follow an optimal sleep hygiene including regular bedtimes, sufficient nighttime sleep, and regular naps as far as possible.

References

1. Peden M, Scurfiled R, Sleet D, Mohan D, Jayder A, Jarawan E. World Report on roaf traffic injury prevention. http://www.who.int/violence_injury_prevention/publications/road_traffic/world_report/summary_en_rev.pdf. 2004. World Health Organization.
2. Arbus L, Tiberge M, Serres A, Rouge D. Drowsiness and traffic accidents. Importance of diagnosis. Neurophysiol Clin. 1991;21:39–43.
3. Akerstedt T. Consensus statement: fatigue and accidents in transport operations. J Sleep Res. 2000;9(4):395.
4. Garbarino S, Nobili L, Beelke M, De Carli F, Ferrillo F. The contributing role of sleepiness in highway vehicle accidents. Sleep. 2001;24(2):203–6.
5. Horne JA, Reyner LA. Sleep related vehicle accidents. Br Med J. 1995;310:565–7.
6. Bundesamt für Statistik. Strassenverkehrsunfälle in der Schweiz 1996. Bern: BFS; 1998.
7. Leger D. The cost of sleep-related accidents: a report for the national commission on sleep disorders research. Sleep. 1994;17:84–93.
8. Arbus L, Tiberge M, Serres A, Rouge D. Somnolence et accidents de la circulation routière. Importance du diagnostic. J Neurophysiol Clin. 1991;21: 39–43.
9. Connor J, Whitlock G, Norton R, Jackson R. The role of driver sleepiness in car crashes: a systematic review of epidemiological studies. Accid Anal Prev. 2001;33(1):31–41.
10. Robb G, Sultana S, Ameratunga S, Jackson R. A systematic review of epidemiological studies investigating risk factors for work-related road traffic crashes and injuries. Inj Prev. 2008;14(1):51–8.
11. Brown BP. Medical conditions, medications, and driving. Can Fam Physician. 1998;44:705–6.

12. Guibert R, Duarte-Franco E, Ciampi A, Potvin L, Loiselle J, Philibert L. Medical conditions and the risk of motor vehicle crashes in men. Arch Fam Med. 1998;7(6):554–8.
13. Sonnen AEH. General principles of assessment and the concept of acceptable risk. Epilepsy and Driving, a European View; a workshop of the International Bureau for Epilepsy, European Association; Eds. Internat. Bureau for Epilepsy, Heemstede, 1997; 11–30.
14. Brown B, Baass K. Seasonal variation in frequencies and rates of highway accidents as a function of severity. Transp Res Rec. 1997;1581:59–65.
15. Horne JA, Baulk SD. Awareness of sleepiness when driving. Psychophysiology. 2004;41(1):161–5.
16. Broughton R, Ghanem Q, Hishikawa Y, Sugita Y, Nevsimalova S, Roth B. Life effects of narcolepsy in 180 patients from North America, Asia and Europe compared to matched controls. Can J Neurol Sci. 1981;8(4):299–304.
17. Broughton RJ, Guberman A, Roberts J. Comparison of the psychosocial effects of epilepsy and narcolepsy/cataplexy: a controlled study. Epilepsia. 1984;25(4):423–33.
18. Aldrich MS. Automobile accidents in patients with sleep disorders. Sleep. 1989;12:487–94.
19. Findley L, Unverzagt M, Guchu R, Fabrizio M, Buckner J, Suratt P. Vigilance and automobile accidents in patients with sleep apnea or narcolepsy. Chest. 1995;108:619–24.
20. Ozaki A, Inoue Y, Nakajima T, Hayashida K, Honda M, Komada Y, et al. Health-related quality of life among drug-naive patients with narcolepsy with cataplexy, narcolepsy without cataplexy, and idiopathic hypersomnia without long sleep time. J Clin Sleep Med. 2008;4(6):572–8.
21. Kotterba S, Müller N, Steiner G, Mayer G. Driving in narcolepsy – analysis by questionnaire among patients (article in German). Somnologie. 2002;6:39–50.
22. George CFG, Boudreau AC, Smiley A. Comparison of simulated driving performance in narcolepsy and sleep apnea patients. Sleep. 1996;19:711–7.
23. Sagaspe P, Taillard J, Chaumet G, Guilleminault C, Coste O, Moore N, et al. Maintenance of wakefulness test a predictor of driving performance in patients with untreated obstructive sleep Apnea. Sleep. 2007;30(3):327–30.
24. De Raedt R, Ponjaert-Kristoffersen I. Can strategic and tactical compensation reduce crash risk in older drivers? Age Ageing. 2000;29(6):517–21.
25. Findley LJ, Suratt PM, Dinges DF. Time-on-task decrements in "steer clear" performance of patients with sleep apnea and narcolepsy. Sleep. 1999;22(6):804–9.
26. Kotterba S, Müller N, Steiner G, Mayer G. Narkolepsie und Fahrtauglichkeit. Narcolepsy and driving. Akt Neurol. 2004;31:273–8.
27. Pakola SJ, Dinges DF, Pack AI. Driving and sleepiness review of regulations and guidelines for commercial and noncommercial drivers with sleep apnea and narcolepsy. Sleep. 1995;18:787–96.
28. Krieger J, McNicholas WT, Levy P, De Backer W, Douglas N, Marrone O, et al. Public health and medicolegal implications of sleep apnoea. Eur Respir J. 2002;20(6):1594–609.
29. Alonderis A, Barbe F, Bonsignore M, Calverley P, De Backer W, Diefenbach K, et al. Medico-legal implications of sleep apnoea syndrome: driving license regulations in Europe. Sleep Med. 2007;9(4):362–75.
30. Aldrich CK, Aldrich MS, Aldrich TK, Aldrich RF. Asleep at the wheel: the physician's role in preventing accidents 'just waiting to happen'. Postgrad Med J. 1986;80:233–40.
31. Horne JA, Reyner LA. Counteracting driver sleepiness: effects of napping, caffeine, and placebo. Psychophysiology. 1996;33(3):306–9.
32. Reyner LA, Horne JA. Evaluation of 'in-car' countermeasures to sleepiness: cold air and radio. Sleep. 1998;21(1):46–50.
33. Philip P, Sagaspe P, Taillard J, Valtat C, Moore N, Akerstedt T, et al. Fatigue, sleepiness, and performance in simulated versus real driving conditions. Sleep. 2005;28(12):1511–6.
34. Mathis J, Seeger R, Kehrer P, Wirtz G. Fahreignung bei Schläfrigkeit: Empfehlungen für Ärzte bei der Betreuung von Patienten mit vermehrter Schläfrigkeit. Schweiz Med Forum, 2007; 328–32.
35. Reyner LA, Horne JA. Falling asleep whilst driving: are drivers aware of prior sleepiness? Int J Leg Med. 1998;111(3):120–3.
36. Mitler MM, Gujavarty KS, Browman CP. Maintenance of wakefulness test: a polysomnographic technique for evaluating treatment efficacy in patients with excessive somnolence. Electroencephalogr Clin Neurophysiol. 1982;53:658–61.
37. Doghramji K, Mitler MM, Sangal RB, Shapiro C, Taylor S, Walsleben J, et al. A normative study of the maintenance of wakefulness test (MWT). Electroencephalogr Clin Neurophysiol. 1997;103(5):554–62.
38. Banks S, Barnes M, Tarquinio N, Pierce RJ, Lack LC, McEvoy RD. The maintenance of wakefulness test in normal healthy subjects. Sleep. 2004;27(4):799–802.
39. Philip P, Sagaspe P, Taillard J, Chaumet G, Bayon V, Coste O, et al. Maintenance of wakefulness test, obstructive sleep apnea syndrome, and driving risk. Ann Neurol. 2008;64(4):410–6.
40. Sagaspe P, Taillard J, Akerstedt T, Bayon V, Espie S, Chaumet G, et al. Extended driving impairs nocturnal driving performances. PLoS ONE. 2008;3(10):e3493.
41. Rogers AE, Aldrich MS, Lin X. A comparison of three different sleep schedules for reducing daytime sleepiness in narcolepsy. Sleep. 2001;24(4):385–91.
42. Gijerde H, Christophersen AS, Morland J. Amphetamine and drugged driving. J Traffic Med. 1992;20:21–6.
43. Hurst PM. Amphetamines and driving. Alcohol Drugs Driving. 1987;3:13–7.
44. Mills KC, Spruill SE, Kanne RW, Parkman KM, Zhang Y. The influence of stimulants, sedatives, and fatigue on tunnel vision: risk factors for driving and piloting. Hum Factors. 2001;43(2):310–27.

45. Halliday R, Naylor H, Brandeis D, Callaway E, Yano L, Herzig K. The effect of D-amphetamine, clonidine, and yohimbine on human information processing. Psychophysiology. 1994;31(4):331–7.
46. de Wit H, Enggasser JL, Richards JB. Acute administration of d-amphetamine decreases impulsivity in healthy volunteers. Neuropsychopharmacology. 2002;27(5):813–25.
47. Mitler MM, Hajdukovic R, Erman M, Koziol JA. Narcolepsy. J Clin Neurophysiol. 1990;7(1):93–118.
48. Mitler MM, Hajdukovic R, Erman MK. Treatment of narcolepsy with methamphetamine. Sleep. 1993;16(4):306–17.
49. Godbout R, Montplaisir J. All-day performance variations in normal and narcoleptic subjects. Sleep. 1986;9(1 Pt 2):200–4.

Part IX

Treatment of Narcolepsy

Treatment of Narcolepsy

G.J. Lammers

Keywords
Narcolepsy • Treatment • Sodium oxybate • Modafinil • Antidepressants • Methylphenidate

Introduction

The core problem of patients suffering from narcolepsy is their inability to remain fully alert or even awake during the day. Paradoxically, narcolepsy patients may also have difficulty remaining asleep during the night. In addition, the strict physiological boundaries of specific components of wake and sleep stages are loosened. This leads to partial isolated expressions, which particularly for REM sleep can result in symptoms such as cataplexy, hypnagogic hallucinations, and sleep paralysis [1, 2].

In human patients, loss of hypocretin is considered the major cause of narcolepsy. Hypocretin deficiency explains the narcoleptic phenotype, since selective loss of hypocretin in experimental animal models induces a similar phenotype [3, 4]. Additional animal studies revealed that the hypocretin system normally sustains wakefulness and orchestrates the proper execution and alternation of sleep-wake stages [5]. Hypocretin probably promotes wakefulness through its excitatory effects on brain regions implicated in arousal. There are particularly dense projections not only to histaminergic neurons of the tuberomammillary nucleus and noradrenergic neurons of the locus ceruleus, but also to dopaminergic neurons in the ventral tegmental area and serotonergic neurons of the dorsal raphe [6, 7]. Excitation of cholinergic neurons in the basal forebrain and the laterodorsal and pedunculopontine tegmental (LDT/PPT) nuclei may also help promote wakefulness [6].

A persistent cholinergic/aminergic imbalance may be required for the development of cataplexy [7, 8]. Monoaminergic nuclei are among the sites receiving the most dense hypocretin projections, and the cholinergic nuclei, to a lesser extent. Hypocretin deficiency may thus induce such an imbalance by reducing aminergic tone disproportionately [7]. In support of this idea, drugs that increase aminergic tone, particularly of the adrenergic system, can reduce cataplexy, perhaps by restoring this balance [6].

Targeting the underlying cause or prevention is the ultimate goal with every disorder. Unfortunately, narcolepsy cannot be cured, and

G.J. Lammers (✉)
Department of Neurology, Leiden University Medical Center, PO Box 9600, 2300 RC Leiden, The Netherlands
e-mail: G.J.Lammers@lumc.nl

C.R. Baumann et al. (eds.), *Narcolepsy: Pathophysiology, Diagnosis, and Treatment*,
DOI 10.1007/978-1-4419-8390-9_37,

hypocretin substitution is not yet practical as the hypocretin peptides do not easily cross the blood–brain barrier. Alternative routes, such as nasal application, have been tried in animal studies. It has been shown that hypocretin enters the brain via the nasal route, but only one study suggests a functional benefit [9]. Therefore, it seems unlikely that this will become a clinical option in the near future. Alternatively, hypocretin agonists that cross the blood–brain barrier are worth considering, but not much progress has been made so far in the development of selective agonists.

Clinical management thus relies on symptomatic treatments. Two treatment modalities have proven to be effective: behavioral modification and pharmacological therapy. As a rule, both are needed to achieve success.

Behavioral Modification

It is very important that patients learn to accept the diagnosis and its consequences. This highly facilitates the implementation of behavioral modifications and decreases the burden of the disease. A supportive social environment (e.g., patient group organizations and support groups) and education of family, schools, and employers by patients and clinicians all improve acceptance.

The most important behavioral advice includes the recommendation to live a regular life, going to bed at the same hour each night, and getting up at the same time each morning. Scheduled daytime naps or short naps just before certain activities demanding a high degree of attention temporarily improve sleepiness in most patients. The optimal frequency, duration, and timing of these naps have to be established on an individual basis [10].

Because narcoleptic patients are probably more sensitive to the sleep-inducing properties of carbohydrates, they should not eat large carbohydrate-rich meals [11]. Alcohol should preferably be avoided for similar reasons.

Despite such measures, there will remain residual complaints in the majority of patients, requiring adjuvant pharmacological treatment.

Pharmacological Treatment

A variety of medications may be used in the treatment of narcolepsy, and no single drug or combination of drugs is effective in all. As most treatments predominantly improve either excessive daytime sleepiness (EDS) or cataplexy, combinations are often needed to control both symptoms. The only available drug that may improve all major symptoms of narcolepsy is sodium oxybate. Because it is a hypnotic, it has the additional advantage that it may improve nocturnal sleep, in contrast to the other drugs.

Before discussing individual drugs, it is important to discuss what may be expected from pharmacological treatment, what the goals should be, and what the arguments are to choose a certain drug or dose. Although randomized, double-blind, placebo-controlled studies are indispensable in the struggle to optimize treatment for any disorder, they are not always helpful for treatment decisions in individual patients.

How to Choose a Drug for an Individual Patient

The treatment goal is of major importance in the judgment of the efficacy of treatment. Sleepiness will never be completely alleviated in any patient, whereas cataplexy may completely disappear in some. Patients must be made aware of this, and this knowledge must guide physicians in trying new drugs or combinations of drugs and in deciding on the right balance between efficacy and side effects.

Ideally, treatment decisions should be driven by generally accepted objective tests to quantify the severity and the individual impact of a symptom. Unfortunately, there is no such test for narcolepsy as a whole or for its various symptoms. Sleepiness can be assessed with subjective and objective tests, but none is generally accepted as a valid indicator of daytime functioning [12]. In fact, it is uncertain whether impaired concentration while awake or sleeping in the daytime is the more important symptom. If inattentiveness is the major problem, then vigilance tests may be more appropriate than sleep tests [13]. Nocturnal sleep, cataplexy, hypnagogic hallucinations, and sleep paralysis all present similar assessment problems. In addition, cataplexy is a challenge to quantify as it varies in frequency, duration, the number of muscles involved, and behavioral consequences, such as avoidance of situations in which attacks may occur. Thus, in the absence of objective tests, history taking is the main instrument to evaluate efficacy and the occurrence of side effects.

Applying the results of clinical trials to individual patients can be difficult. Drug efficacy as assessed in groups is of relatively little importance for individuals because individual differences in efficacy, side effects, and tolerability can be large. Pharmacokinetic aspects (i.e., rapid-onset, fast-acting versus slow-onset, long-acting medications) may be more important than the expected efficacy. The interpretation of pharmacological trials is also hampered by the lack of well-designed studies of older drugs, by a shortage of studies comparing different substances, and by the lack of reliable and relevant objective outcome measures.

Thus, treatment decisions for an individual patient should take into account all these considerations. Moreover, it is best to start with one drug at a time, targeting the most disabling symptoms, and explain that it sometimes takes months before an optimal situation is reached. Last, it is worth remembering that some combinations of drugs such as modafinil and sodium oxybate may have synergistic effects.

Treatment of EDS

Stimulants are the main treatment for EDS [14, 15]. These include dextroamphetamine (5–60 mg/day), methamphetamine (10–50 mg/day), methylphenidate (10–60 mg/day), modafinil (100–400 mg/day), and mazindol (1–6 mg/day). They enhance the release and inhibit the reuptake of catecholamines and to a lesser extent of serotonin in the central nervous system and the periphery [16]. They are also weak inhibitors of monoamine oxidase (MAO). Mazindol has been withdrawn in most countries due to severe side effects (including pulmonary hypertension and valvular regurgitation) with related drugs that suppress appetite, in particular fenfluramines [17]. As some patients respond better to mazindol than any other drug, it may remain an option, provided treatment is closely monitored.

Caffeine may alleviate sleepiness but has a weak effect: the alerting effect of six cups of strong coffee is comparable with that of 5 mg of dexamphetamine [16, 18].

Side effects and tolerance are major drawbacks in the use of stimulants. The most important side effects include irritability, agitation, headache, and peripheral sympathetic stimulation. These are usually dose related. Tolerance develops in about a third of patients, leading to use of high dosages [16, 18]. Some patients tend to increase their dosage because they prefer high alertness at the expense of an overactive mind and body. Still, addiction does not seem to be a problem in narcoleptics, although data are limited [18]. Induction or aggravation of hypertension might be expected, but seems not to be a significant problem when using normal therapeutic doses [19]. Induction of psychosis and hallucinations is rare [16]. Nocturnal sleep may be disturbed in patients who use high doses or who take stimulants in the evening.

Long-acting agents (modafinil, dexamphetamine, and methamphetamine) seem to be tolerated better than the short-acting drugs (methylphenidate). The quick and short-acting ones can be used to good effect when "targeted" at social events or difficult periods during the day. For this reason, combinations of stimulants may be tailored to the circumstances. Unfortunately, there are no studies assessing the advantages or disadvantages of combinations of stimulants.

Modafinil is usually grouped with the stimulants, but it has a different pharmacological mode of action. An alpha-1 agonistic action was presumed initially, but questioned later. Its exact mechanism has not yet been elucidated, but a role as blocker of the dopamine transporter has been suggested [20]. Its efficacy has been studied in large randomized placebo-controlled studies and is probably comparable to that of the stimulants, although direct comparisons are lacking [21, 22]. The advantage of modafinil over the classical stimulants lies in the lower frequency and severity of side effects. Irritability and agitation are less problematic [23]. The possibility of induction of human hepatic cytochrome P450 enzymes by modafinil should be borne in mind. Modafinil increases the metabolism of oral contraceptives, and women of child-bearing age should use an alternate form of contraception or a contraceptive containing 50 μg or higher ethinyl estradiol should be prescribed [15].

Recent studies with sodium oxybate (SO), the sodium salt of gamma-hydroxybutyrate (GHB), have shown that it is effective in reducing EDS, particularly at dosages of 6–9 g per night [24–27]. The effect on EDS was similar to that of modafinil, and side effects were generally mild [28]. However, the combination therapy was even more effective. (For a more detailed description of the role of SO in narcolepsy, see below.)

Selegiline and brofaromine (available in some European countries) may alleviate EDS as well [15].

Treatment of REM Sleep Dissociation Phenomena

Most studies concerning the treatment of the REM dissociation phenomena have focused on cataplexy, but most drugs that improve cataplexy also reduce hypnagogic hallucinations and sleep paralysis. SO and tricyclic antidepressants are the most effective treatments. The different tricyclic antidepressants all inhibit the reuptake of norepinephrine and serotonin, and are potent REM sleep inhibitors. The most commonly used ones are imipramine (10–100 mg/day), protryptiline (2.5–40 mg/day), and clomipramine (10–150 mg/day) [1, 14, 15]. Most authors consider clomipramine to be the treatment of choice [29]. Low dosages of 10–20 mg/day may be remarkably effective and well tolerated. Some patients experience improvement of EDS when treated with protryptiline or clomipramine. As with stimulants, side effects and, to a lesser extent, tolerance are the major drawbacks. Side effects are largely due to the anticholinergic properties; the most frequently reported ones are dry mouth, increased sweating, sexual dysfunction (impotence, delayed orgasm, and erection and ejaculation dysfunction), weight gain, tachycardia, constipation, blurred vision, and urinary retention. These are severe enough to lead to dose reductions or stopping its use. However, in some patients, very low doses may be very effective without causing significant side effects. Tricyclic antidepressants should never be stopped abruptly because of the risk of severe aggravation of cataplexy, which may even lead to status cataplecticus.

Many alternative antidepressants have been studied, especially selective serotonin reuptake inhibitors and more selective noradrenergic reuptake inhibitors such as fluoxetine, zimelidine, viloxazine, femoxitine, fluvoxamine, and paroxetine in a relative higher dosage than the tricyclics [1, 8, 15]. All these substances appear to have anti-cataplectic properties and usually less bothersome side effects than the tricyclics. These substances seem to act mainly via less selective desmethyl metabolites, which are potent adrenergic uptake inhibitors [30].

During recent years, venlafaxine and atomoxetine have become very popular in the treatment of cataplexy, although no randomized placebo-controlled studies are available. In the author's experience, venlafaxine is not clearly superior. Atomoxitine, however, has been effective when other medications failed [31, 32].

SO is the best studied drug and is a very potent inhibitor of cataplexy (see also below) [24, 33–35]. It has never been compared to any antidepressant, so it is difficult to know whether it is really more effective. However, the relatively mild side effect profile makes it a more favorable drug, even independent of the beneficial effect of SO on the other symptoms.

Other alternatives less well studied and probably less potent are mazindol, selegiline, and brofaromine. As with SO, these compounds may improve sleepiness and the REM dissociation phenomena. Therefore, they can sometimes be used as monotherapy in patients with EDS and REM dissociation phenomena, particularly in milder forms.

Several drugs may theoretically be expected to aggravate cataplexy, but the only one for which this is reliably documented is prazosin, an alpha-1 antagonist used to treat hypertension [36].

Treatment of Disturbed Nocturnal Sleep

Disturbed nocturnal sleep can be a major complaint of patients. Unfortunately, treatment options are limited. SO is the only drug with a proven long-term effect on nocturnal sleep [26]. Short-term beneficial effects of benzodiazepines have been described as well [37]. Although nocturnal sleep may be improved with these drugs, improvement of EDS is uncommon.

Sodium Oxybate

This section contains more detailed information when compared to the previous ones. SO treatment is relatively new, and the instructions for its use are somewhat unusual.

Sodium oxybate is the sodium salt of GHB. It is a natural metabolite of GABA and acts as a hypnotic. Although GHB receptors have been identified in the brain, the therapeutic effect is predominantly mediated through the GABA-B receptor [38]. Early studies using low dosages (30 mg/kg twice a night) showed an effect on cataplexy and hypnagogic hallucinations and some improvement of nocturnal sleep and EDS [33]. More recent large, randomized, double-blind, placebo-controlled studies, in which higher dosages were used (up to 9 g/night), showed a beneficial effect on all symptoms (although not significant for all) [24–28, 34, 35]. The strongest effect is on cataplexy, but higher doses seem to reduce EDS as effectively as modafinil [28]. The combination of SO and modafinil has an even better effect on EDS, and it may also allow for the use of lower doses of other stimulants [28, 39]. Available trials were not designed to study the effect on hypnagogic hallucinations and sleep paralysis, but these symptoms tended to improve in all studies.

Efficacy and side effects of SO are both dose dependent. Side effects are usually mild, particularly when compared to those occurring with other drugs. While the most frequent side effect is nausea, the most disabling ones are enuresis and sleep walking. Lowering the dose may solve these problems. Weight loss can occur [40]. Note that while the sedating effects are immediately apparent, it can take several weeks to see the maximal improvements in EDS and cataplexy of a certain dose. In long-term follow-up studies, there is no indication that tolerance develops, and abrupt cessation does not induce rebound cataplexy either [34]. However, long-term clinical experience shows tolerance for the sleep-promoting effects in a substantial proportion of patients, but with remaining efficacy for the other symptoms.

Pharmacological interactions are not known, and combined use with other medications used in the treatment of narcolepsy seems safe. The dose should be reduced in patients with hepatic failure.

Patients starting to use SO need to be given specific instructions: two nocturnal doses are required, and eating must be avoided at least 2 h before the first ingestion to allow optimal absorption. Patients should prepare the doses in two small containers before going to bed and take the first dose only after they have lain down in bed. The second dose must be ingested 3–4 h after the first, and there must be at least 3 h between the second dose and the scheduled wake-up time. Although this may sound complicated, in practice, patients only need to set an alarm for the second dose for a few days to weeks. SO should not be used in conjunction with other sedatives or alcohol. If patients have consumed alcohol in the evening, they should omit one or both doses afterward. In patients with comorbid OSAS, treatment should be closely monitored since SO may worsen OSAS. Co-treatment with CPAP may be indicated [41]. The usual starting dose is 2.25 g twice a night. The dose must be gradually increased keeping in mind that the optimal daytime effects are reached after weeks.

Unfortunately, there is some potential for misuse. GHB is used as a party drug, because it can produce a disinhibited, slightly euphoric state, and it has been used in cases of rape and theft as it can cause amnesia. Continuous (day and night) use of high dosages may lead to dependence and withdrawal symptoms on cessation. Overdose can be fatal, though deaths probably resulted from combinations of GHB with alcohol or other drugs. These risks rightly raise concerns among patients and their physicians, but it is important to realize that when the drug is properly used, it is relatively safe with a low risk of dependence [42].

Treatment of Associated Symptoms/Disorders

Obesity and overweight are common in narcolepsy with cataplexy and should be managed using conventional weight loss techniques. Interestingly, the higher BMI is better explained by hypocretin deficiency than by the sleep phenotype [43]. Since there is no evidence of higher caloric intake, a lower metabolic rate has been proposed as explanation [44, 45]. (See also the chapter "Appetite and Obesity.")

Lack of energy is another associated symptom in up to 60% of patients. It is important to separate this from sleepiness. Treatment with stimulants and/or SO seems to have only limited effect on this complaint [46].

Sleep apnea is more prevalent in narcolepsy patients. Treatment, however, usually does not improve the EDS, and compliance with CPAP and other treatments can be a problem. Whether there is a medical indication for treatment is controversial [47]. Treatment with SO may facilitate the acceptance of CPAP treatment. However, since SO may worsen the course of sleep apnea, it is important in these cases that patients are compliant [41].

Periodic limb movements are more prevalent as well. Treatment must be considered if there is coexistent RLS, especially in severe cases.

There is a high incidence of REM sleep behavior disorder in narcolepsy. If needed, treatment with clonazepam or melatonin can be initiated, which is similar as in idiopathic cases. [48].

Recommendations for the Initiation of Pharmacological Treatment

Pharmacological treatment and behavioral advice should be tailored individually. The recommendations given below should, therefore, only be considered as a general guide to initiate pharmacotherapy.

For patients who mainly suffer from EDS, modafinil is a good first choice. If EDS is relatively mild or mostly situational, methylphenidate as "on demand" treatment may be a good alternative. If modafinil monotherapy is insufficient, then combination therapy with SO or methylphenidate can be considered. Increasing the dose of modafinil beyond 400 mg daily is not recommended when there has been no effect at all with doses up to 400 mg.

SO as first-line treatment is appropriate for patients who predominantly suffer from cataplexy and/or a disturbed nocturnal sleep. OSAS must, however, be excluded first as the explanation for disturbed nocturnal sleep. In case of comorbid OSAS, the combination of CPAP and SO may be considered.

Patients presenting with substantial EDS and cataplexy are also good candidates for first-line SO treatment. If EDS persists, addition of modafinil or methylphenidate may have a beneficial effect. If cataplexy is not completely controlled, a very low dose (10 mg) of clomipramine can be added.

Future Treatments

Nonpharmacological

The regulation of skin temperature is altered in narcolepsy and may worsen daytime vigilance and disrupted nocturnal sleep. Preliminary studies show that skin temperature manipulation improves both daytime vigilance and nocturnal sleep, and further studies in non-laboratory circumstances are needed [49, 50].

Pharmacological

Several new approaches are now under development that may provide new therapeutic options for patients with narcolepsy. Modafinil contains equal amounts of both the r- and s-enantiomers. Armodafinil is a new formulation that contains only the r-enantiomer, has a longer half-life, and has been shown to be effective in narcolepsy patients [51]. Direct comparisons with modafinil are not available, and the drug is currently not available in most European countries [14]. Signaling through the histamine H3 receptor reduces the activity of monoaminergic neurons, and several H3 antagonists are currently being studied for the treatment of EDS with encouraging results in a pilot study [52].

Intravenous immunoglobulins administered close to disease onset might have a beneficial effect, especially on cataplexy [53]. Note, however, that studies were small and not blinded, that possible spontaneous severity fluctuations may have influenced outcome, and that the placebo effect may be large [54]. Only randomized, double-blind, placebo-controlled studies can solve these issues, but there are problems in carrying out such a study: narcolepsy is an uncommon disease, immunoglobulins are expensive, and there is a relative shortage of immunoglobulins.

From a theoretical point of view, the best therapies would correct the underlying hypocretin deficiency. While some small peptides have been developed that act as agonists at the hypocretin receptors, these cannot cross the blood–brain barrier, and no small compounds have yet been identified that would be therapeutically practical. Gene therapy is also promising but has potentially dangerous side effects if genes insert randomly into the patient's DNA. Transplantation of hypocretin neurons might potentially provide a cure [55]. However, the current techniques need to be improved and there is the potential problem of an immune reaction to the graft in view of the autoimmune hypothesis of narcolepsy. Ultimately, a better understanding of the pathologic process that kills the hypocretin neurons should result in effective prevention strategies.

References

1. Overeem S, Mignot E, van Dijk JG, Lammers GJ. Narcolepsy: clinical features, new pathophysiologic insights, and future perspectives. J Clin Neurophysiol. 2001;18(2):78–105.
2. Broughton R, Valley V, Aguirre M, et al. Excessive daytime sleepiness and the pathophysiology of narcolepsy-cataplexy: a laboratory perspective. Sleep. 1986;9:205–15.
3. Chemelli RM, Willie JT, Sinton CM, et al. Narcolepsy in orexin knockout mice: molecular genetics of sleep regulation. Cell. 1999;98:437–51.
4. Nishino S, Ripley B, Overeem S, et al. Hypocretin (orexin) deficiency in human narcolepsy. Lancet. 2000;355:39–40.
5. Sakurai T. The neural circuit of orexin (hypocretin): maintaining sleep and wakefulness. Nat Rev Neurosci. 2007;8(3):171–81.
6. Peyron C, Tighe DK, Den Pol AN, et al. Neurons containing hypocretin (orexin) project to multiple neuronal systems. J Neurosci. 1998;18:9996–10015.
7. Nishino S. Clinical and neurobiological aspects of narcolepsy. Sleep Med. 2007;8(4):373–99.
8. Nishino S, Mignot E. Pharmacological aspects of human and canine narcolepsy. Prog Neurobiol. 1997; 52:27–78.
9. Deadwyler SA, Porrino L, Siegel JM, Hampson RE. Systemic and nasal delivery of orexin-A (Hypocretin-1) reduces the effects of sleep deprivation on cognitive performance in nonhuman primates. J Neurosci. 2007;27(52):14239–47.
10. Mullington J, Broughton R. Scheduled naps in the management of daytime sleepiness in narcolepsy-cataplexy. Sleep. 1993;16:444–56.
11. Bruck D, Armstrong S, Coleman G. Sleepiness after glucose in narcolepsy. J Sleep Res. 1994;3:171–9.
12. Lammers GJ, van Dijk JG. Daytime tests for sleepiness: indications, interpretation, and pitfalls. In: Overeem S, Reading P, editors. Sleep disorders in neurology. A practical approach. Wiley-Blackwell; 2010. p. 28–40.
13. Fronczek R, Middelkoop HA, van Dijk JG, Lammers GJ. Focusing on vigilance instead of sleepiness in the assessment of narcolepsy: high sensitivity of the Sustained Attention to Response Task (SART). Sleep. 2006;29(2):187–91.
14. Wise MS, Arand DL, Auger RR, et al. American Academy of Sleep Medicine. Treatment of narcolepsy and other hypersomnias of central origin. Sleep. 2007;30(12):1712–27.
15. Billiard M, Bassetti C, Dauvilliers Y, et al. EFNS Task Force. EFNS guidelines on management of narcolepsy. Eur J Neurol. 2006;13(10):1035–48.
16. Mitler MM, Aldrich MS, Koob GF, Zarcone VP. Narcolepsy and its treatment with stimulants. ASDA standards of practice. Sleep. 1994;17:352–71.
17. Ryan DH, Bray GA, Helmcke F, et al. Serial echocardiographic and clinical evaluation of valvular regurgi-

tation before, during, and after treatment with fenfluramine or dexfenfluramine and mazindol or phentermine. Obes Res. 1999;7:313–22.
18. Parkes JD, Dahlitz M. Amphetamine prescription. Sleep. 1993;16:201–3.
19. Wallin MT, Mahowald MW. Blood pressure effects of long-term stimulant use in disorders of hypersomnolence. J Sleep Res. 1998;7:209–15.
20. Volkow ND, Fowler JS, Logan J, et al. Effects of modafinil on dopamine and dopamine transporters in the male human brain: clinical implications. JAMA. 2009;301(11):1148–54.
21. Broughton RJ, Fleming JA, George CF, et al. Randomized, double-blind, placebo-controlled crossover trial of modafinil in the treatment of excessive daytime sleepiness in narcolepsy. Neurology. 1997;49: 444–51.
22. US Modafinil in Narcolepsy Multicenter Study Group. Randomized trial of modafinil as a treatment for the excessive daytime somnolence of narcolepsy. Neurology. 2000;54:1166–75.
23. Bastuji H, Jouvet M. Successful treatment of idiopathic hypersomnia and narcolepsy with modafinil. Prog Neuropsychopharmacol Biol Psychiatry. 1988;12:695–700.
24. US Xyrem® Multicenter Study Group. A randomized, double blind, placebo-controlled multicenter trial comparing the effects of three doses of orally administered sodium oxybate with placebo for the treatment of narcolepsy. Sleep. 2002;25:42–9.
25. The Xyrem® International Study Group. A double-blind, placebo-controlled study demonstrates that sodium oxybate is effective for the treatment of excessive daytime sleepiness in narcolepsy. J Clin Sleep Med. 2005;1:391–7.
26. US Xyrem® Multicenter Study Group. A 12-month, open-label, multicenter extension trial of orally administered sodium oxybate for the treatment of narcolepsy. Sleep. 2003;26:31–5.
27. Mamelak M, Black J, Montplaisir J, Ristanovic R. A pilot study on the effects of sodium oxybate on sleep architecture and daytime alertness in narcolepsy. Sleep. 2004;27:1327–34.
28. Black J, Houghton WC. Sodium oxybate improves excessive daytime sleepiness in narcolepsy. Sleep. 2006;29(7):939–46.
29. Parkes D. Introduction to the mechanism of action of different treatments of narcolepsy. Sleep. 1994;17: S93–6.
30. Nishino S, Arrigoni J, Shelton J, et al. Desmethyl metabolites of serotonergic uptake inhibitors are more potent for suppressing canine cataplexy than their parent compounds. Sleep. 1993;16:706–12.
31. Mignot E, Nishino S. Emerging therapies in narcolepsy-cataplexy. Sleep. 2005;28(6):754–63.
32. Niederhofer H. Atomoxetine also effective in patients suffering from narcolepsy? Sleep. 2005;28(9):1189.
33. Lammers GJ, Arends J, Declerck AC, et al. Gammahydroxybutyrate and narcolepsy: a double-blind placebo-controlled study. Sleep. 1993;16:216–20.
34. US Xyrem® Multicenter Study Group. Sodium oxybate demonstrates long-term efficacy for the treatment of cataplexy in patients with narcolepsy. Sleep Med. 2004;5:119–23.
35. Xyrem® International Study Group. Further evidence supporting the use of sodium oxybate for the treatment of cataplexy: a double-blind, placebo-controlled study in 228 patients. Sleep Med. 2005;6:415–21.
36. Guilleminault C, Mignot E, Aldrich M, Quera-Salva MA, Tiberge M, Partinen M. Prazosin contraindicated in patients with narcolepsy. Lancet. 1988;2(8609):511.
37. Thorpy MJ, Snyder M, Aloes FS, et al. Short-term triazolam use improves nocturnal sleep of narcolepstics. Sleep. 1992;15:212–6.
38. Carter LP, Koek W, France CP. Behavioral analyses of GHB: receptor mechanisms. Pharmacol Ther. 2009;121(1):100–14.
39. Broughton R, Mamelak M. Effects of nocturnal gamma-hydroxybutyrate on sleep/waking patterns in narcolepsy-cataplexy. Can J Neurol Sci. 1980;7(1): 23–31.
40. Husain AM, Ristanovic RK, Bogan RK. Weight loss in narcolepsy patients treated with sodium oxybate. Sleep Med. 2009;10(6):661–3.
41. Feldman NT. Clinical perspective: monitoring sodium oxybate-treated narcolepsy patients for the development of sleep-disordered breathing. Sleep Breath. 2010;14(1):77–9.
42. Lammers GJ, Bassetti C, Billiard M, et al. Sodium oxybate is an effective and safe treatment for narcolepsy. Sleep Med. 2010;11(1):105–6.
43. Kok SW, Overeem S, Visscher TL, et al. Hypocretin deficiency in narcoleptic humans is associated with abdominal obesity. Obes Res. 2003;11(9):1147–54.
44. Lammers GJ, Pijl H, Iestra J, et al. Spontaneous food choice in narcolepsy. Sleep. 1996;19(1):75–6.
45. Poli F, Plazzi G, Di Dalmazi G, et al. Body mass index-independent metabolic alterations in narcolepsy with cataplexy. Sleep. 2009;32(11):1491–7.
46. Droogleever Fortuyn HA, Fronczek R, Smitshoek M, et al. Prevalence and correlates of severe experienced fatigue in patients with narcolepsy. Submitted.
47. Sansa G, Iranzo A, Santamaria J. Obstructive sleep apnea in narcolepsy. Sleep Med. 2010;11(1):93–5.
48. Billiard M. REM sleep behavior disorder and narcolepsy. CNS Neurol Disord Drug Targets. 2009;8(4):264–70.
49. Fronczek R, Raymann RJ, Overeem S, et al. Manipulation of skin temperature improves nocturnal sleep in narcolepsy. J Neurol Neurosurg Psychiatry. 2008;79(12):1354–7.
50. Fronczek R, Raymann RJ, Romeijn N, et al. Manipulation of core body and skin temperature improves vigilance and maintenance of wakefulness in narcolepsy. Sleep. 2008;31(2):233–40.
51. Darwish M, Kirby M, Hellriegel ET, Robertson Jr P. Armodafinil and modafinil have substantially different pharmacokinetic profiles despite having the same terminal half-lives: analysis of data from three randomized, single-dose, pharmacokinetic studies. Clin Drug Investig. 2009;29(9):613–23.

52. Lin JS, Dauvilliers Y, Arnulf I, et al. An inverse agonist of the histamine H(3) receptor improves wakefulness in narcolepsy: studies in orexin–/– mice and patients. Neurobiol Dis. 2008;30(1):74–83.
53. Dauvilliers Y. Follow-up of four narcolepsy patients treated with intravenous immunoglobulins. Ann Neurol. 2006;60(1):153.
54. Fronczek R, Verschuuren J, Lammers GJ. Response to intravenous immunoglobulins and placebo in a patient with narcolepsy with cataplexy. J Neurol. 2007; 254(11):1607–8.
55. Arias-Carrión O, Murillo-Rodríguez E. Cell transplantation: a future therapy for narcolepsy? CNS Neurol Disord Drug Targets. 2009;8(4):309–14.

Treatment of Narcolepsy in Children

Michel Lecendreux

Keywords
Pediatrics • Narcolepsy • Modafinil • Sodium oxybate • Immunotherapy • Histamine

Introduction

Narcolepsy is a chronic primary sleep-wake disorder, characterized by excessive daytime sleepiness (EDS), sudden sleep episodes, and attacks of muscle atonia mostly triggered by emotions (cataplexy). Narcolepsy is a lifelong disorder, however, not progressive, and occurrence during childhood is frequent. Similar to the adult phenotype, pediatric narcolepsy shows a large variability in its presentation. Symptoms can start abruptly, sometimes even dramatically, e.g., with the occurrence of sudden and complete cataplexy, or progressively and insidiously, with slow progression of EDS or weight gain over weeks to years. The latter may cause difficulties in recognizing the condition and in making the diagnosis at an early stage.

Therefore, narcolepsy in children is probably under-diagnosed, frequently mistaken for other diseases. Indeed, with cataplexy being the most specific symptom of the disorder, appearing often years after EDS onset, making an appropriate diagnosis is difficult.

The treatment of narcolepsy is essentially symptomatic. At present, there is no perspective in regard to a definitive cure with remission of symptoms. Considering the effect of the disorder on the individual life, its dramatic consequences particularly in children and adolescents, medication is required at a very early stage of the disease, as soon as the diagnosis is confirmed. Given the chronic natural history of narcolepsy, the disease must be diagnosed appropriately before medication-based therapy is initiated. Non-medication-based approaches are always favored in children and adolescents, albeit often in combination with pharmacotherapy, before the initiation of any psychotropic medication.

Non-medication-Based Approaches: Behavior and Lifestyle

A healthy lifestyle and regular waking and sleep routine are strongly recommended for children and adolescents suffering from narcolepsy. It is important to make sure that late afternoon naps (after school) do not interfere with nighttime sleep periods.

M. Lecendreux (✉)
Pediatric Sleep Center and Narcoleptic Reference Center, Hospital Robert Debré, Paris, France
e-mail: michel.lecendreux@rdb.aphp.fr

C.R. Baumann et al. (eds.), *Narcolepsy: Pathophysiology, Diagnosis, and Treatment*,
DOI 10.1007/978-1-4419-8390-9_38, © Springer Science+Business Media, LLC 2011

Although no studies on the effectiveness of naps are available, one or more daytime sleep periods are generally recommended. One to two scheduled routine naps of 20–30 min increase daytime wakefulness and psychomotor performances. These naps should be encouraged, but they are not always accepted in a school environment. At school, we recommend to give the teachers the most detailed information possible on the disease and its effects, and to help searching for adapted solutions. Nutritional advice, regular meal times, and physical activity should also be encouraged at an early stage in children and adolescents, in order to avoid gain weight and to help maintain regular growth.

Counseling or brief psychotherapy is often required to enable the child to accept the loss of their previous healthy state, and to accept the reality of a disabling chronic condition. In many cases, young children will not easily understand that excessive daytime sleepiness can be a true disease, as sleep itself is perceived as a normal condition. Long-term psychotherapy is recommended in children or adolescents presenting with poor self-esteem, lack of social interactions, or morbid thinking, leading in some cases to chronic depression.

Patient associations also play a vital role, allowing young patients to exchange information and advice on managing their disease, interacting with their peers, e.g., in group meetings, Internet forums, or meetings at holiday camps. Social interactions should be encouraged in children and adolescents with narcolepsy in order to ameliorate the acceptance of the disease and to discuss and generate positive attitudes toward the most disabling symptoms.

For both young children under pharmacotherapy and adolescents, training and education are crucial to enhance the benefits of the treatment. The need of being treated on a daily basis represents a challenge for many young children – many of them are reluctant to take daily medication. Behavioral therapy or coaching can be helpful to increase the acceptance of treatment and to help the child perceive its positive aspects. Parents or older siblings and sometimes teachers certainly play a fundamental role in helping the child improve his compliance to medication, especially when the treatment is prescribed several times a day. Most medications are prescribed during the daytime period – they are often given during the main meals. Treatments such as sodium oxybate – although not yet approved in children and adolescents – are given at bedtime and 4 h after sleep onset, requiring specific and operant strategies from the child, and efficient supervision from the parents.

Medication-Based Therapies

In children, narcolepy symptoms often lead to immediate disability, which causes a dramatic decrease in the child's well-being and functioning. Pharmacotherapy is often required soon after the diagnosis is established, in order to relieve the child and to restore a normal or subnormal level of alertness in children at school. Although symptomatic treatment with wakefulness-promoting agents such as modafinil has a significant impact on the disease in children and adolescents, the perspectives and the constraints of a lifelong treatment are a major concern for the narcoleptic child. The hope for a curative treatment is occupying many young narcoleptic patients. In this line, some clinicians and researchers assume that medication should be delivered at the very beginning of the disease, due to the hope that an early and effective therapeutic could counteract the degenerative process that leads to the loss of hypocretin neurons.

Most importantly, it must be underlined that no medication awarded a marketing authorization, and none is available for the treatment of narcolepsy in children and adolescents, which poses a genuine healthcare issue. While there is a lack of established guidelines for the treatment of the disease in children, there is at least a number of best practices that are generally acknowledged, including the preference for monotherapy from disease onset (lack of evidence for treatment combinations), and sufficiently long-term administration of a treatment at effective doses. The recently proposed guidelines for adults [1] are not applicable to children. For children, practices are generally based on empirical data, although few clinical trials have been conducted in recent years.

Treatment of Excessive Daytime Sleepiness

Wakefulness-promoting agents (modafinil) and psychostimulants (methylphenidate) represent the first-line treatments of EDS in children and adolescents.

The mode of action of modafinil is complex and not fully understood, but an alpha-1-adrenergic action, a direct and indirect action on the dopaminergic systems, and an impact on the serotoninergic/GABAergic mechanisms have been suggested to explain its action. Modafinil helps increasing daytime wakefulness at dosages between 100 and 600 mg/day taken in one to two doses in the morning and at noon. The side effects are generally mild and include headache, nervousness, anxiety, nausea, and insomnia at the start of treatment. Long-term dependence or tolerance has not been reported. Because of its enzyme-inducing effect on cytochrome P450, sexually active young women must be advised to use normal-dose contraception containing 50 μg of ethinyl estradiol. The favorable effect of modafinil on excessive daytime sleepiness in narcoleptic children has been reported in a small series of 13 patients [2].

A double-blind randomized study in 26 subjects [3] reported satisfactory tolerance and efficacy of modafinil in 6–16-year-old children who suffered from excessive sleepiness or sleep apnea syndrome. A total of 19 children were enrolled in the modafinil treatment arm, with a satisfactory tolerance particularly on cardiovascular parameters, at dosages ranging from 100 to 400 mg/day.

A European study was recently conducted to assess the tolerance and efficacy of modafinil in a pediatric population (unpublished results). In this open-label study in 92 children (6–16 years) suffering from narcolepsy or obstructive sleep apnea, modafinil was administered at two dosages ranging from 100 to 400 mg per day, according to tolerance and efficacy. The study found an overall improvement of excessive daytime sleepiness in 91% of patients. The most frequent adverse event was headache. The cardiovascular tolerance was satisfactory. Modafinil improved social competence and behavioral disorders in both children suffering from narcolepsy and those with sleepiness because of obstructive sleep apnea.

Both immediate-release and extended-release methylphenidate are considered second-line treatment of excessive daytime sleepiness in narcolepsy in children. These compounds are well known by pediatricians because of their usefulness in the treatment of attention-deficit hyperactivity disorder (ADHD) in children and adolescents. Methylphenidate blocks dopamine and norepinephrine reuptake, but does not inhibit monoaminergic vesicle transporters. The dosages differ between compounds with different pharmacokinetics. For the immediate-release compound, they are between 0.5 and 1 mg/kg/day. The most frequent side effects are loss of appetite, nervousness, tics, headache, and sleep-onset insomnia. The extended-release forms enable a better coverage of daytime requirements and reduce side effects. The pharmacokinetics of the compound plays a major role on the clinical effects of the drug, and several formulations of methylphenidate are available across countries. The combination of extended- and immediate-release forms of methylphenidate is helpful when both long-lasting and immediate effects are required.

The third-line agents include mazindole and amphetamines. Mazindole (available only in a subset of Western countries) is an imidazoline derivative that blocks dopamine and norepinephrine reuptake. It is active against both sleepiness and cataplexy. The dosage ranges from 1 to 4 mg per day, taken in the morning and at noon. Gastrointestinal and cardiac side effects may occur, as well as headache and dizziness. Amphetamines (which are not available in many countries) must be used with caution because of their side effects (e.g., insomnia, irritability, high blood pressure, and psychotic reactions) and because of the risk of tolerance and dependence.

Sodium oxybate (gamma-hydroxybutyrate) has significant benefits on nocturnal sleep, cataplexy, and excessive daytime sleepiness in adult narcolepsy, and is therefore considered a first-line treatment. Although there are encouraging clinical reports on the efficacy and tolerability of sodium oxybate in children and adolescents, the compound warrants further extensive studies in the pediatric population. Sodium oxybate acts on the GABA system, particularly by stimulating the GABA-B receptors, and produces a more physiological sleep than benzodiazepines, as it increases slow-wave sleep and preserves REM sleep. The anti-cataplectic and antisleepiness effects of this molecule are poorly understood. It is conceivable that these effects are due to the improvement of sleep quality. In adults, treatment is initiated with 1.5 g in two doses, taken at night (in view of its very brief half-life of 40–60 min). Thereafter, the dose is progressively increased up to 9 g/night. The most common side effects are headache, dizziness, depression, sleepwalking, enuresis, and terminal insomnia (the patient wakes at around 4–5 a.m. and is unable to go back to sleep). Alcohol consumption is strictly contraindicated. Due to the risk of misuse of this substance, it is subject to strict issue conditions (psychostimulant register). Although safety data on the use of sodium oxybate in children and adolescents are still lacking, good efficacy both on daytime and nighttime symptoms, and good tolerability over more than a year have been observed in our clinic in adolescents presenting with severe narcolepsy with cataplexy.

Treatment of Cataplexy

In many countries, first-line treatment of cataplexy is sodium oxybate. In children and adolescents, this substance has been found to be effective for the treatment of cataplexy [4]. In an open-label pilot study in a small number of children and adolescents suffering from severe narcolepsy ($n=8$), frequency and severity of cataplexy improved in 88% of subjects. Epworth sleepiness score decreased from 19 to 13. Psychostimulant and anti-cataplectic treatments were maintained throughout the study.

Second-line treatment against cataplexy includes selective serotonin reuptake inhibitors (SSRIs) such as fluoxetine (20–60 mg/day) and the norepinephrine and serotonin reuptake inhibitor venlafaxine (75–300 mg). These compounds are equally effective against hypnagogic hallucinations and sleep paralysis. Tricyclic antidepressants have also been suggested, particularly clomipramine, starting at low doses of 10–20 mg and increasing up to 75 mg/day if required. The use of the latter is limited by anticholinergic side effects and potential cardiovascular threats (QT prolongation). After abrupt discontinuation of tricyclics, a rebound effect is frequently experienced. Finally, at the doses indicated for sleepiness, and based on clinical experience, mazindole may also be beneficial in the treatment of cataplexy in adolescents.

Treatment of Dyssomnia

Benzodiazepines and non-benzodiazepine hypnotics are mostly ineffective for dyssomnia, and these compounds are only rarely used in childhood narcolepsy. No clinical trials have been conducted for this indication in children. Furthermore, sleep-promoting agents may increase daytime sleepiness, and tolerance may develop in young patients.

Sodium oxybate reduces the number of nocturnal arousals and increases slow-wave sleep. This treatment may be of particular interest in cases of early-onset dyssomnia, which often occurs in children suffering from narcolepsy. However, targeted studies including polysomnographic assessment await their execution in children and adolescents.

Treatment of Associated Comorbidities

In the absence of sufficient evidence, systematic trials are warranted for the treatment of obstructive sleep apnea and periodic leg movements in sleep in children. Nighttime continuous positive airway pressure treatment may be indicated for children or adolescents suffering from severe obstructive sleep apnea. In children, there is no standard treatment for periodic leg movements in sleep. Dopaminergic agonists appear to be effective in children suffering from RLS [4].

Mood disorders such as depression and ADHD have not been reported in children and adolescents with narcolepsy. There is, however, no doubt that this important clinical aspect deserves more attention, especially when stimulants or antidepressants are prescribed at an young age. Improving mood, wakefulness, and attention may support the child in his daily living, not only at school, but also in many different settings, which in the end may lead to a significant increase in the child's self-esteem. Therefore, ADHD symptoms – which are currently underestimated in narcoleptic children – certainly require more careful investigation from the clinicians.

New Therapeutic Alternatives

Selective histamine (anti-H3) antagonists are already tested in phase II trials in narcoleptic adults, with encouraging results [5]. Atomoxetine is indicated in ADHD and has been the subject of a polysomnographic study in children [6]. This compound is emerging as a first-line alternative to psychostimulant drugs in ADHD and may have a beneficial effect on cataplexy reported in adults [7].

Transdermal methylphenidate patches provide a transdermal release of methylphenidate, enabling sustained release of the active substance and avoiding a rebound effect, which is frequently the cause of sleep-onset insomnia. This new system may help children with problems of administering the drug several times a day and, therefore, might improve long-term treatment compliance.

For associated dyssomnia, melatonin is a possible alternative treatment for children and adolescents, because it has an excellent tolerance and a beneficial chronobiological effect. Currently, small studies with melatonin have been published for children suffering from ADHD. Caution should be applied in the use of this substance in adolescents, because of its potential effect on spermatogenesis. Studies on melatonin in narcolepsy children are lacking to date.

Emerging Therapies

There is considerable evidence (however, no proof) that narcolepsy is caused by an autoimmune attack. Therefore, immunosuppressant treatments may emerge as treatments of choice to limit or inhibit the autoimmune process. Corticosteroids have been tested, with one case reported in the literature [8]. A 3-week treatment protocol at a rate of 1 mg/kg/day did not significantly improve the clinical symptoms in this narcolepsy case. Side effects of corticosteroids must be considered, particularly in children, because of their consequences on growth and metabolism. To our knowledge, no further study using corticotherapy has been published.

Human intravenous immunoglobulins (IVIg) are blood derivatives, with a high immunomodulation power. Tegeline, for instance, has been selected for its satisfactory tolerance and relative safety, providing sufficient immunosuppressant efficacy. Thus, these compounds are used as immunomodulation treatment for various autoimmune diseases such as idiopathic thrombocytopenic purpura in adults and children, Birdshot retinochoroidopathy, adult Guillain–Barré syndrome, and Kawasaki syndrome. The mode of action of IVIg in autoimmune disorders is complex. The targets of immunotherapy depend on the pathogenesis of the disease, with an involvement of both cellular and humoral immunity [9].

For narcolepsy patients, we and others believe that high doses of IVIg may downregulate T-cell functions and pathogenic cytokines as well as unidentified autoantibodies, and interfere with autoantigen recognition through major histocompatibility complex class II molecules such as DQB1*0602.

Intravenous high-dose IvIG treatment in recent-onset childhood narcolepsy with cataplexy was tested by our group [10]. Immunoglobulin therapy was administrated only 4 months after the onset of the first narcoleptic symptoms in a 10-year-old boy, who presented with unexplained weight gain (10 kg over 1 year), excessive daytime sleepiness, sleep attacks, and clear-cut cataplexy. IVIg perfusion at the dose of 1 g/kg/day for two consecutive days was administrated to the child, 3 months after cataplexy onset. Three days after treatment, episodes of cataplexy improved significantly and this effect persisted during the next 3 weeks. Notably, not a single cataplexy or irresistible nap was reported (as observed independently by the parents and qualified as a dramatic improvement by the mother). Thereafter, a mild relapse (recurrence of partial cataplexy) was observed, but this was not severe enough to require conventional anti-cataplectic agent. Because mild side effects had occurred during perfusion administration (transient headache, fever, and flushing), the mother decided not to repeat the IVIg perfusion after 3 weeks. The long-term outcome of the child is similar to that in other narcoleptic adolescents, and requires a combination therapy with a wake-promoting agent (modafinil) and an anti-cataplectic drug (venlafaxine).

Soon after our report, Dauvilliers and colleagues published new evidence on the efficacy of IVIg treatment in four recent-onset narcoleptic patients, including a 10-year-old male and a 12-year-old female patient, who participated in this open-label clinical trial [11]. IVIg therapy was administrated at a dose of 1 mg/kg/day over 2 consecutive days and repeated three times at 4-week intervals. All four subjects completed the protocol and showed clinical improvement in the frequency and severity of cataplexy. The most striking finding was the persisting improvement in cataplexy, hypnagogic hallucinations, and sleep paralysis, even several months after three IVIg perfusions. Several months after treatment, none of the four patients required anti-cataplectic medication.

More recently, a clinical open trial with IVIg has been published by the Italian group of Plazzi. Four children and adolescents participated in a protocol performing IVIg perfusions at a dose of 0.4 g/kg/day for 5 days, repeated for 3 months, and followed by the same single-day dose every month for the following 6 months [12]. This study, however, showed no significant clinical improvement in three of four patients. In one patient, a 15-year-old female, a dramatic improvement in cataplexy was observed. Interestingly, this patient had a higher CSF hypocretin-1 level than the other children, and her symptoms were less severe before treatment administration. The authors discussed that the dramatic improvement monitored in that particular case could also reflect the natural history of narcolepsy with cataplexy, which can sometimes lead to a spontaneous remission of symptoms.

At present, and based on the few studies published in this field, it is not possible to draw conclusions on the efficacy and beneficial long-term effects of IVIg treatment. We believe that IVIg therapy in children with recent-onset (under 6 months) narcolepsy with cataplexy might procure the child a supplementary chance to improve the condition and to decrease the ulterior need for conventional treatments.

Thus, double-blind, randomized, placebo-controlled trials are needed to evaluate the safety and long-term efficacy of IVIg in narcolepsy in children and adolescents with recent symptomatic onset.

Other Alternatives

The presumed role of an infectious trigger in narcolepsy was recently examined by Aran and coworkers [13]. The authors mentioned a methodological limitation of many studies: when the disorder is diagnosed several years after onset, the chance to detect a possible infectious trigger is very poor. Attempts to duplicate (unpublished) findings of elevated antistreptococcal titers in long-standing adult narcolepsy cases were unsuccessful. The hypothesis underlying the present study is that streptococcal infections would be best detected in recent-onset narcoleptics. Therefore, the authors have tested 200 narcoleptic patients with hypocretin deficiency, but focused on recent-onset cases and 200 age-matched healthy controls for markers of immune response to β-hemolytic streptococcus (anti-streptolysin O (ASO); anti-DNAse B (ADB)) and *Helicobacter pylori* (Anti-Hp IgG), two bacterial infections known to trigger autoimmunity. The results supported the hypothesis of a possible infectious trigger in narcolepsy, showing significant higher titers of antistreptococcal antibodies in patients with narcolepsy for both ASO and ADB. These data support the hypothesis that streptococcal infections are probably a significant environmental trigger for narcolepsy and suggest that HLA-positive narcoleptic patients with increased antistreptococcal antibodies, who were diagnosed close to disease onset, could be good candidates for antibiotic treatment administrated alone or in addition with immunotherapy.

Conclusions

The pharmacological treatment of narcolepsy in children is mostly based on empirical data and clinical experience in both adults and children. Treatment guidelines have not yet been established in narcoleptic children. Recently introduced compounds have markedly improved the symptoms and the well-being of young patients with narcolepsy with cataplexy. However, the available treatment options are purely symptomatic, and their use in children is off-label. Therefore, there is need for well-designed clinical trials to improve treatment of narcolepsy in children and adolescents. Non-pharmacological approaches are also helpful and should be promoted, especially in very young children.

At present and in our opinion, narcolepsy with recent onset in children or adolescents should be considered a therapeutic emergency, even though data from clinical studies on immunotherapy are controversial. Future targeted therapies such as immunosuppressive treatment or antibiotics, administrated soon after onset of the disease, will probably provide a real opportunity for children to improve their condition on a long-term basis.

References

1. Billiard M, Bassetti C, Dauvilliers Y, Dolenc-Groselj L, Lammers GJ, Mayer G, et al. EFNS Task Force. EFNS guidelines on management of narcolepsy. Eur J Neurol. 2006;13:1035–48.
2. Ivanenko A, Tauman R, Gozal D. Modafinil in the treatment of excessive daytime sleepiness in children. Sleep Med. 2003;4:579–82.
3. Owens J. Modafinil is well tolerated in children and adolescents with excessive sleepiness and obstructive sleep apnea: a 6-week double-blind study. Oral communication. Minneapolis: APSS; 2007.
4. Konofal E, Arnulf I, Lecendreux M, Mouren MC. Ropinirole in a child with attention-deficit hyperactivity disorder and restless legs syndrome. Pediatr Neurol. 2005;32:350–1.
5. Lin JS, Dauvilliers Y, Arnulf I, Bastuji H, Anaclet C, Parmentier R, Kocher L, Yanagisawa M, Lehert P, Ligneau X, Perrin D, Robert P, Roux M, Lecomte JM, Schwartz JC. An inverse agonist of the histamine H(3) receptor improves wakefulness in narcolepsy: studies in orexin(−/−) mice and patients. Neurobiol Dis. 2007.
6. Sangal RB, Owens J, Allen AJ, Sutton V, Schuh K, Kelsey D. Effects of atomoxetine and methylphenidate on sleep in children with ADHD. Sleep. 2006;29(12):1573–85.
7. Niederhofer H. Atomoxetine also effective in patients suffering from narcolepsy? Sleep. 2005;28:1189.
8. Hecht M, Lin L, Kushida CA, Umetsu DT, Taheri S, Einen M, et al. Report of a case of immunosuppression with prednisone in an 8-year-old boy with an acute onset of hypocretin-deficiency narcolepsy. Sleep. 2003;26:809–10.
9. Dalakas MC. Intravenous immunoglobulin in autoimmune neuromuscular diseases. JAMA. 2004;291: 2367–75.
10. Lecendreux M, Maret S, Bassetti C, Mouren MC, Tafti M. Clinical efficacy of high-dose intravenous immunoglobulins near the onset of narcolepsy in a 10-year-old boy. J Sleep Res. 2003;12:347–8.
11. Dauvilliers Y, Carlander B, Rivier F, Touchon J, Tafti M. Successful management of cataplexy with intravenous immunoglobulins at narcolepsy onset. Ann Neurol. 2004;56:905–8.
12. Plazzi G, Poli F, Franceschini C, Parmeggiani A, Pirazzoli P, Bernardi F, et al. Intravenous high-dose immunoglobulin treatment in recent onset childhood narcolepsy with cataplexy. J Neurol. 2008;255: 1549–54.
13. Aran A, Lin L, Nevsimalova S, Plazzi G, Hong SC, Weiner K, et al. Elevated anti-streptococcal antibodies in patients with recent narcolepsy onset. Sleep. 2009;32:979–83.

Index

C.R. Baumann et al. (eds.), *Narcolepsy: Pathophysiology, Diagnosis, and Treatment*,
DOI 10.1007/978-1-4419-8390-9,

O

Printed in the United States of America